T0188939

Lecture Notes in Artificial Intelligence 13601

Subseries of Lecture Notes in Computer Science

More information about this subseries at https://link.springer.com/bookseries/1244

Poncelet Pascal · Dino Ienco (Eds.)

Discovery Science

25th International Conference, DS 2022
Montpellier, France, October 10–12, 2022
Proceedings

Editors
Poncelet Pascal
University of Montpellier
Montpellier, France

Dino Ienco
INRAE
Montpellier, France

ISSN 0302-9743 ISSN 1611-3349 (electronic)
Lecture Notes in Artificial Intelligence
ISBN 978-3-031-18839-8 ISBN 978-3-031-18840-4 (eBook)
https://doi.org/10.1007/978-3-031-18840-4

LNCS Sublibrary: SL7 – Artificial Intelligence

This Springer imprint is published by the registered company Springer Nature Switzerland AG
The registered company address is: Gewerbestrasse 11, 6330 Cham, Switzerland

Preface

The Discovery Science conference presents a unique combination of latest advances in the development and analysis of methods for discovering scientific knowledge, coming from machine learning, data mining, and intelligent data analysis, with their application in various scientific domains. The 25th International Conference on Discovery Science (DS 2022) was held in Montpellier, France, during October 10–12, 2022. This was the second time the conference was organized as a stand-alone physical event.

For its first 20 editions, DS was co-located with the International Conference on Algorithmic Learning Theory (ALT). In 2018 it was co-located with the 24th International Symposium on Methodologies for Intelligent Systems (ISMIS 2018). DS 2019 was a stand-alone event, whereas DS 2020 and DS 2021 were online-only events.

DS 2022 received 56 international submissions. Each submission was reviewed by at least two Program Committee (PC) members in a single-blind manner. The PC decided to accept 27 regular papers and 12 short papers. This resulted in an acceptance rate of 48% for regular papers.

The conference included three keynote talks. Leman Akoglu (Carnegie Mellon University) contributed a talk titled "Unsupervised Model Selection in Outlier Detection: The Elephant in the Room"; Luca Maria Aiello (IT University of Copenhagen) gave a presentation titled "Coloring Social Relationships"; and Stefan Kramer (University of Mainz) contributed a talk titled "35 Years of 'Scientific Discovery: Computational Explorations of the Creative Processes' – From the Early Days to the State of the Art". Abstracts of the invited talks are included in the front matter of these proceedings. Besides the presentation of regular and short papers in the main program, the conference offered a session titled "Late Breaking Contributions" featuring poster and spotlight presentations of very recent research results on topics related to discovery science.

We are grateful to Springer for their continued long-term support. Springer publishes the conference proceedings, as well as a regular special issue of the Machine Learning journal on discovery science. The latter offers authors a chance to publish significantly extended and reworked versions of their DS conference papers in this prestigious journal, while being open to all submissions on DS conference topics.

This year, Springer (LNCS) supported the best student paper award. For DS 2022, the awardees are Annunziata D'Aversa, Stefano Polimena, Gianvito Pio, and Michelangelo Ceci (for the paper "Leveraging spatio-temporal autocorrelation phenomena to improve the forecasting of the energy consumption in smart grids"). We would like to thank Roberto Interdonato who joined the Program Chairs of the conference for the selection of the best student paper.

On the program side, we would like to thank all the authors of submitted papers and the PC members for their efforts in evaluating the submitted papers, as well as the keynote speakers. On the organization side, we would like to thank all the members of the Organizing Committee, in particular Virginie Feche and Elena Demchenko, for the smooth preparation and organization of all conference associated activities. We are also grateful to the people behind EasyChair for developing the conference organization system that proved to be an essential tool in the paper submission and evaluation process, as well as in the preparation of the Springer proceedings.

The DS 2022 conference was organized under the auspices of several universities and research institutes in Montpellier: the University of Montpellier, the University of Paul Valery, INRAE, Inria, and CIRAD. Significant support, especially through human resources, was also provided by the University of Montpellier and INRAE. Finally, we are indebted to all conference participants, who contributed to making this exciting event a worthwhile endeavor for all involved.

October 2022

Poncelet Pascal
Dino Ienco
Sašo Džeroski

Organization

General Chair

Sašo Džeroski	Jožef Stefan Institute, Slovenia

Program Committee Chairs

Pascal Poncelet	University of Montpellier, France
Dino Ienco	INRAE, France

Publicity Chair

Roberto Interdonato	CIRAD, France

Local Arrangements Chairs

Virginie Feche	University of Montpellier, France
Elena Demchenko	University of Montpellier, France

Program Committee

Reza Akbarinia	Inria, France
Giuseppina Andresini	University of Bari Aldo Moro, Italy
Martin Atzmueller	Osnabrück University, Germany
Elena Battaglia	University of Turin, Italy
Colin Bellinger	National Research Council of Canada, Canada
Robert Bossy	INRAE, France
Alberto Cano	Virginia Commonwealth University, USA
Michelangelo Ceci	University of Bari Aldo Moro, Italy
Mattia Cerrato	University of Mainz, Germany
Bruno Cremilleux	Universite de Caen Normandie, France
Wouter Duivesteijn	Eindhoven University of Technology, The Netherlands
Sašo Džeroski	Jožef Stefan Institute, Slovenia
Johannes Fürnkranz	Johannes Kepler University Linz, Austria
Edith Gabriel	INRAE, France
Sabrina Gaito	University of Milan, Italy
Dragan Gamberger	Rudjer Boskovic Institute, Croatia

Francesco Gullo	UniCredit, Italy
Nadine Hilgert	INRAE, France
Kouichi Hirata	Kyushu Institute of Technology, Japan
Jaakko Hollmén	Aalto University, Finland
Dino Ienco	INRAE, France
Roberto Interdonato	CIRAD, France
Stefan Kramer	Johannes Gutenberg University Mainz, Germany
Vincent Labatut	Université d'Avignon, France
Baptiste Lafabregue	Université de Haute-Alsace, France
Anne Laurent	University of Montpellier, France
Nada Lavrač	Jozef Stefan Institute, Slovenia
Tomislav Lipic	Rudjer Boskovic Institute, Croatia
Gjorgji Madjarov	Ss. Cyril and Methodius University, Macedonia
Giuseppe Manco	ICAR-CNR, Italy
Giuseppe Mangioni	University of Catania, Italy
Bruno Martins	University of Lisbon, Portugal
Elio Masciari	University of Naples Federico II, Italy
Florent Masseglia	Inria, France
Anna Monreale	University of Pisa, Italy
Nisrine Mouhrim	Sidi Mohamed Ben Abdellah University, Morocco
Tsuyoshi Murata	Tokyo Institute of Technology, Japan
Claire Nédellec	INRAE, France
Pance Panov	Jozef Stefan Institute, Slovenia
Ruggero G. Pensa	University of Turin, Italy
Bernhard Pfahringer	University of Waikato, New Zealand
Pascal Poncelet	University of Montpellier, France
Jan Ramon	Inria, France
Chedy Raïssi	Inria, France
Mathieu Roche	CIRAD, France
Arnaud Sallaberry	Université Paul-Valéry Montpellier 3, France
Maximilien Servajean	Université Paul-Valéry Montpellier 3, France
Tomislav Smuc	Rudjer Boskovic Institute, Croatia
Marina Sokolova	University of Ottawa and Institute for Big Data Analytics, Canada
Arnaud Soulet	Université de Tours, France
Andrea Tagarelli	University of Calabria, Italy
Alberto Tonda	INRAE, France
Davide Vega	Uppsala University, Sweden
Herna Viktor	University of Ottawa, Canada
Matteo Zignani	University of Milan, Italy
Albrecht Zimmermann	Universite de Caen Normandie, France

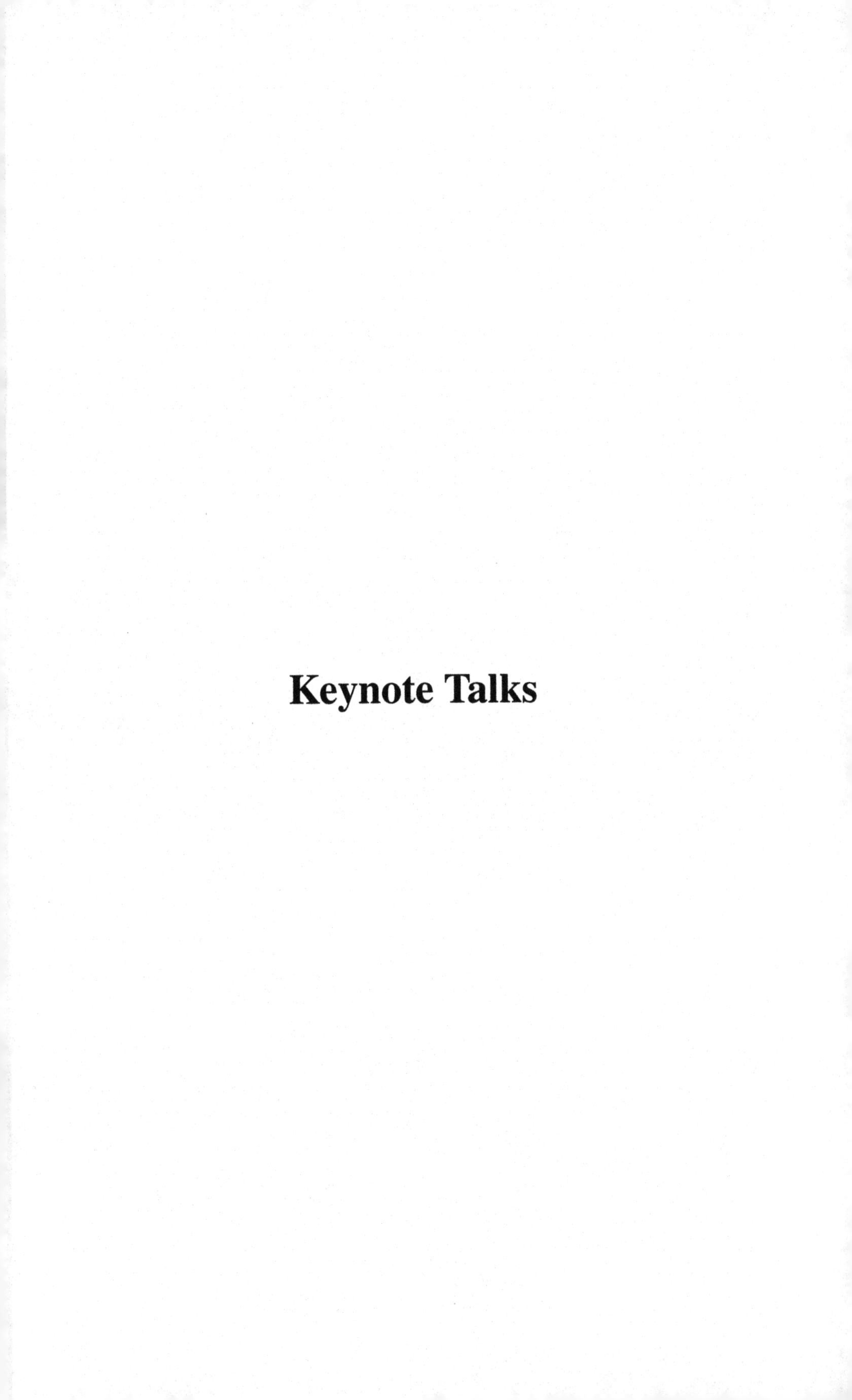

Keynote Talks

Unsupervised Model Selection in Outlier Detection: The Elephant in the Room

Leman Akoglu

Carmegie Mellon University, USA

Outlier mining has a large literature containing numerous detection algorithms, as it finds high-stakes applications in numerous domains including finance, cybersecurity, surveillance, to name a few. However, given a new detection task, it is unclear how to choose an algorithm to use, nor how to set its hyperparameter(s) (HPs) in unsupervised settings. HP tuning is an ever-growing problem with the arrival of many new deep learning based outlier detectors. While they have appealing properties such as task-driven representation learning and end-to-end optimization, deep models come with a long list of HPs. Surprisingly, the issue of model selection in the outlier mining literature has been “the elephant in the room”; a significant factor in unlocking the utmost potential of (deep) models, yet little said or done to systematically tackle the issue.

In this talk, I will first quantitatively demonstrate the HP sensitivity of deep outlier detectors from various families based on a large-scale evaluation study. Next I will present a couple of new directions we have taken to tackle the unsupervised outlier model selection problem, including meta-learning based solutions that transfer “experience” from historical tasks for model selection for a new task, and scalable hyper-ensemble modeling that fully bypasses/obviates model selection.

Coloring Social Relationships

Luca Maria Aiello

IT University of Copenhagen, Denmark

Social relationships are the key determinant of crucial societal outcomes, including diffusion of innovation, productivity, happiness, and life expectancy. To better attain such outcomes at scale, it is therefore paramount to have technologies that can effectively capture the type of social relationships from digital data. NLP researchers have tried to do so from conversational text but mostly focusing on sentiment or topic mining, techniques that fall short on either conciseness or exhaustiveness. We propose a theoretical model of 10 dimensions (colors) of social relationships that is backed by decades of research in social sciences and that captures most of the common relationship types. We trained a deep-learning model to classify text along these ten dimensions, and we reached performance up to 0.98 AUC. By applying this tool on large-scale conversational data, we show that the combination of the predicted dimensions suggests both the types of relationships people entertain and the types of real-world communities they shape. We believe that the ability of capturing interpretable social dimensions from language using AI will help in closing the gap between the oversimplified social constructs that existing social network analysis methods can measure and the multifaceted understanding of social dynamics that has been developed by decades of theoretical research.

35 Years of 'Scientific Discovery: Computational Explorations of the Creative Processes' – From the Early Days to the State of the Art

Stefan Kramer

University of Mainz, Germany

It was 35 years ago, in April 1987, when the first book on computational models of scientific discovery was published: "Scientific Discovery: Computational Explorations of the Creative Processes" by Pat Langley, Herbert Simon, Gary Bradshaw, and Jan Zytkow contained a comprehensive account of systems for discovering quantitive empirical laws as well the discovery of qualitative and structural models, and marked an important milestone in a new branch of AI. Since then, methods for equation discovery, symbolic regression and the automation of science have been developed and refined, with many interesting problems remaining. Currently, deep neural networks (DNNs), representation learning, explainable AI (XAI), graph neural networks (GNNs), and many other technical innovations are bringing new elements into the field. At the same time, progress in the natural and life sciences is increasingly made by (and often requires) methods from AI and ML to produce models with high predictive and explanatory power. In the talk, I will review progress in the field, applications from the natural and life sciences as well as a new test environment, with many options for extensions, that frames machine discovery as a reinforcement learning problem.

Contents

Regression and Limited Data

Model Optimization in Imbalanced Regression

Aníbal Silva[1(✉)], Rita P. Ribeiro[1,2], and Nuno Moniz[1,2]

[1] Faculty of Sciences - University of Porto, Porto, Portugal
up201008538@up.pt, rpribeiro@fc.up.pt
[2] INESC TEC, Porto, Portugal
nmmoniz@inesctec.pt

Abstract. Imbalanced domain learning aims to produce accurate models in predicting instances that, though underrepresented, are of utmost importance for the domain. Research in this field has been mainly focused on classification tasks. Comparatively, the number of studies carried out in the context of regression tasks is negligible. One of the main reasons for this is the lack of loss functions capable of focusing on minimizing the errors of extreme (rare) values. Recently, an evaluation metric was introduced: Squared Error Relevance Area ($SERA$). This metric posits a bigger emphasis on the errors committed at extreme values while also accounting for the performance in the overall target variable domain, thus preventing severe bias. However, its effectiveness as an optimization metric is unknown. In this paper, our goal is to study the impacts of using $SERA$ as an optimization criterion in imbalanced regression tasks. Using gradient boosting algorithms as proof of concept, we perform an experimental study with 36 data sets of different domains and sizes. Results show that models that used $SERA$ as an objective function are practically better than the models produced by their respective standard boosting algorithms at the prediction of extreme values. This confirms that $SERA$ can be embedded as a loss function into optimization-based learning algorithms for imbalanced regression scenarios.

Keywords: Imbalanced regression · Asymmetric loss functions · Model optimization · Boosting

1 Introduction

Supervised learning assumes there is an unknown function mapping a set of independent to one or more dependent variables. Learning algorithms aim to approximate such an unknown function through optimization processes. A key decision rests on choosing which preference criterion, e.g. a loss function, should be used. Such a decision entails critical definitions and assumptions on what should be considered a successful approximation. Most importantly, we should stress that, these decisions are commonly aimed at minimizing the overall error across the entire domain of the target (dependent) variable of a given data set.

P. Pascal and D. Ienco (Eds.): DS 2022, LNAI 13601, pp. 3–21, 2022.
https://doi.org/10.1007/978-3-031-18840-4_1

By assuming that all values are equally important, traditional optimization processes tend to produce models that have a particular focus on the most common values of the target variable. This is not the goal of many real-world applications that configure imbalanced domain learning tasks.

In imbalanced learning, the following holds: *i)* the target variable has a non-uniform or skewed distribution; *ii)* the values across the domain of the target variable are not equally important; and *iii)* the focus is on the rare cases, i.e. values that are poorly represented in the data set. Examples of this type of predictive task spread from classification to regression. They include multiple real-world applications in different areas, such as finance, where the user might be interested in fraud detection, and environmental sciences, to mitigate the occurrence of natural catastrophes, such as floods and hurricanes.

Focusing on Imbalanced Regression, several challenges impose the non-triviality of predicting extreme values. From a supervised learning perspective, these include two main ones: 1) the definition of suitable and non-uniform preferences over a continuous and possibly infinite domain of the target variable; 2) map such preference regarding the extreme values into an evaluation metric that would adequately allow model selection and, possibly, optimization. Regarding the first challenge, a proposal [22, 26] exists that suggests a mapping of the target variable domain into a well-defined space (the relevance space), which gives information about the relevance of a given instance based on its target value. As for the second challenge, while there are some proposals for specially tailored evaluation metrics [8, 13, 27] in an imbalanced regression scenario, only a very few works exist on including such metrics in the optimization process. We focus on this second challenge. In particular, we build on recent work that introduced the Squared Error Relevance Area ($SERA$) [22] metric. This metric allows for errors of equal magnitude to have different impacts depending on the relevance of the target values. Moreover, while it focuses on errors in cases with extreme target values, it also accounts for the errors committed across all the rest of the target values, preventing a severe bias towards the extreme values. However, despite its demonstrated interest in model selection tasks, it is unclear if it is possible to use it directly in optimization processes.

In this work, our main contribution is to show that $SERA$ can be used as an optimization loss function in machine learning algorithms, with the ability to generalize its predictive power for both average and extreme target value instances. Our demonstration efforts consist of empirical evaluation using gradient boosting algorithms and a test bed of 36 data sets. Results show us that, overall, $SERA$ can be used as an optimization loss function. In addition, when these models optimized with $SERA$ under-perform w.r.t. models optimized via standard loss functions (e.g. MSE), the former still have the ability to outperform on extreme values, opening horizons to a broader set of applications in the realm of Imbalance Regression (e.g. Deep Learning).

The paper is organized as follows. In Sect. 2, we provide a review of recent related work regarding imbalanced domain learning. In Sect. 3, we formulate the problem of Imbalanced Regression, introducing $SERA$, the loss function that we will use to optimize our models. In addition, we also provide the details needed to embed this loss function in gradient boosting models, which will be our baselines. In Sect. 4, we demonstrate how *SERA* can be integrated as a custom optimization metric. In Sect. 5, we provide the experimental study and discuss the obtained results. Finally, in Sect. 6 we conclude our work with further research directions.

2 Related Work

The study of imbalanced learning has been advocated over the years, as it poses well-known challenges to standard predictive learning tasks [2]. There are three main strategies to cope with imbalanced domain learning problems: data-level, algorithm-level and hybrid.

Data-level approaches are the most common ones. They allow for any standard machine learning to be used, as they act in a pre-processing stage by changing the data distribution to reduce the imbalance. Generally speaking, we can group them into under-sampling, over-sampling, generation of synthetic examples, or their combination. Even though far more data-level methods have been proposed for classification, few exist for regression (e.g. [3]). An adaptation of the Synthetic Minority Over-sampling Technique (SMOTE) [4], initially proposed for classification, has been made for regression and named SMOTEr [25]. More recently, in the context of Deep Learning, a method to deal with missing data in an imbalanced regression domain was proposed in [28]: Deep Imbalanced Regression (DIR). The specificity of this method lies in the fact that there may be missing values close to a high (low)-representative neighborhood in the target variable distribution. The distribution of the target variable is smoothed across the entire domain, considering a similarity kernel based on statistical properties of the data to estimate missing values. However, there is a caveat from these approaches – they add artificial instances that may not represent the reality with which we are faced or remove common cases that can represent a crucial discriminating aspect for the predictive task. Moreover, the models are not specifically optimized toward predicting those rare cases. Thus, it is not easy to assess whether the change made to the data distribution would effectively map to the intended predictive focus [18].

Regarding algorithm-level approaches, one of the most popular methods is cost-sensitive learning [10,11,24]. These methods use costs to emphasize/relax errors committed by predictions at specific target values. An error committed at a rare or extreme value in imbalanced domains should have a higher cost than a common value. A problem linked to these methods is that assessing the exact cost of a given error is highly domain-dependent and not straightforward [10]. Other methods include the optimization of an asymmetric loss function in standard learning algorithms [9,21]. In the context of regression, few contributions have been made regarding optimization techniques. In [9], the authors focused on the prediction of extreme values by defining a branched asymmetric loss function in the residual space, using the Gradient Boosting algorithm as a training model. Here, the loss function branches into a quadratic function - for values around the mean, so-called normal; and exponential function - for extreme values. However, this loss function has the caveat of a precise definition of normality and thus the requirement of a pre-defined threshold defined in the residual space.

Recently, an ensemble model that consists of embedding SMOTE in several variations of the AdaBoost algorithm was proposed, both in classification [5], and regression tasks [19]. There is, however, a caveat to these models, as mentioned before - they add artificial instances that may not represent the reality with which we are faced. This is especially critical when we are tackling ecological or health domains, where there is no guarantee that the generated instances may be valid observations.

In standard regression tasks, a given model's quality or predictive power is typically assessed by metrics such as the Mean Squared Error (MSE) or the Mean Absolute Error (MAE). These metrics have one property in common: the importance attributed to each observation is uniform, which is not adequate if we are facing a problem of imbalanced regression. Several metrics were proposed for regression with the same goal of assigning uneven importance to instances. The Linear Exponential (LINEX) [27] loss function, which, controlled by a parameter, differentiates over and underestimations. The Relevance-Weighted Root Mean Squared Error ($RW\text{-}RMSE$) [17], a modified version of $RMSE$ which takes into the account the relevance of a given observation. This metric has the caveat of neglecting values which have a low relevance. The utility-based F-measure F^u_β [26], is a function that depends on variations of both the well-known precision and recall, implemented in the context of regression. It relies on the values of relevance and utility assigned to a prediction for a given true value. Nevertheless, both values depend on a given threshold defined in the relevance and utility space. All the metrics mentioned above share the same limitation: they are threshold dependent.

In this work, we follow the same principle as in [9], but with another loss function - Squared Error Relevance Area ($SERA$), a metric recently presented in the context of imbalanced regression [22]. Using the definition of relevance associated with the target variable domain, this metric explicitly gives the notion of asymmetry regarding the loss in different ranges of the target variable. This metric is not dependent on any threshold and thus does not face the problems referred to in the above metrics. Due to such characteristics, it presents the best option for exploring the possibility of model optimization in imbalanced regression tasks.

3 Imbalanced Regression

Consider $\mathcal{D}$ a training set defined as $\mathcal{D} = \{\langle \boldsymbol{x}_i, y_i \rangle\}_{i=1}^N$, where $\boldsymbol{x}_i$ is a feature vector of the feature space $\mathcal{X}$ composed by m independent variables and y_i an instance of the feature space $\mathcal{Y}$ that depends on the feature space $\mathcal{X}$. In a supervised learning setting, our aim is to find the function f that maps the feature space $\mathcal{X}$ onto $\mathcal{Y}$, $f : \mathcal{X} \rightarrow \mathcal{Y}$. Depending on the nature of $\mathcal{Y}$, we can face a classification (if $\mathcal{Y}$ is discrete) or a regression problem (if $\mathcal{Y}$ is continuous). To obtain the best approximation function of f, h, the standard approach in supervised learning is to consider a loss function $\mathcal{L}$, responsible for the optimization of a set of parameters $\boldsymbol{\Theta}$ which tune a model to extract predictions that better describes new instances from the feature space $\mathcal{Y}$.

Here, we will focus on the problem of imbalanced regression, i.e., when the target variable $\mathcal{Y} \in \mathbb{R}$ presents a skewed distribution and the most important values for the prediction task are extreme (rare) values. The most commonly used loss function in regression is the Mean Squared Error (MSE). However, this metric is not adequate for our prediction task. The constant which minimizes MSE is the mean of the target variable, $\bar{y}$, which is counter intuitive for our predictive focus: the extreme values. In an imbalanced regression scenario, an appropriate loss function should search the parameter space $\boldsymbol{\Theta}$ such that it encompasses a good predictive power for both common (around the mean) and uncommon (extremes) instances of our target domain $\mathcal{Y}$. However, this is not a trivial task to accomplish.

3.1 Relevance Function

In this study, we define an extreme value based on the notion of relevance introduced by [26]. The authors define a relevance function $\phi : \mathcal{Y} \rightarrow [0, 1]$ as a continuous function that expresses the application-specific bias concerning the target variable domain $\mathcal{Y}$ by mapping it into a $[0, 1]$ scale of relevance, where 0 and 1 represent the minimum and maximum relevance, respectively. With the assumption that extreme values are the values of interest, authors have also proposed a method that automatically constructs the $\phi(.)$ function. It achieves that by interpolating a set of control points provided by the adjusted boxplot, a non-parametric modification to Tukey's boxplot, proposed by [14]. In particular, this method uses the median and the whiskers as the set of key points to interpolate. In Fig. 1 we depict the result of the automatic mapping of the adjusted boxplot to the relevance space, as introduced by [26], for three different scenarios based on the type of extremes indicated to be of interest: low, high or both (default).

3.2 Squared Error Relevance Area (SERA)

Once we have the domain of the target variable mapped into a relevance space, we now present the asymmetric loss function $SERA$. We will use $SERA$ to improve the predictive power of extreme values in Gradient Boosting algorithms, emphasizing that any model which relies on optimization in the parameter space $\boldsymbol{\Theta}$ could be used to perform this improvement.

Let $\mathcal{D} = \{\langle \boldsymbol{x}_i, y_i \rangle\}_{i=1}^{N}$ be a data set and $\phi : \mathcal{Y} \rightarrow \{0, 1\}$ a relevance function defined for the target variable Y. Considering the subset $\mathcal{D}^t \subseteq \mathcal{D}$ of instances such that $\mathcal{D}^t = \{\langle \boldsymbol{x}_i, y_i \rangle \in \mathcal{D} \,|\, \phi(y_i) \geq t\}$, we can define a Squared Error-Relevance (SER_t) to estimate the error of a given model with respect to a given cutoff t as

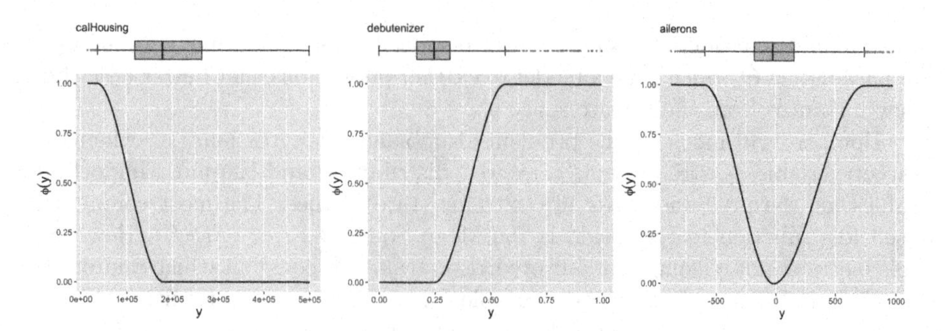

Fig. 1. The adjusted boxplot (top) of a target variable y and its automatically inferred relevance function $\phi(y)$ (bottom) for three data sets with different type of extreme values: low (left), high (middle) and both (right).

$$SER_t = \sum_{y_i \in \mathcal{D}^t} (\hat{y}_i - y_i)^2 \tag{1}$$

This is the sum of squared errors for all the instances such that the relevance of the target value is bounded by a given threshold t. Since this metric only depends on instances such that $\phi(y_i) \geq t$, we will have that, for any given $\delta \in \mathbb{R}^+$, s.t. $t + \delta \leq 1$: $SER_{t+\delta} \leq SER_t$. Finally, its maximum and minimum value are ascertained when $t = 0$ and $t = 1$, respectively.

In the same work, the authors took a step further and integrated this estimate w.r.t. all possible cutoff values (i.e., between 0 and 1). There, they defined this area as the Squared Error Relevance Area ($SERA$), and it is given by

$$SERA = \int_0^1 SER_t \, dt = \int_0^1 \sum_{y_i \in \mathcal{D}^t} (\hat{y}_i - y_i)^2 \, dt \tag{2}$$

This area has some important properties that can help us better understand the performance of models in an imbalanced regression setting. First, it encompasses all the possible relevance thresholds constrained in the definition of SER_t, removing the need to explicitly define a threshold. Secondly, it is a decreasing and monotonic function. Note that we are integrating along with all possible relevance values. Since we know by definition that $\mathcal{D}^{t+\delta} \subseteq \mathcal{D}^t$, the higher the relevance threshold is, the lower will be the number of instances considered, but also more relevant. Thus, on the one hand, values of SER_t which have a high relevance will have a greater contribution to this area when compared with instances where the relevance is small. The squared errors of these latter instances are accounted less times for $SERA$, when compared to the high relevance instances. On the other hand, the area will be smaller at points where the relevance is high (since we are only considering the observations that have a high relevance). In this sense, we are explicitly penalizing high relevance errors, which are usually harder to optimize while keeping the entire data domain. Finally, since this metric is built by integrating SER_t over all the relevance domain, which is convex, convexity is also preserved. Also, this metric must be differentiable. By the same token, given that SER_t is differentiable, so is $SERA$.

4 Optimization Loss Function for Imbalanced Regression

We aim to study the possibility and impact of embedding $SERA$ as an optimization loss function in supervised learning algorithms. Its use in already implemented learning algorithms only requires the proposal of a custom loss function, where the only thing we need to provide is the first and second-order derivatives.

The first-order derivative of $SERA$ is obtained by evaluating the first derivative w.r.t. a given prediction $\hat{y}_j$, as follows

$$\begin{aligned} \frac{\partial SERA}{\partial \hat{y}_j} &= \frac{\partial}{\partial \hat{y}_j} \int_0^1 \sum_{y_i \in \mathcal{D}^t} (\hat{y}_i - y_i)^2 \, dt \\ &= 2 \int_0^1 \sum_{y_i \in \mathcal{D}^t} (\hat{y}_i - y_i) \, \delta_{ij} \, dt \end{aligned} \tag{3}$$

where δ_{ij} is the Kronecker's delta, which takes value of 1 if $i = j$ and 0 otherwise. Since we need to take into the account all the possible relevance values a given observation is encompassed in, we can write the expression above as

$$\frac{\partial SERA}{\partial \hat{y}_j} = 2 \int_0^1 (\hat{y}_j - y_j) \Big|_{y_j \in \mathcal{D}^t} dt \tag{4}$$

The second derivative w.r.t. a given prediction $\hat{y}_j$ is obtained by

$$\frac{\partial^2 SERA}{\partial \hat{y}_j^2} = 2 \int_0^1 \mathbf{1}(y_j \in \mathcal{D}^t) \, dt \tag{5}$$

where $\mathbf{1}(.)$ is an indicator which takes the value of 1 if the argument holds, and 0 otherwise.

In this paper, we will drive our efforts using Gradient Boosting algorithms, namely two well-known variants of Gradient Boosting Regression Trees (GBRT) [12], XGBoost [6] and LGBM [15]. For that, we resort to the implementations found in **R** [20] from the packages `xgboost` [7] and `lightgbm` [23], respectively.

To approximate the values of the two derivatives, we resort to the trapezoidal rule with a uniform grid of T equally spaced intervals between $[0, 1]$. In this approximation, we can expand the summations and deduce the following expressions (see Appendix A for all derivation steps).

$$\begin{aligned} \frac{\partial SERA}{\partial \hat{y}_j} &\approx \frac{1}{T} \left(1 + 2n_j + \mathbf{1}\left(y_j \in \mathcal{D}^{t_T}\right)\right) (\hat{y}_j - y_j), \\ \frac{\partial^2 SERA}{\partial \hat{y}_j^2} &\approx \frac{1}{T} \left(1 + 2n_j + \mathbf{1}\left(y_j \in \mathcal{D}^{t_T}\right)\right) \end{aligned} \tag{6}$$

where $n_j \in [1, T-1]$ is the number of times the instance y_j contributes to $SERA$ derivative. From this, we can infer that, for a given prediction, the first and second-order derivatives will be greater (assuming a greater error for extreme values) as the relevance increases, as there will be a higher contribution from n_j. These derivatives, in addition with $SERA$, were implemented in **R** [20].

From now on, we will designate XGBoost and LGBM models optimized with $SERA$ as XGBoost^{S}and LGBM^S, while models optimized with MSE XGBoost^{M}and LGBM^M, respectively.

Computational Complexity. Another important aspect is the computational complexity introduced by $SERA$. The trapezoidal rule has a computational complexity of $O(T)$, where T is the number of steps taken to discretize an integral. SER_t has a computational complexity of $O(|\mathcal{D}^t|)$, where $|\mathcal{D}^t|$ is the number of instances with relevance higher or equal to a given threshold t. $SERA$ will consider $|\mathcal{D}^{t_0}| + |\mathcal{D}^{t_1}| + ... + |\mathcal{D}^{t_T}|$ instances for all the T steps of the Riemann's sum. In the worst-case scenario, all the target values have a constant and maximum relevance equal to 1. In that case, $|\mathcal{D}|$ is the number of instances accounted for all steps. Thus, $SERA$ will have a computational complexity of $O(T \times |\mathcal{D}|)$. Regarding the computational complexity introduced in XGBoost and LGBM, we only need to take into consideration the additional complexity of the first and second order derivatives. Thus, using the approximation found in Appendix A, the computational complexity will be again $O(T \times |\mathcal{D}|)$.

5 Experimental Study

Our goal is to answer the research question (**Q**) that motivated this work: can $SERA$ be used as an optimization loss function to reduce errors for both extreme and common values?

In this section, we take into consideration a group of 36 regression data sets from several domains with an imbalanced distribution on the target variable. Given these data sets and the models described in Sect. 4, we will start by describing our experimental setup in Sect. 5.1. Namely, the considered data sets with an imbalanced domain, and the grid-search procedure for parameter tuning of models. In Sect. 5.2 we refer to the Bayes Sign Test used to assess the statistical significance of the results. Given the best parameters for each model, we present and discuss the obtained results for all data sets in Sect. 5.3.

5.1 Experimental Setup

To study the effects of using $SERA$ as an optimization loss function, a wide range of data sets from several domains in the context of imbalanced regression is used. These data sets, with their respective main properties, are presented in Table 1. From them, we extracted the number of instances $|\mathcal{D}|$, the number of nominal (Nom) and numerical (Num) variables. In addition, and to give a notion of the imbalance present in the target variable, we resort to the automatic method proposed in [22] that is based on the adjusted boxplot. We calculated the number of instances such that $\phi(y) = 1$, as representative of the number of extreme (rare) target value instances ($|\mathcal{D}_R|$) and the Imbalance Ratio (IR), calculated as $|\mathcal{D}_R|/|\mathcal{D}| \times 100\%$. Finally, we also include the type of imbalance for each target variable as follows: if the adjusted boxplot only presents outliers below or above the respective fence, the type of extremes is low (L) or high (H), respectively, while if it presents outliers below and above the fences, the type is both (B).

Table 1. Data sets description: $|\mathcal{D}|$ — nr of instances, **Nom** — nr. of nominal attributes, **Num** — nr. of numeric attributes, $|\mathcal{D}_R|$ — nr. of extreme (rare) instances, i.e. $\phi(y) = 1$ IR — imbalance ratio and Type — type of extremes.

Id	Dataset	$\|\mathcal{D}\|$	Nom	Num	$\|\mathcal{D}_R\|$	IR	Type
1	diabetes	35	0	3	4	12.90	H
2	triazines	151	0	61	4	2.72	B
3	a7	160	3	9	7	4.58	H
4	autoPrice	165	10	16	3	1.85	L
5	elecLen1	399	0	3	4	1.01	H
6	housingBoston	407	0	14	40	10.90	B
7	forestFires	416	0	13	7	1.71	H
8	wages	429	7	4	1	0.23	B
9	strikes	501	0	7	1	0.20	H
10	mortgage	841	0	16	60	7.68	L
11	treasury	841	0	16	79	10.37	L
12	musicorigin	848	0	118	15	1.80	B
13	airfoild	1203	0	6	11	0.92	H
14	acceleration	1387	3	12	30	2.21	B
15	fuelConsumption	1413	12	26	27	1.95	B
16	availablePower	1443	7	9	75	5.48	B
17	maxTorque	1442	13	20	43	3.07	B
18	debutenizer	1918	0	8	90	4.92	H

Id	Data set	$\|\mathcal{D}\|$	Nom	Num	$\|\mathcal{D}^R\|$	IR	Type
19	space_ga	2487	0	7	21	0.85	B
20	pollen	3080	0	5	32	1.05	B
21	abalone	3343	1	8	374	12.60	B
22	wine	5199	0	12	1022	24.47	H
23	deltaAilerons	5705	0	6	528	10.20	B
24	heat	5922	3	9	39	0.66	B
25	cpuAct	6555	0	22	227	3.59	L
26	kinematics8fh	6556	0	9	50	0.77	B
27	kinematics32fh	6556	0	33	53	0.82	B
28	pumaRobot	6556	0	33	91	1.41	B
29	deltaElevation	7615	0	7	1802	31	H
30	sulfur	8065	0	6	606	8.12	B
31	ailerons	11003	0	41	186	1.72	B
32	elevators	13280	0	18	1598	13.68	B
33	calHousing	16513	0	9	23	0.14	L
34	house8H	18229	0	9	305	1.70	B
35	house16H	18229	0	17	303	1.69	B
36	onlineNewsPopRegr	31716	0	60	2879	9.98	B

To assess the effectiveness of each model, we performed a random partition for each data set, where 80% will be used to tune the models while the remaining 20% to make predictions under the best model configuration found in a given data set. To tune the parameters for each model, we will use a grid-search approach with a 10-fold stratified cross-validation. We define a workflow of a given algorithm j as the tuple $\boldsymbol{W}^{(j)} = (\boldsymbol{M}_j, \boldsymbol{\Theta}^{(j)}) = \{\mathcal{W}_q^{(j)}\}_{q=1}^{e}$, where e is the number of different workflows considered for a given tuple, $\boldsymbol{M}_j$ denote the algorithm used, $\boldsymbol{\Theta}^{(j)}$ the respective set of parameters, which are described in Table 2.

Table 2. Models parameters considered for grid search.

Model	R Package	Parameters
XGBoost	xgboost [7]	nrounds $= \{250, 500\}$
LGBM	lightgbm [15]	max_depth $= \{3, 5, 7\}$ $\eta = \{10^{-3}, 10^{-2}, 10^{-1}\}$

Given the workflows obtained from the grid-search, we start by providing a methodology to answer the question that motivated this work, **Q**. It consists on the following tasks.

T1: For each data set, and for each model in M, we select the workflow that had the lowest score according to $SERA$. This score is calculated by averaging the results obtained by cross-validation on the 80% partition.
T2: Given the best workflows, we compare them using the Bayes Sign Test [1] (Sect. 5.2). The designation we give to Gradient Boosting models is $\mathcal{W}^M$, in case they are optimized using a standard loss function (MSE) and $\mathcal{W}^S$ in case they are optimized using $SERA$.
T3: Next, we train our best workflows for each data set with the partitioned 80% and, with the remaining 20%, we assess the quality of their predictions (Sect. 5.3). This quality will also be evaluated by plotting $SERA$ curves.

5.2 Results on Model Optimization

With the top workflows from each model obtained by **T1**, we can assess the performance of our models in task **T2**. For that, we resort to the Bayes Sign Test. Briefly, this test compares two models on a multi data set scenario by measuring their score difference (a prior probability) for all data sets, returning a probability measure (the posterior) hinting if a model is practically better than another, or if they are equivalent. This equivalence is measured in a given interval and is defined as the Region Of Practical Equivalence (ROPE) [16]. The prior z_i, where i indicates a given data set, is determined by averaging the normalized difference below for all k-folds

$$z_i = \frac{1}{10}\sum_{k=1}^{10} \frac{\mathcal{L}_k(\mathcal{W}^S) - \mathcal{L}_k(\mathcal{W}^M)}{\mathcal{L}_k(\mathcal{W}^M)} , \tag{7}$$

and taking $\mathcal{L}$ as $SERA$ or MSE. After determining this mean difference for all data sets, we feed into the Bayes Sign Test the vector $\boldsymbol{z}$ and a ROPE between $[-1\%, 1\%]$, returning the posterior probability $p(z)$ that a given model is practically better or equivalent than the other.

The results from this evaluation are depicted in Fig. 2 and provide us with two perspectives according to the considered error metrics. Regarding MSE (left column), the standard models are practically better (with a $p(z)$ of 0.92 for XGBoost^{M}and a $p(z)$ of 0.7 for LGBM^M). Concerning $SERA$ (right column), results tell us that both algorithms with our optimization are practically better (with a $p(z)$ of 0.65 for XGBoost^{S}and a $p(z)$ of 0.76 for LGBM^S).

Thus, from a statistical points of view and in a model optimization scenario, there is a clear trade-off between the standard and our models when assessing their scores with different metrics. This was somewhat expected as these metrics have a different predictive focus as it was already mentioned above. Nevertheless, from this test we are able to infer the ability of $SERA$ as a loss function to lower the errors obtained in a problem of Imbalance Regression. With this, we finish

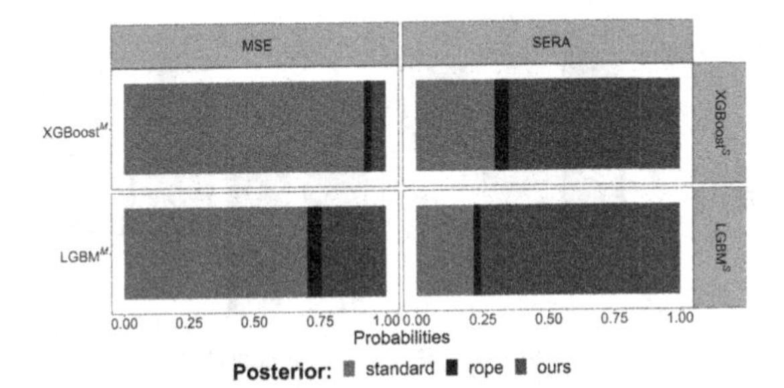

Fig. 2. Comparison between our models optimized with $SERA$, LGBM^{S}and XGBoost^S, against the standard models LGBM^{M}and XGBoost^M. Each color denotes the probability of our implementation (green) or standard (red) being practically better or equivalent (blue) to one another according with the Bayes Sign Test with the ROPE interval $[-1\%, 1\%]$. The left and right column denote the results of the Bayes Sign Test with MSE and $SERA$, respectively. (Color figure online)

our second task **T2** and partially answered our main question **Q**. Next, we aim to show that $SERA$ do in fact improve the predictive power for both common and extreme values in an out-of-sample scenario (i.e., using our test data).

5.3 Results in Out-of-Sample

Using the parameters found in the best workflows obtained for each model and for each data set in task **T1**, we train our models and assess their predictions with the (20%) out-of-sample data. Given these predictions, we calculated $SERA$ and MSE for all the models in the considered data sets (cf. Tables 3 and 4 in Appendix B).

From the obtained results, we initially take into consideration a rank evaluation of our models. For that purpose, and for a given data set, the rank of 1 is assigned to the model which provided the lowest score. Figure 3 depicts the rank distribution for each model over all the considered data sets. In MSE-based ranking (left column), XGBoost^{M}had a median rank of 1, followed by LGBM^{M}with a rank of 2, and finally XGBoost^{S}and LGBM^S, both with a median rank of 3. In $SERA$-based ranking (right column), both XGBoost^{M}and XGBoost^S had a median rank of two, followed by LGBM^{S}with a median rank of 3, and finally LGBM^{M}with a rank of 4. From these ranking distributions, we can take a somewhat expected conclusion: standard models were the top performers when evaluated under MSE, and an unexpected one: under $SERA$, XGBoost^{M}and XGBoost^{S}compete with each other.

Next, we focus on each algorithm independently. While for XGBoost, XGBoost^S had the lowest score in 3 and 18 of the data sets when evaluated by MSE and $SERA$, respectively, for LGBM, LGBM^{S}had a better score in 11 and 26 of the data sets when evaluated by MSE and $SERA$.

Although the standard models had the lowest score in more data sets than ours, it does not mean that, for a given data set, those errors were the lowest

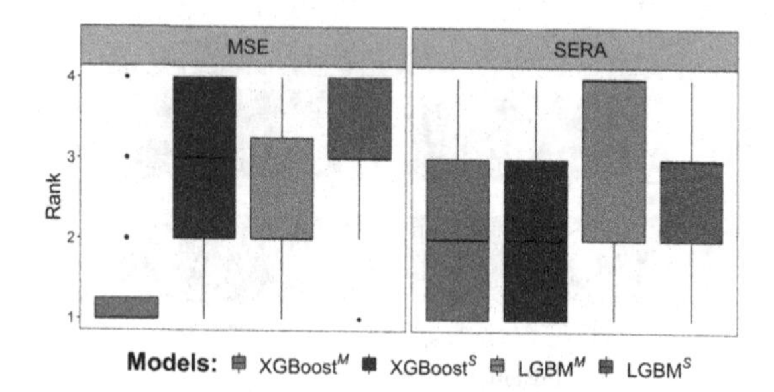

Fig. 3. Rank distribution of models by MSE and $SERA$ results in out-of-sample.

across all the relevance domain. To study this statement, we follow [22] and show $SERA$ curves for six selected data sets. These curves are built by calculating the error SER_t as the relevance threshold t for $\phi(y)$ increases and are shown in Fig. 4.

In addition to $SERA$ curves, we also define a turning point for each data set. That point is the minimum relevance value φ for which a model optimized with $SERA$ has a $SERA$ estimate for all the values with a relevance greater or equal to φ, i.e. $SERA_{\phi(.)\geq\varphi}(\mathcal{M}^S)$, lower than the $SERA$ estimate obtained by the standard model in the same conditions, i.e. for all the values with a relevance greater or equal to φ, i.e. $SERA_{\phi(.)\geq\varphi}(\mathcal{M}^M)$. More formally, and for a specific data set, the turning point is then a threshold ϕ_t obtained by

$$\phi_t = \min\{\varphi \in [0,1] \mid SERA_{\phi(.)\geq\varphi}(\mathcal{M}^S) < SERA_{\phi(.)\geq\varphi}(\mathcal{M}^M)\}. \quad (8)$$

In the curves of Fig. 4, the turning points are represented by dashed lines, and the shadowed regions represent the relevance domain for which the condition above holds.

The plots from the first row of Fig. 4 show us $SERA$ curves where models that were optimized with $SERA$ had the lowest score. As depicted, for house8H data set, XGBoost^{S}had the lowest score in most of the relevance domain (ϕ_t = 0.09). For housingBoston data set, although XGBoost^{S}has the lowest $SERA$, its turning point occurs at a higher threshold ($\phi_t = 0.48$). Finally, we show the curves for musicorigin data set for which LGBM^{S}had the lowest score. Here, the turning point also occurs at a very low relevance ($\phi_t = 0.01$).

Regarding the second row of plots of Fig. 4, we show three data sets where XGBoost^{M}had the lowest $SERA$ score. From these examples, we see that although XGBoost^{M}was the top performer, there are several turning points in the relevance domain for which our models surpassed XGBoost^M ($\phi_t = 0.22$, $\phi_t = 0.23$, $\phi_t = 0.78$ for deltaElevation, space_ga and strikes data sets, respectively).

From this analysis we can conclude that *SERA* can be used as a loss function to reduce errors in both extreme and common values (first row of Fig. 4). Even when models optimized with *SERA* do not provide the best score across the whole relevance domain, they can still perform better for different relevance domains (second row from Fig. 4). With this, we conclude our task **T3** and answered the question that motivated this work **Q**.

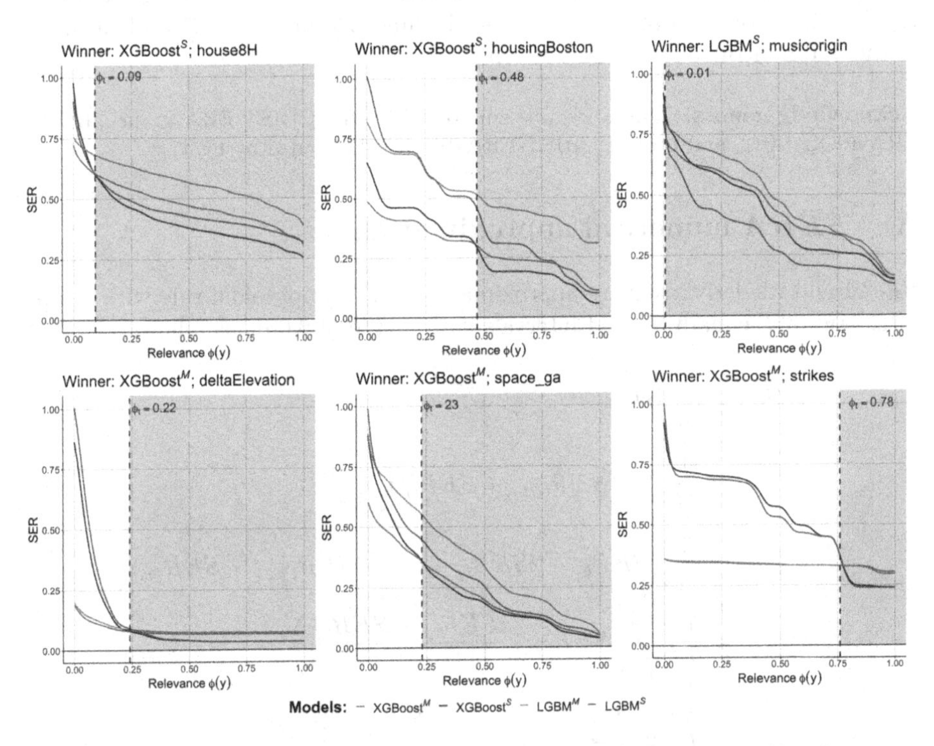

Fig. 4. *SERA* curves for six selected data sets. The first row provides data sets where our models had the lowest error, while the second row provides data sets where XGBoost^M had the lowest error. The highlighted region in each graph depicts the turning point where models optimized with *SERA* started to have a lower error w.r.t. standard models.

6 Conclusions

In this work, we addressed the problem of embedding *SERA* as an optimization loss function to improve the predictive power for both extreme and normal values. We used as our baselines two well-known variations of Gradient Boosting models, XGBoost and LGBM. Results showed that the embedment of *SERA* as an optimization loss function in our algorithms provided the ability of reducing errors for both extreme and normal values. In addition, when underperformed by standard models, our algorithms were able to provide a more accurate prediction in a high relevance domain. For the sake of reproducibility, all the experiments are available in github[1].

Finally, we highlight that hyper-parameter optimization, due to the construction of $SERA$, is not restricted to the considered algorithms, opening the use of this loss function to several domains of Machine Learning (e.g. Deep Learning) in the realm of Imbalanced Regression.

Acknowledgements. This work was supported by the CHIST-ERA grant CHIST-ERA-19-XAI-012, and project CHIST-ERA/0004/2019 funded by FCT.

A *SERA* numerical approximation

$SERA$ and its derivatives are approximated by the trapezoidal rule with a uniform grid of T equally spaced intervals with a step of 0.001, as follows.

$$\begin{aligned}
SERA &= \int_0^1 SER_t \, dt \\
&\approx \frac{1}{2T} \sum_{k=1}^{T} (SER_{t_{k-1}} + SER_{t_k}) \\
&= \frac{1}{2T} (SER_{t_0} + 2SER_{t_1} + ... + 2SER_{t_{T-1}} + SER_{t_T}) \\
&= \frac{1}{T} \left(\sum_{k=1}^{T-1} SER_{t_k} + \frac{SER_{t_0} + SER_{t_T}}{2} \right) \\
&= \frac{1}{T} \left(\frac{1}{2} \sum_{y_i \in \mathcal{D}^{t_0}} (\hat{y}_i - y_i)^2 + \sum_{y_i \in \mathcal{D}^{t_1}} (\hat{y}_i - y_i)^2 + ... \right. \\
&\quad \left. + \sum_{y_i \in \mathcal{D}^{t_{T-1}}} (\hat{y}_i - y_i)^2 + \frac{1}{2} \sum_{y_i \in \mathcal{D}^{t_T}} (\hat{y}_i - y_i)^2 \right)
\end{aligned} \tag{9}$$

[1] https://github.com/anibalsilva1/IRModelOptimization.

Similarly, the derivative of $SERA$ w.r.t. a given prediction $\hat{y}_j$ is obtained by

$$\begin{aligned}
\frac{\partial SERA}{\partial \hat{y}_j} &\approx \frac{1}{T}\frac{\partial}{\partial \hat{y}_j}\Bigg(\frac{1}{2}\sum_{y_i\in\mathcal{D}^{t_0}}(\hat{y}_i-y_i)^2+\sum_{y_i\in\mathcal{D}^{t_1}}(\hat{y}_i-y_i)^2+\ldots\\
&\quad+\sum_{y_i\in\mathcal{D}^{t_{T-1}}}(\hat{y}_i-y_i)^2+\frac{1}{2}\sum_{y_i\in\mathcal{D}^{t_T}}(\hat{y}_i-y_i)^2\Bigg)\\
&=\frac{1}{T}\Bigg(\sum_{y_i\in\mathcal{D}^{t_0}}(\hat{y}_i-y_i)\delta_{ij}+2\sum_{y_i\in\mathcal{D}^{t_1}}(\hat{y}_i-y_i)\delta_{ij}+\ldots\\
&\quad+\sum_{y_i\in\mathcal{D}^{t_{T-1}}}(\hat{y}_i-y_i)\delta_{ij}+\frac{1}{2}\sum_{y_i\in\mathcal{D}^{t_T}}(\hat{y}_i-y_i)\delta_{ij}\Bigg)\\
&=\frac{1}{T}\Bigg((\hat{y}_j-y_j)\Big|_{y_j\in\mathcal{D}^{t_0}}+2(\hat{y}_j-y_j)\Big|_{y_j\in\mathcal{D}^{t_1}}+\ldots\\
&\quad+2(\hat{y}_j-y_j)\Big|_{y_j\in\mathcal{D}^{t_{T-1}}}+(\hat{y}_j-y_j)\Big|_{y_j\in\mathcal{D}^{t_T}}\Bigg)\\
&=\frac{1}{T}\Bigg((\hat{y}_j-y_j)\Big|_{y_j\in\mathcal{D}^{t_0}}+2\sum_{k=1}^{T-1}(\hat{y}_j-y_j)\Big|_{y_j\in\mathcal{D}^{t_k}}+(\hat{y}_j-y_j)\Big|_{y_j\in\mathcal{D}^{t_T}}\Bigg)\\
&=\frac{1}{T}\Bigg(\mathbf{1}\left(y_j\in\mathcal{D}^{t_0}\right)+2\sum_{k=1}^{T-1}\mathbf{1}\left(y_j\in\mathcal{D}^{t_k}\right)+\mathbf{1}\left(y_j\in\mathcal{D}^{t_T}\right)\Bigg)(\hat{y}_j-y_j)
\end{aligned} \tag{10}$$

Note that any given instance y_j will always have at least zero relevance, i.e. $\phi(y_j)\geq 0$, so the first term of Eq. (10) will always be taken into account. Nevertheless, not all the summation terms will be considered for cases where $\phi(y_j)<1$. With this in mind, we define

$$n_j=\sum_{k=1}^{K_j}\mathbf{1}\left(y_j\in\mathcal{D}^{t_k}\right) \tag{11}$$

where n_j is the number of times the instance y_j contributes to $SERA$ derivative, where $K_j\in[1,T-1]$. Equation (10) becomes then

$$\frac{\partial SERA}{\partial \hat{y}_j}\approx\frac{1}{T}\left(1+2n_j+\mathbf{1}\left(y_j\in\mathcal{D}^{t_T}\right)\right)(\hat{y}_j-y_j) \tag{12}$$

In this context, the second derivative for a given prediction $\hat{y}_j$ is obtained by

$$\frac{\partial^2 SERA}{\partial \hat{y}_j^2}\approx\frac{1}{T}\left(1+2n_j+\mathbf{1}\left(y_j\in\mathcal{D}^{t_T}\right)\right) \tag{13}$$

We now proceed to a study on the degree of error committed by using the approximations above (Eqs. (12), (13)) against the use of the trapezoidal rule directly on Eqs. (4) and (5). For that, we resort to the predictions obtained for XGBoost^S. The rationale is the following: 1) for each data set, we compute the first and second derivatives using both methods; 2) we calculate the absolute difference between the results obtained from both methods; 3) we average these differences over all instances.

The results obtained using this evaluation are depicted in the left box plot of Fig. 5.

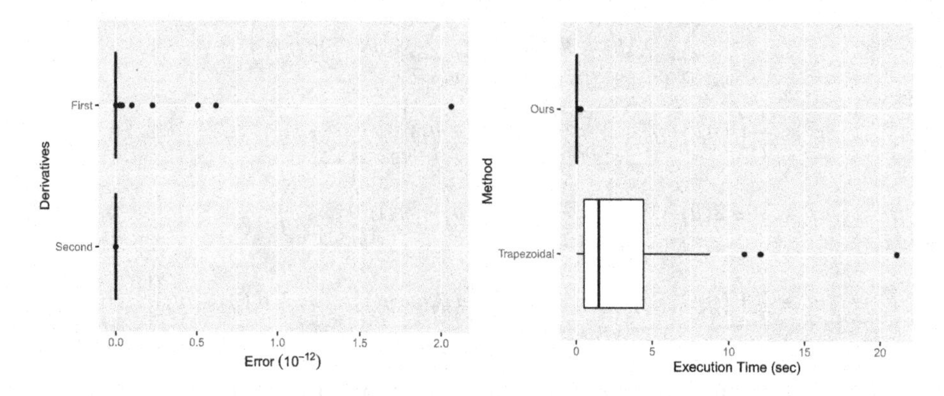

Fig. 5. Left: Absolute error differences for the first and second derivatives between our approximations and the trapezoidal rule. **Right**: Execution time (in seconds) of the first and second derivative for a given data set, under our approximation and the trapezoidal rule.

Results show that both first and second derivative approximations have a minor error ($\approx 10^{-12}$). Sim On the right box plot of Fig. 5, we also show the difference in execution time using both methods for each data set. Here, the execution time is measured as the time taken to evaluate both the first and second derivatives. As we can see, there is a non-negligible difference between our approximation and the trapezoidal rule as the size of a data set increases.

B Tables of Results

In this section, we report the MSE and $SERA$ results obtained in out-of-sample of each data set.

Table 3. *MSE* results in out-of-sample, with the best models per data set in bold. Model (#wins): XGBoost^M (27), XGBoost^S (2), LGBM^M (6), LGBM^S (1).

Id	XGBoost^S	XGBoost^M	LGBM^S	LGBM^M
1	6.41e-01	**3.36e-01**	5.56e-01	5.15e-01
2	**1.00e-02**	1.41e-02	1.42e-02	1.55e-02
3	4.60e+01	4.76e+01	4.67e+01	**4.56e+01**
4	8.64e+07	2.39e+07	6.94e+07	**1.60e+07**
5	9.81e+05	**3.34e+05**	9.18e+05	3.75e+05
6	1.44e+01	**1.08e+01**	2.25e+01	1.84e+01
7	1.68e+04	4.62e+03	**1.82e+02**	4.29e+02
8	2.30e+01	1.78e+01	2.36e+01	**1.69e+01**
9	1.23e+06	4.88e+05	1.32e+06	**4.80e+05**
10	1.98e+00	**2.57e-02**	6.84e-01	7.49e-01
11	3.42e+00	**8.19e-02**	1.28e+00	9.15e-01
12	3.46e-01	**3.00e-01**	3.24e-01	3.10e-01
13	1.20e+01	**2.17e+00**	7.59e+01	1.65e+02
14	7.72e-01	**4.87e-01**	2.13e+00	2.14e+00
15	2.37e-01	**1.11e-01**	8.85e-01	7.62e-01
16	5.27e+01	**2.53e+01**	2.23e+02	2.53e+02
17	4.21e+01	**1.33e+01**	7.23e+02	1.02e+03
18	2.24e-02	**4.66e-03**	2.16e-02	8.45e-03
19	1.46e-02	**9.74e-03**	1.59e-02	1.41e-02
20	2.93e+00	**2.35e+00**	2.72e+00	2.45e+00
21	6.66e+00	**4.87e+00**	5.96e+00	5.59e+00
22	2.02e+00	**3.48e-01**	9.53e-01	7.85e-01
23	4.78e-08	2.67e-08	3.36e-08	**2.47e-08**
24	2.06e+00	**6.77e-01**	4.43e+01	4.18e+01
25	2.00e+01	**4.52e+00**	1.25e+02	8.32e+01
26	2.36e-03	**2.00e-03**	1.01e-02	1.06e-02
27	7.12e-02	**6.56e-02**	1.00e-01	1.13e-01
28	8.44e-05	**6.40e-05**	8.02e-05	6.76e-05
29	7.93e-06	2.01e-06	9.01e-06	**1.99e-06**
30	2.07e-03	**1.93e-03**	2.64e-03	2.89e-03
31	3.89e+02	**2.16e+02**	1.37e+03	1.23e+03
32	5.85e-06	**4.37e-06**	1.65e-05	1.39e-05
33	8.39e+09	**3.38e+09**	1.19e+10	1.47e+10
34	1.49e+09	**1.09e+09**	1.38e+09	1.13e+09
35	1.23e+09	**8.63e+08**	1.12e+09	9.48e+08
36	**9.89e+07**	1.01e+08	1.07e+08	9.95e+07

Table 4. *SERA* results in out-of-sample, with the best models per data set in bold. Model (#wins): XGBoost^M (16), XGBoost^S (14), LGBM^S (5), LGBM^M (1).

id	XGBoost^S	XGBoost^M	LGBM^S	LGBM^M
1	3.86e+00	**1.48e+00**	3.36e+00	2.13e+00
2	**1.47e-01**	2.50e-01	2.03e-01	2.66e-01
3	1.70e+03	1.75e+03	**1.58e+03**	1.61e+03
4	1.18e+09	3.63e+08	9.99e+08	**1.87e+08**
5	3.50e+07	**1.34e+07**	3.59e+07	1.76e+07
6	**6.39e+02**	6.47e+02	9.58e+02	1.15e+03
7	5.82e+05	2.27e+05	**4.66e+03**	1.41e+04
8	**7.06e+02**	9.11e+02	9.06e+02	8.94e+02
9	8.70e+07	**5.47e+07**	8.47e+07	5.48e+07
10	1.90e+02	**1.91e+00**	4.43e+01	4.86e+01
11	4.22e+02	**1.12e+01**	1.32e+02	8.08e+01
12	3.03e+01	3.24e+01	**2.32e+01**	3.57e+01
13	1.07e+03	**2.47e+02**	1.05e+04	1.59e+04
14	5.11e+01	**4.35e+01**	2.03e+02	2.34e+02
15	1.63e+01	**1.46e+01**	9.51e+01	9.17e+01
16	**5.77e+03**	6.72e+03	4.26e+04	5.02e+04
17	3.52e+03	3.09e+03	**1.29e+05**	2.12e+05
18	4.71e+00	1.14e+00	**4.94e+00**	2.59e+00
19	2.38e+00	**2.30e+00**	2.90e+00	3.58e+00
20	**4.84e+02**	6.05e+02	5.47e+02	7.81e+02
21	**1.92e+03**	2.26e+03	2.25e+03	2.86e+03
22	**4.30e+01**	1.54e+02	1.70e+02	5.50e+02
23	1.14e-05	1.65e-05	**1.11e-05**	1.62e-05
24	4.05e+02	**3.48e+02**	2.25e+04	2.63e+04
25	1.01e+04	**1.58e+03**	5.71e+04	4.39e+04
26	**9.31e-01**	1.12e+00	4.27e+00	5.02e+00
27	**2.90e+01**	3.82e+01	5.48e+01	7.93e+01
28	**2.24e-02**	2.82e-02	3.13e-02	4.05e-02
29	1.78e-03	**1.33e-03**	2.04e-03	1.47e-03
30	**2.68e+00**	2.93e+00	3.66e+00	4.61e+00
31	2.77e+05	**2.23e+05**	1.93e+06	2.21e+06
32	**5.84e-03**	6.20e-03	1.87e-02	1.98e-02
33	1.21e+13	**5.64e+12**	1.61e+13	2.36e+13
34	**2.82e+12**	3.25e+12	3.08e+12	3.85e+12
35	**2.24e+12**	2.33e+12	2.30e+12	3.07e+12
36	7.57e+11	7.81e+11	**7.24e+11**	7.46e+11

References

1. Benavoli, A., Corani, G., Demšar, J., Zaffalon, M.: Time for a change: a tutorial for comparing multiple classifiers through bayesian analysis. J. Mach. Learn. Res. **18**(77), 1–36 (2017)
2. Branco, P., Torgo, L., Ribeiro, R.P.: A survey of predictive modeling on imbalanced domains. ACM Comput. Surv. **49**(2), 1–50 (2016). https://doi.org/10.1145/2907070
3. Branco, P., Torgo, L., Ribeiro, R.P.: Pre-processing approaches for imbalanced distributions in regression. Neurocomputing **343**, 76–99 (2019). https://doi.org/10.1016/j.neucom.2018.11.100
4. Chawla, N.V., Bowyer, K.W., Hall, L.O., Kegelmeyer, W.P.: SMOTE: synthetic minority over-sampling technique. J. Artif. Int. Res. **16**(1), 321–357 (2002)
5. Chawla, N.V., Lazarevic, A., Hall, L.O., Bowyer, K.W.: SMOTEBoost: improving prediction of the minority class in boosting. In: Lavrač, N., Gamberger, D., Todorovski, L., Blockeel, H. (eds.) PKDD 2003. LNCS (LNAI), vol. 2838, pp. 107–119. Springer, Heidelberg (2003). https://doi.org/10.1007/978-3-540-39804-2_12
6. Chen, T., Guestrin, C.: XGBoost: a scalable tree boosting system. In: 22nd ACM SIGKDD International Conference on Knowledge Discovery and Data Mining, KDD 2016, pp. 785–794. ACM (2016). https://doi.org/10.1145/2939672.2939785
7. Chen, T., et al.: XGBoost: Extreme Gradient Boosting (2022). https://CRAN.R-project.org/package=xgboost
8. Christoffersen, P.F., Diebold, F.X.: Further results on forecasting and model selection under asymmetric loss. J. Appl. Economet. **11**(5), 561–571 (1996)
9. Ehrig, L., Atzberger, D., Hagedorn, B., Klimke, J., Döllner, J.: Customizable asymmetric loss functions for machine learning-based predictive maintenance. In: 2020 8th International Conference on Condition Monitoring and Diagnosis (CMD), pp. 250–253 (2020). https://doi.org/10.1109/CMD48350.2020.9287246
10. Elkan, C.: The foundations of cost-sensitive learning. In: 17th International Conference on Artificial Intelligence, vol. 1, pp. 973–978 (2001)
11. Fan, W., Stolfo, S.J., Zhang, J., Chan, P.K.: AdaCost: misclassification cost-sensitive boosting. In: 16th International Conference on Machine Learning, pp. 97–105. ICML 1999. Morgan Kaufmann Publishers Inc. (1999)
12. Friedman, J.H.: Greedy function approximation: a gradient boosting machine. Ann. Statist. **29**(5), 1189–1232 (2001). http://www.jstor.org/stable/2699986
13. Granger, C.W.J.: Outline of forecast theory using generalized cost functions. SpanEconRev **1**(2), 161–173 (1999). https://doi.org/10.1007/s101080050007
14. Hubert, M., Vandervieren, E.: An adjusted boxplot for skewed distributions. Comput. Statist. Data Anal. **52**, 5186–5201 (2008). https://doi.org/10.1016/j.csda.2007.11.008
15. Ke, G., et al.: LightGBM: a highly efficient gradient boosting decision tree. In: 31st International Conference on Neural Information Processing Systems, pp. 3149–3157. NIPS 2017, Curran Associates Inc. (2017)
16. Kruschke, J., Liddell, T.: The Bayesian new statistics: two historical trends converge. SSRN Electron. J. (2015). https://doi.org/10.2139/ssrn.2606016
17. Moniz, N.: Prediction and Ranking of Highly Popular Web Content. Ph.D. thesis, Faculty of Sciences, University of Porto (2017)
18. Moniz, N., Monteiro, H.: No free lunch in imbalanced learning. Knowl. Based Syst. **227**, 107222 (2021). https://doi.org/10.1016/j.knosys.2021.107222

19. Moniz, N., Ribeiro, R., Cerqueira, V., Chawla, N.: SMOTEBoost for regression: improving the prediction of extreme values. In: IEEE 5th International Conference on Data Science and Advanced Analytics (DSAA), pp. 150–159 (2018). https://doi.org/10.1109/DSAA.2018.00025
20. R Core Team: R: A Language and Environment for Statistical Computing. R Foundation for Statistical Computing, Vienna, Austria (2020). https://www.R-project.org/
21. Rengasamy, D., Rothwell, B., Figueredo, G.P.: Asymmetric loss functions for deep learning early predictions of remaining useful life in aerospace gas turbine engines. In: International Joint Conference on Neural Networks (IJCNN), pp. 1–7 (2020). https://doi.org/10.1109/IJCNN48605.2020.9207051
22. Ribeiro, R., Moniz, N.: Imbalanced regression and extreme value prediction. Mach. Learn. **109**, 1–33 (2020). https://doi.org/10.1007/s10994-020-05900-9
23. Shi, Y., et al.: LightGBM: Light Gradient Boosting Machine (2022). https://CRAN.R-project.org/package=lightgbm
24. Sun, Y., Kamel, M.S., Wong, A.K., Wang, Y.: Cost-sensitive boosting for classification of imbalanced data. Pattern Recogn. **40**(12), 3358–3378 (2007). https://doi.org/10.1016/j.patcog.2007.04.009
25. Torgo, L., Ribeiro, R.P., Pfahringer, B., Branco, P.: SMOTE for regression. In: Correia, L., Reis, L.P., Cascalho, J. (eds.) EPIA 2013. LNCS (LNAI), vol. 8154, pp. 378–389. Springer, Heidelberg (2013). https://doi.org/10.1007/978-3-642-40669-0_33
26. Torgo, L., Ribeiro, R.: Utility-based regression. In: Kok, J.N., Koronacki, J., Lopez de Mantaras, R., Matwin, S., Mladenič, D., Skowron, A. (eds.) PKDD 2007. LNCS (LNAI), vol. 4702, pp. 597–604. Springer, Heidelberg (2007). https://doi.org/10.1007/978-3-540-74976-9_63
27. Varian, H.R.: A bayesian approach to real estate assessment. Studies in Bayesian Econometric and Statistics in Honor of Leonard J. Savage, pp. 195–208 (1975)
28. Yang, Y., Zha, K., Chen, Y., Wang, H., Katabi, D.: Delving into deep imbalanced regression. CoRR abs/2102.09554 (2021). arXiv:abs/2102.09554

Discovery of Differential Equations Using Probabilistic Grammars

Boštjan Gec[1,2(✉)], Nina Omejc[1,2], Jure Brence[1,2], Sašo Džeroski[1,2], and Ljupčo Todorovski[1,3]

[1] Department of Knowledge Technologies, Jožef Stefan Institute, Ljubljana, Slovenia
bostjan.gec@ijs.si
[2] Jožef Stefan International Postgraduate School, Ljubljana, Slovenia
[3] Faculty of Mathematics and Physics, University of Ljubljana, Ljubljana, Slovenia

Abstract. Ordinary differential equations (ODEs) are a widely used formalism for mathematical modeling of dynamical systems, a task omnipresent in many scientific domains. The paper introduces a novel method for inferring ODEs from data. It extends ProGED, a method for equation discovery that employs probabilistic context-free grammars for constraining the space of candidate equations. The proposed method can discover ODEs from partial observations of dynamical systems, where only a subset of state variables can be observed. The new method's empirical evaluation shows it can reconstruct the ODEs of the well-known Van der Pol oscillator from synthetic simulation data. In terms of reconstruction performance, improved ProGED compares favorably to state-of-the-art methods for inferring ODEs from data.

Keywords: Partial observability · Dynamical systems · System identification · Equation discovery · Symbolic regression · Probabilistic context-free grammars · Ordinary differential equations

1 Introduction

Dynamical systems describe phenomena that evolve through time. The temporal evolution of the system is commonly modeled by the time derivatives of the state variables leading to ordinary differential equations (ODEs) of the form $\dot{\boldsymbol{u}} = f(\boldsymbol{u}, \boldsymbol{\theta})$, where $\boldsymbol{u}$ is the d-dimensional vector of state variables, $\dot{\boldsymbol{u}}$ is the vector of their time derivatives describing the change of their values through time, $f : \mathbb{R}^d \times \mathbb{R}^p \to \mathbb{R}^d$ is a function, referred to as *model structure*, and $\boldsymbol{\theta} \in \mathbb{R}^p$ are constant *model parameters*. The solution of the ODEs is obtained by assuming an initial state $\boldsymbol{u}_0 = \boldsymbol{u}(t_0)$, i.e., the values of the state variables at the initial time point t_0. ODEs are numerically simulated [5] to obtain the state trajectory $\boldsymbol{u}(t)$, i.e., the values of the state variables at time points $t > t_0$.

Mathematical modeling of an observed dynamical system is an inverse problem: given observations of the trajectory $\boldsymbol{u}(t)$, we aim at finding *appropriate* model structure f and model parameters $\boldsymbol{\theta}$. The simulated trajectory should

P. Pascal and D. Ienco (Eds.): DS 2022, LNAI 13601, pp. 22–31, 2022.
https://doi.org/10.1007/978-3-031-18840-4_2

closely match the observed one and simpler model structures are preferred over more complex ones. Classical approaches to system identification assume that the model structure f is provided at input. It is often supposed to be linear (i.e., the right-hand sides of the ODEs are linear combinations of state variables), or it is manually inferred from first principles in the domain of the dynamical system.

Equation discovery (also known as symbolic regression) approaches aim to infer f and $\boldsymbol{\theta}$ from data. While they mostly aim at discovering algebraic equations, some are also used for learning ODEs. The simplest approach would numerically calculate the time derivatives of the state variables $\dot{\boldsymbol{u}}$ and add them to the set of observed variables. Consequently, ODEs can be discovered as algebraic equations with variables corresponding to the time derivatives on the left-hand side. Lagrange [12] and SINDy [2] follow this framework and use linear regression variants to learn ODEs that are linear in the model parameters $\boldsymbol{\theta}$.

This simple approach has two limitations. First, calculating derivatives is numerically unstable [8], leading to high sensitivity to noisy measurements. Second, all the state variables have to be observed. More general approaches can infer ODEs in scenarios with *partial observability*, where not all state variables are observed. L-ODEfind [9] and GPoM [7] can induce polynomial ODEs from partial observations. Pret [10] and ProBMot [3] employ numerical simulation of ODEs and can learn arbitrary ODEs, not necessarily linear in parameters. They use knowledge about modeling dynamics to limit the space of candidate structures.

Recently proposed probabilistic approaches to symbolic regression, e.g., Bayesian scientist [4] and ProGED [1] can not learn ODEs from data. In this paper, we propose two extensions of ProGED for the discovery of ODEs. The first simply calculates the numerical derivatives of the state variables. The second employs methods for numerically simulating ODEs and can be applied to partially observed systems. We evaluate the proposed extensions by reconstructing the ODEs of the Van der Pol (VDP) oscillator from simulated, synthetic data in both full and partial observability scenarios. We compare the performance of ProGED with the performance of SINDy, L-ODEfind, and GPoM.

Section 2 reviews the related work on inferring ODEs from data. We introduce the extensions of ProGED for ODE discovery in Sect. 3. Section 4 reports the results of our comparative analysis of the methods. Section 5 summarizes our contributions and outlines the directions for further research.

2 Related Work

Several methods, based on fast and simple linear regression, have been developed for identifying ODEs that are linear in the parameters. They numerically calculate the time derivatives of the state variables. Lagrange [12] then introduces new higher-order terms with multiplication of the system variables and uses linear regression to find the constant parameters of the polynomial ODEs involving these terms. SINDy [2] extends the Lagrange approach by introducing new terms with custom basis functions. Sparse regression techniques, such as Lasso, help control the complexity of discovered equations and reduce over-fitting.

L-ODEfind [9] is also based on sparse linear regression on terms introduced by polynomial basis functions. It tackles the partial observability problem by rewriting a system of n first-order ODEs to a single n-th order ODE, the $n-1$ unobserved variables being implicitly included and their values numerically approximated as higher-order derivatives. Since the inverse transformation to the first-order ODEs is not uniquely determined, L-ODEfind is useful for predictive modeling, but ultimately inappropriate as a system identification framework.

GPoM [7], Generalized Polynomial Modeling, is another approach that uses higher-order derivatives as a proxy for missing variables. GPoM does not express a model with a single, higher-order ODE like L-ODEfind, but rather uses differential embedding to construct a system of first-order ODEs. GPoM starts with a random set of polynomial ODEs within the user-constrained space and then optimizes the best model's structure with a genetic algorithm. The algorithm evaluates the goodness of fit of candidate models to the original data, for which repetitive, computationally costly numerical simulation is required. The best model is mutated into neighbours by adding or removing monomials.

3 Methods

ProGED discovers algebraic equations following the generate-and-test paradigm. In the generate phase, ProGED addresses the task of structure identification, in which we construct candidate equations. The test phase performs parameter estimation, in which we fit the values of unknown numeric parameters to data. Among a large number of tested equations, ProGED chooses the ones with the lowest error-of-fit. ProGED composes candidate equations from algebraic expressions, sampled from a probabilistic context-free grammar (PCFG).

A context-free grammar (CFG) is defined by the tuple $(\mathscr{T}, \mathscr{N}, \mathscr{R}, S)$. When defining arithmetic expressions, the set of terminal symbols $\mathscr{T}$ consists of symbols representing variables (e.g. x, y), operators or functions (e.g. $+, \cdot, \sin$), and constant parameters (c). The nonterminal symbols in $\mathscr{N}$ do not appear in expressions, but represent higher-level concepts in the language of mathematics, such as polynomials, monomials or terms. The set $\mathscr{R}$ contains production (rewrite) rules $A \rightarrow \alpha_1 \dots \alpha_k$, where $A \in \mathscr{N}$ and $\alpha_i \in \mathscr{N} \cup \mathscr{T}$. A production rule specifies how to replace a particular nonterminal symbol with a string of nonterminal and terminal symbols. In a probabilistic context-free grammar, each production rule is assigned a probability, so that the probabilities of all production rules with the same nonterminal symbol on the left-hand side sum up to 1.

The generation of a random expression from a PCFG begins with the string (starting symbol) S and proceeds by successively applying production rules to the string until only terminal symbols remain. Whenever more than one rule applies, we randomly choose a rule, according to the probabilities. The final result of one instance of the sampling process is an arithmetic expression, which we transform to its canonical form using the symbolic mathematics engine SymPy.

Besides acting as generators of expressions, grammars are a powerful way of encoding background knowledge. Note that a PCFG defines a probability distribution over the space of candidate expressions, which allows the user to impose an inductive bias by manipulating the production probabilities. For example, we can manipulate the complexity of generated equations through the probabilities of recursive productions, or express a bias towards trigonometric functions by raising their respective probabilities. In the absence of background knowledge we can use a universal grammar for generating an arbitrary expression, composed of the four basic operations (+, −, *, /), as well arbitrary functions.

After generation, a candidate equation contains constants, the values of which must be fitted to data. Since the equations are, in general, non-linear in their parameters, a universal optimization algorithm is used to minimize the error-of-fit on the data, which can be computationally demanding, but is more flexible than the approaches based on linear regression. ProGED uses the differential evolution [11] algorithm for numerical optimization.

3.1 Algebraic Equations and Numeric Differentiation

ODEs can be transformed into algebraic equations by numerically calculating the derivatives of the state variables and considering them as the left-hand side of an algebraic equation. In that case, we estimate parameter values by minimizing the difference L between observed and predicted values of the time derivatives of the state variables, $L = \sum_{i=1}^{n} (\dot{u}(t_i) - \hat{\dot{u}}(t_i))^2$, where $\dot{u}(t_i)$ represents the value of the derivative of variable u at time t_i, numerically estimated from observed data, $\hat{\dot{u}}$ represents the corresponding predicted value, obtained by evaluating the candidate equation, and n is the number of time points.

This simple transformation is common among ODE discovery approaches. However, its use is problematic if we deal with sparsely sampled measurements and high levels of noise. Also, this approach is possible only when the measurements of all system variables are readily available.

3.2 Differential Equations and Direct Simulation

To address the limitations of numerical differentiation, we introduce an approach based on simulating differential equations. During each step of parameter estimation, we must compute the error-of-fit of the candidate equation with a given set of parameter values. To obtain this, we solve the initial-value problem by performing a full simulation of the system of ODEs, using the function `odeint` from the SciPy library. We define the error-of-fit as the mean-squared-error of the simulated trajectories, with respect to the true trajectories of observed variables. In other words, we minimize the error $L = \sum_{u \in \mathcal{U}_{\text{obs}}} \sum_{i=1}^{n} (u(t_i) - \hat{u}(t_i))^2$, where $\mathcal{U}_{\text{obs}}$ is the set of all *observed* variables, $u(t_i)$ represents the observed value u at time t_i, $\hat{u}(t_i)$ represents the corresponding simulated value (i.e., the value obtained by simulating the candidate equation), and n is the number of time points.

3.3 Parallel Computation

Most existing approaches to equation discovery are difficult to parallelize. In contrast, ProGED uses a Monte-Carlo algorithm to sample expressions from a PCFG, so each run of the procedure is completely independent of the results of any previous runs. This allows for parallelization of parameter estimation, by far the most computationally demanding step in ProGED, especially when identifying a partially-observed system of ODEs. Accelerating this step is critical for managing the computation time of ProGED. The easy parallelization and the ability to make good use of any available high-performance-computing resources is therefore an important advantage of our approach.

4 Experimental Evaluation

In this section, we compare the ability of different approaches to identify the Van der Pol (VDP) oscillator, a non-conservative oscillator with non-linear damping. It is defined by a second-order ODE or the following system of first-order ODEs:

$$\begin{aligned}\dot{x} &= y \\ \dot{y} &= \mu(1 - x^2)y - x,\end{aligned}$$

where x and y are the state variables, and μ is the model parameter of the oscillator, referred to as the damping constant. In the next two subsections, we introduce the setup for our experiments with this oscillator and present the results.

4.1 Experimental Setup

Data. We generated the data by numerically simulating the VDP ODEs, using the `solve_ivp` function from the SciPy Python library. The damping constant μ was set to 0.5. We chose the initial values of $x(t_0) = -0.2$ and $y(t_0) = -0.8$ and simulated the system for $n-1 = 5000$ equidistant time points in the interval of $[0, 50)$, leading to a time step of 0.01 s. We set the values of the simulation parameters of relative and absolute tolerance to 10^{-12}. The resulting data set includes three columns: time and x, y coordinates of the simulated trajectory.

Methods. We evaluated SINDy, L-ODEfind, GPoM and ProGED in fully observed and partially observed scenarios. For the VDP ODEs, there are two scenarios for partial observability, depending on whether x or y is observed. SINDy is the only evaluated method without support for partial observability. We included it in the comparison as the most widely used equation discovery method to obtain a baseline for the performance in fully observed scenarios. We evaluated ProGED in the fully observed scenario twice – using numerical differentiation and ODEs simulation. All the methods we tested were downloaded from their public Github repositories.

For SINDy, we used the default polynomial library up to the third-degree and a SR3 sparse regression algorithm with the sparsity parameter set to 0.25.

To run L-ODEfind, we first had to add the VDP model in its model library. In the full observability scenario, we set the maximum degree of ODEs to one and the maximum order of the polynomials to three. In the partial observability scenario, the maximum degree of ODEs was set to two. L-ODEfind outputs the best model in a higher dimensional space, which can not be uniquely transformed back to the first order system of ODEs. However, as we used the known VDP model that includes the equation $\dot{x} = y$, a conversion of the second order equation to a system of first order ODEs was obvious, so we performed the conversion by hand, which allowed us to evaluate the method nonetheless.

The GPoM settings besides the ones mentioned in L-ODEfind, required the specification of the number of dimensions that will be created from each observed input variable. When both variables were observed, we kept the dimensions at one, but in partial observability scenarios, we increased the desired dimensions to two, so that one additional variable was constructed by differentiating the observed variable. We set the integration method to fourth-order Runge-Kutta, with 5120 maximum integration steps.

For generating equations in ProGED, we used a probabilistic context-free grammar for polynomial expressions with the following production rules:

$$\begin{aligned} P \rightarrow &\quad P + M \ [0.4] \mid M \ [0.6] \\ M \rightarrow &\quad M * V \ [0.4] \mid c \ [0.6] \\ V \rightarrow &\quad x \ [0.5] \mid y \ [0.5] \end{aligned}$$

where the sets of terminal and non-terminal symbols are $\mathscr{N} = \{P, M, V\}$ and $\mathscr{T} = \{c, x, y, +, *\}$ respectively, with P being the start symbol. Rules for P produce a sum of monomials M that represent products of the system variables V and a constant parameter c. The probabilities are set so that the majority of generated polynomials consist of one to three low order terms.

We employed the parallel computation approach described in Sect. 3.3, generating and testing 1000 systems of ODEs for each experiment. We set the relative and absolute tolerances of simulation to 10^{-6}. We set the maximum number of iterations for differential evolution to 1000, population size to 50, the recombination parameter to 0.88 and the mutation parameter to 0.45. We optimized recombination and mutation parameters using grid search.

Performance Measures. We compare the results of the different methods through three metrics that quantify the accuracy of the reconstructed models, as well as the values of its parameters. The first metric is the *trajectory error*, calculated for a given state variable using the relative-root-mean-square-error $TE_u = \sqrt{\sum_{i=1}^{n}(\hat{u}(t_i) - u(t_i))^2 / \sum_{i=1}^{n}(\overline{u} - u(t_i))^2}$, where $u(t)$, $\hat{u}(t)$ and $\overline{u}$ denote the observed and the simulated values of the state variable u at the time point t (trajectory) and the average value of observations, respectively.

Furthermore, *reconstruction error* (*RE*) measures the difference between the model parameters of the reconstructed ODEs and the parameters of the original ones by root-mean-square error. Finally, ΔM is the difference between the number of terms in the reconstructed ODEs and the original ones. When calculating *RE* and ΔM, we consider only those terms in the reconstructed ODEs with constant parameters above 0.01.

Table 1. Comparison of the trajectory error (**TE**), reconstruction error (**RE**), the number of extra parameters in the discovered equations (**ΔM**) and the time required for computation (**T**) for different methods. For each method, its approach is classified (**type**) as either numerical differentiation ("num.") or simulation ("sim."). Variable observability (**obs.**) is indicated for each experiment, where "XY" indicates full observability, "X" and "Y" denote scenarios where only the variable x/y is observed.

Method	Type	obs.	TEx	TEy	RE	ΔM	T
SINDy	num.	XY	$4.89 \cdot 10^{-4}$	$5.61 \cdot 10^{-4}$	$2.96 \cdot 10^{-5}$	0	1.10s
L-ODEfind	num.	XY	$1.90 \cdot 10^{-3}$	$2.19 \cdot 10^{-3}$	$5.98 \cdot 10^{-5}$	0	1.15 s
		X	$5.80 \cdot 10^{-4}$	$6.90 \cdot 10^{-4}$	$2.88 \cdot 10^{-4}$	0	1.11 s
		Y	1.016	0.995	0.67	8	0.56 s
GPoM	sim.	XY	$3.55 \cdot 10^{-3}$	$4.05 \cdot 10^{-4}$	$2.70 \cdot 10^{-4}$	0	30.44 s
		X	$2.55 \cdot 10^{-3}$	$2.84 \cdot 10^{-3}$	$1.76 \cdot 10^{-4}$	0	46.75 s
		Y	1.196	0.998	0.604	4	38.33 s
ProGED	num.	XY	$1.07 \cdot 10^{-3}$	$2.23 \cdot 10^{-4}$	$2.55 \cdot 10^{-3}$	0	*11 min
	sim.	XY	$1.45 \cdot 10^{-4}$	$1.66 \cdot 10^{-4}$	$9.66 \cdot 10^{-6}$	0	*713 h
	sim.	X	$7.96 \cdot 10^{-2}$	3.16	0.73	0	*500 h
	sim.	Y	0.44	$3.57 \cdot 10^{-2}$	0.32	0	*230 h

*The timings for ProGED represent an estimation of the total computation time, if the experiments were performed in a single thread. In practice, we ran it in parallel, which reduced the real-time computation time by a factor of 500–1,000.

4.2 Results

Table 1 summarizes the results of the experiments. In the fully observed scenario, all methods have excellent performance, achieving *trajectory errors* below 10^{-2} and recovering ODEs with no extraneous terms. We attribute the good results in this scenario to the fact that we used noise-free data and a high temporal resolution, which is ideal for methods that rely on numerical differentiation.

ProGED using simulation of ODEs achieved the best results in the fully observed scenario according to all metrics except computation time, which is orders of magnitude worse as compared to the other methods. This difference arises because ProGED performs parameter estimation for each of the sampled candidate equations, whereas the methods based on linear regression perform structure search and parameter estimation in a single step. The difference is even more significant when employing direct numerical simulation, which is performed at every step of parameter estimation for each candidate equation. Although GPoM also simulates the models in every iteration, it only has a very constrained space of model structures to consider and simulate.

When only the variable x is observed, L-ODEfind and GPoM are able to accurately reconstruct the VDP ODEs. On the other hand, the two methods fail in the scenario where only y is observed. This behavior can be explained by the fact that both methods model the unobserved variable as the derivative of the observed variable. This assumption perfectly matches with VDP ODEs, when x is observed, i.e., under the assumption of $y = \dot{x}$. On the other hand, in the scenario when only y is observed, the methods assume that $x = \dot{y}$, which prevents them from reconstructing the ODEs.

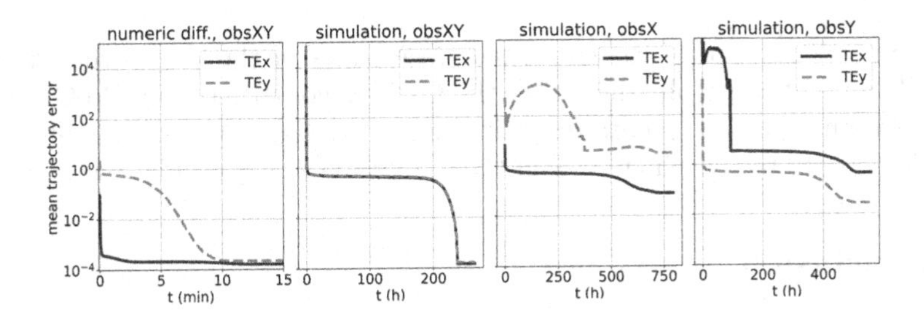

Fig. 1. Learning curves of ProGED in the full observability scenario (first two graphs) and the two partial observability scenarios (last two graphs). The full blue lines correspond to the learning curves for the x equation, while the dashed orange lines correspond to the learning curves for the y equation. The estimation of learning curves is explained in the Appendix, Section a. (Color figure online)

In partially observed scenarios, ProGED is able to reconstruct the ODE of the observed variable fairly well, while having difficulties in fitting the parameters of the equation for the unobserved state variable. The learning curves for ProGED depicted in Fig. 1 provide an explanation. During system identification in a partially-observed scenario, ProGED directly minimizes only the error-of-fit of the observed variable and the corresponding error falls monotonically. In contrast, ProGED optimizes the equation for the unobserved variable only indirectly, resulting in a TE that initially increases, before eventually finding a minimum an order of magnitude above the error of the observed variable.

Tables 1 and 2 in the Appendix[1] report the reconstructed equations. Note that L-ODEfind in the two partially observed scenarios reconstructs second-order equations. To compare them with the original VDP ODEs, it is straightforward to manually convert them back to first order ODEs with the introduction of the first derivative of the observed state variable as a new (unobserved state) variable, e.g., $\dot{x} = y$ or $\dot{y} = x$. Table 2 presents the converted first-order ODEs. Notably, ProGED was always able to identify the correct equation structure.

[1] Appendix: http://kt.ijs.si/~ljupco/ed/ds-2022/appendix.pdf.

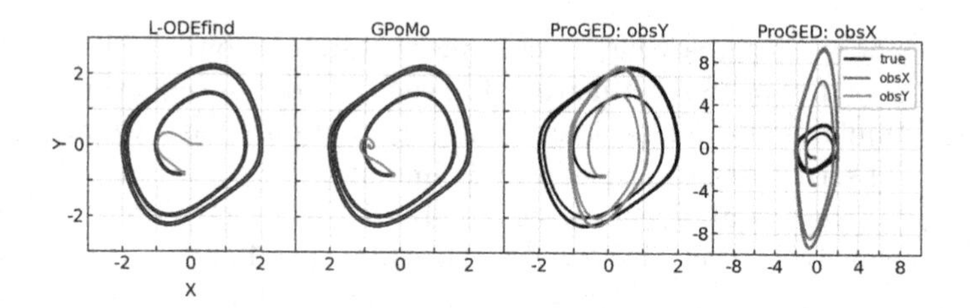

Fig. 2. Simulated trajectories of the ODEs reconstructed by L-ODEfind, GPoM, and ProGED (last two graphs, one for each partial observability scenario). Data trajectories are shown in black, reconstructed trajectories when only one of x or y is observed are shown in blue and orange, respectively. Green dots denote initial values. (Color figure online)

Finally, Fig. 2 depicts the simulated trajectories of the reconstructed equations. The last two graphs show that the ODEs reconstructed by ProGED lead to simulations that are *qualitatively* similar to the observations. The differences are related only to the differences in the amplitude of the oscillations.

5 Conclusion

The two main contributions of the paper are as follows. First, we extend the scope of ProGED, a probabilistic algorithm for symbolic regression, to the task of discovering ODEs from data. One of the two extensions also applies to situations where only some of the state variables of the dynamical system at hand can be observed. Second, we compare the reconstruction abilities of and computational resources needed by ProGED and three other state-of-the-art methods for automated modeling of dynamical systems: SINDy, L-ODEfind, and GPoM.

The comparison shows that using computationally expensive full numerical simulation of ODEs, ProGED can successfully reconstruct equations for the observed variables in both full and partial observability scenarios. While numerical simulation requires vast computational resources, ProGED's parallel implementation significantly reduces the actual response time of the reconstruction experiments. Finally, ProGED is considerably more robust to the selection of the observed state variable. This property is essential in real modeling scenarios where we cannot choose the observations freely.

This paper opens several directions for further research. First, we would like to test the robustness of the compared methods to noisy and sparsely sampled observation data. To this end, artificial noise should be added to the synthetic simulation data prepared for the experiments presented in this paper and additional dynamical systems should be considered. Second, ProGED uses vast computational resources to simulate the candidate differential equations, since the general numeric optimization method for estimating the model parameters requires full simulation for each set of parameter values. One can address this issue by using efficient, surrogate-based optimization methods [6] or more innovative strategies, that combine simulation and numerical differentiation, to address the issue of partial observability.

Acknowledgements. The authors acknowledge the financial support of the Slovenian Research Agency via the research core funding No. P2-0103 and project No. N2-0128.

References

1. Brence, J., Todorovski, L., Džeroski, S.: Probabilistic grammars for equation discovery. Knowl.-Based Syst. **224**, 107077 (2021)
2. Brunton, S.L., Proctor, J.L., Kutz, J.N.: Discovering governing equations from data by sparse identification of nonlinear dynamical systems. In: Proceedings of the National Academy of Sciences, vol. 113, no. 15, pp. 3932–3937 (2016)
3. Čerepnalkoski, D.: Process-based models of dynamical systems: representation and induction. Ph.D. thesis, Jožef Stefan International Postgraduate School, Ljubljana, Slovenia (2013)
4. Guimerà, J., et al.: A Bayesian machine scientist to aid in the solution of challenging scientific problems. Sci. Adv. **6**(5), eaav6971 (2020)
5. Hindmarsh, A.C., et al.: SUNDIALS: suite of nonlinear and differential/algebraic equation solvers. ACM Trans. Math. Softw. (TOMS) **31**(3), 363–396 (2005)
6. Lukšič, Ž, Tanevski, J., Džeroski, S., Todorovski, L.: Meta-model framework for surrogate-based parameter estimation in dynamical systems. IEEE Access **7**, 181829–181841 (2019)
7. Mangiarotti, S., Coudret, R., Drapeau, L., Jarlan, L.: Polynomial search and global modeling: two algorithms for modeling chaos. Phys. Rev. E **86**, 046205 (2012)
8. Ramm, A.G., Smirnova, A.B.: On stable numerical differentiation. Math. Comput. **70**, 1131–1153 (2001)
9. Somacal, A., et al.: Uncovering differential equations from data with hidden variables. Phys. Rev. E **105**, 054209 (2022)
10. Stolle, Reinhard, Bradley, Elizabeth: Communicable knowledge in automated system identification. In: Džeroski, Sašo, Todorovski, Ljupčo (eds.) Computational Discovery of Scientific Knowledge. LNCS (LNAI), vol. 4660, pp. 17–43. Springer, Heidelberg (2007). https://doi.org/10.1007/978-3-540-73920-3_2
11. Storn, R., Price, K.: Differential evolution-a simple and efficient heuristic for global optimization over continuous spaces. J. Global Optim. **11**(4), 341–359 (1997)
12. Todorovski, L., Džeroski, S.: Declarative bias in equation discovery. In: Proceedings of the 14th International Conference on Machine Learning, pp. 376–384 (1997)

Hyperparameter Importance of Quantum Neural Networks Across Small Datasets

Charles Moussa(✉), Jan N. van Rijn, Thomas Bäck, and Vedran Dunjko

LIACS, Leiden University, Niels Bohrweg 1, 2333 Leiden, CA, Netherlands
c.moussa@liacs.leidenuniv.nl

Abstract. As restricted quantum computers are slowly becoming a reality, the search for meaningful first applications intensifies. In this domain, one of the more investigated approaches is the use of a special type of quantum circuit – a so-called quantum neural network – to serve as a basis for a machine learning model. Roughly speaking, as the name suggests, a quantum neural network can play a similar role to a neural network. However, specifically for applications in machine learning contexts, very little is known about suitable circuit architectures, or model hyperparameters one should use to achieve good learning performance. In this work, we apply the functional ANOVA framework to quantum neural networks to analyze which of the hyperparameters were most influential for their predictive performance. We analyze one of the most typically used quantum neural network architectures. We then apply this to 7 open-source datasets from the OpenML-CC18 classification benchmark whose number of features is small enough to fit on quantum hardware with less than 20 qubits. Three main levels of importance were detected from the ranking of hyperparameters obtained with functional ANOVA. Our experiment both confirmed expected patterns and revealed new insights. For instance, setting well the learning rate is deemed the most critical hyperparameter in terms of marginal contribution on all datasets, whereas the particular choice of entangling gates used is considered the least important except on one dataset. This work introduces new methodologies to study quantum machine learning models and provides new insights toward quantum model selection.

Keywords: Hyperparameter importance · Quantum neural networks · Quantum machine learning

1 Introduction

Quantum computers have the capacity to efficiently solve computational problems believed to be intractable for classical computers, such as factoring [42] or simulating quantum systems [12]. However, with the Noisy Intermediate-Scale Quantum era [33], quantum algorithms are confronted with many limitations

P. Pascal and D. Ienco (Eds.): DS 2022, LNAI 13601, pp. 32–46, 2022.
https://doi.org/10.1007/978-3-031-18840-4_3

(e.g., the number of qubits, decoherence, etc.). Consequently, hybrid quantum-classical algorithms were designed to work around some of these constraints while targeting practical applications such as chemistry [27], combinatorial optimization [10], and machine learning [2]. Quantum models can exhibit clear potential in special datasets where we have theoretically provable separations with classical models [18,22,35,46]. More theoretical works also study these models from a generalization perspective [8]. Quantum circuits with adjustable parameters, also called quantum neural networks, have been used to tackle regression [25], classification [14], generative adversarial learning [50], and reinforcement learning tasks [18,44].

However, the value of quantum machine learning on real-world datasets is still to be investigated in any larger-scale systematic fashion [13,32]. Currently, common practices from machine learning, such as large-scale benchmarking, hyperparameter importance, and analysis have been challenging tools to use in the quantum community [39]. Given that there exist many ways to design quantum circuits for machine learning tasks, this gives rise to a hyperparameter optimization problem. However, there is currently limited intuition as to which hyperparameters are important to optimize and which are not. Such insights can lead to much more efficient hyperparameter optimization [5,11,26].

In order to fill this gap, we employ functional ANOVA [16,45], a tool for assessing hyperparameter importance. This follows the methodology of [34,41], who employed this across datasets, allowing for more general results. For this, we selected a subset of several low-dimensional datasets from the OpenML-CC18 benchmark [4], that are matching the current scale of simulations of quantum hardware. We defined a configuration space consisting of ten hyperparameters from an aggregation of quantum computing literature and software. We extend this methodology by an important additional verification step, where we verify the performance of the internal surrogate models. Finally, we perform an extensive experiment to verify whether our conclusions hold in practice. While our main findings are in line with previous intuition on a few hyperparameters and the verification experiments, we also discovered new insights. For instance, setting well the learning rate is deemed the most critical hyperparameter in terms of marginal contribution on all datasets, whereas the particular choice of entangling gates used is considered the least important except on one dataset.

2 Background

In this section, we introduce the necessary background on functional ANOVA, quantum computing, and quantum circuits with adjustable parameters for supervised learning.

2.1 Functional ANOVA

When applying a new machine learning algorithm, it is unknown which hyperparameters to modify in order to get high performances on a task. Several

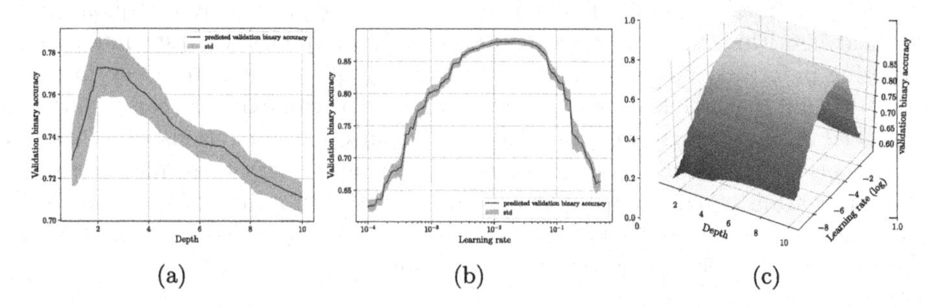

Fig. 1. Examples of marginals for a quantum neural network with validation accuracy as performance on the banknote-authentication dataset. The hyperparameters correspond to the learning rate used during training (a), and the number of layers, also known as depth (b), and their combination (c). The hyperparameter values for learning rate are on a log scale. When considered individually, we see for instance that depth and learning rate should not be set too high for better performances. However, when grouped together, the learning rate seems most influential.

techniques exist that assess hyperparameter importance, such as functional ANOVA [36]. The latter framework can detect the importance of both individual hyperparameters and interaction effects between different subsets of hyperparameters. We first introduce the relevant notation, based on the work by Hutter *et al.* [16].

Let A be a machine learning algorithm that has n hyperparameters with domains $\Theta_1, \ldots, \Theta_n$ and *configuration space* $\boldsymbol{\Theta} = \Theta_1 \times \ldots \times \Theta_n$. An instantiation of A is a vector $\boldsymbol{\theta} = \{\theta_1, \ldots, \theta_n\}$ with $\theta_i \in \Theta_i$ (this is also called a *configuration* of A). A partial instantiation of A is a vector $\boldsymbol{\theta}_U = \{\theta_{i_1}, \ldots, \theta_{i_k}\}$ with a subset $U = \{i_1, \ldots, i_k\} \subseteq N = [n] = \{1, \ldots, n\}$ of the hyperparameters fixed, and the values for other hyperparameters unspecified. Note that $\boldsymbol{\theta}_N = \boldsymbol{\theta}$.

Functional ANOVA is based on the concept of a marginal of a hyperparameter, i.e., how a given value for a hyperparameter performs, averaged over all possible combinations of the other hyperparameters' values. The *marginal performance* $\hat{a}_U(\boldsymbol{\theta}_U)$ is described as the average performance of all complete instantiations $\boldsymbol{\theta}$ that have the same values for hyperparameters that are in $\boldsymbol{\theta}_U$. As an illustration, Fig. 1 shows marginals for two hyperparameters of a quantum neural network and their union. As the number of terms to consider for the marginal can be very large, the authors of [16] used tree-based surrogate regression models to calculate efficiently the average performance. Such a model yields predictions $\hat{y}$ for the performance p of arbitrary hyperparameter settings.

Functional ANOVA determines how much each hyperparameter (and each combination of hyperparameters) contributes to the variance of $\hat{y}$ across the algorithm's hyperparameter space $\boldsymbol{\Theta}$, denoted $\mathbb{V}$. Intuitively, if the marginal has high variance, the hyperparameter is highly important to the performance measure. Such framework has been used for studying the importance of hyperparameters of common machine learning models such as support vector machines, random forests, Adaboost, and residual neural networks [34,41]. We refer to [16] for a complete description and introduce the quantum supervised models considered in this study along with the basics of quantum computing.

2.2 Supervised Learning with Parameterized Quantum Circuits

Basics of Quantum Computing. In quantum computing, computations are carried out by the manipulation of qubits, similarly to classical computing with bits. A system of n qubits is represented by a 2^n-dimensional complex vector in the Hilbert space $\mathcal{H} = (\mathbb{C}^2)^{\otimes n}$. This vector describes the state of the system $|\psi\rangle \in \mathcal{H}$ of unit norm $\langle\psi|\psi\rangle = 1$. The bra-ket notation is used to describe vectors $|\psi\rangle$, their conjugate transpose $\langle\psi|$ and inner-products $\langle\psi|\psi'\rangle$ in $\mathcal{H}$. Single-qubit computational basis states are given by $|0\rangle = (1,0)^T, |1\rangle = (0,1)^T$, and their tensor products describe general computational basis states, e.g., $|10\rangle = |1\rangle \otimes |0\rangle = (0,0,1,0)$.

The quantum state is modified with unitary operations or gates U acting on $\mathcal{H}$. This computation can be represented by a quantum circuit (see Fig. 2). When a gate U acts non-trivially only on a subset $S \subseteq [n]$ of qubits, we denote such operation $U \otimes \mathbb{1}_{[n]\setminus S}$. In this work, we use, the Hadamard gate H, the single-qubit Pauli gates X, Z, Y and their associated rotations R_X, R_Y, R_Z:

$$H = \frac{1}{\sqrt{2}}\begin{pmatrix}1 & 1\\ 1 & -1\end{pmatrix}, Z = \begin{pmatrix}1 & 0\\ 0 & -1\end{pmatrix}, R_Z(w) = \exp\left(-i\frac{w}{2}Z\right),$$
$$Y = \begin{pmatrix}0 & -i\\ i & 0\end{pmatrix}, R_Y(w) = \exp\left(-i\frac{w}{2}Y\right), X = \begin{pmatrix}0 & 1\\ 1 & 0\end{pmatrix}, R_X(w) = \exp\left(-i\frac{w}{2}X\right), \tag{1}$$

The rotation angles are denoted $w \in \mathbb{R}$ and the 2-qubit controlled-Z gate $\updownarrow = \mathrm{diag}(1,1,1,-1)$ as well as the $\sqrt{\mathrm{iSWAP}}$ given by the matrix

$$\frac{1}{\sqrt{2}}\begin{pmatrix}\sqrt{2} & 0 & 0 & 0\\ 0 & 1 & i & 0\\ 0 & i & 1 & 0\\ 0 & 0 & 0 & \sqrt{2}\end{pmatrix}. \tag{2}$$

Measurements are carried out at the end of a quantum circuit to obtain bitstrings. Such measurement operation is described by a Hermitian operator O called an observable. Its spectral decomposition $O = \sum_m \lambda_m P_m$ in terms of eigenvalues λ_m and orthogonal projections P_m defines the outcomes of this measurement, according to the Born rule: a measured state $|\psi\rangle$ gives the outcome λ_m and gets projected onto the state $P_m|\psi\rangle / \sqrt{p(m)}$ with probability $p(m) = \langle\psi| P_m |\psi\rangle = \langle P_m\rangle_\psi$. The expectation value of the observable O with respect to $|\psi\rangle$ is $\mathbb{E}_\psi[O] = \sum_m p(m)\lambda_m = \langle O\rangle_\psi$. We refer to [30] for more basic concepts of quantum computing, and follow with parameterized quantum circuits.

Parameterized Quantum Circuits. A parameterized quantum circuit (also called *ansatz*) can be represented by a quantum circuit with adjustable real-valued parameters $\boldsymbol{\theta}$. The latter is then defined by a unitary $U(\boldsymbol{\theta})$ that acts

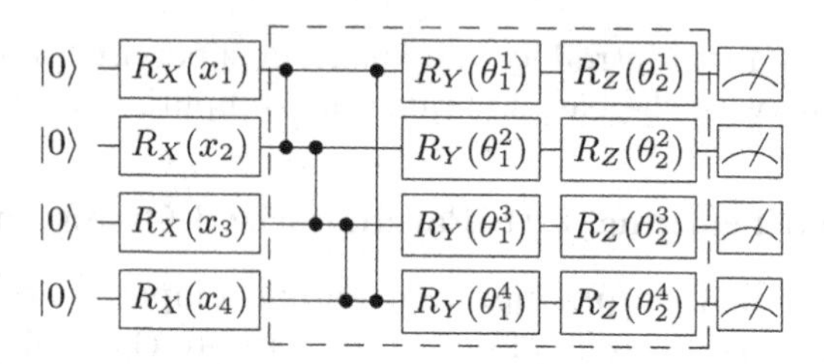

Fig. 2. Parameterized quantum circuit architecture example with 4 qubits and ring connectivity (qubit 1 is connected to 2, 2 to 3, 3 to 4, and 4 to 1 makes a ring). The first layer of R_X is the encoding layer U_{enc}, taking a data instance $x \in \mathbb{R}^4$ as input. It is followed by the entangling part with Ctrl-Z gates. Finally a variational layer U_{var} is applied. Eventually, we do measurements to be converted into predictions for a supervised task. The dashed part can be repeated many times to increase the expressive power of the model.

on a fixed n-qubit state (e.g., $|0^{\otimes n}\rangle$). The ansatz may be constructed using the formulation of the problem at hand (typically the case in chemistry [27] or optimization [10]), or with a problem-independent generic construction. The latter are often designated as *hardware-efficient.*

For a machine learning task, this unitary encodes an input data instance $x \in \mathbb{R}^d$ and is parameterized by a trainable vector $\boldsymbol{\theta}$. Many designs exist but hardware-efficient parameterized quantum circuits [19] with an alternating-layered architecture are often considered in quantum machine learning when no information on the structure of the data is provided. This architecture is depicted in an example presented in Fig. 2 and essentially consists of an alternation of encoding unitaries U_{enc} and variational unitaries U_{var}. In the example, U_{enc} is composed of single-qubit rotations R_X, and U_{var} of single-qubit rotations R_z, R_y and entangling Ctrl-Z gates, represented as ⫶ in Fig. 2, forming the entangling part of the circuit. Such entangling part denoted U_{ent}, can be defined by connectivity between qubits.

These parameterized quantum circuits are similar to neural networks where the circuit architecture is fixed and the gate parameters are adjusted by a classical optimizer such as gradient descent. They have also been named quantum neural networks. The parameterized layer can be repeated multiple times, which increases its *expressive power* like neural networks [43]. The data encoding strategy (such as reusing the encoding layer multiple times in the circuit - a strategy called *data reuploading*) also influences the latter [31,40].

Finally, the user can define the observable(s) and the post-processing method to convert the circuit outputs into a prediction in the case of supervised learning. Commonly, observables based on the single-qubit Z operator are used. When applied on $m \leq n$ qubits, the observable is represented by a $2^m - 1$ square diagonal matrix with $\{-1, 1\}$ values, and is denoted $\mathcal{O} = Z \otimes Z \otimes \cdots \otimes Z$.

Having introduced parameterized quantum circuits, we present the hyperparameters of the models, the configuration space, and the experimental setup for our functional ANOVA-based hyperparameter importance study.

3 Methods

In this section, we describe the network type and its hyperparameters and define the methodology that we follow.

3.1 Hyperparameters and Configuration Space

Many designs have been proposed for parameterized quantum circuits depending on the problem at hand or motivated research questions and contributions. Such propositions can be aggregated and translated into a set of hyperparameters and configuration space for the importance study. As such, we first did an extensive literature review on parameterized quantum circuits for machine learning [2,14,15,17,18,21,23–25,32,38,44,47–50] as well as quantum machine learning software [1,3,7]. This resulted in a list of 10 hyperparameters, presented in Table 1. We choose them so we balance between having well-known hyperparameters that are expected to be important, and less considered ones in the literature. For instance, many works use Adam [20] as the underlying optimizer, and the learning rate should generally be well chosen. On the contrary, the entangling gate used in the parameterized quantum circuit is generally a fixed choice.

From the literature, we expect data encoding strategy/circuit to be important. We choose two main forms for U_{enc}. The first one is the hardware-efficient $\bigotimes_{i=1}^{n} R_X(x_i)$. The second takes the following form from [3,14,17]:

$$U_{\text{enc}}(\boldsymbol{x}) = U_z(\boldsymbol{x})H^{\otimes n} \tag{3}$$

$$U_z(\boldsymbol{x}) = \exp\left(-i\pi\left[\sum_{i=1}^{n} x_i Z_i + \sum_{\substack{j=1,\\ j>i}}^{n} x_i x_j Z_i Z_j\right]\right). \tag{4}$$

Using data-reuploading [31] results in a more expressive model [40], and this was also demonstrated numerically [18,31,44]. Finally, pre-processing of the input is also sometimes used in encoding strategies that directly feed input features into Pauli rotations. It also influences the expressive power of the model [40]. In this work, we choose a usual activation function *tanh* commonly used in neural networks. We do so as its range is $[-1, 1]$, which is the same as the data features during training after the normalization step.

The list of hyperparameters we take into account is non-exhaustive. It can be extended at will, at the cost of more software engineering and budget for running experiments.

3.2 Assessing Hyperparameter Importance

Once the list of hyperparameters and configuration space are decided, we perform the hyperparameter importance analysis with the functional ANOVA framework. Assessing the importance of the hyperparameters boils down to four steps.

Table 1. List of hyperparameters considered for hyperparameter importance for quantum neural network, as we named them in our Tensorflow-Quantum code.

Hyperparameter	Values	Description
Adam learning rate	$[10^{-4}, 0.5]$ (log)	The learning rate with which the quantum neural network starts training. The range was taken from the automated machine learning library Auto-sklearn [11]. We uniformly sample taking the logarithmic scale.
batch size	16,32,64	Number of samples in one batch of Adam used during training
depth	$\{1, 2, \cdots, 10\}$	Number of variational layers defining the circuit
is data_encoding hardware efficient	True, False	Whether we use the hardware-efficient circuit $\bigotimes_{i=1}^{n} R_X(x_i)$ or an IQP circuit defined in Eq. 3 to encode the input data.
use reuploading	True, False	Whether the data encoding layer is used before each variational layer or not.
have less rotations	True, False	If True, only use layers of R_Y, R_Z gates as the variational layer. If False, add a layer of R_X gates.
entangler operation	cz, sqiswap	Which entangling gate to use in U_{ent}
map type	ring, full, pairs	The connectivity used for U_{ent}. The ring connectivity use an entangling gate between consecutive indices $(i, i+1), i \in \{1, \ldots, n\}$ of qubits. The full one uses a gate between each pair of indices $(i, j), i < j$. Pairs connect even consecutive indices first, then odd consecutive ones.
input activation function	linear, tanh	Whether to input $tanh(x_i)$ as rotations or just x_i.
output circuit	2Z, mZ	The observable(s) used as output(s) of the circuit. If 2Z, we use all possible pairs of qubit indices defining $Z \otimes Z$. If mZ, the tensor product acts on all qubits. Note we do not use single-qubit Z observables although they are quite often used in the literature. Indeed, they are provably not using the entire circuit when it is shallow. Hence we decided to use $Z \otimes Z$ instead. Also, a single neuron layer with a sigmoid activation function is used as a final decision layer similar to [38]

Firstly, the models are applied to various datasets by sampling various configurations in a hyperparameter optimization process. The performances or metrics of the models are recorded along. The sampled configurations and performances serve as data for functional ANOVA. As functional ANOVA uses internally tree-based surrogate models, namely random forests [6], we decided to add an extra step with reference to [34]. In the second step, we verify the performance of the internal surrogate models. We cross-evaluate them using regression metrics commonly used in surrogate benchmarks [9]. Surrogates performing badly at this step are then discarded from the importance analysis, as they can deteriorate the quality of the study. Thirdly, the marginal contribution of each hyperparameter over all datasets can be then obtained and used to infer a ranking of their importance. Finally, a verification step similar to [34] is carried out to confirm the inferred ranking previously obtained. We explain such a procedure in the following section.

3.3 Verifying Hyperparameter Importance

When applying the functional ANOVA framework, an extra verification step is added to confirm the output from a more intuitive notion of hyperparameter importance [34]. It is based on the assumption that hyperparameters that perform badly when fixed to a certain value (while other hyperparameters are optimized), will be important to optimize. The authors of [34] proposed to carry out a costly random search procedure fixing one hyperparameter at a time. In order to avoid a bias to the chosen value to which this hyperparameter is fixed, several values are chosen, and the optimization procedure is carried out multiple times. Formally, for each hyperparameter θ_j we measure $y^*_{j,f}$ as the result of a random search for maximizing the metric, fixing θ_j to a given value $f \in F_j, F_j \subseteq \Theta_j$. For categorical θ_j with domain Θ_j, $F_j = \Theta_j$ is used. For numeric θ_j, the authors of [34] use a set of 10 values spread uniformly over θ_j's range. We then compute $y^*_j = \frac{1}{|F_j|}\sum_{f\in F_j} y^*_{j,f}$, representing the score when not optimizing hyperparameter θ_j, averaged over fixing θ_j to various values it can take. Hyperparameters with lower values for y^*_j are assumed to be more important since the performance should deteriorate more when set sub-optimally.

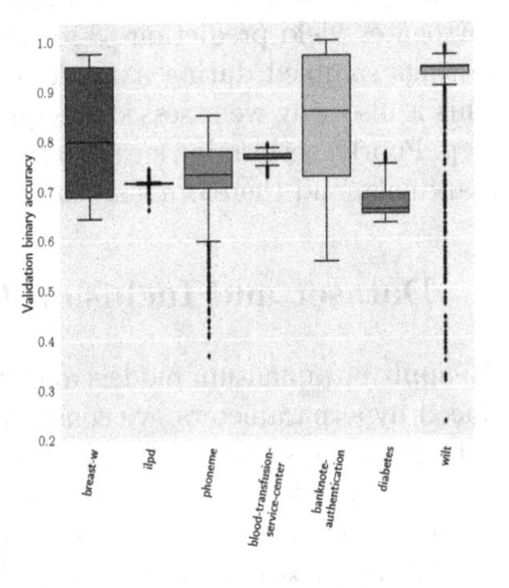

Fig. 3. Performances of 1 000 quantum machine learning models defined by different configurations of hyperparameters over each dataset. The metric of interest in the study is the 10-fold cross-validation accuracy. We take the best-achieved metric per model trained over 100 epochs.

In our study, we extend this framework to be used on the scale of quantum machine learning models. As quantum simulations can be very expensive, we

Table 2. List of datasets used in this study. The number of features is obtained after a usual preprocessing used in machine learning methods, such as one-hot-encoding.

Dataset	OpenML Task ID	Number of features	Number of instances
breast-w	15	9	699
diabetes	37	8	768
phoneme	9952	5	5 404
ilpd	9971	11	583
banknote-authentication	10093	4	1 372
blood-transfusion-service-center	10 101	4	748
wilt	146820	5	4 839

carry out the verification experiment by using the predictions of the surrogate instead of fitting new quantum models during the verification experiment. The surrogates yield predictions $\hat{y}$ for the performance of arbitrary hyperparameter settings sampled during a random search. Hence, they serve to compute $y^*_{j,f}$. This is also why we assessed the quality of the built-in surrogates as the second step. Poorly-performing surrogates can deteriorate the quality of the constructed marginals, and therefore lead to poorly-supported conclusions.

4 Dataset and Inclusion Criteria

To apply our quantum models and study the importance of the previously introduced hyperparameters, we consider classical datasets. Similarly to [34], we use datasets from the OpenML-CC18 benchmark suite [4]. In our study, we consider only the case where the number of qubits available is equal to the number of features, a common setting in the quantum community. As simulating quantum circuits is a costly task, we limit this study to the case where the number of features is less than 20 after preprocessing.[1] Our first step was to identify which datasets fit this criterion. We include all datasets from the OpenML-CC18 that have 20 or fewer features after categorical hyperparameters have been one-hot-encoded, and constant features are removed. Afterwards, the input variables are also scaled to unit variance as a normalization step. The scaling constants are calculated on the training data and applied to the test data.

The final list of datasets is given in Table 2. In total, 7 datasets fitted the criterion considered in this study. For all of them, we picked the OpenML Task ID giving the 10-fold cross-validation task. A quantum model is then applied using the latter procedure, with the aforementioned preprocessing steps.

[1] A 10-fold cross-validation run in our experiment takes on average 262 minutes for 100 epochs with Tensorflow Quantum [7].

5 Results

In this section, we present the results obtained using the hyperparameters and the methodology defined in Sect. 3 with the datasets described in Sect. 4. First, we show the distribution of performances obtained during a random search where configurations are independently sampled for each dataset. Then we carry out the surrogate verification. Finally, we present the functional ANOVA results in terms of hyperparameter importance with marginal contributions and the random search verification per hyperparameter.

5.1 Performance Distributions per Dataset

For each dataset, we sampled independently 1 000 hyperparameter configurations and run the quantum models for 100 epochs as budget. As a performance measure, we recorded the best validation accuracy obtained over 100 epochs. Figure 3 shows the distribution of the 10-fold cross-validation accuracy obtained per dataset. We observe the impact of hyperparameter optimization by the difference between the least performing and the best model configuration. For instance, on the wilt dataset, the best model gets an accuracy close to 1, and the least below 0.25. We can also see that some datasets present a smaller spread of performances. ilpd and blood-transfusion-service-center are in this case. It seems that hyperparameter optimization does not have a real effect, because most hyperparameter configurations give the same result. As such, the surrogates could not differentiate between various configurations. In general, hyperparameter optimization is important for getting high performances per dataset and detecting datasets where the importance study can be applied.

5.2 Surrogate Verification

Functional ANOVA relies on an internal surrogate model to determine the marginal contribution per hyperparameter. If this surrogate model is not accurate, this can have a severe limitation on the conclusions drawn from functional ANOVA. In this experiment, we verify whether the hyperparameters can explain the performances of the models. Table 3 shows the performance of the internal surrogate models. We notice low regression scores for the two datasets (less than 0.75 R2 scores). Hence we remove them from the analysis.

5.3 Marginal Contributions

For functional ANOVA, we used 128 trees for the surrogate model. Figure 4(a,b) shows the marginal contribution of each hyperparameter over the remaining 5 datasets. We distinguish 3 main levels of importance. According to these results, the learning rate, depth, and the data encoding circuit and reuploading strategy are critical. These results are in line with our expectations. The entangler gate, connectivity, and whether we use R_X gates in the variational layer are the least important according to functional ANOVA. Hence, our results reveal new insights into these hyperparameters that are not considered in general.

5.4 Random Search Verification

In line with the work of [34], we perform an additional verification experiment that verifies whether the outcomes of functional ANOVA are in line with our expectations. However, the verification procedure involves an expensive, post-hoc analysis: a random search procedure fixing one hyperparameter at a time. As our quantum simulations are costly, we used the surrogate models fitted on the cur-

Table 3. Performances of the surrogate models built within functional ANOVA over a 10-fold cross-validation procedure. We present the average coefficient of determination (R2), root mean squared error (RMSE), and Spearman's rank correlation coefficient (CC). These are common regression metrics for benchmarking surrogate models on hyperparameters [9]. The surrogates over ilpd and blood-transfusion-service-center obtain low scores (less than .75 R2), hence we remove them from the study.

Dataset	R2 score	RMSE	CC
breast-w	0.8663	0.0436	0.9299
diabetes	0.7839	0.0155	0.8456
phoneme	0.8649	0.0285	0.9282
ilpd	0.1939	0.0040	0.4530
banknote-authentication	0.8579	0.0507	0.9399
blood-transfusion-service-center	0.6104	0.0056	0.8088
wilt	0.7912	0.0515	0.8015

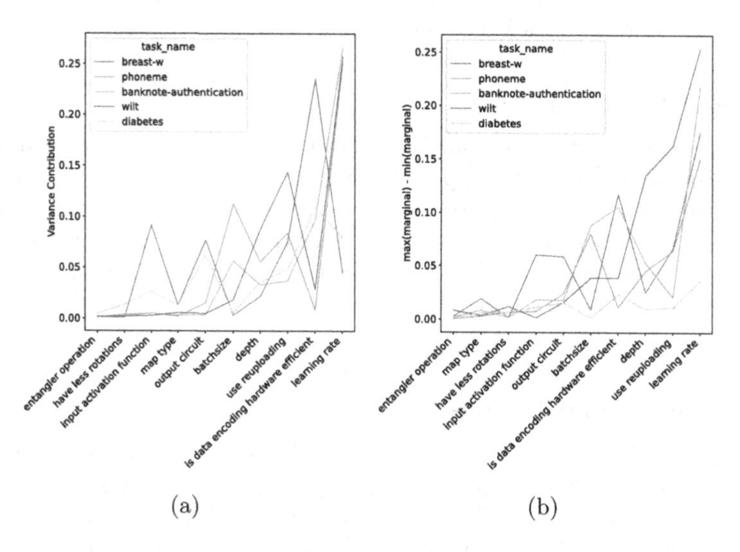

Fig. 4. The marginal contributions per dataset are presented as a) the variance contribution and b) the difference between the minimal and maximal value of the marginal of each hyperparameter. The hyperparameters are sorted from the least to most important using the median. We distinguish from the plot 3 main levels of importance.

rent dataset considered over the 1 000 configurations obtained initially to predict the performances one would obtain when presented with a new configuration.

Figure 5 shows the average rank of each run of random search, labeled with the hyperparameter whose value was fixed to a default value. A high rank implies poor performance compared to the other configurations, meaning that tuning this hyperparameter would have been important. We witness again the 3 levels of importance, with almost the same order obtained. However, the input_activation_function is deemed more important while batch size is less.

More simulations with more datasets may be required to validate the importance. However, we retrieve empirically the importance of well-known hyperparameters while considering less important ones. Hence functional ANOVA becomes an interesting tool for quantum machine learning in practice.

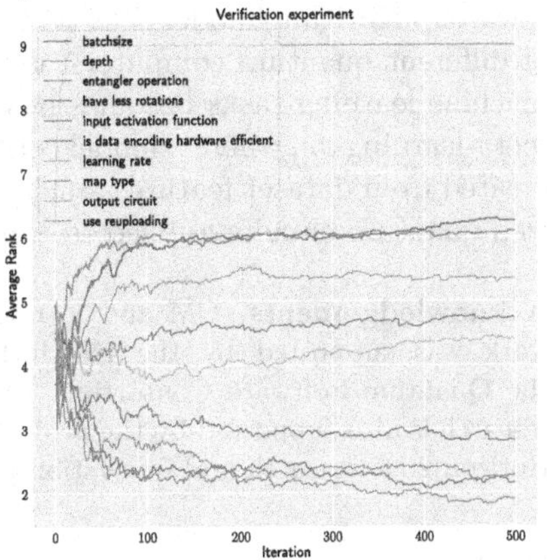

Fig. 5. Verification experiment of the importance of the hyperparameters. A random search procedure up to 500 iterations excluding one parameter at a time is used. A lower curve means the hyperparameter is deemed less important.

6 Conclusion

In this work, we study the importance of hyperparameters related to quantum neural networks for classification using the functional ANOVA framework. Our experiments are carried out over OpenML datasets that match the current scale of quantum hardware simulations (i.e., datasets that have at most 20 features after pre-processing operators have been applied, hence using 20 qubits). We selected and presented the hyperparameters from an aggregation of quantum computing literature and software. Firstly, hyperparameter optimization highlighted datasets where we observed high differences between configurations. This underlines the importance of hyperparameter optimization for these datasets. There were also datasets that showed little difference. These led us to extend the methodology by adding an additional verification step of the internal surrogate performances. From our results, we distinguished 3 main levels of importance. On the one hand, Adam's learning rate, depth, and the data encoding strategy are deemed very important, as we expected. On the other hand, the less considered hyperparameters such as the particular choice of the entangling gate and using 3 rotation types in the variational layer are in the least important group. Hence, our experiment both confirmed expected patterns and revealed new insights for quantum model selection.

For future work, we plan to further investigate methods from the field of automated machine learning to be applied to quantum neural networks [5,11,26]. Indeed, our experiments have shown the importance of hyperparameter optimization, and this should become standard practice and part of the protocols applied within the community. We further envision functional ANOVA to be employed in future works related to quantum machine learning and understanding how to apply quantum models in practice. For instance, it would be interesting to consider quantum data, for which quantum machine learning models may have an advantage. Plus, extending hyperparameter importance to techniques for scaling to a large number of features with the number of qubits, such as dimensionality reduction or divide-and-conquer techniques, can be left for future work. Finally, this type of study can also be extended to different noisy hardware and towards algorithm/model selection and design. If we have access to a cluster of different quantum computers, then choosing which hardware works best for machine learning tasks becomes possible. One could also extend our work with meta-learning [5], where a model configuration is selected based on meta-features created from dataset features. Such types of studies already exist for parameterized quantum circuits applied to combinatorial optimization [28,29,37].

Acknowledgements. CM and VD acknowledge support from TotalEnergies. This work was supported by the Dutch Research Council (NWO/OCW), as part of the Quantum Software Consortium programme (project number 024.003.037). This research is also supported by the project NEASQC funded from the European Union's Horizon 2020 research and innovation programme (grant agreement No. 951821).

References

1. ANIS, M.S., et al.: Qiskit: an open-source framework for quantum computing (2021). https://doi.org/10.5281/zenodo.2573505
2. Benedetti, M., Lloyd, E., Sack, S., Fiorentini, M.: Parameterized quantum circuits as machine learning models. Quantum Sci. Technol. **4**(4), 043001 (2019)
3. Bergholm, V., Izaac, J.A., Schuld, M., Gogolin, C., Killoran, N.: Pennylane: Automatic differentiation of hybrid quantum-classical computations. CoRR abs/1811.04968 (2018)
4. Bischl, B., et al.: OpenML benchmarking suites. In: Proceedings of the Neural Information Processing Systems Track on Datasets and Benchmarks (2021)
5. Brazdil, P., van Rijn, J.N., Soares, C., Vanschoren, J.: Metalearning: Applications to Automated Machine Learning and Data Mining. Springer, 2nd edn. (2022). https://doi.org/10.1007/978-3-030-67024-5
6. Breiman, L.: Random forests. Mach. Learn. **45**(1), 5–32 (2001)
7. Broughton, M., et al.: TensorFlow quantum: a software framework for quantum machine learning. arXiv:2003.02989 (2020)
8. Caro, M.C., Gil-Fuster, E., Meyer, J.J., Eisert, J., Sweke, R.: Encoding-dependent generalization bounds for parametrized quantum circuits. Quantum **5**, 582 (2021)
9. Eggensperger, K., Hutter, F., Hoos, H.H., Leyton-Brown, K.: Efficient benchmarking of hyperparameter optimizers via surrogates. In: Proceedings of the Twenty-Ninth AAAI Conference on Artificial Intelligence, pp. 1114–1120. AAAI Press (2015)
10. Farhi, E., Goldstone, J., Gutmann, S.: A quantum approximate optimization algorithm. arXiv:1411.4028 (2014)
11. Feurer, M., Eggensperger, K., Falkner, S., Lindauer, M., Hutter, F.: Auto-Sklearn 2.0: hands-free automl via meta-learning. arXiv:2007.04074v2 [cs.LG] (2021)

12. Georgescu, I.M., Ashhab, S., Nori, F.: Quantum simulation. Review of modern. Physics **86**, 153–185 (2014)
13. Haug, T., Self, C.N., Kim, M.S.: Large-scale quantum machine learning. CoRR abs/2108.01039 (2021)
14. Havlíček, V., et al.: Supervised learning with quantum-enhanced feature spaces. Nature **567**(7747), 209–212 (2019)
15. Heimann, D., Hohenfeld, H., Wiebe, F., Kirchner, F.: Quantum deep reinforcement learning for robot navigation tasks. CoRR abs/2202.12180 (2022)
16. Hutter, F., Hoos, H., Leyton-Brown, K.: An efficient approach for assessing hyper-parameter importance. In: Proceedings of the 31th International Conference on Machine Learning, ICML 2014. JMLR Workshop and Conference Proceedings, vol. 32, pp. 1130–1144 (2014)
17. Jerbi, S., Fiderer, L.J., Nautrup, H.P., Kübler, J.M., Briegel, H.J., Dunjko, V.: Quantum machine learning beyond kernel methods. CoRR abs/2110.13162 (2021)
18. Jerbi, S., Gyurik, C., Marshall, S., Briegel, H.J., Dunjko, V.: Parametrized quantum policies for reinforcement learning. In: Advances in Neural Information Processing Systems 34, pp. 28362–28375 (2021)
19. Kandala, A., et al.: Hardware-efficient variational quantum eigensolver for small molecules and quantum magnets. Nature **549**(7671), 242–246 (2017)
20. Kingma, D.P., Ba, J.: Adam: a method for stochastic optimization (2015)
21. Liu, J.G., Wang, L.: Differentiable learning of quantum circuit born machines. Phys. Rev. A **98**, 062324 (2018)
22. Liu, Y., Arunachalam, S., Temme, K.: A rigorous and robust quantum speed-up in supervised machine learning. Nat. Phys. **17**(9), 1013–1017 (2021)
23. Marshall, S.C., Gyurik, C., Dunjko, V.: High dimensional quantum machine learning with small quantum computers. CoRR abs/2203.13739 (2022)
24. Mensa, S., Sahin, E., Tacchino, F., Barkoutsos, P.K., Tavernelli, I.: Quantum machine learning framework for virtual screening in drug discovery: a prospective quantum advantage. CoRR abs/2204.04017 (2022)
25. Mitarai, K., Negoro, M., Kitagawa, M., Fujii, K.: Quantum circuit learning. Phys. Rev. A **98**, 032309 (2018)
26. Mohr, F., van Rijn, J.N.: Learning curves for decision making in supervised machine learning - a survey. CoRR abs/2201.12150 (2022)
27. Moll, N., et al.: Quantum optimization using variational algorithms on near-term quantum devices. Quantum Sci. Technol. **3**(3), 030503 (2018)
28. Moussa, C., Calandra, H., Dunjko, V.: To quantum or not to quantum: towards algorithm selection in near-term quantum optimization. Quantum Sci. Technol. **5**(4), 044009 (2020)
29. Moussa, C., Wang, H., Bäck, T., Dunjko, V.: Unsupervised strategies for identifying optimal parameters in quantum approximate optimization algorithm. EPJ Quantum Technol. 9(1) (2022)
30. Nielsen, M.A., Chuang, I.L.: Quantum Computation and Quantum Information: 10th Anniversary. Cambridge University Press, New York (2011)
31. Pérez-Salinas, A., Cervera-Lierta, A., Gil-Fuster, E., Latorre, J.I.: Data re-uploading for a universal quantum classifier. Quantum **4**, 226 (2020)
32. Peters, E., et al.: Machine learning of high dimensional data on a noisy quantum processor. NPJ Quantum Inf. **7**(1), 161 (2021)
33. Preskill, J.: Quantum Computing in the NISQ era and beyond. Quantum **2**, 79 (2018)

34. van Rijn, J.N., Hutter, F.: Hyperparameter importance across datasets. In: Proceedings of the 24th ACM SIGKDD International Conference on Knowledge Discovery & Data Mining, KDD 2018, pp. 2367–2376. ACM (2018)
35. Sajjan, M., et al.: Quantum computing enhanced machine learning for physicochemical applications. CoRR arXiv:2111.00851 (2021)
36. Saltelli, A., Sobol, I.: Sensitivity analysis for nonlinear mathematical models: numerical experience. Matematicheskoe Modelirovanie 7 (1995)
37. Sauvage, F., Sim, S., Kunitsa, A.A., Simon, W.A., Mauri, M., Perdomo-Ortiz, A.: Flip: a flexible initializer for arbitrarily-sized parametrized quantum circuits. CoRR abs/2103.08572 (2021)
38. Schetakis, N., Aghamalyan, D., Boguslavsky, M., Griffin, P.: Binary classifiers for noisy datasets: a comparative study of existing quantum machine learning frameworks and some new approaches. CoRR abs/2111.03372 (2021)
39. Schuld, M., Killoran, N.: Is quantum advantage the right goal for quantum machine learning? Corr abs/2203.01340 (2022)
40. Schuld, M., Sweke, R., Meyer, J.J.: Effect of data encoding on the expressive power of variational quantum-machine-learning models. Phys. Rev. A **103**, 032430 (2021)
41. Sharma, A., van Rijn, J.N., Hutter, F., Müller, A.: Hyperparameter importance for image classification by residual neural networks. In: Kralj Novak, P., Šmuc, T., Džeroski, S. (eds.) DS 2019. LNCS (LNAI), vol. 11828, pp. 112–126. Springer, Cham (2019). https://doi.org/10.1007/978-3-030-33778-0_10
42. Shor, P.W.: Polynomial-time algorithms for prime factorization and discrete logarithms on a quantum computer. SIAM Rev. **41**, 303–332 (1999)
43. Sim, S., Johnson, P.D., Aspuru-Guzik, A.: Expressibility and entangling capability of parameterized quantum circuits for hybrid quantum-classical algorithms. Advanced Quantum Technologies **2**(12), 1900070 (2019)
44. Skolik, A., Jerbi, S., Dunjko, V.: Quantum agents in the gym: a variational quantum algorithm for deep q-learning. CoRR abs/2103.15084 (2021)
45. Sobol, I.M.: Sensitivity estimates for nonlinear mathematical models. Math. Model. Comput. Exp. **1**(4), 407–414 (1993)
46. Sweke, R., Seifert, J., Hangleiter, D., Eisert, J.: On the quantum versus classical learnability of discrete distributions. Quantum **5**, 417 (2021)
47. Wang, H., Gu, J., Ding, Y., Li, Z., Chong, F.T., Pan, D.Z., Han, S.: QuantumNAT: quantum noise-aware training with noise injection, quantization and normalization. CoRR abs/2110.11331 (2021)
48. Wang, H., Li, Z., Gu, J., Ding, Y., Pan, D.Z., Han, S.: QOC: quantum on-chip training with parameter shift and gradient pruning. CoRR abs/2202.13239 (2022)
49. Wossnig, L.: Quantum machine learning for classical data. CoRR abs/2105.03684 (2021)
50. Zoufal, C., Lucchi, A., Woerner, S.: Quantum generative adversarial networks for learning and loading random distributions. NPJ Quantum Inf. **5**(1), 103 (2019)

ImitAL: Learned Active Learning Strategy on Synthetic Data

Julius Gonsior[1(✉)], Maik Thiele[2], and Wolfgang Lehner[1]

[1] Technische Universität Dresden, Dresden, Germany
{julius.gonsior,wolfgang.lehner}@tu-dresden.de
[2] Hochschule für Technik und Wirtschaft Dresden, Dresden, Germany
maik.thiele@htw-dresden.de

Abstract. Active Learning (AL) is a well-known standard method for efficiently obtaining annotated data by first labeling the samples that contain the most information based on a query strategy. In the past, a large variety of such query strategies has been proposed, with each generation of new strategies increasing the runtime and adding more complexity. However, to the best of our knowledge, none of these strategies excels consistently over a large number of datasets from different application domains. Basically, most of the existing AL strategies are a combination of the two simple heuristics *informativeness* and *representativeness*, and the big differences lie in the combination of the often conflicting heuristics. Within this paper, we propose ImitAL, a domain-independent novel query strategy, which encodes AL as a learning-to-rank problem and learns an optimal combination between both heuristics. We train ImitAL on large-scale simulated AL runs on purely synthetic datasets. To show that ImitAL was successfully trained, we perform an extensive evaluation comparing our strategy on 13 different datasets, from a wide range of domains, with 7 other query strategies.

Keywords: Annotation · Active learning · Imitation learning · Learning to rank

1 Introduction

Machine Learning (ML) has found applications across a wide range of domains and impacts (implicitly) nearly every aspect of nowaday's life. Still, one of the most limiting factors of successful application of ML is the absence of labels for a training set. Usually, domain experts that are rare and costly are required to obtain a labeled dataset. Thus, to improve the manual label task is a prime object to improve. For example, the average cost for the common label task of segmenting a single image reliably is 6,40 USD[1].

Reducing the amount of necessary human input into the process of generating labeled training sets is of utmost importance to make ML projects possible and

[1] According to scale.ai as of December 2021.

P. Pascal and D. Ienco (Eds.): DS 2022, LNAI 13601, pp. 47–56, 2022.
https://doi.org/10.1007/978-3-031-18840-4_4

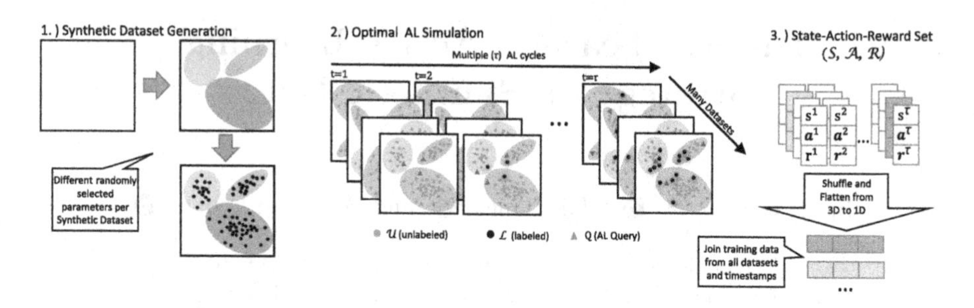

Fig. 1. General overview on the training procedure of ImitAL

scalable. A standard approach to reduce the number of required labels without compromising the quality of the trained ML model, is to exploit Active Learning (AL). The approach consists of an iterative process of selecting exactly those unlabeled samples for labeling by the domain experts that benefit the to-be trained model the most. Given a small initial labeled dataset $\mathcal{L} = \{(x_i, y_i)\}_i^n$ of n samples x_i with the respective labels y_i and a large unlabeled pool $\mathcal{U} = \{x_i\}, x_i \notin \mathcal{L}$, an ML model called *learner* θ is trained on the labeled set. A *query strategy* then subsequently chooses a batch of b unlabeled samples Q, which will be labeled by the human experts and added to the set of labeled data $\mathcal{L}$. This AL cycle repeats τ times until a stopping criterion is met.

The challenge of applying AL is the almost paradoxical problem to be solved: how to decide, which samples are most beneficial to the ML model, without knowing the label of the samples, since this is exactly the task to be learned by the to-be-trained ML model.

During the past years, many different AL query strategies have been proposed, but to our knowledge, none excels consistently over a large number of datasets and from different application domains. By deliberately focusing on domain-independent AL strategies we aim to shed some light onto this problem. Even though various extensive general survey papers [14,15,20] exist, no clearly superior AL strategy has been identified. The results of the individual evaluations in papers with newly proposed AL strategies suggest that current AL strategies highly depend on the underlying dataset domain. Even more interestingly, the naïve baseline of randomly selecting samples often achieves surprisingly competitive results [7,9,11,13].

At its core, the vast majority of AL strategies rely on the same set of two simple heuristics: *informativeness* and *representativeness*. The first favors samples that foremost improve the classification model, whereas the latter favors samples that represent the overall sample distribution in the feature vector space. Most recent AL strategies add more layers of complexity on top of the two heuristics in their purest form, often resulting in excessive runtimes. This renders many AL strategies unusable in large-scale and interactively operating labeling projects, which are exactly those projects that would benefit the most from "optimal" learning strategies.

We are presenting IMITAL, a novel AL strategy, which at its core is a Neural Network (NN) trained on very large simulated AL episodes with the goal to optimally combine the basic AL heuristics informativeness and representativeness. As it is not practically feasible to enumerate all possible real-world datasets as training data in the simulations, we are approximating them by using synthetic datasets instead. The benefit of synthetic datasets is that we can leverage the knowledge about all the labels to construct an optimal AL strategy, which then serves as training basis for IMITAL. Our work falls therefore under the category of "learning AL strategies". According to our knowledge, our approach is, in contrast to similar works [9,11,13], the first one to solely utilize purely synthetical data to train the strategy. We can present a pre-trained, ready-to-apply AL strategy which can be applied without any further necessary transfer-learning or fine-tuning in any domain.

We start in Sect. 2 by presenting our synthetic datasets simulation process, followed by our Imitation Learning (IL) procedure in Sect. 3. In Sect. 4, we are comparing IMITAL with 7 common AL strategies on 13 real-world datasets and conclude in Sect. 5.

2 Simulating AL on Synthetic Training Data

For the IL training procedure of IMITAL we need an *expert* AL strategy, which the neural network behind IMITAL can learn to imitate. In order to capture the characteristics of "all" possible datasets we pursue the idea by generating initially nearly "infinite" synthetic datasets[2] and computing an optimal AL strategy on them, leveraging the information about the known full labels for the synthetic datasets.

We construct the nearly-optimal strategy by selecting a batch of those samples for labeling, which will result in the highest accuracy (in the following called *reward*), if they each were added to the set of labeled samples $\mathcal{L}$. As this process is computationally heavy, we do not consider all possible batches, but perform a *pre-selection* based on a heuristic, which selects a promising and diverse set of the top-k batches. Details of the pre-selection are explained in Sect. 3.

The results of the AL simulation for each AL cycle t for a specific synthetic dataset is a *state-action-reward triple.* The state s is represented as a triple $s = (\mathcal{U}^t, \mathcal{L}^t, \theta^t)$, consisting of the set of unlabeled samples $\mathcal{U}^t$, the set of labeled samples $\mathcal{L}^t$, and the state of the learner model θ^t trained on $\mathcal{L}^t$. The corresponding actions $\boldsymbol{a}_s$ is a set of the pre-selected queries x, whereas the respective rewards $\boldsymbol{r}_s$ for each of these actions is a set of rewards r. The optimal choice $Q_s^t \in \boldsymbol{a}_s$ for the AL cycle t can be easily computed from the given accuracies – the action with the highest future accuracy. This simulation is repeated α-times using different synthetic datasets. The accumulated state-action-reward pairs, denoted as the triple $(\mathcal{S}, \mathcal{A}, \mathcal{R})$, reflect then the input for IL training procedure

[2] For generating the synthetic datasets the algorithm by [4], which is a runtime efficient method for creating a diverse range of synthetic datasets of varying shape and resulting classification hardness, is used.

of the NN of IMITAL. The whole synthetic data training generation and training of IMITAL has to be done only once, afterwards it is applicable to real-world datasets without any further transfer learning or fine-tuning steps.

3 Training a Neural Network by Imitation Learning

The final step of IMITAL is to use the generated state-action-reward triples $(\mathcal{S}, \mathcal{A}, \mathcal{R})$ for training an NN as AL query strategy. Therefore, we are deploying the ML technique IL [12], where demonstrated expert actions are being replicated by the ML model. The training task for IMITAL is to find patterns in the presented actions.

Subsequently (Sect. 3.1) we will first explain the IL learning process, followed by the details of the NN input and output encoding (Sect. 3.2), and lastly, the necessary pre-selection process (Sect. 3.3).

3.1 Imitation Learning

For training IMITAL we use Imitation Learning (IL), where an expert demonstrates an optimal strategy, which the neural network behind IMITAL learns to replicate. We use *behavioral cloning* [12] as a variant of IL, which reduces IL to a regular regression ML problem. The desired outcome is a trained strategy returning the optimal action for a given state. We use the state-action-reward set $(\mathcal{S}, \mathcal{A}, \mathcal{R})$ to extract an optimal strategy $\hat{\pi}$, which we are then demonstrating to the to-be-trained network $\hat{\pi}(s) = \text{argmax}_{x \in \boldsymbol{a}_s, r_x \in \boldsymbol{r}_s}(r_x)$. For a given state s, the action set $\mathcal{A}$ contains all pre-selected actions $\boldsymbol{a}_s$ for this state; the reward set $\mathcal{R}$ contains the respective rewards $\boldsymbol{r}_s$. As the optimal strategy only contains the optimal actions, it can be used to construct the optimal batch by taking the b-highest actions. In other words: we train a network $\widehat{\pi}$ predicting for a given state s and a possible action $x \in \boldsymbol{a}_s$ – which equals labeling the sample represented by this action – the reward $r_x \in \boldsymbol{r}_s$. The expected future accuracy is in our case demonstrated by the true reward $\dot{r}$ function as $\dot{r}(s, x) = r_x$.

3.2 Neural Network Input and Output Encoding

Before using the state-action-reward set $(\mathcal{S}, \mathcal{A}, \mathcal{R})$ to train the network predicting the future accuracy, we first transform it into a fixed sized vector representation using feature encoding, and thus making IMITAL dataset agnostic. NNs are limited by the number of the neurons to either a fixed size input, or when using *recurrent* NN to circumvent this limitation, they often suffer from the case of memory loss where the beginning of the input sequence is forgotten due to exploding or vanishing gradients [6]. The last problem occurs more frequently the larger and more length-varying the input is, which is the case for AL. The raw actions set $\mathcal{A}$ may then contain – depending on the number of unlabeled samples – many possible samples, or just a few, varying again drastically. That

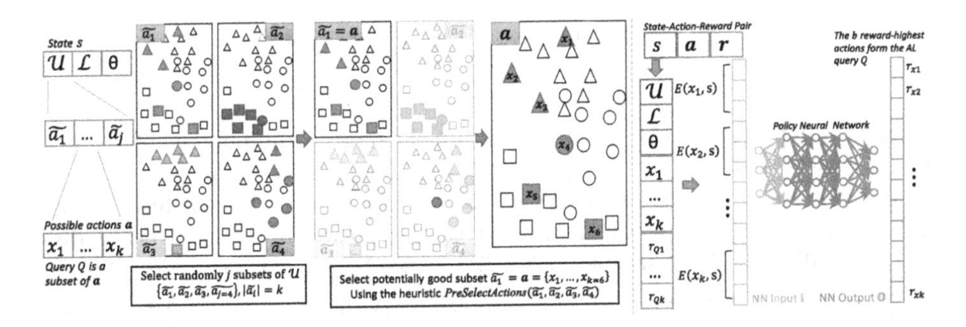

Fig. 2. Pre-selection process and action meaning for IMITAL, example for $j = 4$, $k = 6$, and $b = 3$, and encoding of a state-action-triple

underpins the already mentioned *pre-selection* method, reducing the number of possible actions to a fixed size.

The transformation of the state-action-reward set into a suitable form for the network is called *input* and *output encoding*. Figure 2 displays the general procedure of the encoding to the right. We chose a *listwise* input encoding, where we enter k possible actions $x \in \boldsymbol{a}_s, |\boldsymbol{a}_s| = k$ at once into the network, in contrast to a *pointwise* encoding, where a single action is entered at a time. This has the benefit of enabling the network to compare each possible action relatively to the others, enabling IMITAL to take batch-aware AL query decisions. Batch-awareness is, as thoroughly explained in [8], a beneficial and desirable property of AL query strategies, meaning that the joint diversity of all samples of the final AL batch Q is taken into account. The input of the network is defined by the vector $\mathbb{I}_s = \{E(x, s)|x \in \boldsymbol{a}_s\}$, with E being the encoding of the action x. The output $\mathbb{O} = (\widehat{r}_1, \ldots, \widehat{r}_k)$ of the network consists of exactly $|\boldsymbol{r}_s|$ output neurons, one for each of the predicted accuracies $\widehat{r}_x$ for the respective possible actions x. The amount of output neurons equals therefore the amount of pre-selected actions: $|\boldsymbol{r}_s| = |\boldsymbol{a}_s| = k$. We use a final softmax layer of k output neurons, each per possible action. The b highest output neurons indicate the samples for the unlabeled query Q.

A single action represents an unlabeled sample $x \in \mathcal{U}$. The input encoding function $E(x, s)$ defines on what basis the network can make the AL query strategy decision. We use the state $s = (\mathcal{U}, \mathcal{L}, \theta)$ to calculate the encoding, which is a 5-tuple consisting of multiple parts, the individual functions will be explained in the following paragraphs $E(x, s) = (u_1(x, \theta), u_2(x, \theta), u_3(x, \theta), dl(x, \mathcal{L}), du(x, \mathcal{U}))$. The complete network input vector $\mathbb{I}$ consists then of 5-times k values, an encoded 5-tuple for each unlabeled sample x out of the set of possible actions $\boldsymbol{a}$ $\mathbb{I}_s = \{E(x_1, s), \ldots, E(x_k, s)\}, x \in \boldsymbol{a}_s$ As mentioned in the beginning, a good AL query strategy takes informativeness as well as representativeness into account. Informativeness is derived by $u_i(x, \theta)$, a function computing the uncertainty of the learner θ for the i-th most probable class for the sample $x \in \mathcal{U}$, given the probability of the learner $P_\theta(y|x)$ in classifying x with the label y:

$$u_i(x,\theta) = \begin{cases} P_\theta\Big(\big(\text{argmax}_{y,i} P_\theta(y|x)\big) \Bigm| x\Big), & \text{if } i \leq C \\ 0, & \text{otherwise} \end{cases} \tag{1}$$

argmax$_{_,i}$ denotes the i-th maximum argument, and C the number of classification classes.

For representativeness we compute $dl(x, \mathcal{L})$ and $du(x,\mathcal{U})$, the first denoting the average distance to all labeled samples, the latter the average distance to all unlabeled samples $dl(x, \mathcal{L}) = \frac{1}{|\mathcal{L}|}\sum_{x_l \in \mathcal{L}} d(x, x_l)$, $du(x,\mathcal{U}) = \frac{1}{|\mathcal{U}|}\sum_{x_u \in x} d(x, x_u)$, where $d(x_1, x_2)$ is an arbitrary distance metric between the points x_1 and x_2. We use the Euclidean distance for small feature vector spaces, and recommend using the cosine distance for high-dimensional feature vector space. Both feature encoding functions represent the most raw, unpreprocessed forms of informativeness and representativeness, as the neural network should learn necessary transformations of the feature vector space.

3.3 Pre-selection

Instead of considering all possible actions, we pre-select promising actions, whose individual samples have the largest diversity and whose individual samples are the furthest away from each other, similar to [8]. The pre-selection fulfills two objectives: first and foremost, we can present a fixed amount of actions to the network, and secondly it keeps the runtime of the simulations within a processable range. A positive side effect of the fixed-size input of the network is the low and static runtime of ImitAL, which is almost independent of the size of the dataset. The effect is especially apparent with very large datasets.

We start the pre-selection by drawing randomly j possible actions $\{\widetilde{a}_1, \ldots, \widetilde{a}_j\}$, with each $\widetilde{a}$ being a subset of $\mathcal{U}$. After that we use a heuristic to select the top-k most promising actions $\boldsymbol{a}$ out of the random ones. The pre-selection process is illustrated in Fig. 2 at the left side.

We are using a heuristic to filter out potentially uninteresting actions. By calculating the average distance to the already labeled samples of all the samples in each possible action set $\widetilde{a}$ and select the action set $\boldsymbol{a}$ having the highest average distance: $\boldsymbol{a} = \text{argmax}_{\widetilde{a}} \sum_{x \in \widetilde{a}} dl(x, \mathcal{L})$, where $\widetilde{a}$ contains k unlabeled samples. Thus, we are ensuring that we sample evenly distributed from each region in the sample vector space during the training process. We compute the heuristic for j random possible batches, instead of all possible subsets of $\mathcal{U}$.

4 Evaluation

The goal of ImitAL is to learn a domain-independent AL strategy from synthetic datasets, which combines the strength of both the basic informativeness and the representativeness heuristics. For evaluation, we are comparing ImitAL therefore with 7 AL strategies on 13 real-world datasets from varying domains.

4.1 Experiment Details

The datasets are from the UCI ML Repository [1] with varying sample size, feature size, and application domain, similar to the evaluations of [7,9,13]. As an additional larger dataset the table classification dataset DWTC [2] was also included. For the experiments we started with a single random sample of each class, and ran the AL loop with a batch size b of 5 for 25 cycles, or until all data was labeled. We repeated this for 1,000 times with varying initial labeled samples to generate statistically stable results. As learner model θ a simple NN with 2 hidden layers and 100 neurons each was used. The datasets were split randomly into a 50 % train and 50 % test evaluation set.

As evaluation metric we used the *area-under-the-curve* (AUC) of the learning curve, as has been also done recently in the AL survey by Chan et. al. [20]. This makes it easy to calculate the mean of 1,000 times repeated experiments. Similarly to [5] we are further normalizing the AUC values by the maximum possible AUC value – a rectangle of 100% F1-Scores for each time step – to additionally enable comparisons across datasets.

The training of ImitAL is highly parallelizable, as the generation of the synthetic datasets and the respective AL simulation may run completely in parallel. For a full training of ImitAL with the best parameters we needed 100,000 computation jobs, resulting in a set of 1,000,000 state-action pairs as training data. In total, ~1M CPU-hours were needed for all experiments conducted for this paper, including testing out different NN and IL configurations, and training the final version of ImitAL. For the final version of ImitAL we set the parameter of the simulated AL cycle τ to 10, the pre-sampling parameter k to 20 and j to 10 during training, and 2 during application, as this suffices for a trained ImitAL. The batch size was fixed to a standard value of 5 for the used UCI datasets.

4.2 Comparison with Other Active Learning Strategies

Our evaluation compares 7 AL strategies against our AL strategy, ImitAL. The results are shown in Table 1. Each displayed value is the mean of F1-AUC values for the 1,000 repeated runs. As the percentages are often quite similar, we additionally included the ranks. The displayed percentages are rounded, but the ranks are computed on the complete numbers, which can lead to different ranks for otherwise equally rounded displayed percentages.

We included Least Confidence (LC) and Uncertainty Entropy (Ent) [17], the two most common and basic variants of the informativeness heuristic, where greedily the most uncertain samples based on the classification probability of the learner model are selected for labeling. The Graph Density (GD) strategy [3] was added as a pure representativeness heuristic-based strategy which solely focuses on sampling evenly from the vector space. BatchBALD [8] is a popular AL strategy which works well for computer vision deep neural networks. Querying Informative and Representative Examples (QUIRE) [7] is a computationally

Table 1. F1-AUC-scores (%) for different AL query strategies, mean for 1,000 repeated experiments each, including the ranks and the ranked mean. Empty cells indicate no calculable results within the maximum runtime window of seven days.

	IMITAL	LC	QBC	Ent	Rand	GD	BatchBALD	QUIRE
abalone	21.2 (2)	19.3 (5)	19.6 (4)	17.8 (6)	**21.3 (1)**	15.6 (7)	21.1 (3)	11.2 (8)
adult	**54.5 (1)**	53.5 (4)	54.1 (2)	53.5 (3)	51.8 (5)	47.9 (7)	51.3 (6)	
australian	**83.9 (1)**	83.8 (2)	83.8 (3)	83.8 (2)	83.0 (5)	83.6 (4)	79.8 (6)	71.5 (7)
BREAST	94.0 (3)	**94.4 (1)**	94.4 (2)	**94.4 (1)**	92.8 (4)	91.6 (5)	90.9 (6)	84.6 (7)
DWTC	**69.3 (1)**	65.3 (5)	65.9 (4)	63.4 (6)	67.8 (2)	52.8 (7)	66.1 (3)	50.1 (8)
fertility	88.2 (2)	87.8 (3)	87.7 (4)	87.8 (3)	87.0 (5)	**88.2 (1)**	86.8 (6)	86.8 (7)
flags	**57.5 (1)**	55.5 (6)	55.6 (5)	54.7 (7)	55.8 (4)	56.6 (2)	55.9 (3)	43.7 (8)
german	74.5 (2)	74.2 (4)	74.2 (5)	74.2 (4)	74.3 (3)	**75.5 (1)**	74.1 (6)	71.5 (7)
glass	**68.9 (1)**	67.6 (2)	67.4 (3)	66.6 (6)	67.3 (4)	66.3 (7)	67.1 (5)	40.6 (8)
heart	**79.0 (1)**	78.8 (3)	78.8 (5)	78.8 (3)	78.8 (4)	78.9 (2)	78.3 (6)	71.4 (7)
ionos	**88.9 (1)**	88.6 (2)	88.5 (3)	88.6 (2)	88.0 (5)	88.2 (4)	82.9 (6)	53.5 (7)
wine	**95.2 (1)**	94.8 (2)	94.6 (4)	94.6 (3)	94.4 (5)	94.3 (6)	93.5 (7)	84.9 (8)
zoo	**93.7 (1)**	93.3 (2)	92.9 (6)	93.2 (3)	93.1 (4)	92.8 (7)	93.1 (5)	92.7 (8)
mean %	**74.5 (1)**	73.6 (3)	73.7 (2)	73.2 (5)	73.5 (4)	71.7 (7)	72.4 (6)	58.7 (8)
mean (r)	1.38	3.15	3.85	3.77	3.92	4.62	5.23	7.54

expensive combination of both heuristics, and Query-bycommittee (QBC) [16] a combination of the uncertainty of multiple learner models [10][3].

IMITAL learns a combination of the two heuristics informativeness and representativeness. For the datasets FERTILITY, FLAG, GERMAN, and HEART `GD` is much better than `LC`. This is an indication that on these datasets a pure informativeness heuristic is challenged the most, whereas for the other strategies `LC` still seems to be the safest bet as a general-purpose AL strategy. IMITAL successfully learned to combine the best of both strategies, which can be especially seen by the superior performance on the datasets. `QBC` achieved quite competitive results, but at the cost of almost twice as high running cost than IMITAL due to the expensive retraining of multiple learner models instead of a single one. The good results from the original `QUIRE` and `BatchBALD` paper could not be reproduced by us. Additionally, the runtime of `QUIRE` was so high that not even one AL experiment finished within seven days. The pre-selection of IMITAL with our used parameters means that IMITAL always considers a fixed amount of 40 unlabeled samples during each AL iteration, making it 10 times faster than even the second fastest `LC` strategy, which has to consider all unlabeled samples.

We also performed a significance test to prove that IMITAL is not only by chance but indeed statistically sound better than the competitors. We used a Wilcoxon signed-rank test [19] with a confidence interval of 95% to calculate the proportional win/tie/losses between IMITAL and each competing strategy.

[3] We used for all strategies the implementations from the open-source AL framework ALiPy [18].

For each of the 1,000 starting points, we took the F1-values of all the 25 AL iterations[4] of the two strategies to compare. Our null hypothesis is that the mean of both learning curves is identical. If the null hypothesis holds true we count this experiment repetition as a tie, and otherwise as a win or loss depending on which strategy performed according to the better mean. Due to lack of space we are omitting the table with the results of all datasets, but overall, ImitAL won at least 35% more often compared to each strategy than lost against them. It also has to be noticed that the majority of the direct comparisons resulted in a tie with a total amount of 55%.

5 Conclusion

We presented a novel approach of training a universally applicable AL query strategy on purely synthetic datasets by encoding AL as a listwise learning-to-rank problem. For training, we chose IL, as it is cheap to generate a huge amount of training data when relying on synthetic datasets. Our evaluation showed that ImitAL successfully learned to combine the two basic AL heuristics informativeness and representativeness by outperforming both heuristics and other AL strategies over multiple datasets of varying domains. In the future, we want to include more requirements of large ML projects into the state-encoding of ImitAL to make it more applicable.

Acknowledgements. This research and development project is funded by the German Federal Ministry of Education and Research (BMBF) and the European Social Funds (ESF) within the "Innovations for Tomorrow's Production, Services, and Work" Program (funding number 02L18B561) and implemented by the Project Management Agency Karlsruhe (PTKA). The author is responsible for the content of this publication.

The authors are grateful to the Center for Information Services and High Performance Computing [Zentrum für Informationsdienste und Hochleistungsrechnen (ZIH)] at TU Dresden for providing its facilities for high throughput calculations.

References

1. Dua, D., Graff, C.: UCI machine learning repository (2017)
2. Eberius, J., Braunschweig, K., Hentsch, M., Thiele, M., Ahmadov, A., Lehner, W.: Building the Dresden web table corpus: a classification approach, pp. 41–50, December 2015
3. Ebert, S., Fritz, M., Schiele, B.: Ralf: A reinforced active learning formulation for object class recognition. In: 2012 IEEE Conference on Computer Vision and Pattern Recognition, pp. 3626–3633 (2012). https://doi.org/10.1109/CVPR.2012.6248108

[4] As the exact p-values of the Wilcoxon signed-rank test are only computed for a sample size of up to 25, and for greater values an approximate – in our case not existent – normal distribution has to be assumed, we decided to stop our AL experiments after 25 iterations.

4. Guyon, I.: Design of experiments of the nips 2003 variable selection benchmark. In: NIPS Workshop on Feature Extraction and Feature Selection, vol. 253 (2003)
5. Guyon, I., Cawley, G., Dror, G., Lemaire, V.: Results of the active learning challenge. J. Mach. Learn. Res. Proc. Track **16**, 19–45 (2011)
6. Hochreiter, S., Bengio, Y., Frasconi, P., Schmidhuber, J., et al.: Gradient flow in recurrent nets: the difficulty of learning long-term dependencies (2001)
7. Huang, S.j., Jin, R., Zhou, Z.H.: Active learning by querying informative and representative examples. In: Lafferty, J., Williams, C., Shawe-Taylor, J., Zemel, R., Culotta, A. (eds.) Advances in Neural Information Processing Systems, vol. 23, pp. 892–900. Curran Associates, Inc. (2010)
8. Kirsch, A., v. Amersfoort, J., Gal, Y.: BatchBALD: efficient and diverse batch acquisition for deep bayesian active learning. In: NIPS, vol. 32, pp. 7026–7037. Curran Associates, Inc. (2019)
9. Konyushkova, K., Sznitman, R., Fua, P.: Discovering general-purpose active learning strategies. arXiv preprint arXiv:1810.04114 (2018)
10. Lewis, D.D., Gale, W.A.: A sequential algorithm for training text classifiers. In: Croft, B.W., van Rijsbergen, C.J. (eds.) SIGIR 1994, pp. 3–12. Springer, London (1994). https://doi.org/10.1007/978-1-4471-2099-5_1
11. Liu, M., Buntine, W., Haffari, G.: Learning how to actively learn: a deep imitation learning approach. In: Proceedings of the 56th Annual Meeting of the Association for Computational Linguistics, Melbourne, Australia (Volume 1: Long Papers), pp. 1874–1883. Association for Computational Linguistics, July 2018. https://doi.org/10.18653/v1/P18-1174
12. Michie, D., Camacho, R.: Building symbolic representations of intuitive real-time skills from performance data. In: Machine Intelligence, vol. 13, pp. 385–418. Oxford University Press (1994)
13. Pang, K., Dong, M., Wu, Y., Hospedales, T.: Meta-learning transferable active learning policies by deep reinforcement learning. arXiv preprint arXiv:1806.04798 (2018)
14. Ren, P., et al.: A survey of deep active learning. ACM Comput. Surv. (CSUR) **54**(9), 1–40 (2021)
15. Settles, B.: Active learning literature survey. Computer Sciences Technical Report 1648 (2010)
16. Seung, H.S., Opper, M., Sompolinsky, H.: Query by committee. In: Proceedings of the Fifth Annual Workshop on Computational Learning Theory, New York, NY, USA, pp. 287–294. COLT 1992, Association for Computing Machinery (1992). https://doi.org/10.1145/130385.130417
17. Shannon, C.E.: A mathematical theory of communication. Bell Syst. Tech. J. **27**(3), 379–423 (1948)
18. Tang, Y.P., Li, G.X., Huang, S.J.: ALiPy: active learning in Python. arXiv preprint arXiv:1901.03802 (2019)
19. Wilcoxon, F.: Individual comparisons by ranking methods. Biometrics Bull. **1**(6), 80–83 (1945)
20. Zhan, X., Liu, H., Li, Q., Chan, A.B.: A comparative survey: benchmarking for pool-based active learning. In: IJCAI, pp. 4679–4686, August 2021. https://doi.org/10.24963/ijcai.2021/634, survey Track

Incremental/Continual Learning

Predicting Potential Real-Time Donations in YouTube Live Streaming Services via Continuous-Time Dynamic Graph

Ruidong Jin[1,2], Xin Liu[2(✉)], and Tsuyoshi Murata[1,2]

[1] Tokyo Institute of Technology, Tokyo, Japan
ruidong.jin@net.c.titech.ac.jp, murata@c.titech.ac.jp
[2] National Institute of Advanced Industrial Science and Technology, Tokyo, Japan
xin.liu@aist.go.jp

Abstract. Online live streaming services (e.g., YouTube Live, Twitch) are booming in recent years and gaining popularity in people's cyber life. Real-time gifts paid by viewers in live streaming bring considerable profits and fame to streamers, whereas only a few works are interested in the donation system on live streaming platforms. In this paper, we focus on the real-time donation 'Superchat' on YouTube live platform and build a continuous-time dynamic graph to model the interactions among viewers based on real-time chat messages. Live streaming viewers tend to respond to the superchat immediately, demonstrating the possibility of predicting the real-time donations by analyzing other active viewers and chat messages. We design a temporal graph neural network architecture to dynamically predict the potential viewers who send donations during live streaming. Also, our model can predict the exact periods when superchat appears. Extensive experiments on three live streaming video datasets show our proposed model's effectiveness and robustness compared to baseline methods from other fields.

Keywords: Online live streaming · Real-time donation · Continuous-time dynamic graph · Dynamic node label prediction

1 Introduction

In recent years, online live streaming services have been booming and growing on Social Network Sites (SNS). Due to the advancement of the Internet and wide usage of mobile devices, live streaming services have been considered a convenient and entertaining way to enjoy real-time media. Many video media platforms, such as YouTube Live[1] and Twitch[2], provide live streaming service content covering broadcast news, sports matches, entertainment, video games, and so on [24]. Users prefer these services on live streaming services rather than

[1] https://www.youtube.com/.
[2] https://www.twitch.tv/.

P. Pascal and D. Ienco (Eds.): DS 2022, LNAI 13601, pp. 59–73, 2022.
https://doi.org/10.1007/978-3-031-18840-4_5

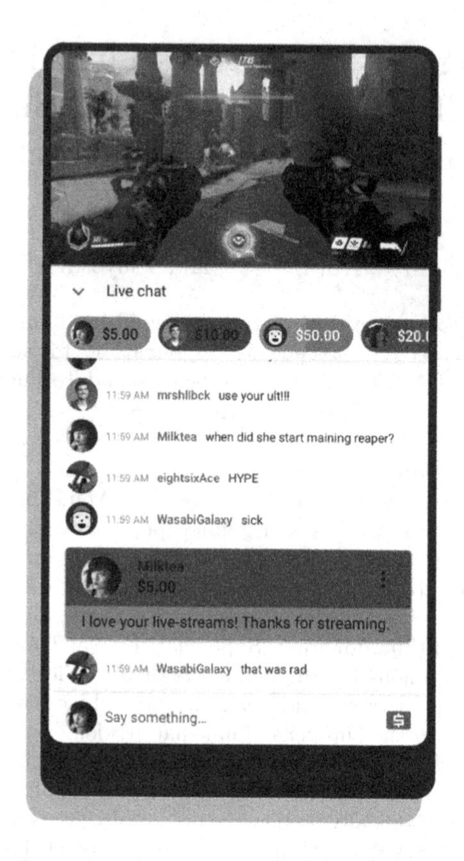

Fig. 1. Superchat messages on YouTube live streaming.

traditional TV due to convenience, better content, low cost, appointment viewing, customized channels, and break-free shows [8,24]. Live streaming service is bringing massive popularity and profits. According to the investigation report provided by Fortune Business Insights[3], the global video streaming market is valued at $372.07 billion in 2021. The market is projected to grow from $473.39 billion in 2022 to $1,690.35 billion by 2029. Twitch, one of the biggest live-streaming platforms, has on average 30 million daily visitors and more than 7 million unique streamers every month. Specifically, the Covid-19 pandemic considerably positively affected the video markets. The increasing adoption of online learning, work from home, and remote patient monitoring in health services has rapidly increased the demand for live streaming services.

The appeal point of live streaming is that viewers can send real-time chat messages to interact with other viewers, and the streamers can also interact with their audience via chat messages. The chat message interaction system shortens the distance between streamers and viewers in cyberspace. Live streaming makes popular streamers become 'celebrities' to their viewers [2]. Also, it makes

[3] https://www.fortunebusinessinsights.com/video-streaming-market-103057.

viewers feel like a part of the community. Some viewers are willing to support their favorite streamers and donate some money during the real-time live streaming. The donation system is called 'superchat' on YouTube Live or 'subscription' on Twitch. Figure 1 demonstrates the superchat donation on YouTube live platform. Superchat is a particular chat message accompanied by an amount of donation. It has a unique effect and will be pinned in the chatbox for a period. A superchat usually costs from $1 to $500. Donation means profit, popularity, and motivation to produce more high-quality content. It is a significant part of streamers' income, especially those who make a living on live streaming. With the development of the live streaming market, there is an urgent need for live streamers to know more about their potential donors and the expected donation income.

Currently, research on live streaming services is still in an early stage. Some researchers study live streaming itself. Computer vision methods are interested in video quality improvement and image recognition in live streaming. Also, live-streaming platforms can be a rich source for data collection (e.g., chat-data analysis). Several works focus on highlight detection [1], sentimental analysis [6], and fraud detection [10]. However, there are few AI-related works on predicting donations in live streaming services.

With the concerns mentioned above, we study the interactions among viewers and real-time chat messages in online live streaming services. Specifically, We focus on the superchat on the YouTube Live platform. Empirically, chat messages are always booming when a superchat appears in live streaming. The streamer would like to appreciate the donation, and other viewers tend to send chat messages to respond to the superchat. Thus, a superchat message is always followed by many response chat messages and usually has unique text content. Based on this fact, we attempt to predict the superchat message by analyzing its relations with other nearby chat messages and its text content. Specifically, we propose to use the continuous-time dynamic graph to model the complicated connections among thousands of viewers and millions of chat messages. Further, we employ Temporal Graph Network (TGN) [17] to analyze the relation between superchat messages and nearby chat messages and finally predict the donations.

Our proposed method predicts the real-time donation and the exact period it appears. We transform the superchat detection problem into a dynamic node label classification problem in continuous-time dynamic graphs, which is discussed in a few prior works. Specifically, unlike the traditional node label classification problem where node labels are static and constant, node labels in this task are dynamic and changing. Solving the dynamic node label classification task can predict the exact time when superchat appears. We also conducted massive experiments to evaluate our proposed approach. The experimental results demonstrate that our proposed approach achieves a 0.902 AUC score in the dynamic node label prediction task. The results are significantly superior to other baselines, including decision tree algorithms, time sequence models, static graph neural networks, and NLP text classification models.

The main contributions of the paper are summarized as follows:

- We research the gift donations and donors in live streaming services and try to predict the virtual gift donations via dynamic graph neural network models. As far as we are concerned, we are the first to focus on this innovative and significant topic of online live streaming donations and donors.
- We represent the live streaming chat messages and interactions among viewers in a continuous-time dynamic graph. Furthermore, we design a temporal graph neural network model to solve the dynamic node label classification problem and predict the potential superchat donations on YouTube live streaming platform.
- We conduct experiments and prove that our proposed approach achieves excellent performance on the dynamic node label classification task. Our approach outperforms baseline methods by a large margin, including traditional machine learning algorithms, sequence learning models, and state-of-the-art dynamic graph neural network methods.

The remainder of the paper is organized as follows. Section 2 summarizes the recent literature on online live streaming services and dynamic graph learning. Section 3 introduces the approaches to generating dynamic graphs from live streaming chat data and predicting potential donations through the temporal graph neural networks. Section 4 reports the experiment results and evaluations, and also demonstrates a case study to prove the feasibility of the proposed model. Finally, Sect. 5 concludes our research.

2 Related Work

We represent the related work from two aspects: online live streaming and dynamic graph learning.

2.1 Online Live Streaming Service

Online live streaming services have attracted increasing attention due to the development of high-speed Internet and mobile devices. Live streaming communities on different platforms are always formed by the various streaming content genres [3]. For example, many YouTube live streamers like to share their daily life and experience. Those streamers are also called ‘vloggers’ [7]. Besides, game streaming channels are much more popular on the twitch platform. Viewers like to watch others play video games to release tension, kill time, and seek common topics with friends [19].

Virtual donation in live streaming is a promising topic and has caught the attention of many researchers. Current research considers the reasons behind virtual donation as that donation represents viewers’ appreciation and approval of the streamer, or the recognition and happiness for shared contents [9]. Besides, the donation information is entirely public on live streaming channels. When viewers donate the streamer, others will notice it. Other viewers tend to be affected by such noticeable actions and are likely to follow the groups and send more donations [13]. Therefore, donations in online streaming services signal a group interaction and an event for viewers to interact with others.

2.2 Dynamic Graph Learning

Graph learning has produced many successes [27]. The techniques of learning embedding vectors on graphs have been widely acknowledged for graph-related downstream tasks such as node classification [20], link prediction [25], and graph classification [26]. The main challenge in graph learning is finding a proper way to encode graph structures, including nodes and edges, into low-dimension hidden embedding vectors. Embedding vectors can be fed to machine learning models and deep learning structures, such as random-walk-based algorithms [14] and graph neural networks [20].

Learning on dynamic graphs has been a heated topic recently. In the early stage, research on dynamic graphs focused on discrete-time dynamic graphs. Discrete-time dynamic graphs consist of a timed sequence of snapshots of the graph [11,18]. The existing static graph methods can be directly applied to it. However, most of the real-life graph-structured data is constantly evolving. The continuous-time dynamic graph is a more general style of the dynamic graph. It consists of a timed list of events, including edge creation or deletion, node creation or deletion, and node or edge status evolution. Only recently, some studies on continuous-time dynamic graphs have been proposed [12,17,21,23].

3 Methodology

This section proposes a method to identify the potential donors and real-time donations in live streaming services. Empirically, chat messages are always booming when a superchat appears in YouTube live streaming. The streamer would like to appreciate the donation, and other viewers tend to send chat messages to respond to the superchat. Moreover, the size of superchat messages is always long, and the words are usually well-organized to attract others' attention. Thus, we hypothesize that a superchat message is always followed by many response chat messages and usually has unique text content. It is therefore possible to identify the superchat message by analyzing its text content and relations with other nearby chat messages.

We propose to use the continuous-time dynamic graph to model the complicated relations among thousands of viewers and millions of chat messages. The nodes represent viewers, and the edges represent the interactions between viewers. Further, we employ Temporal Graph Network (TGN) to analyze the relation between superchat messages and nearby chat messages. TGN is a subclass of neural networks that operates on dynamic graphs. It learns the continuously evolving node representations and finally contributes to predicting potential superchats and donors.

In the following, we first describe the online live streaming dataset used in the research. Next, we introduce an approach to generate a continuous-time dynamic graph to represent the live streaming viewers and their chat messages. Then, we elaborate on how TGN predict the potential live streaming viewers who may send superchat to streamers and the specific period they sent superchats.

Finally, we show the effort of adjusting the model to training on the imbalanced live streaming dataset.

Table 1. Detailed dataset information

Item	Type	Description
timestamp	string	UTC timestamp
body	string	chat message
membership	string	membership status
isSuperchat	boolean	is superchat message
isModerator	boolean	is channel moderator
isVerified	boolean	is verified account
amount(only for superchat)	number	donation amount
currency(only for superchat)	string	currency symbol
significance(only for superchat)	number	donation significance
id	string	anonymized chat id
channelId	string	anonymized viewer id
originVideoId	string	streaming video id
originChannelId	string	streamer channel id

3.1 Dataset

We use a YouTube live streaming dataset *VTuber 1B: Large-scale Live Chat and Moderation Events Dataset*[4]. VTuber 1B is a huge collection of over a billion of live chat messages, superchats, and moderation events (ban and deletion) across hundreds of YouTubers' live streams, especially English streamers and Japanese Streamers. Our research use the chat message data ranging from Mar. 2021 to Apr. 2021, including 484 live streaming channels, over 6,000 streaming videos, over 180 million live chat messages, and over 500,000 superchat messages. The detailed information is listed in Table 1.

We split the original YouTube live streaming dataset into several live streaming videos according to the streaming video ID and streamer channel ID. Each video contains thousands of chat messages from hundreds of viewers in timestamp order. Then, we use a pre-trained Sentence Transformers [15,16] language model to encode all the chat message texts into sentence embedding vectors. Sentence embedding vectors will be fed to the temporal graph network structures later.

3.2 Dynamic Graph Generation

We propose an algorithm to generate continuous-time dynamic graphs to represent viewers' chat messages and interactions in live streaming videos. The graph

[4] https://www.kaggle.com/datasets/uetchy/vtuber-livechat.

Algorithm 1: Generate dynamic graph from a timed sequence of live streaming chat messages

```
   Data: A timed sequence S of chat message.
   Result: A continuous-time dynamic graph G
 1 initialization;
 2 Separate S into several batches ;
 3 for batch ← batches do
 4 |   for msg_1, msg_2 ∈ S[batch] and msg_1 earlier than msg_2 do
 5 |   |   if SequenceMatcher(msg_1, msg_2) > thrs then delete msg_2// Drop
   |   |       duplicated chat messages
 6 |   end
 7 |   for msg ← S[batch] do
 8 |   |   if msg appears for the first time then
 9 |   |   |   Create a new node for msg in G
10 |   |   else
11 |   |   |   Update node embedding in G by the newest sentence embedding
   |   |   |   vector
12 |   |   end
13 |   |   for active_node ← active_list do
14 |   |   |   if cosine_sim(active_node, msg)> thrs then Generate an edge
   |   |   |   between Node active_node and Node msg in G
15 |   |   end
16 |   |   msg → active_list;
   |   |   // Keep msg active for a period
17 |   |   Delete expired nodes in active_list;
   |   |   // Drop inactive nodes
18 |   end
19 end
```

is composed of batches of dynamic graphs along the time. The nodes represent viewers, and the edges represent the interactions between viewers. Node creation/deletion arises when the viewer enters or leaves the streaming channels. Edge changing occurs when the viewer sends new chat messages and interacts with others. Moreover, each node is associated with a feature vector, which will dynamically update according to the chat messages posted by the corresponding viewer.

Algorithm 1 demonstrates how to generate a dynamic graph from a timed sequence of chat messages by exploiting text content, sentence embedding vectors, viewer ID, timestamps, and a label indicating if it is a superchat message. The details are as follows:

1. Preprocess the raw chat messages in live streaming as a dataset. Separate all the chat messages into several batches along the time and filter the duplicated, non-sense, and too short chat messages (lines 1–6).
2. Traverse all the chat messages in the batch. Create a new node for the viewer who sends the chat message for the first time. The newly-created node feature

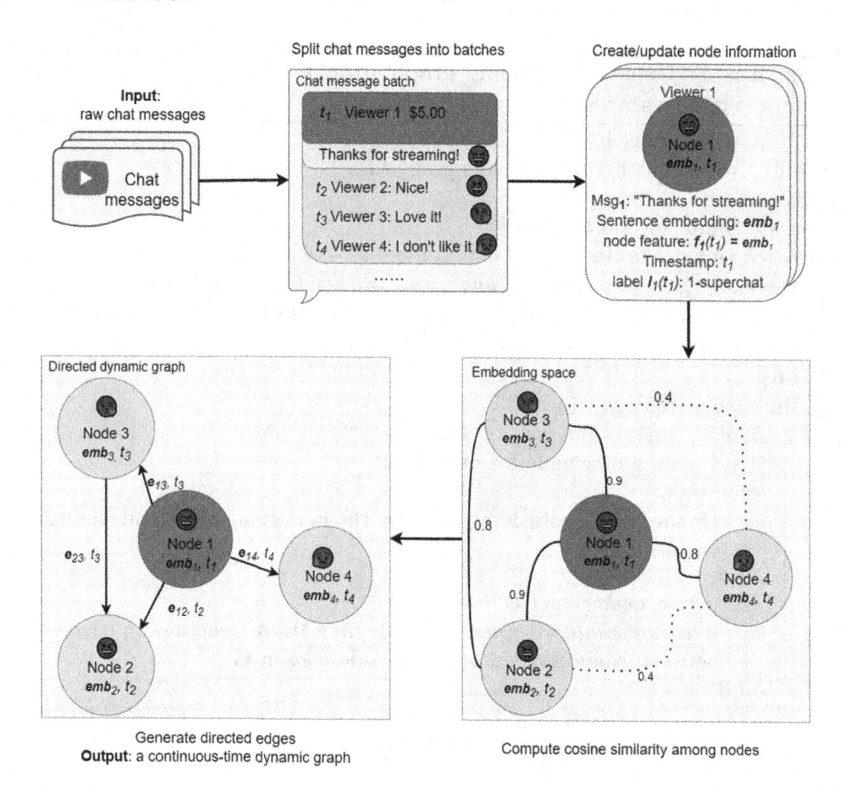

Fig. 2. Generate dynamic graphs from YouTube chat messages

vectors are initialized by the sentence embedding vectors of their first chat messages (lines 7–9).

3. Update node features by the newest sentence embedding vectors the viewer posts (lines 10–12).
4. Compute the cosine similarity between the new node and each node in the active node list. The cosine similarity is used to evaluate the 'distance' of nodes in the embedding space. Generate directed temporal edges for node pairs with cosine similarity higher than the threshold (lines 14–16).
5. Add the new node to the active list for a period (line 16). Drop inactive old nodes from the active node list (line 17). Repeat step 2 to 4 until all the chat messages are visited.

We associate a binary dynamic node label $\mathbf{L}(t) \in \{0, 1\}$ to each node in the dynamic graph. Label 0 means the node does not send any superchat at time t, and label 1 means the node has sent a superchat at time t. Specifically, if a node posts a superchat, the node label will temporarily change from 0 to 1 until the superchat expires.

Figure 2 is an intuitive description of how to generate a dynamic graph from chat messages. The input is a timed-sequence of raw chat messages. Four viewers

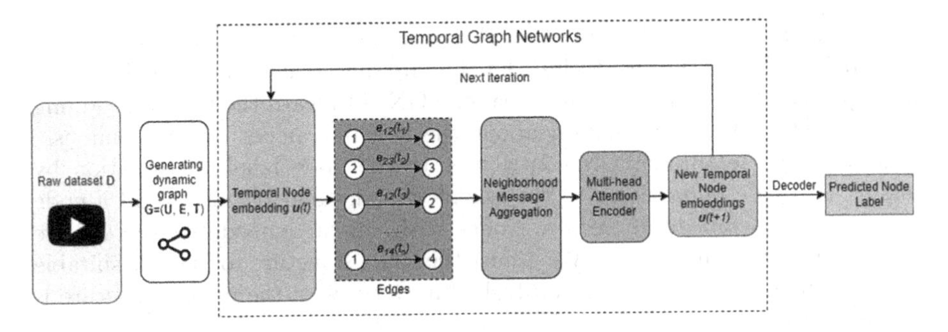

Fig. 3. Overview of the model structure

post chat messages $msg_1 \sim msg_4$ at timestamp $t_1 \sim t_4$, and the first one is a superchat. Four chat messages appear in the same batch, and the chat messages are encoded to sentence embedding vectors $\mathbf{emb}_1 \sim \mathbf{emb}_4$. Node 1∼4 are created to represent the corresponding viewers. The node features are initialized by $\mathbf{emb}_1 \sim \mathbf{emb}_4$. A dynamic node label is associated to each node to identify the superchat and normal messages. The blue color represents superchat nodes, and the green color represents the normal nodes. Then we calculate the cosine similarity and generate edges for the node pairs in the active node list. Node 2 and node 3 have similar opinions toward node 1, while node 4 has the opposite one. Therefore, edge $\mathbf{e}_{12}$ from node 1 to node 2, edge $\mathbf{e}_{23}$ from node 2 to node 3, and edge $\mathbf{e}_{13}$ from node 1 to node 3 are generated. Node 4 responds to node 1 but does not positively correlate to node 2 and node 3. Thus, only an edge $\mathbf{e}_{14}$ from node 1 to node 4 is generated. Edge directions are from old to new nodes because new chat messages are affected by old ones, and messages propagate from old nodes to new nodes.

3.3 Temporal Graph Neural Network

We represent the YouTube live streaming data as a dynamic graph $\mathbf{G} = (\mathbf{U}, \mathbf{E}, \mathbf{T})$. $t \in \mathbf{T}$ is the timestamp. $\mathbf{u}_s(t), \mathbf{u}_d(t) \in \mathbf{U}$ are temporal node features at timestamp t. $\mathbf{E} \subset \mathbf{U} \times \mathbf{U}$ is the edge set. $\mathbf{e}_{sd}(t) \in \mathbf{E}$ is the temporal edge feature between node $\mathbf{u}_s(t)$ and node $\mathbf{u}_d(t)$ at timestamp t. Temporal edge features consist of the chat message length and the cosine similarity of two node embedding vectors. Every node $\mathbf{u}(t) \in \mathbf{U}$ is associated with a temporal node label $l_u(t) \in \{0, 1\}$. $\mathbf{L}$ denotes the dynamic node labels for all the interaction events in $\mathbf{G}$. During model training, $\mathbf{L}$ is split into a training set $\mathbf{L}^{train}$, a validation set $\mathbf{L}^{val}$, and a test set $\mathbf{L}^{test}$. Therefore, our research target can be mathematically represented as following: Given a continuous time dynamic graph ($\mathbf{G}$), temporal node features ($\mathbf{U}$), edge features ($\mathbf{E}$), and a portion of known dynamic node labels ($\mathbf{L}^{\mathbf{train}}$ and $\mathbf{L}^{\mathbf{val}}$), how to learn a mapping $\mathcal{F} : \mathbf{G} = (\mathbf{U}, \mathbf{E}, \mathbf{T}) \rightarrow \mathbf{L}$ to predict the remaining dynamic node labels ($\mathbf{L}^{\mathbf{test}}$).

We introduce a temporal graph neural network model TGN to predict dynamic node labels in the continuous-time dynamic graph obtained in Sect. 3.2.

TGN is an extension of Graph Neural Network(GNN) on dynamic graphs. GNN is a subclass of deep learning techniques that are specifically built to do inference on graph-based data. The primary goal of TGN is to learn the node embedding vectors that contain information about the neighbor nodes and continuously updating node features. TGN contributes to the node label classification by passing the node information to the neighbor nodes and trying to find a node with a similar embedding vector. Nodes with similar embedding vectors are likely to have the same labels. We adjust the TGN structure to make it suitable for the dynamic node classification task. An overview of the model structure is demonstrated in Fig. 3.

TGN model contains three phases: Interaction analysis, Neighbor message aggregator, and Multi-head attention encoder. First, the model splits the time-sequence temporal edge list $\mathbf{E}$ into several batches. Each edge represents an interaction between a pair of nodes. For example, edges $\mathbf{e}_{12}(t_1)$ represents the interaction between node $\mathbf{u}_1(t_1)$ and node $\mathbf{u}_2(t_1)$ that happened at timestamp t_1. Next, the model aggregates temporal node embeddings from neighbor nodes. The multi-head attention encoder updates the temporal node embeddings $\mathbf{u}(t+1)$. $\mathbf{u}(t+1)$ will be the input in the next training batch. Finally, node embeddings are fed to an MLP decoder and the softmax function to output the predicted dynamic node labels.

3.4 Strategies for Data Imbalance

The superchat donations only account for a small part of all the chat messages in the live streaming. Thus, the dataset is extremely imbalanced, severely confusing the model and resulting in poor performance. We take the following strategies to alleviate the data imbalance:

- We refine the dataset by dropping duplicated and meaningless chat messages. Also, we filter the chat messages shorter than a particular size. It reduces the amount of non-superchat samples in the dataset.
- We increase the positive samples by tuning the superchat expiring time and the cosine similarity threshold in Algorithm 1.
- We employ an under-sampling strategy to alleviate the gap between positive and negative samples.
- We apply a cost-sensitivity learning method to self-adjust the penalty factor in loss function during the model training batches.

4 Experiments

We conduct some experiments on the dynamic node label prediction task. We first explain the experimental settings and baselines. Then we discuss the experimental results and model evaluation. Finally, we show a case study to prove our model's feasibility.

Table 2. Statistics of the live streaming dynamic graphs.

Dataset length	Short	Mid	Long
Durations (hrs.)	8.61	47.22	78.49
Node num	6,225	28,582	41,156
Edge num	1,660,813	9,498,600	15,097,110
Positive label num	105,207	258,079	525,964
Positive ratio	6.3%	2.7%	3.4%

4.1 Dataset Description

We prepared three continuous-time dynamic graphs generated from the live streaming dataset mentioned in Sect. 3.1. The detailed statistics are listed in Table 2. Three dynamic graphs represent the different lengths of live streaming videos. 'Short' dataset contains chat messages in an 8-h live streaming video. 'Mid' dataset contains chat messages in a 47-h video compilation of a week. And 'Long' dataset contains chat messages in a 78-h video compilation of two weeks. It is noticeable that the ratios of positive labels are extremely small in all three dynamic graphs. It is reasonable because superchat messages are only a small part out of all chat messages in a real-world situation. Therefore, the experiments will be conducted on the imbalanced dataset.

4.2 Experiment Setup

We randomly split the chat messages in the dataset into training, validation, and test sets. The first 50%/70%/90% of chat messages are counted as the training set, and the remaining are equally separated into validation and test sets. We choose Parametric Rectified Linear Unit (PReLU) as the activation function. The model is trained by the Adam optimizer with a learning rate equal to 0.001. The training runs 20 iterations at most, and an early stopping strategy is implemented if the validation loss does not decrease for 5 iterations.

4.3 Baselines

We consider traditional decision tree methods, sequence-based models, static graph representation learning methods, and NLP text classification methods as baselines. The details are listed below.

1. **GBDT**: A Gradient Boost Decision Tree (GBDT) classifier from the scikit-learn toolkit.
2. **XGBoost**: An ensemble gradient boosting decision tree model from XGBoost library.
3. **LSTM-FCN** [5]: A time sequence model combining long short-term memory (LSTM) networks and fully convolution network (FCN).

Table 3. The results for predicting the hospital-region labels by different approaches.

Model	Training set Ratio								
	50%			70%			90%		
	Short	Mid	Long	Short	Mid	Long	Short	Mid	Long
GBDT	0.500	0.472	0.500	0.552	0.469	0.515	0.489	0.458	0.472
XGBoost	0.500	0.509	0.500	0.553	0.495	0.518	0.513	0.0462	0.499
LSTM-FCN	0.468	0.505	0.507	0.497	0.499	0.506	0.431	0.499	0.500
ALSTM-FCN	0.485	0.505	0.508	0.499	0.499	0.499	0.500	0.501	0.472
GCN	0.500	0.499	0.499	0.500	0.499	0.499	0.500	0.499	0.499
GAT	0.510	0.510	0.510	0.531	0.531	0.531	0.548	0.548	0.548
BERT	0.630	0.560	0.580	0.570	0.570	0.590	**0.620**	0.540	0.670
Our model	**0.654**	**0.610**	**0.784**	**0.620**	**0.610**	**0.795**	0.520	**0.854**	**0.902**

4. **ALSTM-FCN** [5]: An alternative LSTM-FCN with attention layers following the LSTM cells.
5. **GCN** [20]: Graph Convolutional Networks (GCN) on static graphs.
6. **GAT** [22]: Graph Attention Networks (GAT) on static graphs.
7. **BERT** [4]: Bidirectional Encoder Representations from Transformers(BERT) is a transformer-based model for NLP pre-training.

GBDT and XGBoost are gradient boosting decision tree models. We replace the MLP decoder in our model with GBDT and XGBoost to test the performance. LSTM-FCN and ALSTM-FCN achieved state-of-the-art performance on the task of time sequence classification. Every time node features update, the new node features and existing node hidden embeddings will be fed to the models and output new node hidden embeddings, then wait for the next update. GCN and GAT are static graph neural network models. We build a static graph to represent the interactions among viewer nodes, where the edge weight is the frequency of node interactions. The temporal node embeddings are used to predict the dynamic node labels. BERT is a transformer-based machine learning technique for NLP pre-training. We exploit a Japanese BERT pretrained model to encode the chat messages of each viewer and integrate them as long sentences. The BERT for sequence classification model provided by Huggingface is used to classify the long sentences and predict the viewers.

4.4 Evaluation

Table 3 demonstrates the experimental results of our proposed model and the baselines. We evaluate the performance in terms of Area under the ROC Curve(AUC) score to measure the model's prediction quality, regardless of the classification threshold and how the datasets are imbalanced. Our proposed model shows the almost best overall performance. In particular, it achieves the highest AUC score on all three datasets and all the training set ratios. The only

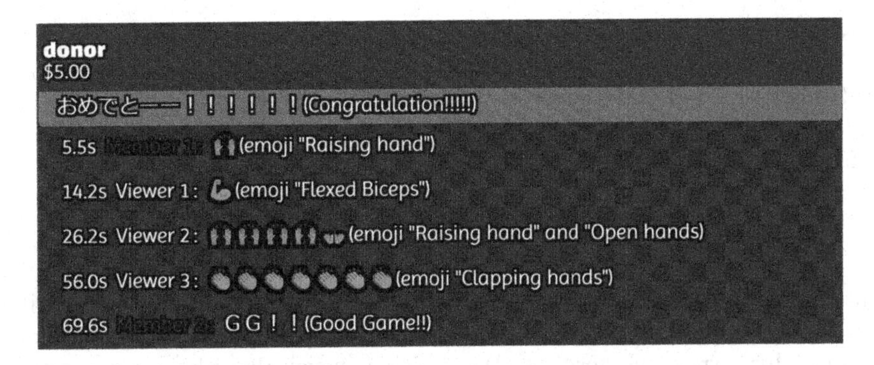

Fig. 4. A case study of a superchat message and following responses. Time before messages is the time offset after the post of superchat message.

exception that our model performs poorer than BERT arises on the short video dataset with a training set ratio equal to 50%. The inferred reason is that the test samples are fewest over all cases. Thus the standard error and variance may influence the test result more than in other cases.

Our proposed model apparently outperforms the other baselines by a large margin. We attribute this to the deliberately designed continuous-time dynamic graph that considers the changing node interactions and the frequent node embedding update. As to the baselines, GBDT and XGBoost are substitutes of the origin MLP decoder. However, the decision tree structure is incompatible with the neural network structure, resulting in worse training efficiency. LSTM-FCN and ALSTM-FCN are time sequence models and can deal with continuous feature updates. However, they cannot fully use the graph structure information, leading to poor performance. GCN, GAT and BERT are not designed to fit temporal data. They lose much temporal information. Furthermore, GCN and GAT are good at exploiting the graph structure and node interactions but cannot deal with the temporal node features. In contrast, BERT analyzes the all-time node features but lacks graph structure information. Regarding the fact that BERT performs better than GCN and GAT, we infer that the node feature update has more influence than the graph structure on the dynamic node label prediction task. In other words, chat messages in live streaming videos are more significant than viewers' interactions.

4.5 Case Study

We demonstrate a case study to validate the feasibility of our proposed model. We check the superchat messages that are correctly predicted in the result of the 'Mid' dataset with the training set ratio equal to 90%. Figure 4 illustrates a $5.00 superchat message and some correlated chat messages within a particular period. The node which posts the superchat message is predicted as positive by our proposed model. The response chat messages are sent from the neighbor nodes. The superchat writes 'Congratulation!!!' in Japanese, and the following messages respond to the superchat with emojis like 'Raising hands,' 'Clapping hands', and text 'GG!'(abbreviation of 'Good game'). All these messages are positively correlated to the superchat message 'Congratulation', proving that our model can accurately identify the superchat message and its correlated messages within a particular period.

5 Conclusion

This work focuses on the chat messages and viewer interactions on the YouTube live streaming platform and presents a model to predict real-time donations in online live streaming services. We design a novel algorithm to generate a continuous-time dynamic graph representing the viewers and chat messages. We exploit a temporal graph neural network structure to predict the potential real-time donations based on the generated dynamic graphs. Our approach has an excellent performance in terms of AUC score, outperforming the baselines, including decision tree classifiers, time-sequence models, static graph neural networks, and NLP text classification models. We apply a case study to prove our model's feasibility by checking the origin chat messages in the prediction results. To the best of our knowledge, we are the first to combine live streaming services and dynamic graph neural network models. Our work contributes to the research of real-time donations in live streaming services, which is an innovative and promising topic.

Acknowledgements. This work is partly supported by JST SPRING (grant number JPMJSP2106), JSPS Grant-in-Aid for Scientific Research (grant number 21K12042, 17H01785), and the New Energy and Industrial Technology Development Organization (grant number JPNP20006).

References

1. Chu, W.-T., Chou, Y.-C.: On broadcasted game video analysis: event detection, highlight detection, and highlight forecast. Multimedia Tools Appl. **76**(7), 9735–9758 (2016). https://doi.org/10.1007/s11042-016-3577-x
2. Fietkiewicz, K.J., Dorsch, I., Scheibe, K., Zimmer, F., Stock, W.G.: Dreaming of stardom and money: micro-celebrities and influencers on live streaming services. In: Social Computing and Social Media. User Experience and Behavior, pp. 240–253 (2018)
3. Hamilton, W.A., Garretson, O., Kerne, A.: Streaming on twitch: fostering participatory communities of play within live mixed media. In: SIGCHI2014, pp. 1315–1324 (2014)
4. Jacob Devlin, Ming-Wei Chang, K.L.K.T.: Bert: pre-training of deep bidirectional transformers for language understanding. arXiv preprint arXiv:1810.04805 (2018)
5. Karim, F., Majumdar, S., Darabi, H., Chen, S.: LSTM fully convolutional networks for time series classification. IEEE Access **6**, 1662–1669 (2018)
6. Kavitha, G., Saveen, B., Imtiaz, N.: Discovering public opinions by performing sentimental analysis on real time Twitter data. In: International Conference on Management of Data 2018, pp. 1–4 (2018)

7. Ladhari, R., Massa, E., Skandrani, H.: Youtube vloggers' popularity and influence: the roles of homophily, emotional attachment, and expertise. J. Retail. Consum. Serv. **54**, 102027 (2020)
8. Lee, M., Choi, H., Cho, D., Lee, H.: Cannibalizing or complementing? The impact of online streaming services on music record sales. Procedia Comput. Sci. 91, 662–671 (2016)
9. Lee, S.E., Choi, M., Kim, S.: They pay for a reason! the determinants of fan's instant sponsorship for content creators. Telematics Inform. **45**, 101286 (2019)
10. Li, Z., et al.: Live-streaming fraud detection: a heterogeneous graph neural network approach. In: KDD2021, pp. 3670–3678 (2021)
11. Liben-Nowell, D., Kleinberg, J.: The link-prediction problem for social networks. J. Am. Soc. Inform. Sci. Technol. **58**(7), 1019–1031 (2007)
12. Nguyen, G.H., Lee, J.B., Rossi, R.A., Ahmed, N.K., Koh, E., Kim, S.: Continuous-time dynamic network embeddings. In: Companion Proceedings of the The Web Conference 2018, pp. 969–976 (2018)
13. Payne, K., Keith, M.J., Schuetzler, R.M., Giboney, J.S.: Examining the learning effects of live streaming video game instruction over twitch. Comput. Hum. Behav. **77**, 95–109 (2017)
14. Perozzi, B., Al-Rfou, R., Skiena, S.: DeepWalk: online learning of social representations. In: KDD2014, pp. 701–710 (2014)
15. Reimers, N., Gurevych, I.: Sentence-BERT: sentence embeddings using siamese BERT-networks. arXiv preprint arXiv:1908.10084 (2019)
16. Reimers, N., Gurevych, I.: Making monolingual sentence embeddings multilingual using knowledge distillation. arXiv preprint arXiv:2004.09813 (2020)
17. Rossi, E., Chamberlain, B., Frasca, F., Eynard, D., Monti, F., Bronstein, M.: Temporal graph networks for deep learning on dynamic graphs. arXiv preprint arXiv:2006.10637 (2020)
18. Sankar, A., Wu, Y., Gou, L., Zhang, W., Yang, H.: DySAT: deep neural representation learning on dynamic graphs via self-attention networks. In: WSDM2020, pp. 519–527 (2020)
19. Sjöblom, M., Hamari, J.: Why do people watch others play video games? An empirical study on the motivations of twitch users. Comput. Hum. Behav. **75**, 985–996 (2017)
20. Thomas N. Kipf, M.W.: Semi-supervised classification with graph convolutional networks. arXiv preprint arXiv:1609.02907v4 (2017)
21. Trivedi, R., Farajtabar, M., Biswal, P., Zha, H.: DyRep: learning representations over dynamic graphs. In: ICLR2019 (2019)
22. Veličković, P., Cucurull, G., Casanova, A., Romero, A., Lio, P., Bengio, Y.: Graph attention networks. arXiv preprint arXiv:1710.10903 (2017)
23. Xu, D., Ruan, C., Korpeoglu, E., Kumar, S., Achan, K.: Inductive representation learning on temporal graphs. arXiv preprint arXiv:2002.07962 (2020)
24. Yang, H., Lee, H.: Exploring user acceptance of streaming media devices: an extended perspective of flow theory. IseB **16**(1), 1–27 (2017). https://doi.org/10.1007/s10257-017-0339-x
25. Zhang, M., Chen, Y.: Link prediction based on graph neural networks. In: NeurIPS2018, pp. 5171–5181. NIPS 2018 (2018)
26. Zhang, M., Cui, Z., Neumann, M., Chen, Y.: An end-to-end deep learning architecture for graph classification. In: AAAI2018, vol. 32, no. 1 (2018)
27. Zhou, J., et al.: Graph neural networks: a review of methods and applications. arXiv preprint arXiv:1812.08434 (2018)

Semi-supervised Change Point Detection Using Active Learning

Arne De Brabandere[1(✉)], Zhenxiang Cao[2], Maarten De Vos[2,3], Alexander Bertrand[2,4], and Jesse Davis[1]

[1] DTAI, Department of Computer Science, KU Leuven, Belgium
{arne.debrabandere,jesse.davis}@cs.kuleuven.be
[2] STADIUS Center for Dynamical Systems, Signal Processing and Data Analytics, Department of Electrical Engineering, KU Leuven, Belgium
{zhenxiang.cao,maarten.devos,alexander.bertrand}@esat.kuleuven.be
[3] Department of Development and Regeneration, KU Leuven, Belgium
[4] Leuven.AI - KU Leuven Institute for AI, Leuven, Belgium

Abstract. The goal of change point detection (CPD) is to find abrupt changes in the underlying state of a time series. Currently, CPD is typically tackled using fully supervised or completely unsupervised approaches. Supervised methods exploit labels to find change points that are as accurate as possible with respect to these labels, but have the drawback that annotating the data is a time-consuming task. In contrast, unsupervised methods avoid the need for labels by making assumptions about how changes in the underlying statistics of the data correlate with changes in a time series' state. However, these assumptions may be incorrect and hence lead to identifying different change points than a user would annotate. In this paper, we propose an approach in between these two extremes and present AL-CPD, an algorithm that combines active and semi-supervised learning to tackle CPD. AL-CPD asks directed queries to obtain labels from the user and uses them to eliminate incorrectly detected change points and to search for new change points. Using an empirical evaluation on both synthetic and real-world datasets, we show that our algorithm finds more accurate change points compared to existing change point detection methods.

Keywords: Change point detection · Active learning · Semi-supervised learning

1 Introduction

Time series are time-ordered sequences that report the observed values of a variable of interest at each time step. The observed values depend on the underlying state of the system, which usually does not remain constant but changes over

A. De Brabandere and Z. Cao—Equal contribution.

P. Pascal and D. Ienco (Eds.): DS 2022, LNAI 13601, pp. 74–88, 2022.
https://doi.org/10.1007/978-3-031-18840-4_6

time. For example, when monitoring a person's physical activity using on-body accelerometers, the state of the system is the activity that is currently being performed, which affects the observed values in the acceleration signals. The problem of change point detection (CPD) is to locate abrupt changes in the underlying state of a time series [1]. In the example of physical activity monitoring, change points occur when the person transitions from one activity to another.

Existing CPD algorithms can be categorised into two groups: supervised and unsupervised approaches. Supervised CPD methods exploit labels to learn where the change points of a time series are located. These methods treat the CPD problem as a multi-class [20] or binary [10,11] classification task. Multi-class methods classify each window of the time series as its corresponding state. Change points are detected when the predicted state changes between two consecutive windows. Binary classification methods learn whether a given location in the time series is a change point or not. The features used as input to supervised methods depend on the application and the type of data that is used. For example, supervised segmentation for transportation mode detection [20] uses application-specific features such as the magnitude of the acceleration measured by an accelerometer, or the speed derived from GPS data. Therefore, these methods are hard to generalise to other datasets. Moreover, they require a sufficient amount of labelled data in order to achieve good detection accuracy, and the resources and time needed for annotating the data may not always be available.

Unsupervised CPD methods can be subdivided into classical model-based approaches and data-driven model-free approaches. Classical approaches such as the cumulative sum (CUSUM) [3] and the generalised likelihood ratio (GLR) [2] use a sliding window approach to estimate the underlying statistical models of adjacent subsequences of the time series. The parameters in the estimated models are assumed to be constant if there is no change point in between. Hence, a change point is detected when the models significantly differ. Some other studies [14,15] further improved the performance of these approaches by estimating the density ratio. The assumption of these methods is that the density ratio of consecutive window pairs remains constant when there is no change point. Approaches such as FLOSS [12] and ESPRESSO [9] rely on changes in temporal shape patterns, whereas AutoPlait [16] detects changes in the parameters of a hidden Markov model learned from the time series. However, all model-based algorithms face the same problem: their final performance heavily depends on whether the actual data follows the assumed parametric model. It is often hard to guarantee this condition in complex real-world datasets. Recently, a variety of unsupervised data-driven learning algorithms have been proposed, which are typically based on (deep) neural networks such as convolutional neural networks (CNN) [17] and graph neural networks (GNN) [22]. However, these methods also make assumptions about the changes in the underlying statistics of the time series. For example, the time-invariant representation (TIRE) framework [8] uses an autoencoder under the assumption that some latent features should remain constant in the absence of a change point.

Despite the wide range of existing algorithms, finding the correct change points of a time series remains a challenging task because the time series may have multiple possible definitions of change points. On the one hand, supervision enables tailoring the method to the problem at hand, but requiring fully annotated data imposes a huge time burden on a user. On the other hand, unsupervised approaches rely on assumptions which may not correspond to the user's intuitions or may not be appropriate for a specific problem. Hence a mismatch can arise between the change points found by the algorithm and the correct ones.

In order to fill this gap, we propose an active, semi-supervised approach to change point detection. By employing active learning, we can focus the labelling effort to specific locations in the time series that will be particularly informative in order to minimize the manual effort. In summary, our contributions are as follows:

1. We propose an active learning approach to CPD (AL-CPD) which asks a small number of directed queries to the user in order to obtain labels. AL-CPD exploits these labels to (1) eliminate incorrectly detected change points and (2) detect new change points in a semi-supervised setting.
2. We perform an empirical evaluation on both synthetic and real-world time series and show that AL-CPD outperforms existing CPD methods.

2 AL-CPD

In change point detection, the goal is to find abrupt transitions in the underlying state of a time series. More specifically, we define the problem as follows:

Given: A set of n time series $x_1, \ldots, x_n$
Find: Locations $t_i^1, \ldots, t_i^{s_i}$ of the change points of each series x_i

The input consists of n sequences $x_1, \ldots, x_n$ where each x_i is a numerical time series that can be univariate or multivariate. Instead of representing the data as a single long time series, our input format can represent time series collected over multiple batches. For example, an activity recognition dataset is typically collected from multiple subjects. Our data format can represent each subject's data as a separate sequence. Each sequence x_i consists of multiple segments that each correspond to an underlying state of the time series. The goal is to find the locations $t_i^1, \ldots, t_i^{s_i}$ of the change points, i.e., the transitions between the segments, for each sequence. Note that CPD can be tackled in an *offline* or *online* setting. Here, we only consider the offline case where all data is collected before running the algorithm.

Our algorithm approaches the offline CPD task by employing an active learning strategy that queries a human annotator in order to intelligently acquire labels that the algorithm can exploit to better identify the relevant change points. Because each query entails a manual effort from the user, the goal is to find good change points using a small number of queries. Designing such an algorithm poses two key challenges. First, given a set of potential change points, which ones should be queried to the user? Because this focuses on a fixed set of change points, this step of the algorithm is concerned with increasing the precision, that is, eliminating false positive change points. Second, how can the acquired labels be used to improve the algorithm used to detect candidate change points? This requires moving from an unsupervised change point detection setting to a semi-supervised setting. The effect of this step is to identify new change points in order to increase the recall, i.e., the fraction of ground truth change points that are found by the algorithm.

2.1 Algorithm Outline

Algorithm 1 shows the main steps of AL-CPD, our proposed change point detection algorithm. Initially, the algorithm has no labels and hence operates in an unsupervised setting. As shown on lines 1–3, we run TIRE [8] on each sequence to find the initial set of candidate change points C. By automatically learning features using an autoencoder (AE), TIRE makes no distributional assumptions about the change points. The AE takes a window of size s as input and learns two types of features: time-invariant features ($\mathbf{f}^{ti}$) which are used for detecting change points, and time-variant features ($\mathbf{f}^{tv}$) which are only used to reconstruct the time series. When no change point is present, the time-invariant features should remain constant. Therefore, the model minimises the dissimilarity between the time-invariant features extracted from adjacent windows using a time-invariant loss function:

$$\mathcal{L}^{ti} = \sum_{t} ||\mathbf{f}^{ti}_{t+1} - \mathbf{f}^{ti}_{t}||_2.$$

where $\mathbf{f}^{ti}_{t}$ and $\mathbf{f}^{ti}_{t+1}$ are the time-invariant features at time t and $t+1$, respectively. After training, TIRE detects candidate change points by finding peaks in the dissimilarity between the time-invariant features of consecutive windows.

On lines 4–16, the active learning phase of our algorithm improves the candidate change points using two steps: (1) selecting candidates (lines 8–11), and (2) finding new candidates (lines 12–15). These steps represent our key algorithmic contributions. The active learning phase asks queries one by one until the number of queries reaches a user-defined query budget b. When this phase terminates, the algorithm returns all selected candidate change points. In the following two subsections, we describe each step in detail.

2.2 Selecting Candidate Change Points

The first step employs an active learning strategy to identify and remove incorrectly detected candidate change points. For this, we train a random forest classifier [4] m that classifies each candidate in C as a correct or incorrect change point and only keep the candidates predicted as correct change points. We use a model instead of TIRE's change point score in order to avoid querying the label for multiple similar candidate change points. While other classification methods could be relevant, we selected random forests due to their computational efficiency and their ability to select relevant features.

Algorithm 1: AL-CPD

Input: Time series $X = \{x_1, \ldots, x_n\}$, window size s, query budget b
Output: Locations of the change points of each time series x_i

```
1  INITIALISATION
2  C = TIRE(X, s)
3  r = ⌊|C| * 0.1⌋
4  ACTIVE LEARNING
5  Q = ∅
6  m = TrainClassifier(C, Q, s)
7  while |Q| < b do
8      q = LeastCertainCandidate(C, m)
9      a = Query(q)
10     Q = Q ∪ {(q, a)}
11     m = TrainClassifier(C, Q, s)
12     if |Q| mod r = 0 then
13         T = {t ∈ C | p_m(t) > 0.9}
14         C = C ∪ Filter(STIRE(X, T, s), C, s)
15     end
16 end
17 return {t ∈ C | p_m(t) > 0.5}
```

In order to train a model, we construct a training set as follows. First, we extract a feature representation from each candidate change point t in C using TSFuse [6]. While there exist feature extraction systems that compute a similar set of features, we employ TSFuse because this feature extraction system has an efficient implementation. In an active learning system, this is important to minimise the time that the user has to wait. Using the *fast* set of transformers listed in [7], we build a feature vector F_1 from the interval $[t-s, t]$ and F_2 from $[t, t+s]$ for each candidate t. We then compute the difference $\varDelta F = F_2 - F_1$ to measure the change in the feature values. Second, because the number of labelled examples is initially small, we use the local and global consistency label spreading algorithm [23] to increase the amount of labelled data.[1] Third, every candidate that has a label or for which the propagated label has a certainty larger than 90% is added to the training set.

We employ an uncertainty sampling active learning strategy to acquire labels. In each iteration of Algorithm 1, the LeastCertainCandidate(C, m) function computes the certainty of each candidate t in C as $|p_m(t) - 0.5|$ where $p_m(t)$ is the probability predicted by the model m. It returns the candidate with the lowest

[1] Because the label propagation algorithm performs poorly when given high-dimensional data, we first reduce the dimensionality of the feature space using a principal component analysis (PCA) transformation (setting the number of components such that the explained variance is at least 0.9) and standardise the PCA components.

certainty as the query q. The Query(q) function obtains the answer a from the user, which is *true* if there is a change point close to queried candidate change point and *false* otherwise. We add each query-answer tuple (q, a) to Q.

2.3 Finding New Candidate Change Points

Whereas the first step focuses on improving the precision by selecting change points, the second step aims to identify new change points in order to improve the recall. Using the unsupervised TIRE approach to find the initial candidate change points may result in some of the true change points being missed. Therefore, we propose a semi-supervised version of TIRE ("STIRE").

A key challenge to adapting TIRE is its time-invariant loss [8], which pushes the feature representations of neighbouring windows to be close to each other, even when labelled data indicates that this should not be the case due to the presence of a ground-truth change point. Therefore, we modify the time-invariant loss function such that it only forces the time-invariant features for consecutive windows to be close to each other in the latent space when the algorithm is confident that no change point occurs. To this end, we assign labels to all input time windows corresponding to their underlying state. The labels vary only at the temporal indices of confident accurate detections, i.e., the candidate change points for which the random forest model of step 1 predicts a probability larger than 0.9. We replace the time-invariant loss with the triplet loss [21] which is defined as follows:

$$\mathcal{L}^{tri} = \sum_{t} \max\{d_p(\mathbf{f}_t^{ti}) - d_n(\mathbf{f}_t^{ti}) + \gamma, 0\},$$

with

$$d_p(\mathbf{f}_t^{ti}) = \begin{cases} ||\mathbf{f}_{t+1}^{ti} - \mathbf{f}_t^{ti}||_2^2 & \text{if } p_m(t) > 0.9 \text{ or } a is true \\ ||\mathbf{f}_t^{ti} - \mathbf{f}_{t-1}^{ti}||_2^2 & \text{otherwise} \end{cases}$$

and

$$d_n(\mathbf{f}_t^{ti}) = ||\mathbf{f}_t^{ti} - \mathbf{f}_N^{ti}||_2^2,$$

where γ represents the pre-defined margin, and $p_m(t)$ is the probability that there is a change point at time t (more specifically a change between time $t-1$ and t) as predicted by the random forest model m. $\mathbf{f}_t^{ti}$ represents the time-invariant features at time t and $\mathbf{f}_N^{ti}$ denotes the time-invariant features of a negative time window sample, i.e., the time-invariant features extracted from a time window with a different label than the current window at time t. We always select this negative time window randomly from all the segments that come before the previous or after the next true positive change point that was selected by the random forest model of step 1 ($p_m(t) > 0.9$). Similar to the original TIRE model, we also include the reconstruction loss:

$$\mathcal{L}^{rec} = \sum_{t} ||\widehat{w}_t - w_t||_2^2,$$

which encourages the encoded features to contain all information needed for reconstructing the current input window w_t at time t, where $\widehat{w}_t$ denotes the reconstructed window. Finally, the reconstruction loss is combined with the triplet loss via a weighted sum:

$$\mathcal{L} = \mathcal{L}^{rec} + \lambda \mathcal{L}^{tri},$$

where λ controls the balance between the two losses.

After training STIRE, we filter the change points by removing all duplicates, i.e., all candidates that were previously identified. We only keep the candidates for which the time distance to any candidate change point in C is larger than the window size s. Since training STIRE can be time-consuming, we only run this step after every r queries, where we set r to 10% of the number of initial change point candidates: $r = \lfloor |C| * 0.1 \rfloor$.

3 Experiments

We evaluate our active change point detection algorithm on both synthetic and real-world time series datasets to answer the following research questions:

Q1: Can AL-CPD detect change points more accurately than existing methods?
Q2: How many labels does AL-CPD need to find accurate change points?
Q3: How much do each of AL-CPD's two components, (1) using a random forest to select candidates, and (2) using a semi-supervised TIRE variant to find new candidates, contribute to its overall performance?
Q4: What is the sensitivity of AL-CPD to its hyperparameter values?

Because we run our experiments on multiple different applications, we are unable to evaluate application-specific supervised methods. Therefore, we only compare our approach to unsupervised baselines:

GLR [2] The Generalised Likelihood Ratio method fits an auto-regressive model on each adjacent window pair of the time series and detects change points by measuring the dissimilarity of the parameters in the AR model.

RuLSIF [15] The Relative unconstrained Least-Squares Importance Fitting method detects change points by estimating the density ratio of each pair of consecutive windows.

KL-CPD [5] The Kernel Learning CPD method optimises a lower bound of test power using an auxiliary generative model. It learns features using a Seq2Seq model and measures the dissimilarity between neighbouring windows using the maximum mean discrepancy.

TIRE [8] The Time-Invariant REpresentation model maps overlapping windows of the time series onto a feature space using an autoencoder. Change points are detected based on the dissimilarity between windows in the learned feature space.

FLOSS [12] The Fast Low-cost Online Semantic Segmentation algorithm uses the Matrix Profile to find changes in the temporal shape patterns of the time series. It requires the number of ground truth segments as input.

3.1 Datasets

We run the experiments on seven datasets, of which four are artificially constructed and three are based on real-life measurements. Table 1 shows the properties of each dataset. The synthetic datasets are similar to those introduced in [8] and [15]. Three of these synthetic datasets are generated based on a 1-dimensional auto-regressive model:

$$s(t) = a_1 s(t-1) + a_2 s(t-2) + \epsilon_t \quad (1)$$

in which the error term ϵ_t follows a Gaussian distribution $\epsilon_t \sim \mathcal{N}(\mu_t, \sigma_t^2)$. In our experiments, we set the initial state in (1) as: $s(1) = s(2) = 0$ and the default values of the parameters are set to the same values as in [8]: $a_1 = 0.6$, $a_2 = -0.5$, $\mu_t = 0$, and $\sigma_t = 1.5$ unless explained otherwise. Each of these datasets consists of 10 randomly generated sequences. In each sequence, 48 change points are inserted along the temporal axis at each $t_n = t_{n-1} + \lfloor \tau_n \rfloor$, with $t_0 = 0$ and $t_n \sim \mathcal{N}(100, 10)$. We introduce 4 types of change points, leading to the following datasets:

Jumping Mean (JM). The Jumping Mean dataset is generated by changing the value of μ_t at each t_n.
Scaling Variance (SV). In the Scaling Variance dataset, the value of σ_t is changed at each t_n.
Changing coefficients (CC). Here, we set $a_2 = 0$ and alternately draw a_1 from two independent uniform distributions every time a change point is crossed.
Gaussian Mixtures (GM). In this dataset, the time samples in the consecutive segments are alternatively sampled from two different Gaussian mixture distributions.

We include three real-world datasets:

Activity Recognition 1 (AR1). The HASC Challenge 2011 dataset [13] consists of human activity recognition data collected by a triaxial accelerometer. Similar to [8], we select the data from one person (subject 671) and use the magnitude of the acceleration as input. Each segment corresponds to one of the following six activities: staying still, walking, jogging, skipping, ascending stairs, and descending stairs.
Activity Recognition 2 (AR2). We collected a second activity recognition dataset from 8 participants. Each participant performed a sequence of activities consisting of standing, walking, jogging, cycling, ascending stairs, and descending stairs. Similar to the AR1 dataset, the magnitude of the acceleration is collected by a triaxial accelerometer.
Bee Dance (BD). The bee dance dataset [18] consists of six sequences of a bee performing a three-stage waggle dance. Each sequence is a three-dimensional time series representing the location in 2D coordinates and angle differences.

All datasets except AR2 are the same datasets as those used in [8]. We did not include the well log dataset as the number of change points for this dataset is too small to evaluate the active learning step of our proposed algorithm.

Table 1. Dataset properties: number of sequences, and the length and number of change points per sequence (min.–max.).

	Sequences	Length	Change points
JM	10	4836–4925	48
SV	10	4834–4918	48
GM	10	4847–4932	48
CC	10	4864–4907	48
AR1	1	39397	36
AR2	8	15003–25103	22
BD	6	602–1124	15–28

3.2 Methodology

Hyperparameter Settings. The two steps of our algorithm rely on a window size s. For each dataset, we use a window size smaller than the expected interval between change points, but long enough to capture the statistics of the segments. We set s to 30 for the synthetic datasets (JM, SV, GM, CC), 300 for the activity recognition datasets (AR1, AR2), and 15 for the bee dance dataset (BD).

In addition to the window size, our algorithm has several parameters that are independent of the dataset. For the random forest model, we use the implementation of scikit-learn [19] with the default hyperparameter settings. For (S)TIRE, we learn both time-domain and frequency-domain features. Each auto-encoder learns 3 features: 2 time-invariant features and 1 time-variant feature. In the loss function, the values of γ and λ are set to 0.1 and 0.001, respectively. We train the networks for 200 epochs using the Adam optimiser.

Evaluation. We compare the detected change points to the ground truth change points by computing the precision, recall, and F1 score. These metrics are defined as follows:

$$\text{precision} = \frac{TP}{TP + FP} \quad \text{recall} = \frac{TP}{TP + FN} \quad \text{F1 score} = 2 \cdot \frac{\text{precision} \cdot \text{recall}}{\text{precision} + \text{recall}}$$

The number of true positives (TP) is computed as the number of predicted change points that are within a distance s from one of the ground truth change points. Any predicted change point that is further than s from all ground truth change points is counted as a false positive (FP). The number of false negatives (FN) is the number of ground truth change points that have a distance larger than s to any predicted change point.

Typically, evaluating active learning methods involves reporting the performance in terms of the number of examples labelled by an annotator. In our setting, each example corresponds to one of the candidate change points. However, the number of candidate change points identified by AL-CPD varies as more labels are acquired. Therefore, we report the performance relative to the amount of work a human would have to do to fully annotate the data. This would require partitioning each sequence into non-overlapping windows of size s and then labelling each window as either containing a change point or not. We refer to the total labelling effort as the number of potential change points in a dataset:

$$P = \sum_{i=1}^{n} \left\lfloor \frac{N_i}{s} \right\rfloor$$

where n ranges over sequences and N_i is the length of the ith sequence. The number of potential change points P for each dataset is as follows:

	JM	SV	GM	CC	AR1	AR2	BD
P	1622	1617	1622	1621	131	516	328

When running active learning, AL-CPD receives a *true* answer if its queried candidate change point t is within s samples from at least one ground truth change point. Otherwise, it receives a *false* answer.

3.3 Q1: Comparison to Existing Change Point Detection Algorithms

Table 2 shows the precision, recall, and F1 score for the baselines and AL-CPD. We run AL-CPD for three different query budgets that correspond to 5%, 10% and 20% of all potential change points. In terms of the F1 score, AL-CPD substantially outperforms the baselines on all datasets after querying 20% of the potential change point locations. On the four synthetic datasets, querying only 5% of the potential change points leads to better results compared to the baselines. Because of the well-defined underlying process, the synthetic datasets contain many similar change points. Hence, learning the definition of a change point requires fewer labelled examples (i.e., fewer queries) compared to the more complex real-world datasets.

GLR, RuLSIF, and KL-CPD achieve a perfect recall on all datasets except GM and BD for RuLSIF. However, these methods find many false positives, which results in a low precision. Hence, for the 20% query budget, the improved performance of AL-CPD over these baselines can be attributed to the better precision. FLOSS performs worse than all other baselines and AL-CPD in terms of all evaluation metrics. Compared to TIRE, our algorithm achieves a better precision on all datasets and a better recall on 5 out of 7 datasets after querying 10% of the potential change point locations.

Note that for the AR1 dataset, AL-CPD scores zero for all metrics for the 5% query budget. This occurs because none of the queried candidate windows contained a change point. Hence, the learned random forest predicts that no other windows contain a change point, leading to all candidate change points being discarded.

Table 2. Precision, recall, and F1 score of each baseline and AL-CPD after querying 5%, 10%, and 20% of all possible change point locations. For each dataset, we highlight the best-performing baseline in bold and annotate each baseline outperformed by AL-CPD after querying 20%, 10%, and 5% of the change point locations with |, ||, and |||, respectively.

Precision

	Baselines					AL-CPD		
	GLR	RuLSIF	KL-CPD	FLOSS	TIRE	5%	10%	20%
JM	0.584$^{\vert\vert\vert}$	0.590$^{\vert\vert\vert}$	0.573$^{\vert\vert\vert}$	0.548$^{\vert\vert\vert}$	0.861$^{\vert\vert\vert}$	0.923	0.945	**0.990**
SV	0.585$^{\vert\vert\vert}$	0.589$^{\vert\vert\vert}$	0.581$^{\vert\vert\vert}$	0.556$^{\vert\vert\vert}$	0.702$^{\vert\vert\vert}$	0.756	0.810	**0.908**
GM	0.578$^{\vert\vert\vert}$	0.596$^{\vert\vert\vert}$	0.582$^{\vert\vert\vert}$	0.565$^{\vert\vert\vert}$	0.906$^{\vert\vert\vert}$	0.982	**1.000**	**1.000**
CC	0.583$^{\vert\vert\vert}$	0.057$^{\vert\vert\vert}$	0.574$^{\vert\vert\vert}$	0.571$^{\vert\vert\vert}$	0.738$^{\vert\vert\vert}$	0.790	0.836	**0.954**
AR1	0.354$^{\vert\vert}$	0.366$^{\vert\vert}$	0.382$^{\vert\vert}$	0.389$^{\vert\vert}$	0.500$^{\vert\vert}$	0.000	0.889	**0.941**
AR2	0.485$^{\vert\vert\vert}$	0.532$^{\vert\vert\vert}$	0.523$^{\vert\vert\vert}$	0.415$^{\vert\vert\vert}$	0.630$^{\vert\vert\vert}$	0.826	0.856	**0.973**
BD	0.670$^{\vert\vert\vert}$	0.679$^{\vert\vert\vert}$	0.682$^{\vert\vert\vert}$	0.474$^{\vert\vert\vert}$	0.741$^{\vert\vert\vert}$	0.809	0.833	**0.869**

Recall

	Baselines					AL-CPD		
	GLR	RuLSIF	KL-CPD	FLOSS	TIRE	5%	10%	20%
JM	**1.000**	**1.000**	**1.000**	0.548$^{\vert\vert\vert}$	0.944$^{\vert\vert\vert}$	0.962	0.977	0.979
SV	**1.000**	**1.000**	**1.000**	0.556$^{\vert\vert\vert}$	0.852$^{\vert\vert\vert}$	0.919	0.933	0.940
GM	**1.000**	0.994	**1.000**	0.565$^{\vert\vert\vert}$	0.987$^{\vert\vert}$	0.985	0.994	0.994
CC	**1.000**	**1.000**	**1.000**	0.571$^{\vert\vert\vert}$	0.783$^{\vert\vert\vert}$	0.831	0.875	0.875
AR1	**1.000**	**1.000**	**1.000**	0.556$^{\vert}$	0.861	0.000	0.361	0.611
AR2	**1.000**	**1.000**	**1.000**	0.591$^{\vert\vert}$	0.830	0.551	0.722	0.807
BD	**1.000**	0.852	**1.000**	0.453$^{\vert\vert\vert}$	0.717$^{\vert\vert\vert}$	0.764	0.775	0.852

F1 score

	Baselines					AL-CPD		
	GLR	RuLSIF	KL-CPD	FLOSS	TIRE	5%	10%	20%
JM	0.738$^{\vert\vert\vert}$	0.742$^{\vert\vert\vert}$	0.729$^{\vert\vert\vert}$	0.548$^{\vert\vert\vert}$	0.900$^{\vert\vert\vert}$	0.942	0.961	**0.985**
SV	0.738$^{\vert\vert\vert}$	0.741$^{\vert\vert\vert}$	0.735$^{\vert\vert\vert}$	0.556$^{\vert\vert\vert}$	0.770$^{\vert\vert\vert}$	0.829	0.867	**0.923**
GM	0.733$^{\vert\vert\vert}$	0.745$^{\vert\vert\vert}$	0.736$^{\vert\vert\vert}$	0.565$^{\vert\vert\vert}$	0.945$^{\vert\vert\vert}$	0.983	**0.997**	**0.997**
CC	0.736$^{\vert\vert\vert}$	0.108$^{\vert\vert\vert}$	0.729$^{\vert\vert\vert}$	0.571$^{\vert\vert\vert}$	0.760$^{\vert\vert\vert}$	0.810	0.855	**0.912**
AR1	0.523$^{\vert}$	0.536$^{\vert}$	0.553$^{\vert}$	0.458$^{\vert\vert}$	0.633$^{\vert}$	0.000	0.514	**0.741**
AR2	0.649$^{\vert\vert\vert}$	0.692$^{\vert\vert}$	0.684$^{\vert\vert}$	0.486$^{\vert\vert\vert}$	0.709$^{\vert\vert}$	0.650	0.774	**0.880**
BD	0.799$^{\vert}$	0.752$^{\vert\vert\vert}$	0.806$^{\vert}$	0.453$^{\vert\vert\vert}$	0.726$^{\vert\vert\vert}$	0.779	0.793	**0.857**

3.4 Q2: Labelling Effort of AL-CPD

We investigate the effect of the number of acquired labels on AL-CPD's performance. Specifically, we evaluate how many queries are required to obtain an F1 score that is larger than a chosen percentage of the final F1 score achieved when using an unlimited query budget.

Table 3 shows the percentage of change point locations needed to achieve an F1 score of at least 80%, 90% and 95% of the final F1 score. Achieving an F1 score of at least 80% of the final F1 score requires labelling between 0.1% and 21.4% of all potential change point locations. In other words, the user saves between 78.6% and 99.9% of the labelling effort compared to manually labelling each window. Even to achieve an F1-score of 95% of the final one, AL-CPD still reduces the effort compared to completely labelling the data by at least 68.7%.

Table 3. Percentage of change point locations that the user has to label in order to obtain an F1 score of at least 80%, 90% and 95% of the final F1 score.

	JM	SV	GM	CC	AR1	AR2	BD
80%	0.1%	1.9%	0.7%	0.2%	21.4%	7.2%	3.7%
90%	0.1%	10.9%	0.7%	9.3%	30.5%	16.1%	10.7%
95%	5.1%	19.3%	0.9%	15.7%	31.3%	22.9%	25.0%

3.5 Q3: Contribution of Each Component of AL-CPD

Our algorithm has two components: (1) selecting candidates by training a classifier, and (2) finding new candidates by training TIRE in a semi-supervised setting. For research question **Q3**, we analyse the effect of each component on the algorithm's performance. To do so, we perform an ablation study that compares the AL-CPD algorithm to two variants:

1. The **A** variant includes only component 1 of our algorithm, i.e., the active learning step for selecting candidates.
2. The **S** variant includes only component 2 of our algorithm, i.e., the semi-supervised setting of TIRE for finding new candidates.

In order to compare our algorithm to the two variants, we evaluate the area under the learning curve (ALC) of the precision, recall, and F1 score. The ALC is defined as follows:

$$\text{ALC} = \frac{1}{n}\sum_{i=0}^{n} e(i)$$

where n is the total number of queries and $e(i)$ is the evaluation metric (i.e., precision, recall, or F1 score) computed after the ith query.

Table 4 shows that the **A** variant results in a better precision than the **S** variant on most datasets. However, because the candidate selection component of the **A** variant removes some of the true positive change points, this variant has the lowest recall. By searching for new candidates, the **S** variant improves the recall on all datasets. In terms of the F1-score, the full AL-CPD algorithm outperforms both **A** and **S** on four datasets. For the other three datasets, the **S** variant outperforms AL-CPD due to a better recall.

Table 4. Area under the learning curve of the precision, recall, and F1 score for each variant. The performance of the best variant is highlighted in bold.

	Precision			Recall			F1 score		
	A	S	AL-CPD	A	S	AL-CPD	A	S	AL-CPD
JM	**0.973**	0.961	0.968	0.941	**0.974**	0.972	0.956	0.967	**0.970**
SV	0.885	0.832	**0.895**	0.847	0.926	**0.927**	0.862	0.875	**0.909**
GM	0.974	**0.977**	0.974	0.965	**0.991**	0.971	0.973	**0.984**	0.976
CC	0.910	0.849	**0.914**	0.767	**0.874**	0.857	0.829	0.860	**0.883**
AR1	0.757	0.759	**0.798**	0.515	**0.868**	0.560	0.592	**0.802**	0.637
AR2	0.923	0.840	**0.933**	0.700	**0.861**	0.760	0.782	**0.844**	0.827
BD	0.864	0.836	**0.877**	0.701	0.805	**0.821**	0.769	0.819	**0.845**

3.6 Q4: Sensitivity Analysis

For research question **Q4**, we analyse the hyperparameter sensitivity of AL-CPD. Our algorithm has three main hyperparameters: the window size s and two hyperparameters specific to STIRE: the balance between the reconstruction and triplet loss λ and the margin γ.

Figure 1 compares the ALC of the F1 score for three different values of each hyperparameter. For the window size s, we multiply the default window sizes by a factor 0.5, 1 and 1.5 and report the average ALC of the F1-score over all datasets. On average, shorter window sizes decrease the F1 score. This is expected since capturing the characteristics of a segment requires a sufficiently long portion of the time series. Longer windows do not further improve the F1 score, and may even decrease the performance when exceeding the distance between consecutive change points. The performance of our algorithm is robust w.r.t. the reconstruction and triplet loss λ and the margin γ, since the ALC of the F1 score is almost not affected by the values of these hyperparameters.

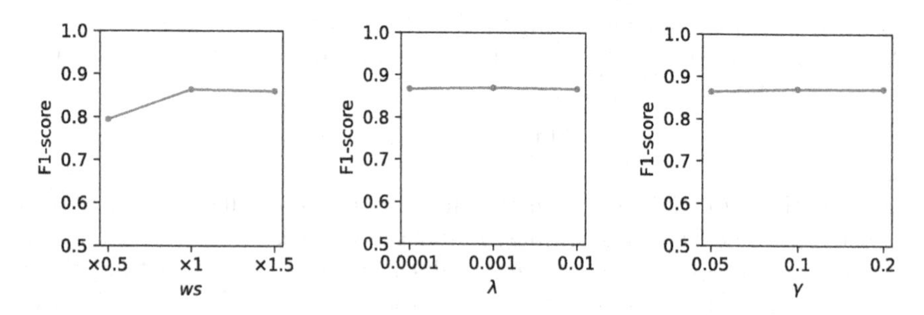

Fig. 1. ALC of the F1 score for three different values of the window size s, the balance between the reconstruction and triplet loss, and the margin γ. The ALC is averaged over all datasets.

4 Conclusion

This paper presented AL-CPD, a change point detection algorithm that combines active and semi-supervised learning. Instead of only relying on assumptions about the changes in the underlying statistics of the given time series, AL-CPD asks directed queries to the user in order to obtain labels. Our algorithm exploits these labels to eliminate incorrectly detected change points and to search for new change points. In an empirical evaluation, we compared the performance of AL-CPD to existing unsupervised CPD methods and showed that AL-CPD is able to find more accurate change points with a query budget of at most 20% of all potential change points.

Acknowledgements. This work is supported by the Research Foundation Flanders (FWO) under TBM grant number T004716N, by the Flemish government under the "Onderzoeksprogramma Artificiële Intelligentie (AI) Vlaanderen" programme, and by VLAIO ICON-AI CONSCIOUS (HBC.2020.2795).

References

1. Aminikhanghahi, S., Cook, D.J.: A survey of methods for time series change point detection. Knowl. Inf. Syst. **51**(2), 339–367 (2017)
2. Appel, U., Brandt, A.V.: Adaptive sequential segmentation of piecewise stationary time series. Inf. Sci. **29**(1), 27–56 (1983)
3. Basseville, M., Nikiforov, I.V., et al.: Detection of abrupt changes: theory and application, vol. 104, prentice Hall Englewood Cliffs (1993)
4. Breiman, L.: Random for. Mach. Learn. **45**(1), 5–32 (2001)
5. Chang, W.C., Li, C.L., Yang, Y., Póczos, B.: Kernel change-point detection with auxiliary deep generative models. arXiv preprint arXiv:1901.06077 (2019)
6. De Brabandere, A., Op De Beéck, T., Hendrickx, K., Meert, W., Davis, J.: TSFuse: Automated feature construction for multiple time series data. Mach. Learn. (2022). https://doi.org/10.1007/s10994-021-06096-2
7. De Brabandere, A., Robberechts, P., Op De Beéck, T., Davis, J.: Automating feature construction for multi-view time series data. In: ECMLPKDD Workshop on Automating Data Science (2019)
8. De Ryck, T., De Vos, M., Bertrand, A.: Change point detection in time series data using autoencoders with a time-invariant representation. IEEE Trans. Signal Process. **69**, 3513–3524 (2021)
9. Deldari, S., Smith, D.V., Sadri, A., Salim, F.: Espresso: entropy and shape aware time-series segmentation for processing heterogeneous sensor data. Proc. ACM on Interact. Mobile, Wearable Ubiquitous Technol. **4**(3), 1–24 (2020)
10. Desobry, F., Davy, M., Doncarli, C.: An online kernel change detection algorithm. IEEE Trans. Signal Process. **53**(8), 2961–2974 (2005)
11. Feuz, K.D., Cook, D.J., Rosasco, C., Robertson, K., Schmitter-Edgecombe, M.: Automated detection of activity transitions for prompting. IEEE Trans. Human-Mach. Syst. **45**(5), 575–585 (2014)
12. Gharghabi, S., et al.: Domain agnostic online semantic segmentation for multi-dimensional time series. Data Min. Knowl. Disc. **33**(1), 96–130 (2019)

13. Kawaguchi, N.,et al.: HASC2011corpus: Towards the common ground of human activity recognition. In: Proceedings of the 13th International Conference on Ubiquitous Computing, pp. 571–572 (2011)
14. Kawahara, Y., Sugiyama, M.: Sequential change-point detection based on direct density-ratio estimation. Stat. Analysis Data Mining: ASA Data Sci. J. **5**(2), 114–127 (2012)
15. Liu, S., Yamada, M., Collier, N., Sugiyama, M.: Change-point detection in time-series data by relative density-ratio estimation. Neural Netw. **43**, 72–83 (2013)
16. Matsubara, Y., Sakurai, Y., Faloutsos, C.: Autoplait: Automatic mining of co-evolving time sequences. In: Proceedings of the 2014 ACM SIGMOD International Conference on Management of data, pp. 193–204 (2014)
17. Munir, M., Siddiqui, S.A., Dengel, A., Ahmed, S.: Deepant: a deep learning approach for unsupervised anomaly detection in time series. Ieee Access **7**, 1991–2005 (2018)
18. Oh, S.M., Rehg, J.M., Balch, T., Dellaert, F.: Learning and inferring motion patterns using parametric segmental switching linear dynamic systems. Int. J. Comput. Vision **77**(1), 103–124 (2008)
19. Pedregossa, F., et al.: Scikit-learn: Machine learning in Python. J. Mach. Learn. Res. **12**, 2825–2830 (2011)
20. Reddy, S., Mun, M., Burke, J., Estrin, D., Hansen, M., Srivastava, M.: Using mobile phones to determine transportation modes. ACM Trans. Sensor Netw. (TOSN) **6**(2), 1–27 (2010)
21. Schroff, F., Kalenichenko, D., Philbin, J.: Facenet: A unified embedding for face recognition and clustering. In: 2015 IEEE Conference on Computer Vision and Pattern Recognition (CVPR), pp. 815–823 (2015). https://doi.org/10.1109/CVPR.2015.7298682
22. Zhang, R., Hao, Y., Yu, D., Chang, W.C., Lai, G., Yang, Y.: Correlation-aware unsupervised change-point detection via graph neural networks (2020)
23. Zhou, D., Bousquet, O., Lal, T., Weston, J., Schölkopf, B.: Learning with local and global consistency. In: Advances in neural information processing systems vol. 16 (2003)

Adaptive Neural Networks for Online Domain Incremental Continual Learning

Nuwan Gunasekara(✉), Heitor Gomes, Albert Bifet, and Bernhard Pfahringer

AI Institute, University of Waikato, Hamilton, New Zealand
ng98@students.waikato.ac.nz

Abstract. Continual Learning (CL) poses a significant challenge to Neural Network (NN)s, where the data distribution changes from one task to another. In Online domain incremental continual learning (OD-ICL), this distribution change happens in the input space without affecting the label distribution. In order to adapt to such changes, the model being trained risks forgetting previously learned knowledge (stability). On the other hand, enforcing that the model preserves past knowledge will cause it to fail to learn new concepts (plasticity). We propose Online Domain Incremental Networks (ODIN), a novel method to alleviate catastrophic forgetting by automatically detecting the end of a task using concept drift detection. As a consequence, ODIN does not require the specification of task ids. ODIN maintains a pool of NNs, each trained on a single task and frozen for further updates. A Task Predictor (TP) is trained to select the most suitable NN from the frozen pool for prediction. We compare ODIN against popular regularization and replay methods. It outperforms regularization methods and achieves comparable predictive performance to replay methods.

Keyword: Online domain incremental continual learning

1 Introduction

Though modern Neural Network (NN)s have shown great success in image classification and natural language processing, they assume training data to be Independent and Identically Distributed (IID). Due to this assumption, once confronted with a distribution shift in the input data, the model may undergo costly retraining to preserve old knowledge while adjusting to the new distribution. Without retraining, an NN receiving non-IID data forgets its past knowledge when confronted with a distribution shift. This phenomenon is known as "catastrophic forgetting" in the literature [9,16].

Continual Learning (CL) attempts to minimize this catastrophic forgetting in NNs via replay and regularization methods [16]. Though current replay methods outperform regularization methods in terms of performance, they may not be suitable for situations with memory and privacy constraints on the replay buffer

P. Pascal and D. Ienco (Eds.): DS 2022, LNAI 13601, pp. 89–103, 2022.
https://doi.org/10.1007/978-3-031-18840-4_7

[2,16]. Even though offline CL methods have been proposed, current research mainly focuses on online methods to solve catastrophic forgetting in NNs. This allows one to develop continually learning agents which are adaptive but also resilient to catastrophic forgetting.

Online domain incremental continual learning (ODICL) focuses on online CL models, which learn from one input distribution to another with minimum catastrophic forgetting. Here the class distribution remains the same. There are many practical applications of this scenario in the modern IoT world. For example, one can use an ODICL approach to avoid costly retraining of an X-ray image classification model after a distribution shift in the incoming data due to some hardware changes in the X-ray machine [19]. The same scenario can be valid for many NN models that rely on hardware sensor inputs. Also, on specific ODICL settings, replay approaches may be less preferred due to constraints on having a replay buffer. Mainly these are privacy constraints on the replay buffer [2,16].

Considering the practical importance of non-replay ODICL, this work proposes an ODICL method that alleviates catastrophic forgetting in NNs. It is superior to regularization methods and competitive to replay methods. Here a Convolutional Neural Network (CNN) is trained online. Once confronted with a concept shift, it freezes a copy of the current CNN. Task Predictor (TP) is trained to pick the best CNN from the frozen pool for prediction. This approach is further extended to automatically detect concept shifts in incoming data using a Task Detector (TD) instead of relying on an external task id signal. Experiment results reveal that both the proposed methods, with and without automatic TD, surpass the current popular regularization methods and have competitive performance compared to replay methods.

The main contributions of this paper are the following:

1. Online Domain Incremental Networks (ODIN): we introduce a novel method to alleviate catastrophic forgetting for Online domain incremental continual learning without using instance replay. Here, a frozen copy of the training CNN is saved in a pool at the end of each task. A Task Predictor is trained to predict the best frozen CNN for evaluation for a given instance. The experiment results reveal that ODIN yields better accuracy than regularization and competitive performance to replay baselines. Furthermore, an in-depth investigation is done to better understand the effectiveness of different TPs on three ODICL datasets.
2. Instead of relying on an external task id signal during training, ODIN uses an automatic Task Detector mechanism to detect tasks in the incoming data. ADaptive sliding WINdow (ADWIN) is used to detect drifts in CNN's loss. An incremental drift in the loss is determined as the end of a task. Furthermore, incremental or decremental drifts in CNN's loss detected by ADWIN allow ODIN to dynamically increase or decrease the learning rate. To the best of our knowledge, this automatic Task Detector with Dynamic Learning-Rate (DL) adjustment for ODICL has not been proposed before.

The rest of the paper is organized as follows. The following section presents the current developments in Online domain incremental continual learning, including some practical use cases. The next section then presents the proposed ODIN for ODICL. The experiments section explains the experimental setup where the proposed method is compared against popular ODICL methods on three datasets. It also provides insights into the effectiveness of different Task Predictors. The final section provides conclusions and directions for future research.

2 Related Work

The literature has thoroughly documented that an NN receiving non-IID data forgets past knowledge when confronted with a concept shift [9,16]. Continual Learning attempts to continually learn with minimal forgetting of past concepts [9,16]. In ODICL, this learning happens online, and the data stream comprises different concepts (distributions) with the same label distribution [16].

CL algorithms use two popular approaches to avoid catastrophic forgetting in NNs: regularization and replay. Regularization algorithms like Elastic Weight Consolidation (EWC) [9] and Learning without Forgetting (LwF) [13] adjust the weights of the network in such a way that it minimizes the overwriting of the weights for the old concept. EWC uses a quadratic penalty to regularize updating the network parameters related to the past concept. It uses the Fisher Information Matrix's diagonal to approximate the importance of the parameters [9]. EWC has some shortcomings: 1) the Fisher Information Matrix needs to be stored for each task, 2) it requires an extra pass over each task's data at the end of the training [16]. Though different versions of EWC address these concerns [16], [5] seems suitable for online CL by keeping a single Fisher Information Matrix calculated by a moving average. LwF uses knowledge distillation to preserve knowledge from past tasks. Here, the model related to the old task is kept separate, and a separate model is trained on the current task. When the LwF receives data for a new task (X_{new}, Y_{new}), it computes the output (Y_{old}) from the old model for new data X_{new}. During training, assuming that $\hat{Y_{old}}$ and $\hat{Y_{new}}$ are predicted values for X_{new} from the old model and new model, LwF attempts to minimize the loss: $\alpha L_{KD}(Y_{old}, \hat{Y_{old}}) + L_{CE}(Y_{new}, \hat{Y_{new}}) + R$ [16]. Here L_{KD} is the distillation loss for the old model, and α is the hyper-parameter controlling the strength of the old model against the new one. L_{CE} is the cross-entropy loss for the new task. R is the general regularization term. Due to this strong relation between old and new tasks, it may perform poorly in situations where there is a huge difference between old and new task distributions [16].

Replay methods present a mix of instances from the old and current concepts to the NN based on a given policy while training. This reduces the forgetting as the training instances from the old concepts avoid complete overwriting of past concepts' weights. GDumb [18], Experience Replay (ER) [6], and Maximally Interfered Retrieval (MIR) [1] are some of the most popular CL replay methods. GDumb attempts to maintain a class-balanced memory buffer using instances from the stream. At the end of the task, it trains the model using the buffered instances. ER uses reservoir sampling [20] to sample instances from the stream to fill the buffer. Reservoir sampling ensures that every instance in the stream has the same probability of being selected to fill the buffer. ER uses random sampling to retrieve instances from the memory buffer. Despite its simplicity, ER has shown competitive performance in ODICL [16]. Five (three buffer and two non-buffer) tricks have been proposed by [4] to improve the accuracy of ER in the Online Class Incremental Continual Learning (OCICL) setting. The buffer tricks are independent buffer augmentation, balanced reservoir sampling, and loss-aware reservoir sampling. The two non-buffer tricks are bias control and exponential learning rate decay. Except for bias control which controls the bias of newly learned classes, these tricks can be used in ODICL to improve the performance of a replay method. MIR uses the same reservoir sampling as ER to fill the memory buffer. However, when retrieving instances from the buffer, it first does a virtual parameter update using the incoming min-batch. Then it selects the top k randomly sampled instances with the most significant loss increases by the virtual parameter update for training. In the online implementation in [16], this virtual update is done on a copy of the NN. Replay Using Memory Indexing (REMIND) [7] takes this approach to another level by storing the internal representations of the instances by the initial frozen part of the network and using a randomly selected set of these internal representations to train the last unfrozen layers of the network. Here, REMIND can store more instances' representations using internal low-dimensional features. In general, these replay approaches are motivated by how the hippocampus in the brain stores and replays high-level representations of the memories to the neocortex to learn from them [7]. The empirical survey by [16] suggests that ER and MIR perform better on ODICL than other online CL methods.

Recent research has focused on using ODICL methods to avoid costly retraining in practical situations where the model is confronted with a concept shift. ODICL has been used in X-ray image classification to avoid costly retraining on distribution shifts due to unforeseen shifts in hardware's physical properties [19]. Also, it has been used to mitigate bias in facial expression and action unit recognition across different demographic groups [8]. Furthermore, ODICL was used to counter retraining on concept shifts for multi-variate sequential data of critical care patient recordings [2]. The authors highlight some replay methods' infeasibility due to strong privacy requirements in clinical settings. This concern is further highlighted in the empirical study in [16].

Most current ODICL methods rely on an explicit end-of-task signal during training. EWC and LwF use this signal to optimize weights, while replay methods can use it to update their replay buffer. However, GDumb , ER, and MIR do not rely on this signal for replay buffer updates. Though [16] defines ODICL as training without the end of the task signal. Implementations such as [8] and [2] use the end of the task signal to employ CL methods such as EWC and LwF. However, on the other hand, the implementation in [19] assumes a gradual distribution shift in the input data distribution where instances from both the new and old tasks can appear in the stream for a certain period.

ODIN comes in two versions. One assumes the presence of an end-of-task signal at training, whereas the other proposes an automated task detection method. When a concept shift is detected, the proposed method freezes a copy of the training NN and adds it into a pool. A predictor is trained to choose the best network from the frozen pool for a given evaluation instance. As the method avoids a replay buffer, it is a good candidate for settings with higher privacy requirements.

3 Online Domain Incremental Networks

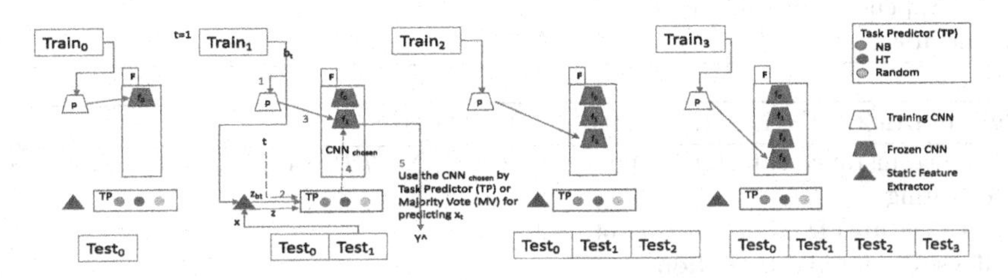

Fig. 1. Proposed ODIN: 1) train network p with incoming mini-batch b_t for t^{th} task, 2) train TP using extracted features and task id, 3) freeze a copy of p at the end of task t 4) at prediction, if enabled, TP predicts CNN$_{chosen}$ via extracted x features 5) predict using CNN$_{chosen}$ or Majority Vote.

In ODICL, the training set is composed of multiple concepts of non-IID data where each concept has a different input distribution with the same label distribution [16]. The goal of the learning algorithm is to minimize catastrophic forgetting of past concepts while performing well on the current concept [7,16]. The initial version of ODIN assumes the availability of the task id at training, which signals the end of a concept to the learning model. However, this information is not available to the model during evaluation. The refined version of ODIN is extended to detect the end of a concept automatically. So, ODIN can be applied to situations where the external task id signal is unavailable at training.

We propose an Online Domain Incremental Networks (ODIN), where CNN p is trained on each concept t with a given Task Predictor (TP). The TPs could be Naive Bayes (NB), or Hoeffding Tree (HT). The TP is trained on mini-batch b_t using extracted features from a feature extractor. Feature extractors extract features from high-dimensional data. Hence, it allows one to use simple learning algorithms on high-dimensional data [10]. Usually, a pre-trained network is used as a feature extractor [10], and its last layer features are used to train the TP. At the end of each task's training, a copy of p is frozen and added to the frozen pool F. Algorithm 1, along with Fig. 1, further explains this training approach.

In ODIN, there are two vote aggregation methods for prediction: Weighted Voting (WV) or votes from the best CNN (CNN$_{best}$). Weighted Voting uses the TP's probabilities for each frozen CNN as weights. In the CNN$_{best}$ case,

Algorithm 1. ODIN TRAINING ALGORITHM

Input: p: training CNN, F: pool of frozen CNNs, T: task set, X_t: training set for task t, TP: Task Predictor

1: Initialize pool $F = \{\}$
2: **for** all task $t \in T$ **do**
3:     **for** all mini-batch b_t in training set X_t for task t **do**
4:         Train p with the computed the loss L_{b_t} for mini-batch b_t
5:         **if** task predictor TP is Naive Bayes or Hoeffding Tree **then**
6:             $z \leftarrow$ extract features from mini-batch b_t via feature extractor
7:             TRAIN $TP(z, t)$
8:         **end if**
9:     **end for**
10:     Append a copy of p to F
11: **end for**

Algorithm 2. ODIN PREDICTION ALGORITHM

Input: x_t: instance of task t, F: pool of frozen CNNs, TP: Task Predictor, useWeightedVoting

1: $z \leftarrow$ features from instance x_t of task t
2: **if** useWeightedVoting **then**
3:     **if** TP is Majority Vote **then**
4:         $votes \leftarrow 1/|F| \sum_{f=1}^{|F|} \text{PREDICT}_f(x_t)$
5:     **else**
6:         $votes \leftarrow 1/|F| \sum_{f=1}^{|F|} \text{PREDICT}_{TP}(z)_f \times \text{PREDICT}_f(x_t)$
7:     **end if**
8: **else**
9:     **if** TP is Random **then**
10:         Select CNN_{chosen} randomly from pool F
11:     **else**
12:         $\text{CNN}_{chosen} \leftarrow \arg\max_{f \in F} \text{PREDICT}_{TP}(z)$
13:     **end if**
14:     $votes \leftarrow \text{PREDICT}_{\text{CNN}_{chosen}}(x_t)$
15: **end if**

Output: $votes$

Algorithm 3. DYNAMIC LEARNING-RATE

Input: lr_0: learning rate at start, d: decay factor$(0 < d < 1)$, n: instances seen since last drift, $upwardDrift$: whether the estimated loss going up

1: **if** $upwardDrift$ **then**
2:     $lr \leftarrow lr_0 * (1 + d^n)$
3: **else**
4:     $lr \leftarrow lr_0 * (d^n)$
5: **end if**

Output: lr

it is either selected randomly from the F pool or the one predicted by TP. Algorithm 2 further explains this.

Generally, NN's loss distribution changes when the underlying input distribution changes as the network weights need to be readjusted to match the new distribution. Once a drift in the loss is detected, 1) it would be helpful to learn following the direction of the loss, where the network learns faster if there is an upward drift in the loss, and it learns slower if the drift in the loss is decreasing. Usually, in ODICL, NN's loss gradually decreases for a given task with non-IID training instances. Hence, 2) it would be helpful to reduce the magnitude of the learning for the incoming instances further away from the task's start so that the later instances of the same task do not disturb the learned weights too much.

Here we use the drift detector ADWIN [3] to monitor the loss of p. ADWIN uses exponential histograms for memory efficiency and discards the buffer related to the previous concept once a drift is detected. It also provides an estimation for the mean of the current items in the buffer. Once ADWIN detects a drift in the loss, one can compare the current estimated loss against the previous estimated loss to identify the direction of the loss. This helps determine whether to increase or decrease the learning rate (point 1). Also, the number of instances seen after the drift can be used to decrease the magnitude of the learning rate (point 2).

Algorithm 4. ODIN TRAINING ALGORITHM WITH DYNAMIC LEARNING-RATE

Input: p: training CNN, F: pool of frozen CNNs, T: task set, X_t: training set for task t, TP: Task Predictor, lr_0: learning rate at start, d: learning rate decay factor

```
1: Initialize pool F = {}
2: Initialize n = 0
3: Initialize upwardDrift = true
4: for all task t ∈ T do
5:     for all mini-batch b_t in training set X_t for task t do
6:         Compute the loss L_{b_t} of mini-batch b_t for p
7:         Update ADWIN_p with L_{b_t}
8:         if ADWIN_p detects change then
9:             n ← 0
10:            if change is upward then
11:                upwardDrift ← true
12:            else
13:                upwardDrift ← false
14:            end if
15:        else
16:            n ← n + 1
17:        end if
18:        if task predictor TP is Naive Bayes or Hoeffding Tree then
19:            z ← extract features from mini-batch b_t via feature extractor
20:            TRAIN TP(z, t)
21:        end if
22:        lr ← DYNAMIC LEARNING-RATE(lr_0, d, n, upwardDrift)
23:        train p with loss L_{b_t} and lr
24:    end for
25:    Append a copy of p to F
26: end for
```

The easiest way to manage the learning of a NN is to adjust the learning rate. In order to learn faster for the upward drifts in the loss detected by ADWIN, ODIN increases the learning rate. For the downwards drifts it decreases the learning rate to prevent against large changes of presumably already well-adjusted weights. Also, to decrease the magnitude of the learning in either direction, it uses a decaying factor d^n where d is $0 < d < 1$ and n is the number of instances seen since the last drift. This decaying factor is discussed in [4]. However, they continually decrease the learning rate from the start of learning in their work. Hence it forces NN not to learn too much from instances of later tasks. On the other hand, this Dynamic Learning-Rate (DL) in ODIN allows p to best adjust to the current task. Algorithm 3 and Algorithm 4 explain ODIN's Dynamic Learning-Rate adjustment mechanism.

Algorithm 5. ODIN TRAINING ALGORITHM WITH DYNAMIC LEARNING-RATE AND AUTOMATIC TASK DETECTOR

Input: p: training CNN, F: pool of frozen CNNs, T: task set, X_t: training set for task t, TP: Task Predictor, lr_0: learning rate at start, d: learning rate decay factor

1: Initialize pool $F = \{\}$
2: Initialize $taskId = 0$
3: Initialize $n = 0$
4: Initialize $upwardDrift = true$
5: **for** all task $t \in T$ **do**
6:     **for** all mini-batch b_t in training set X_t for task t **do**
7:         Compute the loss L_{b_t} of mini-batch b_t for p
8:         Update ADWIN$_p$ with L_{b_t}
9:         **if** ADWIN$_p$ detects change **then**
10:             $n \leftarrow 0$
11:             **if** change is upward **then**
12:                 $upwardDrift \leftarrow true$
13:                 $taskId \leftarrow taskId + 1$
14:                 Append a copy of p to F
15:             **else**
16:                 $upwardDrift \leftarrow false$
17:             **end if**
18:         **else**
19:             $n \leftarrow n + 1$
20:         **end if**
21:         **if** task predictor TP is Naive Bayes or Hoeffding Tree **then**
22:             $z \leftarrow$ extract features from mini-batch b_t via feature extractor
23:             TRAIN $TP(z, taskId)$
24:         **end if**
25:         $lr \leftarrow$ DYNAMIC LEARNING-RATE(lr_0, d, n, $upwardDrift$)
26:         train p with loss L_{b_t} and lr
27:     **end for**
28: **end for**

Some of the proposed ODICL algorithms rely on an explicit end of the task signal (task id) to identify the start of a new task. The initial ODIN version also relies on explicit task ids to distinguish different tasks for training. This reliance on an explicit task id may preclude one from employing current ODICL algorithms in real-life settings where it may be challenging to identify such a signal explicitly.

One can assume the upward drift in p's loss detected by ADWIN is due to the distribution shift in the underlying input features. Hence, ODIN determines the end of the task when an upward drift is detected. Line 12–15 in Algorithm 5 explains this automatic Task Detector (TD) in ODIN. Line 25 of Algorithm 5 further integrates DL with this automatic TD. With automatic TD, if the new task is similar to the past task, ADWIN might not detect an upward drift in the loss, as the learning on the new task can improve the prediction of the previous task due to backward knowledge transfer [16]. Hence, detected task ids may not align with actual task ids. Therefore in automatic task detection, the current training network p is included in the F pool only for prediction. In the experiments, the effectiveness of different versions of ODIN were compared against popular regularization and replay baselines.

4 Experiments

The experiments attempt to understand the effectiveness of ODIN against popular online CL baselines. Also, they attempt to identify the effectiveness of Dynamic Learning-Rate adjustments. Furthermore, experiments attempt to identify the effectiveness of the Task Predictor. Lastly, they attempt to identify the effectiveness of ODIN with an automatic Task Detector against online CL baselines that do not use the external end of task signal.

The experiments were done on three datasets: CORe50 [14], RotatedCIFAR10, and RotatedMNIST. With RotatedCIFAR10 and RotatedMNIST, 90° rotations (0°, 90°, 180°, -90°) of the original images from CIFAR10 [11] and MNIST [12] were considered separate tasks. There were four tasks in each of those two datasets. With CORe50, 11 distinct sessions (8 indoor and 3 outdoor) of the same object were considered as separate tasks: tasks 0–2,4–8 indoor, 3,9, and 10 outdoor. Here the 10 object categories were considered as the class labels. Though it uses the same dataset as in [14] for ODICL, task separation is more natural than the random separation in [14]. Also, our CORe50 version had a separate evaluation set for each task rather than a mixed evaluation set, as in [14]. This allows one to better understand forgetting in the ODICL setting.

In the Experiments, different versions of ODIN were compared against regularization baselines: LwF, EWC, and replay baselines: ER and MIR. For ER, an extended buffer version was considered in the experiments. Instead of randomly replacing an item from the buffer, we replace an instance from the most represented task's most represented class. It is referenced as ER_{TbCb} in this paper. This ER_{TbCb} is a further extension of [4], where we attempt to balance the buffer with regard to both task and class. This extended ER_{TbCb} was considered so that

Table 1. Datasets

Dataset	# tasks	Instances per train/test task	# classes	Channels, H, W
CORe50	11	2000/1000	10	3, 32, 32
RotatedCIFAR10	4	50000/10000	10	3, 32, 32
RotatedMNIST	4	60000/10000	10	1, 28, 28

ODIN without automatic TD can be compared against a good replay method that utilizes the external end of task signal. All the replay methods used a 1k instance buffer. Also, all the methods use a simple CNN (33450 parameters) with four convolution layers. Two types of TPs were used in the experiments for ODIN: NB[1], and HT(see footnote 1) . Quantized ResNet-18 was used as the feature extractor, and flattened last layer features were used to train the TPs: NB and HT. ResNet was chosen considering [10], where HT was trained on extracted features by the ResNet feature extractor for images. Also, three types of vote aggregation methods were considered in the experiments: Majority Vote (MV), Weighted Voting (WV), and the use of just CNN_{best} just by itself. In the experiments, we also considered a hypothetical scenario of ODIN, where the task id is available at evaluation and is used to determine the correct frozen CNN. This is presented as the "known_{tid}" in the results. It indicates achievable performance if task prediction is perfect. In the results for ODIN, TP_{WV} represents Task Predictor with Weighted Voting, $\text{TP}_{No\text{WV}}$ represents: Task Predictor without Weighted Voting, $\text{TP}^{\text{DL}}_{\text{WV}}$ represents: Task Predictor with Weighted Voting and Dynamic Learning-Rate, $\text{TP}^{\text{DL}}_{No\text{WV}}$ represents: Task Predictor without Weighted Voting and with Dynamic Learning-Rate, $\text{TP}^{\text{TD}}_{\text{WV}}$ represents: Task Predictor with Weighted Voting and automatic Task Detector and, $\text{TP}^{\text{DLTD}}_{\text{WV}}$ represents: Task Predictor with Weighted Voting, Dynamic Learning-Rate and automatic Task Detector[2]. ODIN Dynamic Learning-Rate used the same learning rate decay factor (0.999995) as in [4]'s continuous learning rate decay.

All experiments were run using the Avalanche [15] CL platform. The online buffer implementations of ER and MIR were from [16]. Average accuracy and forgetting defined in [16] were used in the evaluation. All experiments were run three times, and relevant averages and standard deviations were considered in the evaluation.

Table 2 contains the average accuracy and forgetting after training on the last task for each method that uses task ids. Considering the average accuracy ranks, ODIN $\text{NB}^{\text{DL}}_{\text{WV}}$ produces the best results. It also has very little forgetting considering average forgetting ranks. Extended ER_{TbCb} has better average accuracy but lags a bit behind ODIN $\text{NB}^{\text{DL}}_{\text{WV}}$ when considering the average accuracy ranks. Except for ODIN random, all ODIN versions achieve better accuracy than the regularization baselines EWC and LwF. Both of the regularization baselines

[1] Source code github.com/nuwangunasekara/ODIN uses online NB and HT [17].

[2] In the legend of the plots, superscripts and subscripts are in lowercase letters.

Table 2. Average accuracy and forgetting after training on the last task (use end of the task signal)

Dataset	LwF	EWC	ER_{TbCb}	ODIN							
				$known^{*}_{tid}$	random	MV	HTwv	NBwv	NB^{DL}_{WV}	$NB_{No}wv$	$NB^{DL}_{No}wv$
Accuracy											
CORe50	0.44 ± 0.02	0.47 ± 0.03	0.65 ± 0.01	0.69 ± 0.01	0.44 ± 0.00	0.54 ± 0.02	0.63 ± 0.05	0.62 ± 0.01	**0.66** ± 0.01	0.62 ± 0.01	0.65 ± 0.01
RotatedCIFAR10	0.44 ± 0.01	0.42 ± 0.01	0.42 ± 0.02	0.49 ± 0.01	0.37 ± 0.02	**0.45** ± 0.01	**0.45** ± 0.01	0.43 ± 0.02	0.44 ± 0.01	0.40 ± 0.00	0.41 ± 0.00
RotatedMNIST	0.66 ± 0.01	0.52 ± 0.01	**0.84** ± 0.01	0.97 ± 0.00	0.48 ± 0.01	0.65 ± 0.02	0.52 ± 0.01	0.79 ± 0.00	0.80 ± 0.00	0.78 ± 0.01	0.78 ± 0.00
Avg	0.51 ± 0.01	0.47 ± 0.02	**0.64** ± 0.01	0.72 ± 0.01	0.43 ± 0.01	0.55 ± 0.02	0.53 ± 0.02	0.62 ± 0.01	0.63 ± 0.01	0.60 ± 0.00	0.62 ± 0.00
Avg Rank	6.00	7.67	3.67		10.00	5.00	4.67	4.33	**2.33**	6.67	4.67
Forgetting											
CORe50	0.19 ± 0.04	0.15 ± 0.05	-0.01 ± 0.02	0.00 ± 0.00	-0.01 ± 0.01	**-0.03** ± 0.01	0.03 ± 0.04	0.01 ± 0.01	0.01 ± 0.00	0.00 ± 0.01	0.01 ± 0.00
RotatedCIFAR10	0.01 ± 0.02	0.07 ± 0.02	0.04 ± 0.00	0.00 ± 0.00	0.02 ± 0.01	**-0.03** ± 0.01	0.04 ± 0.02	-0.02 ± 0.00	-0.02 ± 0.01	0.00 ± 0.00	0.00 ± 0.00
RotatedMNIST	0.24 ± 0.01	0.60 ± 0.02	0.16 ± 0.01	0.00 ± 0.00	0.20 ± 0.02	0.12 ± 0.02	0.60 ± 0.02	**0.12** ± 0.00	0.12 ± 0.01	0.12 ± 0.00	0.13 ± 0.00
Avg	0.14 ± 0.02	0.27 ± 0.03	0.06 ± 0.01	0.00 ± 0.00	0.07 ± 0.01	**0.02** ± 0.02	0.22 ± 0.02	0.04 ± 0.01	0.03 ± 0.00	0.04 ± 0.00	0.04 ± 0.00
Avg Rank	8.00	9.67	5.67		5.33	**1.67**	8.67	4.00	2.67	4.33	5.00

! ER_{TbCb} is ER with an extended task-balanced and class-balanced online buffer.
* $known_{tid}$ is a hypothetical scenario where task id is known at evaluation.

Table 3. Average accuracy and forgetting after training on the last task (do not use end of the task signal)

Dataset	ODIN		ER	MIR
	NB^{TD}_{WV}	NB^{DLTD}_{WV}		
Accuracy				
CORe50	0.47 ± 0.04	0.52 ± 0.03	0.61 ± 0.03	**0.62** ± 0.03
RotatedCIFAR10	0.42 ± 0.01	**0.44** ± 0.00	0.27 ± 0.01	0.28 ± 0.01
RotatedMNIST	0.71 ± 0.03	0.69 ± 0.04	**0.78** ± 0.03	0.77 ± 0.01
Avg	0.53 ± 0.02	0.55 ± 0.02	**0.56** ± 0.02	**0.56** ± 0.02
Avg Rank	3.00	2.67	2.33	**2.00**
Forgetting				
CORe50	0.18 ± 0.03	0.17 ± 0.03	-0.09 ± 0.01	**-0.10** ± 0.04
RotatedCIFAR10	0.00 ± 0.01	**0.00** ± 0.00	0.01 ± 0.01	0.01 ± 0.01
RotatedMNIST	0.22 ± 0.04	0.25 ± 0.04	**0.06** ± 0.02	0.07 ± 0.02
Avg	0.14 ± 0.03	0.14 ± 0.03	-0.01 ± 0.02	**-0.01** ± 0.02
Avg Rank	2.67	3.00	2.33	**2.00**

have quite a high forgetting rate. ODIN NB_{WV} yields better results than ODIN $NB_{No}WV$ with less average forgetting. This shows that Weighted Voting boosts accuracy. Also, ODIN NB_{WV} yields better accuracy than ODIN HT_{WV} with less forgetting. This shows that NB is a better Task Predictor compared to HT. This is further explored in later experiments. Both NB^{DL}_{WV} and $NB^{DL}_{No}WV$ yields superior accuracy compared to NB_{WV} and $NB_{No}WV$. This suggests that Dynamic Learning-Rate improves the overall accuracy. In general, a good Task Predictor, Weighted Voting, and Dynamic Learning-Rate improve ODIN accuracy. Considering the hypothetical ODIN $known_{tid}$ scenario, it is evident that just selecting the correct frozen CNN is sufficient to outperform current baselines by a considerable margin. ODIN $known_{tid}$ also has zero average forgetting across all datasets after training on the last task. This suggests further improvements to the Task Predictors can result in good accuracy gains.

Table 3 only compares ODICL methods that do not use task ids: ODIN NB^{TD}_{WV}, ODIN NB^{DLTD}_{WV}, ER , and MIR, for a fairer comparison. Here, the ODIN

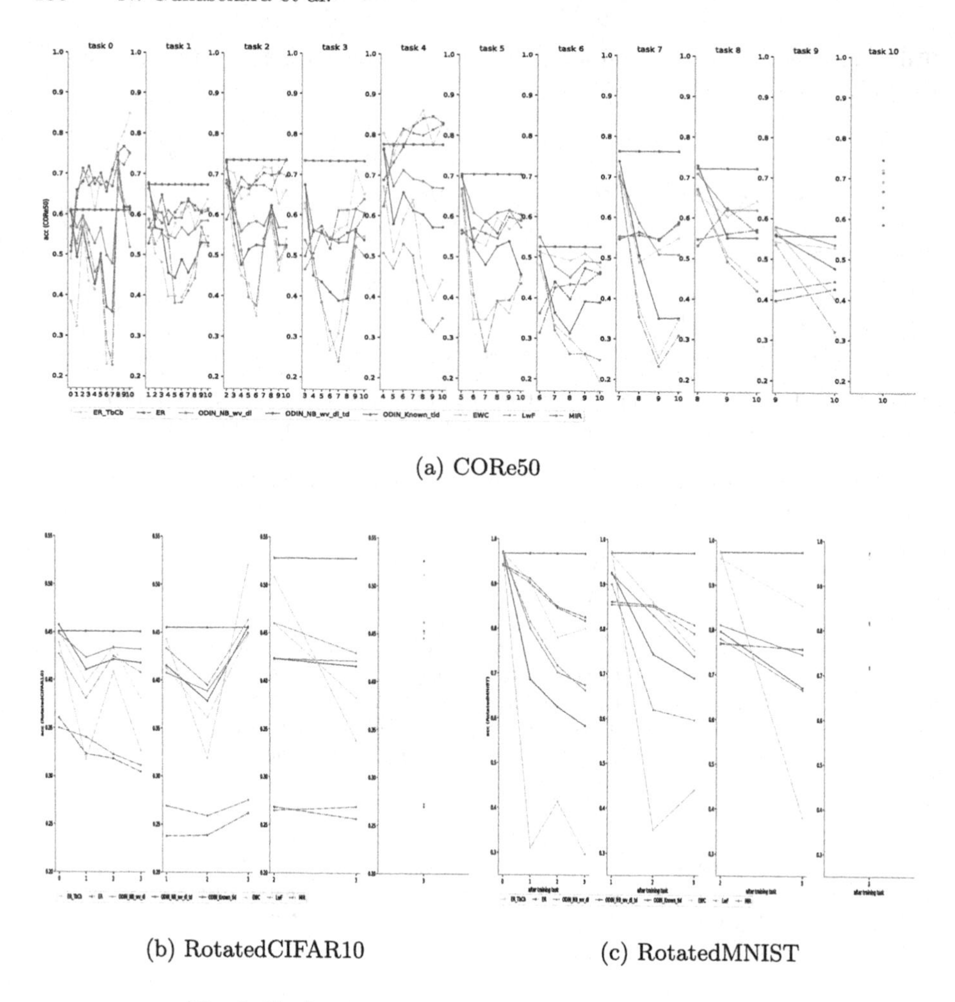

(a) CORe50

(b) RotatedCIFAR10

(c) RotatedMNIST

Fig. 2. Evaluation accuracy after training on each task

versions use ADWIN as a Task Detector. Also, ER and MIR can be included in the same category as they do not rely on an external task id signal. From the results for this category, it is evident that replay methods slightly outperform ODIN $\text{NB}_{\text{WV}}^{\text{DLTD}}$. Also, replay methods have better forgetting in this category. However, compared to the ODIN methods, they seem to perform quite badly on RotatedCIFAR10. Maybe being task aware gives ODIN methods an edge against replay methods on RotatedCIFAR10. Here also, one can see the positive effect of DL on ODIN's accuracy when comparing $\text{NB}_{\text{WV}}^{\text{DLTD}}$ with $\text{NB}_{\text{WV}}^{\text{TD}}$. Considering the results in Tables 2 and 3, it is evident that ODIN $\text{NB}_{\text{WV}}^{\text{DL}}$ performs better than ODIN $\text{NB}_{\text{WV}}^{\text{DLTD}}$. This highlights the importance of a good Task Detector. Furthermore, when comparing the two tables, ER_{TbCb} performs better than ER. This highlights the importance of ER to be task aware. In general, when one considers the results of both tables, it is evident that being task aware gives an edge to an ODICL method. Considering the results in both tables, one can conclude that NB_{WV}^{DLTD} performs well in all the datasets.

To get a deeper understanding of each method's predictive performance on old tasks after training on a new task, Fig. 2a, Fig. 2b, and Fig. 2c plot the accuracy of old tasks after training on a new task for selected methods on CORe50, RotatedCIFAR10, and RotatedMNIST datasets. Here top two ODIN methods from each category (with and without the end of the task signal): ODIN NB_{WV}^{DL}, ODIN NB_{WV}^{DLTD} were compared against the baselines that use the end of the task signal: EWC, LwF, ER_{TbCb}, and baselines that do not need the end of the task signal: ER and MIR. Hypothetical ODIN $known_{tid}$ is also included in the plots to better understand the upper bound of ODIN's Task Predictor. From Fig. 2a, it is evident that replay methods perform quite well on past tasks. Especially task-aware ER_{TbCb}. However, their performance has degraded for recent tasks. On the other hand, ODIN NB_{WV}^{DL} has relatively stable performance across all tasks. Hence on average, ODIN NB_{WV}^{DL} performs well on CORe50. This explains its good average accuracy on CORe50 in Table 2. Also, ODIN NB_{WV}^{DLTD} has a similar accuracy pattern to ODIN NB_{WV}^{DL}. But with less performance. Regularization baselines are quite poor on this dataset. They also have a very high variance. However, as per Fig. 2b, LwF performs well as ODIN versions on RotatedCIFAR10. Nevertheless, the replay methods ER and MIR perform poorly on that dataset except for ER_{TbCb}. It seems that the learning model needs to be aware of the task identities to perform well on RotatedCIFAR10. As per Fig. 2c, replay baselines generally perform well on RotatedMNIST. However, except for ER_{TbCb}, the performance gap between ODIN NB_{WV}^{DL} and other replay methods (ER and MIR) seem to narrow for recent tasks. ODIN NB_{WV}^{DL} performed better on the last task than ER and MIR on this dataset. This shows ODIN NB_{WV}^{DL}'s ability to perform well on current tasks as well as on past tasks. In all three plots, ODIN with hypothetical TP $known_{tid}$ never forgets after training on a new task. However, it does not improve as well. This explains 0.0 average forgetting for ODIN $known_{tid}$ in Table 2.

To further understand the TP's effectiveness, the predicted task id by each TP was compared against the actual task id in non-auto-TD mode against all datasets. This comparison was made for all evaluation instances after training on the last task. Figure 3 shows the ROC curves for the predicted task id and the relevant AUC scores for each TP on each dataset. According to the figure, it is clear that NB is a better Task Predictor for all datasets. This further strengthens the overall strong NB results in Table 2. Figure 3b further explains the effectiveness of NB as a TP when predicting each task for a given dataset. From the per-task ROC curves and AUC scores in Figs. 3b and 3d, it is clear that NB performs similarly on all the tasks for a given dataset. Nevertheless, it does perform slightly better on certain tasks. This is evident in CORe50, with NB performing slightly better for tasks 3,4,5,9, and 10. This generally uniform predictive capability of NB makes it a better Task Predictor than HT.

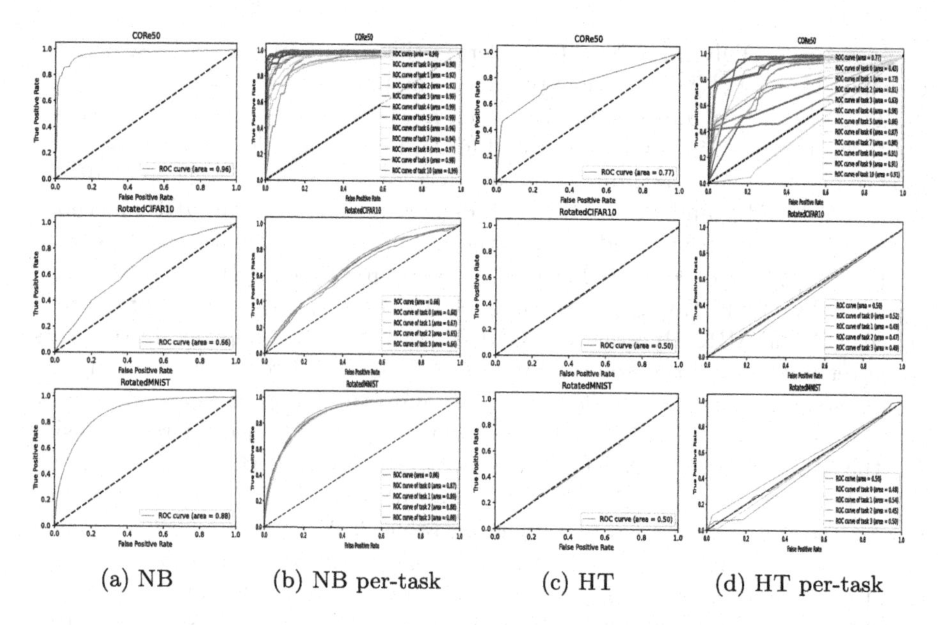

Fig. 3. Effectiveness of TPs: a & c) micro-average ROC curves for predicted task id and AUC scores, b & d) per-task ROC curves and AUC scores, for TPs NB and HT

5 Conclusion

The proposed ODIN produces competitive results for ODICL in comparison to regularization-based approaches. ODIN with and without automatic Task Detector produces competitive results compared to current popular ODICL baselines without requiring an instance buffer. This makes ODIN a suitable replacement for regularization methods in the ODICL setting. Better Task Predictors, more effective Dynamic Learning-Rate mechanisms, and more responsive Task Detector mechanisms could further improve ODIN's performance in ODICL.

References

1. Aljundi, R., et al.: Online continual learning with maximally interfered retrieval. arXiv preprint arXiv:1908.04742 (2019)
2. Armstrong, J., Clifton, D.: Continual learning of longitudinal health records. arXiv preprint arXiv:2112.11944 (2021)
3. Bifet, A., Gavalda, R.: Learning from time-changing data with adaptive windowing. In: Proceedings of the 2007 SIAM International Conference on Data Mining, pp. 443–448. SIAM (2007)
4. Buzzega, P., Boschini, M., Porrello, A., Calderara, S.: Rethinking experience replay: a bag of tricks for continual learning. In: 2020 25th International Conference on Pattern Recognition (ICPR), pp. 2180–2187. IEEE (2021)
5. Chaudhry, A., Dokania, P.K., Ajanthan, T., Torr, P.H.: Riemannian walk for incremental learning: understanding forgetting and intransigence. In: Proceedings of the European Conference on Computer Vision (ECCV), pp. 532–547 (2018)

6. Chaudhry, A., et al.: On tiny episodic memories in continual learning. arXiv preprint arXiv:1902.10486 (2019)
7. Hayes, T.L., Kafle, K., Shrestha, R., Acharya, M., Kanan, C.: REMIND your neural network to prevent catastrophic forgetting. In: Vedaldi, A., Bischof, H., Brox, T., Frahm, J.-M. (eds.) ECCV 2020. LNCS, vol. 12353, pp. 466–483. Springer, Cham (2020). https://doi.org/10.1007/978-3-030-58598-3_28
8. Kara, O., Churamani, N., Gunes, H.: Towards fair affective robotics: continual learning for mitigating bias in facial expression and action unit recognition. arXiv preprint arXiv:2103.09233 (2021)
9. Kirkpatrick, J., et al.: Overcoming catastrophic forgetting in neural networks. Proc. National Acad. Sci. **114**(13), 3521–3526 (2017)
10. Korycki, Ł, Krawczyk, B.: Streaming decision trees for lifelong learning. In: Oliver, N., Pérez-Cruz, F., Kramer, S., Read, J., Lozano, J.A. (eds.) ECML PKDD 2021. LNCS (LNAI), vol. 12975, pp. 502–518. Springer, Cham (2021). https://doi.org/10.1007/978-3-030-86486-6_31
11. Krizhevsky, A., Hinton, G., et al.: Learning multiple layers of features from tiny images (2009)
12. LeCun, Y., Bottou, L., Bengio, Y., Haffner, P.: Gradient-based learning applied to document recognition. Proc. IEEE **86**(11), 2278–2324 (1998)
13. Li, Z., Hoiem, D.: Learning without forgetting. IEEE Trans. Pattern Anal. Mach. Intell. **40**(12), 2935–2947 (2017)
14. Lomonaco, V., Maltoni, D.: Core50: a new dataset and benchmark for continuous object recognition. In: Conference on Robot Learning, pp. 17–26. PMLR (2017)
15. Lomonaco, V., et al.: Avalanche: an end-to-end library for continual learning. In: Proceedings of the IEEE/CVF Conference on Computer Vision and Pattern Recognition, pp. 3600–3610 (2021)
16. Mai, Z., Li, R., Jeong, J., Quispe, D., Kim, H., Sanner, S.: Online continual learning in image classification: an empirical survey. Neurocomputing **469**, 28–51 (2022). https://doi.org/10.1016/j.neucom.2021.10.021, https://www.sciencedirect.com/science/article/pii/S0925231221014995
17. Montiel, J., Read, J., Bifet, A., Abdessalem, T.: Scikit-Multiflow: a multi-output streaming framework. J. Mach. Learn. Res. **19**(72), 1–5 (2018). http://jmlr.org/papers/v19/18-251.html
18. Prabhu, A., Torr, P.H.S., Dokania, P.K.: GDumb: a simple approach that questions our progress in continual learning. In: Vedaldi, A., Bischof, H., Brox, T., Frahm, J.-M. (eds.) ECCV 2020. LNCS, vol. 12347, pp. 524–540. Springer, Cham (2020). https://doi.org/10.1007/978-3-030-58536-5_31
19. Srivastava, S., Yaqub, M., Nandakumar, K., Ge, Z., Mahapatra, D.: Continual domain incremental learning for chest X-Ray classification in low-resource clinical settings. In: Albarqouni, S., et al. (eds.) Domain Adaptation and Representation Transfer, and Affordable Healthcare and AI for Resource Diverse Global Health, pp. 226–238. Springer International Publishing, Cham (2021). https://doi.org/10.1007/978-3-030-87722-4_21
20. Vitter, J.S.: Random sampling with a reservoir. ACM Trans. Math. Softw. (TOMS) **11**(1), 37–57 (1985)

Incremental Update of Locally Optimal Classification Rules

Van Quoc Phuong Huynh(✉), Florian Beck, and Johannes Fürnkranz

FAW Institute, Johannes Kepler University, Linz, Austria
{vqphuynh,fbeck,juffi}@faw.jku.at

Abstract. Incremental learning is a traditional topic that has particularly gained importance in the wake of big data and stream mining. Discrete symbolic representations do not easily allow for gradual refinements of the learned concept. While the problem is less severe for incremental induction of decision trees, it is much harder for incremental rule learning in that there are hardly any incremental rule learning algorithms which are really successful. In this paper, we introduce iLORD algorithm, an adaptation of a recently proposed rule learning algorithm LORD, which aims at finding the best rule for each individual example, to an incremental learning setting. After being initialized with a first batch of training examples, iLORD relies on efficient data structures to summarize the information contained in the training examples, which can be quickly updated and allows to retrieve the best rule for each incoming example. The behavior of iLORD is evaluated with different parameterizations, and compared to other best-known incremental symbolic learning algorithms such as HOEFFDINGTREE and VFDR.

Keywords: Rule learning · Incremental classification · Stream mining · Machine learning · Data mining

1 Introduction

Incremental learning is a traditional topic in machine learning, which has particularly gained importance in the wake of big data and stream mining [1]. While some algorithms such as neural networks are naturally updated in an incremental or batch-incremental fashion, algorithms for learning symbolic representations such as decision trees [11] or rules [6] typically operate in a single-batch setting, where all of the data are assumed to be present at training time.

The main reason for this is that discrete symbolic representations do not easily allow for gradual refinements of the learned concept. In the case of incremental learning of decision trees, the problem is less severe, as the hierarchical structure of the tree guarantees that incremental updates in the lower levels of the tree do not affect the upper levels, which can thus stabilize over time. Thus, even simple algorithms such as ID4 [12], which essentially decide to re-learn a node in a decision tree once it turns out that an alternative split is better, can eventually converge towards a stable concept, at

P. Pascal and D. Ienco (Eds.): DS 2022, LNAI 13601, pp. 104–113, 2022.
https://doi.org/10.1007/978-3-031-18840-4_8

least in the upper levels of the tree. The best-known incremental tree learning algorithm, HOEFFDINGTREE [4], may be viewed as refinement of this idea, where Hoeffding bounds are used in order to determine the minimum number of examples necessary in order to reliably determine the best attribute at each level of the tree.

Incremental rule learning is even harder than incremental induction of decision trees, because there is no hierarchical structure that can be incrementally refined. Whereas a decision tree always covers the entire sample space and each example will always be covered by exactly one path through the tree (which corresponds to a single rule), rule sets are far more erratic: Generalizing one rule may conflict with other previously learned rules, or yield them obsolete. Conversely, a specialization of a rule may leave regions of the instance space uncovered, which have previously been covered by this rule. In brief, every modification of a single rule may impact all other rules. For these reasons, not many successful incremental rule learning algorithms can be found in the literature [6]. The best-known algorithm is VFDR (Very Fast Decision Rules [10]), which also uses Hoeffding bounds for deciding when to update a rule, i.e., when the next addition to add to a rule can be clearly determined. Uncovered examples are checked with the default rule, which can also spawn new rules if necessary. Various variants of this algorithm have been investigated [10], and, for some time, VFDR is also available in the MOA stream mining library [1].

However, updating rules with single conditions may be quite slow, as may be the creation of new rules with single conditions. We strive for a different approach, which, at its core, aims at efficiently retrieving or re-generating the best rule for every example. The resulting algorithm, iLORD, is based on a previously developed rule learning algorithm LORD [7], which maintains an efficient structure for summarizing the data, which can be incrementally updated. From this structure, we can greedily approximate the best rule for a given example. Depending on the available update time, such updates can be performed after every example, or more efficiently, in mini-batches.

In the remainder of the paper, we briefly mention the basic idea of the LORD algorithm in Sect. 2, upon which our approach iLORD is based to extend towards an incremental algorithm in Sect. 3. We then experimentally evaluate iLORD algorithm in Sect. 4 with different batch sizes for initializing and updating the statistics and rule set, and subsequently show that iLORD outperforms its main competitors in term of accuracy, the incremental symbolic learning algorithms HOEFFDINGTREE and VFDR. Finally, we finish the paper with conclusion and some future work in Sect. 5.

2 The LORD Algorithm

The iLORD algorithm builds upon our prior work on LORD (Locally Optimal Rules Discoverer) [7], our efficient implementation of locally optimal rule discovery. It extends LORD with the capability for incremental updates that allow it to be used in streaming scenarios. They key idea of LORD is to strive for finding a (locally) optimal rule for every training example. One of the key obstacles for such an approach is, of course, that learning a single rule is quite expensive, so that the costs for learning a separate rule for each individual training example are prohibitive when done naïvely. LORD solves this problem by first summarizing the dataset in efficient data structures, PPC-trees and

N-lists [3], which can efficiently summarize the count of examples supporting a conjunctive expression and thus be used to quickly find a locally optimal rule for a given example, without further consultation of the dataset. In principle, LORD can use most common rule learning heuristics for guiding its greedy search, e.g. the m-estimate [2] which has been shown to perform very well in a broad empirical comparison of various rule learning heuristics [9]. The large rule set learnt by LORD is indexed via a prefix tree structure, FPO-trees [8], of rule bodies that speeds up the search for covering rules of examples. In prior work, we have shown that the approach can be efficiently used for classification datasets with millions of examples, where other modern rule learning algorithms come to their limits [7].

3 Incremental LORD

In this section, we discuss how the LORD algorithm mentioned in the previous section can be adapted to enable incremental learning. The key idea is that the underlying data structures are updated incrementally. This means that the new example is sorted into the PPC-tree, and that the statistics of all nodes are updated. If necessary, a new node is added to the tree, in which case the PPC-codes of nodes must be recalculated and the N-lists also have to be updated with the newly added nodes. Finally, the current rule set is maintained, but for every new example it is checked whether a better rule can be found on the fly. These updates can be performed for every example, or more efficiently, for a mini-batch of examples.

3.1 Incremental Updates

Incremental updates for the current data structures is shown in Algorithm 1. It takes a mini-batch of (labeled) examples B, and uses them to update the current PPC-tree, the corresponding N-lists, and the current rule set R. iLORD starts by retrieving all selectors that cover the current examples e (typically one selector for each feature value of e) and sorts them down the $ppcTree$. If the current path has already been observed in previous examples, the corresponding counts are incremented (line 8), otherwise a new node will be inserted into the tree (lines 10–11). To facilitate efficient updates, the N-list structure of individual selectors consists of pointers to the PPC-Nodes in the PPC-tree. This avoids traversing the tree to regenerate an N-list for each individual selector when the tree is updated with new examples. However, traversing the PPC-tree to reassign PP-codes is still needed if new PPC-nodes are inserted into the tree (lines 16–19). The references to these new PPC-nodes are then inserted into the corresponding Nlists.

In lines 22–32, true positive ($r.tp$) and false positive ($r.fp$) counts of all rules in R that cover the current example e are updated; and based on that, their heuristic values are re-calculated in lines 33–35. With the idea of learning from failures, in lines 37–43, the rule set is updated with new rules learned from examples that have been classified incorrectly. This focus on misclassifications helps to reduce the running time significantly while still highly approximating the classification performance that could have been achieved when updating the rule set after each new example. For every misclassified example, a locally optimal rule is learned by calling the corresponding function of LORD (line 38). Newly learned rules are only added if they are better than (have a higher heuristic value, or equal heuristic value with a larger true positive count) all rules in the current rule set that cover the example and predict the right class.

Algorithm 1. Update PPC-trees, N-lists and rule set

```
1: function UPDATE(B, R, ppcTree, Nlists)
2:     // Update ppcTree and N-lists
3:     for each example e ∈ B do
4:         S_e ← sorted selector set for e
5:         N ← root node of ppcTree
6:         for each selector s ∈ S_e do
7:             if ∃N' ∈ N.successors with N'.s = s then
8:                 N'.freq ← N'.freq + 1
9:             else
10:                add new child node N'
11:                N'.freq ← 1; N'.s ← s
12:            end if
13:            N ← N'
14:        end for
15:    end for
16:    if new nodes have been added then
17:        Update PPC-codes of ppcTree
18:        Insert new nodes to the corresponding N-lists
19:    end if
20:    // Update rule set R
21:    S ← ∅
22:    for each example e ∈ B do
23:        e.CR ← COVERINGRULES(R, e)
24:        for each rule r ∈ e.CR do
25:            if (r.head = e.class) then
26:                r.tp ← r.tp + 1
27:            else
28:                r.fp ← r.fp + 1
29:            end if
30:            S ← S ∪ r
31:        end for
32:    end for
33:    for each rule r ∈ S do
34:        re-calculate heuristic value h(.) of r
35:    end for
36:    // Learn rules for misclassified examples
37:    for each incorrectly classified example e ∈ B do
38:        r ← the best rule learned from e and Nlists
39:        FCR ← {r ∈ e.CR : r.head = e.class}
40:        if r is better than all rules in FCR then
41:            R ← R ∪ r
42:        end if
43:    end for
44: end function
```

3.2 Overall Algorithm

The above incremental update function, Algorithm 1, is integrated into the overall iLORD algorithm, an incremental version of LORD. iLORD learns and updates a rule set, and classifies every example from a data stream. Like LORD, it also requires a heuristic to evaluate rules, but it needs two more parameters, the size of a starting batch i and the size of the mini-batch b for the incremental updates.

Algorithm 2 iLORD algorithm

Input: data stream DS, initialization size i, heuristic $h(.)$, batch size b
Output: final rule set R, series of accumulated classification performance P

```
1: P ← ∅; p ← 0
2: count ← 0; c_majority ← class_list[0]
3: // Initialize with first i examples.
4: for each example e from the data stream DS do
5:     classify e with c_majority
6:     update p; P ← P ∪ p
7:     update c_majority with the class of e
8:     D0 ← D0 ∪ e; count ← count + 1
9:     if count = i then
10:        break
11:    end if
12: end for
13: R, ppcTree, Nlists ← LORD(D0, h(.))
14: // Process the stream
15: B ← ∅
16: for each example e from DS do
17:     // Classify the current example
18:     e.CR ← COVERINGRULES(R, e)
19:     classify e with the best rule from e.CR
20:     obtain e.class
21:     update p; P ← P ∪ p
22:     // Update in mini-batches of b examples.
23:     B ← B ∪ e
24:     if |B| = b then
25:         UPDATE(B, R, ppcTree, Nlists)
26:         B ← ∅
27:     end if
28: end for
29: return R, P
```

Algorithm 2 shows the execution of iLORD. During its initialization phase (lines 1–12), iLORD collects an initial batch of i examples. It also dynamically tracks the majority class, and uses this for prediction in this phase. After collecting enough examples, iLORD calls LORD to initialize the PPC-tree, N-list structures, and to learn a rule set. In the LORD algorithm, the PPC-tree is no longer used after N-list structures are generated, so the tree is freed for memory saving. In the iLORD version, the N-list structures need to be incrementally updated w.r.t. new incoming examples, so the PPC-tree structure is kept and updated with every new example, as discussed in Sect. 3.1. iLORD uses its current rule set to classify examples in the same way as LORD does (lines 18–19), and updates classification performance based on the true labels (line 20–21). Finally, in lines 23–27, the rule set, PPC-tree and N-list structures are updated by the UPDATE function (Algorithm 1) when a mini-batch with b examples has been collected.

4 Experiments

In this section, we report on the experimental evaluation of the iLORD algorithm. For evaluation, we selected four datasets from the UCI repository of machine learning databases [5], as shown in Table 1. As we noticed some order-dependent effects that equally affected all the studied algorithms, we randomly shuffled the datasets.

Table 1. Datasets used in experiments

#	Datasets	# Exs.	# Attr.	Attr. types	Missing values
1	*adult*	48,842	14	mix	yes
2	*airlines*	539,395	8	mix	yes
3	*connect-4*	67,557	42	categorical	no
4	*covertype*	581,012	55	mix	no

We have compared iLORD with two other symbolic streaming classification algorithms, HOEFFDINGTREE [4], aka very fast decision trees, as well as the very fast decision rules algorithm VFDR [10]. We used the implementation of HOEFFDINGTREE available in the MOA framework [1], and the one of VFDR for the WEKA machine-learning platform, available at https://github.com/oowekyala/vfdr-weka. All three algorithms were implemented in Java. Experiments were run on Intel(R) Core(TM) i3-10110U CPU @2.10 GHz and less than 1 GB available memory.

Unless reported otherwise, all algorithms were used with their default settings, which, for iLORD algorithm, we set to be $m = 0.1$ for the m-estimate heuristic, a batch size $b = 100$, and an initial window of size $i = 5000$ for datasets containing numeric attributes. For datasets with only categorical attributes, iLORD does not need to discretize attributes, so that a smaller number of first examples is sufficient to initialize the rule set. Therefore, in the case of *connect-4*, we set $i = 500$.

For each algorithm and dataset, we obtain a series of evaluation values, one for each example, which reflect the accumulated accuracy over all examples seen up to the point.

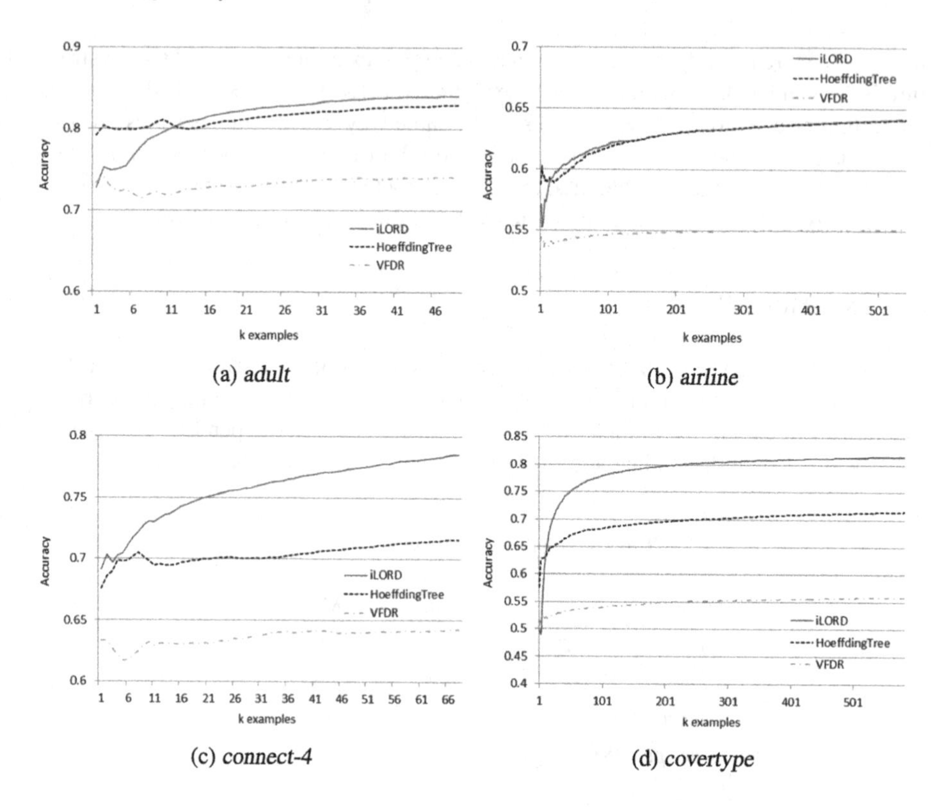

(a) *adult* (b) *airline* (c) *connect-4* (d) *covertype*

Fig. 1. Classification performance series of the algorithms on four datasets

4.1 Comparison to HOEFFDINGTREE and VFDR

Figure 1 shows classification accuracy series of the three algorithms on the four datasets. The accuracies are accumulated over all predictions as shown in Algorithm 2 (lines 20–21), where the number of predictions (in thousands of examples) is shown on the x-axis. In all tested datasets, the classification accuracy of iLORD is substantially higher than VFDR, and in three of the four datasets also considerably higher than the one of HOEFFDINGTREE, while being a close match on the fourth. For *adult* (Fig. 1a) and *covertype* (Fig. 1d), one can also clearly see the initially poor performance of iLORD is because of its majority-class-based classification for the first i examples. However, it quickly recovers and soon surpasses the accuracy of the others. For *airline* (Fig. 1b), iLORD maintains a clear advantage for the first ca. 150k examples, but then HOEFFDINGTREE catches up and both perform approximately the same.

However, the good classification performance of iLORD is achieved at the expense of a higher run-time than the other algorithms, which can be seen in Table 2. The running time of VFDR is always between iLORD and HOEFFDINGTREE for all four datasets. Obviously, the run-time and the accuracy of iLORD also depends on its parameter settings that we will analyze in the next section.

Table 2. Running time (seconds) of the algorithms

#	Datasets	iLORD	HOEFFDINGTREE	VFDR
1	*adult*	7.6	1.3	4.6
2	*airlines*	297.2	109.6	157.6
3	*connect-4*	85.4	1.4	6.0
4	*covertype*	2200.5	24.5	843.6

Table 3. Accuracy vs. running time (seconds) of iLORD on *adult* dataset for different initial window sizes i and batch sizes b

Init size i	Batch size b	Accuracy	Run time
500	1	0.8331	40.4
	10	0.8333	18.7
	100	0.8332	6.1
	1000	0.8335	5.0
	No update	0.8108	0.5
1000	1	0.8385	53.3
	10	0.8384	20.4
	100	0.8385	6.6
	1000	0.8381	5.4
	No update	0.8191	1.1
5000	1	0.8398	104.3
	10	0.8400	26.7
	100	0.8399	7.6
	1000	0.8401	6.6
	No update	0.8337	2.1

4.2 Sensitivity to Parameter Settings

We briefly evaluated the influence of the two parameters, the size of the initial window i and the batch size b. The choice of the learning heuristic, in our case captured in the parameter m of the m-estimate, also has a strong influence on the result, but this issue is not pertinent to a stream setting, and has been analyzed in prior work [7].

Table 3 shows the results on the *adult* dataset, for three different sizes of the initial window ($i = \{500, 1000, 5000\}$), four different versions of the batch size ($b = \{1, 10, 100, 1000\}$), and in the last line the accuracy of the model that is learned from the initial batch only and never updated. A few things become apparent here: First, with respect to the initial batch size, the initial model which is never updated is clearly more accurate for larger sizes of the initialization window. For all initial batch sizes, the incremental updates continue to improve the model, but less for the model that has been learned from a larger initial window. Nevertheless, the latter achieves a higher overall accuracy. Second, with respect to the batch size, the method is not very sensitive with respect to the final accuracy (which is what is shown in the table). However, the run-time is clearly different in that more frequent updates are clearly more costly, and also the convergence rate may differ.

The latter point can be seen from Fig. 2, which shows the effects of varying the batch size (left) or the effects of varying the initial batch size (right) over the course of the classification process. While different batch sizes perform quite similar (and all of them clearly better than the version without updates at the bottom), there are clear differences visible from different choices of the initial window, which are primarily due to the fact that the algorithm only dynamically adjusts and predicts the majority class in this initial phase. However, it can also be seen that the largest initial window size (7000 examples) eventually results in the highest classification. The main reason for this is presumably that this window is used for computing a discretization of the numerical attributes, which is never changed through-out the process. The integration of an on-line discretization method is clearly a subject for future work.

With respect to run-times, there are clear, unsurprising differences between different batch sizes: updates after every example ($b = 1$) are most expensive, but time can be saved when choosing larger mini-batch sizes. As discussed above, the observed accuracy does not differ much in these cases. Similarly, larger initial windows ($i = 5000$) generally result in higher accuracy, but also somewhat higher run-times. However, the run-time increase is particularly observable for smaller mini-batch sizes, whereas the run-times of the algorithm do not seem to be particularly sensitive to higher mini-batch sizes ($b = \{100, 1000\}$). Note that, the run-time for training the initial model depends on only the initial window size i and is much smaller than the run-time for updates from new batches, e.g. in the same experiments shown in Table 3, training the initial models with $i = 500, 1000, 5000$ takes 0.3, 0.37, 1.42 s respectively.

Thus, overall, the best configuration for iLORD seems to be to choose an initial window that should be large enough to increase accuracy, and a typical mini-batch size, such as $b = 100$, in order to increase efficiency.

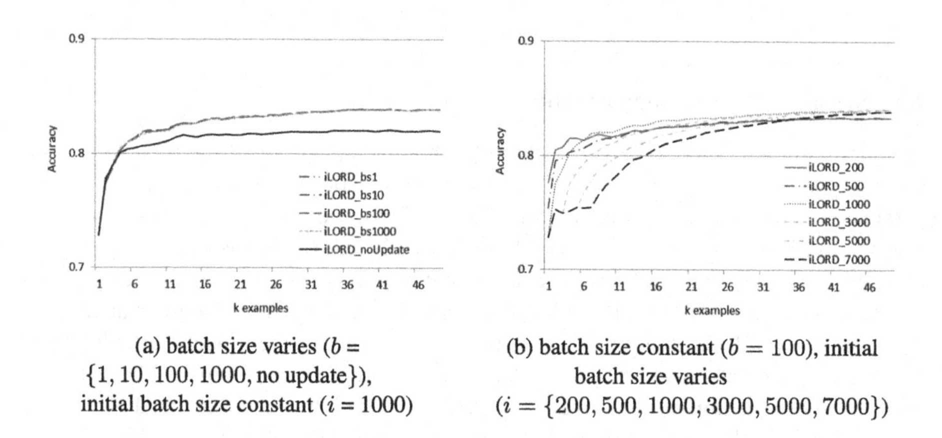

(a) batch size varies ($b = \{1, 10, 100, 1000, \text{no update}\}$), initial batch size constant ($i = 1000$)

(b) batch size constant ($b = 100$), initial batch size varies ($i = \{200, 500, 1000, 3000, 5000, 7000\}$)

Fig. 2. Influence of parameter settings on *adult* dataset.

5 Conclusion

In this work, we have proposed iLORD, an adaptation of the rule learning algorithm LORD for incremental learning from a stream of examples. Its key idea is different than previous work, as it does not aim at incrementally updating a small set of rules, but it keeps a large pool of locally optimal rules and continues to add new rules that are locally optimal for misclassified examples.

As a result, iLORD is somewhat slower than its very fast competitors, but it achieves a higher accuracy. The speed of the algorithm is clearly affected by the cost of the rule set updates, which are independent of the batch size. An interesting property of the algorithm is that its run-time can clearly benefit from performing updates in larger mini-batches similar to other batch-incremental algorithms such as neural networks.

We also observed that a larger initialization window results in a better final performance. This is presumably mostly due to the fact that our algorithm computes a fixed discretization for the numerical attributes from this window, which is never changed in the remaining process. In this initialization phase we currently always make majority class predictions. These weakness should be addressed in future work.

References

1. Bifet, A., Holmes, G., Kirkby, R., Pfahringer, B.: MOA: massive online analysis. J. Mach. Learn. Res. **11**, 1601–1604 (2010)
2. Cestnik, B.: Estimating probabilities: a crucial task in machine learning. In: Aiello, L. (ed.) Proceedings of the 9th European Conference on Artificial Intelligence (ECAI-90), Stockholm, Sweden, pp. 147–150. Pitman (1990)
3. Deng, Z.H., Lv, S.L.: PrePost+: an efficient N-lists-based algorithm for mining frequent itemsets via children-parent equivalence pruning. Expert Syst. Appl. **42**(13), 5424–5432 (2015)
4. Domingos, P., Hulten, G.: Mining high-speed data streams. In: Proceedings of the 6th ACM SIGKDD International Conference on Knowledge Discovery and Data Mining (KDD), pp. 71–80 (2000)
5. Dua, D., Graff, C.: UCI machine learning repository (2017). http://archive.ics.uci.edu/ml
6. Fürnkranz, J., Gamberger, D., Lavrač, N.: Foundations of Rule Learning. Springer, Heidelberg (2012). https://doi.org/10.1007/978-3-540-75197-7
7. Huynh, V.Q.P., Beck, F., Fürnkranz, J.: Efficient learning of large sets of locally optimal classification rules. Mach. Learn. (2023). accepted with minor revisions
8. Huynh, V.Q.P., Küng, J.: FPO tree and DP3 algorithm for distributed parallel frequent itemsets mining. Expert Syst. Appl. **140**, 112874 (2020)
9. Janssen, F., Fürnkranz, J.: On the quest for optimal rule learning heuristics. Mach. Learn. **78**(3), 343–379 (2010)
10. Kosina, P., Gama, J.: Very fast decision rules for classification in data streams. Data Min. Knowl. Disc. **29**(1), 168–202 (2013). https://doi.org/10.1007/s10618-013-0340-z
11. Murthy, S.K.: Automatic construction of decision trees from data: a multi-disciplinary survey. Data Min. Knowl. Disc. **2**(4), 345–389 (1998)
12. Schlimmer, J.C., Fisher, D.H.: A case study of incremental concept induction. In: Kehler, T. (ed.) Proceedings of the 5th National Conference on Artificial Intelligence (AAAI), Volume 1: Science, Philadelphia, PA, USA, pp. 496–501. Morgan Kaufmann (1986)

Policy Evaluation with Delayed, Aggregated Anonymous Feedback

Guilherme Dinis Junior(✉), Sindri Magnússon, and Jaakko Hollmén

Stockholm University, Stockholm, Sweden
{guilherme,sindri.magnusson,jaakko.hollmen}@dsv.su.se

Abstract. In reinforcement learning, an agent makes decisions to maximize rewards in an environment. Rewards are an integral part of the reinforcement learning as they guide the agent towards its learning objective. However, having consistent rewards can be infeasible in certain scenarios, due to either cost, the nature of the problem or other constraints. In this paper, we investigate the problem of delayed, aggregated, and anonymous rewards. We propose and analyze two strategies for conducting policy evaluation under cumulative periodic rewards, and study them by making use of simulation environments. Our findings indicate that both strategies can achieve similar sample efficiency as when we have consistent rewards.

Keywords: Reinforcement learning · Markov Decision Process (MDP) · Reward estimation

1 Introduction

The idea behind reinforcement learning (RL) is that an agent learns how to make decisions that maximize rewards through a policy π, by interacting with the environment [16]. RL is suitable for sequential decision making problems, where the choice of an action affects future states and actions, and there is a wide range of applications for it. For example, within the transportation sector, RL has been used for optimizing multiple objectives on ride ordering [11,17,18], and car pooling [7]; in recommender systems, it has been successfully applied to problems with large discrete item spaces [4,5], and to generate feeds of digital content [19,20].

The growth in applications of RL compels us to reflect on some of the fundamental assumptions and limitations in the theory, in order to broaden its application. One assumption is that the rewards given at each step transition are tied to the action taken by the policy in that state. Temporal difference (TD) learning methods for instance, which are core methods of learning from logged data, rely on having rewards for every transition step [15]. In practice though, rewards can be observed with delay and they can also be observed as an aggregate from multiple actions, i.e. cumulative. In marketing and advertising, for instance, a person can be exposed to several impressions before they make

P. Pascal and D. Ienco (Eds.): DS 2022, LNAI 13601, pp. 114–123, 2022.
https://doi.org/10.1007/978-3-031-18840-4_9

purchase. Similarly, we can imagine distributed learning settings with low powered devices that connect to a network at time intervals to send data - there are savings in sending a summary of the rewards observed since the last connection instead of a stream of values corresponding to every action. Problems where rewards are only observed with delay and, when so, they are aggregated over the last P actions, have been explored in bandits but are under-explored in RL literature.

We formulate the problem of learning from delayed, aggregated and anonymous feedback (DAAF) as one of delayed credit attribution. Our aim is to explore ways to conduct policy evaluation under such conditions for MDPs with discrete state and action spaces. Our research questions are as follows: **RQ1** - to what extent can standard policy evaluation work when feedback is delayed, aggregated and anonymous? and **RQ2** - how can the dynamics of a DAAF trajectory be leveraged to estimate the value function of a policy?

We investigate policy evaluation under DAAF without any intervention, and propose two approaches to address policy estimation. In our experiments, we demonstrate empirically our proposed approaches' ability to achieve sample efficiency from data compared to having non-delayed and non-aggregated rewards per action. Our main contributions are (1) an investigation of DAAF with existing policy evaluation methods; (2) the proposal of two approaches to conduct policy evaluation under DAAF with reasonable sample efficiency; and (3) a sequence learning problem, alphabet sequence (ABC), where actions are proportional to states, designed to for its simplicity and complexity scaling.

To the best our of knowledge, our work is the first to explore this problem in the RL setting. The rest of this paper is organized as follows: in Sect. 2, we present work related to the subject matter. Then, in Sects. 3 and 4, we formulate the problem and present our proposed solutions. Our methodology is described in Sect. 5, followed by our findings in Sect. 6. In Sect. 7 we discuss the results along with thoughts on future work, leaving our final remarks to Sect. 8.

2 Related Work

Though we investigate DAAF, we are not the first to research reward dynamics in RL. We highlight three areas that are relevant to our work: Non-markovian Rewards, Credit Assignment, and Inverse reinforcement learning (IRL). On the subject of non-markovian rewards, in [8] the authors explore the problem of maximizing rewards when an agent can actively choose whether to observe the reward for an action at a cost. Their proposed solution relies on estimating the gain of a reward observation, and pay the cost whenever the gain is higher. While their setting has absent rewards, the distinction here is that in our setting rewards are delayed, aggregated and anonymous; the agent cannot decide when to observe a reward; and there is no explicit cost. A more closely related line of work is that explored in [2] and [6], where rewards are delayed, aggregated, anonymous. In both cases, there is an adversarial component described as the reward for every action being split over a time horizon d. In our setting, the rewards for a state-action pair materialize fully on aggregate, i.e. they are not split over multiple time steps. The work in [13] has the closest setting to ours, the only distinction being that, like the other papers presented so far, their focus is on the multi-armed bandits and learning a policy as opposed to policy evaluation. We therefore extend the work in [13] to the RL setting, with attention to policy evaluation instead of control.

Turning to credit assignment, when doing policy evaluation we are given an existing policy π, and our goal is to estimate the value of using it in an environment. The value and state-value functions of a policy π, $V_{\pi}(s)$ and $Q_{\pi}(s, a)$ respectively, can be learned by using credit assignment methods, such as Monte Carlo and SARSA [15]. These methods learn how to attribute credit for future outcomes to the states and/or actions in a given trajectory, and much of the research in this area concerns mechanisms to do accurate credit attribution. For instance, in [12], the authors propose a counterfactual approach to discerning credit assignment to actions of a policy using hindsight data, while the authors in [3] propose learning a credit weighting function based on states. In either case, the credit assignment methods expect rewards for policy actions, and our study differs in that we address the task of credit assignment, in the context of policy evaluation, with DAAF. Our problem is more closely related to the single-step structural credit assignment problem of multi-agent systems, where given the a global reward from the environment, we wish to determine the contribution of each agent to the outcome [1].

Finally, we have the area of inverse reinforcement learning, the subject of which is deriving a reward function from observations. It is typically employed to extract knowledge from experts to then design RL agents, a task known as learning from imitation. In [14], this problem is formulated as a supervised learning task, where feature maps representing state-action pairs are used to learn rewards that can maximize the similarity of a learned policy to that of an expert policy or trajectory. Later works have extended this problem, with either novel formulations or extensions to new conditions. In [10], the authors generate a mixture of reward functions for different behavior clusters of policies in order to extract rules from learned agents. Their reward is defined as a linear combination of features and feature reward weights. Unlike the traditional IRL setting, we consider problems where the observation of rewards is limited, instead of fully absent, and the observed rewards are delayed, aggregated and anonymous. As such, one could conceptually use IRL techniques to extract a reward function for our use case. Our approach differs in that we seek to leverage the rewards available and their structure to constrain the reward function we estimate.

3 Preliminaries

We consider the typical RL setting, where an agent has been trained to maximize rewards in an unknown environment. The environment is modeled as an MDP with a set of states $\mathbf{S}$, actions $\mathbf{A}$, transition dynamic $\mathbf{T}(S, A, S')$ and rewards $(S, A) \rightarrow R$. A state S_t encapsulates information about the environment at time step t. An action A_t is chosen by the agent given the state $\pi(A_t|S_t)$ - which is a density function indicating the probability of action A given that we are in state S - upon which the environment transitions into a new state S_{t+1} and the agent receives a reward R_t. Thus the agent follows the learning trajectory $S_1, A_1, R_1, S_2, A_2, R_2, ..., S_T, A_T, R_T$ where T is the length of the episode, which is finite for episodic tasks and infinite for continuous tasks. The goal of the agent is to maximize future cumulative rewards, i.e. the sum of rewards from future states $\sum_{t=1}^{T} \gamma^t R_t$, where $\gamma \in [0, 1]$ is a discount factor that can be chosen to favor near-term rewards ($\gamma < 1$) or give all rewards equal weight ($\gamma = 1$). For this work we limit our analysis to finite horizon tasks, an assume $\gamma = 1$ throughout.

One of the key problems in RL is to estimate the state-value function $V_\pi(S)$ or action-value function $Q_\pi(S, A)$ for a policy π. The state-value function $V_\pi(S)$ tells us the maximum returns from being in state S and using π to make decisions starting from that state. The action-value function $Q_\pi(S, A)$ tells us the maximum returns from being in state S, taking a specific action A, and from there using the policy π to make decisions. Knowing the state or action value functions allows us to compare policies, and determine which ones provide better returns for a given problem.

4 Policy Evaluation with DAAF

In this paper, we study policy evaluation when the reward is observed with some delay. Additionally, rewards are observed on aggregate, i.e. rewards observed at time t correspond to the sum of rewards for the last P steps. In our experiments, we make P a constant, but our solution easily extends to P coming from any discrete and bounded distribution $P \sim \tau(t)$. Mathematically, with a constant P, if we denote by R_t^o the reward signal observed by the agent at time t then we have:

$$R_t^o = \begin{cases} \sum_{i=t-P+1}^{P} \gamma^i R_i & \text{if (t mod P = 0)} \\ \emptyset & \text{otherwise} \end{cases} \quad (1)$$

where $\emptyset$ is the absence of a reward, and γ is the discount factor. And since we consider undiscounted episodic tasks ($\gamma = 1$) with deterministic rewards, then whenever the feedback is observed we get: $R_t^o = \sum_{i=1}^{P} R(s, a)_p$. One known method for policy evaluation is SARSA [16], and it uses the update function: $Q(S, A) \leftarrow Q(S, A) + \alpha * (R + \gamma * Q(S', A') - Q(S, A))$. Without a reward, the update function of the algorithm cannot be executed. And with non-markovian rewards, as is the case of DAAF, our value updates can make inaccurate attributions of credit to state-action pairs, which in turn leads to incorrect policy estimation. To address the problem of misattributed credited, we propose two solutions: zero impute missing rewards (ZI-M) and linear estimation of stateaction rewards (LEAST). Both of them rely on using existing policy evaluation algorithms without any direct modifications, and changing only the trajectory data. We describe them next.

Zero Impute Missing Rewards. as the name suggests, is an approach whereby we simply make the assumption that an absent reward corresponds to a reward value of zero. This enables us use standard policy evaluation algorithms such as SARSA, without any changes. Note that the rewards used for estimating $Q_\pi(s,a)$ can still differ from the true rewards. However, this method maintains the trajectory observed, and the returns observed from a given starting state are thus closer to the true returns when the discount factor $\gamma = 1$.

Linear Estimation of State-Action Rewards. LEAST is our second approach, and it comprises of estimating the values of the rewards for each state-action pair, thus replacing both the aggregated anonymous and missing rewards in the trajectory data with their estimate, $\hat{R}_t$. To estimate the true reward for a state-action pair, we first take note of the structure of the problem. The aggregated anonymous rewards are observed at fixed time step intervals, P. We denote by $R(s,a)$ the average reward from the state-action pair (s,a). Then our goal is to use the data observed from the aggregated rewards to estimate $R(s,a)$ for all pairs (s,a). This is naturally formulated as a least-squares problem. Mathematically, we can denote by the vector of $R(s,a)$ for all pairs (s,a), i.e., $x = [R(s,a)]_{(s,a)\in S\times A}$. We can define a matrix and vector $B \in \mathbb{R}^{N\times(S*A)}$ and $c \in \mathbb{R}^N$ such that our estimated reward is the solution to the least squares problem $\min_x ||Bx - c||_2^2$ [9]. Each row in the matrix B is constructed from a single reward period window P, and each column corresponds to the number of observations of state-pairs within that same window. Our factors to be learned, x, are the average rewards for each state-action pair, and c is the DAAF observed in the window. To illustrate this structure, assume we have an MDP with two states and one action, and a reward of 10 for every action in any state. Assuming our reward period $P = 2$, and that over two transitions we observe both states, an entry row for our regression estimation would be $A_i = [1,1]$ with each 1 indicating a single observation for both $(s = 0, a = 0)$ and $(s = 1, a = 0)$; and $b_i = 20$ the undiscounted DAAF observed. Our formulation is similar to that described by the authors in [14] for the task of learning a reward function for a policy that mimics given trajectory data through supervised learning. When the approximation of $R(s,a)$ is accurate, LEAST can provide a more reliable observation of returns to estimate $Q_\pi(s,a)$.

5 Methodology

We follow a quantitative approach to answer our research questions. We conduct simulations comparing our solutions against a baseline approach, and make use of multiple runs to generate statistics from the results. To monitor convergence, we first compute a solution to the action-value function using dynamic programming. We then use the approximated action-value function $Q_{dp}(s,a)$ as a reference when carrying out policy evaluation with the SARSA algorithm to compute the error of the estimated Q_π. In our experiments we compare four different methods: (1) the case of full rewards (FR) - no delays nor aggregation

(2) skip missing rewards (S-M) - the naïve approach of ignoring value updates when the reward is missing and taking DAAF as the reward for the current state-action if available, and our proposed approaches (3) ZI-M and (4) LEAST.

Environments - To establish some generality, we study two MDPs with different properties: Grid World and ABC. *Grid World* is a relatively known maze problem, described in detail in [16]. The agent is placed at the starting point of a grid sized $H \times L$, and their objective is to reach the exit. There are many variations of this problem, and we chose the simplest version, where the goal is to find the exit in as few steps as possible. We get the following MDP: S is the position of the player on the grid; A one of four options: up, down, left, right; $T(S, A, S')$ is the transition function from one state to the next, depending on the chosen action; and $S, A \rightarrow R$ is -1 for every transition, -100 for falling into a cliff, and 0 for reaching the exit. In Fig. 1, we have examples of two Grid World maps.

(a) gridworld_01 (b) gridworld_05

Fig. 1. Example Grid World levels used in our experiments - S is the starting position, G is the goal (exit), and X are cliffs that send that agent back to the starting position. An agent can go up, down, left or right.

ABC is a problem we propose for its simplicity and complexity scaling. We have a sequence of states: $A, B, C, D...N$, and the goal of an agent is to learn to choose each state in the right order. Given a state in position i, the agent should choose the next state in $i + 1$. If they choose correctly, the agent advances and there is no penalty, and otherwise there is a penalty. It results in the following MDP: S indicates the current state in the sequence; A a state to choose next; $T(S, A, S')$ is the transition function from one state to the next, depending on the chosen action; and $S, A \rightarrow R$ is defined by a function $D(p^{\pi}_{a+1}, p_a) + c$ - where p_a is the current state and p^{π}_{a+1} is the position of the agent's chosen next state, and c is a constant penalty for every action. The distance function D penalizes actions based on how far they are from the right choice[1]. ABC is an episodic task, where a terminal state is reached when the agent selects the final state in the sequence. We note that, more generally, the ABC problem can be unbounded or have any arbitrary sequence of states e.g. going from 1 to 10000. The idea is that the game advances in one direction only, from one state to the next in sequence. And, equally important, as the sequence length grows, so does the

[1] E.g. if $N = 5$, in state 4 the agent ought to select action 4; selecting 3 yields a penalty of $-(N - 4 + 3)$ and selecting 5 yields a penalty of -1.

possible number of actions to choose from. The data needed to learn a policy grows as a function of $S * A$, and since $|S| = |A|$, this growth is quadratic.

Data Collection - In all of our experiments, we evaluate a random policy, i.e. a policy that can chose any action with equal probability, both on the Grid World and ABC environments. We use the random policies to generate trajectory data for policy evaluation with the SARSA Algorithm. Both each environment, we use different configurations - e.g. shorter and longer sequences for ABC, smaller and larger grids for Grid World - and we experiment with different reward periods, P. During evaluation, we measure root mean-squared-error (RMSE) against an action-value function Q_{dp} obtained using dynamic programming. To compute a measure of variance, we run policy evaluation 100 times for each configuration.

6 Results

We selectively analyse a few results, starting with ABC. In Fig. 2, we have the RMSE plots for level 16 of ABC across four different reward periods. First, we can observe what happens when no intervention is made (S-M) under DAAF - the estimated value gets worse over time, despite an initial improvement. Turning to our proposed methods, ZI-M and LEAST, both of them converge to a values relatively close to what one would get with the full rewards. LEAST appears to obtain slightly more accurate estimates than ZI-M, and these patterns are also observed in Grid World configurations for levels 1–6 with the exception of level 5 (Fig. 3). We further compared our proposed solutions by measuring Spearman's rank correlation of the final Q_{π} values across configurations against the baseline Q_{dp}. The S-M method performed worst, overall, on the ABC configurations compared to Grid World, and it had higher variance. LEAST's estimated values more strongly correlated to the baseline Q_{dp} than ZI-M (results are omitted due to space constraints). Summaries of the results for RMSE of ABC and Grid World are in Tables 1 and 2, respectively, for select levels[2]

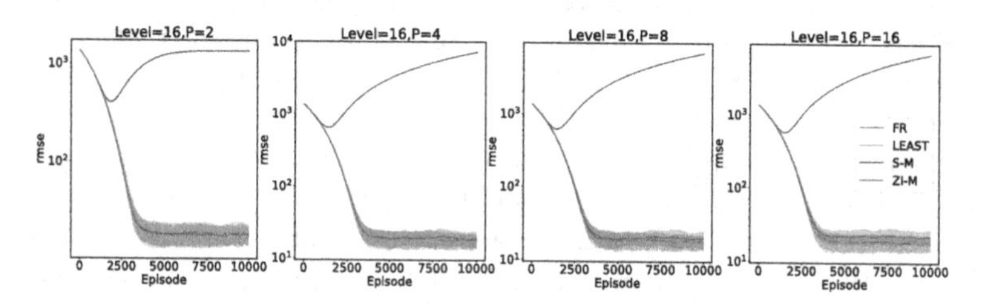

Fig. 2. ABC, Level = 16 with increasing reward period (P). The y-axis is on a logarithmic scale. Filled area is the standard deviation.

[2] Complete results can be found at https://github.com/dsv-data-science/rl-daaf.git.

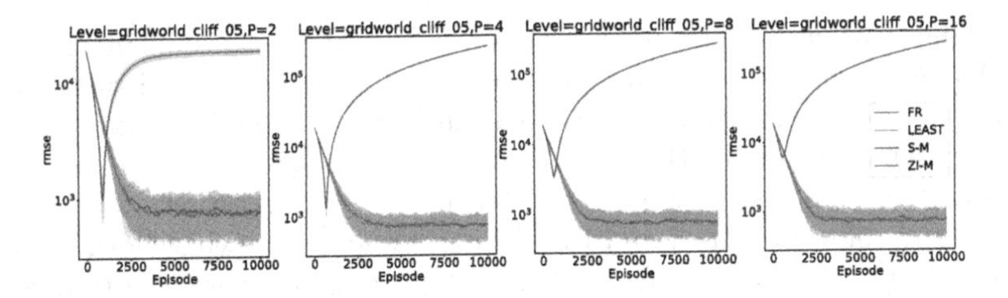

Fig. 3. Grid World, Map = *gridworld_05* with increasing reward period (P). The y-axis is on a logarithmic scale. Filled area is the standard deviation.

Table 1. RMSE for ABC problem configurations. We report the mean and standard deviation from 100 runs. The best results are highlighted.

Level/Map	P	S-M	ZI-M	LEAST
8	2	185.377 ± 6.437	$\mathbf{6.111 \pm 1.704}$	6.147 ± 1.638
8	4	3852.374 ± 47.991	6.494 ± 1.366	$\mathbf{5.854 \pm 1.429}$
8	8	2687.982 ± 41.446	7.736 ± 1.761	$\mathbf{5.927 \pm 1.436}$
8	16	2498.938 ± 34.716	9.369 ± 1.825	$\mathbf{5.927 \pm 1.301}$
24	2	3791.951 ± 30.037	$\mathbf{35.215 \pm 8.889}$	35.84 ± 7.414
24	4	8978.408 ± 47.414	35.883 ± 8.86	$\mathbf{34.621 \pm 9.044}$
24	8	8685.247 ± 38.652	38.318 ± 9.237	$\mathbf{33.698 \pm 7.316}$
24	16	8506.618 ± 30.93	40.295 ± 9.685	$\mathbf{36.907 \pm 8.98}$

Table 2. RMSE for Grid World configurations. We report the mean and standard deviation from 100 runs. The best results are highlighted.

Level/Map	P	S-M	ZI-M	LEAST
01	2	146.72 ± 30.606	23.908 ± 7.277	$\mathbf{18.358 \pm 7.056}$
01	4	15464.607 ± 301.918	29.085 ± 7.457	$\mathbf{18.786 \pm 7.001}$
01	8	10844.884 ± 297.05	34.247 ± 7.233	$\mathbf{18.786 \pm 8.061}$
01	16	4669.388 ± 237.253	36.61 ± 9.242	$\mathbf{18.107 \pm 7.386}$
05	2	18195.32 ± 1118.722	$\mathbf{779.158 \pm 280.31}$	806.638 ± 350.586
05	4	255804.832 ± 3890.733	768.801 ± 287.025	$\mathbf{739.775 \pm 272.95}$
05	8	265567.7 ± 4103.359	$\mathbf{756.606 \pm 280.82}$	826.515 ± 337.681
05	16	279066.206 ± 5158.172	$\mathbf{759.32 \pm 312.452}$	762.568 ± 308.587

7 Discussion and Future Work

We conducted policy evaluation using the SARSA algorithm without any alterations to the trajectory data to answer our first research question. In our simulations, the estimated policy diverged from the baseline Q_{dp}. This limitation was more pronounced in the ABC problem than for Grid World. One possible factor is the higher variance of rewards per action in the ABC environment compared to Grid World, where rewards are consistent for any step except when the agent falls into a cliff. Another factor is the cardinality of actions $|A|$, since in ABC $|S| = |A|$, which increases the chances of misattribution of credit over longer reward period windows. To answer our second research question, we compared our proposed solutions, ZI-M and LEAST, and found that while both achieved comparable results across configurations of both environments, the former had higher error in the estimate of the Q_π in most cases, except Grid World level 5. In ABC, as the reward period increased, the error for ZI-M followed. We speculate that this is due to the bias introduced from using zero or the cumulative reward. One limitation of LEAST is that it requires some initial observations of every (s, a) pair before estimating the average reward $R(s, a)$. With a random policy, and a constrained enough environment, collecting those observations is feasible. Under more constrained settings, e.g. with little exploration, it can be more practical to opt for ZI-M until enough data is available to estimate the average reward. Despite their differences, both proposed solutions converged with similar sample efficiency as FR on the tested environments. A natural extension of the work presented here is the study DAAF in discounted tasks. A second line of extension is to adapt the solutions proposed to problems with large or infinite state spaces, where function approximation is employed. And finally, our reseached as focused on the task of policy evaluation, and this problem framing and solution can be extended to control tasks.

8 Summary and Conclusions

In this paper, we presented a novel problem of conducting policy evaluation DAAF that comes at intervals. Through simulations using the well known environment of Grid World and a proposed environment called ABC, we demonstrated the inadequacy of using algorithms such as SARSA with missing rewards data. We proposed two solutions that alter the trajectory data, both of which were found to be sample efficient and effective for policy evaluation in undiscounted episodic tasks. Our findings indicate that estimating a policy is still feasible with DAAF. Future work will investigate how to expand our proposed methods to more common settings, e.g. with discounting, to further expand the applicability of RL in real world settings.

References

1. Agogino, A.K., Tumer, K.: Unifying temporal and structural credit assignment problems, pp. 980–987. AAMAS 2004. IEEE Computer Society, USA, July 2004
2. Cesa-Bianchi, N., Gentile, C., Mansour, Y.: Nonstochastic bandits with composite anonymous feedback, pp. 750–773. PMLR, July 2018. ISSN: 2640–3498
3. Chelu, V., Borsa, D., Precup, D., Hasselt, H.P.V.: Selective credit assignment. arXiv preprint arXiv:2202.09699 (2022)
4. Chen, H., et al.: Large-scale interactive recommendation with tree-structured policy gradient, vol. 33(1), pp. 3312–3320 (2019)

5. Chen, M., Beutel, A., Covington, P., Jain, S., Belletti, F., Chi, E.H.: Top-k off-policy correction for a REINFORCE recommender system, pp. 456–464. WSDM 2019. Association for Computing Machinery (2019)
6. Garg, S., Akash, A.K.: Stochastic bandits with delayed composite anonymous feedback, October 2019. arXiv:1910.01161
7. Jindal, I., Qin, Z.T., Chen, X., Nokleby, M., Ye, J.: Optimizing taxi carpool policies via reinforcement learning and spatio-temporal mining, pp. 1417–1426 (2018)
8. Krueger, D., Leike, J., Evans, O., Salvatier, J.: Active Reinforcement Learning: observing Rewards at a Cost, November 2020. arXiv:2011.06709
9. Lawson, C.L., Hanson, R.J.: Least-squares approximation, pp. 963–964. John Wiley and Sons Ltd., GBR, January 2003
10. Lee, K., Rucker, M., Scherer, W.T., Beling, P.A., Gerber, M.S., Kang, H.: Agent-based model construction using inverse reinforcement learning, pp. 1–12. WSC 2017. IEEE Press (2017)
11. Li, M., et al.: Efficient ridesharing order dispatching with mean field multi-agent reinforcement learning, pp. 983–994. WWW 2019. Association for Computing Machinery (2019)
12. Mesnard, T., et al.: Counterfactual credit assignment in model-free reinforcement learning, pp. 7654–7664. PMLR, ISSN: 2640-3498 (2021)
13. Pike-Burke, C., Agrawal, S., Szepesvari, C., Grunewalder, S.: Bandits with delayed, aggregated anonymous feedback, June 2018. arXiv:1709.06853
14. Ratliff, N.D., Bagnell, J.A., Zinkevich, M.A.: Maximum margin planning, pp. 729–736. ICML 2006. Association for Computing Machinery, New York, NY, USA (2006)
15. Sutton, R.S.: Learning to predict by the methods of temporal differences. Mach. Lang. **3**(1), 9–44 (1988). https://doi.org/10.1007/BF00115009
16. Sutton, R.S., Barto, A.G.: Reinforcement Learning: an Introduction. Adaptive Computation and Machine Learning Series. The MIT Press, Cambridge, Massachusetts, second edition edn. (2018)
17. Wang, Z., Qin, Z., Tang, X., Ye, J., Zhu, H.: Deep reinforcement learning with knowledge transfer for online rides order dispatching, pp. 617–626. ISSN: 2374-8486
18. Xu, Z., et al.: large-scale order dispatch in on-demand ride-hailing platforms: a learning and planning approach, pp. 905–913. KDD 2018. Association for Computing Machinery (2018)
19. Zhao, Y., Zhou, Y.H., Ou, M., Xu, H., Li, N.: Maximizing cumulative user engagement in sequential recommendation: an online optimization perspective, pp. 2784–2792. KDD 2020. Association for Computing Machinery, New York, NY, USA, August 2020
20. Zou, L., Xia, L., Ding, Z., Song, J., Liu, W., Yin, D.: Reinforcement learning to optimize long-term user engagement in recommender systems, pp. 2810–2818. KDD 2019. Association for Computing Machinery, New York, NY, USA, July 2019

Spatial and Temporal Analysis

Spatial Cross-Validation for Globally Distributed Data

Rita Beigaitė[1(✉)], Michael Mechenich[1], and Indrė Žliobaitė[1,2]

[1] Department of Computer Science, University of Helsinki, Helsinki, Finland
{rita.beigaite,indre.zliobaite}@helsinki.fi

[2] Department of Geosciences and Geography, University of Helsinki, Helsinki, Finland

Abstract. Increasing amounts of large scale georeferenced data produced by Earth observation missions present new challenges for training and testing machine-learned predictive models. Most of this data is spatially auto-correlated, which violates the classical i.i.d. assumption (identically and independently distributed data) commonly used in machine learning. One of the largest challenges in relation to spatial auto-correlation is how to generate testing sets that are sufficiently independent of the training data. In the geoscience and ecological literature, spatially stratified cross-validation is increasingly used as an alternative to standard random cross-validation. Spatial cross-validation, however, is not yet widely studied in the machine learning setting, and theoretical and empirical support is largely lacking. Our study aims at formally introducing spatial cross-validation to the machine learning community. We present experiments on data sets from two different domains (mammalian ecology and agriculture), which include globally distributed multi-target data, and show how standard cross-validation may lead to over-optimistic evaluation. We propose how to use tailored spatial cross-validation in this context to achieve more realistic assessment of performance and prudent model selection.

Keywords: Spatial cross-validation · Geospatial data · Model evaluation

1 Introduction

Cross-validation is a widely-used procedure for model selection and performance evaluation. It is expected to yield a reliable choice of model when the training sample is independent from, and distributed identically to, the validation sample [3].

Earth observation data are often spatially structured, such that observations geographically near each other are more similar than observations separated by greater distances [16]. Spatial auto-correlation is often strongly present in such data [10]. This property of geographic data violates the assumption of identical and independent distribution (i.i.d.), and leads to potential leakage of information from training to validation folds in the standard cross-validation setting.

P. Pascal and D. Ienco (Eds.): DS 2022, LNAI 13601, pp. 127–140, 2022.
https://doi.org/10.1007/978-3-031-18840-4_10

Spatial cross-validation is increasingly used as a strategy to make the cross-validation folds more independent from each other in ecology and geoscience studies [14,18,19]. This strategy has not been extensively analyzed in computer science either empirically or theoretically. The lack of theoretical backing has caused discussion in the ecological literature concerning whether spatial cross-validation is a proper method for estimating model performance [15,23]. However, this discussion is more about whether spatial cross-validation is the right tool to evaluate the accuracy of produced maps, rather than the generalization ability of machine-learned models. Evaluation of map accuracy is one of the goals in ecological modelling studies. By contrast, in machine learning the goal is to build a model which generalizes well over the observed patterns. A notable difference between machine learning and typical ecological studies that use spatial cross-validation is the size of data sets. In ecological studies the data is often small and it is usually concentrated around a specific region or a country, while the machine learning literature increasingly focuses on very large scale data sets. While the prevalence of Earth observation data in machine learning is increasing, typically, only the classical random cross-validation is used to evaluate the ability of the model to predict on unseen data.

In this paper, we introduce and analyze the spatial cross-validation task setting from the machine learning perspective where the data sets contain large numbers of observations which are globally distributed. Such data is becoming increasingly available in machine learning with rapid technological improvements in Earth observations [13].

Our aim is to introduce spatial cross-validation to the machine learning community, and empirically show in what ways a standard random cross-validation can fail to indicate that a model is over-fitting and provide an over-optimistic estimate of model performance. We present recommendations on how to use spatial k-fold cross-validation. We suggest using spatial cross-validation for globally distributed and spatially auto-correlated data whenever the size of the data set allows it.

The rest of the paper is organized as follows. After summarizing related work on spatial cross-validation in ecological studies in Sect. 2, we formally define spatial cross-validation in Sect. 3. We describe our experimental setup for evaluating spatial cross-validation in Sect. 4. In Sect. 4.3 we discuss the results, and draw conclusions in Sect. 5.

2 Related Work

In ecological studies, several strategies have been proposed for spatial cross-validation [19]. One of these strategies is geographic blocking [1,4,17,19]. In this approach, the data set is divided spatially into distinct geographic subsets, each spatially isolated from the others.

The simplest way to divide observations spatially is to overlay rectilinear blocks of a specified width and height on the mapped data set [22]. These blocks divide the study area vertically and/or horizontally, and may be used as folds in spatial cross-validation. Note that this blocking may be done in environmental space rather than in geographic space, for instance, by clustering observations based on the environmental conditions represented, then using each cluster as a fold in cross-validation [19,22]. It is often recommended to use each block as a separate fold, to allow more data for model training [19].

Another existing strategy for spatially-informed cross-validation is to leave a margin between the training and validation data, by removing data within a buffer of a pre-defined radius around each validation point. This can be done as a spatial leave-one-out (LOO) [14], spatial leave-pair-out [2], or spatial k-fold [18] cross-validation. One of the challenges of using buffered cross-validation is selecting the optimal distance for the buffer radius [21]. Often the buffer radius is decided by measuring spatial auto-correlation using the range of the variogram for the target variable or model residuals [14,18]. In practice, the choice is most often empirical.

While studies that implement spatial cross-validation exist, we are not aware of any analyses showing whether, how, and why spatial cross-validation is expected to outperform random cross-validation in the machine learning model evaluation and selection process.

In this article, we investigate a spatial k-fold cross-validation method that is computationally more efficient than LOO, and which does not require manual spatial or environmental blocking, for which the blocking strategy must be tailored to the specifics of the given data set.

3 Spatial k-Fold Cross-Validation

Spatial cross-validation may be considered a special case of a modified cross-validation approach [8] in which the data set is *trimmed*, and a number of points are discarded. Conceptually related approaches have been used in ecological [2,14,18–20] and remote sensing [12] contexts. In the case of spatial k-fold cross-validation, training points are removed within a so-called *buffer* radius of distance r from validation points. A visual example is provided in Fig. 1.

With Algorithm 1, we formally introduce the procedure of spatial k-fold cross-validation. First, the data set is randomly split into training and validation sets as in classical random k-fold cross-validation. Then, in each iteration, all training points closer than a chosen distance r to validation points are removed from the training set, and are not used in either training or validation. This is expected to remove potential data leakage. Finally, the reduced training sets are used in the same way as in regular random cross-validation.

The distance r can be determined by measuring at what distance spatial auto-correlation notably weakens. The challenge is that in environmental data,

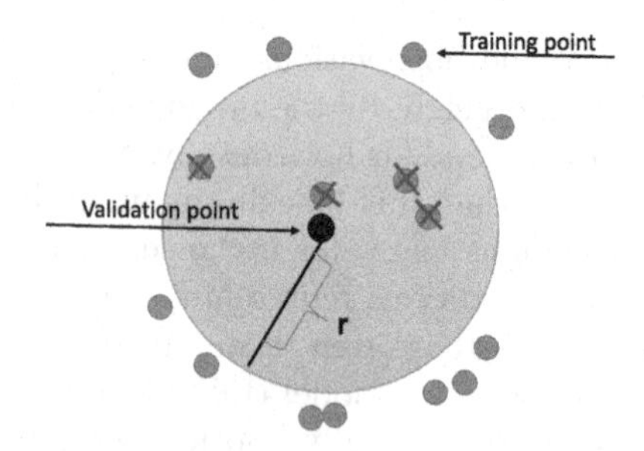

Fig. 1. An example of removing training points (blue colour) close to the validation point (black colour) within the buffer radius r (Color figure online)

auto-correlation can be present between observations as far as several thousand kilometers apart[1] [11]. If we were to choose a buffer of thousands of kilometers, it would remove most of the training points. In addition, in a multi-target setting, each target can have a different strength of spatial auto-correlation. Furthermore, for some variables auto-correlation can decrease with distance and start to increase again. Therefore, we recommend using a buffering strategy around validation points, and suggest experimenting with buffers of several radius distances, selecting the maximum distance which still retains sufficient data points for training.

In the following section, we compare the performance of the classical random and the spatial cross-validation that we recommend.

4 Evaluation of Performance

We conduct an empirical evaluation of spatial k-fold cross-validation on two real-world data sets[2] with globally distributed data. We choose a prediction task setting of multi-target regression where targets are dependent on each other and require a complex model to capture these relationships.

4.1 Data Sets

The first data set is from the field of mammalian ecology. The modeling task here is to predict the composition of vegetation from measures of the teeth of large, herbivorous mammals. The targets are 13 vegetation fractions that come from

[1] The distance between two georeferenced points can be calculated using the *Harvesine* distance formula, which gives shortest-path spherical distances between two points from their longitude and latitude coordinates [7].

[2] We made the data sets and our code publicly available at https://github.com/ritabei/Spatial-cross-validation.

Algorithm 1. Spatial k-fold cross-validation

Require: data set $(d_1, d_2, \dots d_p) \in \mathcal{D}$, where p is the size of the data set
Require: number of folds k
Require: buffer radius r
Require: model $\mathcal{M}$
Ensure: cross-val error e
  for $i = 1$ to k **do**
    Split $\mathcal{D}$ into $\mathcal{D}_i^{train}$ and $\mathcal{D}_i^{val}$ for the *i-th* split
    for $j = 1$ to n **do** ▷ for each instance of the train set $\mathcal{D}_i^{train}$
      for $l = 1$ to m **do** ▷ for each instance of the validation set $\mathcal{D}_i^{val}$
        if distance(d_j^{train}, d_l^{val}) $\leq$ r **then**
          $D_i^{train} \leftarrow D_i^{train} \setminus d_j^{train}$ ▷ removing points too close
          $n \leftarrow n - 1$ ▷ decreasing the number of iterations
        end if
      end for
    end for
    Train the model $\mathcal{M}$ using reduced training set D_i^{train}
    Compute the training error e_i
  end for
  $e \leftarrow \frac{1}{k}\sum_{i=1}^{n} e_i$ ▷ computing the average of errors

---

the MODIS [6] land cover product (MCD12C1, for the year 2001). The features are nine different functional dental traits of mammals originating from [9,24]. The size of this data set is 28224 observations. These observations are distributed over a 50 × 50 km grid with only one (or none) observation per georeferenced grid cell.

The second data set we derive from the global soil profile data [5]. Here, the task is to predict the soil texture, i.e., the proportions of salt, silt and sand from other soil chemical and physical properties (nine features in total). This data set has 10279 observations. In contrast to the mammalian ecology data set, this data set is not systematically gridded, and several observations can be recorded at the same or very close coordinates.

4.2 Experimental Design

Comparison of Testing and Cross-Validation Errors. First, as we expect each continent to present a unique environmental profile, to which our models must generalize and for which we require an estimate of model performance, we subset the data from one continent at a time and treat it as an independent test set. Then, we build a predictive model and estimate its cross-validation error on all the remaining continents. Finally, we compare the estimated cross-validation error with the test error of the held-out continent. We repeat this for each continent for which observations are present, for both of the data sets. For instance, we take the points from South America as the testing set and conduct model assessment via cross-validation on the data points from North America, Europe, Asia, and Africa. An example of the first cross-validation fold when South America is the testing subset is provided in Fig. 2.

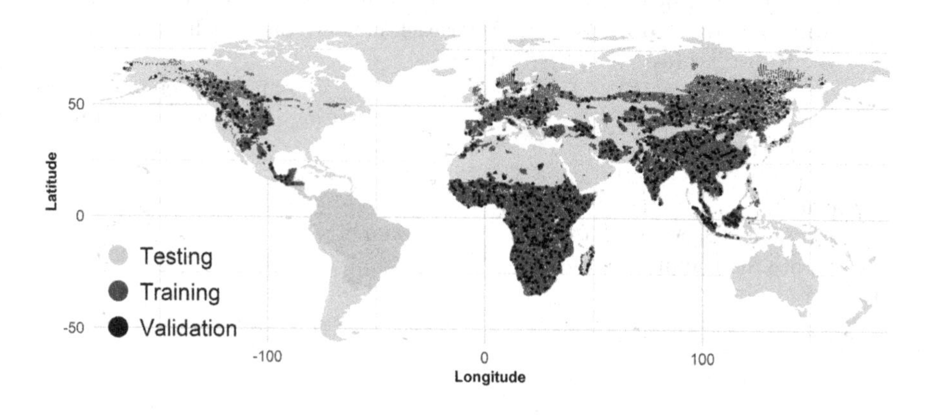

(a) Random cross-validation

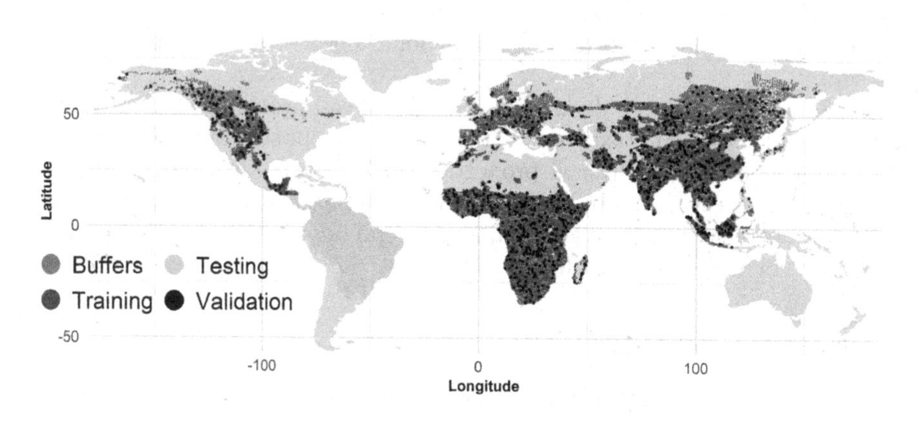

(b) Spatial cross-validation (50 km buffer)

Fig. 2. An example of one cross-validation fold for mammalian ecology data when South America is a testing set. Note that for better visibility, the amount of validation points is reduced and their size is increased. An example of one cross-validation fold for mammalian ecology data when South America is a testing set. Note that for better visibility, the amount of validation points is reduced and their size is increased

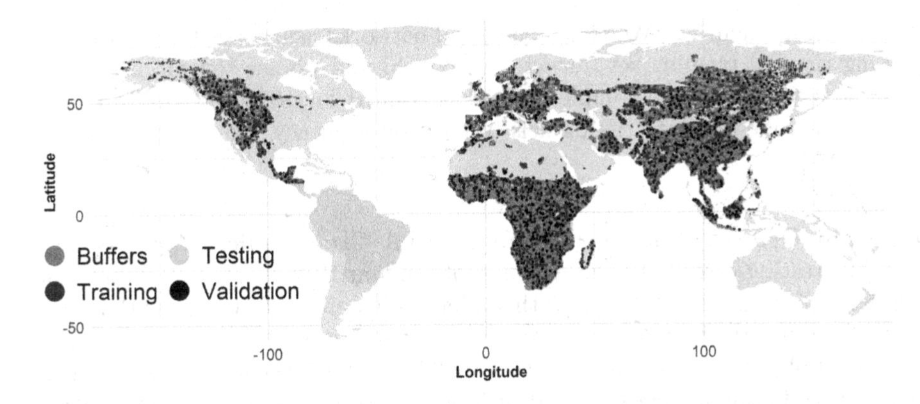

(c) Spatial cross-validation (100 km buffer)

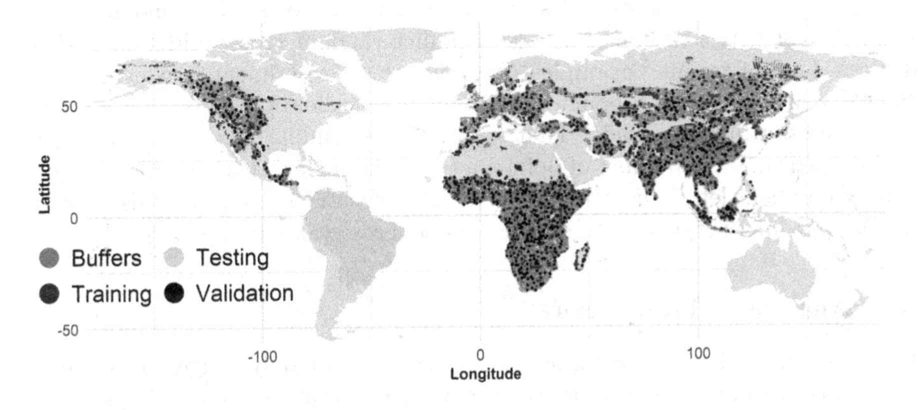

(d) Spatial cross-validation (150 km buffer)

Fig. 2. (*continued*)

Predictive Models. For each testing iteration we build two neural network (NN) models: one intentionally over-complex and one relatively simple. The former model (for both data sets) consists of an input layer, two hidden layers of 100 neurons which are activated by the *sigmoid* function, and an output layer activated by the *softmax* function. The *softmax* function is used to capture compositional dependencies in the data. The simple model has the same structure apart from the hidden layers. It has only one hidden layer: for the mammalian ecology data set the hidden layer consists of 10 neurons, and for the soil data set it consists of 8 neurons. We train the neural networks using the *Adam* optimizer passing the full training set during each epoch, and we use the *mean absolute error* loss function. The complex model is trained for 10000 epochs to ensure over-fitting. We evaluate the models' prediction accuracy using mean absolute error (MAE).

Random Cross-Validation Versus Spatial Cross-Validation. We compare the testing error with the estimate of random 10-fold cross-validation and spatial 10-fold cross-validation with buffer radius sizes of 50 km, 100 km, and 150 km around the validation points of the random cross-validation. The buffer increment of 50 km is motivated by the 50×50 km grid of the mammalian ecology data set. Buffering farther than 150 km distance reduces the data set excessively. Note that in this comparison random cross-validation has a potential advantage: the greater number of points used in the training process.

In order to check how results are affected when we deprive random cross-validation of this advantage, we also conduct random cross-validation with a decreased set of training examples. We conduct an experiment in which we use the same number of data points in random cross-validation folds as in reduced spatial cross-validation folds. That is, we repeat simple cross-validation four times. Each time the number of points in each training fold is randomly reduced. This experiment let us see whether the smaller size of the training folds leads to a more pessimistic estimate of simple random cross-validation error.

4.3 Analysis of Performance

In ecological studies, the empirical variogram is commonly suggested for measuring and visualizing spatial auto-correlation in variables. In a variogram, the average squared difference between pairs of observations within separation distance bins are plotted as a function of distance. If, for a given variable, pairs of observations near each other have small squared differences, the variable is positively spatially auto-correlated. Pairwise differences generally increase with distance to a *sill*, where the trend levels. The distance to the sill is termed the *range*; this is the distance over which spatial auto-correlation is detectable.

Variograms of the target variables of each data set are shown in Fig. 3. We can observe that some of the variables have in general very small variance, while for some variables variance increases rapidly with distance and can decrease again. Such plots suggest these variables do not exhibit clear and consistent ranges and sills, and thus do not provide a means for choosing a buffer distance for these globally distributed data sets.

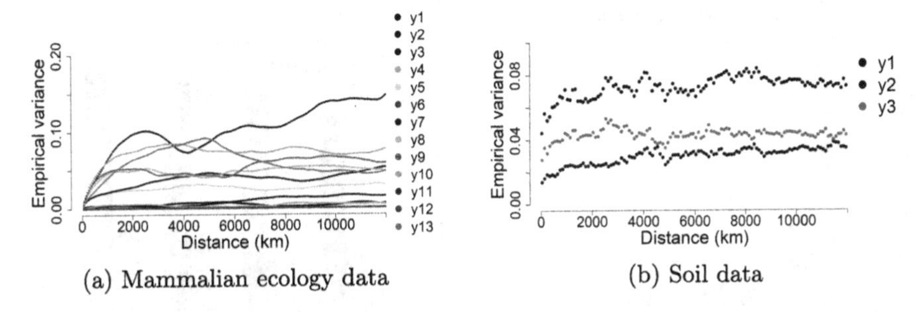

(a) Mammalian ecology data (b) Soil data

Fig. 3. Variance in targets as a function of distance

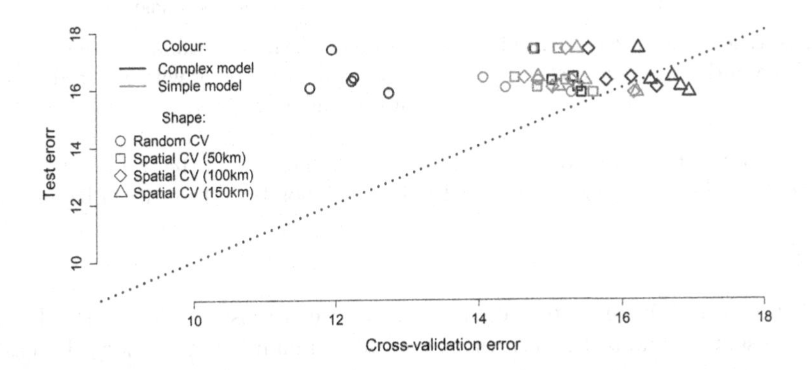

Fig. 4. Soil data. Comparison of cross-validation (CV) and test errors on each continent. The dotted diagonal line indicates where cross-validation error would be equal to the test error

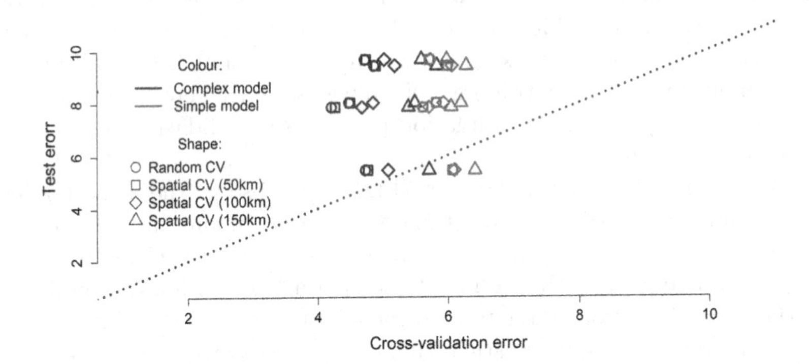

Fig. 5. Mammalian ecology data. Comparison of cross-validation (CV) and test errors on each continent.The dotted diagonal line indicates where cross-validation error would be equal to the test error

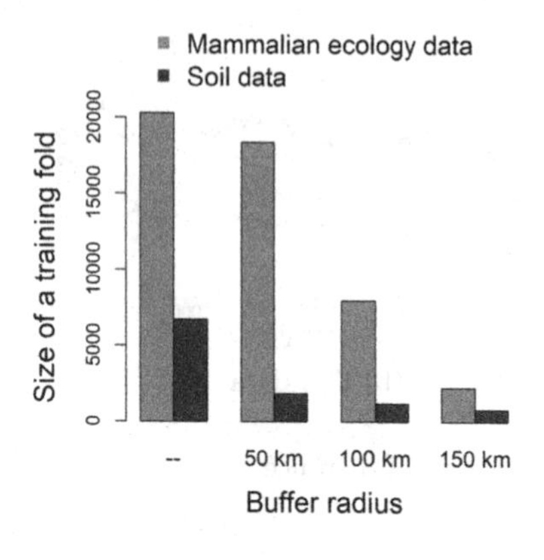

(a) Decrease in the size of a training fold when the buffer size increases

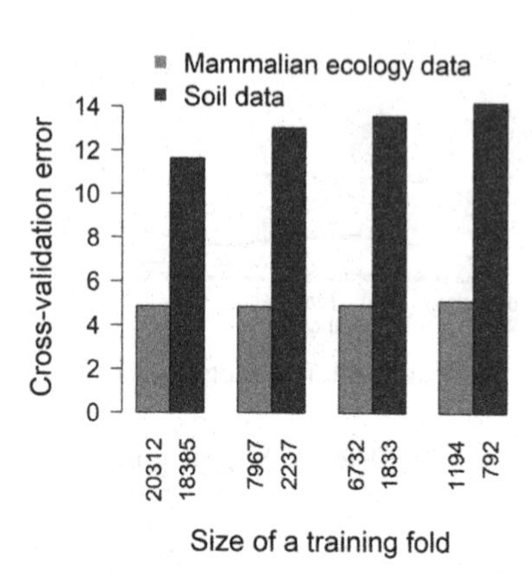

(b) Increase in the estimate of the random cross-validation error with reduced size of a training fold

Fig. 6. A link between random cross-validation error and size of the training fold. The order of bars (from left to right) in the (b) plot corresponds to the order of bars in plot (a)

Figures 4 and 5 illustrate a comparison between cross-validation (both random and spatial) and testing errors for the soil and mammalian ecology data sets, respectively. In the ideal case we would expect cross-validation error to be equal to the test error on unseen data and lie on (or very close to) the dotted diagonal line of the plot. For the soil data set the random cross-validation estimate of the complex models is far from this line. It indicates highly over-optimistic estimation of the models' performance. The reason could be that the soil data set includes observations which are very close or at the same coordinates. Thus, with random cross-validation, examples which are very similar to the validation set are memorised during the training. When we use spatial cross-validation and exclude the points close to the validation points, cross-validation estimates get closer to the dotted line.

In the mammalian ecology data set (Fig. 5) we observe the same tendency of over-optimistic cross-validation estimates of both random and spatial cross-validation. However, the spatial cross-validation estimates are closer to the true testing error. The greater the radius of the spatial cross-validation buffer, the closer the cross-validation estimate is to the true testing error.

We also observe over-pessimistic estimates of spatial cross-validation for the soil data with a 150 km radius buffer. This happens due to significantly reduced training set. In Fig. 6a we can observe that with increasing radius of the buffer, the training folds of both data sets are reduced considerably. This decrease in the number of points in the training set leads to higher random cross-validation errors for the soil data set (Fig. 6b). Nevertheless, despite being over-pessimistic,

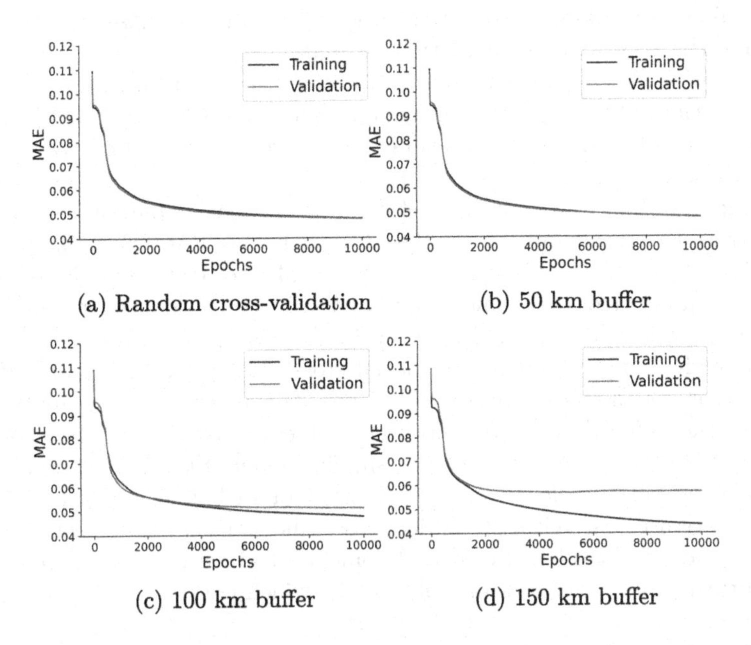

Fig. 7. An example of validation and training errors of mammalian ecology data set when model is over-complex and starts to over-fit

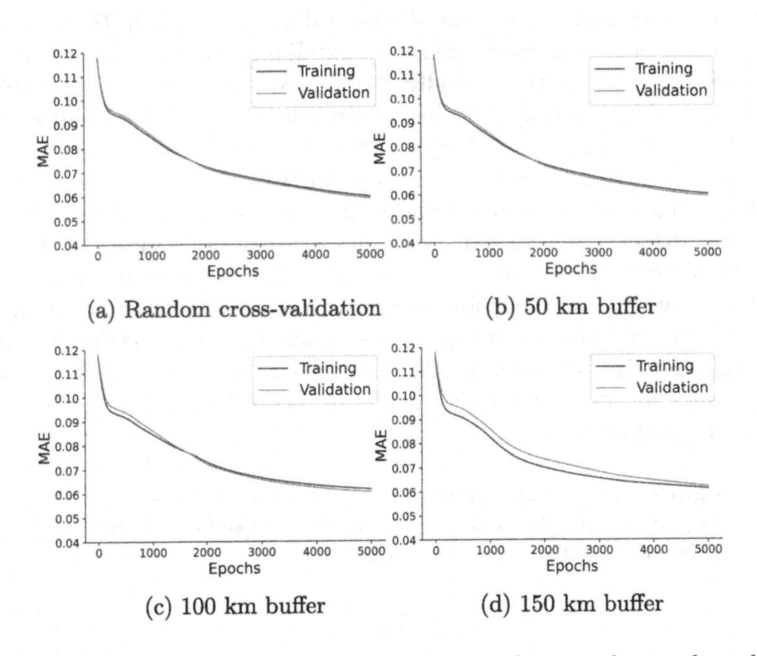

Fig. 8. An example of validation and training errors of mammalian ecology data set when model is simple

the spatial cross-validation estimate for the soil data is still closer to the testing error than the random cross-validation estimate.

We observe that for the mammalian ecology data set this increase in the cross-validation error (Fig. 6b) is negligible. One of the reasons could be that the original number of data points is higher. Therefore, we still have a sufficient amount of data for training the model.

Examples of the training process of mammalian ecology data in Figs. 7 and 8 demonstrate that random cross-validation does not indicate when the model is over-fitting. When the validation fold is very similar to the training folds, validation and training errors are almost equal. Even if the model is over-complex for the data no indication of over-fitting can be observed (Fig. 7a). With increased distance of the buffer radius in the spatial cross-validation, we can observe validation error starting to increase (Fig. 7c,d). When spatial cross-validation is used with the buffer distance of 150 km, the plots clearly indicate over-training.

When we look at the plots of the simple model (Fig. 8), we observe that the validation error is close to the training error in both random and spatial cross-validation cases. Without spatial cross-validation, it would seem that the complex model is an adequate fit and could be trained on even more epochs. This, in turn, would lead to choice of a model which performs poorly on unseen data.

5 Conclusions

We empirically examined how spatial cross-validation can help to improve the model selection process in the presence of spatial auto-correlation. We have carried out experiments on two globally distributed data sets from the fields of mammalian ecology and agriculture. Our results confirm that random cross-validation fails to indicate when models start to over-fit complex data, and its estimation of the error tends to be over-optimistic.

Spatial cross-validation proved to be an adequate modification addressing these issues. However, such modification has drawbacks. Removal of the points close to the validation set increases computational complexity and decreases the size of the training data. Therefore, we recommend using it with caution when the initial data size is small as it can lead to an over-pessimistic estimate of model performance. For further research, it is important to investigate the choice of the buffer distance for global data sets as the auto-correlation can still be present thousands of kilometers away.

Acknowledgments. We thank Tang Hui for the initial pre-processing of the mammalian ecology data set. Research leading to these results was supported by the Academy of Finland (grants no. 314803 and 341623).

References

1. Adams, M.D., Massey, F., Chastko, K., Cupini, C.: Spatial modelling of particulate matter air pollution sensor measurements collected by community scientists while cycling, land use regression with spatial cross-validation, and applications of machine learning for data correction. Atmos. Environ. **230**, 117479 (2020)
2. Airola, A., et al.: The spatial leave-pair-out cross-validation method for reliable auc estimation of spatial classifiers. Data Min. Knowl. Disc. **33**(3), 730–747 (2019)
3. Arlot, S., Celisse, A.: A survey of cross-validation procedures for model selection. Stat. Surv. **4**, 40–79 (2010)
4. Bahn, V., McGill, B.J.: Testing the predictive performance of distribution models. Oikos **122**(3), 321–331 (2013)
5. Batjes, N.: Harmonized soil profile data for applications at global and continental scales: updates to the wise database. Soil Use Manag. **25**(2), 124–127 (2009)
6. Channan, S., Collins, K., Emanuel, W.: Global mosaics of the standard modis land cover type data. University of Maryland and the Pacific Northwest National Laboratory, College Park, Maryland, USA 30 (2014)
7. Chopde, N.R., Nichat, M.: Landmark based shortest path detection by using a* and haversine formula. Int. J. Innov. Res. Comput. Commun. Eng. **1**(2), 298–302 (2013)
8. Feluch, W., Koronacki, J.: A note on modified cross-validation in density estimation. Comput. Stat. Data Analysis **13**(2), 143–151 (1992)
9. Galbrun, E., Tang, H., Fortelius, M., Žliobaitė, I.: Computational biomes: The ecometrics of large mammal teeth. Palaeontol. Electron. **21**(21.1. 3A), 1–31 (2018)
10. Getis, A.: A history of the concept of spatial autocorrelation: a geographer's perspective. Geogr. Anal. **40**(3), 297–309 (2008)
11. Hijmans, R.J.: Cross-validation of species distribution models: removing spatial sorting bias and calibration with a null model. Ecology **93**(3), 679–688 (2012)
12. Karasiak, N., Dejoux, J.-F., Monteil, C., Sheeren, D.: Spatial dependence between training and test sets: another pitfall of classification accuracy assessment in remote sensing. Mach. Learn. **111** 1–26 (2021). https://doi.org/10.1007/s10994-021-05972-1
13. Lary, D., et al.: Machine learning applications for earth observation. In: Mathieu, P.-P., Aubrecht, C. (eds.) Earth Observation Open Science and Innovation. ISRS, vol. 15, pp. 165–218. Springer, Cham (2018). https://doi.org/10.1007/978-3-319-65633-5_8
14. Le Rest, K., Pinaud, D., Monestiez, P., Chadoeuf, J., Bretagnolle, V.: Spatial leave-one-out cross-validation for variable selection in the presence of spatial autocorrelation. Glob. Ecol. Biogeogr. **23**(7), 811–820 (2014)
15. Meyer, H., Pebesma, E.: Machine learning-based global maps of ecological variables and the challenge of assessing them. Nat. Commun. **13**(1), 1–4 (2022)
16. Miller, H.J.: Tobler's first law and spatial analysis. Ann. Assoc. Am. Geogr. **94**(2), 284–289 (2004)
17. Ploton, P., et al.: Spatial validation reveals poor predictive performance of large-scale ecological mapping models. Nat. Commun. **11**(1), 1–11 (2020)
18. Pohjankukka, J., Pahikkala, T., Nevalainen, P., Heikkonen, J.: Estimating the prediction performance of spatial models via spatial k-fold cross validation. Int. J. Geogr. Inf. Sci. **31**(10), 2001–2019 (2017)
19. Roberts, D.R., et al.: Cross-validation strategies for data with temporal, spatial, hierarchical, or phylogenetic structure. Ecography **40**(8), 913–929 (2017)

20. Schratz, P., Muenchow, J., Iturritxa, E., Richter, J., Brenning, A.: Hyperparameter tuning and performance assessment of statistical and machine-learning algorithms using spatial data. Ecol. Model. **406**, 109–120 (2019)
21. Trachsel, M., Telford, R.J.: Estimating unbiased transfer-function performances in spatially structured environments. Climate of the Past **12**(5), 1215–1223 (2016)
22. Valavi, R., Elith, J., Lahoz-Monfort, J.J., Guillera-Arroita, G.: blockCV: an R package for generating spatially or environmentally separated folds for k-fold cross-validation of species distribution models. Methods Ecol. Evol. **10**(2), 225–232 (2019)
23. Wadoux, A.M.C., Heuvelink, G.B., De Bruin, S., Brus, D.J.: Spatial cross-validation is not the right way to evaluate map accuracy. Ecol. Model. **457**, 109692 (2021)
24. Žliobaitė, I., et al.: Herbivore teeth predict climatic limits in kenyan ecosystems. Proc. Natl. Acad. Sci. **113**(45), 12751–12756 (2016)

Leveraging Spatio-Temporal Autocorrelation to Improve the Forecasting of the Energy Consumption in Smart Grids

Annunziata D'Aversa[1], Stefano Polimena[1], Gianvito Pio[1,2](✉), and Michelangelo Ceci[1,2,3]

[1] Department of Computer Science, University of Bari Aldo Moro, Bari, Italy
{annunziata.daversa,stefano.polimena,gianvito.pio, michelangelo.ceci}@uniba.it

[2] Big Data Lab, CINI Consortium, Rome, Italy

[3] Department of Knowledge Technologies, Jozef Stefan Institute, Ljubljana, Slovenia

Abstract. Smart grids are networks that distribute electricity by relying on advanced communication technologies, sensor measurements, and predictive methods, to quickly adapt the network behavior to different possible scenarios. In this context, the adoption of machine learning approaches to forecast the customer energy consumption is essential to optimize network planning operations, avoid unnecessary energy production, and minimize power shortages. However, classical forecasting methods are not able to take into account spatial and temporal autocorrelation phenomena, naturally introduced by the spatial proximity of consumers, and by the seasonality of the energy consumption trends.
In this paper, we investigate the adoption of several solutions to take into account spatio-temporal autocorrelation phenomena. Specifically, we investigate the contribution provided by the explicit representation of temporal information related to historical measurements using multiple strategies, as well as that of simultaneously predicting multiple future consumption measurements in a multi-step predictive setting. Finally, we investigate the effectiveness of injecting descriptive features to make the learning methods aware of the spatial closeness among the consumers.
The experimental evaluation performed on a real-world electrical network demonstrated the positive contribution of making the models aware of spatio-temporal autocorrelation phenomena, and proved the overall superiority of models based on the multi-step predictive setting.

Keywords: Energy forecasting · Multi-step prediction · Spatio-temporal autocorrelation

1 Introduction

The infrastructures for the energy distribution are continuously subject to evolutions, mainly because of the generally increasing energy demand, as well as

P. Pascal and D. Ienco (Eds.): DS 2022, LNAI 13601, pp. 141–156, 2022.
https://doi.org/10.1007/978-3-031-18840-4_11

of the introduction of new technologies, such as renewable power plants and car charging stations. The need of managing complex scenarios led to the definition of the so-called *smart grids*, that are distribution networks that exploit sensor measurements, advanced communication technologies and predictive components, to quickly adapt the network behavior to multiple possible situations. In this context, the accurate forecasting of the customer energy consumption is fundamental, not only to optimize the planning of network maintenance operations over the long term, but also to properly tune the production of energy from fossil fuel power stations. Indeed, producing energy from fossil sources generally leads to high CO_2 emissions, and the overproduction may also lead to the need of additional resources for storage. On the other hand, the underestimation of the energy consumption may compromise the system reliability, since an excessive demand could easily degenerate into a blackout. For these reasons, it is of paramount importance to predict the energy consumption in the network.

Machine learning methods can fruitfully be adopted to support this task, since they are able to exploit historical data, temporal trends and other consumer characteristics to build accurate predictive models. In general, the temporal dimension plays a central role for this task. Indeed, we can expect to observe cyclical behaviors, for example, along the months of each year (i.e., a generally higher consumption during summer and winter, mainly due to heating/cooling systems, rather than during spring and autumn).

The temporal dimension can generally introduce autocorrelation phenomena, known as the correlation of a signal with a *delayed copy* of itself as a function of delay, or the similarity between observations as a function of the time lag between them [4]. Analogously, the spatial closeness can influence the measurements: the Tobler's first law of geography [17] states that "everything is related to everything else, but near things are more related than distant things". In this specific context, spatially close consumers may exhibit a similar behavior, mainly because they live in similar climatic conditions. Although considering temporal and spatial autocorrelation phenomena should generally lead to a higher accuracy of the learned models [15], they have not yet been fully exploited in the context of the prediction of the energy consumption. Indeed, in the literature we can find only few works that investigated their contribution for the forecasting of the energy consumption, which are based on classical ARIMA models [6,12]. On the other hand, their positive effect on the accuracy of the learned predictive models has been observed in the context of the energy production from photovoltaic power plants [5]. However, the challenges arising while aiming to predict the energy production and the energy consumption are different: while the former task is much more dependent on physical factors, such as weather conditions, in the latter, the prediction is mainly dependent on the behavior of consumers. Therefore, it is expected that the temporal dimension is more influential on the prediction of the energy consumption than for the prediction of the energy production.

In this paper, we propose a method for the forecasting of the monthly energy consumption of the consumers of a smart grid on a yearly horizon. The proposed approach is able to properly capture and model both temporal and spatial autocorrelation phenomena. Different strategies are proposed for both the temporal and the spatial dimensions, each of which is able to properly model specific temporal/spatial characteristics and relationships among different measurements. Finally, we investigate the possibility to predict the 12 monthly measurements of the considered yearly horizon simultaneously, in a multi-step predictive setting, that, as we will emphasize in Sect. 2, is able to implicitly model the temporal relationships among the measurements at different time points, for both descriptive and target variables.

The rest of the paper is organized as follows. In Sect. 2, we briefly discuss existing related work. In Sect. 3, we describe the proposed approach for the forecasting of the energy consumption in smart grids, taking into account both temporal and spatial autocorrelation phenomena. In Sect. 4 we describe our experiments on a real-world energy distribution network. Finally, in Sect. 5, we draw some conclusions and outline possible future work.

2 Related Work

In the literature, we can find several works that propose methods for the prediction of the energy consumption, at different spatial and temporal scales: from high and very localized geographical resolutions (e.g., hourly measurements of a single sensor) to coarser temporal resolutions (e.g., days, months, years) and/or covering a large geographic area (e.g., a region or a country). Existing approaches can also be categorized as *single-step* methods, that aim to predict the value of a target attribute for a single future time step, and *multi-step* methods, that aim to predict the value of a target attribute for multiple steps ahead. In [16], the authors described different strategies that can be adopted to solve the latter task, including *recursive*, *direct* and *Multi-Input Multi-Output (MIMO)* strategies. The *recursive* strategy exploits an approach based on self learning, that iterates a single-step ahead predictive model to obtain the desired forecasts: after estimating the next value of the sequence, it is fed back as a descriptive variable for the subsequent prediction. The *direct* strategy is based on learning a set of independent predictive models, where the i-th model is able to return a prediction for the i-th time points in the future. Note that both *recursive* and *direct* strategies are actually single-step approaches that are applied multiple times to obtain a multi-step ahead prediction. On the other hand, the *MIMO* strategy aims to learn one global model that returns a vector of predictions, also possibly taking into account the existence of dependencies between future values, that in principle may be beneficial in terms of forecasting accuracy [3].

In [2], the authors proposed a deep learning architecture to forecast the customer energy consumption for the next month, using the measurements of the previous 12 months and other information such as the target month and the category of the customer (e.g., residential, business, etc.). Among the considered deep learning models, LSTM achieved the lowest mean absolute error.

In [18] the authors compared the performance of different methods, such as Linear Regression, Regression Trees and Multivariate Adaptive Regression Spline (MARS), for the prediction of the next month energy consumption using climate data and the characteristics of the buildings (e.g., size of living area, number of rooms, etc.). The authors also aggregated the individual consumptions to predict the monthly consumption for groups of buildings. Results showed that MARS was the best model for individual households, while regression trees outperformed the competitors for the prediction of the consumption of the groups.

In [10], the authors adopted the *direct* strategy to predict the electric load 10 d ahead using ARIMA and LSTM. The models were evaluated on three electrical networks and the results showed a general superiority of LSTM.

Despite several studies have been proposed for energy consumption forecasting, only a few of them investigated the possible contribution coming from spatial and temporal autocorrelation phenomena. An attempt in this direction has been done in [6,12], where the authors considered spatial autocorrelation phenomena for the forecasting of the regional electricity consumption. In these works, a spatial ARMA model (SAR-ARMA) and a spatial ARIMA model (ARIMA-Sp) were proposed. However, auto-regressive approaches usually train a model based on the target variable only, and are not able to take into account additional features and possible dependencies between them and the target variable.

In [9], the authors proposed a deep neural network, called LSTNet, which combines convolutional neural networks to capture short-term patterns and LSTM or GRU for long-term patterns. To overcome the issue caused by the vanishing gradient, which affects the possibility to properly capture long-term interdependencies, the authors proposed the introduction of a recurrent-skip layer or an attention mechanism. Similarly, in [14], the authors proposed TPA-LSTM, an attention-recurrent neural network that allows the model to learn interdependencies among multiple variables across all previous time-steps.

The consideration of the spatial and of the temporal dimensions gained a general interest for other tasks related to time-series forecasting, even if not specifically focused on the prediction of the energy consumption. In particular, neural network architectures that simultaneously consider both temporal and spatial dimensions have been recently proposed. A relevant example is Graph WaveNet [19], a spatio-temporal graph convolutional network for multi-step forecasting, tailored for the prediction of traffic conditions at different locations. It uses dilated convolution networks to capture temporal dependencies and a self-adaptive adjacency matrix to capture spatial correlations. Another relevant example applied in the same domain is GMAN [20], which exploits a graph multi-attention network, with spatial and temporal attention mechanisms. Since it can be considered as one of the most recent approaches for multi-step prediction, that also consider spatio-temporal aspects, it will be considered as a state-of-the-art competitor in our experimental evaluation (see Sect. 4).

3 The Proposed Method

In this section, we describe our approach to forecast the monthly energy consumption of consumers on a yearly horizon. Therefore, the goal is to predict, for each consumer, 12 energy consumption values, i.e., one for each month of the subsequent year. As mentioned in Sect. 1, predicting such values is useful for planning network maintenance operations, as well as for tuning the energy production from fossil sources.

In the following subsections, we report the details of the proposed strategies to take into account the temporal and the spatial autocorrelation phenomena. After properly representing the temporal and the spatial dimensions, different standard regression models can be learned on top. At the end of the following subsection, we also briefly introduce the considered regressors and their extension to the multi-step predictive setting proposed in this paper.

3.1 Modeling the Temporal Autocorrelation

We propose different strategies to take into account the temporal autocorrelation, exploiting historical data about consumptions. We investigate two forecasting settings, namely, single-step (SS), where the 12 predictions are obtained by a *recursive* approach, and multi-step (MS), falling in the *MIMO* category, which goal is that of learning a global predictive model that returns the whole vector of 12 predictions. More formally, considering a time series of length w of energy consumptions for the consumer c, the SS setting consists in the exploitation of the historical measurements up to the time-step t-1 to predict the next time-step $y_{c,t}$. Through the recursive strategy, the predicted value $y_{c,t}$ is considered as a real measurement for the forecast of the energy consumption $y_{c,t+1}$, and so on up to predict $y_{c,t+11}$ (see the left part of Fig. 1).

Note that the adopted *recursive* strategy exhibits both advantages and disadvantages with respect to the *direct* strategy. Among the strong points, we can mention that the number of training instances increases (roughly by a factor of w), thanks to the fact that the measurement at a given month is considered multiple times, in different positions of the w-dimensional training time series (see, for example, the measurement related to Dec 2018 in the left part of Fig. 1). On the other hand, this aspect introduces the disadvantage of losing the temporal semantics of each descriptive feature, namely, each feature does not represent the same month of the year for all the training instances. This means that the model learned in this setting cannot easily detect and exploit seasonality phenomena. Another disadvantage is that, since it relies on a self-training approach, forecasting errors at the initial time-steps may be propagated to subsequent time-steps [13]. In order to alleviate the first issue, keeping the advantages of the recursive strategy, we explicitly represent the temporal information through additional features. In this respect, we propose two alternative settings:

- **SS-DT** (Described Target time-step), that introduces two additional descriptive features, namely the year j_t and the month m_t of the target value to predict $y_{c,t}$;

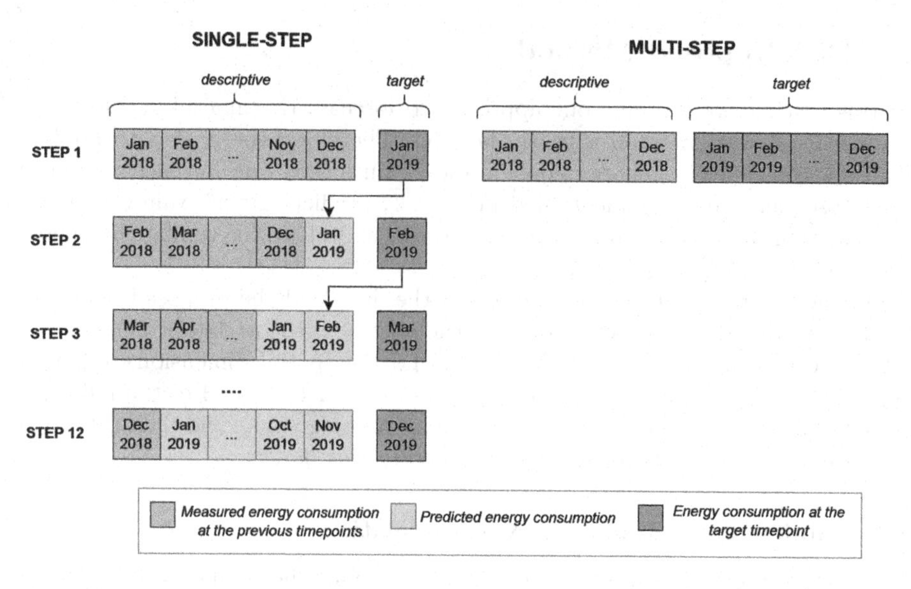

Fig. 1. A graphical representation of the single-step (SS) and multi-step (MS) learning settings. In the SS setting, the prediction for the i-th step is added to the descriptive variables for the prediction of the $(i+1)$-th step, while in the MS setting a global method able to simultaneously predict the value for all the 12 steps is learned.

- **SS-DTP** (Described Target and Previous time-steps), that introduces the year j_t and the month m_t of the target value to predict $y_{c,t}$, as well as the years $j_{t-1}, j_{t-2}, \ldots, j_{t-w}$ and the months $m_{t-1}, m_{t-2}, \ldots, m_{t-w}$ of the considered w previous observations.

It is noteworthy that, although SS-DT and SS-DTP explicitly represent the information about the year and the month associated with a given descriptive feature, the absolute value of a month does not properly represent the temporal cyclicity. In other words, December (12) 2018 may appear very distant to January (1) 2019, while it is actually temporarily close. To alleviate this issue, we resort to directional statistics that allow considering the *temporal position* of the target month, as well as that of the months historical data refer to (only in the case of SS-DTP). At this purpose, we use directional statistics that *envelope* the probability density function around the circumference of a unit circle representing the months of the year (see Fig. 2). More specifically, we compute the radial closeness between two months m_1 and m_2, represented as integer values in the interval $[1;12]$, on the unit circumference as $2\pi - d_r(m_1, m_2)$, where:

$$d_r(m_1, m_2) = min\left(\frac{2\pi}{12} \cdot |m_1 - m_2|, 2\pi - \frac{2\pi}{12} \cdot |m_1 - m_2|\right) \tag{1}$$

is the radial distance between m_1 and m_2 on the acute angle (see Fig. 2 for an example of radial distance computed between February and May).

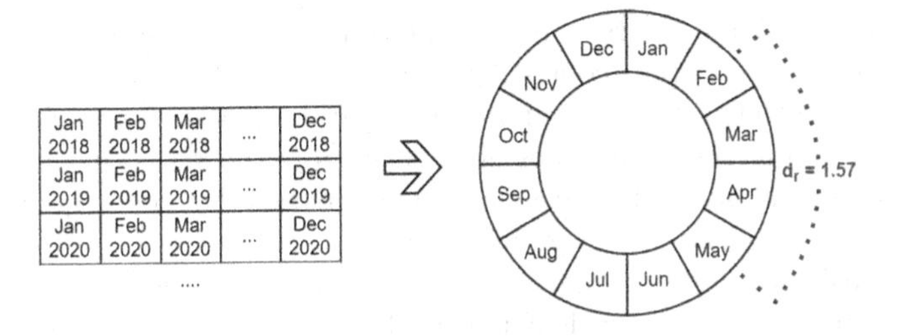

Fig. 2. Representation of the month of the year on the circumference of a unit circle. In the example, the radial distance between February and May is computed as $d(2,5) = min(2\pi/12 \cdot |2-5|, 2\pi - 2\pi/12 \cdot |2-5|) = min(1.57, 4.71) = 1.57$.

In our case, we compute the radial closeness between a given month in the descriptive attributes and the month of the target time-step to predict. Henceforth, the settings that exploit this radial closeness will be distinguished through a **C** (cyclical), appended to the name of the setting.

As regards the MS setting, we adopt the MIMO strategy to forecast 12 time-steps $y_{c,t}, \ldots, y_{c,t+11}$ for the consumer c at the same time. In this setting, we consider as input features the monthly energy consumption of the previous year (i.e., of the previous $w = 12$ months) and the year of the target time-step (see the right part of Fig. 1). Unlike the SS setting, MS does not need additional features to represent temporal relationships. Indeed, it is implicitly able to capture potential temporal dependencies, since the i-th feature always represents the i-th month of the year. On the other hand, while the recursive SS setting may be more suited when the training data is limited, MS preserves the dependencies also between the predicted values, and avoids the propagation of errors typical of the recursive SS strategy.

Note that, however, not all the regression methods can be easily extended to work in this setting. In our system, we adopt three different regressors, namely, Linear Regression, Regression Trees and Random Forests, also because of their ability to produce accurate models also when the available training data is poor.

Linear Regression methods aim to identify a linear model with coefficients $q = (q_1, q_2 \ldots, q_p)$, where p corresponds to the number of descriptive features plus 1 (the intercept), that minimizes the residual sum of squares between the observed target values in the training set, and the predictions provided by the linear approximation. For multi-step prediction, in our case, since we need to predict the consumption for the 12 subsequent months, identifying a predictive linear model corresponds to finding a matrix of coefficients $Q \in \mathbb{R}^{p \times 12}$ such that $\frac{1}{N}\sum_{i=1}^{N} ||u_i^\top Q - v_i^\top||_2^2$ is minimized, where $u_i \in \mathbb{R}^p$ is the vector of the descriptive features of the i-th training instance concatenated with a 1 (to take into account the intercept), N is the number of training instances, and $v_i \in \mathbb{R}^{12}$ is the vector of target values for the 12 subsequent months for the training instance u_i.

Learning methods for the construction of **Regression Trees** and ensemble thereof (e.g., **Random Forests**) are usually based on top-down induction procedures. Starting from the root node containing all the training instances, at each iteration, the *best* split, consisting of a descriptive feature and a threshold, is identified such that it well discriminates/separates the instances falling in the resulting children nodes. Leaf nodes of the tree store the actual predictions. The identification of the best split relies on some heuristics that, for regression tasks, are usually based on the reduction of the variance.

The extension of these approaches to solve multi-step tasks consists in storing multiple output values in the leaf nodes (12 in our case), and in a modified heuristics able to globally consider the contribution of the split towards the proper prediction of all the target values. Specifically, we adopt the *arithmetic mean of the variance reduction* computed over all the target time-steps.

3.2 Modeling the Spatial Autocorrelation

As mentioned in Sect. 1, taking into account the spatial autocorrelation in the construction of the predictive models may be beneficial in terms of accuracy, since spatially close consumers could exhibit a similar behavior, mainly due to similar climatic conditions. We evaluate the contribution coming from the adoption of two different spatial statistics [5]: the Local Indicator of Spatial Association (LISA) [1] and the Principal Coordinates of Neighbor Matrices (PCNM) [7].

According to [1], *i)* a LISA for a given observation must give an indication of the extent of significant spatial clustering of similar values around that observation, and *ii)* the sum of LISAs for all observations must be proportional to a global indicator of spatial association. In our case, given the set of n consumers, we first compute a neighborhood matrix $A \in \{0,1\}^{n \times n}$ as:

$$A[c_a, c_b] = \begin{cases} 1 \; if dist(c_a, c_b) < maxDist \\ 0 \; otherwise \end{cases} \tag{2}$$

where c_a and c_b are the a-th and the b-th consumers (with $1 \leq a \leq n$ and $1 \leq b \leq n$), $dist(c_a, c_b)$ is the geodesic distance between consumers, and $maxDist$ is a user-defined threshold on the maximum distance to consider the spatial autocorrelation phenomena among consumers as relevant. The matrix A is then normalized so that the sum of each row equals to 1[1], as follows:

$$A'[c_a, c_b] = \frac{1}{max(\sum_{i=1}^{n} A[c_a, c_i], 1)} A[c_a, c_b] \tag{3}$$

Using the matrix A', we can estimate the contribution of the neighborhood on each descriptive feature. Specifically, we first compute the z-score normalization for each descriptive feature x of each consumer c_a as:

$$x'_{c_a} = \frac{x_{c_a} - \mu_{x,c_a}}{\sigma_{x,c_a}}, \tag{4}$$

[1] Some rows in the normalized matrix can have a sum of 0, when the corresponding consumer has no other consumers falling in its neighborhood, according to $maxDist$.

where μ_{x,c_a} and σ_{x,c_a} are the average and the standard deviation of the descriptive variable x for the consumer c_a. Using the normalized value x'_{c_a}, we compute the spatial indicator I_{x,c_a} for the variable x of the consumer c_a as:

$$I_{x,c_a} = x'_{c_a} \cdot \sum_{i=1}^{n} (A'[c_a, c_i] \cdot x'_{c_i}) \tag{5}$$

The computed spatial indicators, one for each feature, can finally be added as additional descriptive features. Therefore, this solution leads to the introduction of w additional features, that represent the initial descriptive features influenced by the spatial closeness with other consumers.

A different approach to consider the spatial autocorrelation, as mentioned before, is represented by the PCNM. It allows us to extract additional, separate, spatial descriptive attributes, starting from the closeness among consumers. Its computation consists of the following main steps:

1. Compute a truncated squared distance matrix, as follows:

$$D^* = \begin{cases} dist(c_a, c_b)^2 \ if dist(c_a, c_b) \leq maxDist \\ 4 \cdot maxDist \ otherwise \end{cases} \tag{6}$$

where $maxDist$ is a user-defined threshold.

2. Perform the Principal Coordinate Analysis (PCoA) [8] on D^*. This analysis consists in the diagonalization of Δ, where:

$$\Delta = -\frac{1}{2}\left(I - \frac{1 \cdot 1^\top}{n}\right) D^* \left(I - \frac{1 \cdot 1^\top}{n}\right) \tag{7}$$

with I be the identity matrix, and 1 be a vector of 1s. After diagonalization, the principal coordinates are obtained by scaling each eigenvector of Δ by the square root of its correspondent eigenvalue. Note that the eigenvalues can be either positive or negative. Eigenvectors associated with high positive (resp., negative) eigenvalues represent a high positive (resp., negative) autocorrelation. Since we are interested in considering only positive spatial autocorrelation phenomena (i.e., spatially close consumers with similar behaviors, rather than spatially distant consumers with similar behaviors), only eigenvectors corresponding to positive eigenvalues are kept and used as spatial descriptors.

Henceforth, the settings that exploit the spatial dimension will be distinguished through **LISA** or **PCNM**, appended to their name. In Fig. 3, a graphical overview of all the proposed learning setting is provided, where the temporal, the spatial or both temporal and spatial dimensions are considered.

4 Experiments

In this section, we describe the considered real-world dataset and the experimental setting. Then, we show and discuss the obtained results.

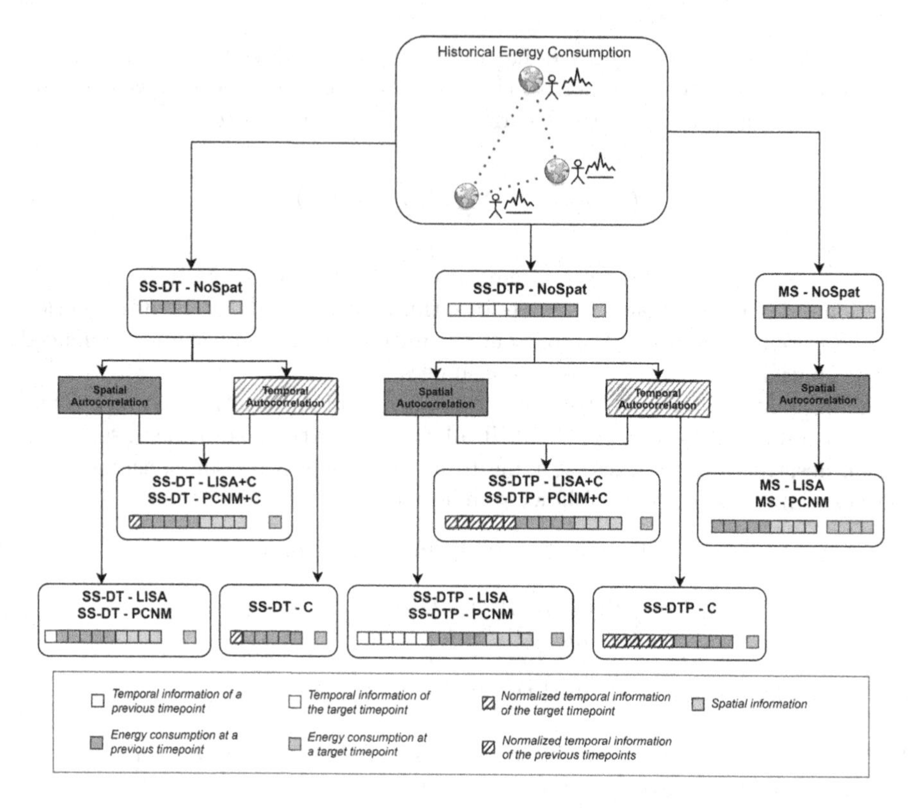

Fig. 3. A graphical overview of all the proposed learning setting is provided, where the temporal, the spatial or both temporal and spatial dimensions are considered.

4.1 Experimental Setting

We considered a dataset of an electrical network of a small city in the South of Italy consisting of 159 customers. Each customer is associated with the geographic coordinates (latitude and longitude) of the energy substation he/she is connected to in the network. The dataset consists of energy consumption data (in kWh) collected every month for a period of 10 years, i.e., from 2010 to 2019. Following a cross-validation setting for time-series, we iteratively consider each year from 2012 to 2019 as target year (see the quantitative information of the dataset in Table 1), with the goal of predicting the energy consumption for all the months of the target year, for all the customers of the network.

We performed the experiments with all the settings proposed in Fig. 3, to properly assess the contribution coming from the specific strategy adopted to take into account temporal and/or spatial autocorrelation phenomena. For LISA, we computed 12 indexes, one for each descriptive variable representing previous consumptions. For PCNM, we extracted 15 eigenvectors, following the experimental results reported in [5]. For both, the threshold $maxDist$ was set to 0.3 km, which is adequate in the context of a small city. As regressors, as introduced

Table 1. Quantitative information of each fold of the considered dataset.

Fold	Training period	Testing period	SS training instances	MS training instances
1	2010–2011	2012	1,908	159
2	2010–2012	2013	3,816	318
3	2010–2013	2014	5,724	477
4	2010–2014	2015	7,632	636
5	2010–2015	2016	9,540	795
6	2010–2016	2017	11,448	954
7	2010–2017	2018	13,356	1,113
8	2010–2018	2019	15,264	1,272

in Sect. 3.1, we considered Linear Regression (**LR**), Regression Trees (**RT**) and Random Forests (**RF**), available in *scikit-learn*. All the regressors were run with the default values for their parameters, except for the regression trees, for which we performed a grid search for the pruning parameter *ccp_alpha* $\in \{0.2, 0.5, 1.0\}$. In Sect. 4.2, we report the best obtained results (i.e., with *ccp_alpha* $= 1.0$).

As state-of-the-art competitor, we considered **GMAN** [20], a recently proposed neural network that is able to capture both spatial and temporal dimensions, through attention mechanisms, and of performing multi-step predictions. We adapted GMAN so that the temporal embedding encodes the month of each time-step, instead of the day and the hour, as in its original implementation. We also optimized its user-defined threshold ϵ on the spatial closeness, considering $\epsilon = 0.1$ (as suggested in [20]) and $\epsilon = 0.05$. In Sect. 4.2, we report only the best obtained results (i.e., with $\epsilon = 0.05$). Note that GMAN also performs a tuning phase on a validation set. Therefore, for this method, the results on the first fold are not available, since it requires data of an additional year as validation set.

As evaluation measure, we adopted the Relative Squared Error (RSE), which, contrary to other common measures like the RMSE, allows us to evaluate the predictive accuracy with respect to a simple predictor based on the average: a RSE close to 0.0 (resp. 1.0) means that the model has a perfect predictive accuracy (resp., equivalent to that of the simple average predictor), while a RSE over 1.0 means that the model is worse than the simple predictor. Formally, $RSE = \frac{\sum_t (r^t - \widetilde{r}^t)^2}{\sum_t (r^t - \overline{r})^2}$, where r^t and $\widetilde{r}^t$ are the true and the predicted values, respectively, for the t-th time-step, and $\overline{r}$ is the average value in the dataset.

4.2 Results and Discussion

In Table 2, we show the RSE result for each testing fold (target year), obtained by the considered regressors in the proposed settings, and by the competitor GMAN. We recall that the results of the first fold (2012) for GMAN are not available because it requires an additional year of data for its validation phase. Moreover, we do not report the results obtained in some settings of the LR (i.e., SS-DTP

NoSpat, LISA and PCNM), since it was not able to fit a proper model (i.e., RSE > 10) with the small amount of available training data for the first fold.

Looking at Table 2, we can make several observations. First, for the years 2012 and 2013, the RSE values appear quite high. This is due to the scarce availability of training data for these folds (see Table 1). An exception is represented by the results obtained by MS, especially in the settings MS+PCNM and MS+LISA, that achieved good results also for these years. This may be due to the fact that the poor availability of historical data has been compensated by the captured dependencies among different time-steps and by the exploitation of the spatial information. Note that MS+PCNM appeared to be the setting that provided the best results overall for most of the years. Focusing on the regressors, the adoption of RF generally provided the best results in most of the settings, and when learned from the MS+PCNM setting, it led to the best absolute results. Note that, as emphasized in Sect. 3.1, learning methods for the induction of multi-step RTs and RF simultaneously optimize the construction of the model by considering all the time-steps. The capability of RF to reduce the variance in the predictions with respect to RT provided further improvements.

Looking at the results obtained by the considered state-of-the-art competitor GMAN, we can notice that, besides not being able to make predictions for the year 2012, the obtained RSE for the 2013 is very high, and quite close to the average baseline for the 2014. The RSE values become more acceptable for the subsequent years, but still higher than those achieved by the approaches proposed in this paper. These results prove that the approaches proposed in this paper to capture temporal and spatial autocorrelation phenomena are very effective with respect to those adopted by GMAN, and confirm the limitation of deep neural network architectures when the available training data is poor.

Overall, the strongest contribution appears to come from the MS setting. This observation is also clear from the average results shown in Fig. 4, where we can easily observe that the charts related to MS generally appear the lowest ones (i.e., with the lowest RSE), independently on the regressor. This confirms that the temporal dimension (and, especially, temporal autocorrelation phenomena) is fundamental for the prediction of the energy consumption in smart grids, and that capturing dependencies between different target time-steps provides higher advantages than explicitly representing the temporal information in the descriptive attributes, as done in the ST-DTP setting, and than adopting the radial temporal closeness (C). We further stress this aspect by observing the line charts in Fig. 5, where we plot the average RSE per month obtained by the best configurations according to Fig. 4 for each pair of setting (MS, SS-DT, SS-DTP) and regressor. From Fig. 5, we can observe that GMAN generally achieved an average high RSE, and that the MS setting led to more stable errors over the months of the year. This is due to its capability of capturing possible dependencies among the months of the year, and to avoid the propagation of errors introduced by recursive approaches. An interesting case is observable in the period April-May, where the highest prediction errors are made by almost all the approaches, probably due to the abrupt climatic changes that often happen

Table 2. Results in terms of RSE for each testing fold. The best result for each regressor (sub-table) and fold (column) is emphasized in bold, while the best result overall for each fold (column) is emphasized in bold with a gray background.

			2012	2013	2014	2015	2016	2017	2018	2019
GMAN			-	9.640	0.840	0.650	0.370	0.280	0.270	0.359
Linear Regression	SS-DT	NoSpat	0.364	0.719	0.263	0.302	0.219	0.245	0.130	0.242
		LISA	0.366	0.726	0.264	0.305	0.221	0.251	0.133	0.243
		PCNM	0.392	0.757	0.279	0.301	0.225	0.247	0.131	0.252
		C	0.363	0.715	0.266	0.298	0.219	**0.241**	0.130	0.240
		LISA+C	0.364	0.720	0.266	0.302	0.221	0.247	0.133	0.241
		PCNM+C	0.391	0.752	0.281	0.298	0.224	0.243	0.130	0.249
	SS-DTP	NoSpat	-	**0.712**	0.260	0.289	0.214	0.249	**0.127**	**0.235**
		LISA	-	0.718	**0.258**	0.293	0.214	0.253	0.129	0.236
		PCNM	-	0.750	0.274	**0.288**	0.219	0.252	0.128	0.245
		C	0.402	**0.712**	0.261	0.290	0.214	0.250	**0.127**	0.236
		LISA+C	**0.359**	0.717	0.260	0.292	0.214	0.254	0.129	0.236
		PCNM+C	0.382	0.750	0.274	0.289	0.219	0.252	0.128	0.245
	MS	NoSpat	0.384	0.792	0.324	0.312	**0.205**	0.298	0.132	0.276
		LISA	0.417	0.862	0.332	0.330	0.235	0.333	0.136	0.284
		PCNM	0.394	0.826	0.350	0.319	0.214	0.302	0.134	0.282
Regression Trees	SS-DT	NoSpat	0.737	**0.805**	0.464	0.569	0.750	0.672	0.317	0.405
		LISA	0.774	1.866	0.540	0.430	0.619	0.915	0.295	0.448
		PCNM	0.820	0.885	0.428	0.412	0.432	0.688	0.302	0.374
		C	0.462	1.386	0.414	0.456	0.447	0.444	0.341	0.445
		LISA+C	0.437	1.429	0.527	0.472	0.469	0.684	0.351	0.390
		PCNM+C	0.492	1.634	0.446	0.521	0.380	0.766	0.392	0.366
	SS-DTP	NoSpat	0.624	1.361	0.500	0.631	0.480	0.377	0.336	0.475
		LISA	0.504	1.273	0.608	0.464	0.901	0.833	**0.274**	0.413
		PCNM	0.998	1.411	**0.397**	1.577	0.842	0.866	0.440	0.587
		C	0.851	1.072	0.556	0.689	0.694	0.562	0.332	0.504
		LISA+C	0.880	1.462	0.475	0.571	0.498	0.627	0.391	0.388
		PCNM+C	0.794	0.854	0.520	0.779	0.575	0.455	0.295	0.587
	MS	NoSpat	**0.364**	1.026	0.425	0.463	0.349	0.436	0.630	0.448
		LISA	0.443	1.096	0.732	0.502	0.307	**0.366**	0.323	0.435
		PCNM	0.460	0.984	0.454	**0.390**	**0.303**	0.467	0.337	**0.348**
Random Forests	SS-DT	NoSpat	0.300	0.893	0.336	0.307	0.197	0.572	0.132	0.251
		LISA	0.296	0.915	0.345	0.302	0.215	0.570	**0.127**	0.251
		PCNM	0.336	0.855	0.344	0.305	**0.188**	0.564	0.130	0.256
		C	0.320	0.912	0.326	0.310	0.211	0.553	0.135	0.244
		LISA+C	0.305	0.902	0.330	0.311	0.226	0.565	0.133	0.249
		PCNM+C	0.332	0.889	0.315	0.304	0.200	0.555	0.133	0.244
	SS-DTP	NoSpat	0.297	0.882	0.312	0.293	0.197	0.587	0.134	0.248
		LISA	0.302	0.876	0.333	0.306	0.217	0.595	0.131	0.251
		PCNM	0.331	0.869	0.324	0.297	0.196	0.573	0.133	0.248
		C	0.320	0.883	0.316	0.296	0.193	0.588	0.135	0.246
		LISA+C	0.303	0.904	0.327	0.294	0.208	0.592	0.128	0.247
		PCNM+C	0.325	0.855	0.320	0.298	0.199	0.568	0.132	0.249
	MS	NoSpat	0.262	0.578	0.254	0.277	0.195	**0.219**	0.148	0.234
		LISA	0.263	**0.520**	0.291	0.286	0.200	0.226	0.147	**0.229**
		PCNM	**0.259**	0.534	**0.253**	**0.270**	0.197	**0.219**	0.148	0.236

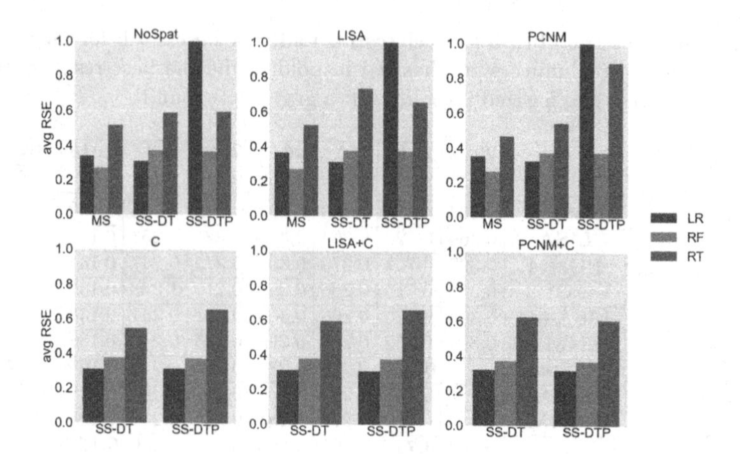

Fig. 4. Results in terms of average RSE. For readability, the results of LR in the upper part are graphically truncated to 1.0 (but they are actually around 1.5).

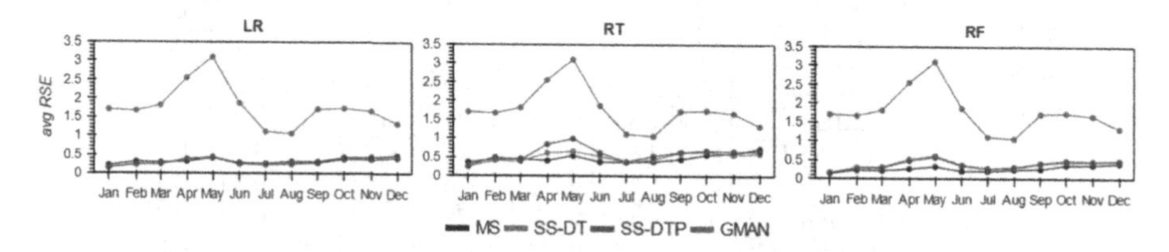

Fig. 5. RSE results averaged over the years for each month, obtained by the best configurations (see Fig. 4) for each pair of setting (MS, SS-DT, SS-DTP) and regressor.

in the South of Italy during such a period. On the other hand, the settings based on MS are able to provide accurate predictions also in these cases.

5 Conclusion

In this paper, we proposed different approaches to take into account temporal and spatial autocorrelation phenomena while learning forecasting models for the prediction of the energy consumption in smart grids. For the temporal dimension, we investigated the contribution of the explicit representation of temporal information related to historical measurements, also through the temporal radial closeness, and that of predicting the value for multiple future time-steps simultaneously. For the spatial dimension, we investigated the contribution coming from the injection of LISA indexes and eigenvectors computed through the PCNM.

The experiments proved the overall superiority of models learned in the multi-step predictive setting, and the positive contribution coming from the PCNM, also when the available training data are scarce. The learned models also significantly outperformed the considered state-of-the-art competitor GMAN, which is based on a multi-attention neural network architecture.

For future work, we will consider the adoption of the proposed strategies for short-term predictions, in a nowcasting environment, and the integration of transfer learning techniques [11] to further improve the predictive accuracy when the available data related to a specific geographic area are poor.

References

1. Anselin, L.: Local indicators of spatial association – LISA. Geogr. Anal. **27**(2), 93–115 (1995)
2. Berriel, R.F., Lopes, A.T., Rodrigues, A., Varejao, F.M., Oliveira-Santos, T.: Monthly energy consumption forecast: a deep learning approach. In: 2017 International Joint Conference on Neural Networks (IJCNN), pp. 4283–4290. IEEE (2017)
3. Bontempi, G., Ben Taieb, S.: Conditionally dependent strategies for multiple-step-ahead prediction in local learning. Int. J. Forecast **27**(3), 689–699 (2011)
4. Box, G.E.P., Jenkins, G.M., Reinsel, G.C., Ljung, G.M.: Time Series Analysis: Forecasting and control, 5th edn. Wiley (2015)
5. Ceci, M., Corizzo, R., Fumarola, F., Malerba, D., Rashkovska, A.: Predictive modeling of PV energy production: how to set up the learning task for a better prediction? IEEE Trans. Ind. Inf. **13**(3), 956–966 (2016)
6. Cabral, D.A., Legey, J., et al.: Electricity consumption forecasting in Brazil: a spatial econometrics approach. Energy **126**, 124–131 (2017)
7. Dray, S., Legendre, P., Peres-Neto, P.R.: Spatial modelling: a comprehensive framework for principal coordinate analysis of neighbour matrices (PCNM). Ecol. Model. **196**(3–4), 483–493 (2006)
8. Gower, J.C.: Some distance properties of latent root and vector methods used in multivariate analysis. Biometrika **53**(3–4), 325–338 (1966)
9. Lai, G., Chang, W.C., Yang, Y., Liu, H.: Modeling long-and short-term temporal patterns with deep neural networks. In: The 41st International ACM SIGIR Conference on Research & Development in Information Retrieval, pp. 95–104 (2018)
10. Masum, S., Liu, Y., Chiverton, J.: Multi-step time series forecasting of electric load using machine learning models. In: Rutkowski, L., Scherer, R., Korytkowski, M., Pedrycz, W., Tadeusiewicz, R., Zurada, J.M. (eds.) ICAISC 2018. LNCS (LNAI), vol. 10841, pp. 148–159. Springer, Cham (2018). https://doi.org/10.1007/978-3-319-91253-0_15
11. Mignone, P., Pio, G.: Positive unlabeled link prediction via transfer learning for gene network reconstruction. In: Ceci, M., Japkowicz, N., Liu, J., Papadopoulos, G.A., Raś, Z.W. (eds.) ISMIS 2018. LNCS (LNAI), vol. 11177, pp. 13–23. Springer, Cham (2018). https://doi.org/10.1007/978-3-030-01851-1_2
12. Ohtsuka, Y., Oga, T., Kakamu, K.: Forecasting electricity demand in Japan: a Bayesian spatial autoregressive ARMA approach. Comput. Stat. Data Anal. **54**(11), 2721–2735 (2010)
13. Serafino, F., Pio, G., Ceci, M.: Ensemble learning for multi-type classification in heterogeneous networks. IEEE Trans. Knowl. Data Eng. **30**(12), 2326–2339 (2018)
14. Shih, S.-Y., Sun, F.-K., Lee, H.: Temporal pattern attention for multivariate time series forecasting. Mach. Learn. **108**(8), 1421–1441 (2019). https://doi.org/10.1007/s10994-019-05815-0
15. Stojanova, D., Ceci, M., Appice, A., Džeroski, S.: Network regression with predictive clustering trees. In: Gunopulos, D., Hofmann, T., Malerba, D., Vazirgiannis, M. (eds.) ECML PKDD 2011. LNCS (LNAI), vol. 6913, pp. 333–348. Springer, Heidelberg (2011). https://doi.org/10.1007/978-3-642-23808-6_22

16. Taieb, S.B., Bontempi, G., Atiya, A.F., Sorjamaa, A.: A review and comparison of strategies for multi-step ahead time series forecasting based on the NN5 forecasting competition. Expert Syst. Appl. **39**(8), 7067–7083 (2012)
17. Tobler, W.R.: A computer movie simulating urban growth in the detroit region. Econ. Geogr. **46**(sup1), 234–240 (1970)
18. Williams, K.T., Gomez, J.D.: Predicting future monthly residential energy consumption using building characteristics and climate data: a statistical learning approach. Energy Build. **128**, 1–11 (2016)
19. Wu, Z., Pan, S., Long, G., Jiang, J., Zhang, C.: Graph WaveNet for deep spatial-temporal graph modeling. arXiv preprint arXiv:1906.00121 (2019)
20. Zheng, C., Fan, X., Wang, C., Qi, J.: GMAN: a graph multi-attention network for traffic prediction. In: AAAI 2020, vol. 34 , no. 01, pp. 1234–1241 (2020)

Elastic Product Quantization for Time Series

Pieter Robberechts[(✉)], Wannes Meert, and Jesse Davis

Department of Computer Science, Leuven.AI, KU Leuven, B-3000 Leuven, Belgium
{pieter.robberechts,wannes.meert,jesse.davis}@kuleuven.be

Abstract. Analyzing numerous or long time series is difficult in practice due to the high storage costs and computational requirements. Therefore, techniques have been proposed to generate compact similarity-preserving representations of time series, enabling real-time similarity search on large in-memory data collections. However, the existing techniques are not ideally suited for assessing similarity when sequences are locally out of phase. In this paper, we propose the use of product quantization for efficient similarity-based comparison of time series under time warping. The idea is to first compress the data by partitioning the time series into equal length sub-sequences which are represented by a short code. The distance between two time series can then be efficiently approximated by pre-computed elastic distances between their codes. The partitioning into sub-sequences forces unwanted alignments, which we address with a pre-alignment step using the maximal overlap discrete wavelet transform (MODWT). To demonstrate the efficiency and accuracy of our method, we perform an extensive experimental evaluation on benchmark datasets in nearest neighbors classification and clustering applications. Overall, the proposed solution emerges as a highly efficient (both in terms of memory usage and computation time) replacement for elastic measures in time series applications.

1 Introduction

Data mining applications on large time series collections are constrained by the computational cost of similarity comparisons between pairs of series and memory constraints on the processing device. The general approach to overcome these constraints is to first apply a transformation that produces a compact representation of the time series that retains it's main characteristics. Many techniques have been proposed to generate such representations for time series analysis, including techniques based on Discrete Fourier Transform (DFT) [6], Discrete Wavelet Transform (DWT) [3], Singular Value Decomposition (SVD) [2], and segmentation [9].

Most of these techniques are based on the Euclidean distance as the metric for similarity. However, there are some cases where the Euclidean distance, and lock-step measures in general, may not be entirely adequate for estimating

P. Pascal and D. Ienco (Eds.): DS 2022, LNAI 13601, pp. 157–172, 2022.
https://doi.org/10.1007/978-3-031-18840-4_12

similarity [18]. The reason is that the Euclidean distance is sensitive to distortions along the time axis. To avoid this problem, similarity models should allow some elastic shifting of the time dimension to detect similar shapes that are not locally aligned. This is resolved by elastic measures such as Dynamic Time Warping (DTW) [22]. The ability to accommodate temporal aberrations comes at a cost however, as the standard dynamic programming approach for computing the DTW measure has a quadratic computational complexity.

To accommodate efficient *approximate* nearest neighbor (NN) search in large datasets under time warping, Zhang et al. [29] proposed to adapt the technique of product quantization (PQ) [8]. In the standard feature-vector case, PQ is extremely performant for approximate NN search based on Euclidean distance. Its core idea is to (1) partition the vectors into disjoint subspaces, (2) cluster each subspace independently to learn a codebook of centroids, (3) re-represent each vector by a short code composed of its indices in the codebook, and (4) efficiently conduct the search over these codes using look-up tables. This has the dual benefits of greatly shrinking the memory footprint of the training set while simultaneously dramatically reducing the number of computations needed.

Unfortunately, naively combining the conventional PQ with DTW results in both missed and unwanted alignments. First, PQ segments a time series and the optimal alignment can cross segments. Second, applying DTW on a segment forces an unwanted alignment at the beginning and end of a segment. Zhang et al. [29] got round these challenges with a *filter-and-refine* post-processing step that calculates the exact DTW distances between the best candidate time series to filter out the erroneous alignments. However, this *increases* the memory footprint of NN-DTW (as both the original training set and a compact representation have to be retained) and does not solve the problem of false dismissals due to missed alignments. Instead, we propose a pre-alignment step using the maximal overlap discrete wavelet transform (MODWT). The resulting method is an approximate DTW method that is fast in its own right while it can still benefit from previous advances in speeding-up DTW, such as constraint bands and pruning strategies.

The contributions of this paper are as follows: (1) We bridge the gap between PQ and DTW, introducing a pre-alignment step to minimize the effect of segmentation that is part of PQ on DTW; (2) We show how our method is compatible with and speeds up tasks such as nearest neighbours and clustering; (3) We demonstrate empirically the utility of our approach by comparing it to the most common distance measures on the ubiquitous UCR benchmarks.

2 Background

A large body of literature is available on dynamic time warping and product quantization. In this section, we restrict our presentation to the notations and concepts used in the rest of the paper.

2.1 Dynamic Time Warping

Dynamic Time Warping (DTW) computes the distance between two time series A and B after optimal alignment [22]. The alignment is computed using (1) dynamic programming with

$$\text{dtw\_dist}[i,j] = (A_i - B_j)^2 + \min \begin{cases} \text{dtw\_dist}[i-1,j-1] \\ \text{dtw\_dist}[i,j-1] \\ \text{dtw\_dist}[i-1,j] \end{cases}$$

where i and j are indices for A and B, and (2) the minimum cost path in the matrix dtw_dist$[i,j]$. The value $DTW(A,B) = \text{dtw\_dist}[\text{length}(A), \text{length}(B)]$ is the distance between series A and B. DTW is particularly useful for comparing the shapes of time series, as it compensates for subtle variations such as shifts, compression and expansion.

The standard dynamic programming approach for computing the DTW measure has a quadratic computational complexity. Four common approaches exist to enable scaling to large datasets: constraint bands or warping windows [22], lower-bound pruning [10–12,20,24,27], pruning warping alignments [25], and DTW approximations [23,26]. While proposed methods across these four approaches reduce the theoretical complexity of DTW down from quadratic, they incur other costs. Constraining bands decreases accuracy, especially in domains with large distortions. Lower-bound pruning is increasingly efficient for larger datasets, but cannot be applied for tasks such as clustering since it requires the complete pair-wise distance matrix. Pruning warping alignments is a valuable addition but has limited impact when many suitable warping paths exist. Approximate methods introduce an additional complexity that requires more memory and loses the computational simplicity of the original DTW algorithm making it often slow in practice [28]. In contrast, our proposed approach is fast in its own right while maintaining compatibility with the aforementioned techniques.

2.2 Product Quantization

Product Quantization (PQ) [8] is a well-known approach for approximate nearest neighbors search for standard feature-vector data using the Euclidean distance. It confers two big advantages. One, it can dramatically compress the size of the training set, which enables storing large datasets in main memory. Two, it enables quickly computing the approximate Euclidean distance between a test example and each training example.

PQ compresses the training data by partitioning each feature vector used to describe a training example into M equal sized groups, termed *subspaces*. It then learns a *codebook* for each subspace. Typically, this is done by running k-means clustering on each subspace which only considers the features assigned to that subspace. Then the values of all features in the subspace are replaced by a single v-bit code representing the id of the cluster centroid c_k that the example is assigned to in the current subspace. Hence, each example is re-represented by M v-bit code words. This mapping is termed the *quantizer*.

At test time, finding a test example's nearest neighbor using the squared Euclidean distance can be done efficiently by using table look-ups and addition. For a test example, a look-up table is constructed for each subspace. This table stores the squared Euclidean distance to each of the K cluster centroids in that subspace. Then the approximate distance to each training example is computed using these look-up tables and the nearest example is returned.

3 Approximate Dynamic Time Warping with Product Quantization

This section introduces the DTW with Product Quantization (PQDTW) approach [29]. First, we discuss how each component of the original PQ method can be adapted to a DTW context. This encompasses learning the codebook, encoding the data and computing approximate distances between codes. Then, we extend the base method to compensate for the alignment loss caused by partitioning the time series. Finally, we explain how PQDTW can be used in NN search and clustering.

3.1 Training Phase

The training phase comprises the learning of a codebook using DTW Barycenter Averaging (DBA) k-means [19]. Let us consider a training set of time series $X = [x_1, x_2, \ldots, x_N] \in \mathbb{R}^{N \times D}$ (i.e., N time series of length D). Each time series in the dataset is first partitioned in M sub-sequences, each of length D/M. Subsequently, a sub-codebook for each $m \in \{1, \ldots, M\}$ is computed: $C^m = \{c_k^m\}_{k=1}^K$, with centroids $c_k^m \in \mathbb{R}^{D/M}$. Each C_m is obtained by running the DBA k-means clustering over the mth part of the training sequences, where K is the number of clusters. The $K \times M$ centroids obtained by k-means represent the most commonly occurring patterns in the training time series' subspaces.

Two other pre-processing steps can be performed during the training phase to speed-up the encoding of time series and computing the symmetric distances between codes: the construction of the Keogh envelopes [10] of all centroids and the computation of a distance look-up table with the DTW distance between each pair of centroids in a subspace. These steps are explained in the next sections.

The training step has to be performed once to optimize it for a specific type of data, but can be reused to speed up computations on future examples from the same domain.

3.2 Encoding Time Series

Using the codebook, we can represent any time series as a short code. The idea is to (1) partition a time series into sub-sequences (2) independently encode each subsequence to an identifier, and (3) re-represent the time series as a concatenation of the identifiers. A given time series $x \in \mathbb{R}^D$ is therefore mapped as:

$$x \to \left[q_1(x_1, \ldots x_{D/M}), \ldots, q_M(x_{D-D/M+1}, \ldots, x_D)\right],$$

where $q_m : \mathbb{R}^{D/M} \to \{1, \ldots, K\}$ is the quantizer associated with the mth subspace that maps a sub-sequence

$$x^m = \left(x_{m\times(D/M))}, \ldots x_{(m+1)\times(D/M)}\right)$$

to the identifier of the nearest centroid c_k^m in the codebook. Formally, this search is defined as:

$$q_m(x^m) = arg\,min_{k\in\{1,\ldots,K\}} \text{DTW}\left(x^m,\ c_k^m\right).$$

Practically, this search is performed by linearly comparing a D/M-dimensional sub-sequence to K centroids (i.e., a NN-DTW query), which has a computational complexity of $O(K \times (D/M)^2)$ with standard DTW. Since the sub-sequences are quantized separately using M distinct quantizers, the overall computational complexity of encoding a time series is $O(K \times D^2/M)$.

The quadratic complexity of the dynamic programming approach to DTW makes the NN-DTW queries required to encode a time series highly computationally demanding. In regular NN-DTW, cheap-to-compute lower bounds are a key strategy to combat this by pruning the expensive DTW computations of unpromising nearest neighbour candidates [20]. This involves computing an enclosing envelope around the query (i.e., the sub-sequence to be encoded) which is reused to compute the actual lower bound between the query and each test time series. This would be inefficient in our use case, since it would require the construction of the envelope every time the PQ is used to encode a time series. Therefore, we reverse the query/data role in the lower bound search [20]. This enables computing the envelopes only once around the codebook, which can be done during the training phase.

Many DTW lower bounds have been proposed, including LB Kim [11], LB Keogh [10], LB Improved [12], LB New [24] and LB Enhanced [27]; as well as cascading lower bounds that start with a looser (and computationally cheap) one and progress towards tighter lower bounds [20]. In the experimental evaluation of this paper, we use a cascading lower bound of LB Kim and the reversed LB Keogh, which provides an effective trade-off between speed and tightness for small window sizes (i.e., as we obtain after partitioning the time series) [27]. Nevertheless, other bounds might be more effective depending on the time series' properties. Given the pre-computed upper and lower envelopes at training time, the cost of computing these bounds is only $O(D/M)$.

While this encoding is the most costly part of the PQDTW algorithm, it can be executed offline in many applications. For example, in NN search, the dataset can be encoded during the training phase and the costs can be amortized over multiple subsequent queries.

3.3 Computing Distances Between Time Series

Consider two time series x and $y \in \mathbb{R}^D$ and a trained product quantizer q. The original PQ paper [8] proposes two methods to estimate the distance between these two time series: symmetric and asymmetric distance computation.

Symmetric Distance. This method computes the distance between the PQ-codes of x and y. Therefore, both time series are first encoded by the product quantizer. Secondly, a distance score is computed by fetching the centroid distances from q's pre-computed distance table. This means that the distances between the centroids c_i^m and c_j^m of x and y in each subspace m need to be aggregated into one distance d as:

$$\hat{d}(x,y) = d(q(x),\ q(y)) = \sqrt{\sum_{m=1}^{M} d(c_i^m,\ c_j^m)^2}.$$

The distances $d(c_i^m,\ c_j^m)^2$ between each pair of centroids in a subspace are pre-computed during the training phase and stored in a M-by-K-by-K look-up table. Hence, symmetric distance computation is very efficient, taking only $O(M)$ table look-ups and additions.

Asymmetric distance. This method encodes only one of the two time series and estimates the distance between the PQ code of x and the original series y as

$$\hat{d}(x,y) = d(q(x),\ y) = \sqrt{\sum_{m=1}^{M} d(c_i^m,\ y^m)^2}.$$

The distances $d(c_i^m,\ y^m)^2$ between the M centroids of x and the M subspaces of y have to be computed on-the-fly. Hence, this method is inefficient to compute the distances between a single pair of time series. However, when computing the distances between a query time series y and a database with many time series $X = \{x_n\})_{n=1}^N$, it becomes efficient to first construct a distance look-up table for each pair $(y^m,\ c_i^m)$ with $m \in \{1,\ldots,M\}$ and $i \in \{1,\ldots,K\}$. The computation of this look-up table takes $O(D \times K)$ DTW computations, and is performed just once per query. Subsequently, the distance computation itself takes only $O(M)$ table look-ups and additions.

Whether symmetric or asymmetric is the most appropriate distance measure depends on the application, the amount of data and the required accuracy.

3.4 Memory Cost

Due to the conversion of time series to short codes, product quantization allows large time series collections to be processed in memory. Since these codes consist of M integers ranging from 1 to K, the parameters M and K control the main memory cost of our approach. Typically, K is set as 256, such that each code can be represented by $8M$ bits (i.e., each integer in the code is represented by 8 bits). If a D-dimensional time series is represented in single-precision floating-point format (i.e., $32 \times D$ bits), a PQ-code with $K = 256$ compresses the original series by a factor $32D/8M = 4D/M$. For example, time series of length 140 can be represented $80\times$ more efficiently by PQ-codes with 7 subspaces. Larger values for M lead to faster performance and a higher memory cost, but the effect on representation error is domain-dependent.

In addition to the data memory cost, our method requires a small amount of additional memory for storing the codebook ($32 \times D \times K$ bits), pre-computed distance look-up table ($32 \times K^2 \times M$ bits) and Keogh envelopes ($2 \times 32 \times D \times K$ bits). In total, this corresponds to $32 \times K \times (3 \times D + K \times M)$, which is negligible in relation to the data memory cost. For our previous example, with $D = 140$, $K = 256$ and $M = 7$, the total cost is limited to 2.3 MB.

3.5 Pre-alignment of Subspaces

When partitioning time-series into equal length sub-sequences, the endpoints of these sub-sequences might not be aligned well. This problem is illustrated in Fig. 1. The middle row of the plot shows the subspaces obtained by dividing two similar time series of the Trace dataset [4] in four equal partitions. Notice that the distinctive peak near the first split point falls in different subspaces for both series. Because it is not possible to warp across segments, the effect is that the location of the split point will tend to contribute disproportionately to the estimated similarity, resulting in a higher approximate distance. In this section we introduce a pre-alignment step to deal with this problem.

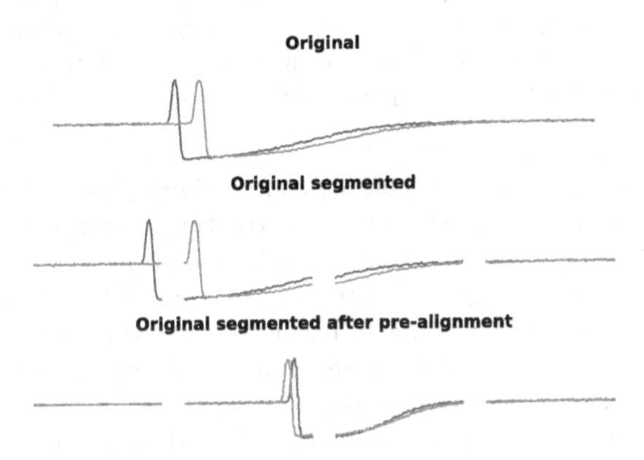

Fig. 1. Segmentation of two time series of the Trace [4] data set (top) in four subspaces using fixed-length subspaces (middle) and our own MODWT-based method (bottom).

The idea is to identify local structures in the time series data and segment the time series at the boundaries of these structures. For this, we use the Maximal Overlap Discrete Wavelet Transform (MODWT), as proposed by Hong et al. [7]. Via convolution of a raw time series $x \in \mathbb{R}^D$ and the basis functions (Haar wavelet) of the MODWT, we obtain the scale coefficients $c_{j,i}$, where j is the level of the decomposition $\in \{1, \ldots, J\}$ and $i \in \{1, \ldots, D\}$. These coefficients are proportional to the mean of the raw time series data. The scale coefficients of the MODWT have a length D that is the same as that of the raw time series.

Next, time segment points are extracted as the points at which the signs of the differences between the time series data and scale coefficients change, as shown in Fig. 2. Since the complexity of MODWT is only $O(J \times D)$, this segmentation step does not increase our method's overall complexity.

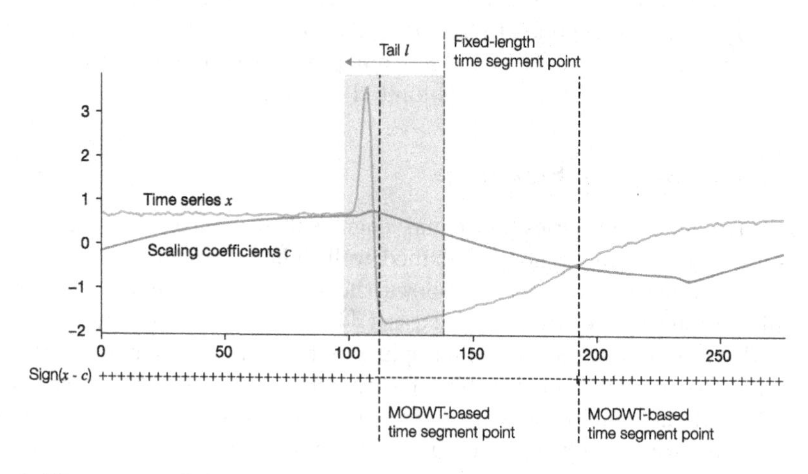

Fig. 2. Illustration of the segmentation procedure based on the pre-alignment of time series. MODWT-based time segment points are extracted as the points at which the signs of the differences between the time series data and scale coefficients change. If the tail of one of the fixed length segment points contains one of these MODWT-based segment points, the MODWT-based point is used instead.

Since one time series archive typically contains series with different patterns, partitioning the time series based on local structures implies that the number of subspaces can vary between sequences. If this happens, the distances between two such series cannot be approximated by the sum of the distances between their subspaces. Therefore, a degree of fuzziness is needed to cut the time series such that each series is split in the same number of segments of similar length while avoiding splits of distinctive local structures. This is achieved by specifying a tail, t, measured backwards from each original fixed-length split point l, within which the cut should be applied; thus the cut will fall between $l - t$ and l. If the MODWT method identifies split points in this period, the right-most point is used to split the series; otherwise, the l remains the original split point. Hence, we obtain subspaces varying in length between l and $l + t$. This is illustrated in Fig. 2. Finally, we re-interpolate the obtained segments to have the same length $l + t$ [15], which is required to be able to pre-compute the Keogh envelopes.

4 Data Mining Applications

Similarity comparisons between pairs of series are a core subroutine in most time series data mining approaches. In this section, we discuss how PQDTW can be incorporated in 1-NN and hierarchical clustering.

4.1 NN Search with PQ Approximates

A large body of empirical research has shown that NN-DTW is the method of choice for most time series classification problems [5,18,27]. However, being a lazy learner, the main drawback is its time and space complexity. The entire training set has to be stored and the classification time is a function of the size of this training set. Therefore, NN-DTW still has severe tractability issues in some applications. This is especially true for resource-constrained devices such as wearables.

NN search with PQ is both fast (only the query has to be encoded online and only M additions are required per distance calculation) and reduces significantly the memory requirements for storing the training data. It proceeds as follows: Given a query time series $y \in \mathbb{R}^D$ and a database of N time series $X = \{x_n\}_{n=1}^N$ where each $x_n \in \mathbb{R}^D$. At training time, a PQ is first trained on (a subset of) the database time series, which are subsequently encoded (Sect. 3.2) using the trained PQ. At prediction time, the approximate distance is computed between the query vector y and each encoded database time series. In most cases, one should use the asymmetric version, which obtains a lower distance distortion. When the training set is large, the $O(D \times K)$ DTW computations required to compute the asymmetric distance look-up table is relatively small. The only exceptions are queries in small databases (albeit, techniques other than PQ are more appropriate in such cases) or applications with many time-critical queries.

The linear scan with PQDTW is fast compared to the state-of-the-art NN-DTW methods [20], but still slow for a large number of N. To handle million-scale search, a search system with inverted indexing was developed in the original PQ paper [8].

4.2 Clustering with PQ Approximates

While all of the conventional clustering approaches rely on similarity comparisons in which DTW can be substituted by PQDTW, we focus on the hierarchical algorithms in this paper. These have great visualization power in time series clustering and do not require the number of clusters as an additional parameter. However, at the same time, hierarchical clustering does not scale to large datasets, because it requires the computation of the full pairwise distance matrix. Therefore, lower-bound pruning cannot be applied.

For constructing a pairwise distance matrix, asymmetric distance computation is an expensive operation since it involves the computation of the full DTW distance matrix between the subspaces of each time series in the dataset and the codebook (i.e., $N \times K \times M$ DTW computations). This is only acceptable if the number of time series is a lot bigger than the number of centroids in the codebooks.

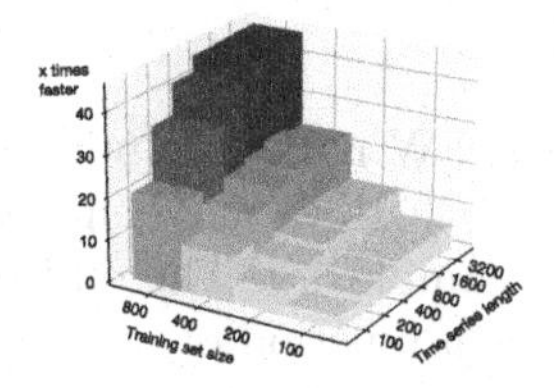

(a) Comparison of the runtime of PQDTW (no pre-alignment) with DTW

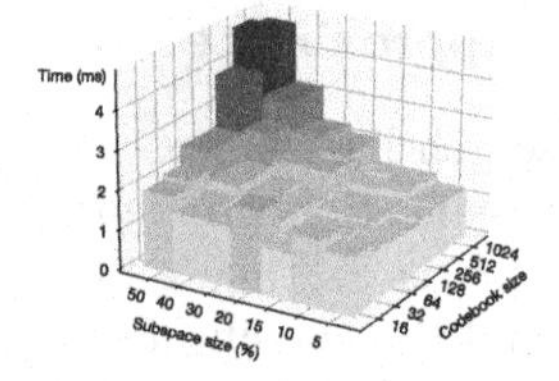

(b) Effect of the subspace and codebook size on the runtime of PQDTW

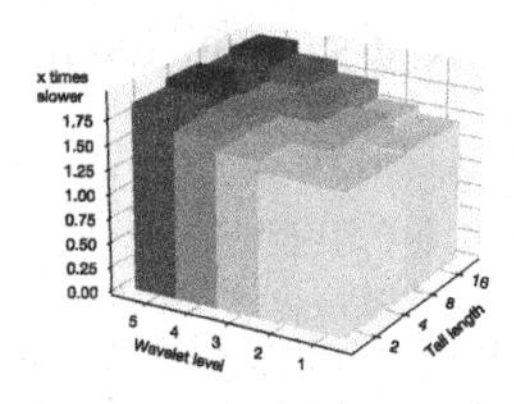

(c) Effect of pre-alignment on the runtime of PQDTW

Fig. 3. Empirical evaluation of the time complexity of PQDTW on a random walks dataset.

Using symmetric distance computation, two similar time series have a high likelihood to be mapped to the same centroids, resulting in an approximate distance of zero. While the resulting errors on the estimated distance are small, this might be problematic in clustering applications where the ranking of distances is important. This is solved by partially replacing the estimated distance when subspaces are encoded to the same code. As an efficient and elegant replacement value, we propose the Keogh lower bound. Given a subquantizer q_m and two subspaces x^m and y^m, the distance value would be $max(lb(x^m, q_m(y^m)), lb(q_m(x^m), y^m))$. This bound is guaranteed to be between 0 and the exact distance.

5 Experimental Settings

This section describes the experimental settings for the evaluation of PQDTW.

Platform. We ran all experiments on a set of identical computing servers[1] using a single core per run. In order to reduce the variance in runtime caused by other processes outside our control and the variance caused by the random selection of centroids in the DBA k-means step of the PQDTW encoding step, we executed each method five times with different seeds and report the mean accuracy and median run time.

Baselines. We compare PQDTW against the most common and state-of-the-art distance measures for time series: Euclidean distance (ED), dynamic time warping (DTW) [22], window-constrained dynamic time warping (cDTW) [22], and shape-based distance (SBD) [17]. For DTW we use the PrunedDTW [25] technique to prune unpromising alignments. For constrained DTW, we consider window sizes of 5 and 10%, as well as the window size which leads to the minimal 1NN classification error on the training set. We denote the window size with a suffix (e.g., cDTW10) and using cDTWX for the optimal one. SBD is a state-of-the-art shape-based distance measure, achieving similar results to cDTW and

[1] Intel Core i7-2600 CPU @ 3.40 GHz; 15 Gb of memory; Ubuntu GNU/Linux 18.04.

DTW while being orders of magnitude faster. In addition, we compare against SAX [13], which is perhaps the most studied symbolic representation for time series. We use an alphabet size $\alpha = 4$, and segments of length $l = 0.2 * L$ (where L is length of the time series) [16]. Finally, we compare against standard PQ using the Euclidean distance (i.e., a version of PQDTW without pre-alignment that uses ED instead of DTW), denoted PQ_{ED}.

Implementation. We implemented PQDTW, ED, DTW, and SBD under the same framework, in C(ython) [14], for a consistent evaluation in terms of both accuracy and efficiency. For repeatability purposes, we make all source code available.[2] For SAX, we use the Cython implementation available in `tslearn`.[3]

Parameter Settings. A disadvantage of the PQDTW approach is that we have many hyper-parameters to tune. We use a default codebook size of 256 (or all time series in the training set if there are less examples) and symmetric distance computation. To determine the optimal subspace size, wavelet level, tail and quantization window, we use the Tree-structured Parzen Estimator algorithm [1] which we ran for 12 h on each dataset. This hyper-parameter tuning is a one-time effort. We use 5-fold cross validation on the training set with a test set of 25% and evaluate the 1NN classification error. This results in multiple Pareto optimal solutions with respect to runtime and accuracy. We report the results for the most accurate solution on the training set.

Statistical Analysis. We analyze the results of every pairwise comparison of algorithms over multiple datasets using the Friedman test followed by the post-hoc Nemenyi test. We report statistical significant results with a 95% confidence level.

6 Experimental Results

The goal of this evaluation is to demonstrate the efficiency and accuracy of PQDTW in classification and clustering applications. This evaluation will first demonstrate the empirical time complexity of PQDTW in comparison to DTW and evaluate the effect of parameter settings on a synthetically generated random walk dataset. Second, we benchmark PQDTW for time series classification and clustering on 48 UCR datasets.[4]

6.1 Empirical Time Complexity

We begin with an evaluation of the empirical time complexity on a random walk dataset. Although these random walks are not ideal to evaluate the PQDTW

[2] https://github.com/probberechts/PQDTW.

[3] We use *tslearn v0.5.0.5*. See https://tslearn.readthedocs.io.

[4] Only the datasets available since 2018 [4] were used to keep the runtime of the experiments manageable, while achieving a maximal overlap with existing research.

Table 1. Comparison of PQDTW against other distance measures for 1NN and hierarchical complete linkage clustering. The column "Mean difference" contains the mean and standard deviation of the relative difference in classification error (1NN) and rand index (clustering) between PQDTW and the corresponding measure, whereas "Speedup" indicates the factor by which PQDTW speeds up the runtime.

	1NN		**Clustering**	
	Mean error difference	Speedup	Mean ARI difference	Speedup
ED	0.017(066)	×14.00	0.013(099)	×2.64
DTW	−0.014(064)*	×25.01	0.004(122)	×225.20
cDTW5	−0.036(044)*	×12.91	0.008(097)	×32.83
cDTW10	−0.029(052)*	×15.81	−0.002(110)	×59.01
cDTWX	−0.037(050)*	×14.15	−0.003(106)	×50.45
SBD	−0.021(056)*	×6.45	−0.011(105)	×47.18
SAX	0.293(199)+	×190.63	0.043(140)	×884.77
PQ_{ED}	0.038(071)+	×0.83	−0.006(057)	×0.75

* PQDTW performs worse ($p < 0.05$); + PQDTW performs better ($p < 0.05$)

algorithm due to lack of common structures that can be aligned, they allow us to do reproducible experiments on a large set of time series collections of varying sizes and time series lengths. The results show a significant speedup of PQDTW (subspace size = 20%, no pre-alignment) over DTW, which improves relatively for longer time series (Fig. 3a). For computing the pairwise distance matrix of 100 time series, PQDTW is between 2.9 times (length 100) and 5.6 times faster (length 3200). Interestingly, due to lower bound pruning, the average computation time of PQDTW per pair of time series decreases a lot if the number of time series grows. Therefore, for a collection of 800 time series of length 3200, PQDTW is already 45.8 times faster.

The parameters that affect the speed of PQDTW most are the subspace size and codebook size (Fig. 3b). In accordance with the theoretical time complexity $O(K \times D^2/M)$, the runtime increases linearly when less subpaces or a larger codebook is used.

Finally, the pre-alignment step has a minor effect on the runtime (Fig. 3c), which is mainly determined by the level of the wavelet decomposition. Increasing the tail length does not have a significant effect.

6.2 1NN Classification

Table 1 reports the classification error and runtime of PQDTW against the state-of-the-art distance measures. For DTW and cDTW, we use the Keogh lower bound for early stopping. The statistical test suggests that there is no significant difference between PQDTW and ED. PQDTW performs at least as well in 23 datasets. All other distance measures that operate on the raw data outperform PQDTW with statistical significance. However, Fig. 4 shows that the difference in accuracy between PQDTW and cDTWX (i.e., the best performing measure) is small in all cases, while PQDTW is 14x faster on average. Additionally, PQDTW compresses the training data by a factor varying between 26.2 and 2622.4, depending on the dataset and PQDTW's parameter settings. From this experiment, we can conclude that (1) PQDTW is competitive with ED, but is much faster and requires far less space; and that (2) PQDTW outperforms SAX and PQ_{ED}, the baseline dimensionality reduction techniques based on ED.

6.3 Hierarchical Clustering

We use agglomerative hierarchical clustering with single, average, and complete linkage criteria. To evaluate the obtained clustering, we compute a threshold that cuts the produced dendrogram at the minimum height such that k clusters are formed, with k corresponding to the number of classes in the dataset. Subsequently, we compute the Rand Index (RI) [21] over the test set using the class labels as the ground truth clustering. The major difference in performance among hierarchical methods is the linkage criterion and not the distance measure [17]. Since the complete linkage criterion gave the best results, we only report these in Table 1. There are no significant differences among all distance measures that we evaluated. Figure 4b shows that the differences in RI between PQDTW and cDTWX are indeed small. Since the full distance matrix has to be computed to obtain a hierarchical clustering and lower bound pruning cannot be applied, the gain in performance is larger compared to 1NN. Our approach is one order of magnitude faster than cDTW and SBD, and two orders of magnitude faster than DTW.

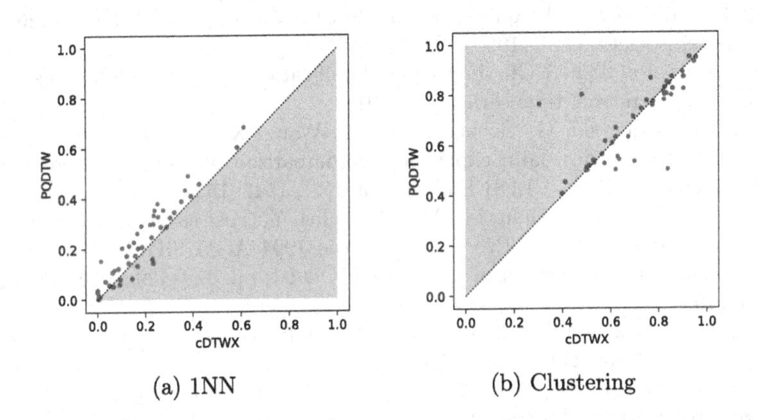

(a) 1NN (b) Clustering

Fig. 4. Comparison of (a) the 1NN classification error and (b) the rand index for hierarchical complete linkage clustering with PQDTW and cDTWX over 48 UCR datasets. Circles in the green area indicate datasets for which PQDTW performs better than cDTWx. (Color figure online)

7 Conclusions

This work presented PQDTW, a generalization of the product quantization algorithm for Euclidean distance to DTW. By exploiting prior knowledge about the data through quantization and compensating for subtle variations along the time axis through DTW, PQDTW learns a tighter approximation of the original time series than other piecewise approximation schemes proposed earlier. Overall, the results suggest that PQDTW is a strong candidate for time series data analysis applications in online settings and in situations where computation time and storage demands are an issue.

Acknowledgements. This work was partially supported by iBOF/21/075, the KU Leuven Research Fund (C14/17/070), VLAIO ICON-AI Conscious, and the Flemish Government under the "Onderzoeksprogramma Artificiële Intelligentie (AI) Vlaanderen" program.

References

1. Akiba, T., Sano, S., Yanase, T., Ohta, T., Koyama, M.: Optuna: a next-generation hyperparameter optimization framework. In: Proceedings of the 25th ACM SIGKDD International Conference Knowledge Discovery and Data Mining (2019)
2. Chan, F.P., Fu, A.C., Yu, C.: Haar wavelets for efficient similarity search of time-series: with and without time warping. IEEE Trans. Knowl. Data Eng. **15**(3), 686–705 (2003)
3. Chan, K.P., Fu, A.W.C.: Efficient time series matching by wavelets. In: Proceedings 15th International Conference on Data Engineering. ICDE 99, p. 126. IEEE Computer Society, USA (1999)
4. Dau, H.A., et al.: The UCR time series classification archive (2018). https://www.cs.ucr.edu/~eamonn/time_series_data_2018/
5. Ding, H., Trajcevski, G., Scheuermann, P., Wang, X., Keogh, E.: Querying and mining of time series data: experimental comparison of representations and distance measures. Proc. VLDB Endowment **1**(2), 1542–1552 (2008)
6. Faloutsos, C., Ranganathan, M., Manolopoulos, Y.: Fast subsequence matching in time-series databases. In: Proceedings of the 1994 ACM SIGMOD International Conference on Management of Data. SIGMOD 94, pp. 419–429. ACM Press, New York (1994)
7. Hong, J.Y., Park, S.H., Baek, J.G.: SSDTW: shape segment dynamic time warping. Expert Syst. Appl. **150**, 113291 (2020)
8. Jegou, H., Douze, M., Schmid, C.: Product quantization for nearest neighbor search. IEEE Trans. Pattern Anal. Mach. Intell. **33**(1), 117–128 (2010)
9. Keogh, E., Chakrabarti, K., Pazzani, M., Mehrotra, S.: Dimensionality reduction for fast similarity search in large time series databases. Knowl. Inf. Syst. **3**(3), 263–286 (2001)
10. Keogh, E., Ratanamahatana, C.A.: Exact indexing of dynamic time warping. Knowl. Inf. Syst. **7**(3), 358–386 (2005)
11. Kim, S.W., Park, S., Chu, W.W.: An index-based approach for similarity search supporting time warping in large sequence databases. In: Proceedings 17th International Conference on Data Engineering, pp. 607–614. IEEE (2001)

12. Lemire, D.: Faster retrieval with a two-pass dynamic-time-warping lower bound. Pattern Recognit. **42**(9), 2169–2180 (2009)
13. Lin, J., Keogh, E., Wei, L., Lonardi, S.: Experiencing sax: a novel symbolic representation of time series. Data Min. Knowl. Disc. **15**(2), 107–144 (2007)
14. Meert, W., Hendrickx, K., Van Craenendonck, T., Robberechts, P.: DTAIDistance (2022). https://doi.org/10.5281/zenodo.3981067https://github.com/wannesm/dtaidistance
15. Mueen, A., Keogh, E.: Extracting optimal performance from dynamic time warping. In: Proceedings of the 22nd ACM SIGKDD International Conference on Knowledge Discovery and Data Mining. KDD 2016, pp. 2129–2130. ACM Press, New York (2016)
16. Nguyen, T.L., Gsponer, S., Ifrim, G.: Time series classification by sequence learning in all-subsequence space. In: 2017 IEEE 33rd International Conference on Data Engineering (ICDE). ICDE 1, pp. 947–958(2017)
17. Paparrizos, J., Gravano, L.: k-Shape: efficient and accurate clustering of time series. In: Proceedings of the 2015 ACM SIGMOD International Conference on Management of Data, pp. 1855–1870 (2015)
18. Paparrizos, J., Liu, C., Elmore, A.J., Franklin, M.J.: Debunking four long-standing misconceptions of time-series distance measures. In: Proceedings of the 2020 ACM SIGMOD International Conference on Management of Data. SIGMOD 20, pp. 1887–1905. ACM Press, New York (2020). https://doi.org/10.1145/3318464.3389760
19. Petitjean, F., Ketterlin, A., Gançarski, P.: A global averaging method for dynamic time warping, with applications to clustering. Pattern Recognit. **44**(3), 678–693 (2011)
20. Rakthanmanon, T., et al.: Searching and mining trillions of time series subsequences under dynamic time warping. In: Proceedings of the 18th ACM SIGKDD International Conference on Knowledge Discovery and Data Mining, pp. 262–270. ACM Press, New York (2012)
21. Rand, W.M.: Objective criteria for the evaluation of clustering methods. J. Am. Stat. Assoc. **66**(336), 846–850 (1971). https://www.jstor.org/stable/2284239
22. Sakoe, H., Chiba, S.: Dynamic programming algorithm optimization for spoken word recognition. IEEE Trans. Signal Process. **26**(1), 43–49 (1978)
23. Salvador, S., Chan, P.: Toward accurate dynamic time warping in linear time and space. Intell. Data Anal. **11**(5), 561–580 (2007)
24. Shen, Y., Chen, Y., Keogh, E., Jin, H.: Accelerating time series searching with large uniform scaling. In: Proceedings of the 2018 SIAM International Conference on Data Mining. SIAM Publications, pp. 234–242 (2018)
25. Silva, D.F., Batista, G.E.A.P.A.: Speeding up all-pairwise dynamic time warping matrix calculation. In: Proceedings of the 2016 SIAM International Conference on Data Mining, pp. 837–845. SIAM Publications (2016)
26. Spiegel, S., Jain, B.J., Albayrak, S.: Fast time series classification under lucky time warping distance. In: Proceedings of the 29th Annual ACM Symposium on Applied Computing, pp. 71–78 (2014)
27. Tan, C.W., Petitjean, F., Webb, G.I.: Elastic bands across the path: a new framework and method to lower bound DTW. In: Proceedings of the 2019 SIAM International Conference on Data Mining, pp. 522–530. SIAM (2019)
28. Wu, R., Keogh, E.J.: FastDTW is approximate and generally slower than the algorithm it approximates. IEEE Trans. Knowl. Data Eng. (2020)

29. Zhang, H., Dong, Y., Li, J., Xu, D.: Dynamic time warping under product quantization, with applications to time series data similarity search. IEEE IoT-J, 1 (2021). https://doi.org/10.1109/JIOT.2021.3132017

Stress Detection from Wearable Sensor Data Using Gramian Angular Fields and CNN

Michela Quadrini[1(✉)], Sebastian Daberdaku[2], Alessandro Blanda[2], Antonino Capuccio[2], Luca Bellanova[2], and Gianluca Gerard[2]

[1] School of Science and Technology, University of Camerino, Via Madonna Delle Carceri 9, 62032 Camerino, MC, Italy
michela.quadrini@unicam.it

[2] Sorint.Tek, Via Zanica 17, 24050 Grassobbio, BG, Italy
{sdaberdaku,ablanda,acapuccio,lbellanova,ggerard}@latek.it

Abstract. Stress is a body reaction that is one of the principal causes of many physical and mental disorders, including cardiovascular disease and depression. Developing robust methods for rapid and accurate stress detection plays an important role in improving people's life quality and wellness. Prior research shows that analyzing physiological signals collected from wearable sensors is a reliable predictor of stress. For stress detection, methods based on machine learning techniques have been defined in the literature. However, they require hand-crafted features to be effective. Deep learning-based approaches overcome these limitations.

In this work, we introduce STREDWES, a method for stress detection that analyzes biosignals obtained from wearable sensor data. STREDWES extracts signal fragments using a sliding windows approach and converts them into Gramian Angular Fields images. These images are then classified using a Convolutional Neural Network, a deep learning algorithm. We apply our method to a publicly available dataset. The analysis of the performance values shows that our method outperforms other state-of-the-art competitors.

Keywords: Convolutional neural network · Gramian angular field · Biosignal · Stress detection

1 Introduction

Psychological stress is a body reaction, defined as "the non-specific response of the body to any demand upon it" [1]. Its effects and symptoms impact humans both physically and emotionally, playing a significant role in overall behavior, wellbeing, and potentially personal and professional successes [11]. Moreover, stress is one of the principal causes of many health problems and mental diseases. According to the British Health and Safety Executive, stress accounted for 50% of all work-related ill health cases in 2020/21 and this rate is increased of

P. Pascal and D. Ienco (Eds.): DS 2022, LNAI 13601, pp. 173–183, 2022.
https://doi.org/10.1007/978-3-031-18840-4_13

13% since 2015/16 [5]. Therefore, developing robust methods for the rapid and accurate detection of stress impacts on quality of life: detecting and managing stress before it turns into a more severe problem is crucial.

Assessments, like the Perceived Stress Scale (PSS) [8], are the most commonly used to detect stress. However, these techniques are time-consuming and unreliable: questionnaires often contain a set of questions designed by psychologists and could be psychologically invasive. Therefore, the challenge is to find a way to detect users' stress states reliably, automatically, and non-invasively. Since stress is a physiological response to a stimulus, multiple physiological signals can be employed to monitor such physiological reactions. Their analysis can detect the presence of stress as a binary variable (stress/no stress). The recent increase in usage of wearable devices, such as smartphones and smartwatches, permits tracking steps and monitoring other physical activities of their users non-invasively. Plarre *et al.* [13] and Hovsepian *et al.* [6] introduced stress detection systems employing biosignals, such as an electrocardiogram (ECG) and respiration (RESP). In this context, machine learning and deep learning methodologies achieve promising results. Uddin and Canava proposed an approach based on a convolutional neural network (CNN) and random forest [20]. In [2], various physiological signals such as peripheral pulse volume (BVP), ECG, and electromyography (EMG) are used as input to a multichannel CNN for improving the classification accuracy of the different affective states.

The analysis of the approaches in the literature and the definition of stress as a non-specific response of the body suggests that combining several features would provide better performance in stress detection. To the best of our knowledge, the signal set encoding into a single image has not been investigated yet for improving stress detection. Nevertheless, the output image formalizes the temporal correlations among time points of different signals where a measurement is taken. Motivated by the results obtained for the load-carrying weight and posture classifications in [9], we introduce STREDWES, a method for stress detection based on the wearable sensors data and a CNN architecture as a data classifier. The data obtained from different wearable sensors determine a set of signals that, in our approach, will be converted into a multichannel image encoding their temporal correlations.

After preprocessing the data (consisting of resampling, outlier removal, and normalization), the proposed method determines a dataset of samples that are signal fragments obtained using the sliding window approach converted into an image corresponding to the Gramian Angular Fields (GAF) [22]. Then, the images are classified by a CNN architecture in *stress*, *neutral*, and *amusement*.

We applied the described approach to a public dataset, Wearable Stress and Affect Detection (WESAD) [19]. WESAD is a publicly available dataset containing data recorded from both a wrist-(Empatica E4) and a chest-worn (RespiBAN) device. The performances obtained with the proposed approach on the WESAD are compared with other classical machine learning algorithms [19] and with the multimodal-multisensory sequential fusion model (MMSF) [10]. The comparison shows that our approach outperforms the competitors.

2 Materials and Methods

In this work, we propose a methodology based on CNN that combines a set of signals to stress detection. We apply our approach on the WESAD dataset.

2.1 Dataset

WESAD is a public multimodal dataset containing laboratory-recorded data of 15 subjects. The wearable sensors used to record data detected blood volume pulse (BVP), ECG, electrodermal activity (EDA), electromyogram (EMG), respiration (RESP), body temperature (TEMP), and three-axis acceleration (ACC). In particular, the dataset contains high-resolution physiological (ECG, EDA, EMG, RESP, and TEMP) and motion (ACC) data sampled 700 Hz from a chest-worn device (RespiBan) and lower resolution data from a wrist-worn device (Empatica E4). The data consists of 14 time series of about 2 h each, the total time of the experiment. These time series represent two major stimuli: amusement and stressful condition. These two conditions were interchanged between different subjects to avoid effects of order. Moreover, the experiments also measured baseline and two meditation periods. Figure 1 summarises the protocol.

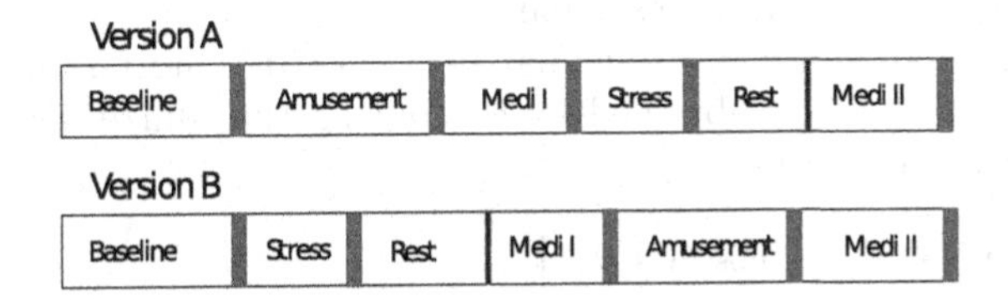

Fig. 1. The two different versions of the study protocol.

2.2 Preprocessing

Resampling. The dataset contains data sampled at 700 Hz and lower resolution data. Therefore, we first standardize the sampling step of all signals to simplify further processing. The signals sampled 700 Hz were resampled 35 Hz by downsampling by an integer factor applying a lowpass filter (finite impulse response of order 5). The Fourier method upsampled the signals at 35 Hz.

Outliers Removal. Some signals show anomalous peaks, probably due to instrumental error. Therefore, we eliminated those anomalies from each time series by applying a Hampel filter. The method takes sliding windows of 1 minute as input. It computes the mean μ and standard deviation σ of values related to the interval. The observations that exceed the threshold of 3σ from the mean of the corresponding window are considered outliers (Pearson's rule) and are replaced with the closest value in chronological order. This approach allows us to replace the outliers with values of the dataset without introducing high frequencies.

2.3 Sample Construction

Representation. To represent the temporal correlations among time points, we associate an image to signals in a time window by employing the GAF representation [22]. The matrix represents time series in a polar coordinate system: each element is the cosine of the summation of angles. To build the GAF matrix, we first rescale the observations of the time series. Therefore, let $X = \{x_1, x_2, \ldots, x_n\}$ be the considered time series with n components. Each of them are rescaled to the interval $[-1, 1]$ by applying the mean normalization:

$$\tilde{x}_j = \frac{(x_i - \max(X)) + ((x_i - \min(X))}{\max(X) - \min(X)} \quad . \tag{1}$$

Hence, the scaled series is represented by $\tilde{X} = \{\tilde{x}_1, \tilde{x}_2, \ldots, \tilde{x}_n\}$. This is transformed to a polar coordinates system by encoding the value as the angular cosine and the time stamp as the radius:

$$\begin{cases} \theta_i = \arccos(\tilde{x}_i), & \tilde{x}_i \in \tilde{X} \\ r_i = \frac{t_i}{n}, & \text{with } i \in \{1, \ldots, N\} \end{cases} \tag{2}$$

where t_i is the time stamp and n is the number of samples used to regularize the span of the polar coordinate system.

Finally, Gramian summation angular field (GASF) and Gramian difference angular field can be easily obtained by computing the sum/difference between the points of the time series

$$\text{GASF} = \begin{bmatrix} \cos(\theta_1 + \theta_1) & \ldots & \cos(\theta_1 + \theta_n) \\ \vdots & \ddots & \vdots \\ \cos(\theta_n + \theta_1) & \ldots & \cos(\theta_n + \theta_n) \end{bmatrix} = \tilde{X}^T \cdot \tilde{X} - \sqrt{I - \tilde{X}^2}^T \cdot \sqrt{I - \tilde{X}^2} \ , \tag{3}$$

$$\text{GADF} = \begin{bmatrix} \sin(\theta_1 - \theta_1) & \ldots & \sin(\theta_1 - \theta_n) \\ \vdots & \ddots & \vdots \\ \sin(\theta_n - \theta_1) & \ldots & \sin(\theta_n - \theta_n) \end{bmatrix} = \sqrt{I - \tilde{X}^2}^T \cdot \tilde{X} - \tilde{X}^T \cdot \sqrt{I - \tilde{X}^2} \ , \tag{4}$$

where I is the unit row vector $[1, 1, \ldots, 1]$.

Figure 2 shows GADF and GASF images of a signal window from WESAD, respectively. Note that Eqs. 3 – 4 produce a 1D matrix as an output of the encoding process. The matrix represents a heatmap whose values range from 0 (blue) to 1 (red). As a next step, we applied the RGB colour map to the image, thus resulting in a three channel matrix (interested readers can refer to [21, 22]).

Dataset Entry. The dataset entries consist of images (corresponding to Gramian Angular Fields Matrices) determined from multimodal signal fragments. To obtain them, there are several variables to consider, such as the method to compute the GAF images (summation or difference), the time window

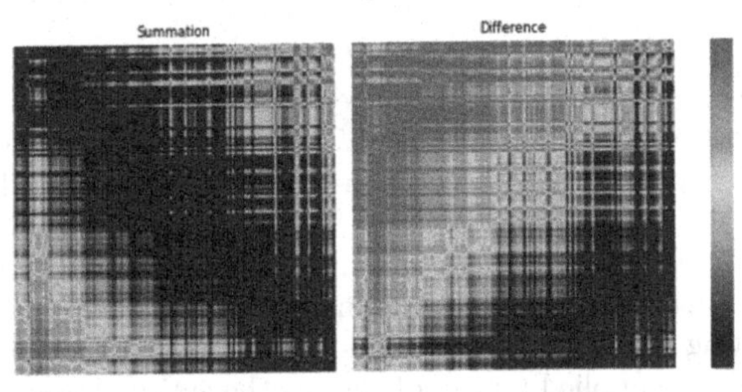

Fig. 2. GADF and GASF images of a single signal window in WESAD, respectively. (Color figure online)

Table 1. Description of the employed hyperparameters.

Hyperparameter	Description
Batch size	Batch dimension
GAF method	Method to compute the GAF images
Image size	The dimension of GAF images
Learning rate	Learning rate of the optimization algorithm
Optimizer	Optimization algorithms
Time step	The step of the sliding window
Window size	The dimension of the sliding window

length, and the image size. The used hyperparameters with their description are listed in Table 1.

Precomputing images for all the possible combinations of the parameters is expensive in terms of computation time and space. Therefore, we define a sample generator, *ImgGenerator*, to give as input to the CNN that will generate the samples online one batch at a time. The offsets of sliding windows are precomputed, ensuring that each window does not span over more than one emotional state. Such offsets are also randomly shuffled at the end of each training epoch.

The sliding windows, and consequently the GAF images, are extracted "on the fly" when the neural network requests a new batch from the generator. Moreover, fixed the d size of the GAF image, the samples produced by the ImgGenerator are matrices of dimensions $(d, d, 14)$, where 14, i.e. the number of channels of the image, corresponds to the number of time series available for each subject. Such samples depend on the hyperparameters' choice. To select the best performing hyperparameters, we considered the classification accuracy and sparse categorical cross-entropy of the validation set as the loss function. For each configuration of hyperparameter values, a neural network was trained on the samples related to these parameters for a given number of epochs. The obtained loss function and the accuracy over the generated samples were evaluated.

2.4 Convolutional Neural Network

CNN is one of the most successful deep learning architectures for its capacity to analyze spatial information. It has been designed to process multiple data types, especially two-dimensional images. The basic structure of CNNs consists of an input layer, convolution layers, non-linear layers, pooling layers, and an output layer. The input of the CNN is formalized as a tensor characterized by a shape. After passing through a convolutional layer, the image becomes abstracted to a feature map. Such feature maps are obtained at each convolution layer by computing convolutions between local patches and weight vectors called filters. The filters are applied repeatedly across the entire dataset since identical patterns can appear regardless of the location in the data. In this way, the efficiency of the training is improved by reducing the number of parameters to learn. Then, the non-linear properties of feature maps are increased by non-linear layers. Moreover, each pooling layer performs maximum or average subsampling of non-overlapping regions in feature maps. This non-overlapping subsampling permits the CNNs to aggregate local features to identify more complex ones.

3 Results

3.1 Implementation

Our approach is implemented in Python using the Tensorflow package [4]. The implementation uses the library Keras [3] for developing both the sample generator and Convolutional Neural Network. The neural network is composed of two 2D convolutional layers with 32 filters, each followed by a MaxPooling2D layer (with pool size = (2, 2)) and a layer of Dropout (with 0.25 dropout rate), and finally from two fully connected layers with a Dropout layer equals to 0.5. The activation function used is the ReLU for all layers except the last one that uses the softmax function. Adam is the optimization algorithm with the learning rate set as a definable parameter during the creation of the network. All experiments were performed with the Amazon SageMaker Service using one "ml.g4dn.xlarge" instance. Figure 3 outlines the approach, while the implementation is available from the corresponding author upon reasonable request.

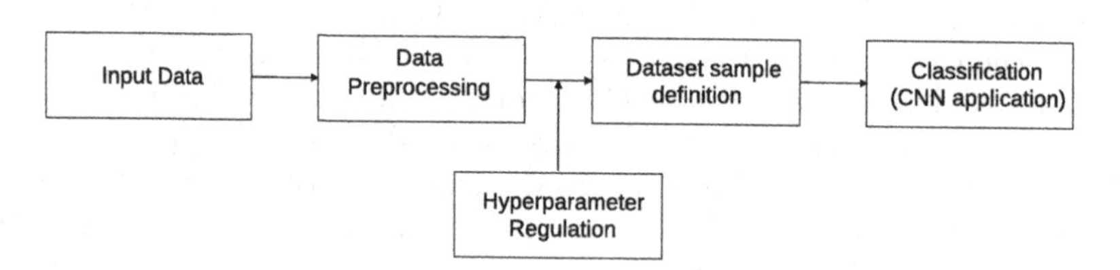

Fig. 3. General architecture of our approach.

3.2 Experiments

The 15 subjects of the dataset were split into two disjoint sets (train and test), according to the sex of 12 and 3 people, respectively. To train our model,

- *batch size*, tested integer values from 4 to 512 in logarithmic scale;
- *GAF method*, tested method: "summation", "difference";
- *image size*, tested integer values from 16 to 128 in logarithmic scale;
- *learning rate*, tested real value from $1 \cdot e^{-7}$ to $1 \cdot e^{-1}$ in logarithmic scale;
- *optimizer*, tested methods: "Adam", "AMSGrad" and "SGD";
- *time step*, tested integer value from 1 to 30 in auto scale[1];
- *window size*, tested integer value from 30 to 300 in logarithmic scale.

For each combination of hyperparameters, the network is trained for up to 50 epochs. To avoid overfitting, the method stops the training by employing the early stopping method if the loss function on the train set does not improve for five consecutive epochs. At the end of the training, it selects the net weights corresponding to the epoch with the best result on the validation loss. The Bayesian hyperparameter search is repeated for 100 iterations. The method achieves the maximum accuracy, 98.79%, for the hyperparameters reported in Table 2.

Table 2. Selected hyperparameters

Hyperparameter	Best value
Time step	1
Optimizer	AMSGrad
Learning rate	0.0012
Batch size	82
Window size	293
Image size	105
GAF method	Difference

Due to the limited data sample we use Cross-validation, a resampling procedure, to evaluate machine learning models. This procedure has a single parameter, called k, which indicates the number of groups into which a given sample of data must be divided. We randomly split the data into five groups composed of 3 subjects each. In each procedure step, a group forms the test dataset, while the remaining groups set up a training dataset. Given the results obtained from the hyperparameter search, we fixed the values of some of them (time step equals 1, GAF method is 'difference', and the choice optimizer is 'AMSGrad') and

[1] the tuning algorithm chooses the best scale for the hyperparameter exploration among linear, logarithmic, and reverse logarithmic.

updated the range of the others. We performed a Bayesian search considering the following intervals of hyperparameters:

- *batch size*, integer values from 4 to 82 by using auto scale;
- *image size*, integer values from 48 to 105 by using auto scale;
- *learning rate*, real values from $9.884e^{-4}$ to $4.235e^{-3}$, by using auto scale;
- *window size*, integer values from 145 to 300 by using auto scale.

As in the previous hyperparameter search, we trained the network for a maximum of 50 epochs by considering the Bayesian search. However, the method stops the training in advance by the early stopping method if the loss function on the validation set does not improve for five consecutive epochs to avoid overfitting. We obtained an average accuracy of the 92.58% and a standard deviation of 10.09% by considering the hyperparameter values reported in Table 3.

Table 3. Selected hyperparameters with 5-Fold Cross-Validation

Hyperparameter	Best value
Batch size	43
Image size	83
Learning rate	0.0016
Window size	294

Moreover, the high variance value indicates that the split train–validation significantly affects the performance of the results. This aspect could be a limit of the model because it could not generalize well when applied to new subjects. To overcome this limitation, we consider the Leave-One-Subject-Out Cross-Validation (LOSOCV), i.e., an approach that utilizes each subject as a "test" set and the remaining 14 as a "training" set. In this way, we train the model for a maximum of 10 epochs. For each LOSOCV iteration, the hyperparameter search was performed on the 14 training subjects with the 5-fold CV described earlier. The model stops the training when the first local minimum of the training loss is reached and selects the weights at that minimum. We obtain an average accuracy of the 91.16% and a standard deviation of 0.09%. Finally, we compare the performance of STREDWES with other approaches based on classical machine learning algorithms [19] and the multimodal-multisensory sequential fusion model (MMSF) [10] defined in the literature that take as input the set of physiological signals recorded in WESAD. Table 4 shows the accuracy and F1 score of each approach.

Table 4. Accuracy and F1-score of the considered approaches.

Method	Accuracy (%)	Weighted F1-score (%)
K-nearest	56.14	48.70
Decision Tree	63.56	58.05
Random Forest	74.97	64.08
AdaBoost	79.57	68.85
Linear discriminant analysis	75.80	71.56
MMSF	85.00	86.00
STREDWES	**92.97**	**92.31**

4 Conclusions and Future Work

In this work, we have introduced STREDWES, a method for stress detection based on the wearable sensors data that takes advantage of temporal correlations of the signal fragments obtained using the sliding windows approach, and CNN as a classifier. We have applied our method to a public dataset, WESAD, and we have analyzed the performance and effectiveness of our model against the competitors. The results show that our method outperforms the competitors.

As a future work, we intend to apply our method to other public datasets, like SWELL [7]. Moreover, we will investigate the role of the length of the sliding windows by considering other approaches based on entropy, like [14], that have obtained valuable results in the scenario of the prediction of protein-protein interaction sites. Another important future direction is to explore how to extract and represent the correlations that characterize the sliding windows. Possible representations are images like Markov Transition Field (MTF) [21], simplicial complexes by investigating the topological interpretation of the temporal relations as in [12]. Other representations to consider are arc-annotated sequences for the analysis and comparison of signals exploiting tools like [18] and strings, which allow applying techniques from formal methods to identify patterns [17]. Motivated by the previous results in [15,16], we also intend to apply deep learning techniques such as graph convolution networks or recurrent neural networks.

Acknowledgements. MQ is supported by the "GNCS - INdAM". The authors are grateful to Simone Cardis @ Sorint.Tek for his helpful insights and to Zerina Koplikaj @ Sorint.Tek for proofreading the manuscript.

References

1. Bara, C.P., Papakostas, M., Mihalcea, R.: A deep learning approach towards multimodal stress detection. In: AffCon@ AAAI, pp. 67–81 (2020)
2. Chakraborty, S., Aich, S., Joo, M.I., Sain, M., Kim, H.C.: A multichannel convolutional neural network architecture for the detection of the state of mind using physiological signals from wearable devices. J. Healthc. Eng. **2019**, 5397814 (2019)

3. Chollet, F., et al.: Keras (2015). https://keras.io
4. Girija, S.S.: Tensorflow: large-scale machine learning on heterogeneous distributed systems. Software available from tensorflow. org **39**(9), (2016)
5. Health and Safety Executive: HSE on work-related stress (2021). http://www.hse.gov.uk/statistics/causdis/-ffstress/index.htm. Accessed 7 Mar 2022
6. Hovsepian, K., Al'Absi, M., Ertin, E., Kamarck, T., Nakajima, M., Kumar, S.: cStress: towards a gold standard for continuous stress assessment in the mobile environment. In: Proceedings of the 2015 ACM International Joint Conference on Pervasive and Ubiquitous Computing, pp. 493–504 (2015)
7. Koldijk, S., Sappelli, M., Verberne, S., Neerincx, M.A., Kraaij, W.: The swell knowledge work dataset for stress and user modeling research. In: Proceedings of the 16th International Conference on Multimodal Interaction, pp. 291–298 (2014)
8. Lee, E.H.: Review of the psychometric evidence of the perceived stress scale. Asian Nurs. Res. **6**(4), 121–127 (2012)
9. Lee, H., Yang, K., Kim, N., Ahn, C.R.: Detecting excessive load-carrying tasks using a deep learning network with a Gramian angular field. Autom. Constr. **120**, 103390 (2020)
10. Lin, J., Pan, S., Lee, C.S., Oviatt, S.: An explainable deep fusion network for affect recognition using physiological signals. In: Proceedings of the 28th ACM International Conference on Information and Knowledge Management, pp. 2069–2072 (2019)
11. McEwen, B.S.: Protective and damaging effects of stress mediators. N. Engl. J. Med. **338**(3), 171–179 (1998)
12. Piangerelli, M., Maestri, S., Merelli, E.: Visualising 2-simplex formation in metabolic reactions. J. Mol. Graph. Model. **97**, 107576 (2020)
13. Plarre, K., et al.: Continuous inference of psychological stress from sensory measurements collected in the natural environment. In: Proceedings of the 10th ACM/IEEE International Conference on Information Processing in Sensor Networks, pp. 97–108. IEEE (2011)
14. Quadrini, M., Cavallin, M., Daberdaku, S., Ferrari, C.: ProSPs: protein sites prediction based on sequence fragments. In: International Conference on Machine Learning, Optimization, and Data Science, pp. 568–580. Springer, Cham (2021). https://doi.org/10.1007/978-3-030-95467-3_41
15. Quadrini, M., Daberdaku, S., Ferrari, C.: Hierarchical representation and graph convolutional networks for the prediction of protein–protein interaction sites. In: Nicosia, G., et al. (eds.) LOD 2020. LNCS, vol. 12566, pp. 409–420. Springer, Cham (2020). https://doi.org/10.1007/978-3-030-64580-9_34
16. Quadrini, M., Daberdaku, S., Ferrari, C.: Hierarchical representation for PPI sites prediction. BMC Bioinf. **23**(1), 1–34 (2022)
17. Quadrini, M., Merelli, E., Piergallini, R.: Loop grammars to identify RNA structural patterns. In: BIOINFORMATICS, pp. 302–309 (2019)
18. Quadrini, M., Tesei, L., Merelli, E.: ASPRAlign: a tool for the alignment of RNA secondary structures with arbitrary pseudoknots. Bioinformatics **36**(11), 3578–3579 (2020)
19. Schmidt, P., Reiss, A., Duerichen, R., Marberger, C., Van Laerhoven, K.: Introducing WESAD, a multimodal dataset for wearable stress and affect detection. In: Proceedings of the 20th ACM International Conference on Multimodal Interaction, pp. 400–408 (2018)

20. Uddin, M.T., Canavan, S.: Synthesizing physiological and motion data for stress and meditation detection. In: 2019 8th International Conference on Affective Computing and Intelligent Interaction Workshops and Demos (ACIIW), pp. 244–247. IEEE (2019)
21. Wang, Z., Oates, T.: Encoding time series as images for visual inspection and classification using tiled convolutional neural networks. In: Workshops at the Twenty-Ninth AAAI Conference on Artificial Intelligence (2015)
22. Wang, Z., Oates, T.: Imaging time-series to improve classification and imputation. In: Twenty-Fourth International Joint Conference on Artificial Intelligence (2015)

Multi-attribute Transformers for Sequence Prediction in Business Process Management

Gonzalo Rivera Lazo(✉) and Ricardo Ñanculef

Universidad Técnica Federico Santa María, Valparaíso, Chile
gfrivera@alumnos.inf.utfsm.cl, jnancu@inf.utfsm.cl

Abstract. Leveraging event logs to predict the evolution of an ongoing process is a challenging task in business process management (BPM). During the last years, sequence prediction models based on recurrent neural nets have demonstrated promise in this task attracting considerable interest from the community. Meanwhile, Transformer-based models and other architectures substituting recurrence with attention have become state-of-the-art in other sequence modeling tasks, especially in natural language processing. This paper investigates models based on the Transformer to predict operational business processes. In contrast to recent studies, we propose Multi-attribute Transformers, which exploit activities, resources, and time stamps for prediction, exploring different architectures to encode and integrate this information into the model. We also present multi-task variants of these models, which can predict the next activity of an ongoing process, when it will occur, and which resource it will trigger. Finally, we thoroughly evaluated these models in real datasets. In particular, we found that Multi-attribute Transformers can outperform Transformers that only use information about previous activities of the process. Moreover, our methods are competitive or better than existing multi-attribute recurrent models, allow significantly more parallelism during training and inference, and lead to more transparent/accountable predictions through the attention weights matrices.

Keywords: Deep learning · Transformer · Attention · Multi-attribute · Next event prediction · Business process management

1 Introduction

Recently, digital transformation has gained popularity in almost all industries and sectors. An essential resource to bring this transformation into an organisation is an *information system* that supports internal and external processes. Events performed by users who interact with this system can be recorded, generating *event logs* that represent the actual execution of a business process.

This research has been supported by the Scotiabank-USM alliance to promote the development and communication of computer science.

P. Pascal and D. Ienco (Eds.): DS 2022, LNAI 13601, pp. 184–194, 2022.
https://doi.org/10.1007/978-3-031-18840-4_14

Many business process mining tools exist to extract relevant information from these logs, such as identifying bottlenecks and decision points. However, these techniques typically do not provide *operational support*, i.e., the capacity to give users recommendations during the execution of a process. To this end, statistical and machine learning methods which predict the next activity of an ongoing and incomplete process have started to be investigated. Still, the task of predicting the time until the next event and the resources responsible for the running trace have received little attention.

While deep learning methods like recurrent neural networks have outperformed traditional machine learning and statistical methods in ongoing process monitoring tasks, they also involve a higher computational complexity. Allowing for significantly more parallelism during training, attention-based architectures such as Transformers look like a promising way to circumvent this problem. Unfortunately, current methods to train these models on business process data do not contemplate attributes other than activities. However, previous research on predictive process monitoring suggests information such as employed resources and time stamps are valuable information for prediction and decision making. On the one hand, predictive models using this context information often perform better than those which don't. On the other hand, giving this information as part of the model's predictions strengthens operational support by providing a more comprehensive picture of the ongoing process. Many other applications, including time-series forecasting and neural machine translation, have benefited from context and multiple input sources [5,16].

In this paper, we investigate Transformer based models for operational business processes prediction that can exploit activities, resources, and time stamps to make better and more informed predictions. In addition, we present multi-task variants of these models which address the task of predicting not just the next activity but other attributes attached to it, including the resource it could trigger and its expected timestamp. Finally, we thoroughly compare the proposed and existing models in three real datasets, observing that multi-attribute Transformers are competitive or better than existing recurrent models.

The remainder of the paper is structured as follows. First, Sect. 2 provides basic definitions and formalizes the problem. Then, Sect. 3 presents related work relevant to our proposal. Section 4 details the proposed method. Section 5 describes the experimental setup, a baselines comparison, and discusses the results. Finally, Sect. 6 summarizes the findings and contributions, concluding this paper and outlining future work.

2 Definitions and Problem Statement

A business process is a series of coordinated activities carried out at a particular moment in a technical or organizational context. These operations could employ one or more organizational resources and involve internal and external entities.

Definitions. We define an *event* e_i as the execution of an activity related to a business process in the organization. An event can be characterized by

m attribute values $e_i = \{a_i^{(1)}, a_i^{(2)}, \dots a_i^{(m)}\}$ such as activity ID, resource, start timestamp, and completion timestamp. Besides, we define a *case* $c_j = \langle e_{j,1}, e_{j,2}, \dots e_{j,l_j} \rangle$ as an entire instance of a business process, that is, a sequence of $l_j = l(c_j)$ events sorted chronologically and linked by a unique identifier usually called *Case ID*. Finally, we define an *event log* $E = \{c_1, c_2, \dots c_p\}$ as a collection of p cases produced by the execution of different business processes instances and recorded into an information system.

Problem. Giving an incomplete case $c^{(1:h)}$ representing an ongoing process from which we known only h activities, we are asked to return the most probable event following the sequence e_{h+1}, with all its m attribute values. The *prefix length* h can vary from case to case from a minimum of h_* to $l(c) - 1$ where $l(c)$ is the (unknown) length of c. To address this task, we are given an event log E representing the past execution of processes at the organization.

3 Related Work

Traditional statistical methods as Hidden Markov models (HMM) [10], Bayesian techniques [11] and Annotated transition systems (ATS) [1] have been proposed to solve predictive process monitoring. Classical machine learning classification methods such as Support Vector Machines (SVM) and Random Forest have also been used in process prediction [19]. Unfortunately, these models often have failures supporting long sequences.

A systematic literature review by Verenich et al. [21], which tested various models, concluded that Recurrent Neural Networks outperformed other methods on at least 13 of 16 datasets. Recently, another systematic literature review [15] has shown that deep models have higher accuracy than traditional machine learning methods in sequence modeling. This research also stated that attribute encoding is fundamental when feeding models with multiple attribute types.

In predictive process monitoring, Evermann et al. [6] used an embedding technique for categorical attributes and a shallow LSTM. The method of Tax et al. [18] shares a similar architecture but applies directly one-hot encoding. Camargo et al. [4] also evaluated the benefit of including event attributes (as a triplet of activities, resources, and timestamps) into the model through different concatenation strategies. Lin et al. [13] combined only categorical attributes with an LSTM encoder-decoder architecture and proposed aggregating a Modulator layer between them, outperforming the other methods that predict the next activity. The attention mechanism proposed by Jalayer et al. [8] generates a context vector by weighting information pieces from the input sequence. Later, Jalayer et al. [9] expanded this method to accept multiple categorical variables using a hierarchical attention approach, and Wickramanayake et al. [22] proposed to extract explanations from this model, visualizing the attention weights corresponding to a given output.

Lastly, the ProcessTransformer presented by Bukhsh et al. (2021) [3] uses a minimal pre-processing step and encodes only the activity sequence into a

Transformer to perform three tasks separately: predicting the next activity, the next timestamp, and the remaining time of an incomplete trace. The POP-ON method proposed in [14] only uses the activity sequence in the Multi-head Attention layer and concatenates all the other attributes. Besides, it applies an embedding technique to encode the attributes. The method by Philipp et al. [17] encodes activity sequences using only sin and cos positional embedding. Therefore, attention-based models with attributes other than the activities do have been explored. However, previous works have not presented results on predicting the resource or the triplet (activity, resource & timestamp) following an incomplete sequence.

4 Proposed Architectures

We predict the evolution of a business process instance by training a neural net f that consists of two main parts: an *encoder* and a *decoder*. The encoder maps an incomplete multi-attribute case $c^{(1:h)} = \langle e_1, e_2, e_3, ..., e_h \rangle$, with $e_i = \{a_i^{(1)}, a_i^{(2)}, ..., a_i^{(m)}\}$, into a continuous representation $E(c)$. The decoder transforms this representation into m probability distributions $\hat{y}_1, \ldots, \hat{y}_m$ where $\hat{y}_k$ leads to the most likely values of $a_{h+1}^{(k)}$, the next event's k-th attribute.

We present and evaluate different Transformer-based architectures to implement the encoder and the decoder. The encoder's architectures differ in the way to integrate continuous and categorical attributes into the model. The decoder's architectures differ in the way to handle the prediction task: as multiple independent tasks or as a multi-task prediction problem with shared parameters.

4.1 Encoder Architectures

Baseline. To progressively evaluate the effect of adding attributes, our baseline model obtains $E(c)$ using a Transformer encoder that represents the events in a case $c^{(1:h)}$ using activity IDs only. The input to the model is thus a sequence of h tokens or categorical values $\langle a_1, a_2, \ldots a_h \rangle$. The encoder starts by processing this sequence through an embedding layer equipped with positional encoding [20]. The embedding layer includes a learnable matrix that maps an activity ID into a continuous feature vector of dimension d_k. A sinusoidal encoding vector of the same dimension is added to the embedding to make the representation of each activity sensitive to its position in the input case. Then, an attention-based block that consists of two layers: a trainable multi-head self-attention mechanism and a trainable position-wise feed-forward net. Each layer includes residual connections [7] and applies Layer Normalization [2].

One Encoder. Our second approach, "One-Tr", represents an input case combining different categorical and continuous attributes that describe the events in the case. Attributes often available in information systems and studied in previous works include activities, timestamps, and resources.

To combine categorical attributes such as resources and activities, we adopt an *early fusion* approach. First, we concatenate the raw token sequences and then feed the resulting sequence into an encoder. For instance, if $s_a = \langle a_1, a_2, \dots a_h \rangle$ denotes the sequence of activity IDs and $s_r = \langle r_1, r_2, \dots, r_h \rangle$ denotes the sequence of resource IDs, the input to the encoder is

$$s_c = s_a \oplus s_r := \langle a_1, a_2, \dots a_h, r_1, r_2, \dots, r_h \rangle . \tag{1}$$

The encoder has the same architecture as in the *Baseline* approach. This time, however, the embedding layer learns continuous representations for all the categorical attributes. Furthermore, as the attention block on top of the embedding layer can simultaneously access and combines any subset of positions of their input sequence, this block can learn relationships between different event attributes (e.g. activities and resources) in a flexible and data-driven way.

To integrate continuous attributes into the model, such as features derived from timestamps, we adopt an *late fusion* approach, i.e. it concatenates the output of the Transformer encoder—in this case—with the output of a simple feed-forward net that encodes continuous data. We found this method to be the simplest and most effective for handling attributes of different data types: categorical attributes as activity IDs and resources significantly benefit from deep pre-processing before being combined with continuous data as timestamps.

Multiple Encoders. Our last architecture also explores the *late fusion* approach to combine categorical attributes with each other, such as resources and activity IDs. This method called "Multi-Tr" uses a different Transformer encoder to embed different token sequences. As in the previous approach, the model can exploit continuous data as timestamps by concatenating the encoders' output with the output of a feed-forward net that pre-processes this data.

In natural business processes, activities and resources are often related to each other. However, using independent encoders, the Transformer's self-attention block cannot combine features corresponding to different attributes. We evaluate the benefit of using a *Modulator layer* to circumvent this limitation.

Modulator. A Modulator layer [13] receives a set of representations $\{r_1, \dots, r_t\}$ and expands each r_i by computing the outer products $r_i \circ r_j, \forall j \neq i$,

$$\tilde{r}_i = r_i \oplus r_i \circ r_1 \oplus r_i \circ r_2 \cdots \oplus r_i \circ r_t . \tag{2}$$

We evaluate using this layer to allow our models to learn inter-dependencies between activities and resources after processing them with independent Transformer encoders and before concatenating them with continuous attributes.

4.2 Simplified Decoder Architectures

The sequence of tensors produced by the last encoder's block in the baseline approach or by the different fusion operations in the proposed architectures can be transformed into a flat representation $E(c) \in \mathbb{R}^d$ applying Global Average Pooling [20]. We explore two ways to use this vector to predict the next activity and other event attributes, such as resources and expected timestamps.

Specialized Layers. The simplest approach consists in using different sub-nets to predict different attributes. For simplicity, we use a classic three-layer neural net with Softmax activation for categorical attributes and linear activation for continuous attributes. If $a_{h+1}^{(k)}$ denotes an attribute of interest, each net learns a conditionally independent probability distribution

$$p(a_{h+1}^{(k)}|E(c^{(1:h)})) \approx p(a_{h+1}^{(k)}|c^{(1:h)})\,. \tag{3}$$

Multi-task Approach. Previous art has found that sharing information through common latent representations may be beneficial for improving the performance of multiple related tasks. Arguably, $E(c)$ already captures dependencies among input attributes but not among target attributes. In addition, error signals in a neural net are back-propagated from the output layers to the input layers in a process that may suffer attenuation. Thus, we explore the benefit of using a shared hidden representation immediately before the output layers for each task.

5 Experiments and Discussion

We evaluate the proposed method on three real-life event logs widely used in predictive process monitoring literature. This section presents the pre-processing pipelines, datasets, and the evaluation method. At the end of the section, we discuss our experimental results.

Datasets and Preprocessing. Helpdesk[1] is a dataset extracted from an Italian ticketing management system between 2010 to 2014. *BPIC 2012*[2] is a log of online loan applications in a Dutch financial institute from october 2011 to march 2012. Finally, *BPIC 2017*[3] is a dataset provided by the same company in *BPIC 2012* but contains more heterogeneous samples. Statistics are presented in Table 1.

For each dataset, we sorted events chronologically and kept the first 80% of the data for training and the remainder 20% for testing. Also, we added an end token to each sequence. Then, we augmented the number of samples by applying a k-prefix generator function, which maps a trace to the list of all its prefixes starting from length ml = 1 to the maximum individual case length. This procedure attempts to replicate a real-world monitoring environment in which we have to predict the evolution of ongoing processes, i.e., incomplete cases. Finally, the resulting sequences were padded to a fixed length. Each sequence of length k leads to a prediction task where the input is a sequence of length $k-1$ and the target is the last event.

It is worth noting that some authors exclude short sequences from the datasets. For instance, [13] only retains sequences containing at least 5 steps of events in *BPIC 2017* and *BPIC 2012*. For *Helpdesk*, [13] only works with sequences of length larger than 3. We preferred to preserve the original datasets to favor a clearer comparison with most previous studies. Following [4], we

[1] https://data.4tu.nl/repository/uuid:0c60edf1-6f83-4e75-9367-4c63b3e9d5bb.
[2] https://doi.org/10.4121/uuid:3926db30-f712-4394-aebc-75976070e91f.
[3] https://data.4tu.nl/articles/BPI_Challenge_2017/12696884.

Table 1. Statistics of datasets used for evaluation. Case length/duration in days.

			Unique activities	Case length		Case duration	
Dataset	Cases	Events		Max.	Avg.	Max.	Avg.
Helpdesk	4,580	21,348	14	15	4.6	60	40.69
BPI 2012	13,087	262,200	36	175	20.0	137.5	8.6
BPI 2017	31,509	1,202,267	26	180	38.1	286	21.9

transformed timestamps into three time-related features: time between the previous and current event, the time between the next-to-last event and the current event, and the time passed since the case was initiated.

Evaluation Setup. As in previous studies, we use accuracy to evaluate the performance of the different methods in predicting categorical attributes, i.e., next activity and resource. In contrast, we used MAE (Mean average error) to assess the ability of the methods to predict continuous (time-related) attributes. In any case, we report the average of these metrics among all the processed sequences.

As in a few previous works, evaluation imposes a minimum prefix length $\text{ml} > 1$, i.e., incomplete cases with prefix length $k < \text{ml} = 1$ are not considered for assessing the performance of the method. In our experiments, we use $\text{ml} = 1$ to replicate the needs of real-world scenarios. Nevertheless, when we report results from other works, we also include the minimum prefix length used by the authors.

Neural Net Setup. For training the neural net, we used batch sizes of 16, 32, and 128 in *Helpdesk*, *BPIC 2012* and *BPIC 2017* respectively. We used a learning rate of 0.001 and 10 epochs for all datasets. Further, every model was trained and tested five times and the average performance was calculated.

Results. Table 2 summarizes the performance of different models trained to predict the next activity. Results on *Helpdesk* and *BPIC 2012* show that using Transformer encoders to combine activity and resource attributes outperforms other approaches if we set a minimum prefix length of $\text{ml} = 1$ for evaluation. Arguably, the advantage of [9] in *BPIC 2017* is due to the use of two more attributes: transaction life and location. Table 2 shows also that the proposed methods outperform scores reported for LSTM-based methods in [18], and [8], even if these studies restrict evaluation to $\text{ml} = 2$. As a minimum prefix length increases, the input sequences carry more information about the process. Therefore, we should expect an increase in the model's performance.

Regarding our methods, results in Table 3 confirm that those that consider the sequences of activities and resources are the ones that obtain the greater accuracies. In addition, results show that the use of the Modulator layer did not significantly enhance prediction, suggesting that the Multi-Attribute Transformers on their own can perform prediction accurately. Furthermore, in these datasets adding the timestamp as an input increase the model instability, as is illustrated in Fig. 1.

Table 2. Baselines comparison for the next activity with inputs as Activity (A), Resource (R), Timestamp (T) and Categorical attributes (C).

Method	Input	Data split	Prefix (ml)	Accuracy		
				Helpdesk	BPIC2012	BPIC2017
LSTM [4]	A, R, T	70/30	1	0.789	0.786	–
ProcessTransformer [3]	A	60/20/20	1	0.856	–	–
HAM-net bi-LSTM [9]	C	80/20	1	0.844	0.868	**0.929**
One-Tr (this paper)	A, R	80/20	1	0.922	**0.884**	0.823
Multi-Tr (this paper)	A, R	80/20	1	**0.924**	0.892	0.769
LSTM [18]	A, T	66/33	2	0.712	–	–
Attention bi-LSTM [8]	A	80/20	2	0.833	0.816	–
MM-Pred LSTM [13]	C	70/20/10	4	0.916	0.974	0.974

Figure 1 also shows the performance of our models as we vary the minimum prefix length ml. Results are mixed. In *Helpdesk*, we observe the trend we expected: accuracy increases as ml increases. However, performance remains relatively stable in *BPIC 2012* while in *BPIC 2017*, there is no trend whatsoever. To understand these results, we must note that increasing ml decreases the total number of test instances on which models are evaluated because there are fewer suffixes to predict.

Table 3 shows that the best results we achieve in the task of predicting the next resource are 88%, 78% and 76% from *Helpdesk*, *BPIC 2012* and *BPIC 2017* respectively. To the best of our knowledge, these are the first results demonstrating the application of artificial intelligence on this task. Results on *Helpdesk* and *BPIC 2012* also suggest that training the models to predict the next activity and resource jointly can slightly improve their performance in predicting the next activity. Furthermore, Multi-Task architectures benefit from timestamps features.

Finally, 3 shows a slight decrease in the accuracies of the two categorical prediction tasks when we jointly train the model for the three tasks. On the other hand, results suggest that the task of predicting the next timestamp does not benefit from the shared or specialized hidden representation within a Multi-task model.

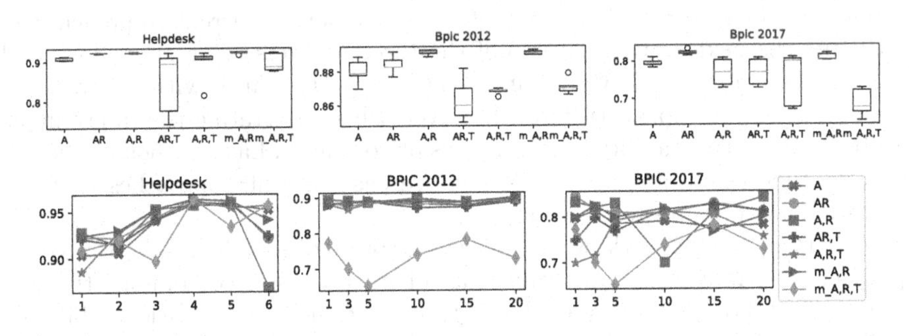

Fig. 1. First row: Accuracies on the next activity task by model on each dataset. Second row: Accuracies on the next activity task by trace length.

Table 3. Accuracies of our methods according the task or Multi-task on each datasets

Next Activity				
Architecture	Input	Accuracy		
		Helpdesk	BPIC 2012	BPIC 2017
One-Tr	A	90.9	88.0	79.6
One-Tr	A, R	92.2	88.4	**82.3**
Multi-Tr	A, R	**92.4**	**89.2**	76.9
One-Tr	A, R, T	84.9	86.3	76.9
Multi-Tr	A, R, T	89.4	86.8	75.3
Tr-Mod	A, R	92.2	89.1	80.9
Tr-Mod	A, R, T	89.7	87.1	68.6

Next Activity & Resource							
Architecture	Input	Accuracy|Accuracy					
		Helpdesk		BPIC 2012		BPIC 2017	
		A	R	A	R	A	R
One-Tr	A, R	91.9	87.9	86.3	76.7	76.9	73.5
One-Tr	A, R,T	76.8	74.9	82.9	75.5	71.3	62.8
S One-Tr	A, R,T	90.8	87.9	80.6	74.2	64.1	61.6
Multi-Tr	A, R	92.3	87.9	88.7	77.2	72.2	75.0
Multi-Tr	A, R, T	**92.5**	87.6	**89.4**	77.1	57.7	68.6
S Multi-Tr	A, R, T	92.4	87.6	88.7	**78.6**	**82.0**	**76.3**
Multi-Tr-Mod	A, R	**92.5**	87.6	88.6	78.2	81.9	75.3
Multi-Tr-Mod	A, R, T	91.5	**88.1**	88.3	77.2	80.9	74.2

Next Activity, Resource & Timestamp										
Architecture	Input	Accuracy|Accuracy|MAE								
		Helpdesk			BPIC 2012			BPIC 2017		
		A	R	T	A	R	T	A	R	T
One-Tr	A, R,T	88.4	72.4	86.5	84.7	74.7	**6.55**	**76.9**	**73.8**	**7.73**
Multi-Tr	A, R, T	90.6	82.3	78.0	88.3	**76.5**	6.68	55.9	55.0	8.02
S Multi-Tr	A, R, T	**90.7**	**84.2**	**75.7**	**88.6**	76.3	6.67	36.5	39.0	7.92

6 Conclusions and Final Remarks

This paper presented a method to approach Multi-task predictions in a predictive monitoring environment using Multi-attribute Transformers. In contrast with other methods, we considered as input not only the activity sequences but the resource and timestamp sequences because they can provide valuable information to the model. To do so, we first leverage the Transformer attention method to create continuous representations of the categorical attributes. Then, we join these representations with continuous features extracted from timestamps. Finally, a simplified decoder learns to perform one or multiple prediction tasks on the multi-attribute representation.

The results revealed that our method outperforms current approaches in predicting the next activity on two of the three datasets using single-task and multi-task architectures. We obtained the best performance with the variants that considered multiple attributes. However, when integrating the timestamps, we observed more instability in the model's performance. Lastly, when predicting the next timestamp our method do not overpass current approaches that use dedicated architectures for the task.

As part of ongoing work, we are expanding our models following a Seq-to-Seq approach. Specifically, we are evaluating different ways of combining attention layers as in Libovick et al. (2018) [12] to exploit the representations obtained by the Transformer encoders on each attribute. Finally, we plan to evaluate Multi-attribute Transformers in areas like biology and autonomous navigation.

References

1. Aalst, W.V., Schonenberg, M., Song, M.: Time prediction based on process mining. Inf. Syst. **36**(2), 450–475 (2011)
2. Ba, J.L., Kiros, J.R., Hinton, G.E.: Layer normalization. arXiv:1607.06450 (2016)
3. Bukhsh, Z.A., Saeed, A., Dijkman, R.M.: Processtransformer: predictive business process monitoring with transformer network. arXiv:2104.00721 (2021)
4. Camargo, M., Dumas, M., González-Rojas, O.: Learning accurate LSTM models of business processes. In: Hildebrandt, T., van Dongen, B.F., Röglinger, M., Mendling, J. (eds.) BPM 2019. LNCS, vol. 11675, pp. 286–302. Springer, Cham (2019). https://doi.org/10.1007/978-3-030-26619-6_19
5. Du, S., Li, T., Horng, S.J.: Time series forecasting using sequence-to-sequence deep learning framework. In: Proceeding Parallel Architectures Algorithms Programming (PAAP), pp. 171–176. IEEE (2018)
6. Evermann, J., Rehse, J.R., Fettke, P.: A deep learning approach for predicting process behaviour at runtime. In: Business Process Management Workshops, pp. 327–338 (2017)
7. He, K., Zhang, X., Ren, S., Sun, J.: Deep residual learning for image recognition. In: Proceedings of the IEEE Conference on Computer Vision and Pattern Recognition (CVPR), pp. 770–778. IEEE (2016)
8. Jalayer, A., Kahani, M., Beheshti, A., Pourmasoumi, A., Motahari-Nezhad, H.R.: Attention mechanism in predictive business process monitoring. In: Enterprise Distributed Object Computing Conference (EDOC), pp. 181–186. IEEE (2020)
9. Jalayer, A., Kahani, M., Pourmasoumi, A., Beheshti, A.: HAM-Net: predictive business process monitoring with a hierarchical attention mechanism. KBS **236**, 107722 (2022)
10. Lakshmanan, G.T., Shamsi, D., Doganata, Y.N., Unuvar, M., Khalaf, R.: A markov prediction model for data-driven semi-structured business processes. KAIS **42**(1), 97–126 (2013)
11. Letham, B., Rudin, C., McCormick, T.H., Madigan, D.: Interpretable classifiers using rules and bayesian analysis: building a better stroke prediction model (2015)
12. Libovický, J., Helcl, J., Mareček, D.: Input combination strategies for multi-source transformer decoder. In: ACL (2018)
13. Lin, L., Wen, L., Wang, J.: MM-Pred: a deep predictive model for multi-attribute event sequence, pp. 118–126 (2019)
14. Moon, J., Park, G., Jeong, J.: POP-ON: prediction of process using one-way language model based on NLP approach. Appl. Sci. **11**(2), 864 (2021)
15. Neu, D.A., Lahann, J., Fettke, P.: A systematic literature review on state-of-the-art deep learning methods for process prediction. AIR **55**(2), 801–827 (2021)
16. Nishimura, Y., Sudoh, K., Neubig, G., Nakamura, S.: Multi-source neural machine translation with missing data. CoRR, pp. 92–99 (2018)
17. Philipp, P., Jacob, R., Robert, S., Beyerer, J.: Predictive analysis of business processes using neural networks with attention mechanism. International Conference on Artificial Intelligence in Information and Communication (ICAIIC), pp. 225–230 (2020)
18. Tax, N., Verenich, I., Rosa, M.L., Dumas, M.: Predictive business process monitoring with LSTM neural networks. In: Advanced Information Systems Engineering (AISE), pp. 477–492 (2017)

19. Teinemaa, I., Dumas, M., Maggi, F.M., Di Francescomarino, C.: Predictive business process monitoring with structured and unstructured data. In: La Rosa, M., Loos, P., Pastor, O. (eds.) BPM 2016. LNCS, vol. 9850, pp. 401–417. Springer, Cham (2016). https://doi.org/10.1007/978-3-319-45348-4_23
20. Vaswani, A., et al.: Attention is all you need. In: NIPS, vol. 30 (2017)
21. Verenich, I., Dumas, M., Rosa, M.L., Maggi, F.M., Teinemaa, I.: Survey and cross-benchmark comparison of remaining time prediction methods in business process monitoring. ACM TIST **10**(4), 1–34 (2019)
22. Wickramanayake, B., He, Z., Ouyang, C., Moreira, C., Xu, Y., Sindhgatta, R.: Building interpretable models for business process prediction using shared and specialised attention mechanisms. Knowl.-Based Syst. **248**, 108773 (2022)

Social Media Analysis

Data-Driven Prediction of Athletes' Performance Based on Their Social Media Presence

Frank Dreyer, Jannik Greif, Kolja Günther, Myra Spiliopoulou, and Uli Niemann(✉)

Faculty of Computer Science, Otto von Guericke University, Magdeburg, Germany
{frank.dreyer,jannik.greif,kolja.guenther}@st.ovgu.de,
{myra,uli.niemann}@ovgu.de

Abstract. It is well known in the sports industry that the performance of athletes is strongly influenced by physiological and psychological factors. In recent years, many researchers have analysed whether athlete-generated social media content can be used as proxies for such performance factors, with some promising results. In this study, we investigated whether such proxies are useful features for a machine learning model to predict athletes' performance in subsequent competitions. We extracted millions of tweets that NBA basketball players posted themselves or were tagged in and derived features reflecting players' mood, social media behaviour, and sleep quality before games. Using these and other social media-unrelated features, we performed statistical tests to examine whether the features significantly improve the accuracy of a random forest model for predicting players' BPM scores in upcoming games. The results show that, in particular, the number of tweets a player is tagged in prior to a game significantly improves the predictions of the model. Our findings provide insights for practitioners on the effects of social media on athlete performance that can be used prospectively for mental health awareness training and optimisation of pre-game routines.

Keywords: Machine learning · Athletic performance · Social media · Twitter · Sentiment analysis · Predictive significance

1 Introduction

With the growing presence of social media in all areas of life, allowing people from around the world to react to current events in real time, an increasingly controversial discussion can be noticed: Today more than ever, public figures are exposed to the reactions of millions of people observing and commenting on every step in their life that becomes public.

Athletes, who use social media not only to communicate with peers and fans but also to promote themselves, are no exception to this circumstance. To date there is plenty of anecdotal evidence that the media has the potential to affect the performance of athletes to a great extent. In their work, von Ott and Puymbroeck [17] describe several cases in which athletes' performance in both

P. Pascal and D. Ienco (Eds.): DS 2022, LNAI 13601, pp. 197–211, 2022.
https://doi.org/10.1007/978-3-031-18840-4_15

team and individual sports changed dramatically after being exposed to media criticism. The authors conclude that there is strong evidence that the media affects the performance of athletes.

The influence social media can have on an individual's mood are confirmed by athletes themselves. In an interview, 8-time NBA all-star Vince Carter explains how it is like to be constantly exposed to social media criticism [6]: *"It's an emotional rollercoaster [...]. We as athletes have social media at our fingertips at any time and of course if you're playing well, you go look at your mentions. If you're not playing well, you go look at your mentions and now you have diehard opponent fans saying whatever they want and sometimes we tend to get caught up in what's being said from these persons [...]."*

Such examples give rise to the question if social media content can be used to predict the performance of athletes in upcoming competitions. In many domains, social media content has become the new source of intelligence. Accordingly, it has been successfully used for a wide range of predictive modelling tasks, such as stock price prediction, election results forecasting and even disease outbreak prediction [19]. A model predicting athletic competition performance based on social media posts could also be of great value for various stakeholders in the sports industry. Consider coaches as an example. Their job is to prepare athletes for upcoming competitions so that they have the best possible chance of winning. This preparation is not only a simple matter of improving athletes' physical skills through training. Perhaps more importantly, it also involves strengthening athletes' belief in their own abilities and their mental resilience. To achieve the latter, coaches must be aware about all factors that affect athletes' psychological functioning. This includes both intrapersonal factors (e.g. self-motivation) as well as interpersonal factors (e.g. social support) [11]. Social media content could be a great resource to uncover the satisfaction of such psychological performance factors. An according performance prediction model could then guide coaches if athletes need further mental or physical training to excel in the competition.

This paper addresses the following research question: *Do features derived from athlete-related social media posts lead to an improvement in accuracy of a machine learning model predicting athletes' performance in subsequent competitions?* To answer this research question, we gathered tweets NBA players posted themselves or were tagged in before games. From these tweets, we distilled various features that reflect athlete interaction on social media from different angles. We considered the quantity of the tweets, their temporal information as well as their sentiment. These features were then used in combination with other social media-unrelated features to study their predictive significance on athletic performance. This involved a permutation test that is based on random forests.

The remainder of this paper is organised as follows: In Sect. 2 we summarise existing research that analysed the relationship between social media and athletic performance. On this basis we describe our methodology to answer our proposed research question in Sect. 3. Our findings are presented and discussed in Sect. 4 and 5, respectively. We conclude our paper in Sect. 6.

2 Related Work

To date there have been few studies that examined the relationship between athletes' social media interaction and their performance. In general researchers approach the topic in one of two ways. They either consider the content of social media posts produced by athletes as a proxy for their mood and behaviour and analyse the effects of this proxy on their performance or they consider social media activity as something that distracts athletes from focusing on performing well. In the following we will elaborate on both perspectives.

2.1 Social Media as a Mood and Behaviour Detection Proxy

In the psychology field there is a consensus that an individual's ability to perform a certain task is greatly affected by his or her mood. While a positive mood is often associated with better concentration, motivation, creativity, and cooperation, a negative mood leads to the exact opposite, consuming many attention resources and recovery efforts [24]. Some researchers make use of this mood-performance relationship and consider social media posts as a way how athletes verbalise their feelings. To extract the mood expressed in the posts, sentiment analysis models are used to capture the polarity of a post in a single number. This sentiment score is then related to the performance of athletes in upcoming competitions. For example, Xu and Yu [24] use sentiment analysis to capture the pre-game mood of NBA players from tweets they posted before games and show that there is a positive linear relationship between the approximated mood and the adjusted Plus/Minus game performance metric of the players. Similarly, Grüttner et al. [8] conduct a statistical test to compare the average first serve fault of ATP and WTA tennis athletes achieved between matches where they had a negative vs. a positive pre-match mood. In contrast to Xu and Yu [24] however they do not find a significant difference between the two groups of interest.

Lim et al. [13] go one step further. Backed by an extensive literature review they claim that there is a positive inverted U-shaped relationship between humility and athletic performance. To investigate this hypothesis, they train a linear regression model to predict NFL players Fantasy Football points in upcoming games based on how arrogant or humble the players appear before the game. Similar to the other mentioned researchers before they approximate humility by the social media content the athletes produce before games. Their results strongly suggest that there is indeed an inverted U-shaped relationship between humility and athletic performance.

2.2 Social Media as a Distraction Factor

In the social psychology field, the Distraction-Conflict Theory (DCT) [1] provides a theoretical attempt to explain the causes of impaired performance levels. According to DCT the mere presence of others can provoke an attentional conflict in an individual performing a certain task which in turn leads to elevated drive and probably impaired performance executing the task. Many researchers use DCT to explain the causes of performance drops among athletes by considering social media as a distractor for athletes. For example, Hayes et al. [9] apply DCT by performing semi-structured interviews with elite Australian athletes to understand the elements of social media athletes perceive to be distracting during competitions. The results suggest that there are five distracting elements, including obligation to respond, susceptibility to unwanted commentary, pressure to build and maintain an athlete brand as well as competitor content and mood management. Another study by Grüttner et al. [8] use DCT to justify that high social media usage of athletes before a competition negatively impacts their performance in both a cognitive and motoric way. The authors give two reasons why high social media usage represents a distractor for athletes. Firstly, the time and focus athletes spend on posting messages limit their capacity focusing on the preparation for the next competition. Secondly, the athletes' awareness that other social media users react to their produced content or post messages related to them may trigger internal distractions. To proof this theory, they conduct a statistical test to check if the difference in average first serve fault of tennis athletes between matches where they posted a large vs. a small number of tweets before the match is significant. Here, they assume that the quantity of posts an athlete generates is a good measure for his social media activity before a competition. Lim et al. [13] use the same connection between post quantity and social media activity in their regression analysis in the context of NFL stars. Similarly, Watkins et al. [22] use the iPhone screen time function to measure the number of hours college athletes spend on social media apps per week and relate the corresponding on-screen time to their competition performance after adjusting for confounding factors. All studies come to the same conclusion, that there is significant evidence that heavy social media usage hinders athletic performance.

Other researchers link social media activity to poor sleep quality. Watkins et al. [22] for example assign college athletes to moderate, active, or super active social media users based on their iPhone screen time and perform an ANCOVA to compare the difference in sleep quality among the three groups. Their results show a significant difference between the groups and that sleep quality tends to decrease with increasing social media activity. Jones et al. [12] consider late-night tweeting, i.e. tweets posted in the middle of the night, as a proxy for sleep deprivation. The authors use t-tests to assess how late night tweeting affects various next-day game statistics of NBA players, including shooting percentage, points scored and rebounds. According to their findings it appears that late night tweeting significantly deteriorates NBA players next-day game performance.

3 Methodology

The studies discussed in the previous section show that social media content can be exploited in various ways to construct features that capture the mindset and well-being of athletes before competitions. In this study we assessed whether such features significantly contribute to the accuracy of a machine learning model predicting athletes' competition performance.

3.1 Data Selection

To answer our research question we focused our analysis on NBA basketball players. This choice was made since the NBA provides well-established and easily accessible performance metrics, and basketball players have already been studied in the context of social media [12,22,24]. Furthermore, we gathered social media posts from Twitter, a platform that is extensively used by NBA players to communicate with peers and fans. Figure 1 depicts the inclusion and exclusion criteria for the NBA dataset and the Twitter dataset.

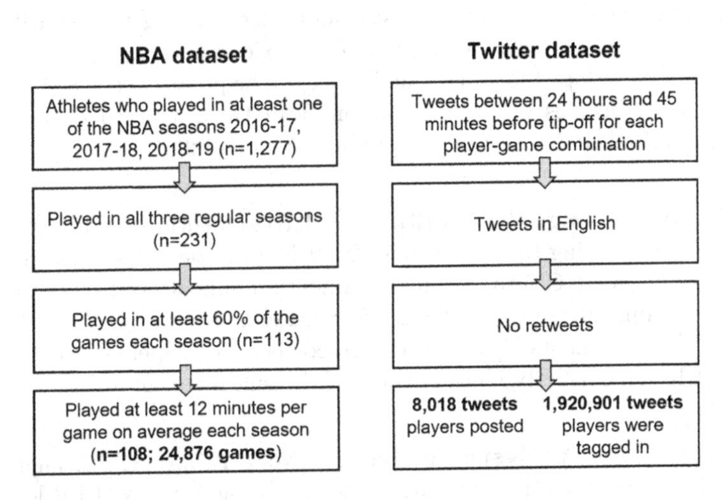

Fig. 1. Inclusion/exclusion criteria for the NBA dataset and the Twitter dataset.

NBA Data: All NBA-related data were collected from basketball-reference.com, a website providing historical basketball statistics from various US and European leagues. We only considered NBA players with a Twitter account and gathered various statistics from games they participated in between the seasons 2016–2019. To avoid potential bias due to the COVID-19 pandemic, we excluded the more recent seasons 2019–2021. Additionally, to account for the effects of long-term injuries we only considered players who obtained playing time in at least 60% of the games each season. Because performance metrics tend to be unreliable when playing time is limited, we only included players with at least one quarter (12 min) of playing time and excluded all games in which the player was on court for less than 5 min. After applying these constraints we had a total of 108 players and 24,876 games in which these players actively participated.

Twitter Data: As social media interaction involves both generating and consuming content, we considered both tweets posted by the players themselves but also tweets produced by others in which the players were mentioned. Particularly, for each game we extracted all player-related tweets that were posted within 24 h before tip-off. Since NBA league policies prohibit players and coaches from using social media from 45 min before a game starts until post-game interviews are completed, we set 45 min before tip-off as an upper limit for tweet extraction for each game. Furthermore, we only considered tweets in English and excluded retweets. These constraints resulted in a total of 8,018 tweets players posted by themselves and 1,920,901 tweets players were tagged in.

3.2 Data Preparation

We preprocessed the textual information of the tweets to bring them into the desired format for a sentiment analysis model determining the polarity of the tweets. To this end, we used VADER [10], a lexicon and rule-based sentiment analysis model developed for social media texts. The polarity scores of the tweets were then used besides other information from the extracted data to create features for a final dataset ready for predictive modelling.

Tweet Preprocessing: We substituted all URLs, mentions and hashtags with placeholders. One thing to note is that VADER only considers empathic uppercasing (e.g. "AMAZING") to capture sentiment amplification [10]. As word elongations (e.g. "amaaaaazing") and gaps between characters (e.g. "D O P E") also intensify word sentiments [7] we decided to correct and uppercase such words so that VADER can correctly identify the shift in sentiment intensity.

Tweet Sentiment Analysis: As the effectiveness of lexicon-based sentiment analysis models is highly context-specific and depends on the words that are present in the lexicon [7] we decided to extend the VADER sentiment lexicon with terms that are frequently used in basketball-related tweets. Therefore, we created a list of words that appeared in at least 0.05% of the tweets and excluded words that were already present in the VADER lexicon. We then manually traversed the list and excluded all terms that did not carry any sentiment or were ambiguous in terms of polarity. For the remaining 101 terms we let 10 annotators rate their polarity, using the same ordinal scale from −4 to 4 of the VADER lexicon. To ensure that participants evaluate each word in a basketball context, 10 randomly selected tweets were added to each word in which the word appeared and presented to the participants. We also familiarised the participants with the meaning of abbreviations like "MVP" (*Most Valuable Player*) or "GOAT" (*Greatest Of All Time*). We averaged over all participants' ratings to obtain a single sentiment score for each word. Table 1 lists example words that were added to the VADER sentiment lexicon.

Table 1. Example words added to VADER lexicon.

Word	Sentiment	
	Mean	SD
goat	3.8	0.42
dpoy	3.8	0.63
mvp	3.6	0.70
lit	3.0	0.67
dope	2.8	0.92
underrated	2.3	0.67
deserved	1.9	1.37
clutch	1.0	1.49
overrated	−2.6	0.70
clown	−2.8	1.03
punk	−3.0	0.82
garbage	−3.5	0.70

The VADER sentiment analysis model with its extended lexicon was then used to determine the polarity of the individual tweets in a range between -1 (very negative sentiment) to 1 (very positive sentiment). The VADER sentiment scores of some example tweets are displayed in Table 2.

Table 2. VADER sentiment scores of example tweets.

Tweet	Sentiment
@RealStevenAdams RESIGN !!	−0.456
Thank you, @swish41 you are my hero! @dallasmavs	0.750
@Pacers @yungsmoove21 Congratulations Thad!!! I hope you're a Pacer forever we love you!! @yungsmoove21	0.921

Final Dataset: With the extracted tweet sentiments we had all information needed to create a final dataset ready for predictive modelling. To allow for a better comparability about the predictive performance of variables we decided to include both social media-related as well as unrelated predictors to the dataset and only considered information that is present before a game starts.

With regard to the social media-related variables we mainly referred to the findings from previous researchers that dealt with the topic (see Sect. 2). Similar to Xu and Yu [24] and Grüttner et al. [8] we used the average sentiment of tweets a player posted before tip-off as a proxy for his mood before the game. Like Grüttner et al. [8] and Lim et al. [13] we used the number of tweets a player posted before tip-off as a measure for his social media activity. Furthermore, with regard to "late night tweeting" [12] we considered tweets players posted at night before the game as a potential indicator for sleep deprivation. To do so we created a binary variable and flagged all games in which a player posted a tweet during normal bedtime (11 p.m. to 7 a.m.) [12] within the time zone of the player's team.

To our knowledge there has not been any research so far that statistically analysed the influence of social media posts athletes were tagged in on their performance. Nevertheless, we believe that there is strong evidence that such posts may also be of use to predict athletic performance. As Hayes el al. [9] indicate, athletes are easily distracted by negative posts addressed to them as such posts give them undesired feelings [9]. For that reason, we included the proportion of negative tweets players were tagged in as a measure for the severity of negative feedback to our set of features. Hayes et al. [9] further note that athletes may also be distracted by the feeling of being compelled to respond to messages addressed to them and that athletes feel guilty if they cannot reply to all messages. As this feeling of guilt may become worse the more posts an athlete is tagged in prior to a competition, we included this variable as a measure for "obligation to respond" [9].

Besides social media-related and unrelated predictors we had to decide for a target that captures the overall performance of an NBA player for a particular game as accurately as possible, considering both offensive and defensive effort. We chose Box Plus Minus (BPM), a metric that uses a player's box score information, position, and the overall performance of the team to estimate the player's contribution in points above league average per 100 possessions played [15]. Table 3 summarises the variables that formed our final dataset.

Table 3. Variables of the final dataset.

	Variable	Summary
(I)	**Social media-related features**	
1	***posted_count***: Number of tweets player posted within 24h before tip-off	$\bar{x}: 0.3$, $s: 1.1$
2	***posted_sentiment***: Mean sentiment of tweets player posted within 24h before tip-off	$\bar{x}: 0.3$, $s: 0.4$, 83% missing
3	***late_night_tweeting***: Flag if player posted a tweet during normal sleeping hours (11pm-7am) in the night before the game	T:2.6%, F:97.4%
4	***tagged_count***: Number of tweets player was tagged within 24h before tip-off	$\bar{x}: 77.1$, $s: 416.9$
5	***tagged_prop_negative***: Proportion of negative tweets player was tagged within 24h before tip-off	$\bar{x}: 0.1$, $s: 0.2$
(II)	**Social media-unrelated features**	
6	***player***: Twitter name of player	108 players
7	***age***: Player age in years	$\bar{x}: 27.4$, $s: 4.0$
8	***tenure***: Years past since player started playing for his current team	$\bar{x}: 3.3$, $s: 2.7$
9	***salary***: Salary of player in Million USD	$\bar{x}: 11.5$, $s: 9.0$
10	***position***: Position of player	SF:14%, PF:18%, PG:22%, SG:22%, C:23%
11	***team***: Team of player	30 teams
12	***opponent_team***: Opponent team	30 teams
13	***homegame***: Flag if homegame for player's team	home:50.4%, away:49.6%
14	***season_type***: Game in regular season or playoffs	regular: 92.5%, playoffs: 7.5%
15	***missing_games***: Number of previous consecutive games player missed, e.g. due to injuries	$\bar{x}: 0.1$, $s: 0.8$
16	***past_BPM***: Player's past 10-game exponential moving average BPM score	$\bar{x}: 0.6$, $s: 4.2$
17	***past_win_percentage***: Team's past 10-game exponential moving average winning percentage	$\bar{x}: 0.5$, $s: 0.2$
(III)	**Target variable**	
18	***BPM***: Player's Box Plus Minus (BPM) score	$\bar{x}: 0.7$, $s: 8.4$

3.3 Predictive Significance Analysis

The formed dataset provided all necessary data to assess whether the features derived from the tweets lead to a significant improvement in accuracy of a model predicting player's BPM score in upcoming games. Unlike other researchers before [13,24] we decided against a linear model in that regard, as its coefficient estimates are prone to be biased if the functional form is inappropriately chosen. Instead, we chose a random forest [3], that, in contrast to parametric models like linear regression, naturally adapts to non-linearities and interactions in the data without any prior knowledge about the data distribution. Besides this advantage, latest research found statistical properties of random forests that turned out to be of use for our analysis. As such, Mentch and Hooker [14] demonstrated that predictions from subsampled random forests can be viewed as incomplete, infinite-order U-statistics that are asymptotically normal so long as the subsample size grows slowly relative to the training set size. The authors made further use of these findings and developed a formal statistical test to assess whether a feature or a set of features make a significant contribution to the prediction for at least one test observation. This test, though valid, becomes computationally prohibitive for test set sizes N_t larger than 20–30 as the test statistic requires the estimation of an $N_t \times N_t$ covariance matrix. Recently, Coleman et al. [5] developed a permutation-style variant of this test that eliminates the need for covariance estimation and thus retains the same computational complexity as the original random forest procedure regardless of the number of test points. The procedure estimates the predictive significance of a subset of variables X by training two random forests: One original forest RF_{orig} that is trained on all features and one reduced forest RF_{red} that is trained on all features where X is randomly permuted to remove any dependence of X to the target variable. The difference in mean squared error (MSE) between the two forests, i.e. $MSE(RF_{red}) - MSE(RF_{orig})$, is then evaluated on a test set as a measure for the importance of X for the prediction of RF_{orig}. To determine the significance of this difference, a permutation distribution is created to approximate a null distribution by repeatedly permuting the predictions between the forests and recomputing the MSE difference. The p-value is then estimated by evaluating the relative frequency of permutations that resulted in a difference as extreme as the observed MSE difference.

We adopted the testing framework by Coleman et al. [5] also in our analysis setting using a significance level of $\alpha = 0.05$ and performed two types of tests: One group test where we assessed the predictive significance of the set of social media derived features (cf. first category in Table 3) as a whole and one marginal test where we assessed the predictive significance of each of these features individually. For all tests we applied 1000 permutations and used 90% of the data for training and the remaining part for testing. To perform the marginal test for *posted_sentiment* we decided to exclude all records where players did not post any tweet before the corresponding game in order to correctly identify effects between player's approximated mood and their BPM score.

For all tests we used the same hyperparameter configuration to train the random forests. In doing so, we have chosen a setting that, in the best case, satisfies the constraints imposed by the test procedure and provides unbiased test results. Consequently, we chose a relatively small subsample size of $n^{0.6}$, where n corresponds to the training set size, as this value also provided robust test results in various experiments conducted by Coleman et al. [5]. By a similar reasoning we set the ensemble size to 500 trees. Since Strobl et al. [21] demonstrated that random forests based on CART trees tend to overestimate the importance of variables the more cut points they offer, we decided to use conditional inference trees as base learners to obtain unbiased predictive significance estimates. Furthermore, we set the minimal node size to 5, the default for regression problems. Finally, we tuned *mtry*, the number of randomly drawn candidate variables for each split, using 10-fold cross-validation with MSE as evaluation measure. This resulted in an optimal value of $mtry = 6$.

3.4 Implementation Details

To collect the tweets, we used *academictwitteR* [2], an R package that provides an interface to access the Twitter Academic Research Product Track v2 API endpoint. Preprocessing of the tweets was done using the R package *textclean* [18]. For efficiency reasons, the random forests needed for the predictive significance tests were trained with the R package *ranger* [23].

4 Results

In our tests, the RF_{orig} achieved a MSE of approximately 61.7 ($RMSE \approx 7.86$, $R^2 \approx 0.13$). The results of the group test, displayed in Fig. 2 (a), suggest that the social media-related features (cf. the first five features in Table 3) make a significant contribution to the prediction of the random forest ($p < 0.001$).

By looking more closely at the marginal tests of the individual features (see Fig. 2 (b)), only *tagged_count*, i.e. the number of tweets in which the player was tagged before tip-off, showed a significant MSE difference between the original and reduced forest ($p \approx 0.002$). The other social media-related features *posted_count* ($p \approx 0.104$), *tagged_prop_negative* ($p \approx 0.346$), *late_night_tweeting* ($p \approx 0.388$) as well as *posted_sentiment* ($p \approx 0.537$) did not significantly contribute to the predictions of the random forest.

In terms of social media-unrelated features the aggregated performance of the player from past games turned out to be the most important feature for the prediction overall, followed by the salary of the player and his position (all $p < 0.001$). Also, *homegame* significantly contributed to the prediction of the random forest ($p \approx 0.007$).

Figure 3 depicts the bivariate relationships of the significant continuous and discrete features to the BPM target. It should be noted that *tagged_count* is represented in logarithmic scale. Interestingly, the relationship between *tagged_count* and BPM appears to be positive. This contradicts our hypothesis

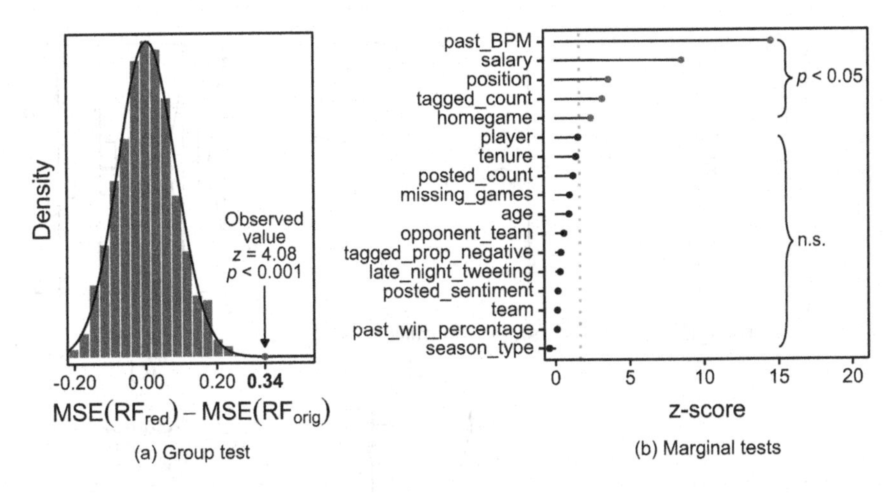

Fig. 2. Results of the group test (a) and marginal tests (b). Social media-related features are highlighted in pink. (Color figure online)

that athletes' performance deteriorates the more messages they receive because they are distracted by the guilt of not being able to respond to all messages.

5 Discussion

The results of our study indicate that features derived from social media posts ultimately lead to a better performing random forest predicting the BPM score of NBA players in upcoming games. However, by having a closer look on the predictive significance of the individual features, only the number of tweets a player was mentioned in before games significantly reduced the MSE of the random forest. With our testing procedure we could not replicate the findings from other studies that support a positive relation between the average sentiment of tweets athletes posted and performance [24], a negative relation between the quantity of tweets athletes produced and performance [8,12,13] as well as a negative relation between night tweeting and performance [12]. Even if athletes' performance depends on their mood, social media activity as well as their sleep quality our findings suggest that the effects are either negligible or our approximations from tweets too inaccurate to predict athletic performance effectively. It should be noted, however, that the random forests in our testing procedure take into account potential confounding factors such as age, tenure, and past performance. Other studies, such as that of Grüttner et al. [8], compared only means using t-tests without adjusting for confounders and thus may have more easily obtained significant results.

Nevertheless, there may be limitations in our study that could have led to inaccuracy of such proxies based on tweets. We believe that one source of error could come from the underlying sentiment analysis model we used to determine the polarity of the tweets. Like other studies before [8,24] we used a lexicon-based

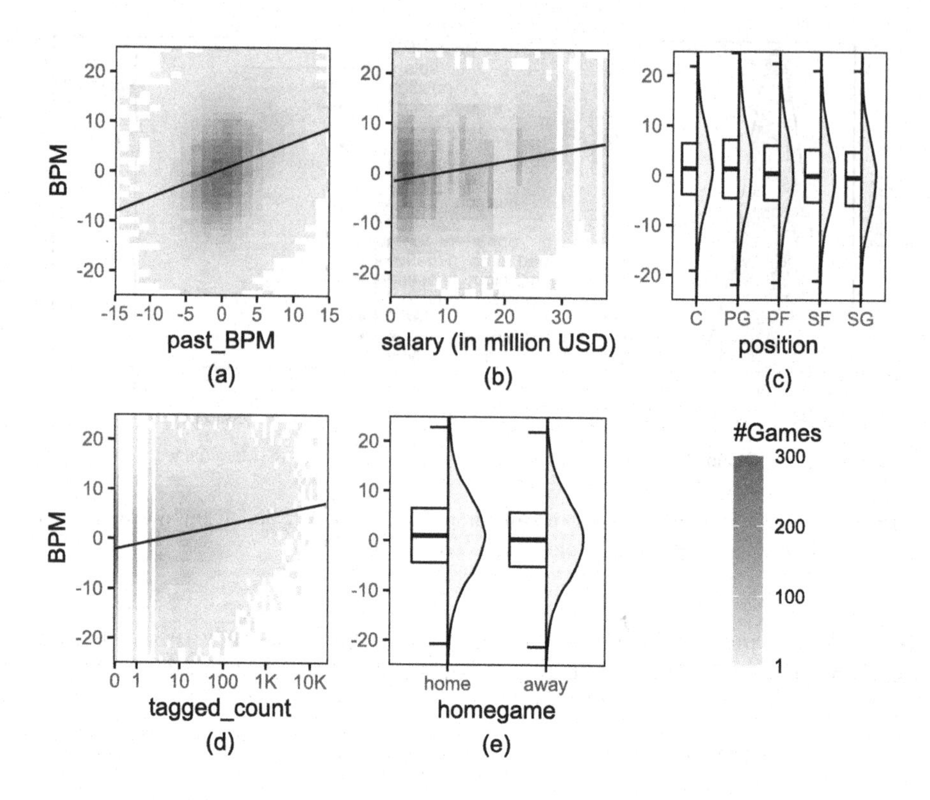

Fig. 3. Relationship between each significant feature (x-axis) and BPM (y-axis).

model in that regard. Such approaches have the advantage of being relatively easy to interpret and efficient to use, but do not consider that words can mean different things in different contexts. This inflexibility often leads to incorrect sentiment predictions. To give an example consider the word "killer". Without any context, one would naturally associate the word with a negative feeling, which is also reflected in the VADER sentiment lexicon, which assigns the word a polarity of −3.3. Now consider the tweet: "@russwest44 should be the most respected NBA player of all time. This guy's game mentality is killer". In this case the sentiment of the word obviously flips to a positive meaning. However, due to the context-unspecific nature of lexicon-based approaches VADER assigns a negative sentiment to the entire tweet of −0.228.

Furthermore, we assumed that each tweet is equally important and free of bias. In terms of tweets the players posted themselves though, more recent tweets may better reflect the true mood of athletes before game time. It is also unclear whether athletes are actually responsible for all of their social media content as nowadays many celebrities employ agencies that maintain their social media profile for them. Similarly, it is unreasonable to believe that athletes actually read all of the up to thousands of tweets in which they are tagged before tip-off. Tweets from teammates, opponent players, family members and friends may have a higher chance of being actually seen by the athletes.

Another limitation concerns our holdout evaluation, which does not take into account the sequential nature of basketball data, since information from future observations is available to influence the predictions of past observations, which may have resulted in overly optimistic estimates. One solution is to resample the data so that only past observations are used to predict future observations, e.g., with leave-one-out cross-validation, blocked cross-validation [20], or leave-future-out cross-validation [4].

Our study provides many opportunities for future research. Firstly, it may be reasonable to repeat our analysis with a more sophisticated sentiment analysis model that is particularly built for basketball-related social media content and capable of detecting context-specific sentiment shifts. In that regard, future research could for example use "distant supervision" [7] to automate the sentiment annotation of basketball-related tweets by making use of emojis or hashtags contained in the tweets. Following this, a transformer-based language model pretrained on tweets like BERTweet [16] could be fine-tuned on the sentiment classification task. Secondly, it may also be interesting to investigate if the expressed emotions in social media text, such as joy, anger, excitement, and fear are useful predictors for athletic performance. Thirdly, future research could also incorporate the time-dependent aspect of social media posts into predictive modelling and train a sequence model to predict athletes' performance in upcoming competitions. Fourthly, our analysis was conducted for only one performance metric, BPM. It would also be interesting to investigate whether our analysis results also apply to other statistics, such as Player Efficiency Rating (PER), Adjusted Plus Minus (APM) or simply field gold percentage (FG%). Fifthly, given that only one competition, basketball, was considered, it might also be of interest to extend the analysis to other competitions. Even within the same competition, one could examine whether there are gender-specific differences by comparing analytical results between the corresponding men's and women's leagues, such as the NBA and WNBA. Lastly, considering that the inclusion of the variable *tagged_count* significantly reduced the MSE of the random forest, it might be interesting to further investigate the influence of the amount of social media content in which athletes are tagged on their performance. In this context, one could also include social media content from other platforms such as Facebook in the analysis.

6 Conclusion

In this study, we investigated the potential of athlete-related social media posts to predict athletes' performance in upcoming competitions. To do this, we extracted tweets NBA players posted themselves or were tagged by others. From these, we derived features that reflect athlete mood, social media behaviour, and sleep quality before games. Using these and other features, we performed a statistical test to investigate whether the MSE of a random forest predicting players' BPM score in upcoming games significantly decreases when the model can utilise the features. The results of this test show that, in particular, the number of tweets NBA players receive before games contributes significantly to the prediction. Contrary to some previous studies, our results neither support a relationship between the average sentiment or number of tweets posted by athletes and their performance nor a relationship between night tweeting and performance. Further research is needed to rule out the possibility that this is due to the inaccuracy of the sentiment analysis model or to limiting assumptions we made about athletes' social media interaction behaviour (e.g. athletes may not be responsible for content posted from their account).

References

1. Baron, R.S.: Distraction-conflict theory: progress and problems. Adv. Exp. Soc. Psychol. **19**, 1–40 (1986). https://doi.org/10.1016/S0065-2601(08)60211-7
2. Barrie, C., Chun-ting Ho, J.: academictwitteR: an R package to access the twitter academic research product track v2 API endpoint. J. Open Source Softw. **6**(62), 3272 (2021). https://doi.org/10.21105/joss.03272
3. Breiman, L.: Random forest. Mach. Learn. **45**(1), 5–32 (2001). https://doi.org/10.1023/A:1010933404324
4. Bürkner, P.C., Gabry, J., Vehtari, A.: Approximate leave-future-out cross-validation for bayesian time series models. J. Stat. Comput. Simul. **90**(14), 2499–2523 (2020). https://doi.org/10.1080/00949655.2020.1783262
5. Coleman, T., Peng, W., Mentch, L.: Scalable and efficient hypothesis testing with random forests (2019). https://doi.org/10.48550/arXiv.1904.07830
6. ESPN: vince carter addresses the negative effects of social media on athletes (2020). https://www.youtube.com/watch?v=1cX5_2YadU4. Accessed 03 Mar 2022
7. Giachanou, A., Crestani, F.: Like it or not: a survey of twitter sentiment analysis methods. ACM Comput. Surv. (CSUR) **49**(2), 1–41 (2016). https://doi.org/10.1145/2938640
8. Grüttner, A., Vitisvorakarn, M., Wambsganss, T., Rietsche, R., Back, A.: The new window to athletes' soul-what social media tells us about athletes' performances. In: Proceeding of Hawaii International Conference on System Sciences (HICSS), pp. 2479–2488 (2020). https://doi.org/10.24251/HICSS.2020.303
9. Hayes, M., Filo, K., Geurin, A., Riot, C.: An exploration of the distractions inherent to social media use among athletes. Sport Manage. Rev. **23**(5), 852–868 (2020). https://doi.org/10.1016/j.smr.2019.12.006
10. Hutto, C., Gilbert, E.: VADER: a parsimonious rule-based model for sentiment analysis of social media text. In: Proceeding of AAAI Conference on Web and Social Media, vol. 8, pp. 216–225 (2014). https://www.aaai.org/ocs/index.php/ICWSM/ICWSM14/paper/view/8109/8122
11. Iso-Ahola, S.E.: Intrapersonal and interpersonal factors in athletic performance. Scandinavian J. Med. Sci. Sports **5**(4), 191–199 (1995). https://doi.org/10.1111/j.1600-0838.1995.tb00035.x
12. Jones, J.J., Kirschen, G.W., Kancharla, S., Hale, L.: Association between late-night tweeting and next-day game performance among professional basketball players. Sleep Health **5**(1), 68–71 (2019). https://doi.org/10.1016/j.sleh.2018.09.005

13. Lim, J.H., Donovan, L.A.N., Kaufman, P., Ishida, C.: Professional athletes' social media use and player performance: evidence from the national football league. Int. J. Sport Commun. **14**(1), 1–27 (2020). https://doi.org/10.1123/ijsc.2020-0055
14. Mentch, L., Hooker, G.: Quantifying uncertainty in random forests via confidence intervals and hypothesis tests. J. Mach. Learn. Res. **17**(1), 841–881 (2016)
15. Myers, D.: About Box Plus/Minus (BPM) (2020). https://www.basketball-reference.com/about/bpm2.html. Accessed 12 Mar 2022
16. Nguyen, D.Q., Vu, T., Nguyen, A.T.: BERTweet: a pre-trained language model for english tweets (2020). https://arxiv.org/abs/2005.10200
17. von Ott, K., Puymbroeck, M.V.: Does the media impact athletic performance. Sport J. **9**(3), (2006)
18. Rinker, T.W.: Textclean: text cleaning tools. Buffalo, New York (2018). https://github.com/trinker/textclean, version 0.9.3
19. Rousidis, D., Koukaras, P., Tjortjis, C.: Social media prediction: a literature review. Multimedia Tools Appl. **79**(9), 6279–6311 (2020). https://doi.org/10.1007/s11042-019-08291-9
20. Snijders, T.A.: On cross-validation for predictor evaluation in time series. In: On Model Uncertainty and its Statistical Implications, pp. 56–69. Springer (1988). https://doi.org/10.1007/978-3-642-61564-1_4
21. Strobl, C., Boulesteix, A.L., Zeileis, A., Hothorn, T.: Bias in random forest variable importance measures: illustrations, sources and a solution. BMC Bioinformatics **8**(1), 1–21 (2007). https://doi.org/10.1186/1471-2105-8-25
22. Watkins, R.A., Sugimoto, D., Hunt, D.L., Oldham, J.R., Stracciolini, A.: The impact of social media use on sleep quality and performance among collegiate athletes. Orthop. J. Sports Med. **9**(7_suppl3) (2021). https://doi.org/10.1177/2325967121S00087
23. Wright, M.N., Ziegler, A.: ranger: a fast implementation of random forests for high dimensional data in C++ and R. arXiv preprint arXiv:1508.04409 (2015). https://arxiv.org/abs/1508.04409
24. Xu, C., Yu, Y.: Measuring NBA players' mood by mining athlete-generated content. In: Proceeding of Hawaii International Conference on System Sciences (HICSS), pp. 1706–1713. IEEE (2015). https://doi.org/10.1109/HICSS.2015.205

Link Prediction with Text in Online Social Networks: The Role of Textual Content on High-Resolution Temporal Data

Manuel Dileo(✉), Cheick Tidiane Ba, Matteo Zignani, and Sabrina Gaito

Computer Science Department, Universitá degli Studi di Milano, Milan, Italy
{manuel.dileo,cheick.ba,matteo.zignani,sabrina.gaito}@unimi.it

Abstract. Machine learning-based solutions for link prediction in Online Social Networks (OSNs) have been the subject of many research efforts. While most of them are mainly focused on the global and local properties of the graph structure surrounding links, a few take also into account additional contextual information, such as the textual content produced by OSN accounts. In this paper we cope with the latter solutions to i) evaluate the role of textual data in enhancing performances in the link prediction task on OSN; and ii) identify strengths and weaknesses of different machine learning approaches when dealing with properties extracted from text. We conducted the evaluation of several tools, from well-established methods such as logistic regression or ensemble methods to more recent deep learning architectures for graph representation learning, on a novel dataset gathered from an emerging blockchain online social network. This dataset represents a valuable playground for link prediction evaluation since it offers high-resolution temporal data on link creation and textual data for each account. Our findings show that the combination of structural and textual features enhances the prediction performance of traditional models. Deep learning architectures outperform the traditional ones and they can also benefit from the addition of textual features. However, some textual attributes can also reduce the prediction power of some deep architectures. In general, deep learning models are promising solutions even for the link prediction task with textual content but may suffer the introduction of structured properties inferred from the text.

Keywords: Online social network · Link prediction · Graph neural networks · Temporal dataset

1 Introduction

In network science, link prediction is one of the most powerful tools, successfully applied in different settings, such as predicting network evolution in online social networks, protein-to-protein interactions, or predicting links in knowledge graphs. The most popular link prediction approaches employ structural information, such as node similarity and centrality measures, to yield the prediction;

P. Pascal and D. Ienco (Eds.): DS 2022, LNAI 13601, pp. 212–226, 2022.
https://doi.org/10.1007/978-3-031-18840-4_16

while more recent approaches rely on graph embedding and graph neural networks to improve prediction performances [10]. When it comes down to link prediction in online social networks, current works have successfully leveraged structural features. However, the role of textual information on link formation remains an open question. This is an important issue to solve as the information derived from text may improve prediction and give insight into the mechanisms leading the link formation process. Indeed, text is crucial in online social networks, being one of the main driving forces of user engagement; and advertisement on content is the main source of revenue for these platforms. Yet, we still have a limited understanding of the impact of text on link formation. One of the main reasons is that it is hard to obtain appropriate data for the task: current research either lacks text information or high-resolution temporal data on network growth. Also, there is a lack of studies on the proper methodology to include the additional information so as to enhance performances in network-specific tasks such as node classification, link prediction, or even community detection.

To this aim, in this work, we performed link prediction with textual information on a temporal attributed network. As a case study, we rely on Steemit, a blockchain-based online social network, that allows the retrieval of high-resolution temporal information that we can use to construct an attributed temporal network. Specifically, we focused on and gathered temporal data about "follow" relationships between users and text content produced by users - posts and comments. As for the availability of suitable and rich data for tackling network-specific tasks, blockchain-based platforms - Web3 platforms - are a great opportunity for researchers in different fields thanks to the huge volume of high-resolution data stored in the supporting blockchains. Indeed, by the nature of blockchains, data are publicly available, validated, and affordable by interfacing with the blockchain's API. Moreover, each piece of information is timestamped since each blockchain block has a validation timestamp; and each block reported multi-faceted interactions and content - social, economic, financial, and textual. So, these data sources have all the features to face tasks and issues related to modern techno-social networks and to support detailed and in-depth analysis of users' traits. Starting from this kind of temporal and heterogeneous data, in this work, we define a methodology to include text information to perform link prediction. Indeed, nodes - users - in the graph are characterized by a set of textual features capturing the statistical properties of their textual content and the topics they treat. Then, we investigate the impact of these text features on the link prediction task, identifying strengths and weaknesses of different machine learning approaches when dealing with properties extracted from text. We conducted the evaluation of several tools, from well-established methods such as logistic regression or ensemble methods to more recent deep learning architectures for graph representation learning, such as graph neural networks and graph autoencoders. In the evaluation, we have taken into account different settings from a full knowledge of structural and textual information to structural features only or subsets of them.

The outcome of the prediction task shows that the combination of structural and textual features improves prediction performance on traditional models. Moreover, we also provide insights on the important text characteristics for link prediction. Through an exhaustive comparison of the models, we show that GNNs outperform the other models in terms of prediction performance, and how textual features increase their power. However, some textual attributes can also reduce the prediction power of some deep architectures. In general, even for the link prediction task with textual content, deep learning models are promising solutions but may suffer the introduction of structured properties inferred from text. Finally, we discuss potential extensions for this work.

The paper is organized as follows. Section 2 provides a brief introduction to the nature of blockchain-based online social networks and a review of works related to link prediction with textual data. In Sect. 3 we describe the construction of the temporal attributed network, the models for predicting links, and how structural and textual features are extracted. In Sect. 4 we provide a description of the dataset, while Sects. 5 and 6 report the main findings of link prediction and a discussion about strengths and weaknesses of the different models and settings.

2 Background

Link Prediction with Text. Link prediction is meaningful for solving numerous issues. Its main objective is to estimate network evolution by inferring the likelihood that pairs of nodes have to either form links or not in the future. Kumar *et al.* [10] reviewed several approaches to link prediction from classical to recent network embedding and deep learning techniques. Some works, such as [2], perform link prediction in dynamic networks. Despite being intensively analyzed, various questions are still open, and many studies try to adapt the prediction problem to recent developments for supplying the newest research gaps. Among the works on link prediction that use text information, [13] relies on text user attributes to model user profile data, using Latent Dirichlet Allocation - LDA - to model topics; here, link prediction is only based on the resulting topic distributions, and not on the network structure. Other works, like [15], have improved prediction performance by fusing a network generated from users' posts with the original "follow" network, but they do not consider content-based features. Overall, relying on text seems to improve prediction performance: however, these approaches have been tested only on static networks. Moreover, there is limited understanding of which text-based features should be used.

Blockchain-Based OSNs. Recently, numerous discussions about the privacy protection problem and the compliance with the regulation arose due to many cyber-attack scandals that highlighted the weaknesses and the deficiencies of the early OSNs centralized structure. In response, several proposals exploiting decentralized technologies have emerged. Among them, we find blockchain-based social networks (BOSNs). Essentially, the blockchain provides a cryptocurrency system and a data storage and validation layer. An interesting case study is Steemit, a notable example of BOSN. Steemit relies on a cryptocurrency-based reward system, to involve users in the growth of the platform, while also stimulating network activities. Also, the Steem blockchain supports two distinct cryptocurrency exchanges, namely the STEEM and the Steem based Dollar (SBD). As every action is stored on a blockchain, these platforms provide a detailed data source of network activity, covering not only the social side, like users' follows, comments, and votes, but also the economic sphere, for example, users' cryptocurrency exchanges. Such characteristics have made BOSNs, especially Steemit, the subject of several recent studies. The most recent advancements are illustrated in a recent survey [5]. Another interesting point is that, unlike in main online social media, due to privacy reasons, Steemit does not allow to create a profile page where users can store personal information such as bio, gender, or location. Hence, Steemit lacks user attributes and the extraction of node features from the textual content is crucial to try to enhance structural link prediction.

3 Methodology

In online social networks, users post content for other users. Given the high amount of content, users can follow each other: when user A starts following another user B, user A starts receiving updates on the B's posts. This allows users to not miss other users' content. Alongside this information, we also have the user-generated content i.e. posts and comments on posts, so that we can understand the impact of user content on the formation of "follow" links. Here, we aim at answering the following research question: what is the impact of textual features on link prediction tasks? In this section, we describe the link prediction models used in this work and how to model data from "follow" operations and user-generated content to generate structural and text-based features used to perform link prediction.

3.1 Graph Construction and Sequence-Based Framework

The first step is to construct the "follow" graphs by retrieving the "follow" relationships. Then, the nodes in the "follow" graphs can be enriched with textual attributes.

"Follow" links and text information can be modeled as an attributed temporal directed graph $\mathcal{G} = (V, E, T, X)$, where V is the set of users, links $(u, v, t) \in E$ denote a directed "follow" link from user u to user v at time t (the time in which user u starts to follow user v), and X is a $|V| \times f$ matrix of node attributes, with f the dimension of attribute vectors. Given a time interval $[t_0, t_1]$, the snapshot graph $\mathcal{G}_{[t_0,t_1]}$ represents the directed graph, where for each link $e = (u, v, t) \in E$, we have that $t \in [t_0, t_1]$.

Given a graph interval snapshot $\mathcal{G}_{[t_0,t_1]}$, the purpose of link prediction is to predict which edges will appear at a successive interval snapshot $\mathcal{G}_{[t_1,t_2]}$. It can be treated as a binary classification task, where we assign label 1 if the link is predicted to form in the following time interval, 0 otherwise.

The main idea is to realize a sequence-based framework so that the evaluation of the link prediction algorithms can be assessed on a successively built dataset. To this aim, we rely on the experimental setting for temporal link prediction presented in [11]. Given a time interval $[t_0, t_1]$, a train set with links in $\mathcal{G}_{[t_0,t_1]}$ can be created and their status can be predicted in the time interval $[t_1, t_2]$. Whereas for test set, links are extracted in $\mathcal{G}_{[t_0,t_2]}$ and their status predicted in $[t_2, t_3]$. Given two graph $\mathcal{G}_{t-1}$ and $\mathcal{G}_t$, where $t-1$ and t are following intervals (for instance, $[t_0, t_1]$ and $[t_1, t_2]$):

- $\mathcal{G}_{t-1}$ is used to compute the well-known state-of-the-art structural features, the textual features, and to retrieve the list of edges and their relative nodes.
- $\mathcal{G}_t$ is obtained as an induced sub-graph constrained around the nodes of G_{t-1}. This limitation makes it possible to yield an effective understanding of how a graph and its connections evolve. Then, only the edges closed in t and not in $t-1$ are considered to form the **positive set**. Simultaneously, starting from the same seed of nodes, also a set of randomly extracted non-existing edges are considered to form the **negative set**. The final dataset results in the combination of the positive and the negative sets and for each item a binary label y is added to indicate if that item is an existing edge or not.

Selecting a subset of edges at random from the original complete set is one of the most common methods to perform test set sampling. Despite this strategy may conduct to over-optimistic results, there are evaluation measures for which subsampling negatives from the test set has no negative effects [17].

The length of the time intervals should be chosen to ensure a sufficient number of snapshots for analysis, a sufficient number of training and test examples for the algorithms, while ensuring the duration between two successive intervals is not too long. This work considers two intervals of two and one month, respectively.

3.2 Learning Algorithms for Link Prediction in Temporal OSNs

Link prediction models refer to predictors that can identify pairs of nodes that will either form a link or not in the future. Two main groups of models were employed to perform the predictions: *traditional* (i.e. well-known in literature) *supervised models* and *graph neural networks*. The first ones are feature-based models (i.e. they can work only with feature vectors extracted for each pair of nodes), while the latter can work directly on graph-structured data.

The traditional supervised models used in this analysis are Logistic Regression (LR), Support Vector Classification (SVC), Multilayer Perceptron (MLP), Random Forest (RF), and Gradient Boosting (GB); their performance evaluation is performed with F1 measure. Below the graph neural networks will be briefly described.

Graph Neural Networks. (GNNs) are a family of neural networks that can operate naturally on graph-structured data [16]. By extracting and utilizing features from the underlying graph, GNNs can make more accurate predictions about entities in these interactions, as compared to models that consider individual entities in isolation.

GNNs learn to map individual nodes to fixed-size real-valued vectors called embeddings. The learned embeddings summarize the structural information of the network taking into consideration also the attributes of the nodes. Then, those vectorial representations can be used to solve different useful problems on graphs (e.g. link prediction). Different GNN variants are distinguished by how these representations are computed, but the general idea is to extend convolution to graphs. To this end, GNNs construct polynomial filters on graphs [4].

Focusing on a particular node v and a 1-hop localized convolution, we can think of this operation as arising of two steps: the aggregation over immediate neighbor features $x_u(u \in \mathcal{N}(v))$ and the combination with the node's own feature x_v. Starting from this idea, different kinds of "aggregation" and "combinations" steps can be considered to build new types of layers. The GNN layers used in this work are:

- *Graph Convolutional Networks (GCNConv).* GCNConv computes the aggregation step between neighbor features by averaging them and the combination step by summing up the neighbor contribution and the node's own feature [9];
- *Graph Sample and Aggregate (SAGEConv).* SAGEConv can be seen as a variation of GCNConv in which different kinds of aggregations can be computed. For instance, focusing on a single node v, the dimension-wise maximum between the embeddings of the neighbors of v can be considered. The combination step is computed by concatenating the embeddings [7];
- *Graph Attention Networks (GATConv).* GATConv introduces an attention mechanism [12] in the computation of the aggregation step. The key benefit of this introduction is to allow for implicitly specifying different importance values to different neighbors of a node. This is achieved by multiplying the embeddings of the neighbors of v by attention weights generated by the attention mechanism at each convolutional step [14].

A graph neural network can be built by stacking layers one after the other with non-linearities, much like a standard CNN. Given a pair of nodes (x, y), the embeddings $h(x)$ and $h(y)$ can be obtained through a GNN and then the probability p that x and y form a link can be estimated as $p = \sigma(h(x)h(y))$ where σ is a sigmoid function applied to the dot product between the node embeddings.

Figure 1 shows the proposed GNN architecture to solve link prediction tasks. There is an input layer with a number of neurons equal to the number of features, two graph convolutional layers that perform the nodes embedding, a dot product to compute the links embedding and then a sigmoid function to output values in $[0, 1]$. The first hidden layer maps points from a space with a number of dimensions equal to the number of features to a space with a certain number of dimensions while the second maps points to a space with two dimensions. In contrast to CNNs, it is typically discouraged to add more than two or three graph convolutional layers to a GNN because it can cause a problem known as over-smoothing [3]. For a discussion related to the computational complexity of GNN models you can refer to [16]. It should be notice that training our GNN model is not computationally more expensive than training a logistic regression model.

We have also tested the *Graph Autoencoder* (GAE) approach. GAE is the natural extension of auto-encoder in the realm of graph computation. This model makes use of latent variables and is capable of learning interpretable latent representations for undirected graphs. Given the adjacency matrix A of a graph (with diagonal elements set to one) and the node features matrix X, GAE uses a two-layer GCN as encoder to calculate the embeddings matrix $Z = \mathrm{GCN}(X, A)$ and then a simple inner product as decoder to reconstruct the adjacency matrix $\hat{A} = \sigma(ZZ^{\intercal})$ where σ is a non-linear function. The GAE architecture used in the experiments is the one presented in [8].

For performance evaluation, since the outputs are the results of the application of a sigmoid function on link embeddings, the area under the receiver operating characteristic curve (AUROC) score will be used to present their performance. Another reason to choose the AUROC score is the robustness of ROC curves and their associated areas to class imbalance, a problem that affects the link prediction domain [17].

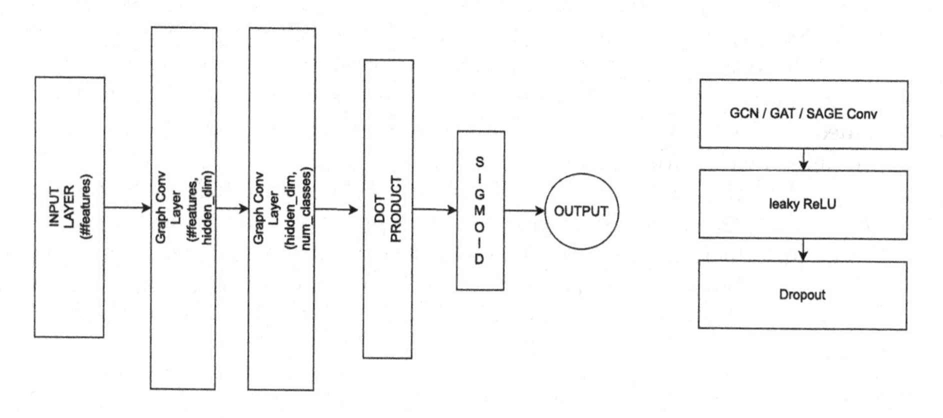

Fig. 1. The proposed GNN architecture to solve link prediction tasks. Left diagram: complete architecture overview. We used two graph convolutional layers to obtain node embeddings and the dot product followed by a sigmoid function to obtain the link predictions. Right diagram: single graph conv. layer overview. We tested GCN, GAT, and SAGE as graph conv. operators, leaky ReLu as activation function and dropout.

3.3 Features for Link Prediction

The extracted features can be split into two types: *structural* and *textual* features. The first ones summarize the key aspects of the user behavior in terms

Table 1. Structural features. The first column reports the category of the feature and the second column details the features belonging to a category.

Type	Structural features
Local similarity metrics	Common neighbors
	Adamic Adar
	Resource Allocation
	Preferential Attachment
	Jaccard's Coefficient
Centrality-based methods	PageRank
	Katz Index
	Local Random Walk
	Neighbors' degree
	In-degree centrality
	Out-degree centrality

of connections with the other users and they regard the topological structure of the social network; the latter is related to the user behavior in terms of published content on the social platform. Textual features can be also distinguished in *text-based statistics* and *user similarity*. All these distinctions are useful to understand which combination of features can give more accurate results when tackling link prediction.

Structural Features. Given a pair of nodes, we compute some of the most used similarity metrics [11]. Specifically, we choose some of the so-called "local" similarity metrics, like Common Neighbors, Jaccard's coefficient, Adamic/Adar's coefficient, Resource Allocation, and Preferential Attachment index. Alongside them, we rely on some centrality-based methods like Local Random Walk, Katz centrality, PageRank, average neighbor degree, in-degree centrality and out-degree centrality, so that for each link we have a pair of measures, one for each vertex. Table 1 summarizes the structural features used in this work.

Textual Features. For each user, we consider two types of features derived from text: *i) text-based statistics* and *ii) user similarity*. Text-based statistics are computed on the corpus formed by user's posts, comments and tags, in the considered time interval. Specifically, we compute the number of posts and comments, the number of tags, the average and standard deviation of the length of the content produced. Whereas for user similarity, we describe the similarity between user content, in terms of language and topics of interest. For topic similarity, we rely on topic modeling with Latent Dirichlet Allocation (LDA), as in [13]. Therefore, given an author and a document, we compute a topic vector;

Table 2. Textual features. The first column reports the category of the feature and the second column details the features belonging to a category. In parentheses, the name of the feature.

Type	Textual features
Text-based statistics	Number of posts (num_post_from, num_post_to)
	Number of comments (num_comment_from, num_comment_to)
	Number of tags (num_tag_from, num_tag_to)
	Average content length (avg_post_length_from, avg_post_length_to)
	Content length standard deviation (std_length_from, std_length_to)
User similarity	Tag Jaccard's coefficient
	Topic distances (Euclidean, Chebyshev, Cosine, Jaccard)

then, to represent a user interest, we average all of its topic vectors. So, given two users and their topic vectors, we can compute a series of similarity/distance measures to capture how close their interests are. We selected cosine similarity, Euclidean distance and Jaccard distance. On the same note, we consider Jaccard's coefficient, to capture language similarity, specifically between the set of tags used by authors. Table 2 summarizes the textual features used in this work.

4 Dataset

Users on Steemit can perform many different actions, called operations. These operations track users' activities with a temporal precision of 3 s. Every operation can be retrieved with a specific API. Through the API, data from June 3, 2016, up to January 21, 2021, have been collected. The starting date is the day the "follow" operation has been made available on Steemit.

Two type of information have been gathered: *a*) the "follow" relationships, available in the `custom_json` transactions; and *b*) posts, comments and their tags, available in the `comment` transactions. Dataset was processed according to the methodology presented in Sect. 3 to generate the attributed graph.

We create the training set with links in $\mathcal{G}_{[t_0,t_1]}$, and predict their status in the following time interval $[t_1, t_2]$. Whereas for the test set, we extract links in $\mathcal{G}_{[t_0,t_2]}$ and predict their status in the next interval, $[t_2, t_3]$. We consider two intervals: period 1 from June 3, 2016, to August 2, 2016 ($[t_0, t_1]$), period 2 from June 3, 2016, to September 2, 2016 ($[t_0, t_2]$); while the time interval $[t_2, t_3]$ refers to period from September 3, 2016, to October 2, 2016. We reduce the intervals by one month after the first period, to have a similar amount of links, as the network started to grow.

The main properties of the resulting graphs are summarized in Table 3.

Since data refers to the early months of Steemit, there is an explosion of new links in the next period for the first two months. The periods considered see a rapid

Table 3. Main properties of the Steemit "follow" graph $\mathcal{G}$ for period 1 $[t_0, t_1]$ - from June 3, 2016, to August 2, 2016 - and the following period 2 $[t_0, t_2]$ - from June 3, 2016, to September 2, 2016.

	$\mathcal{G}_{[t_0,t_1]}$	$\mathcal{G}_{[t_0,t_2]}$
Number of nodes	7,400	20,849
Number of edges	33,920	323,228
Density	0.0006	0.007
Min/Max in degree	0/466	0/2,735
Min/Max out degree	0/643	0/11,824
Avg degree	9.17	31.01
Std degree	25.90	206.43
Strongly connected components	4,972	8,266
Largest SCC	2,313	12,505
New links in the next period	74,228	138,604

network evolution [1] and this makes them interesting intervals to analyze. Note that the maximum in-degree is much smaller than the maximum out-degree. This can be explained by the design of the platform. Following other users is free, does not require confirmation by the followed user and can bring some visibility returns while attracting users is hard and requires a lot of effort [6].

As for textual information, overall, we obtain 327,151 posts, 756,239 comments, and an average number of tags equal to 1.88. The number of contents per user is shown in Fig. 2. Both distributions have high variance and show the heterogeneity of users in terms of the amount of the published content. Some nodes spend a lot of effort posting content and text-based statistics can identify them. They might be influencers, gurus, enthusiasts, and bots, all of whom are typically involved in a lot of relationships with other users.

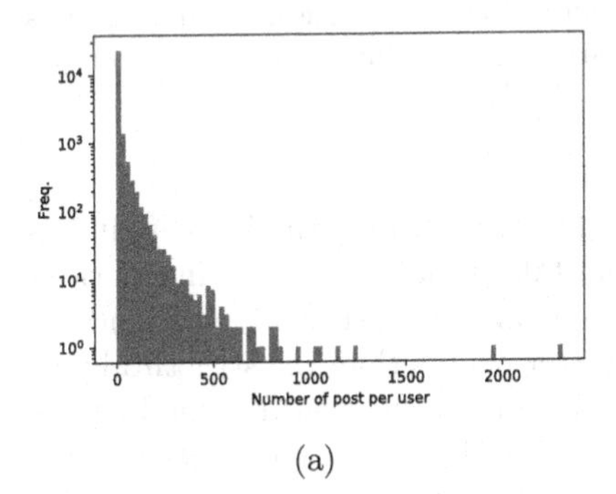

(a)

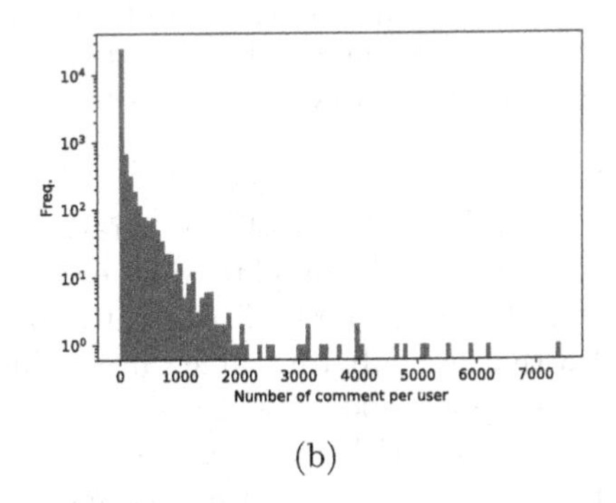

(b)

Fig. 2. Distribution of the number of contents per user from June 3, 2016, to September 2, 2016. In **(a)** the number of posts per user, and in **(b)** the number of comments per user, respectively.

5 Results

We conducted the experiments using the models described in Sect. 3.2 and performing hyperparameter optimization via grid search. The combinations of structural, textual, structural and textual features have been considered. Code is available on a Github repository[1].

5.1 Results for Traditional Models

Figure 3 shows the F1-scores of the Logistic Regression with the different combinations of features for the test set. It shows also the results of a Dummy classifier that generates predictions uniformly at random from $\{0, 1\}$. We choose to present the results of the Logistic Regression as it is the model that has achieved the best performances among the ones considered in this work.

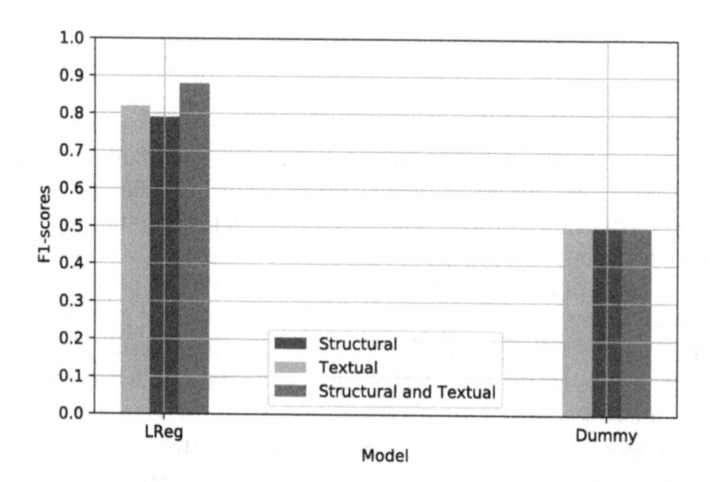

Fig. 3. F1-scores of Logistic Regression (LReg) and Dummy Classifier (Dummy) models for link prediction on the test set, using structural features (blue), textual features (orange), structural and textual features (green). The combination of structural and textual features leads to an increase in F1-score. (Color figure online)

The combination of structural and textual features leads to an increase in F1-score. Due to the rapid network evolution that takes place in the two time periods considered, the best LR configuration uses a strong regularization term and the LR that use only textual features achieves better results than the structural one.

We show the most important 20 structural or textual features for Logistic Regression, with their feature importance, in Fig. 4. We can notice that some textual features are considered more important than the most important structural features. Among the text features, we can see that Jaccard's coefficient

[1] https://github.com/manuel-dileo/link-prediction-with-text.

between the set of tags written by users emerges: this feature is a simple, yet effective way to capture if two authors talk about the same topics so having a few interests in common.

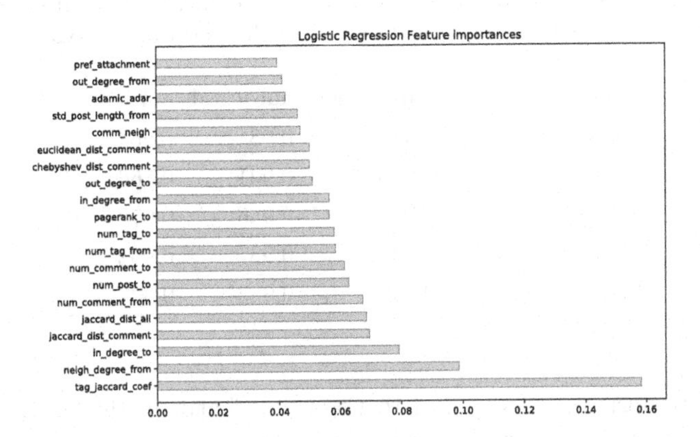

Fig. 4. Feature importance of the top 20 structural or textual important features for Logistic Regression on link prediction task. Importance values are based on the coefficients of the features in the decision function of the Logistic Regression. We observe a blend of both types of features. The most important features are at the bottom of the plot.

5.2 Results for Graph Neural Networks

For GNNs, we conducted the experiments using the architecture proposed in Sect. 3.2. We have used 200 as the number of epochs and batch learning. The following combinations of features have been considered:

- **Constant**. A single constant and equal value for all the nodes as a feature. It means not using any feature.
- **Struct**. Structural features. Specifically, PageRank, in and out-degree, and average neighbor degree have been used. Note that the structure of the graph is always taken into consideration when GNNs are used, so adding structural features acts as a feature augmentation technique.
- **TextStat**. Text-based statistics as node features.
- **TextStat+Topics**. Combination of text-based statistics and user interest vector as node features.
- **StructText**. Combination of structural and textual features.

Table 4 shows the AUROC of GNNs with different combinations of features on the train, validation, and test set. Overall, the results of Graph Neural Network models on link prediction tasks outperform those obtained using traditional supervised models. The use of text-based statistics as node features leads to an

Table 4. Area under the receiver operating characteristic curve of GNNs using no features (Constant), structural features (Struct), text-based statistics features (TextStat), text-statistics and user interest features (TextStat+Topics), structural and text features (StructText), for link prediction on train, validation and test set. The use of text features can lead to an increase in AUROC score but the structure of the network is crucial. Structural feature augmentation makes the performance worse.

Model	auroc_train	auroc_val	auroc_test
Constant	0.937	0.93	0.964
Struct	0.946	0.931	0.766
TextStat	0.94	0.936	0.973
TextStat+Topics	0.943	0.93	0.962
StructText	0.943	0.933	0.902

increase in performance compared to not using features. However, the performance gain is low so the structure is crucial to understand the network evolution. Note also that the addition of user interest vectors as node features does not enhance the performance; hence, not every addition of textual features leads to an increase in performance. The best configuration of hyperparameters has two layers with hidden size of four and two neurons, and Adam as the optimizer, with learning rate and weight decay respectively equal to 0.025 and $5 \cdot 10^{-5}$. Another interesting point is the addition of manually engineered structural features as node attributes. Structural feature augmentation makes the performance worse. This problem arises because the periods considered, as shown in Table 3, see a rapid network evolution; hence, centrality measures are not able to summarize in an effective way the structural information.

Table 5 shows the AUROC scores of GAEs with the different combinations of features on the train, validation, and test set. They describe a behavior very similar to what is presented for GNNs but with worse performances on the test set.

Table 5. Area under the receiver operating characteristic curve of GAEs using no features (Constant), structural features (Struct), text-based statistics features (TextStat), text-statistics and user interest features (TextStat+Topics), structural and text features (StructText), for link prediction on train, validation and test set. The use of text-based statistics as node features leads to an increase in AUROC score.

Model	auroc_train	auroc_val	auroc_test
Constant	0.9034	0.904	0.877
Struct	0.925	0.926	0.868
TextStat	0.917	0.918	0.881
TextStat+Topics	0.90	0.90	0.844
StructText	0.895	0.895	0.867

6 Discussion

In this work, we investigated the role of text on link formation, an important task as text could improve prediction and give insight into the link formation process. To this end, we performed link prediction with text on a temporal attributed network. We relied on Steemit, a blockchain-based online social network, that allows the retrieval of high-resolution temporal information but lacks user attributes due to data control and privacy reasons. We have provided a methodology to use text information alongside traditional structural information and a temporal framework to train and test the models.

First, we showed that the combination of structural and textual features improved prediction performance in terms of F1 score on the traditional supervised models. Then, we showed that some textual features are considered more important than the most important structural features. This is important as we tested on two time intervals where the network changes a lot, hence a dominance of the structural features in terms of importance could have led to poor performance.

GNNs reach an AUROC score of 0.97 working naturally on graph-structured data and using textual information as node features. Textual features enhance the performance of a GNN that works without node features while if the features are augmented through structural information, such as centrality indices, the performance in terms of AUROC score decreases. However, not every addition of textual features leads to an increase in prediction performance; hence, understanding which features extract from textual content and performing a feature selection step, based on the network being studied, is important. In general, deep learning models are promising solutions even for the link prediction task with textual content but may suffer from the introduction of structured properties inferred from text.

Future works will include an extension of the considered features, more precisely economical features. We also plan to test more complex GNNs models. We plan to incorporate several modern deep learning modules into a GNN layer (e.g. Batch Normalization) as well as the addition of dense layers before the graph convolutions. Another interesting point is being able to consider edge features. We want to analyze if graph neural networks only perform prediction, in a black-box way, or whether they also gives us a better understanding of the network being studied. Finally, we will focus on the analysis of different temporal snapshots, based on network evolution and growth: we will study the performance when training and testing snapshots are similar and when they show significant differences.

References

1. Ba, C.T., Zignani, M., Gaito, S.: The role of cryptocurrency in the dynamics of blockchain-based social networks: the case of steemit. Plos One **17**(6), 1–22 (2022). https://doi.org/10.1371/journal.pone.0267612

2. Barracchia, E., Pio, G., Bifet, A., Gomes, H.M., Pfahringer, B., Ceci, M.: Lp-robin: link prediction in dynamic networks exploiting incremental node embedding. Inf. Sci. **606** (2022). https://doi.org/10.1016/j.ins.2022.05.079
3. Chen, D., Lin, Y., Li, W., Li, P., Zhou, J., Sun, X.: Measuring and relieving the over-smoothing problem for graph neural networks from the topological view (2019). 10.48550/ARXIV.1909.03211, arxiv.org/abs/1909.03211
4. Defferrard, M., Bresson, X., Vandergheynst, P.: Convolutional neural networks on graphs with fast localized spectral filtering. **29**, (2016)
5. Guidi, B.: An overview of blockchain online social media from the technical point of view. Appl. Sci. **11**(21), 9880 (2021)
6. Guidi, B., Michienzi, A., Ricci, L.: A graph-based socioeconomic analysis of steemit. IEEE Trans. Comput. Soc. Syst. 1–12 (2020). https://doi.org/10.1109/TCSS.2020.3042745
7. Hamilton, W.L., Ying, R., Leskovec, J.: Inductive representation learning on large graphs (2018)
8. Kipf, T.N., Welling, M.: Variational graph auto-encoders (2016)
9. Kipf, T.N., Welling, M.: Semi-supervised classification with graph convolutional networks (2017)
10. Kumar, A., Singh, S.S., Singh, K., Biswas, B.: Link prediction techniques, applications, and performance: A survey. Physica A-stat. Mech. Appl. **553**, 124289 (2020)
11. Liu, Q., et al.: Network growth and link prediction through an empirical lens. Proceedings of the 2016 Internet Measurement Conference (2016)
12. Niu, Z., Zhong, G., Yu, H.: A review on the attention mechanism of deep learning. Neurocomputing **452** 48–62 (2021). https://doi.org/10.1016/j.neucom.2021.03.091
13. Parimi, R., Caragea, D.: Predicting friendship links in social networks using a topic modeling approach. In: Pacific-Asia Conference on Knowledge Discovery and Data Mining (PAKDD), pp.75–86 (2011)
14. Veličković, P., Cucurull, G., Casanova, A., Romero, A., Lió, P., Bengio, Y.: Graph attention networks (2018)
15. Wang, Z., Liang, J., Li, R.: Exploiting user-to-user topic inclusion degree for link prediction in social-information networks. Expert Syst. Appl. **108**, 143–158 (2018)
16. Wu, Z., Pan, S., Chen, F., Long, G., Zhang, C., Yu, P.S.: A comprehensive survey on graph neural networks. IEEE Trans. Neural Netw. Learn. Syst. **32**(1), 4–24 (2021). https://doi.org/10.1109/tnnls.2020.2978386, https://doi.org/10.1109%2Ftnnls.2020.2978386
17. Yang, Y., Lichtenwalter, R.N., Chawla, N.V.: Evaluating link prediction methods. Knowl. Inf. Syst. **45**(3), 751–782 (2014). https://doi.org/10.1007/s10115-014-0789-0, https://doi.org/10.1007%2Fs10115-014-0789-0

Weakly Supervised Named Entity Recognition for Carbon Storage Using Deep Neural Networks

René Gómez Londoño[1,3], Sylvain Wlodarczyk[1], Molood Arman[1,2,3], Francesca Bugiotti[2,3](✉), and Nacéra Bennacer Seghouani[2,3]

[1] Services Pétroliers Schlumberger, 34000 Montpellier, France
{swlodarczyk,marman2}@slb.com
[2] Paris-Saclay University, CNRS, LISN, 91405 Orsay, France
{francesca.bugiotti,nacera.seghouani}@lri.fr
[3] CentraleSupélec, Paris-Saclay University, 91405 Orsay, France
rene.gomez@student-cs.fr

Abstract. Applying Transfer-Learning based on pre-trained language models has become popular in Natural Language Processing. In this paper, we present a weakly supervised Named Entity Recognition system that uses a pre-trained BERT model and applies two consecutive fine tuning steps. We aim to reduce the amount of human labour required for annotating data by proposing a framework which starts by creating a data set that uses lexicons and pattern recognition on documents. This first noisy data set is used in the first fine tuning step. Then, we apply a second fine tuning step on a small manually refined subset of data. We apply and compare our system with the standard fine tuning BERT approach on large amount of old scanned document. Those documents are North Sea Oil & Gas reports and the knowledge extraction would be used to assess the possibility of future carbon sequestration. Furthermore, we empirically demonstrate the flexibility of our framework showing that it can be applied to entity-identifications in other domains.

Keywords: Natural language processing · Named entity recognition · Deep neural networks · Stratigraphy

1 Introduction

Carbon sequestration in the North Sea is a way to reduce the global warming to below 1.5 °C. Several Northern European countries are currently engaging in solutions to store carbon under the North Sea in old Oil & Gas reservoirs. One of the difficulty in carbon storage is to entirely reassess the ancient reservoirs by interpreting many documents such as end of well reports, or core laboratory reports written during the long life cycle of the reservoir. Those documents are very heterogeneous and many of them are accessible only thanks to OCR techniques that do not provide clean data. In this case of study, the geologists study the rock strata and categorize them given the information embedded in those documents. Multiple analyses are performed in the domain of stratigraphy, that is the study of the physical and temporal relationships between rock layers or strata.

P. Pascal and D. Ienco (Eds.): DS 2022, LNAI 13601, pp. 227–242, 2022.
https://doi.org/10.1007/978-3-031-18840-4_17

For running this analysis, a source of information that is fundamental but generally underused is the set of geological well reports accumulated and produced during the whole history of a reservoir. Before the digital transformation of Oil & Gas industry, these analyses were run on a manually-converted subset of these documents. Nowadays, thanks to cloud computing and new technologies, it could be possible to handle a large amount of heterogeneous data and exploit a valuable source of historical information. Also, from the computational point of view, the analysis becomes more complex to evaluate, and analysis needs all useful data to be considered.

Those documents are underused because the geologists and the petrophysicists need to convert the information manually into structured tables. Usually, from these structured tables, they can populate the numerical models. These documents do not follow a given structure, and old documents are often written by typewriters and are accessible thanks to OCR techniques that do not provide clean data.

Name Entity Recognition (NER) [7] identifies the mentioned entities in unstructured texts and classifies them into target categories. Extracting the correct entities in the domain of the stratigraphy is capital information to evaluate a reservoir. Referring to our context, we can select as classification categories the period, the age, the era, the formation, etc. In the literature the performance of language models based on the Deep Neural Network (DNN) transformers architecture has produced interesting results in information extraction for many specific domains. The problem, however, is to provide the network with the necessary amount of labelled data required for the training phase. A recent state-of-the-art method for NER is to fine-tune a pre-trained BERT model using a labelled dataset with the corresponding entities we want to identify.

In our approach, we create this labelled dataset with a weakly supervised approach by using lexicons and labelling functions. This labelled dataset can be very large but also noisy as it comes from scanned documents and weak supervision. The hyperparameters of this first stage will be adapted to the "noisy" nature of the dataset. We then manually correct a very small subset of the noisy dataset and apply a second fine-tuning step with adapted hyperparameters. By comparing the results with a one-step fine-tuning approach, including the manually corrected dataset, we show that this workflow improves the results of precision by two (2) percentage points and recall by five (5) points. Increasing 5 points in recall means gaining a huge amount of information as we have massive data to process. We propose and test three language models with a human-reviewed data set. We present results for three Name Entity Recognition models, including a light version and compare with the state-of-the-art fine-tuned BERT model. Our results show a precision of 90%, recall of 96%, and F1 score of 93%. We finally provide some recommendations to apply our approach in other domains.

This paper is structured as follows. In Sect. 2 we describe the objectives, and we identify the main contributions of our approach. In Sect. 3 we introduce the fundamentals of our research focusing on the concepts related to Name Entity Recognition. In Sect. 4, we detail our methodology. In Sect. 5 we present the evaluation of the methodology.

In Sect. 6 we discuss related work, and we compare this research to the existing literature. Finally, in Sect. 7 we draw conclusions and some limitations and open challenges that remain subject for future work.

2 Overview

The objective of this research is to build a Named Entity Recognition system using Deep Neural Networks with a weakly supervised training process. To avoid complex feature engineering or continuous labelling and extraction work from the domain experts, we use a deep neural network-based approach. In the context of interest, training data is not available and annotating data is a labour-intensive task for geologists. To overcome such an obstacle, we decided to rely on a distant supervision approach to create noisy labels using external resources like regular expressions and dictionaries. It is a common scenario for a geologist to extract information from a report using regular expressions. Each regular expression identifies an entity and defines a sequence of characters that is used as a search pattern in each report. Multiple chunks of text could match the given search pattern, even text that is not a valid entity. The geologist might not realize this mismatch and erroneous entities are commonly identified (False-Positive). Such matches in NLP tools can produce alignment errors in the labels. As a second scenario, suppose instead using dictionaries related to the energy domain. The matching process should be straightforward and precise. Even in this scenario, False-Positives are commonly produced because of polysemy: words in the entity dictionary might be used in another context with a different meaning.

These two cases demonstrate that additional effort is required for cleaning the results by using pure text matching to extract the final entities. This would drastically hurt the system's scalability. To solve this problem, we use training data to build a deep neural network model that produces clean results and helps us by the generalization capacity of language models to detect unseen entities based on the contextual representation of their tokens Table 1.

The problem we introduced is studied in our domain but is common to many domains [14,21]. In Fig. 1 we show an example where NER is presented as a

Table 1. Sequence to Sequence Task Classification.

Tokens	BIO	BILOU
Diego	B-PER	B-PER
Armando	I-PER	I-PER
Maradonna	I-PER	I-PER
was	O	O
born	O	O
in	O	O
1960	B-DATE	U-DATE

Tokens-entities:
Diego Armando Maradona PERSON
was born in **1960** DATE .

sequence classification task. Specifically, we treat it as a sequence-to-sequence problem: given a token sequence (a sentence) as input, we produce the corresponding sequence of labels as output.

The approach is flexible enough to incorporate new target entities without labour-intense human annotation and sufficiently robust to reduce the necessity of result post-processing. The methodology is composed of the following steps:

1. The first step is the creation of a noisy training set for Named Entity Recognition. Given a set of documents, we aim to facilitate the text extraction task to generate a noisy training set on large data sets using dictionaries and regular expressions. The goal is to build an approach that can be run on distributed processing frameworks.
2. Given a noisy training set, we aim to use transfer learning to evaluate different DNN models incorporating contextual representations and using training techniques to avoid learning the noisy labels.
3. Given a set of pre-trained language models we want to evaluate the performances using a test set reviewed by human annotators. The evaluation shall be done having precision, recall and F1 score as metrics adapted for sequence evaluation.

2.1 Contributions

Given the described challenges, the methodology steps, and the technical constraints, our research achieves the following contributions:

1. The definition of a Named Entity Recognition System, establishing a baseline for future model benchmarking.
2. The implementation of a distributed framework enables data labelling using NER annotation schemas (like BIO and BILOU).
3. The implementation of a detailed two fine-tuning process of a pre-trained BERT model using in the first step, a large and noisy dataset created automatically and in the second step a small and clean human reviewed dataset. The hyperparameters are adapted in each step to fit the specific nature of each training data.
4. The evaluation of the approach utilizing sequence evaluation criteria from CoNLL (precision, recall, and F1 score adapted for text sequences) against human-reviewed data sets.

Furthermore, the same pipeline can be applied to other domains without a huge effort by changing the dictionaries and regular expressions.

3 Background

The main task of this project is to generate a framework to facilitate noisy data set creation, model training, and evaluation for a Named Entity Recognition system. For this purpose in our domain, we focus on a set of entities whose identification is a recurring challenge, given the nature of the geological reports.

The well was drilled beyond the required TD in order to investigate potential in the Carboniferous. At -4042ft TVDSS (4178ft MDKB) a strong reverse drill break (from 160ft/hr to 40ft/hr) was encountered. The well was drilled on to -4058ft TVDSS (4194ft MDKB) and a checkshot survey run to verify with seismic whether it was likely that the Carboniferous had been penetrated. Integration of the well results with seismic confirmed that the amplitude anomaly at 0.54 seconds had been penetrated and that the well was at, or very close to, the Top Carboniferous.

As the commitment depth had been exceeded, no shows had been seen in the well (not even a background of 1ppm) and the Carboniferous was thought to have been tagged, it was decided to TD the well (Figure 2).

1.3 Well data summary

Well no.	112/29-1
Surface location	Lat. 54° 06' 11.5758"N Lon. 04° 19' 20.7060"W UTM (3 deg west) 5995942.7N 413528.8E
Seismic line	JS-IOM92-05
RKB to MSL	136ft
Water depth	110ft
Classification	Exploration
Prim. objective	Triassic Sherwood Sandstone Group
Drill. contractor	Global Marine
Drilling rig	Glomar Adriatic XI
Spud date	13/05/96
TD date	21/05/96
Completion date	24/05/96
Total depth	4194ft MDKB (-4058ft TVDSS)
TD formation	?Carboniferous
Status	P&A dry hole

Fig. 1. An example of an end of well report scanned and converted to pdf format. We manually highlighted the various entities we would like to identify such as the DEPTH_INTERVAL, the FORMATION, the WELL_ID and the AGE.

An example of a well-report is shown in Fig. 1. The text present in the document is very noisy and difficult to interpret, even for a human reader. Documents of this format are written at the end of the drilling process of each well. The document contains critical information to assess a reservoir. When the interpretation of the reservoir is performed during the drilling process, the interpretation of the reported data is handled in real-time by humans. When we need to reassess reservoirs, for example, for evaluating carbon capture storage capabilities, wells were drilled decades ago, and the geologist cannot reread them to assign the information to thousands of wells. That is why we need to create models to perform the task automatically.

The text annotation pipeline uses external resources, matching lexicons in dictionaries, and regular expression patterns. The proposed approach avoids complicated pre/post-processing to provide positive examples for training.

To define the scope of this project, we selected a variant of useful entities to study similar scenarios like the ones proposed by [23]. In the following part of the section, we present each entity and the challenges that we commonly find in its identification process.

Defined Named Entities. An effective analysis must include entities that are: evident from the model, highly noisy, characterized by a limited number of possible instances and finally, entities that could be easily confused between them. Thanks to our methodology we expect to have good accuracy in all of them, but we also aim to detect which are the type of entities that remain challenging to define the future work in this project. The list of entities we are focused on in this presentation are:

(1) **Well Identifier** End of wells reports describe all the studies for one particular well. For instance, `30/2a-8` is a typical well identifier (WELL_ID entity) in the nomenclature of the north sea region. Regular expressions are flexible enough to detect those entities, but we will also detect many noisy labels. For this kind of entity, we want to avoid post-processing operations, improve the quality of the results and generalise the identification (i.e., the USA uses different nomenclature for well identification).
(2) **Period, age & epoch.** The geologic time scale is the "calendar" for events in Earth's history. It subdivides all time into named units of abstract time called eons, eras, periods, epochs, and ages. AGE and PERIOD entities are almost well-defined dictionaries, we expect high-performance detecting them. The EPOCH entity has a specific challenge as it comes from a dictionary containing both unique names and general terms (i.e., `early`, `late`, `lower`, etc.). We aim that in the sentence `the drilling process started late`, the word `late` will not be identified as an EPOCH.
(3) **Formation.** A geological formation consists of a certain amount of rock strata with comparable geological properties. This FORMATION entity is complex, with names ranging from rivers, areas, parks, towns or regions.
(4) **Depth interval and interval.** Depth intervals represent the boundaries of the formations. They usually follow a pattern of `number unit` to/-/and `number unit measure_reference`. The unit could be `feet` or `meters`, with their variations (i.e., `ft`,',", `mt` or `m`). MEASURE_REFERENCE is the reference point or type of the depth (i.e., `True Vertical Depth (TVD)`, `Measure Depth (MD)`, etc.). We also introduced a more relaxed entity, the INTERVAL that follows a similar pattern to the depth interval but without unit and MEASURE_REFERENCE. Since it is a flexible entity, it leads to False Positives, but it helps the model to identify some depth intervals that would be lost otherwise.

4 Methodology

In this section we describe our methodology from the labels generation to the training process of the DNN. Afterwards, we present a more in-depth study for the DNN's training process and finally explain how we use pre-trained language models to accomplish our downstream task.

An overview of the methodology is presented in Fig. 2. It involves multiples stages, starting with the data set creation and finishing with the model training and evaluation.

(1) The lack of labelled training data has limited the development of NLP tools. We use distant supervision resources (dictionaries & regular expressions) to

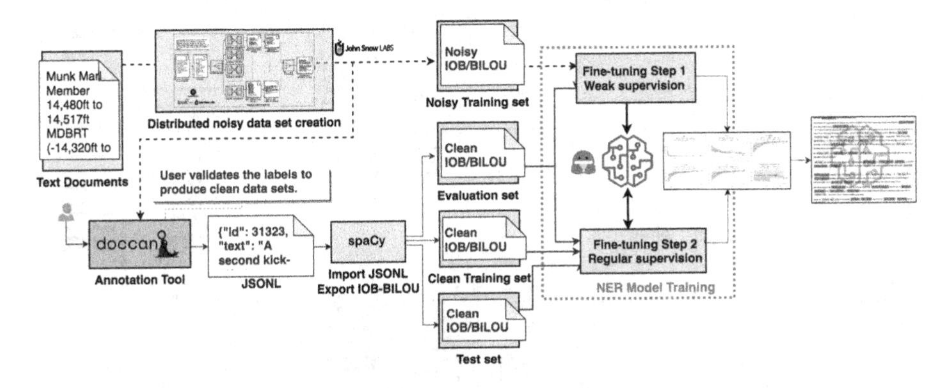

Fig. 2. Implementation pipeline.

create labelled data in a semi-automatic way. This removes the need for intense manual data labelling. The problem is that we get not only True-Positive but also False-Positive examples.

(2) Since we are going to use a noisy data set, we clean part of the labels with an annotation tool to generate a proper evaluation and test set. Notice that we don't annotate from scratch but review the semi-automatic generated labels. We just correct enough examples to control the training process and evaluate the final results.
(3) We used noisy samples to train the model with most of the default parameters, varying the batch size and learning rate. Each batch contains a random number of clean and noisy examples.
(4) According to [23] and [1], using the recommended parameters should be enough. Still, we monitor the training process with a small clean evaluation set to detect in which case the noisy examples start to be learnt by the neural network. In theory, we should see fluctuating loss and performance metrics for the evaluation set.
(5) The output model is selected on the basis of the sequence evaluation performance.

4.1 Noisy Data Set Creation

One of the driver elements in our methodology is the data set creation. Without labelled data we follow a weak supervision approach using dictionaries and regular expressions. Our data set creation pipeline is detailed in Fig. 3.

We remove newline characters and normalize the text to avoid rare characters produced by the OCR system. We then tokenize the text and run a sentence detector model.

The matcher component finds the corresponding chunks where the dictionary or regular expressions match the specific sentence. Lexicons were collected from different internal applications where stratigraphic units are used to describe

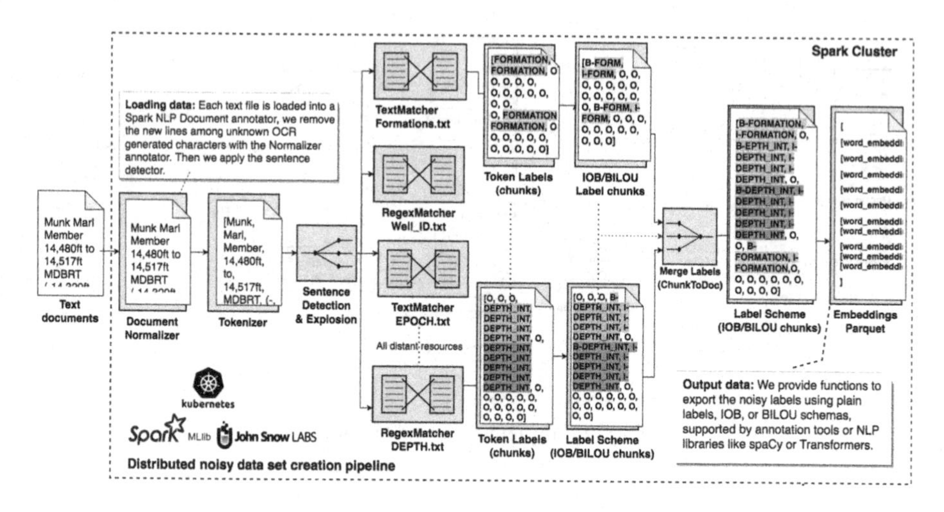

Fig. 3. Spark NLP implementation pipeline

well logs. However, public information like Wikipedia's taxonomies or specific knowledge bases is commonly used as data sources in such applications.

The matches are then converted into token-level labels. When we have overlapped labels, we have to keep the longest match. Here it is crucial to keep the text alignment with the labels, always keeping one label for each token.

Finally, we built an exporter to save BIO/BILOU files.

4.2 Overcoming Noisy Labels Effect

In this section we present our steps to train the DNN avoiding the noise overfitting.

First of all, to reduce some of the negative effects of label errors, we use language models, which means that not only the entity influences the learning process but also the context in which it appears. Under this scenario, noisy examples are harder to be learnt. Not all of them follow one common usually perfect pattern as clean examples do. As additional bias they occur in similar contexts: the representation is then not as close as the clean examples. Batch size and learning rate are fundamental hyperparameters in our context as already stated by [23,25]. We deeply rely on the straightforward approach explored by [23]. The authors demonstrated that larger batch sizes are better to overcome the effect of noisy data labels. The authors argue that the negative impact of uncorrelated or less correlated noise types is diminishing since updates caused by noisy samples are overwhelmed by gradient updates from clean samples. [25] got similar results, observing that DNN trained on noisy labelled datasets with a high learning rate do not memorise noisy labels.

Two-Step Fine-Tuning. [23] suggests that learning with big batch size is enough to mitigate the noise effect. We follow this approach using also a clean evaluation set. This enables to monitor the training evolution to ensure the best hyperparameters configuration for removing noise. Finally, we select the batch size that presents the most consistent behaviour during training. To avoid noise overfitting, we might need to use an early stopping of the training process. We consider this as an adaptation step towards all our domain-specific language where we learn simple entities and patterns. However, if the model does not learn noisy labels it might also be having lousy performance in the difficult to learn patterns, or confusing similar classes like INTERVAL and DEPTH_INTERVAL. In such a case, we plan to run a second fine-tuning stage with regular supervision. It is, having a small training set with clean examples and using traditional hyperparameters to refine the details that might be missing during the first fine-tuning stage. Moreover, we want to evaluate if applying this methodology, we can change the behaviour in the polysemy problem. We expect to influence the algorithm and to see some changes in the predictions for words like *lower* and *late* in the EPOCH entity as the primary example. Additionally, since we are in a transfer learning setting, we use some clean and reliable negative examples to execute a second fine-tuning stage. We intend to evaluate if this helps the model improving the details that might be excluded during the first fine-tuning phase. In this second step we must avoid the forgetting problem [22]. We do this with following strategy: regardless of the errors, we won't target any particular entity but randomly select examples to learn the details. We want to keep the clean training set small, with a similar size to the validation set. Such training set has examples from all the previously learned entities. We are not incorporating a new named-entity or a completely different context. This two-step fine-tuning strategy works even better in more complicated scenarios, where the original training data is not available. Hence, we don't expect any drawback from using it in this more convenient environment.

5 Evaluation

The architecture of the system is provided in Fig. 2. For the project implementation, we use PySpark in a Kubernetes cluster deployed on Google Cloud Platform. For the data set creation, we used a cluster with 16 GB in the driver node and 4 workers with 8 GB RAM each. The training process was done in one single node with 64 GB RAM without GPU.

Specifically, we use Spark NLP for weak data labelling and train the models using Transformers (PyTorch version). The number of resources assigned to the project varied according to the cluster state or the executed task. Our normal configuration for the cluster was with 32 GB of memory in the driver node and four executors with 8 GB each.

The output from the lexicons and regular expressions were cleaned and cross-validated by two engineers using Doccano. Complex examples were verified with domain experts. The reports are publicly available on the Oil and Gas Authority website [16]. The pre-processing code and the OCR were performed by Schlumberger and are not publicly available. The training process was done using the

Table 2. Data sets for training.

Entity	Noisy set	Clean set	Eval set	Test set
WELL_ID	15754	125	151	345
FORMATION	18424	159	167	381
INTERVAL	9218	93	83	189
EPOCH	19366	166	156	360
AGE	11243	130	118	280
PERIOD	7416	79	87	166
DEPTH_INT	4258	40	56	92
TOTAL	85679	792	818	1813

public available HuggingFace Transformers training process with the described hyperparameters. We track our experiments using Weights&Biases (W&B).

Data sets. We collected examples from one thousand different geological reports with more than seven million tokens. We executed the automated noisy data labelling pipeline and we got more than 125,000 sentences with approximately 227,000 entities. However, we did not use the entire data set for our proof of concept. We randomly selected sentences to create the training and evaluation sets. For the noisy training, clean training, and test set we selected respectively 50000, 500, and 1000 sentences.

We present the entities and the number of instances in Table 2.

Evaluation Results. Across all the experiments we use seqeval [18], a framework for sequence labelling evaluation following the CoNLL-2000 shared task data guidelines. Instead of evaluating token by token, the sequence is evaluated based on complete detected named entities. The framework also takes into account class imbalance, ignoring, for instance, the tokens that are not entities labelled as O. We focused our experiments in testing several models using different batch sizes and learning rates as described in the methodology section, evaluating its effect in the fine-tuning steps. For other hyper-parameters, we used the recommended values suggested in [9], with sequences of maximum 128 tokens. Note that we use BERT-Base-Cased like models because we have a lot of capitalized names or upper case codes in our documents.

Figure 4 shows the results over three models: BERT [9] and the HuggingFace distilled version of BERT and RoBERTa [24].

We could see that most of the time, the distilled version of RoBERTa is outperformed by the BERT and the distilled BERT model in all metrics. Hence we decided to focus on the BERT and distilled BERT model.

We present the performance of the two selected models in Table 3.

As explained in Sect. 3 the high performance in entities like DEPTH_INTERVAL and PERIOD were expected, since these entities are consistent with dictionaries

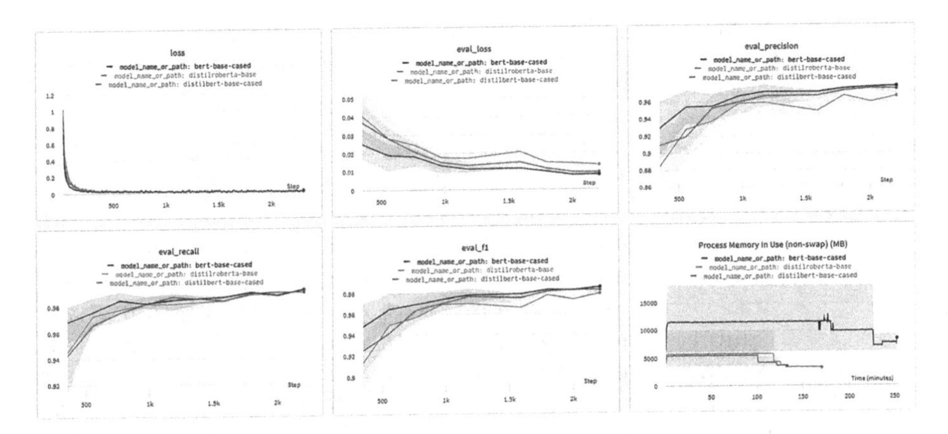

Fig. 4. Benchmark of three pre-trained models.

Table 3. Results for test set. DistilBERT and BERT with a BatchSize of 64. First and second fine-tuning results

Named entity	d-BERT-64 St 1			d-BERT-64 St 2			BERT-64 St 1			BERT-64 St 2			Supp
	P	R	F1	P	R	F1	P	R	F1	P	R	F1	
DEPTH_INT	0.99	0.99	0.99	0.98	0.98	0.98	0.98	0.98	0.98	0.95	0.98	0.96	92
FORMATION	0.90	0.86	0.88	0.84	0.90	0.87	0.92	0.87	0.89	0.85	0.91	0.88	381
WELL_ID	0.46	0.48	0.47	0.90	0.96	0.93	0.46	0.48	0.47	0.91	0.96	0.94	345
AGE	0.97	0.97	0.97	0.96	0.97	0.97	0.96	0.97	0.97	0.96	0.97	0.97	280
PERIOD	0.98	0.99	0.98	0.99	0.99	0.99	0.99	0.99	0.99	0.99	0.99	0.99	166
INTERVAL	0.92	0.97	0.95	0.93	0.97	0.95	0.93	0.95	0.94	0.92	0.96	0.94	189
EPOCH	0.89	0.98	0.93	0.89	0.97	0.93	0.91	0.99	0.94	0.90	0.98	0.94	360

or well-defined patterns. It helps us to evaluate that we are not degrading the performance in the well-known consistent cases. Furthermore, with them, we evaluate the performance in other entities like INTERVAL or FORMATION, where the former is a pattern similar to other non-entities tokens present in the text, and the latter comes from incomplete dictionaries. The WELL_ID is the hardest entity to learn since they have an inconsistent pattern that matches other tokens (i.e., section numbers and coordinates, which can also appear without context).

To validate the advantage of using the two-step fine-tuning approach, we learned a single-step fine-tuning BERT model and the equivalent distilled BERT model using the combination of the noisy and the clean training set as a unique training data set. We present the performances of these models in Table 4.

Result Discussion. The two steps training method presents a slightly better precision (2% points improvement) than the single-step fine-tuning BERT model. Furthermore, the two steps model training has, as expected, a better recall (up

Table 4. Results for test set. BERT with a batch size of 64. BERT Stage 1 and Stage 2 are the two fine-tuned results, whereas stage 2 is the final result. BERT Single-Step is the single-step fine-tuned BERT model

Bert version	Named entity	BERT Stage 1			BERT stage 2			BERT single-Step			Supp
		P	R	F1	P	R	F1	P	R	F1	
Distilled Bert	DEPTH_INT	0.99	0.99	0.99	0.98	0.98	0.98	0.96	0.98	0.97	92
	FORMATION	0.9	0.86	0.88	0.84	0.9	0.87	0.90	0.86	0.88	381
	WELL_ID	0.46	0.48	0.47	0.9	0.96	0.93	0.62	0.64	0.63	345
	AGE	0.97	0.97	0.97	0.96	0.97	0.97	0.97	0.98	0.97	280
	PERIOD	0.98	0.99	0.98	0.99	0.99	0.99	0.99	0.99	0.99	166
	INTERVAL	0.92	0.97	0.95	0.93	0.97	0.95	0.93	0.96	0.94	189
	EPOCH	0.89	0.98	0.93	0.89	0.97	0.93	0.91	0.99	0.94	360
	Micro avg	**0.84**	**0.86**	**0.85**	**0.91**	**0.96**	**0.93**	**0.87**	**0.89**	**0.88**	**1813**
Bert	DEPTH_INT	0.98	0.98	0.98	0.95	0.98	0.96	0.92	0.98	0.95	92
	FORMATION	0.92	0.87	0.89	0.85	0.91	0.88	0.90	0.87	0.89	381
	WELL_ID	0.46	0.48	0.47	0.91	0.96	0.94	0.72	0.74	0.73	345
	AGE	0.96	0.97	0.97	0.96	0.97	0.97	0.97	0.97	0.97	280
	PERIOD	0.99	0.99	0.99	0.99	0.99	0.99	0.99	0.99	0.99	166
	INTERVAL	0.93	0.95	0.94	0.92	0.96	0.94	0.95	0.96	0.96	189
	EPOCH	0.91	0.99	0.94	0.9	0.98	0.94	0.91	0.99	0.94	360
	Micro avg	**0.85**	**0.86**	**0.85**	**0.91**	**0.96**	**0.94**	**0.89**	**0.91**	**0.90**	**1813**

to 5% points) given the fact that the longer training time for the single-step fine-tuning BERT model reduces its flexibility to identify new entities.

We see that the second fine-tuning step improves the accuracy and precision of the models. First, as shown in Table 3 the high performance was maintained for the consistent entities, which was expected since the second training set contains clean examples for all the entities. In other words, we introduced examples of all entities avoiding the catastrophic forgetting problem.

Furthermore, the training set in the second step was focused on providing cleaner examples for the WELL_ID and AGE. Therefore, it makes sense that it helped the model predict multi-token WELL_ID. The second fine-tuning step catches the full WELL_ID with proper boundaries, as shown in Table 5.

In a second round of analysis, we also evaluated the generalization capacity of the models by testing non-existing ages such as `Sylvanian` or `Renotian`. In the sentence "`The late Sylvanian is...`", the token `late` was identified as an EPOCH with a probability of 99% and `Sylvanian` as an AGE with a probability of 70%. Notice that the model never saw `Sylvanian` as an example before, but it appears with a similar structure to other AGE names, and in the same sentence (context) there is an EPOCH (the word `late`), hence the model classified it as an AGE. Nevertheless, it is only 70% confident about the prediction (since it has never been seen before). In the sentence "`I was late for class`", the token `late` was NOT identified as an EPOCH by the BERT_ST2 model: it is the same token as a valid EPOCH, but the context is not valid; therefore it has another

Table 5. Example of multi-token WELL_ID. The model fails to catch the full multi-token WELL_ID with only the first fine-tuned step but succeeds with the second fine-tuned step.

Token	BERT ST1	BERT ST2
Well	B-WELL_ID	B-WELL_ID
13_22a	B-WELL_ID	I-WELL_ID
–	B-WELL_ID	I-WELL_ID
C29X	O	I-WELL_ID
wellsite	O	O
Geological	O	O

meaning. It shows that the second training step has a great potential to improve the capacity to remove the False Positives introduced by words with multiple meanings.

6 Related Work

In this section, we illustrate related work starting from introducing the works that generally studied Named-entity recognition. In the second part of this section, we will analyze the approaches that treat NER using pre-trained word representations. Finally, we will analyse the approaches used in Oil & Gas Industry and other domains.

Named-entity recognition systems have been studied and developed for decades. Nevertheless, the methods using deep neural networks (DNN) have only been introduced in the last decade [14], with recent special improvements given the new capabilities with pre-trained models and transfer learning [14]. **Models with pre-trained word representations** The widely used approach based on DNNs for NER was proposed in [6]. This model applies a Convolutional Network Architecture to the token sequence. Posterior works typically change the encoding part, which ranges from char-based, word-based, and encoding additional features. Examples include predefined word representation like word2vec, GloVe, or BERT or the explicit inclusion of suffixes and prefixes. In this context [12] work focused on changing the CNN with a bidirectional LSTM encoder. They do not perform any pre-processing; they do not take into account morphological information from characters or words. Instead, all features are learned by a CNN, achieving SOTA results. Other approaches [17] take advantage of the usage of a large semantic database and implement distant supervision: the relation classifier is trained using textual features.

Some models are based on general word embeddings, that are fine-tuned for NER. The original work, illustrated in [19], presented an F1 score of 92.2 over the CoNLL 2003 test set. [5] improves this result to 92.6 by using Cross-View Training (CVT). The semi-supervised learning algorithm improves the representations of a Bi-LSTM sentence encoder utilising a mix of labelled and unlabelled data. Zalando Research has also made a great effort in providing SOTA models, getting an F1 score of 92.86 over the same data set [2]. Using the BERT base model(i.e. using the pre-trained embeddings) gives an F1 score of 91. Fine-tuning the same model for NER, however, improves this score to 96.4. In 2019 the pooled version of the approach improved this score to 93.18 [2]. **Energy Industry** a NER for geosciences trained for the Chinese language has been proposed by [20]. They use a generative model, building a data set from seed terms without labelled data with good results. Another system from geoscience is the Portuguese NER [7]. It defines the target entities for the Brazilian sedimentary basins. They used a conventional approach with three different embeddings configurations tested using a BiLSTM-CRF architecture. Some other approaches are focused on unsupervised clustering-based technique to match attributes of a large number of heterogeneous sources as also proposed in [3] to identify entities.

NER in Other Domains. NER is well studied in specific domains like medical data, neuroscience, or scientific data. Bio-NER for the biomedicine field has named entities related to RNA, protein, cell type, cell line, and DNA with different shared tasks. Similarly to the general field, up to 2018 BiLSTM-CRF [13]

Noisy Labels. Label noise has always been an existing problem in machine learning, due to the potential negative impact it has over classification as also stated in [10].

Since weakly supervised learning is gaining a huge attraction, dealing with noise in Deep Neural Networks has become a highly active research field for representation learning [8]. Most works focused on generating and aggregating synthetic noise to well-known data sets [23]. [1] identifies three different approaches to mitigate the effect of noisy labels as widely described in [4,11,15].

7 Conclusion

Named Entity Recognition is the first fundamental step for Information Extraction and Knowledge Base creation. The main objective of our research was to build a NER System for the Oil & Gas industry. However, instead of creating one model for some specific entities in this domain, We aimed to explore a methodology/framework that facilitates the creation of a Named Entity Recognition system based on noisy data labels. The methodology is flexible enough to incorporate new target entities without labour-intense human annotation and sufficiently robust to enhance generalization. We create labels using distant supervision resources like dictionaries and regular expressions. Distant supervision introduces noisy labels, translating mainly into False Positives in the training set. To mitigate the effect of noisy labels, we followed a method with three key elements: (1) Distributed processing - to enable the labelling of bigger data sets than the ones we could have obtained with manual annotation. (2) Transfer learning with pre-trained language models - to learn bidirectional context representations in our domain-specific corpus (3) SOTA training techniques -

to avoid over-fitting the noisy examples. Furthermore, we proposed a two-step fine-tuning approach that showed to be effective in improving the prediction capacity in hard-to-learn named entities. We apply this model to many domain documents from the north-sea and create a knowledge graph that would be used to feed a model.

A similar approach could be applied in other domains where many documents are available. In such scenarios, distant supervision enables extracting thousands of sentences with entities. Even with noise, bigger data sets and the proposed training process will help the model to capture the regular context where entities occur, helping to remove false positives even in domains with polysemy challenges. As future work, we would like to explore the effect of the size of the clean data set on the model performance following our approach. This will allow us to provide clear recommendations on how much data has to be cleaned for the second fine-tuning step. Moreover, as an extension of our work, we can consider replacing the regular expressions and dictionary approach with labelling and transformation functions like in Snorkel [21].

Acknowledgements. We are grateful to the Oil & Gas Authority that provided the access to wells reports used in our research (under the Oil and Gas Authority Licence [16]).

References

1. Abid, A., Zou, J.Y.: Improving training on noisy stuctured labels. CoRR (2020)
2. Akbik, A., Bergmann, T., Vollgraf, R.: Pooled contextualized embeddings for named entity recognition. In: Conference of the North American Chapter of the Association for Computational Linguistics: Human Language Technologies, pp. 724–728 (2019)
3. Arman, M., Wlodarczyk, S., Bennacer Seghouani, N., Bugiotti, F.: PROCLAIM: an unsupervised approach to discover domain-specific attribute matchings from heterogeneous sources. In: Herbaut, N., La Rosa, M. (eds.) CAiSE 2020. LNBIP, vol. 386, pp. 14–28. Springer, Cham (2020). https://doi.org/10.1007/978-3-030-58135-0_2
4. Bahri, D., Jiang, H., Gupta, M.R.: Deep k-nn for noisy labels. CoRR (2020)
5. Clark, K., Luong, M.-T., Manning, C.D., Le, Q.V:. Semi-supervised sequence modeling with cross-view training. CoRR (2018)
6. Collobert, R., Weston, J., Bottou, L., Karlen, M., Kavukcuoglu, K., Kuksa, P.P.: Natural language processing (almost) from scratch. CoRR (2011)
7. Consoli, B., Santos, J., Gomes, D., Cordeiro, F., Vieira, R., Moreira,V.: Embeddings for named entity recognition in geoscience Portuguese literature. In: Proceedings of The 12th Language Resources and Evaluation Conference, pp. 4625–4630, Marseille, France, 2020. European Language Resources Association
8. Deng, Z., Dong, Y., Pang, T., Su, H., Zhu, J.: Adversarial distributional training for robust deep learning. CoRR (2020)
9. Devlin, J., Chang, M.-W., Lee, K., Toutanova, K.: BERT: pre-training of deep bidirectional transformers for language understanding. CoRR, abs/1810.04805 (2018)
10. Frenay, B., Verleysen, M.: Classification in the presence of label noise: a survey. IEEE Trans. Neural Netw. Learn. Syst. **25**(5), 845–869 (2014)

11. Ghosh, A., Kumar, H., Sastry, P.S.: Robust loss functions under label noise for deep neural networks. AAAI'17, pp. 1919–1925. AAAI Press (2017)
12. Huang, Z., Xu, W., Yu, K.: Bidirectional LSTM-CRF models for sequence tagging. CoRR (2015)
13. Khan, M.R., Ziyadi, M., Abdelhady, M.: Mt-bioner: Multi-task learning for biomedical named entity recognition using deep bidirectional transformers. CoRR (2020)
14. Li, J., Sun, A., Han, J., Li, C.: A survey on deep learning for named entity recognition. CoRR (2018)
15. Li, J., Wong, Y., Zhao, Q., Kankanhalli, M.S.: Learning to learn from noisy labeled data. 2019 IEEE/CVF Conference on Computer Vision and Pattern Recognition (CVPR), pp. 5046–5054 (2019)
16. Licence. Oil and Gas Authority Licence (2022) Accessed Jan 2022. https://www.ogauthority.co.uk/media/5850/oga-open-user-licence_210619v2.pdf/
17. Mintz, M., Bills, S., Snow, R., Jurafsky, D.: Distant supervision for relation extraction without labeled data. In: Proceedings of the Joint Conference of the 47th Annual Meeting of the ACL and the 4th International Joint Conference on Natural Language Processing of the AFNLP: Volume 2 - Volume 2, ACL '09, pp. 1003–1011, USA, 2009. Association for Computational Linguistics
18. Nakayama, H.: seqeval: A python framework for sequence labeling evaluation (2018). https://github.com/chakki-works/seqeval
19. Peters, M.E.,et al.: Deep contextualized word representations, CoRR (2018)
20. Qiu, Q., Xie, Z., Liang, W., Tao, L.: Gner: a generative model for geological named entity recognition without labeled data using deep learning. Earth Space Sci. **6**, 931–946 (2019)
21. Ratner, A., Bach, S.H., Ehrenberg, H., Fries, J., Sen, W., Ré, C.: Snorkel. Proc. VLDB Endowment **11**(3), 269–282 (2017)
22. Robins, A.V.: Catastrophic forgetting, rehearsal and pseudorehearsal. Connect. Sci. **7**, 123–146 (1995)
23. Rolnick, D., Veit, A., Belongie, S.J., Shavit, N:. Deep learning is robust to massive label noise. CoRR (2017)
24. Sanh, V., Debut, L., Chaumond, J., Wolf, T.: Distilbert, a distilled version of bert: smaller, faster, cheaper and lighter. ArXiv, abs/1910.01108 (2019)
25. Tanaka, D., Ikami, D., Yamasaki, T., Aizawa, K.: Joint optimization framework for learning with noisy labels. CoRR (2018)

Predicting User Dropout from Their Online Learning Behavior

Parisa Shayan[1(✉)], Menno van Zaanen[2], and Martin Atzmueller[3,4]

[1] Tilburg University, Tilburg, The Netherlands
p.shayan@tilburguniversity.edu
[2] South African Centre for Digital Language Resources, Potchefstroom, South Africa
menno.vanzaanen@nwu.ac.za
[3] Semantic Information Systems Group, Osnabrück University, Osnabrück, Germany
[4] German Research Center for Artificial Intelligence (DFKI), Osnabrück, Germany
martin.atzmueller@uni-osnabrueck.de

Abstract. The Covid-19 pandemic, which required more people to work and learn remotely, emphasized the benefits of online learning. However, these online learning environments, which are typically used on an individual basis, can make it difficult for many to finish courses effectively. At the same time, online learning allows for the monitoring of users, which may help to identify learners who are struggling. In this article, we present the results of a set of experiments focusing on the early prediction of user drop out, based on data from the New Heroes Academy, a learning center providing online courses.

For measuring the impact of user behavior over time with respect to user drop out, we build a range of random forest classifiers. Each classifier uses all features, but the feature values are calculated from the day a user starts a course up to a particular day. The target describes whether the user will finish the course or not. Our experimental results (using 10-fold cross-validation) show that the classifiers provide good results (over 90% accuracy from day three with somewhat lower results for the classifiers for day one and two). In particular, the time-based and action-based features have a major impact on the performance, whereas the start-based feature is only important early on (i. e., during day one).

Keywords: User modeling · Data mining · Dropout prediction

1 Introduction

The COVID-19 pandemic has led to a shift towards online learning [4,7]. The key benefit of online learning is related to providing content to the community through open access platforms which can have economic and educational benefits e. g., reduced market entry time, international partnerships, increased user engagement and satisfaction, growth in the learning curve of users, plus rich evaluation and feedback [1,5].

P. Pascal and D. Ienco (Eds.): DS 2022, LNAI 13601, pp. 243–252, 2022.
https://doi.org/10.1007/978-3-031-18840-4_18

However, even though online learning may have major benefits, it also comes with several challenges, e. g., startup budget requirements, organizational and individual preparation, team effort and development, technical support and crisis management [1]. In addition, one of the main challenges is user participation [9,14]. Not all users of online learning platforms make it to the end of a course successfully. Therefore, in this article we aim to identify such users by investigating user behavior in an online learning platform. Here, we utilize a real-world data set from the New Heroes Academy[1], a learning center providing online courses.

The main purpose of this study is to predict, as quickly as possible, whether a user will finish a course or not. The earlier we know whether a users will drop out, the more time there is to implement possible interventions. These interventions may help to keep users engaged during the course and hence to prevent dropout, increasing the chance they finish the course.
Therefore, this study leads us to target the following research questions:

1. Which features are most influential when predicting user's course completion?
2. How do these change over time (for days)?

Our main contributions are summarized as follows:

1. We present research that aims to predict user dropout as quickly as possible.
2. We analyze and demonstrate the impact of behavioral features on user dropout through a predictive model.

2 Background

In general, prediction of user behavior is a major focus of the research in the research areas of Learning Analytics and Knowledge (LAK) and Educational Data Mining (EDM). LAK mostly measures, collects, analyzes, and reports data on student learning progress, whereas EDM builds and explores learner models, adapting the learning environment using advanced data mining techniques [12].

[3,10] applied data mining techniques to users' demographic and behavioral data. According to their findings, neural networks, Naive Bayes, decision tree, and random forest classifiers are comparable with logistic regression to predict user retention. [15] focused on finding a way to increase user retention rates. For this purpose, they employed Bayesian algorithms, support vector machines (SVM), and decision trees. Their results indicated that Bayesian analysis was more accurate in predicting user retention than other algorithms. On a similar note, [6] examined the weekly dropout classification using SVMs and cumulative features (e. g., number of interactions, number of views per page of the course). Here, the SVM approach reached the higher accuracy. Likewise, [13] analyzed MOOCs using SVM and k-means as classifiers and experimented with different cases. In their results, k-means always lagged behind SVM. According to their findings the best results were obtained after the first 100 interactions and the first seven days after the users' first interaction with the system.

[1] https://www.newheroes.com.

3 Methodology

We are interested in discovering the features that describe which users dropout during a course and how the importance of these features may change over time. In other words, we measure the impact of different types of features over time for early prediction of user dropout. To investigate this, we build a range of random forest classifiers (one for each day) such that the feature values are computed from the day a user starts a course up to that particular day. In each classifier, we extract four groups of features (e. g., time-based, action-based, course-based, and start-based features) to build a model of early prediction for user dropout using a random forest classifier. The target thus defines whether the user will finish the course or not. We evaluate on several metrics using 10-fold cross-validation.

Relating this to previous studies, [3,10] focused on prediction user retention. [15] compared different algorithms to explore the best method to predict user retention whilst in the current study we intend to predict user dropout (in contrast to user retention) as quickly as possible such that we analyze the impact of different type of features over time. While [13] examined a small number of time-based and action-based features to predict user dropout, here we extract a larger range of features. We consider 16 unique action-base features, seven course-based features as well as time-based and start-based features to create an early predictive model of user dropout while taking a time aspect (on a daily basis) into account.

3.1 Data Set

The New Heroes Academy is an online training platform that allows for both online and blended learning (i. e., a combination of online and classroom). The focus here is on the online learning aspect. The New Heroes Academy offers their services to both private and business clients, their common denominator is that the services are subscription-based. Users will earn a certificate once they finish a course. The users finish a course when they complete the required elements (described below) related to that course.

In the current study, we are using a data set that was collected over the period 2016–2020 from the New Heroes Academy website. It consists of 7086 unique users and 198 courses with 17,323 unique user/course pair (13,555 finished courses and 3,768 unfinished courses). To decide on which users we consider not finishing their course, we considered the 90% quantile, which indicates that most users finish their courses within 400 days. As such, we take 400 days as a threshold. Those users taking more than 400 days are considered dropouts. Obviously, the length of a course may have an influence on how long users take to finish a course. The average course consists of 20.71 elements with a standard deviation of 29.40. On average, a user needs to perform 48.34 actions to finish a course. This takes 268.40 days on average with the standard deviation of 620.09.

3.2 Features

In order to describe the interaction of a user with a course, we identify four different types of features. First, *time-based features* can be found in Table 1. These are calculated on a day by day basis. These features are computed over the number of days from the day the user started the course.

Table 1. List of time-based features.

TotalNumberOfActions	Total number of actions a user has performed in the course thus far
AvgNumberOfActions	Average number of actions a user has performed per day
StdDevNumberOfActions	Standard deviation of number of actions a user has performed per day

Second, we identify *action-based features*. These features describe the number of the different types of actions a user has performed up to the time of measurement (i. e., per day). The New Heroes Academy data set identifies 16 different types of actions that are listed in Table 2. These features describe the (total) number of actions per type of action the user has performed within the time period being measured (i. e., number of days under consideration).

Third, we believe that the day a user starts a course may have an influence on the course completion. To represent this type of information, we use *start-based* features. For instance, people starting during school holidays may have allocated time during that period to work specifically on the course. We experimented with additional features (e. g., whether they started in a weekend or not). The preliminary results showed that these features often co-correlated and had no real impact. Therefore, we only considered the week of the year in which a user has started a course. This feature is shown in Table 3.

Finally, the setup of a course may have an influence on the course completion. For instance, extremely long courses may be more difficult to complete as they require longer attention spans.

We can describe the length of a course by counting the number of course elements. Course elements are tasks that a user has to do in order to complete a course. The elements are classified in seven groups (i. e., "Open task", "Goal", "Questionnaire", "Instruction wizard", "Video task", "Textual explanation", and "Video response training (vrt)"). In addition to measuring the overall course length, we count how many of which type of element are found in the course as some elements may be more difficult or take more time to complete than others. We call this group of features *course-based* features and they are described in Table 4. Note that these properties are not related to any user properties. They remain constant for each course.

Table 2. List of action-based features.

AudioPaused	Total number of clicks on the audio pause button
AudioStarted	Total number of clicks on the audio play button
BookmarkCreated	Total number of clicks on the bookmark label
BookmarkViewed	Total number of clicks on "more information" in the bookmark overview
ElementDone	Total number of learning elements done in the course
ElementOpened	Total number of learning elements opened in the course
ElementInProgress	Total number of answered learning elements in the course
JourneyStarted	Total number of learning elements started in the course
JourneyVisited	Total number of learning elements visited in the course
JourneyAssignedToGoal	Total number of learning goals assigned to a course[a]
StuckButtonUsed	Total number of clicks on the stuck button (to get a hint)
SupporterInvited	Total number of learning support invitations sent[b]
SupporterLinkOpened	Total number of learning support invitation opened by the user
ToggleTranscript	Total number of times a user requested the transcript of a video[c]
VideoPaused	Total number of clicks on the pause button of a video
VideoStarted	Total number of clicks on the play button of a video

[a] Before beginning a course, users set learning goals for their personal development.
[b] Users can invite a supporter for a learning element.
[c] The user can click a red button below a video to expand the transcript of that video.

Table 3. List of start-based features.

Week	Number of the week in the year when the course was started

Table 4. List of course-based features.

LengthCourse	Number of elements in the course
Goal	Number of goal elements in the course
InstructionWizard	Number of instruction wizards in the course
OpenTask	Number of open tasks in the course
Questionnaire	Number of questionnaires in the course
TextualExplanation	Number of texts in the course
VideoAssignment	Number of video assignments in the course
VideoResponseTraining (VRT)	Number of video response training in the course

3.3 Pre-processing

To compute the number of actions a user has performed in a course after a particular number of days (say n days), we add all actions that user has performed from day 1 (the day the user started the course) to day n. This is the basis for the time-based, and action-based features. The start-based feature depends on when the user starts the course and the course-based features are computed per course. We use this collection of information for all classifiers such that we will have as many instances as there are user/course combinations in the entire data set. The binary target indicates whether the user finished the course or not. Note that the target does not describe whether the user has finished the course after the n days that form the basis for the information extraction, but it describes whether the user finished the course at some point (potentially in the future).

3.4 Predictive Model

As we aim to build a model that can accurately predict whether users finish the course (or not) as soon as possible. This means that the model should provide high accuracy, but does this using information from the least number of days allowing for sufficient time for potential interventions.

For this, we build classifiers for a range of days: $j = \{1, 2, \ldots, n\}$ days. We expect that the performance of the classifiers using information of more days will have a higher classification performance (as more information is available). At the same time, we would like to have a model that makes the correct decisions as early on in the process. We experiment with classifiers using information from one to twenty days (where day twenty means using the information from day one up to and including day twenty).

3.5 Evaluation

In order to evaluate the quality of our classification models we provide the results of experiments using 10-fold cross-validation for all classifiers. In our experimentation, we provide the following measures: accuracy (i. e., the percentage of correctly classified instances), precision, recall, the F1-score, and AUC (Area Under the Curve).

Furthermore, we compute the accuracy of a majority class classifier as a simple baseline for comparison. Also, (in the respective graph visualization) we provide the standard deviations for the performance metrics to show the variation of the results between the folds. In addition to the classifier performance, we provide information on the feature importance of all the features for all the classifiers. This allows us to investigate which features are most important for classification purposes and, in particular, we can furthermore see how this feature importance changes over time.

4 Results

We first show results of the feature importance within the different classifiers. We focus on the following four different categories of features: time-based, action-based, course-base, and start-based to see the impact of the above features on user dropout over time. Next, we present the performance results of the predictive modeling approach.

4.1 Predictive Model

To measure the impact of user behavior over time, we build a range of random forest classifiers using the above features. Figure 1 shows the feature importance from the day users start the course (day one to day twenty). As aforementioned, the feature vector consists of four types of features e. g., time-based features, action-based, start-based, and course-based features. In general, the weights for the features are quite stable with only minimal changes over time. Only for day one the Week feature is important (as very limited other information is available).

What is striking, is the dominance of action-based features. For instance, the high rate of total number of learning elements done by users in the course (ElementDone) is the most important feature for all days. Furthermore, the high importance of time-based features on user dropout stands out. For example, the standard deviation of actions users have performed in day (StdDevNumberOfActions) one and three have a major influence on user dropout. Furthermore, the course-based features, in specific, Instruction wizard between day two to day five have a large influence on the classifiers. Also, the number of elements in the course (LengthCourse) seems to affect user dropout.

Overall, these results indicate that the ElementDone, Instruction wizard for all days, the Week and StdDevNumberOfActions between day one and day six features have a large impact on the prediction performance. Additionally, some features (LengthCourse as well as TotalNumberOfActions and AvgNumberOfActions) have a decent impact and their influence remains stable over time.

4.2 Evaluation

Figure 2 provides a summary of our experimental results on the time-based analysis. In addition, the respective standard deviations are incorporated into the graph in order to depict the variation of the folds as well.

According to the graph, the majority class baseline accuracy for all days is 78%. What is striking, is that the accuracy for day one and day two is already between 87% and 89% and the performance of the classifiers increases (i. e., over 90% accuracy), from day three. Furthermore, we computed the AUC which starts at 81% (for day one) and increases to 86% (for day twenty). What is also interesting is a consistent value for precision (92%) for almost all days. In contrast, one can observe an increasing trend for recall and F1-score starting with 92% at day one and reaching to 95% and 98% at day twenty accordingly. The accuracy shows a gradual increase over time (after day two).

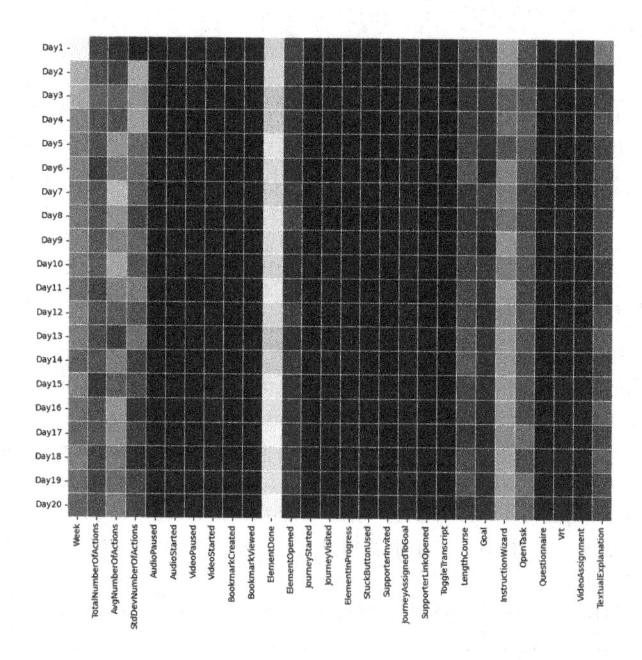

Fig. 1. Distribution of features in order of importance using 10-fold cross-validation with Random Forest classifiers (by day). The more important features are represented by lighter colors whilst the less important features are demonstrated by darker colors. (Color figure online)

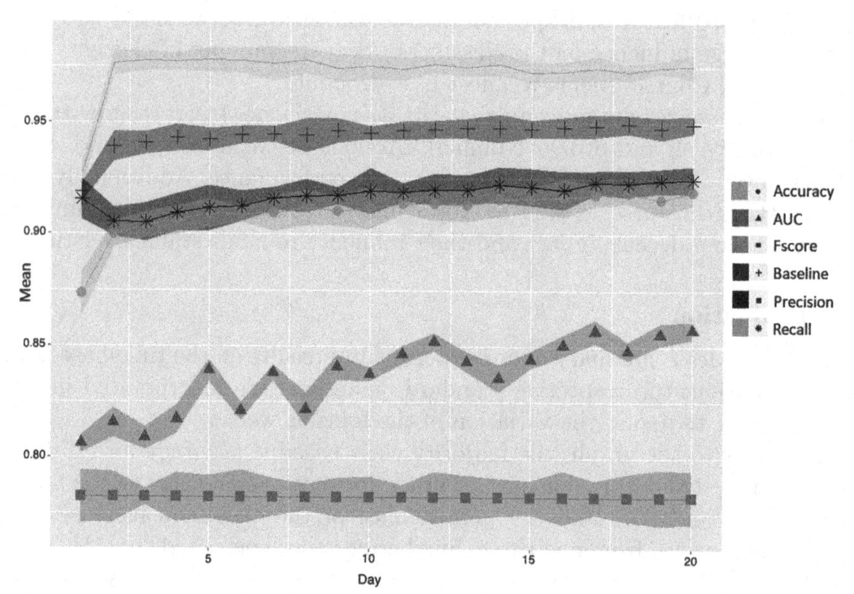

Fig. 2. Plot for majority class baseline, accuracy, precision, recall, F1-score, and AUC (by day) using 10-fold cross-validation plus standard deviation to show the folds variation.

5 Discussion

In this study, to measure the influence of users' behavior on dropout, we constructed a range of random forest classifiers using four different feature groups over time. At the same time, we computed the features' values from the day each user starts a specific course up to a specific day. What was striking, was the feature importance of action-based features (i. e., the total number of learning elements done by users in the course) and time-based features (i. e., standard deviation of actions user has performed) on user dropout. The former for all days and the latter for the first and the third day had a large impact on user dropout. Furthermore, the course-based features (i. e., Instruction wizard) was important from day two onward. Also, the number of elements in the course as well as the average number of elements in the course could affect the user dropout (so there is a difference between long and short courses). Meanwhile, what was interesting, the start-based features (the week number of the year in which the user started the course) was only important during the early period, i. e., day one.

With respect to our evaluation, we provided a majority class baseline which showed an accuracy of 78%. The random forest results indicated 90% accuracy from the third day with sightly lower results for the classifiers for the first and second days which provided high performance for our experimental results. In addition, the AUC value between 81% and 86% implies that the classifiers from day one to day twenty perform quite well.

On essentially all metrics, one can see an upward trend. Recall and precision start from 92% at day one then rose to a high point and peaked into 98% at day twenty. The high precision means that over 90% of our results in predicting user dropout are relevant. The recall score refers to the high rate of total relevant results that classified correctly by our predictive model. The F1-score, defined as a harmonic mean of recall and precision values, shows high scores i. e., from 91% at day one to a high point of 95% at day twenty.

6 Conclusions

The main goal of the current study was to create an early prediction model of user dropout. The experimental results proved that the classifiers provide good results from day three and then the prediction performance remains relatively stable (with only minor increases). The most obvious findings to emerge from this study highlight the importance of time-based and action-based features, whilst the start-based features on day one and the course-based features from day two onward influence the performance of the classifiers.

Overall, the results show at least three major aspects. First, action-based and course-based features, e. g., the number of elements and instruction wizard have been performed by users, had a major influence on user dropout for almost all days. Second, start-based and time-based features, e. g., the number of week in the year in which a user has started a course and the number of actions has been performed from day one to day five, have an effect on the user dropout.

Third, the prediction performance remains relatively stable over time for the rest of course-based and time-based features, e. g., the number and average length of elements in the course as well as total and average number of actions user has performed in the course (per day).

The study contributes to understanding of an early predictive model for user dropout, however, the results open up several questions in need of further investigation. A further study could, for example, assess the behavioral features impact using action-based measurement assuming that users take approximately the same number of actions to finish a course. In addition, an interesting future direction could analyze/discover subgroups regarding their dropout behavior, e. g., [2,8] towards extended user modeling and analysis approaches, c. f., [11].

References

1. Appanna, S.: A review of benefits and limitations of online learning in the context of the student, the instructor and the tenured faculty. Int. J. E-Learn. **7**, 5–22 (2008)
2. Atzmueller, M.: Subgroup discovery. WIREs Data Min. Knowl. Discovery **5**(1), 35–49 (2015)
3. Dekker, G., Pechenizkiy, M., Vleeshouwers, J.: Predicting students drop out: a case study. In: Proceedings of the Educational Data Mining, pp. 41–50 (2009)
4. Heng, K., Sol, K.: Online learning during COVID-19: key challenges and suggestions to enhance effectiveness. Cambodian Education Forum (2020)
5. Kassymova, G., Issaliyeva, S., Aigerim, K.: E-learning and its benefits for students. Pedagogics and Psychology, pp. 249–255 (2019)
6. Kloft, M., Stiehler, F., Zheng, Z., Pinkwart, N.: Predicting MOOC dropout over weeks using machine learning methods. In: Proceedings of the Workshop on Analysis of Large Scale Social Interaction in MOOCs, pp. 60–65. ACL, Doha, Qatar (2014)
7. Lemay, D.J., Bazelais, P., Doleck, T.: Transition to online learning during the COVID-19 pandemic. Comput. Hum. Behav. Reports **4**, 100130 (2021)
8. Lemmerich, F., Ifl, M., Puppe, F.: Identifying influence factors on students success by subgroup discovery. In: Proceedings of Educational Data Mining (2010)
9. Malinen, S.: Understanding user participation in online communities: a systematic literature review of empirical studies. Comput. Hum. Behav. **46**, 228–238 (2015)
10. Pittman, K.L.: Comparison of data mining techniques used to predict student retention. In: NSUWorks (2008)
11. Shayan, P., Rondinelli, R., Zaanen, M., Atzmueller, M.: Descriptive network modeling and analysis for investigating user acceptance in a learning management system context. In: Proceedings of the ABIS, pp. 7–13. ACM, Boston, MA, USA (2019)
12. Shayan, P., van Zaanen, M.: Predicting student performance from their behavior in learning management systems. Int. J. Inf. Educ. Technol. **9**(5), 337–341 (2019)
13. Vitiello, M., Walk, S., Chang, V., Hernández, R., Helic, D., Gütl, C.: MOOC dropouts: a multi-system classifier. In: EC-TEL (2017)
14. Williams, B.: Participation in on-line courses - how essential is it? Educational Technol. Soc. **7**, 1–8 (2004)
15. Zhang, Y., Oussena, S., Clark, T., Kim, H.: Use data mining to improve student retention in higher education - a case study. In: ICEIS (2010)

Efficient Multivariate Data Fusion for Misinformation Detection During High Impact Events

Lucas P. Damasceno[1(✉)], Allison Shafer[2], Nathalie Japkowicz[2], Charles C. Cavalcante[1], and Zois Boukouvalas[2]

[1] Federal University of Ceará, Fortaleza, CE 60455-760, Brazil
lucaspdamasceno@alu.ufc.br, charles@gtel.ufc.br
[2] American University, Washington, DC 20016, USA
{as8273a,japkowicz,boukouva}@american.edu

Abstract. With the evolution of social media, cyberspace has become the de-facto medium for users to communicate during high-impact events such as natural disasters, terrorist attacks, and periods of political unrest. However, during such high-impact events, misinformation on social media can rapidly spread, affecting decision-making and creating social unrest. Identifying the spread of misinformation during high-impact events is a significant data challenge, given the variety of data associated with social media posts. Recent machine learning advances have shown promise for detecting misinformation, however, there are still key limitations that make this a significant challenge. These limitations include the effective and efficient modeling of the underlying non-linear associations of multi-modal data as well as the explainability of a system geared at the detection of misinformation. This paper presents a novel multivariate data fusion framework based on pre-trained deep learning features and a well-structured and parameter-free joint blind source separation method named independent vector analysis, that can reliably respond to this set of limitations. We present the mathematical formulation of the new data fusion algorithm, demonstrate its effectiveness, and present multiple explainability case studies using a popular multi-modal dataset that consists of tweets during several high-impact events.

Keywords: Misinformation detection · Data fusion · Independent vector analysis · Multi-modal learning · Deep learning

1 Introduction

With the evolution of social media technologies, there has been a fundamental change in how information is accessed, shared, and propagated. Propagation of information, particularly misinformation, becomes especially important during high impact events such as pandemics [23,33], natural disasters [1,2], terrorist attacks, periods of political transition or unrest, and financial instability.

P. Pascal and D. Ienco (Eds.): DS 2022, LNAI 13601, pp. 253–268, 2022.
https://doi.org/10.1007/978-3-031-18840-4_19

Recent machine learning advances have shown promise for detecting misinformation [32]; however, the problem remains a significant challenge due to several key limitations [7]. One such limitation is related to the use of multi-modal data, i.e., information collected about the same phenomenon using different modalities. Use of multi-modal data has not been fully leveraged in intelligent systems, which traditionally use a single modality, typically text [28] or images [13]. Machine learning algorithms must be able to understand content holistically to become more effective in detecting misinformation. While it is easy to see that multimodality has great potential in eliminating ambiguity, this is still a significant challenge for many machine learning models.

Early fusion methods provide effective solutions for multi-modal learning since joint representations of input features from different modalities are created before attempting to classify the content, enabling enhanced detection of posts with malicious content [3,6]. However, in most studies, the joint representations are obtained by simply concatenating the individual representations or by implicitly modeling the mutual relationships across the modalities [30]. These techniques are limited since they miss the opportunity to fully exploit the relationship between the different modalities. What is desirable in multi-modal learning, is the ability to *explicitly* learn the mutual relationships among the modalities by letting multiple sources of information adaptively interact while generating the joint representations.

Another key limitation in many multi-modal fusion models is *explainability*, i.e., ability of the model to summarize the causes of its decisions in an efficient manner, illuminate various connections between high-level features and low dimensional joint representations [28], and hence gain the trust of its users [21,26]. Need for explainability becomes even more pronounced in high-impact events, as it is key for an analyst to understand the significance of predictions and suggest mitigations. In addition, during such events, explainability is extremely important in the context of bias and ethical use of artificial intelligence, since understanding the reasons behind certain predictions will enable users to identify potential discrimination against certain groups and demographics.

In this work, we present a novel multi-modal data fusion framework geared at detecting misinformation. Our framework is based on a flexible and computationally efficient multi-modal fusion algorithm named independent vector analysis by multivariate entropy maximization with kernels, IVA-M-EMK, that effectively captures complex, non-linear relationships among textual and visual modalities. Through this work, we make several contributions. First, we present the mathematical formulation of a new data fusion algorithm for the detection of misinformation and theoretically justify the importance of explicitly modeling the non-linear relationships across different modalities. Second, by using a popular multi-modal dataset, we numerically demonstrate how our method enhances the performance of the detection of misinformation by exploiting the underlying complementary information contained in text and image pre-trained deep learning-based features. Finally, through numerical experiments and by using a popular interpretability method, the local interpretable model-agnostic explanations (LIME) [31], we discuss essential aspects of explainability associated with our data fusion method.

2 Materials and Methods

2.1 Dataset

For this study, we use datasets from the MediaEval2016 Image Verification Corpus[1] [9], which include separate labeled training and test tweet text and multimedia datasets. The original data include 15,630 tweet records and 399 multimedia records in the training dataset and 2,177 tweet records and 117 multimedia records in the test dataset. We choose the 2016 vintage datasets over the 2015 vintage datasets because the 2016 training dataset consists of tweets that revolve around a set of events about that are completely different from the set of events discussed from the tweets in the test dataset. Each tweet record is labeled as being "fake" or "real". Tweets labeled as "fake" include any post that shares multimedia content that does not faithfully represent the event that it refers to. This includes content from a past event reposted as being captured for a currently unfolding similar event, context that has been purposely manipulated, or multimedia content published with a false claim about the depicted event [9].

From the datasets, for the tweeted text data we use the tweet ID, tweet text, image ID, and label fields, and create an event field using the data in the associated image ID field. To accompany the tweet data, we use the provided multimedia datasets. The training and test multimedia datasets included image and video data. For the purpose of this work, we exclude the video data and use only the image data. Each tweet text record includes at least one image ID. When more than one image ID is associated with a tweet, we choose only the first image ID listed to represent the image associated with the tweet. We remove tweet records from the dataset that lack an associated event that cannot be derived from an image ID or lack an image ID.

To prepare the data for use in feature extraction, we clean the text data by removing emoji characters, stop words, URLs, Twitter handles, time stamps, and select punctuation. To normalize the text data we lowercase the text, reduce multi-spaces to one space, lemmatize the text, and keep only words that are greater than two characters long. Additionally, we remove tweets that were more than 512 tokens long after cleaning to prepare for future processes. We then identify tweets that use English or a similar language using the Langid Python package, and remove tweet records that are not English or similar in language to English per the International Organization for Standardization (ISO) code for languages. Additionally, for the training dataset, we remove records that are denoted as being retweeted. For the image data, we pre-process each image by resizing the images to 224×224 pixels and normalizing them.

We use the final working training and test datasets for text and image feature creation, as explained in future sections of this paper. When feature creation

[1] https://github.com/MKLab-ITI/image-verification-corpus.

processes result in null records, we remove any null records produced. The final training dataset after feature creation consists of 9,140 tweet records associated with 352 different images and representing 15 unique events. Of the training data tweets, 5,127 are considered fake and 4,013 are considered real. Five of the 15 events represented in the training dataset include both real and fake tweets, while ten of the events include only fake tweets. The final test dataset, after feature creation, consists of 796 tweet records associated with 92 different images and representing 23 unique events. The events represented in the training data and testing datasets are disjoint. Of the test data tweets, 467 are considered fake and 329 are considered real. Seven of the 23 events have both real and fake tweets associated with them, one event has only real tweets associated with it, and fifteen of the events only include fake tweets.

2.2 High-Level Feature Extraction

From the final, raw training and test text and image datasets, we extract text and image features to use in order to classify tweet records as real or fake. We utilize multiple methods to create the features and evaluate each feature's impact on classification accuracy. For the text features, we utilize Word2Vec [27][2] trained on the Google News corpus[3] to generate word embeddings[4]. We run the pre-processed image tensors through the pre-trained VGG-16 model to extract the first fully connected layer with 4,096 hidden units produced by the VGG-16 model to utilize as the image features.[5]

We select the Word2Vec embeddings as our text dataset and the VGG-16 fully connected layer features as our image dataset because they yield the best classification results during model evaluations. We continue our experiments with the training and test datasets for our text data, which consist of a 300-dimensional Word2Vec embedding vector for each tweet record where each tweet is represented by the average word embedding vectors of the words that make up the tweet. Additionally, separate training and test datasets for our image data consist of the 4,096-dimensional fully connected layer from the VGG-16 model for each image associated with an individual tweet record.

2.3 Multi-modal Data Fusion Framework Based on Independent Vector Analysis

We formulate the problem of joint feature generation for detection of unreliable posts as a joint blind source separation (JBSS) problem. In particular, let $\mathbf{X}^{[k]} \in \mathbb{R}^{d \times V}$ is the kth observation matrix from kth modality, where d denotes the number of initial high-level feature vectors in the kth modality and V denotes the total number of tweets. The noiseless JBSS model is given by

[2] We also evaluated features created using Bidirectional Encoder Representations from Transformers, or BERT [18].

[3] https://code.google.com/archive/p/word2vec/.

[4] Additionally, we evaluated Word2Vec trained using our own data.

[5] We also analyzed using the 'avgpool' layer from a pre-trained ResNet-18 model.

$$\mathbf{X}^{[k]} = \mathbf{A}^{[k]}\mathbf{S}^{[k]}, \quad k = 1, ..., K, \tag{1}$$

where $\mathbf{A}^{[k]} \in \mathbb{R}^{d \times N}$ is the kth mixing matrix, and $\mathbf{S}^{[k]} \in \mathbb{R}^{N \times V}$ are latent variable estimates, i.e., k^{th} set of source estimates, which in our setting, correspond to the features. The estimates of the features span the joint low dimensional representation space and will be used to train a machine learning algorithm for the detection of misinformation. It is worth noting that when $K = 1$, (1) it reduces to a simple BSS problem with one modality and the most popular way to achieve BSS is by using independent component analysis (ICA) [4,15,24,28].

IVA provides a smart connection across multiple datasets through the definition of a *source component vector* (SCV), which enables one to take full statistical information across the multi-modal datasets. Using the random vector notation (as opposed to the one written using observations in (1)), we write $\mathbf{x}^{[k]} = \mathbf{A}^{[k]}\mathbf{s}^{[k]}$, $k = 1, ..., K$, where $\mathbf{A}^{[k]} \in \mathbb{R}^{N \times N}$, $k = 1, ..., K$ are invertible mixing matrices and $\mathbf{s}^{[k]} = [s_1^{[k]}, ..., s_N^{[k]}]^\top$ is the vector of features for the kth dataset and $(\cdot)^\top$ denotes the transpose of a vector/matrix. In the IVA model, dependence across corresponding components of $\mathbf{s}^{[k]}$ is taken into account through the SCV which is obtained by vertically concatenating the nth source from each of the K dataset as, $\mathbf{s}_n = [s_n^{[1]}, ..., s_n^{[K]}]^\top$. The goal in IVA is to estimate K demixing matrices to yield source estimates $\mathbf{y}^{[k]} = \mathbf{W}^{[k]}\mathbf{x}^{[k]}$, such that each SCV is maximally independent of all other SCVs. We note that while we consider the noiseless JBSS model, in real world applications the effect of noise is taken into account through dimension reduction, such as principal component analysis (PCA). Thus, we start with an over-determined problem where $d > N$ and use PCA to project the data to a lower dimensional space where $d = N$. This simple step is critical for multi-modal data fusion since each modality might exhibit different levels of noise and thus identifying the optimal signal subspace would help improve generalization abilities of the solution.

The IVA optimization parameter is defined as a set of demixing matrices $\mathbf{W}^{[1]}, \ldots, \mathbf{W}^{[K]}$, which can be collected into a three dimensional array $\mathcal{W} \in \mathbb{R}^{N \times N \times K}$ and can be estimated through the minimization of the IVA objective function given by

$$J_{\text{IVA}}(\mathcal{W}) = \sum_{n=1}^{N} H(\mathbf{y}_n) - \sum_{k=1}^{K} \log \left| \det \left(\mathbf{W}^{[k]} \right) \right| + C. \tag{2}$$

Here $H(\mathbf{y}_n)$ denotes the (differential)[6] entropy of the estimated nth SCV that serves as the term for modeling the complex relationships among the different modalities. By definition, the term $H(\mathbf{y}_n)$ can be written as $\sum_{k=1}^{K} H(y_n^k) - I(\mathbf{y}_n)$, where $I(\mathbf{y}_n)$ denotes the mutual information within the nth SCV.

Therefore, it can be observed that minimization with respect to each demixing matrix $\mathbf{W}^{[k]}$ of (2) automatically increases the mutual information within the components of an SCV, revealing how IVA exploits statistical dependence

[6] We consider continuous-valued random variables and in the sequel, refer to differential entropy as simply entropy for simplicity.

across different modalities. Hence, as observed in (2), the ability to *explicitly* learn the mutual relationships among the multiple modalities depends on the development of *flexible* and *efficient* models for differential entropy and their estimation.

2.4 Effective Density Model for Capturing Multi-modal Associations

The key factor in the explicit modeling of the non-linear relationships across different modalities is the estimation of the true underlying probability density function (PDF) of each estimated SCV. It is clear that minimizing (2) is not a straightforward task since there is no access to the true underlying PDF of each estimated SCV. To mathematically demonstrate this, if $\hat{p}(\mathbf{y}_n)$ denotes the PDF of the nth estimated SCV then its entropy can be expressed as

$$H(\mathbf{y}_n) = -f(p(\mathbf{y}_n), \hat{p}(\mathbf{y}_n)) - E\{\log \hat{p}(\mathbf{y}_n)\}, \tag{3}$$

where $f(p(\mathbf{y}_n), \hat{p}(\mathbf{y}_n))$ denotes the Kullback-Leibler (relative entropy) distance between the density of the nth estimated SCV and the true density of $\mathbf{y}_n$. From (3), we can achieve perfect source estimation as long as the assumed model PDF matches the true latent multivariate density of the nth SCV, i.e., $f(p(\mathbf{y}_n), \hat{p}(\mathbf{y}_n)) = 0$. As demonstrated in [11,17], PDF estimators based on the maximum entropy principle can successfully match multivariate latent sources from a wide range of distributions. The maximum entropy distribution for each $\mathbf{y}_n$ is given by

$$\hat{p}(\mathbf{y}_n) = \exp\left\{-1 + \sum_{m=0}^{M} \lambda_m r_m(\mathbf{y}_n)\right\}, \tag{4}$$

where the Lagrange multipliers λ_m are chosen such that the M number of moment constraints are satisfied.

Thus, the development of *flexible* and *efficient* models for entropy, their estimation using the maximum entropy principle, and their effective integration into (2), requires that we address the following three key issues:

1. Lagrangian multipliers evaluation and choice of constraints: We evaluate the Lagrangian multipliers by the Newton iteration scheme using local and global constraints. The estimation of the Lagrange multipliers highly depends on the proper selection of the constraints in order to provide information about the underlying statistical properties of the data. Failing on this will result in high complexity and poor data characterization.

Following a similar strategy as in [17,20], we jointly use global and local constraints in order to provide flexible multivariate density estimation while keeping the complexity low. Therefore, we use $\mathbf{1}, \mathbf{y}_n, \mathbf{y}_n^2, \mathbf{y}_n/(\mathbf{1}+\mathbf{y}_n^2)$ as the global constraints, since they provide information on the PDF's overall statistics, such as the mean, variance, and higher order statistics (HOS). For the local constraint we use the Gaussian kernel given by,

$$q(\mathbf{y}_n) = \frac{1}{\sqrt{|\boldsymbol{\Sigma}_n|(2\pi)^K}} \exp\left(-\frac{1}{2}(\mathbf{y}_n - \boldsymbol{\mu}_n)^\top \boldsymbol{\Sigma}_n^{-1}(\mathbf{y}_n - \boldsymbol{\mu}_n)\right), \tag{5}$$

where $\boldsymbol{\mu}_n$ denotes the mean vector, $\boldsymbol{\Sigma}_n$ denotes the covariance matrix, and $|\cdot|$ denotes the determinant. The Gaussian kernel provides localized information about the PDF.

It is important to mention that when we add the Gaussian kernel to the multidimensional framework, integration becomes challenging due to the fact that the Gaussian kernel has an infinite support.

2. Multi-dimensional integration during the estimation of the Lagrange multipliers: Multi-dimensional integration is one of the main challenges in our estimation problem. To overcome this problem, we use an efficient multi-dimensional integration technique that is based on Quasi-Monte Carlo (QMC) methods. QMC have shown to be efficient in terms of their rate of convergence and achieve a convergence rate of order $O((\log V)^K/V)$ [19]. Following the steps in [17], we generate a sequence of quasi-random points [29] and using this sequence we approximate the multi-dimensional integrals in a Monte Carlo method manner [19].

3. Efficient multivariate density estimation technique based on IVA by multivariate entropy maximization with kernels (IVA-M-EMK): Once the Lagrange multipliers have been estimated, and we have a full characterization of the underlying PDF and, therefore, a full characterization of the entropy for each estimated SCV, IVA-M-EMK provides estimates of the demixing matrices by minimizing (2). The gradient of (2) with respect to each row vector $\mathbf{w}_n^{[k]}$ of $\mathbf{W}^{[k]}$ is given by

$$\frac{\partial J_{\text{IVA}}}{\partial \mathbf{w}_n^{[k]}} = E\left\{\phi_n^{[k]}(\mathbf{y}_n)\,\mathbf{x}^{[k]}\right\} - \frac{\mathbf{h}_n^{[k]}}{\left(\mathbf{h}_n^{[k]}\right)^\top \mathbf{w}_n^{[k]}}, \tag{6}$$

where $\mathbf{h}_n^{[k]}$ is perpendicular to all row vectors of $\mathbf{W}^{[k]}$ except of $\mathbf{w}_n^{[k]}$ and $\phi_n^{[k]}(\mathbf{y}_n) = -\sum_{i=0}^{M} \lambda_i \frac{\partial r_i(\mathbf{y}_n)}{\partial y_n^{[k]}}$. The estimation of $\mathbf{W}^{[k]}$ is performed with respect to each row vector $\mathbf{w}_n^{[k]}$, $n = 1, \ldots, N$ independently. This is due to the fact that in (6), each gradient direction depends directly on the corresponding estimated source PDF. The sub-optimal gradient directions can lead to slower or sub-optimal convergence, or, in extreme cases, divergence of the source separation algorithm. Following the idea in [11], we perform the optimization routine in a Riemannian manifold rather than a classical Euclidean space since this provides important convergence advantages. We define the domain of our cost function to be the unit sphere in $\mathbb{R}^N$ and project (6) onto the tangent hyperplane of the unit sphere at the point $\mathbf{w}_n^{[k]}$. Since the IVA-M-EMK cost function depends on the number of moment constraints chosen for each SCV, non-monotonic behavior is expected between two consecutive iterations.

2.5 Classification Procedure

The classification process consists of four stages. As mentioned in Sect. 2.1, our dataset is separated into training and testing, where each tweet is represented by text as well as visual content. With this in mind, in the first stage we form our set of tweets in the following way. We denote with $\mathbf{X}_{\text{train}}^{[1]} \in \mathbb{R}^{d_1 \times V_{\text{train}}}$ and $\mathbf{X}_{\text{train}}^{[2]} \in \mathbb{R}^{d_2 \times V_{\text{train}}}$ the training observation matrices for each modality where d_1 denotes the number of initial high-level feature vectors in the textual modality, d_2 denotes the number of initial high-level feature vectors in the visual modality, and V_{train} denotes the number of training tweets. Similarly, $\mathbf{X}_{\text{test}}^{[1]} \in \mathbb{R}^{d_1 \times V_{\text{test}}}$ and $\mathbf{X}_{\text{test}}^{[2]} \in \mathbb{R}^{d_2 \times V_{\text{test}}}$ denote the corresponding testing observation matrices. In the second stage, the mean from each dataset is removed so they are centered and PCA is applied to each $\mathbf{X}_{\text{train}}^{[k]}$, for $k = 1, 2$. For the PCA step, we use an order N, which, in our setting, denotes the number of features from each modality. Then for each $k = 1, 2$, we obtain $\hat{\mathbf{X}}_{\text{train}}^{[k]} \in \mathbb{R}^{N \times V_{\text{test}}}$ and by vertically concatenating each $\hat{\mathbf{X}}_{\text{train}}^{[k]}$ we form a three dimensional array $\hat{\mathbf{X}}_{\text{train}} \in \mathbb{R}^{N \times V_{\text{train}} \times 2}$. In the third stage, we perform IVA on $\hat{\mathbf{X}}_{\text{train}}$, and since we have two modalities, IVA provides two demixing matrices $\mathbf{W}^{[1]} \in \mathbb{R}^{N \times N}$ and $\mathbf{W}^{[2]} \in \mathbb{R}^{N \times N}$. Then, using the estimated demixing matrices we generate $\mathbf{Y}_{\text{train}}^{[1]} = \mathbf{W}^{[1]} \left(\hat{\mathbf{X}}_{\text{train}}^{[1]}\right)^{\top}$ and $\mathbf{Y}_{\text{train}}^{[2]} = \mathbf{W}^{[2]} \left(\hat{\mathbf{X}}_{\text{train}}^{[2]}\right)^{\top}$. The training dataset $\mathbf{Y}_{\text{train}}$ is formed by either concatenating, averaging, or max pooling the estimated SCVs which can be obtained by concatenating the estimated sources from $\mathbf{Y}_{\text{train}}^{[1]}$ and $\mathbf{Y}_{\text{train}}^{[2]}$. Note that $\mathbf{Y}_{\text{train}}$ contains all the extracted features from the multi-modal data and it will be used for training the classification model. The testing dataset is generated by removing the training mean from each multi-modal testing dataset and using the generated PCA transformations from the training phase. The demixing matrices from the training phase are used to transform the testing datasets as follows, $\mathbf{Y}_{\text{test}}^{[1]} = \mathbf{W}^{[1]} \left(\hat{\mathbf{X}}_{\text{test}}^{[1]}\right)^{\top}$ and $\mathbf{Y}_{\text{test}}^{[2]} = \mathbf{W}^{[2]} \left(\hat{\mathbf{X}}_{\text{test}}^{[2]}\right)^{\top}$, where $\mathbf{Y}_{\text{test}}^{[1]} \in \mathbb{R}^{N \times V_{\text{test}}}$, $\mathbf{Y}_{\text{test}}^{[2]} \in \mathbb{R}^{N \times V_{\text{test}}}$. Finally, the testing dataset $\mathbf{Y}_{\text{test}}$ is formed by either concatenating, averaging, or max pooling the estimated SCVs which can be obtained by concatenating the estimated sources from $\mathbf{Y}_{\text{test}}^{[1]}$ and $\mathbf{Y}_{\text{test}}^{[2]}$. In the fourth stage, we train the classification model using $(\mathbf{Y}_{\text{train}})^{\top}$. The specific form of the classification model is unimportant. However, to demonstrate a concrete example, we use Support Vector Machines (SVMs), which have shown reliable performance in a variety of applications, especially with smaller size datasets [16,28]. Once the classification model has been trained, we evaluate its performance using the unseen dataset, $(\mathbf{Y}_{\text{test}})^{\top}$. For all experiments, hyper-parameter optimization and model training and testing is done using a grid search cross-validation with five folds scheme. The entire process was repeated five times (with shuffling before each iteration) to generate well converged statistics.

3 Results and Discussion

3.1 Classification Performance

For all of the experiments, we measure classification performance by employing the F1-score and reporting its macro averaged version. Moreover, we report the total CPU time of the training and testing phases and measure it in seconds.

For the first set of our experiments, we compare the classification performance of three different classification models; one trained using only the high-level textual features, one trained using the high-level visual features, and one trained using the high-level textual features and the high-level visual features concatenated together. From Table 1(a), we see that if we train a classifier with just the high-level textual features, we obtain a classification performance of 40.04%, while if we train by just using the high-level image features, we obtain an F1-score of 65.78%. If we concatenate the high-level text and image features, we obtain a classification performance of 77.59%. This result demonstrates that training a classifier using both modalities yields better classification performance. However, such an approach comes with significant challenges. As we can observe from Table 1(a), concatenating the two modalities results in feature vectors of dimension 4,396, and thus affecting the efficiency of the machine learning algorithm. In addition, without exploiting the complementary information among multiple modalities, discovering the features of greatest importance and how they interact with each other becomes impossible. IVA-M-EMK can address both challenges since it enables simultaneous study of multiple modalities by explicitly exploiting alignments of data fragments where there is a common underlying feature space. This can been seen from Table 1(b), where IVA-M-EMK with $N = 100$ and averaging the SCVs leads to high classification accuracy and superior improvement in terms of the CPU execution time. Moreover, Table 1(b) shows two additional methods to combine the estimated SCVs after IVA-M-EMK has been applied with $N = 100$. Due to the fact, that the "Average" method yields the highest F1-score and lowest CPU execution time, for the rest of our experiments we adopt the "Average" method in order to combine the estimated SCVs.

Table 1. Classification performance in terms of F1-score for different classification scenarios. Total CPU execution time is measured in seconds.

Methods	F1	CPU time
Concatenate	**77.59%**	$\mathbf{1.7 \times 10^4}$
Text	40.04%	2.7×10^3
Image	65.78%	1.03×10^4

(a) Regular approach

Combination	F1	CPU time
Concatenate	73.77%	2.4×10^3
Maximum	73.11%	1.5×10^3
Average	**77.45%**	$\mathbf{1.3 \times 10^3}$

(b) IVA-M-EMK

For the rest of our experiments, we demonstrate how effective modeling of the underlying multi-modal associations using IVA-M-EMK yields superior classification performance in an efficient manner. Figure 1, shows the effectiveness of the IVA-M-EMK algorithm by comparing its performance as a function of the number of features with three widely used IVA algorithms as well as canonical correlation analysis (CCA) [22], in terms of the CPU time and of the F1-score. It is worth mentioning that CCA does not explicitly impose an underlying density model for the joint features, but it implicitly seeks for a pair of vectors with maximum correlation coefficient. On the other hand, different IVA algorithms *explicitly* model the underlying associations by assuming a probability density for the underlying SCVs. In particular, IVA-Laplacian (IVA-L) [25] is an algorithm that takes higher-order statistics (HOS) into account and assumes a Laplacian distribution for the underlying source component vectors. IVA-Gaussian (IVA-G) [5] exploits linear dependencies but does not take HOS into account. Finally, IVA-A-GGD [10] is a more general IVA implementation where both second and higher-order statistics are taken into account. This algorithm assumes a multivariate generalized Gaussian distribution (MGGD) for the underlying sources, and through the estimation of its parameters, multivariate Gaussian and Laplacian distributions become special cases.

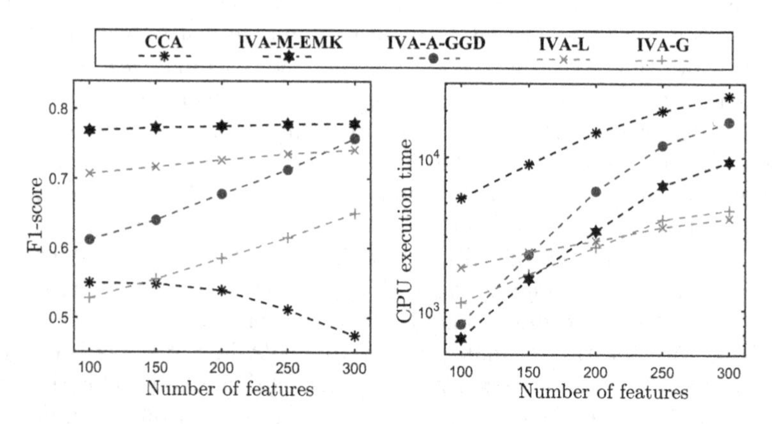

Fig. 1. Performance comparison in terms of F1 score and average CPU time for different number of features when all training samples are used.

From Fig. 1, we see that as the number of features increase, IVA-M-EMK, IVA-A-GGD, and IVA-L provide a desirable performance, followed by IVA-G. Conversely, as the number of features increases, CCA provides the worst performance due to its model simplicity. Overall, IVA-M-EMK performs the best among the five algorithms due to its ability to successfully match multivariate latent sources from a wide range of distributions, even for a small number of features. More importantly, as the number of features increases, the F1-score shows stable behavior. In terms of the CPU time, among the IVA algorithms that use a simple underlying density model, as the number of features increases, IVA-L and IVA-G provide the best performance. For IVA-G, this is due to the assumption of Gaussian distribution for the underlying sources, which simplifies the gradient of the IVA objective function and makes the Hessian positive definite, thus, enabling second-order algorithms to improve the speed of convergence. On the other hand, as expected, IVA-M-EMK and IVA-A-GGD are more computationally expensive; however, IVA-M-EMK has the lowest CPU execution time when the number of features stays low. Note that even if CCA is a much simpler model than the other IVA algorithms, the model mismatch yields poor convergence performance and significantly increases the number of iterations that the algorithm needs to converge.

For the last set of experiments, we compare IVA-M-EMK with several IVA algorithms, CCA, and a fusion approach that is based on PCA in terms of the F1-score and the CPU execution time as a function of the number of training samples. For the PCA approach, PCA is applied to each $\mathbf{X}_{\text{train}}^{[k]}$, for $k = 1, 2$ and the resulting low dimensional features from both modalities are averaged. Since the F1-score for IVA-M-EMK is invariant to the increase in features, we select to use $N = 100$ features for this experiment. From Fig. 2, we see that all IVA algorithms except IVA-G and CCA provide a desirable performance revealing the flexibility of their underlying density models. The fusion approach based on PCA also provides good performance in terms of F1 score and CPU execution time. Overall, the high F1-score along with the low and stable CPU execution time as a function of training tweets, makes IVA-M-EMK an ideal fusion approach for misinformation detection during high-impact events. Finally, we note that the classification results of our approach are on par with results obtained in similar studies such as [8].

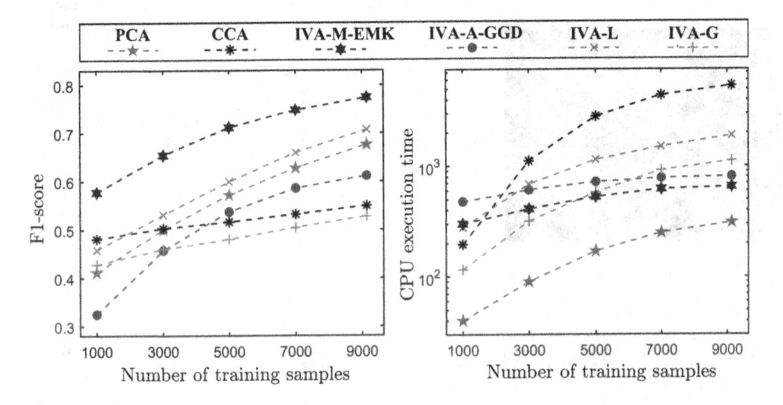

Fig. 2. Performance comparison in terms of F1 score and average CPU time for different number of training samples when $N = 100$.

3.2 Explainability

An important aspect that needs to be addressed when multi-modal fusion algorithms are developed is *explainability*. As a reminder, explainability can be broadly defined as the ability of the model to summarize the causes of its decisions in an efficient manner and thus gain the trust of its users. For our work, we use three different scenarios and several examples from our testing dataset to address different explainability aspects for IVA-M-EMK. We also use these scenarios to discover how different modalities interact with each other and what their impact in the explainability of the classification model is. To support our experiments, we use a popular interpretability tool, LIME [31], which produces local explanations for classifier decisions.

For the first scenario, we examine the impact of $\mathbf{W}^{[1]}$ on the explainability of our system which has been estimated by also taking the visual content into account. Therefore, the training and testing sets are generated using only textual features, where $\mathbf{Y}^{[1]}_{\text{train}} = \left(\hat{\mathbf{X}}^{[1]}_{\text{train}}\right)^{\top}$ and $\mathbf{Y}^{[1]}_{\text{test}} = \left(\hat{\mathbf{X}}^{[1]}_{\text{test}}\right)^{\top}$. In order to take the visual content into account, we introduce $\mathbf{W}^{[1]}$ into the classifier in the following way $\mathbf{Y}^{[1]}_{\text{train}} = \mathbf{W}^{[1]}\left(\hat{\mathbf{X}}^{[1]}_{\text{train}}\right)^{\top}$ and $\mathbf{Y}^{[1]}_{\text{test}} = \mathbf{W}^{[1]}\left(\hat{\mathbf{X}}^{[1]}_{\text{test}}\right)^{\top}$. Figure 3, shows the impact of the visual content through $\mathbf{W}^{[1]}$ on the fake news detector. In the first column, we have the textual and visual tweet content. The second column represents the LIME explanation for the fake news classifier without taking the visual content into account, and the third column takes the visual content into account. In the first row, we have a fake tweet showing the Eiffel Tower lit up

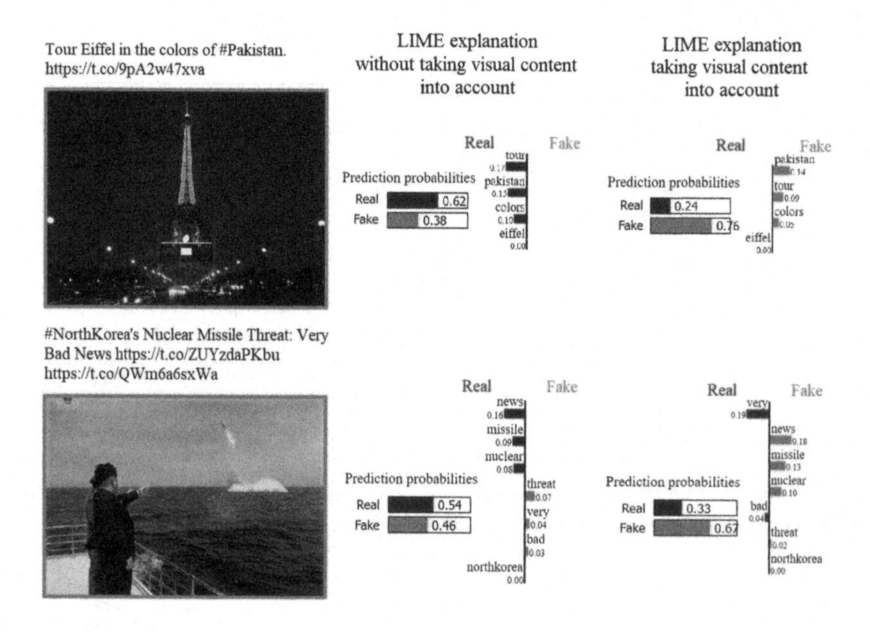

Fig. 3. Impact of the visual content through $\mathbf{W}^{[1]}$ on the classifier.

in the colors of Pakistans national flag[7] suggesting to show solidarity during the Lahore attacks suicide bombing, but in fact, the image emerged to be from the Rugby World Cup. Without the visual features we would not be able to detect this type of misinformation. In the second row, we have also a fake tweet presenting a threat of a nuclear missile attack. The image shows a fake missile coming out of the water, which makes words like missile and nuclear have their weights represented on the fake side. On the other hand, without the visual content showing the fake missile the classifier classifies the tweet as real.

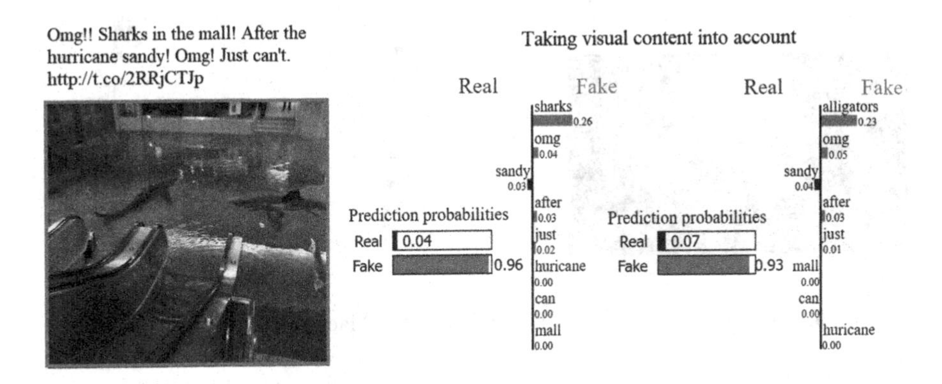

Fig. 4. Similarity between shark and alligator.

For the second scenario, we examine how the classifier behaves when replacing a critical word for the classification decision with an unseen word similar to the original one. Figure 4 presents a fake tweet where sharks were supposedly seen inside a shopping mall after Hurricane Sandy. We can see through the LIME explanation that even though 'alligators' is an unseen word, we are able to accurately classify the tweet, showing the efficiency of our classifier that captures the similarity between the two words since they are animals that can be found in the water.

For the third scenario, we examine the weakness of the proposed approach focusing on the cases where the IVA-based classifier failed to classify a tweet correctly. Figure 5, presents a similar framework to the first setting, where we explore the impact of the insertion of the visual content into the classifier. In the first row, we have a fake tweet showing people united in solidarity with the victims of the Paris attack. The image contains a lot of black pixels and visual content that is hard to verify, even for humans. Through visual inspection, we repeatedly observe that images containing many black pixels yield a misclassified case. Conversely, when the classifier takes only the textual content into account, it correctly classifies the tweet. In the second row, we have a real tweet

[7] https://www.independent.co.uk/news/world/asia/lahore-attack-photo-showing-eiffel-tower-lit-up-in-colours-of-pakistan-flag-is-from-2007-rugby-world-cup-a6959231.html.

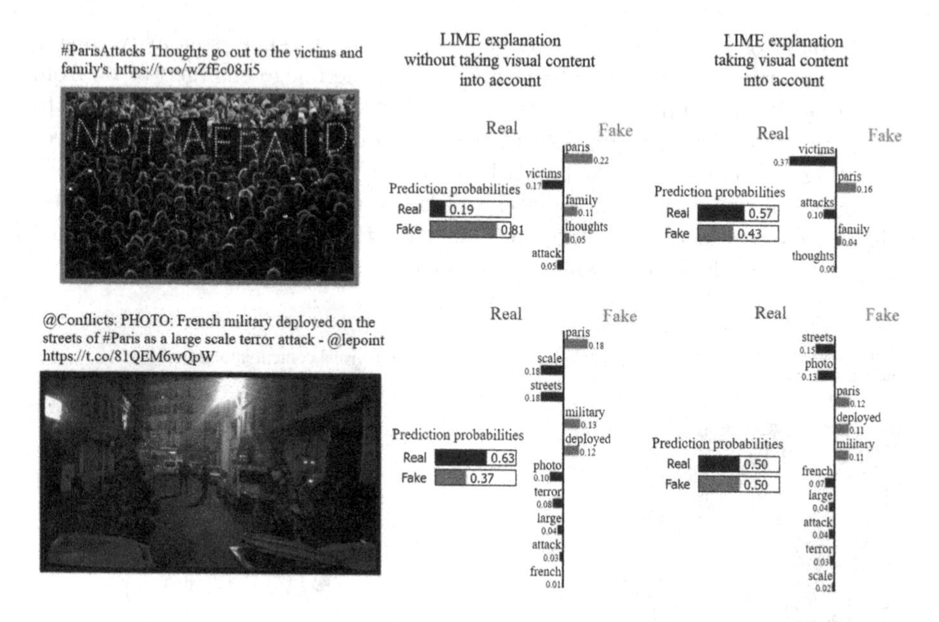

Fig. 5. Weakness cases. Shows the impact of the black pixels on the classifier.

showing military members on a street in Paris after the terrorist attacks. As in the previous case, the image presents many black pixels; in addition, the lightest part of the image is the street, which makes the weight of the word 'street' higher, rendering the classifier unable to identify the correct tweet label.

4 Conclusion

The success of the proposed method raises several interesting questions in terms of the number of modalities and quantitative ways to measure explainability to explore in future work. For this proposed approach, we considered a framework with two modalities represented by the text and visual content. In the future, we are interested in incorporating additional modalities in our study, since as we have demonstrated in [12], the multivariate data fusion model provides enhanced detection performance as the number of modalities increases. Last, as a future direction in terms of explainability, we propose to create formal settings where humans can evaluate whether a set of extracted features have human-identifiable semantic coherence. These quantitative methods have been similarly used for measuring semantic meaning in inferred topics [14]. By developing human-based evaluation metrics, we will not only be able to assess the IVA joint representation space, but more importantly, we will be able to identify potential biases related to specific characteristics of the collected social media posts, enabling us to correct our model before it is deployed at scale.

References

1. The Washington Post (2018). https://rebrand.ly/ieeovv
2. Newsweek (2019). https://rebrand.ly/z6t52a
3. Hateful memes challenge and data set for research on harmful multimodal content. https://ai.facebook.com/blog/hateful-memes-challenge-and-data-set/
4. Adalı, T., Anderson, M., Fu, G.S.: Diversity in Independent Component and Vector Analyses: Identifiability, algorithms, and applications in medical imaging. IEEE Sig. Process. Mag. **31**(3), 18–33 (2014)
5. Anderson, M., Adalı, T., Li, X.L.: Joint blind source separation with multivariate gaussian model: algorithms and performance analysis. Sig. Process. IEEE Trans. **60**(4), 1672–1683 (2012). https://doi.org/10.1109/TSP.2011.2181836
6. Baltrušaitis, T., Ahuja, C., Morency, L.P.: Multimodal machine learning: a survey and taxonomy. IEEE Trans. Pattern Anal. Mach. Intell. **41**(2), 423–443 (2018)
7. BBC: Social media firms fail to act on covid-19 fake news. www.bbc.com/news/technology-52903680, June 2020
8. Boididou, C., Papadopoulos, S., Zampoglou, M., Apostolidis, L., Papadopoulou, O., Kompatsiaris, I.: Detection and visualization of misleading content on twitter. Int. J. Multimedia Inf. Retrieval **7** (2018). https://doi.org/10.1007/s13735-017-0143-x
9. Boididou, C., Papadopoulos, S., Zampoglou, M., Apostolidis, L., Papadopoulou, O., Kompatsiaris, Y.: Detection and visualization of misleading content on twitter. Int. J. Multimedia Inf. Retrieval **7**(1), 71–86 (2018). https://doi.org/10.1007/s13735-017-0143-x
10. Boukouvalas, Z., Fu, G.S., Adalı, T.: An efficient multivariate generalized gaussian distribution estimator: Application to IVA. In: 2015 49th Annual Conference on Information Sciences and Systems (CISS), pp. 1–4. IEEE (2015)
11. Boukouvalas, Z., Levin-Schwartz, Y., Mowakeaa, R., Fu, G.S., Adalı, T.: Independent component analysis using semi-parametric density estimation via entropy maximization. In: 2018 IEEE Statistical Signal Processing Workshop (SSP), pp. 403–407. IEEE (2018)
12. Boukouvalas, Z., Puerto, M., Elton, D.C., Chung, P.W., Fuge, M.D.: Independent vector analysis for molecular data fusion: Application to property prediction and knowledge discovery of energetic materials. In: 2020 28th European Signal Processing Conference (EUSIPCO), pp. 1030–1034. IEEE (2021)
13. Cao, J., Qi, P., Sheng, Q., Yang, T., Guo, J., Li, J.: Exploring the role of visual content in fake news detection. In: Shu, K., Wang, S., Lee, D., Liu, H. (eds.) Disinformation, Misinformation, and Fake News in Social Media. LNSN, pp. 141–161. Springer, Cham (2020). https://doi.org/10.1007/978-3-030-42699-6_8
14. Chang, J., Gerrish, S., Wang, C., Boyd-Graber, J.L., Blei, D.M.: Reading tea leaves: how humans interpret topic models. In: Advances in Neural Information Processing Systems, pp. 288–296 (2009)
15. Comon, P., Jutten, C.: Handbook of Blind Source Separation: Independent Component Analysis and Applications. Academic Press, Cambridge (2010)
16. Cortes, C., Vapnik, V.: Support-vector networks. Mach. Learn. **20**(3), 273–297 (1995). https://doi.org/10.1023/A:1022627411411
17. Damasceno, L.P., Cavalcante, C.C., Adalı, T., Boukouvalas, Z.: Independent vector analysis using semi-parametric density estimation via multivariate entropy maximization. In: ICASSP 2021–2021 IEEE International Conference on Acoustics, Speech and Signal Processing (ICASSP), pp. 3715–3719. IEEE (2021)

18. Devlin, J., Chang, M., Lee, K., Toutanova, K.: BERT: pre-training of deep bidirectional transformers for language understanding. CoRR abs/1810.04805 (2018). arxiv.org/abs/1810.04805
19. Dick, J., Kuo, F.Y., Sloan, I.H.: High-dimensional integration: the quasi-monte Carlo way. Acta Numerica **22**, 133–288 (2013). https://doi.org/10.1017/S0962492913000044
20. Fu, G., Boukouvalas, Z., Adali, T.: Density estimation by entropy maximization with kernels. In: 2015 IEEE International Conference on Acoustics, Speech and Signal Processing (ICASSP), pp. 1896–1900, April 2015. https://doi.org/10.1109/ICASSP.2015.7178300
21. Hansen, L.K., Rieger, L.: Interpretability in intelligent systems – a new concept? In: Samek, W., Montavon, G., Vedaldi, A., Hansen, L.K., Müller, K.-R. (eds.) Explainable AI: Interpreting, Explaining and Visualizing Deep Learning. LNCS (LNAI), vol. 11700, pp. 41–49. Springer, Cham (2019). https://doi.org/10.1007/978-3-030-28954-6_3
22. Hardoon, D.R., Szedmak, S., Shawe-Taylor, J.: Canonical correlation analysis: an overview with application to learning methods. Neural Comput. **16**(12), 2639–2664 (2004)
23. Hiten Patel, M.: Fake news about covid-19 is spreading faster than virus. https://wexnermedical.osu.edu/blog/fake-news-about-covid-19, April 2020
24. Hyvärinen, A., Karhunen, J., Ojá, E.: Independent Component Analysis, vol. 46. Wiley, Hoboken (2004)
25. Kim, T., Eltoft, T., Lee, T.-W.: Independent vector analysis: an extension of ICA to multivariate components. In: Rosca, J., Erdogmus, D., Príncipe, J.C., Haykin, S. (eds.) ICA 2006. LNCS, vol. 3889, pp. 165–172. Springer, Heidelberg (2006). https://doi.org/10.1007/11679363_21
26. Linardatos, P., Papastefanopoulos, V., Kotsiantis, S.: Explainable AI: a review of machine learning interpretability methods. Entropy **23**(1), 18 (2020)
27. Mikolov, T., Chen, K., Corrado, G.S., Dean, J.: Efficient estimation of word representations in vector space abs/1301.3781
28. Moroney, C., et al.: The case for latent variable vs deep learning methods in misinformation detection: an application to covid-19. In: Soares, C., Torgo, L. (eds.) DS 2021. LNCS (LNAI), vol. 12986, pp. 422–432. Springer, Cham (2021). https://doi.org/10.1007/978-3-030-88942-5_33
29. Niederreiter, H.: Random Number Generation and Quasi-Monte Carlo Methods. Society for Industrial and Applied Mathematics, USA (1992)
30. Ramachandram, D., Taylor, G.W.: Deep multimodal learning: a survey on recent advances and trends. IEEE Sig. Process. Mag. **34**(6), 96–108 (2017)
31. Ribeiro, M.T., Singh, S., Guestrin, C.: "Why should i trust you?": Explaining the predictions of any classifier arxiv.org/abs/1602.04938
32. Sharma, K., Qian, F., Jiang, H., Ruchansky, N., Zhang, M., Liu, Y.: Combating fake news: a survey on identification and mitigation techniques. ACM Trans. Intell. Syst. Technol. (TIST) **10**(3), 1–42 (2019)
33. Suciu, P.: Covid-19 conspiracy theories continue to spread and thrive on social media. www.forbes.com/sites/petersuciu/2020/04/24/covid-19-conspiracy-theories-continue-to-spread-and-thrive-on-social-media/#e1a9e8b10076, April 2020

Fairness and Outlier Detection

MQ-OFL: Multi-sensitive Queue-based Online Fair Learning

Farnaz Sadeghi and Herna Viktor(✉)

School of Electrical Engineering and Computer Science,
University of Ottawa, Ottawa, Canada
{fsade079,hviktor}@uottawa.ca

Abstract. Recently, there has been growing interest in fairness considerations in Artificial Intelligence (AI) and AI-based systems, as the decisions made by AI applications may negatively impact individuals and communities with ethical or legal consequences. Indeed, it is crucial to ensure that decisions based on AI-based systems do not reflect discriminatory behavior toward certain individuals or groups. The development of approaches to handle these concerns is an active area of research. However, most existing methods process the data in offline settings and are not directly suitable for online learning from evolving data streams. Further, these techniques fail to take the effects of data skew, or so-called class imbalance, on fairness-aware learning into account. In addition, recent fairness-aware online learning supervised learning approaches focus on one sensitive attribute only, which may lead to subgroup discrimination. In a fair classification, the equality of fairness metrics across multiple overlapping groups must be considered simultaneously. In this paper, we address the combined problem of fairness-aware online learning from imbalanced evolving streams, while considering multiple sensitive attributes. We introduce the Multi-Sensitive Queue-based Online Fair Learning (MQ-OFL) algorithm, an online fairness-aware approach, which maintains valid and fair models over evolving stream. MQ-OFL changes the training distribution in an online fashion based on both stream imbalance and discriminatory behavior of the model evaluated over the historical stream. We compare our MQ-OFL method with state-of-art studies on real-world data sets, and present comparative insights on the performance.

Keywords: Fairness-aware classification · Multi sensitive attribute · Data stream · Imbalanced data

1 Introduction

AI applications have became a necessity to deliver all sorts of decisions, such as screening of job applications, loan credit approval, allocation of health resources, and autonomous driving. However, unconsciously, these automated data-driven systems may lead to discrimination against particular groups of people sharing one or more sensitive attributes (e.g., marital status, age, gender or sex, and

P. Pascal and D. Ienco (Eds.): DS 2022, LNAI 13601, pp. 271–285, 2022.
https://doi.org/10.1007/978-3-031-18840-4_20

ethnicity) [16]. As a recent example, Howard et. al. [9] discusses applications of how bias in the real world can breach into AI systems, such as bias in face recognition applications, voice recognition, and search engines. As a result, a number of studies have been proposed to address this concern. A common theme amongst all these prior works is the assumption of fairness as a static problem, which means the inappropriate discriminate correlations (e.g., marital status or age) is implicitly modeled as a constant and static property. Learning from data stream assumes that new instances arrive continuously and that their properties may change over time due to a phenomenon known as concept drift [7]. Frequently, in a supervised learning setting, such streams are subject to data skew, i.e., class imbalanced, with a disproportion of the number of examples of the different classes [2]. Additionally, the vast majority of the algorithmic fairness literature focused on the simplest case where there are only one sensitive attribute [16]. To the best of our knowledge, this is the first work jointly considers non-stationary imbalanced data distributions where there are more than one sensitive attribute.

The contribution of this paper is three-fold. First, we define a new problem of fairness-aware learning in imbalanced data streams with more than one protected group. Then, we propose a discrimination-aware pre-processing method to handle the trade-off between fairness and accuracy. Second, we introduce an approach to pre-process multi-sensitive attributes that satisfies fairness constraints. Thirdly, our experimental evaluation verify the capability of the proposed model in online settings and for application-driven fairness-aware learning.

The remainder of the paper is organized as follows. Background knowledge and related studies regarding fairness-aware learning are reviewed in Sect. 2, respectively. We introduce our MQ-FOL framework in Sect. 3, followed by an experimental evaluation in Sect. 4. Section 5 concludes the paper.

2 Background

This section details related work and defines key concepts.

2.1 Related Work

A number of research approaches have been proposed to address the problem of bias and discrimination in machine learning systems. They may be categorized into three main groups, namely pre-processing approaches, in-processing approaches and post-processing approaches, based on whether they mitigate bias at the data level, the algorithm design or the output of model, respectively [16]. The first strategy, works under the assumption that in order to learn a fair classifier, the training data should be discrimination-free [11]. In-processing techniques modify and change learning algorithms to limit discrimination [16]. The last category, consists of either adjusting the decision boundary [6] of a model or directly changing the prediction labels. Fairness in an online setting requires simultaneously take the evolution of underlying data distribution into consideration. The Fairness-Aware Hoeffding Tree (FAHT) [23], addresses discrimination by incorporating discrimination-awareness into the model induction process. This is accomplished by introducing the fair information gain splitting criterion, which is able to maintain a moderate predictive performance with low discrimination scores over the course of the stream. Another work [22], proposed a Fairness-Enhancing and concept-Adapting Tree (FEAT) with embedded fair-enhancing splitting criterion. A strength of this approach is the ability of change detection and concept forgetting to handle discriminated and non-stationary data stream. However, these two methods do not consider imbalanced data. Specifically, none of the current online fairness-aware techniques are able to simultaneously handle more than one sensitive attribute. Our work situates in this highly under-explored research direction by including multi-protected groups to provide fair online decision making. Next, we discuss definitions and key concepts in the fairness-aware domain.

2.2 Fairness Definitions

We assume an attribute S, referred as a sensitive attribute with a special value $s \in dom(SA)$ which is a sensitive value that defines the discriminated group. Also, we assume that Z is a binary attribute: $dom(SA) = \{z, \bar{z}\}$. As an example, we use $Z =$ "$maritalstatus$" as the sensitive attribute and $\bar{z} =$ "$single$" as the sensitive value (protected group) with $z =$ "$married''$ (non-protected group). We also consider the class is binary with values {0 as rejected, 1 as granted}.

Fairness definitions fall under different types such as individual, group and the subgroups [16]. Individual notions of fairness means "similar individuals" have to be "treated similarly", while Group notions refers to treating different groups equally and subgroup fairness focus on the best properties of the group and individual notions of fairness. It chooses a group fairness notion and finds whether this metric satisfies a large collection of subgroups [12,13]. Here we define Statistical Parity (S.P.) which is a group notion for fairness. It measures whether the probability of having positive outcome is the same for both protected and non-protected groups. Also, we can refer to it as difference in the probability of a random individual drawn from non-protected group to be predicted as granted (positive) and the probability of a random individual drawn from the protected group to be predicted as granted:

$$S.P. = P(f(x) = y^+ \mid \bar{z}) - P(f(x) = y^+ \mid z) \tag{1}$$

Here, z represents the protected group (e.g., female) and $\bar{z}$ refers to the unprotected group (e.g., male), while y^+ denotes positive predictive outcomes (e.g., an individual was selected for employment). The SP values lie in the $[-1, 1]$ range, with 0 meaning the decision does not depend on the sensitive value (meaning fair), 1 meaning that the protected group is discriminated, and -1 that the non-protected group is discriminated. The next section introduces the concept of Gerrymandering, where we consider more than one sensitive attribute.

2.3 Gerrymandering

The most straightforward setting in fairness is the independent case, with only one sensitive attribute, which can take multiple values, e.g., age only. The presence of multiple sensitive attributes (e.g., ethnicity and age simultaneously) leads to non-equivalent definitions of group fairness. For example, consider a model restricted to S.P. between subgroups defined by ethnicity. Simultaneously, the model can be constrained to S.P. between subgroups defined by gender. We term fairness in this situation independent group fairness. On the other hand, one can consider all subgroups defined by intersections of sensitive attributes (e.g., ethnicity and gender, ethnicity and age, age and gender, and so on), leading to intersectional group fairness. A given algorithm can be independently group fair, e.g., when considering age and gender in isolation, but not intersectionally group fair, e.g., when considering intersections of age and gender groups. For example, [3], showed how facial recognition software had a particularly poor performance for black women. This phenomenon, called fairness gerrymandering, has been studied by [12], where the authors specifically focus on ethnicity and gender.

As shown in Fig. 1 [12], imagine a setting with two binary features, corresponding to ethnicity (say blue and green) and gender (say men and women), both of which are distributed independently and uniformly at random in a population. Consider a classifier that labels an example positive if and only if it corresponds to a blue man, or a green woman. Then the classifier will appear to be equitable when one considers either protected attribute alone, in the sense that it labels both men and women as positive 50 percent of the time, and labels both blue and green individuals as positive 50 percent of the time. However, as pointed out by [12] if one considers any conjunction of the two attributes (such as blue women), then it is apparent that the classifier maximally violates the statistical parity fairness constraint. Similar examples for classification are easily constructed. We remark that the issue raised by this toy example is not merely hypothetical. To avoid this issue, we would like to satisfy a fairness constraint for more than one protected group defined by multiple sensitive attributes.

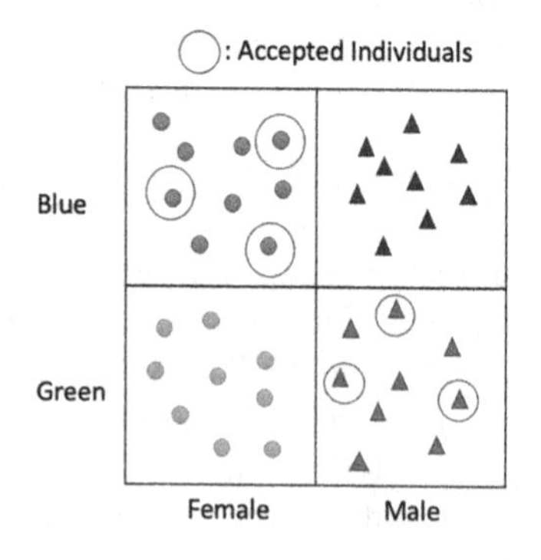

Fig. 1. Gerrymandering illustration (from [12])

Next, we turn our attention to fairness-aware learning from evolving, imbalanced streams that contains multiple sensitive attributes.

2.4 Imbalanced and Drifted Data Stream

Recall that our study assumes that the underlying stream distribution is non-stationary, that is, the characteristics of the stream might change with time leading changes in the joint distribution so that the decision boundary might change over time for two instances i, j, it might hold that $P_i(x, y) \neq P_j(x, y)$, a phenomenon called concept drift [7]. Numerous algorithms have been developed in order to detect and handle concept drift [19]. For instance, Hoeffding's inequality-based Drift Detection Method (HDDM) proposed by [19] employs Hoeffding's inequality [8] to set an upper bound to the level of difference between error rates. That is, using Hoeffding's inequality, triggers a warning level to indicate a drift may have occurred. The threshold used to trigger the warning level is a relaxed condition of the threshold used for the drift level. The data accumulated between the warning level and the drift levels are used as the training set for updating a learning model. We employ HDDM in our work.

Apart from the occurrence of concept drifts, recall that we also assume that the stream is imbalanced with the majority class occurring more often than the minority class, which usually makes the minority class to be overlooked. In two-class problems the minority (underrepresented) class is usually referred to as the positive class, whereas the majority class is considered to be the negative one. There have been several proposals for coping with imbalanced data sets [2] where the main goal is to correctly classify minority examples. In this paper, we employ the Online-OMCQ framework [5], which learns from evolving streams using an incremental, online approach. In Online-OMCQ, we combine batch-based and instance-based learning, as will be discussed in the next section.

3 MQ-OFL Framework

Our MQ-OFL approach, as shown in Fig. 2, consists of three stages: customized queue construction, prediction and fair online learning. Our method includes s class imbalance monitoring and balancing step, that keeps track of the class ratios over the stream and adjusts the proportion of classes by assigning them into related class label queues. In addition, the queue-based system is provided to train customized classifiers based on each sensitive attribute and the subgroups. Each arriving instance is evaluate both based on the sensitive attributes and the class label, while testing for concept drift. In the case where we have concept drift, we employ the previously introduced HDDM algorithm. By employing a fairness aware post-processing method, the decision boundary is adjusted to ensure that the classifier does not incur discrimination.

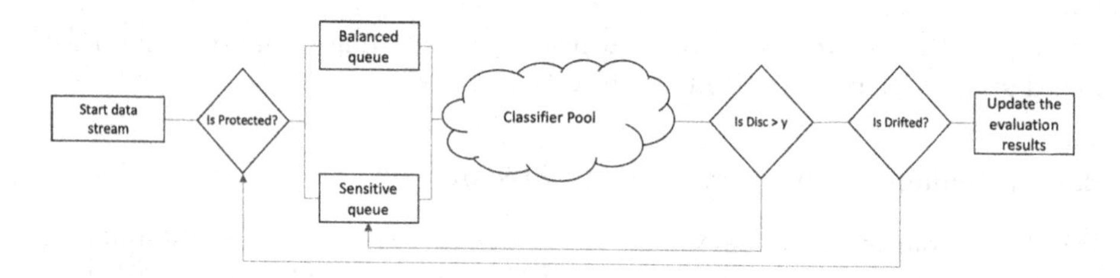

Fig. 2. High-level overview of queue-based fairness aware methodology

3.1 Balanced and Fairness-Aware Pre-processing

In our work, we model each individual (person) as being described by a tuple $((x, \bar{x}), y)$, where $x \in X$ denotes a vector of protected attributes, $\bar{x} \in \bar{X}$ denotes a vector of unprotected attributes, and $y \in 0, 1$ denotes a label. We assume that points (X, y) are drawn i.i.d. from an unknown distribution P. Let D be a binary classifier where $D(X) \in 0, 1$ denote the (possibly randomized) decision induced by D on individual (X, y).

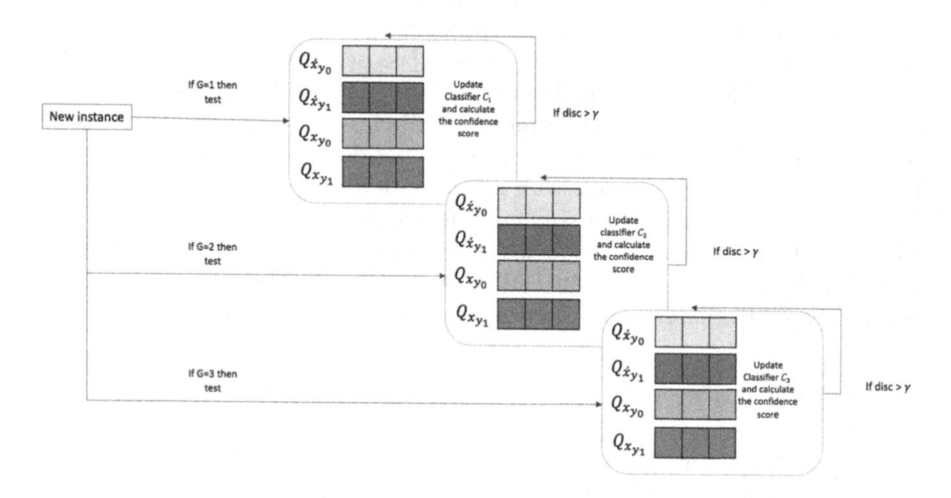

Fig. 3. Example of updating sensitive queues and forming batches

Figure 3 illustrates how $Queue_{F}air$ works when we have two sensitive attributes. Each arriving instance (x_t, y_t) is evaluated based on belonging to a group $G = \{g_1, g_2, g_3\}$ and added to the related queue of equal length $q^t_{C_k} = L$, where c_k is the class label. Together, the queues of same group form a batch. Each of the training batches will customize a separate classifier.

3.2 Classifier Pool

As shown in Fig. 3, each protected group has its own classifier trained based on the importance of that sensitive attribute. First, a candidate classifier pool is established by training a number of base classifiers with related batches to achieve better accuracy and fairness. Generally, base classifiers can be generated on the sample-level, feature-level or algorithm-level, in which classification models are produced from various sample subsets, feature subsets, or by learning algorithms, respectively. We used this idea to combine sample- and feature-level techniques to select samples based on features from the original dataset as a training subset. Second, the classifier with the highest confidence score for arriving instance will choose to predict the label. A confidence score is calculated as an evaluation standard; it shows the probability of the instance being detected correctly by the classifier and it is given as a percentage. The scores are taken on the prediction precision of each classifier for each arriving instance.

3.3 Decision Boundary Adjustment

Adjustment of the decision boundary for discrimination elimination has been investigated in the literature [16]. The authors in [6], proposed an approach to achieving fairness by shifting the decision boundary (SDB) for the protected group in the static datasets. They illustrate that SDB may be combined with a member of the family of learning algorithms that produce a measure of confidence in its prediction. It follows that, since we are dealing with streaming, we do not have access to all predictions through the stream to adjust the boundary accurately. Thus, following [10], we estimate the number of instances n_t which are needed in order to mitigate discrimination at time point t by:

$$n_t = \lfloor \sum_{i=1}^{t} 1 \cdot \mathbb{I}[x_i \in z] \cdot \frac{\sum_{i=1}^{t} 1 \cdot \mathbb{I}[f_i(x_i) = y^+ \mid x_i \in \bar{z}, y^+]}{\sum_{i=1}^{t} 1 \cdot \mathbb{I}[x_i \in \bar{z}, y_i^+]} - \sum_{i=1}^{t} 1 \cdot \mathbb{I}[f_i(x_i) = y^+ \mid x_i \in z, y^+] \rfloor \quad (2)$$

We consider a window size of M to keep misclassified instances together with their confidence scores in descending order. In case we have discrimination, the top n_t number in the window will be adapted at the boundary.

Algorithm 1 summarize our methodology. When online learning starts, at each time step, each instance (x_t, y_t) arrives at time t then receives the predicted label from the classifier with the highest confidence score from classifier pool with label set $Y = \{0, 1\}$. When training starts, the queue sizes for all current protected and non-protected groups assessed; if there are full queues, the classifier is able to update the learning model; i.e., the training process utilizes a balanced set of G and y consisting of the most recent data related to them. It follows that both the batch size p and the sizes of the individual queues are highly domain-dependent; these values are set by inspection. Next, the confidence scores and the Classifier pool are updated, and the HDDM drift detection algorithm is initiated. Finally, we evaluate the discrimination level; if it is higher than a pre-defined α, we employ boundary adjustment. To this end, we use a sliding window model of a pre-defined size M. In particular, we maintain a sliding window of size M for each sensitive attribute to allow for boundary adjustment for different classifiers based on each discriminated group. Finally, the evaluation metrics are updated and the learning process continues.

Algorithm 1. MQ-OFL Methodology

Require: A Discriminated Data Stream D,
1: **while** stream.has_more_instances() at each time step t **do**
2: x_i^t, y_i^t = get.next_instance()
3: y_i^t_predict = Classifier_Highest_Confidence_Score.Predict(x_i^t)
4: **for** $i \in G$ **do**
5: **if** $x_i^t \in g_i$ **then**
6: $IncrementCounter_{g_i}$;
7: $Q_{g_i}^t =_{g_i}^{(t-1)} .append(x_i^t)$;
8: **else**
9: $IncrementCounter_{C_{other}}$;
10: $Q_{C_{other}}^t =_{C_{other}}^{(t-1)} .append(i)$;
11: **end if**
12: **if** $Q_{g_i} == L$ and $Q_{C_{other}} == L$ **then**
13: $Training\_set_{classifier_i} = Q_{g_i} + Q_{c_i}$
14: $Classifier_i.Incremental.Update(Training\_set)$
15: **end if**
16: **end for**
17: Update classifier pool and confidence scores
18: **if** HDDM detects drift **then**
19: Fill the queues with new data
20: **end if**
21: **if** Discrimination level $\geq \alpha$ **then**
22: Adjust the boundary
23: **end ifreturn** G_mean, F_Measure,Statistical_Parity, Model
24: **end while**

4 Experimental Evaluation

In this section, we conduct experiments to evaluate the accuracy and fairness of the MQ-OFL framework. To this end, we first investigate the enhanced discrimination reduction capability of the proposed fair reprocessing. We also show a comprehensive quantitative evaluation to verify the ability of class imbalance and concept adaption or our method. All experiments were conducted on a MacBook Pro with a Dual-Core Intel Core i5 processor, CPU @ 3.1 GHz processor, 8.0 GB RAM on the Mac Catalina Operating System (OS), and the *Name Withheld* Cloud with 10 Core CPUs. Our code was implemented using the Scikit-Learn [20] and Scikit-Multiflow [17] packages in Python version 3.8.2. The framework's implementation and all the code for the experiments will be made available in GitHub upon publication. The Hoeffding Tree (HT) [4] and Hoeffding adaptive tree (HAT) [1] classifiers were used as our base classifier in the model. HTs are incremental decision trees for data stream classification that use Hoeffding's bound to commence online learning. HAT is an extension of HT that adaptively learns from data streams that change over time without needing a fixed-size sliding window. We evaluate MQ-OFL against two recent state-of-the-art fairness-aware stream classifiers FAHT [23] and FEAT [22]. Recall that

FAHT method solves the discrimination problem by introducing a new splitting criterion, called fair information gain (FIG), that jointly considers the fairness gain and information gain of the introduction of an attribute split. On the other hand, FEAT embedded fair-enhancing splitting criterion and includes change detection and concept forgetting to handle discriminated and non-stationary data streams.

4.1 Datasets

Our experimental study is based on the datasets used in the recent works in this research direction [10,16,22]. The following datasets are shown in Table 1: *Adult* [14], the *COMPAS* dataset [15] of criminal recidivism, the *Default* dataset [21] and the *Bank* dataset [18].

There are 48,843 instances in the *Adult* dataset and each instance is described by 14 employment and demographic attributes. Following the state-of-the-art, we conduct experiments on these datasets by setting "gender" and "ethnicity" as the sensitive attributes with female and black being the sensitive value and an annual income of more than 50K as the target class, i.e., the positive classification. The *Bank* dataset comprises 41,188 samples with 20 features and a binary label, indicating whether clients have subscribed to a term deposit. For this dataset, ages less than 25 and more than 60 years and marital status being single/married are considered sensitive. The *CreditCardDefault* dataset considers age (same as *Bank* dataset) and gender as sensitive attributes. We select a subset of the *COMPAS* dataset previously used for fairness experiments, which comprises 5,320 samples with five features (age category, gender, ethnicity, priors count and charge degree) and a binary label indicating whether the defendant re-offended within two years.

The reader should notice that all datasets include more than one sensitive attribute (indicated in the Table 1) which make them useful for evaluating our MQ-OFL method. It must be noted that the selected features aim to facilitate fairness experiments comparable to previous approaches, rather than only focusing on high predictive accuracy.

Table 1. Characteristics of data streams used in experiments.

Dataset	#Instances	#Attributes	Sen.Attr	Imbalanced ratio	Class label
Adult Cen	48843	14	Gender/Ethnicity	$1:3$	$\leq 50K$ or $\geq 50K$
Bank	41188	16	Marital Status/Age	$1:7.5$	Subscription (yes/no)
Default	30000	24	Gender/Age	$1:3.5$	Default Payment (yes/no)
COMPAS	13610	4	Gender/Ethnicity	$1:1.1$	Re-offended (yes/no)

4.2 Evaluation Metrics

We evaluate whether a classifier D is satisfying statistical fairness constraint based on statistical parity (S.P.). This fairness metric is defined with respect to a set of protected groups G if we have more than one sensitive attribute. Each $g : X \rightarrow \{0, 1\} \in G$ has the semantics that $g(x) = 1$ indicates that an individual with protected features x is in group g. Definition 3 (S.P. Subgroup Fairness) [13]; refers to classifier D, distribution P, collection of group indicators G, and parameter $\gamma \in [0, 1]$. For each $g \in G$, define:

$$\alpha_{SP}(g, P) = Pr_p[g(x) = 1], \\ \beta_{SP}(g, D, P) = \|SP(D) - SP(D, g)\| \tag{3}$$

where $SP(D) = Pr_{P,D}[D(X) = 1]$ and $SP(D, g) = Pr_{P,D}[D(X) = 1 \| g(x) = 1]$ denote the overall acceptance rate of D and the acceptance rate of D on group g respectively. We conclude that D satisfies γ-statistical parity (SP) Fairness with respect to P and G if for every $g \in G$:

$$\alpha_{SP}(g, P).\beta_{SP}(g, D, P) \leq \gamma. \tag{4}$$

We refer to SP(D) as the S.P. base rate. For S.P. fairness, if the algorithm D fails to satisfy the γ-fairness condition, then we conclude that D is γ- unfair with respect to P and G. We call any subgroup g which witnesses this unfairness an γ-unfair certificate for (D, P).

Our learning procedure is supervised and is known as first-test-then-train or prequential evaluation [7]. The performance measures we used are the F-measure and geometric mean (G-mean). The F-measure [2] refers to the harmonic mean of two metrics, recall and precision. We used a balanced value, which implies that precision and recall are assumed to carry equal weights in the metric. The F-measure is macro-averaged over the sum of F1-scores over all classes, which assigns equal weights to the existing classes. Additionally, we employed the G-mean [7] value that is the geometric mean of the recall rates of majority and minority classes in the imbalanced data set. The G-mean value is higher only when the classification accuracies of the majority sample and the minority sample are high; therefore, the G-mean value can accurately the classification effect of unbalanced data sets.

4.3 Experimental Results

First, we investigated the value of S.P. as the indicator of fairness violation and vary the γ measure over time to show the maximum unfairness for each subgroup with gerrymandering. Unfortunately, inter-sectional fairness is not statistically estimable in most cases as most intersections are empty. As a remedy, [12] propose max-violation fairness constraints over $G_{gerrymandering}$, where each group is weighed by group size, defined by:

$$max_{g \in G_{gerrymandering}} \frac{\|g\|}{n} \tag{5}$$

Subsequently, following [12], the empty groups are removed, and small groups have relatively low influence unless there is a very large fairness violation. It is also of interest to compare the subgroup fairness achieved by the subgroup customized classifier. We depict the change of γ over the time in Fig. 4.

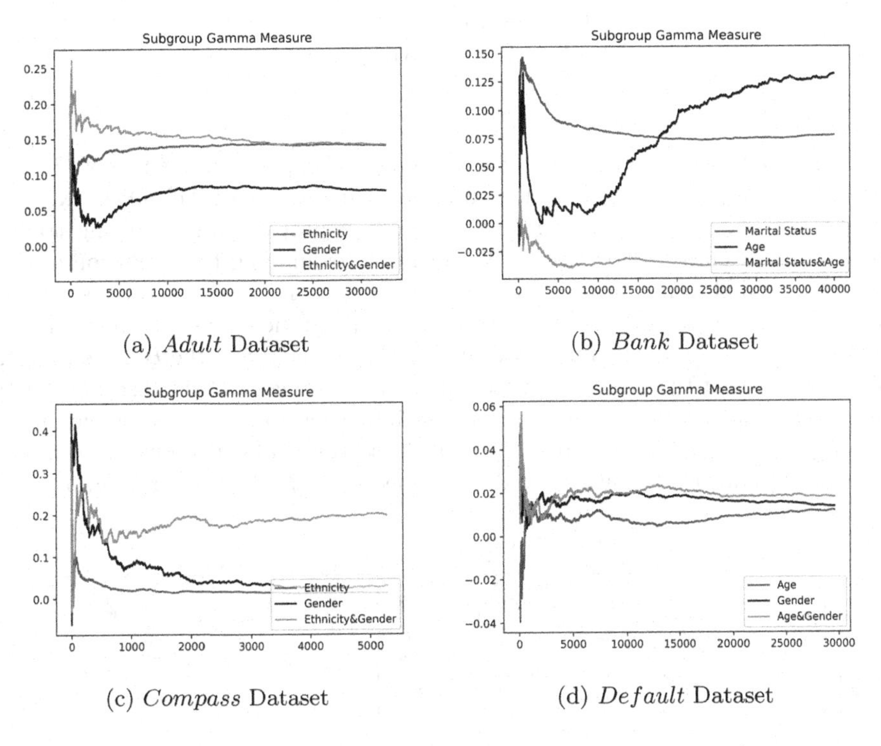

(a) *Adult* Dataset (b) *Bank* Dataset

(c) *Compass* Dataset (d) *Default* Dataset

Fig. 4. Maximum $\gamma_{gerrymandering}$ (lower values are better)

By referring to Fig. 4, we notice that the combination of Ethnicity and Gender has the highest value when we consider the Adult and Compass datasets. That is, Ethnicity and Gender is the subgroup with the higher fairness violation. On the other hand, the combination of Marital-Status and Age in the Bank dataset exhibits the lowest level of unfairness. Further, in the Default stream, the combination of Marital-Status and Gender in the Default dataset has the highest fairness violation amongst all subgroups.

Next, we also present the maximum gerrymandering value for all datasets used in the paper. As shown in Table 2, the value of γ_{Bank} is between -0.02 and 0.13 indicating the highest variation among subgroups. The *Default* dataset reached the lowest value of γ equal to 0.011. We find that Adult is $\gamma_{0.05}$-fair, Bank is $\gamma_{0.03}$-fair, COMPAS is $\gamma_{0.013}$-fair and Default is $\gamma_{0.011}$-fair. The results depicts that our method is empirically necessary to avoid fairness gerrymandering.

Table 2. Gamma measure in each subgroup G

Data	γ_{g1}	γ_{g2}	γ_{g3}
Adult	0.05	0.14	0.16
Bank	0.07	0.131	–0.029
COMPAS	0.013	0.04	0.19
Default	0.011	0.014	0.019

Next, we turn our attention to the second set of experiments, where we investigate accuracy-driven and fairness-oriented capabilities of MQ-OFL. Kearns et al. [12], had shown that varying the input α provides an appealing trade-off between accuracy and fairness. We begin by examining the evolution of the accuracy and discrimination of the model. Based on experimental results we determine the best α and accuracy trade-off. For instance, we show the values of α ranging from 5.0 to 22.5 for *Adult* and between 0 and 2.5 in the *Bank* data set. This implies that the fair customized classifiers aid to control the discrimination propagation and manages to push the discrimination to a low level while maintaining a high prediction capability. Note that the trade-off between accuracy and discrimination can be achieved, by inspection, by adjusting α (Fig. 5).

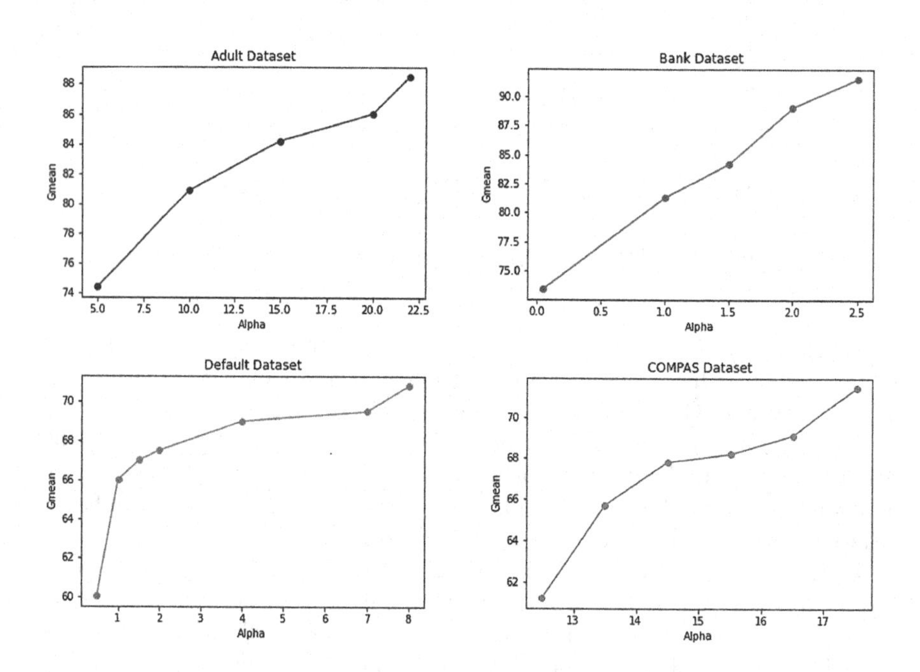

Fig. 5. Accuracy and discrimination trade-off

Finally, we present our results when contrasting different fairness approaches. The HAT and HT classifiers were used as our baseline without considering class imbalance and discrimination. Specifically, Table 3 presents the results of our comparative study contrasting the MQ-OFL, FAHT and FEAT algorithms. Table 3 presents the results of our comparative study when contrasting these three algorithms. In these results, it is clear that our model is capable of diminishing the discrimination to a lower level while maintaining comparable accuracies across all data sets. In addition, the MQ-OFL algorithm produced the highest values in terms of G-mean and lowest value for discrimination for data streams. The same observation holds for the F-measure, where again, MQ-OFL produced higher results, especially handling class imbalance helps to improve this measure.

Table 3. Accuracy-vs-discrimination of learning methods.

	Adult			Bank		
Method	Discrimination	G-mean	F-measure	Discrimination	G-mean	F-measure
MQ-OFL	**15.12**	**83.05**	**80.69**	**1.34**	83.16	**85.11**
FEAT	16.23	81.83	75.69	1.70	79.50	76.95
FAHT	17.40	79.07	71.66	2.65	77.07	73.48
HT	22.60	83.91	72.90	8.10	80.45	71.65
HAT	22.30	**84.07**	76.86	7.80	85.06	72.92
	COMPAS			Default		
Method	Discrimination	G-mean	F-measure	Discrimination	G-mean	F-measure
MQ-OFL	**13.50**	65.74	**63.93**	**1.03**	**66.10**	**64.17**
FEAT	15.80	61.62	59.92	1.83	62.58	44.90
FAHT	16.31	60.05	59.84	2.04	53.30	40.81
HT	21.30	65.38	62.18	8.34	65.10	58.23
HAT	19.70	**66.44**	63.03	8.91	**66.07**	59.08

When assessing the overall accuracy and discrimination level, the reader will notice that MQ-OFL is able to mitigate unfair outcomes and maintains the highest performance in terms of G-mean and F-measure for all datasets, followed by HAT. In terms of discrimination, MQ-OFL appears fairest when considering two sensitive attributes, across all datasets, with keeping the least discrimination score. This is following by FEAT, which was designed for enhanced fairness-aware learning with add-on concept drift adaptation ability to handle non-stationary discriminated data streams. In particular, MQ-OFL achieves a discrimination score of 15.12% and 1.34% at the cost of a slight 1.02% and 2.10% accuracy reduction on the *Adult* and *Bank* dataset, respectively. The results suggest that our MQ-OFL method that seamlessly integrates the fair data merit into classifiers, results into a model that is both accuracy-driven and fairness-driven. MQ-OFL achieves the best discrimination reduction while HT and HAT gives the worst fairness results. This is not surprising; due to the exclusively accuracy-oriented tree construction and the intrinsic discrimination bias of the historic data, a lack of fairness tree can be induced during the construction of the HT and HAT. Therefore, although HAT provides a better prediction performance, it may lead to an unfair model.

5 Conclusion

Current approaches to confront the lack of fairness in AI decision-making systems, mostly consider fairness as a static problem. We introduced an approach for fairness-aware stream classification with the class imbalance and concept drifts which is able to maintain a moderate predictive performance with low discrimination scores over the course of the stream. Moreover, our MQ-OFL method facilitated two or more sensitive attribute by customizing the classifier for each protected group. Our experimental evaluation showed that our approach outperforms other methods in a variety of datasets w.r.t. both predictive performance and fairness preservation. Our class-imbalance-oriented approach effectively learns both sensitive attributes while achieving good predictive performance for both minority and majority classes.

Future studies will include the development of in-processing classifiers. Unfortunately, the number of benchmarking datasets for fairness-aware learning from evolving streams is quite limited; we plan to extend this repository through the creation of additional datasets. Other interesting next steps are extending our work to multi-class classification and evaluating the MQ-OFL method with other fairness metrics to further explore the behavior of the subgroups over the stream.

References

1. Bifet, A., Gavaldà, R.: Adaptive learning from evolving data streams. In: Adams, N.M., Robardet, C., Siebes, A., Boulicaut, J.-F. (eds.) IDA 2009. LNCS, vol. 5772, pp. 249–260. Springer, Heidelberg (2009). https://doi.org/10.1007/978-3-642-03915-7_22
2. Branco, P., Torgo, L., Ribeiro, R.: A survey of predictive modeling on imbalanced domains. ACM Comput. Surv. (CSUR) **49**(2), 1–50 (2016)
3. Buolamwini, J., Gebru, T.: Gender shades: intersectional accuracy disparities in commercial gender classification. In: Conference on Fairness, Accountability and Transparency, pp. 77–91. PMLR (2018)
4. Domingos, P., Hulten, G.: Mining high-speed data streams. In: Proceedings of the sixth ACM SIGKDD International Conference on Knowledge Discovery and Data Mining, pp. 71–80 (2000)
5. Sadeghi, F., Viktor, H.L.: Online-mc-queue: learning from imbalanced multi-class streams. In: Proceedings of the Third International Workshop on Learning with Imbalanced Domains: Theory and Applications, pp. 21–34. PMLR (2021)
6. Fish, B., Kun, J., Lelkes, A.: A confidence-based approach for balancing fairness and accuracy. In: Proceedings of the 2016 SIAM International Conference on Data Mining, pp. 144–152. Society for Industrial and Applied Mathematics (2016)

7. Gomes, H.M., Read, J., Bifet, A., Barddal, J.P., Gama, J.: Machine learning for streaming data: state of the art, challenges, and opportunities. ACM SIGKDD Explor. Newsl. **21**(2), 6–22 (2019)
8. Hoeffding, W.: Probability inequalities for sums of bounded random variables. In: Fisher, N.I., Sen, P.K. (eds.) The collected works of Wassily Hoeffding, Springer, New York, pp. 409–426 (1994). https://doi.org/10.1007/978-1-4612-0865-5_26
9. Howard, A., Borenstein, J.: The ugly truth about ourselves and our robot creations: the problem of bias and social inequity. Sci. Eng. Ethics **24**(5), 1521–1536 (2018)
10. Iosifidis, V., Ntoutsi, E.: FABBOO - online fairness-aware learning under class imbalance. In: Appice, A., Tsoumakas, G., Manolopoulos, Y., Matwin, S. (eds.) DS 2020. LNCS (LNAI), vol. 12323, pp. 159–174. Springer, Cham (2020). https://doi.org/10.1007/978-3-030-61527-7_11
11. Iosifidis, V., Tran, T.N.H., Ntoutsi, E.: Fairness-enhancing interventions in stream classification. In: Hartmann, S., Küng, J., Chakravarthy, S., Anderst-Kotsis, G., Tjoa, A.M., Khalil, I. (eds.) DEXA 2019. LNCS, vol. 11706, pp. 261–276. Springer, Cham (2019). https://doi.org/10.1007/978-3-030-27615-7_20
12. Kearns, M., Neel, S., Roth, A., Wu, Z.: Preventing fairness gerrymandering: auditing and learning for subgroup fairness. In: International Conference on Machine Learning, pp. 2564–2572. PMLR (2018)
13. Kearns, M., Neel, S., Roth, A., Wu, Z.: An empirical study of rich subgroup fairness for machine learning. In: Proceedings of the Conference on Fairness, Accountability, and Transparency, pp. 100–109 (2019)
14. Kohavi, R., Becker, B.: Census income data set (1996). https://archive-beta.ics.uci.edu/ml/datasets/adult
15. Larson, J., Mattu, S., Kirchner, L., Angwin, J.: Propublica compas risk assessment data set (2016). https://github.com/propublica/compas-analysis
16. Mehrabi, N., Morstatter, F., Saxena, N., Lerman, K., Galstyan, A.: A survey on bias and fairness in machine learning. ACM Comput. Surv. (CSUR) **54**(6), 1–35 (2021)
17. Montiel, J., Read, J., Bifet, A., Abdessalem, T.: Scikit-multiflow: a multi-output streaming framework. J. Mach. Learn. Res. **19**(72), 1–5 (2018)
18. Moro, S., Cortez, P., Rita, P.: Bank marketing data set (2014). https://archive.ics.uci.edu/ml/datasets/bank+marketing
19. Ortíz Díaz, A., et al.: Fast adapting ensemble: a new algorithm for mining data streams with concept drift. Sci. World J. (2015)
20. Pedregosa, F., et al.: Scikit-learn: machine learning in python. J. Mach. Learn. Res. **12**, 2825–2830 (2011)
21. Yeh, I.C., Lien, C.H.: The comparisons of data mining techniques for the predictive accuracy of probability of default of credit card clients. Exp. Syst. Appl. **36**(2), 2473–2480. (2009). https://archive.ics.uci.edu/ml/datasets/default+of+credit+card+clients
22. Zhang, W., Bifet, A.: Feat: a fairness-enhancing and concept-adapting decision tree classifier. In: International Conference on Discovery Science, pp. 175–189. Springer, Cham (2020)
23. Zhang, W., Ntoutsi, E.: Faht: an adaptive fairness-aware decision tree classifier. arXiv preprint (2019). arXiv:1907.07237

Multi-fairness Under Class-Imbalance

Arjun Roy[1,2](✉), Vasileios Iosifidis[2], and Eirini Ntoutsi[1,3]

[1] Institute of Computer Science, Free University of Berlin, Berlin, Germany
arjun.roy@fu-berlin.de
[2] L3S Research Center, Leibniz University Hannover, Hanover, Germany
iosifidis@l3s.de
[3] Research Institute CODE, University of the Bundeswehr, Munich, Neubiberg, Germany
eirini.ntoutsi@unibw.de

Abstract. Recent studies showed that datasets used in fairness-aware machine learning for multiple protected attributes (referred to as *multi-discrimination* hereafter) are often imbalanced. The *class-imbalance* problem is more severe for the protected group in the critical minority class (e.g., *female +, non-white +, etc.*). Still, existing methods focus only on the overall error-discrimination trade-off, ignoring the imbalance problem, and thus they amplify the prevalent bias in the minority classes. To solve the combined problem of *multi-discrimination* and class-imbalance we introduce a new fairness measure, *Multi-Max Mistreatment* (*MMM*), which considers both (multi-attribute) protected group and class membership of instances to measure discrimination. To solve the combined problem, we propose Multi-Fair Boosting Post Pareto (MFBPP) a boosting approach that incorporates *MMM*-costs in the distribution update and post-training, selects the optimal trade-off among accurate, class-balanced, and *fair* solutions. The experimental results show the superiority of our approach against state-of-the-art methods in producing the best balanced performance across groups and classes and the best accuracy for the protected groups in the minority class.

Keywords: Multi-discrimination · Class-imbalance · Boosting

1 Introduction

There are growing concerns about the potential discrimination and unfairness of Machine Learning (ML) models in areas of high societal impact like recidivism, job hiring and loan credit. Over the last years a growing body of works has been proposed to address the problem of fairness and algorithmic discrimination [21]. The vast majority of fairness-aware ML approaches however, assumes that discrimination is due to a single protected attribute e.g., only race or only gender

P. Pascal and D. Ienco (Eds.): DS 2022, LNAI 13601, pp. 286–301, 2022.
https://doi.org/10.1007/978-3-031-18840-4_21

(referred hereafter as *mono-discrimination*). In reality though, the roots of discrimination can be ascribed to *multiple protected attributes* (referred hereafter as *multi-discrimination*[1]), e.g., a combination of race, gender and age [17].

The problem of multi-discrimination has attracted attention recently and several approaches to multi-fairness have been proposed [1,12,18,24,25]. However, none of the existing *multi-discrimination* methods considers class-imbalance and the problem arising out of it. Studies [9,11,15] showed that many datasets used in fairness-aware ML research are *class-imbalanced*, i.e., they contain a disproportionately larger amount of instances from the majority class (typically called negative "–" class) comparing to the minority class (typically called positive "+" class). The imbalance is even more pronounced in protected groups like female (vis-a-vis male), non-white (vis-a-vis white) etc. Table 1 highlights the problem in three real-world datasets, which are widely used to evaluate fair-ML algorithms [15]. The Class Imbalance Ratio (*CIR*) is the (+/–) ratio in the whole dataset. For the minority '+' class, we also show the Group Imbalance Ratio (*GIR*) which is the ratio between the protected and non-protected groups, for different protected attributes. As seen in Table 1, within the minority '+' class there exist extreme imbalance between the protected and non-protected group. Thus, within the entire data these protected groups have very less '+' examples. The degree of imbalance varies from attribute to attribute. Thus, giving an uniform and equal importance to tackle discrimination for all the protected attributes may not be sufficient. In these circumstances, a classifier can be highly accurate even by completely ignoring these protected '+' examples. On the other hand, a fair classifier whose working principle is to minimize the difference between performance of the two groups, can have high error rate on both protected and non-protected '+' examples (i.e., predicting them as '–'). Such a situation may result in an acceptable drop in accuracy (in lieu of fairness), however, may lead to heavy under-performance in the positive outcome of some protected groups.

Table 1. Overview of class imbalance ratio (CIR) and protected:non-protected group imbalance ratio in the minority "+" class (GIR) for different protected attributes.

Data	n	Minority (+) class	CIR (+:-)	GIR (Prot. : Non-prot) in "+" class
Adult	45K	>50k	1:3	Race: (1:6), Sex: (1:2)
Bank	40K	*subscription*	1:8.9	Marital (1:3), Age (1:23)
Credit	30K	*default pay.*	1.4:3	Sex (1:1.5), Age (1:6), Marital (1:1.5)

State-of-the-art *multi-discrimination* methods [1,12,18,19,24,25] focus only on error-discrimination trade-off, but ignore this precise imbalance problem.

[1] Through the paper we use the terms "multi-discrimination" and "multi-fairness" interchangeably.

Also, the evaluation strategy presently used ignores to report on this issue of performance of the worst performing group in the minority class. Thus, we need a holistic algorithm approach along with a thorough evaluation mechanism to measure that analyses the performance based on overall error, *multi-discrimination*, imbalance, and protected groups in the minority class. In this work, we target the combined problem of *multi-discrimination* and class-imbalance. Our main contributions are as follows:

i) We extend the definition of multi-group [24] fairness to introduce the notion of *Multi-Max Mistreatment* (*MMM*)[2] that evaluates discrimination for multiple protected attributes and across different classes.
ii) We formulate the *multi-discrimination* under class-imbalance problem as a multi-faceted problem of finding a *MMM*-fair classifier that achieves low overall error, and minimizes performance differences across the classes and groups, to overcome the problem of underrepresented protected groups in the minority (+) class.
iii) We propose Multi-Fair Boosting Post Pareto (MFBPP) algorithm, an in-processing boosting-based approach coupled with a post-processing Pareto Front selection to solve the multiple problems in-hand.
iv) We demonstrate an all round evaluation based on accuracy, imbalance, *multi-discrimination*, and accuracy of protected groups in the positive class to show the superiority of our MFBPP against various state-of-the art approaches w.r.t. *multi-discrimination* under class-imbalance.
v) We offer a flexible alternative of our model to provide solutions per user needs based on user preferences.

The rest of the paper is organized as follows: Related work is summarized in Sect. 2. In Sect. 3 we introduce basic notation and our *Multi-Max Mistreatment* (*MMM*) fairness measure. Our boosting-based method towards an *MMM*-fair classifier is presented in Sect. 3.1 and the experimental evaluation in Sect. 5. We conclude this work in Sect. 6 where we also point to open directions.

2 Related Work

In the following, we summarize related work referring to *multi-discrimination*, and imbalanced learning. Notions built around intersectional discrimination [5, 13] is the most common practice to measure *multi-discrimination*. However, such measures suffer from the drawback of clarity in subgroup definition [6] and scarcity in subgroup distribution [13]. Recently, works [12,24] towards the more operational *multi-discrimination* measure concerning disjoint groups defined by multiple protected attributes came into light. However, they do not take into account ground truth or class membership which is important to consider in presence of class-imbalance. Our introduced *MMM* notion, overcomes the issue by considering both class and multi-group membership of the instances to measure *multi-discrimination*.

[2] The term '*multi*' here refers to both multiple attributes and multiple classes.

A few existing approaches [1,12,18,19,24,25] in supervised learning can handle *multi-discrimination.* [25] introduces fairness-related convex-concave constraints to a logistic regression classifier (FairCons). [1] imposes a set of linear fairness constraints on an exponentiated-gradient reduction method (FairLearn). [12] tackles the fairness-accuracy trade-off by minimizing mutual information between the learning error and the vectorized multiple protected attributes (MI-Fair). [24] applies a Bayes-optimal group-fair classifier (W-ERM) to identify the most-dicriminated group. Fairness-aware learning as a *mini-max* theory has been already used in the literature [18], searching for a Pareto efficient solution of a multi-objective problem (MiniMax). Recently it has been shown that skewed class distributions can affect the discriminatory behaviour of a model [9–11] in the *mono-discrimination* set-up. None of the existing *multi-discrimination* methods considers class-imbalance. Boosting-based approaches have shown their effectiveness in tackling class-imbalance [3,23], fairness [8,11], and multi-class [2] problems. [11] tackles both fairness and class-imbalance but for a single protected attribute (AdaFair).

Our proposed MFBPP considers both *multi-discrimination* and class-imbalance to overcome the limitation of multiple underrepresented groups while delivering accurate solutions across the classes.

3 Basics and *Multi-Max Mistreatment* (*MMM*) Fairness

We assume a dataset $D = (u^{(i)}, s^{(i)}, y^{(i)}) \sim P$ of n instances drawn from the i.i.d distribution P over the domain $U \times S \times Y$, where U is the subspace of *non-protected attributes*, S is the subspace of *protected attributes*, and Y is the class attribute. For simplicity, we assume a binary problem: $Y \in \{+, -\}$ with '+' being the minority (+) class [20]. U and S together define the feature space $X = U \times S$, so $x^{(i)} = (u^{(i)}, s^{(i)})$.

Let the protected subspace consist of k protected attributes: $\{S_1, S_2, \cdots, S_k\}$. Each protected attribute is considered to be binary: $\forall_{j=1,\cdots,k} S_j \in \{g_j, \overline{g_j}\}$ and where g_j and $\overline{g_j}$ represent the *protected group* and the *non-protected group*, respectively w.r.t. protected attribute S_j. Each group g_j ($\overline{g_j}$) w.r.t. a protected attribute S_j can be further subdivided based on class information into: *protected positive* g_{j+}, *protected negative* g_{j-}, *non-protected positive* $\overline{g_{j+}}$ and *non-protected negative* $\overline{g_{j-}}$.

To measure mistreatment in *mono-discrimination* cases, [25] introduced the notion of *Disparate Mistreatment* for a protected feature j as:

$$DM_j = |\delta FNR_j| + |\delta FPR_j| \tag{1}$$

where δFNR_j (δFPR_j) is the discrimination w.r.t. S_j in the positive '+' class (respectively, negative '−' class) defined as:

$$\delta FNR_j = ER(g_{j+}) - ER(\overline{g_{j+}})$$

$$\delta FPR_j = ER(g_{j-}) - ER(\overline{g_{j-}})$$

3.1 *Multi-Max Mistreatment*(*MMM*) Measure

The Disparate Mistreatment measure (c.f., Eq. 1) fails to focus on per-class discrimination due to the summation operation. To ensure fair treatment *across all classes*, for a protected attribute S_j, we measure mistreatment as $\max(|\delta FNR_j|, |\delta FPR_j|)$ where the 'max' operator enforces focus on each of the classes. Moreover, we want to ensure fair treatment *across all protected attributes* $S = \{S_1, \cdots, S_k\}$. Our goal is therefore, to focus on the most discriminated group defined based on a protected attribute and a class. To this end, we introduce a new *multi-discrimination* notion, called *Multi-Max Mistreatment* (*MMM*), that measures the *maximum* discrimination among the protected attributes and for the different classes.

Definition 1. *The* Multi-Max Mistreatment(MMM_S) *due to multiple-protected attributes* $S = \{S_1, \cdots, S_k\}$ *across all classes* $Y = \{+, -\}$, *is defined as:*

$$\mathrm{MMM}_S = \max_{S_j \in S} \left(\max(|\delta FNR_j|, |\delta FPR_j|) \right) \tag{2}$$

where δFNR_j and δFPR_j measure the mistreatment due to S_j in the (+) and (−) class, respectively.

Definition 2. *Given a tolerance threshold* μ, *a classifier* $f(\cdot)$ *is* MMM-*fair iff the maximum mistreatment w.r.t all the protected attributes* $S_j \in S$ *across all classes is less than* μ *i.e.,* $\mathrm{MMM}_S \leq \mu$.

In the ideal case, $\mu = 0$ which signifies no discrimination w.r.t. any protected attribute and in any class.

4 Multi-Fairness-Aware Learning

Our goal is to learn a *MMM*-fair classifier: $f(\cdot) : X \rightarrow Y$ that achieves *equal low* error rates for all the groups ($g_j/\overline{g_j}$, $j = 1, \cdots, k$) in both the classes (+/−). To this end, we first formulate clear objectives (Sect. 4.1) and then, propose a sequential learner approach to find $f(\cdot)$ (Sect. 4.2).

4.1 Multi-discrimination-Free Learning Under Class-Imbalance

We define three objectives for the *MMM*-fair classifier $f(\cdot)$: low overall error (O_1), similar (low) error rates across all classes (O_2), and mitigation of discriminatory outcomes for all protected attributes (O_3).

Objective O_1 targets overall error and is defined as minimizing the classification loss (0–1 loss):

$$O_1 : \; L(f) = \frac{1}{n} \sum_{(x_i, y_i) \in D} |y_i - f(x_i)| \tag{3}$$

where $f(x_i)$ is the predicted and y_i is the true class of x_i.

Objective O_2 explicitly targets class-imbalance by ensuring balanced performance across both classes. Motivated by [7], we define a balanced loss function to minimize the performance differences between the two classes:

$$O_2: \; B(f) = |\frac{1}{|D_+|} \sum_{(x_i,y_i)\in D_+} |y_i - f(x_i)| - \frac{1}{|D_-|} \sum_{(x_i,y_i)\in D_-} |y_i - f(x_i)|| \tag{4}$$

where $D_Y \subset D$, $Y \in \{+,-\}$ denotes the set of instances belonging to class Y.

O_3 is the *multi-discrimination* objective aiming to mitigate discrimination due to multiple protected attributes $S_j \in S$ and across both classes. We call it MMM_S loss, as on optimization it aims to mitigate MMM_S (c.f. Definition 1):

$$O_3: \; \Phi(f) = \max_{S_j\in S}\big(\max_{Y\in\{+,-\}}\big(|\frac{1}{|g_{jY}|} \sum_{(x_i,y_i)\in g_{jY}} |y_i - f(x_i)| - \frac{1}{|\overline{g_{jY}}|} \sum_{(x_i,y_i)\in \overline{g_{jY}}} |y_i - f(x_i)|\,|\big)\big) \tag{5}$$

where $|g_{jY}|$ is the cardinality of group g_j in class $Y \in \{+,-\}$.

The objectives O_3 and O_2 ensure similar performance across all the protected/non-protected groups and the $(+/-)$ classes respectively, thus minimizing the performance bias against the underrepresented protected groups in the minority $(+)$ class, while the objective O_1 would help establish low error rate overall.

4.2 The *MMM*-Fair Boosting Post Pareto (MFBPP) Algorithm

Our goal is to develop a classifier that takes into consideration the above three objectives, eventually solving the problem of *multi-discrimination* under class-imbalance. Boosting-based [22] approaches have been promising in tackling class-imbalance [3,23] and discrimination [8,11]. However, they have also been criticised for being vulnerable in the presence of noise or outliers. As outliers are more likely to be missclassified, boosting may overshoot over the iterations the weights of those instances [16]. Thus, the ensemble obtained at the end of a predefined number of boosting rounds may produce inferior outcomes than an ensemble produced in an earlier round.

Inspired by the literature, we propose a boosting-based learner that *in-training* modifies the distribution weights to incorporate our objective goals. The new weighting puts more attention to the instances from the protected groups in minority class (as they are frequently missclassified) and might therefore, aggravate the weight overshooting problem. To overcome this drawback, we deploy a *post-training* step to select the best solution (partial ensemble).

In-training: *MMM*-Boosted Weight Distribution Update. Let T be the number of boosting rounds. In each round t $(1 \leq t \leq T)$ we train a weak learner (a decision stump) based on the current instance weight distribution D_t. In the first round, all instances receive the same weight: $D_1(x_i) = \frac{1}{n}$. In a later round $0 < t+1 \leq T-1$, the weight distribution is updated as follows:

$$D_{t+1}(x_i) = \frac{D_t(x_i)\exp(-\alpha_t sign(y_i h_t(x_i))) fc_t(x_i)}{Z_t} \tag{6}$$

where as in AdaBoost $\alpha_t = \frac{1}{2} \ln \frac{1-\sum_n D_t(x_i)}{\sum_n D_t(x_i)}$ is the weight of the weak learner h_t, $sign(y_i h_t(x_i))$ returns -1 if $h_t(x_i) \neq y_i$ and 1 otherwise, and Z_t is the normalization factor which ensures that D_{t+1} is a probability distribution. The term $fc_t(x_i)$ is our modification, which corresponds to the *multi-discrimination* cost (MMM-cost) for a misclassified instance x_i defined as:

$$fc_t(x_i) = \begin{cases} \max_{1 \leq j \leq k}(cdc_{ij}^{(t)}), & \textit{if}\; h_t(x_i) \neq y_i \\ 1, \textit{otherwise} \end{cases} \tag{7}$$

where $cdc_{ij}^{(t)}$ is the discrimination weight of instance x_i at round t concerning protected attribute S_j, which depends on the group membership of x_i w.r.t S_j. It is defined as:

$$cdc_{ij}^{(t)} = \begin{cases} 1 + |\delta FNR_j^{1:t}|, & \textit{if}\; (\delta FNR_j^{1:t} \geq 0 \;\wedge x_i \in g_{j+}) \vee (\delta FNR_j^{1:t} \leq 0 \;\wedge x_i \in \overline{g_{j+}}); \\ 1 + |\delta FPR_j^{1:t}|, & \textit{if}\; (\delta FPR_j^{1:t} \geq 0 \;\wedge x_i \in g_{j-}) \vee (\delta FPR_j^{1:t} \leq 0 \;\wedge x_i \in \overline{g_{j-}}); \end{cases} \tag{8}$$

where $\delta FNR_j^{1:t}$ and $\delta FPR_j^{1:t}$ are the cumulative discrimination of the partial ensemble $H_t(x_i) = \sum_{l=1}^{t} \alpha_l h_l(x_i)$ for S_j as in [11].

In each boosting round t we evaluate the partial ensemble H_t and collect the solution vector $\boldsymbol{f_t} = [o_1, o_2, o_3]_t$, where $o_i = O_i(t)$ is a solution point of H_t for the respective objective O_i. In total, T solution vectors are collected. The sequential training stops when the maximum number of iterations T is reached.

Post-training: Selecting Pareto Optimal Solution. Our goal is to find the optimal round $t^* \leq T$ to output the partial-ensemble with the best (O_1, O_2, O_3) objectives trade-off:

$$H_{t^*} = \sum_{l=1}^{t^*} \alpha_l h_l$$

This is achieved in two steps: First, out of all T solutions we select the set of non-dominating optimal solutions. Next, we find the best trade-off solution among the shortlisted ones to get the corresponding optimal t^*.

1. Pareto Front Computation: Among all solution vectors $\boldsymbol{f_t}$, $t = 1, \cdots, T$ collected over the boosting rounds, we find the Pareto Front ($\mathbb{PF}$), i.e., the non-dominated set of Pareto optimal solutions. A solution $\boldsymbol{f_{t'}}$ is said to be dominated by a solution $\boldsymbol{f_t}$ if 1) $O_i(t) \leq O_i(t')\ \forall i \in \{1, 2, 3\}$, and 2) $\exists i \in \{1, 2, 3\}\ O_i(t) < O_i(t')$.

2. Pseudo-Weight Calculation and Choice of Best Solution: To choose the best solution we use the pseudo-weight algorithm [4] that calculates the relative distance of each solution from the worst (maximum value) solution for each objective. The pseudo-weight w_{ti} for $o_i \in \boldsymbol{f_t}$ is given by:

$$w_{ti} = \frac{(o_i^{\max} - o_{ti})/(o_i^{\max} - o_i^{\min})}{\sum_{i=1}^{3}(o_i^{\max} - o_{ti})/(o_i^{\max} - o_i^{\min})} \tag{9}$$

where $o_i^{\max}$ ($o_i^{\min}$) is the maximum or worst (minimum or best) objective value achieved in any of the rounds. This way, for each solution $\boldsymbol{f}_t = [o_1, o_2, o_3]_t$ we compute the corresponding pseudo-weight vector $\boldsymbol{w}_t = [w_{t1}, w_{t2}, w_{t3}]$. Next, we select the solution with the least relative weighted sum as the best trade-off solution w.r.t all the objectives:

$$\boldsymbol{f}_t^* = argmin_t\{(1 - \boldsymbol{w}_t) \cdot \boldsymbol{f}_t\} = argmin_t\{\sum_{i=1}^{3}(1 - w_{ti})o_{ti}\} \quad (10)$$

where $(1 - \boldsymbol{w}_t)$ is the required transformation as the pseudo-weights vector by its nature assigns bigger weight w_{ti} to a smaller objective solution value o_{ti}.

5 Experiments

We evaluate MFBPP performance against state-of-the-art approaches (Sect. 5.2). To show the utility of our *MMM*-cost (Eq. 7) in tackling balanced error (O_2), we plot balanced loss $B(f)$ with varying *MMM* tolerance thresholds μ (Sect. 5.3). Further, we show the changes in the dataset distribution over training and the effectiveness of our approach in promoting underrepresented protected groups (Sect. 5.3). We plot the O_1, O_2, O_3 losses over the rounds to justify the need for post-training selection. At last, in Sect. 5.4 we show the flexibility of MFBPP to intake user preferences for post-training selection.

5.1 Experimental Settings

Baselines: We compare against four state-of-the-art fairness-aware methods:
FairCons [25]: Tackles *multi-discrimination* fairness-related convex-concave constraints,
FairLearn [1]: imposes a set of linear fairness constraints on an exponentiated-gradient reduction method to tackle *multi-discrimination*,
MiniMax [18]: tackles *multi-discrimination* as a mini-max game while searching for a Pareto efficient solution of a multi-objective problem,
W-ERM [24]: applies a Bayes-optimal group-fair classifier to consider algorithmic fairness across multiple overlapping groups simultaneously to tackle the *multi-discrimination* trade-off,
MI-Fair [12]: minimizes mutual information between the learning error and the vectorized multiple protected attributes to tackle *multi-discrimination*, and
AdaFair [11]: uses *mono-discrimination* based boosting algorithm along with summed accuracy and class-imbalance loss to tackle *mono-discrimination* and class-imbalance.

In order to understand the effect of the post-training part as well as the effect of the $\mathbb{PF}$ selection in the post-training part, we also include in the experiments two variations of MFBPP:

- *MFB* that completely discards the post-training part and
- *MFBP* that uses post-training but does not use the Pareto Front $\mathbb{PF}$ set for the final selection but rather selects from all solutions.

Datasets. We report on three imbalanced real-world datasets (c.f., Table 1). Additionally, we also report on Compas [14] (CIR: 1 : 1.2) to show the usability of our method also for class-balanced scenarios. The protected attributes and protected groups studied in the experiments are *Sex* (g_j = "female"), *Race* (g_j = "non-white"), *Marital status/Mari* (g_j = "married"), *Age* (g_j = "$\leq 25\& \geq 60$").

Evaluation Measures. For O_1, we report on accuracy (*Acc*), for O_2 on geometric mean (*G.M*) and for O_3 we report on the proposed MMM-fairness, as well as on the *DM* for each protected attribute. Additionally, we report on the accuracy of the worst performing protected group in the minority (+) class (Wg_+).

Experimental Setup: We set the number of weak learners to $T = 500$. We follow the same evaluation setup as in [11,25] by splitting each dataset randomly into train (50%) and test (50%) and report on the average of 10 random splits.

5.2 Evaluation Results

The discriminatory and predictive performance evaluation of the different approaches is shown in Fig. 1 and Table 2, respectively.

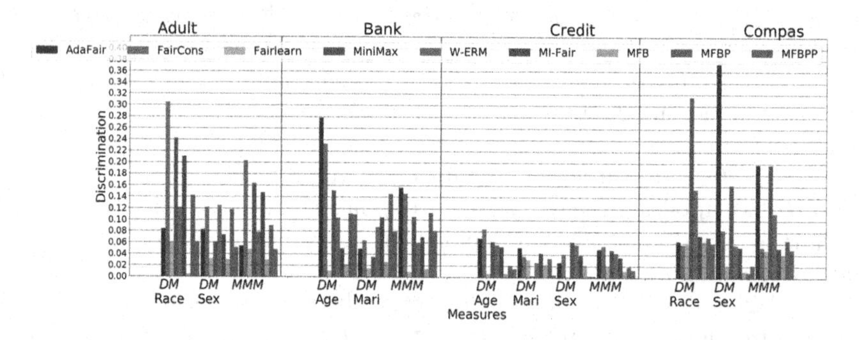

Fig. 1. Discrimination performance: For each dataset, the overall *MMM* score and the DM scores for each protected attribute are shown (lower values are better).

Table 2. Predictive performance evaluation. W_{g+} is the accuracy of the worst performing protected group in the minority (+) class

	Adult			Bank			Credit			Compas		
	Acc	Wg_+	*G.M*	*Acc*	Wg_+	*G.M*	*Acc*	Wg_+	*G.M*	*Acc*	Wg_+	*G.M*
AdaFair	0.84	0.63	0.76	0.88	0.62	0.76	0.81	0.33	0.57	0.65	0.50	0.64
FairCons	0.85	0.43	0.75	**0.91**	0.33	0.59	0.81	0.28	0.55	0.67	0.53	0.66
FairLearn	0.83	0.54	0.73	0.88	0.21	0.46	0.79	0.21	0.45	0.65	0.55	0.64
MiniMax	**0.86**	0.49	0.76	0.90	0.45	0.66	**0.82**	0.37	0.60	**0.68**	0.49	0.67
W-ERM	0.85	0.52	0.75	0.90	0.29	0.56	0.81	0.22	0.47	0.66	0.47	0.64
MI-Fair	0.84	0.65	0.72	0.89	0.69	0.82	0.80	0.59	0.68	0.68	0.60	**0.68**
MFB	0.69	0.64	0.74	0.36	0.28	0.41	0.71	0.68	0.70	0.64	0.63	0.63
MFBP	0.85	0.77	**0.84**	0.85	**0.75**	**0.85**	0.71	0.64	0.69	0.67	0.60	0.65
MFBPP	0.81	**0.79**	0.81	0.81	0.72	0.80	0.74	**0.65**	**0.70**	0.66	**0.63**	0.66

Multi-discrimination: From Fig. 1 we notice that our MFB outperforms all the approaches in all the dataset. MFBPP comes second outperforming the baseline competitors in mitigating *multi-discrimination* (i.e., objective O_3) by producing the lower *MMM* discrimination values in three datasets (Adult: **0.05**, Compas: **0.04**, Credit: **0.01**), while falling behind FairLearn in one dataset (Bank: 0.08). However, in Table 2 we notice that MFB severely underperforms in all the predictive evaluation measures, thus failing to provide a good trade-off between O_1, O_2, and O_3. The closest competitor to us w.r.t. fairness is FairLearn, which however achieves low discrimination by consistently ignoring the minority class (A closer look to Table 2, shows that FairLearn achieves the lowest *G.M* for all four datasets). Approaches like FairCons, MiniMax, and W-ERM result in different levels of discrimination for the different protected attributes and overall high *MMM* values. MI-Fair have mixed outcome with high discrimination in Adult, but performed at par with MFBPP in Bank and Compas data. AdaFair trained on one protected attribute (for Adult: sex, for Bank: marital status, for Credit: sex, for Compas: race) does not mitigate discrimination for other protected attributes and consequently also results in high *MMM* values esp. for Bank and Compas. In case of Adult and Credit datasets, AdaFair, albeit trained for *mono-discrimination* it seems to tackle *multi-discrimination*; the reason is the strong correlation between the protected attributes as revealed by chi-square test with ρ-*value* ≈ 0.

Underrepresented Protected Groups (g_{j+}): In Table 2 we notice that MFBPP and MFBP both outperform the other approaches on Wg_+ by far ([5% − 21% ↑]). Thus, our proposed methods overcome the issue of bias due to the imbalanced distribution of protected groups (c.f Table 1), ensuring high predictive accuracy for any g_{j+}. Note that all the other approaches that even after mitigating *multi-discrimination* fail on this task. MI-Fair emerges as the best among the baseline competitor in all the four datasets behind our proposed methods MFBPP, MFBP, and MFB in Adult, Credit, and Compas datasets, while outperforming only MFB in Bank data.

Balanced Performance: Table 2 shows that our MFPB and MFBP outperform the baselines in $G.M$ in the range $[4\% - 11\%] \uparrow$ for the imbalanced Adult, Bank, and Credit datasets, while being marginally behind MI-Fair, and Minimax in the balanced Compas dataset. We can easily notice that our Acc and $G.M$ values are close to each other for all the datasets with $Acc/G.m \approx 1$. This indicates we achieve $B(f) \approx 0$ (O_2), in all the datasets. Our closest competitors here are MI-Fair, AdaFair and MiniMax. AdaFair explicitly targets class-imbalance for mono-discrimination. MiniMax, and MI-Fair indirectly tackles the problem as they aims at minimizing error for all groups. For other baselines, $Acc/G.m >> 1$, indicating substantial performance differences between the classes.

Overall Accuracy: MFBPP is marginally compromised on the overall Acc in Adult, Credit, and Bank datasets ($[12\% - 5\%] \downarrow$). MiniMax emerged as the winner here, accomplishing the best accuracy in Adult, Credit, and Compas dataset. This is the trade-off we pay to ensure nearly equal performance for all (protected/non-protected) groups across all the classes.

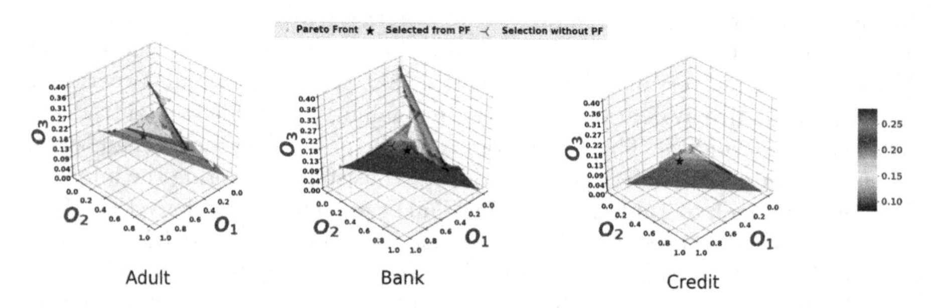

Fig. 2. Visualization of the Pareto front and the selected trade-off solution in the complete solution surface.

Summary: MFBPP provides the best holistic outcome in overall trade-off, outperforms the baselines in mitigating *multi-discrimination*, produces the best predictive performance on underrepresented protected groups ($\forall_j g_{j+}$) in minority class, and equal performance across all classes, while maintaining comparable high accuracy against the baselines. MFB produces the most-fair outcomes but suffers in predictive performance. It overshoots the weight and increases overall error to gain fairness. MFBP solves the overshooting problem but gets outperformed by MFBPP in the fairness task. In Fig. 2 we see that the solution surface after the training-MFB phase of our algorithm is very wide spread in the O_1, O_2, O_3 objective space. By computing $\mathbb{PF}$ as in MFBPP, we narrow down the search space. Using the pseudo-weights, we pick a solution each time close to the origin in the objective space (which is desired). Hence, we are always able to deliver a good trade-off solution without any hyper-parameter tuning. FairLearn also tackled the *multi-discrimination* problem consistently well, but by underperforming in the minority class. MiniMax lacks in *multi-discrimination* convergence but produces the most overall accurate (O_1) predictions. FairCons has difficulty in finding the optimal parameters leading to its poor *multi-discrimination* performance. W-ERM apart from Bank dataset (the most imbalanced), always delivers comparable trade-offs. However, the method is very slow. MI-Fair can be argued as the closest competitor in overall trade-off delivering balanced and accurate performance with low discrimination in three out of the four datasets under study.

5.3 Internal Analysis

This section aims to analyse MFBPP's ability to produce state-of-the-art balanced performance while dealing with *multi-discrimination*. In particular, we try to find answers for three significant points: i) How MFBPP ensures high accuracy for the underrepresented protected groups in the imbalanced minority (+) class? ii) How the overshooting problem affects and, is the post-processing step really required? iii) Does the *multi-discrimination* cost (Eq. 7) also tackle the balanced loss and, what happens if we relax cost by varying the *MMM* threshold μ (Definition 2)? Here we focus the study using only the imbalanced data (Table 1).

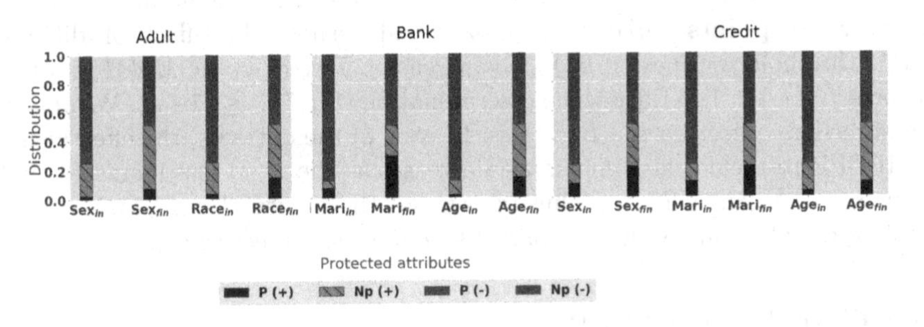

Fig. 3. Changes in instance weight distribution. For each dataset and protected attribute S_j, we depict the initial distribution S_{in} and the final one S_{fin}.

Answer to point (i): We analyse the changes in weight distribution of the various groups from its initial (*ini*) distribution (actual data representation), to boosted weight till the finally selected partial ensemble point (*fin*) in Fig. 3. For any protected attribute S_j, P and Np refer to the respective protected and non-protected groups. Thus, $P(+)$ translates as the protected group in the minority (+) class (g_{j+}). We notice that *ini* weights of each $P(+)$ in every dataset is largely underrepresented. But, in the *fin* weights each $P(+)$ group is boosted significantly. MFBPP increases the weight of the underrepresented groups, thus

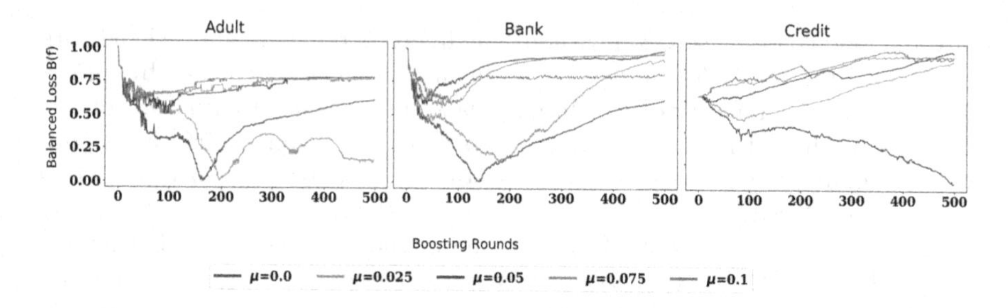

Fig. 4. B(f) loss over boosting rounds with varying *MMM* thresholds μ.

changing the decision boundary to produce highly accurate and unbiased results for all the groups even in case of high imbalance.

Answer to points (ii): We have already shown the effectiveness of our *MMM* cost (Eq. 7) in mitigating *multi-discrimination*. Now to understand its effect on balanced loss we monitor $B(f)$ over the boosting rounds (Fig. 4) for different MMM-tolerance thresholds μ. We see that when $\mu = 0$, the shape of the $B(f)$ loss curve is parabolic for Adult and Bank datasets. In Credit data, the loss continues to descent till the final round. The parabolic curve supports our intuition of the possibility of overshooting the weights due to the possible repeated boosting of noisy instances, whereas a consistently descending loss curve for Credit data shows the uncertainty involved in estimating the optimal size of the ensemble. These results justify the necessity of the post-training selection part.

Answer to points (iii): In Fig. 4 we also show the effect of different *MMM* threshold μ values on $B(f)$. By increasing μ we relax the *MMM* boost i.e. we have $fc_t(x_i) = 1$ in Eq. 6 when discrimination (Eq. 7) is $\leq (1+\mu)$. We observe the immediate effect on the $B(f)$ loss. In each of the datasets, the effectiveness of MFBPP to tackle class imbalance decreases as the $B(f)$ loss increases while we increase the threshold μ. Thus, showcases the ability of our *MMM* cost in tackling the O_2 along with our *multi-discrimination* objective O_3.

5.4 Flexibility of MFBPP

Thus far, we use the pseudo-weight method (Eq. 10) to select the best solution among the ($\mathbb{PF}$) solutions. If information on user preferences exists, in the form of a user-preference vector $\boldsymbol{u} = [u_1, u_2, u_3]$: $u_1 + u_2 + u_3 = 1$, it can be used to select the best solution according to user needs. In this case, we choose the solution $\boldsymbol{f}_{t^*}$ whose corresponding pseudo-weight $\boldsymbol{w}_{t^*}$ is closest according to L1 distance, to the preference vector $\boldsymbol{u}$. To evaluate the effect of such an approach, we mimic four different users and provide their preference vector $\boldsymbol{u}$ as an additional input to MFBPP. In particular, we assume the following users: i) $\boldsymbol{u} = [0.33, 0.33, 0.33]$ indicating *equal preference* to all O_i, ii) $\boldsymbol{u} = [0, 0, 1]$, iii) $\boldsymbol{u} = [0, 1, 0]$, iv) $\boldsymbol{u} = [1, 0, 0]$, indicating preference only for O_i if $u_i = 1$ (Fig. 5).

As expected, the output changes noticeably with changes with $\boldsymbol{u}$. For all datasets the most accurate classifier (for $\boldsymbol{u} = [1, 0, 0]$) delivers Acc at par if not

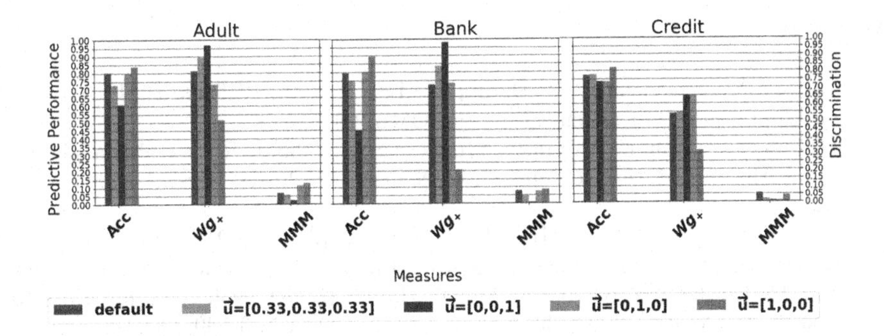

Fig. 5. Performance evaluation for different user preference vectors $\boldsymbol{u}$

better than the state of the art, whereas the fairest (for $\boldsymbol{u} = [1, 0, 0]$) produces state of the art fair predictions. With preference $\boldsymbol{u} = [0.33, 0.33, 0.33]$ the classifier consistently produces good trade-off solutions, however, the default version of MFBPP (without $\boldsymbol{u}$) produces better trade-offs.

6 Conclusions and Outlook

In this work we claimed that *multi-discrimination* under class-imbalance is an important multi-faceted problem of finding low overall error, while minimizing performance differences across the classes and groups. Existing multi-discrimination approaches consider only error-discrimination trade-off, and ignore class-imbalance. This way, they achieve *multi-discrimination* by underperforming in the minority (+) class, especially for the underrepresented protected groups. To this end, we propose the *Multi-Max Mistreatment* fairness measure (*MMM*) and a *MMM*-fair boosting post Pareto classifier (MFBPP) to ensure *MMM*-fairness. Our experiments show the superiority of our method in mitigating *multi-discrimination*, producing best balanced performance across groups and classes along with best accuracy for underrepresented protected groups in the minority (+) class, without a significant compromise on overall accuracy. Further, our method is flexible to user needs as it can select the best solution trade-off according to user preferences. In future, we want study the *multi-discrimination* under class-imbalance problems in the more challenging multi-class and multi-label set-up, where the complexity is much harder.

Acknowledgements. The work of the first author is supported by the Volkswagen Foundation under the call "Artificial Intelligence and the Society of the Future" (the BIAS project). We are sincerely thankful to the invaluable suggestion of Prof. Niloy Ganguly from L3S Research Center, in shaping up the paper to its current form. Most of the work was carried out while the last author was affiliated with Freie Universität Berlin, Germany.

References

1. Agarwal, A., Beygelzimer, A., Dudík, M., Langford, J., Wallach, H.M.: A reductions approach to fair classification. In: ICML (2018)
2. Brukhim, N., Hazan, E., Moran, S., Mukherjee, I., Schapire, R.E.: Multiclass boosting and the cost of weak learning. Adv. Neural. Inf. Process. Syst. **34**, 3057–3067 (2021)
3. Chawla, N.V., Lazarevic, A., Hall, L.O., Bowyer, K.W.: SMOTEBoost: improving prediction of the minority class in boosting. In: Lavrač, N., Gamberger, D., Todorovski, L., Blockeel, H. (eds.) PKDD 2003. LNCS (LNAI), vol. 2838, pp. 107–119. Springer, Heidelberg (2003). https://doi.org/10.1007/978-3-540-39804-2_12
4. Deb, K.: Multi-objective Optimization Using Evolutionary Algorithms, vol. 16. Wiley, New York (2001)
5. Foulds, J.R., Islam, R., Keya, K.N., Pan, S.: An intersectional definition of fairness. In: ICDE, pp. 1918–1921 (2020)
6. Fredman, S.: Intersectional discrimination in EU gender equality and non-discrimination law. European Commission, Brussels, UK (2016)
7. García, V., Mollineda, R.A., Sánchez, J.S.: A new performance evaluation method for two-class imbalanced problems. In: da Vitoria Lobo, N., et al. (eds.) SSPR /SPR 2008. LNCS, vol. 5342, pp. 917–925. Springer, Heidelberg (2008). https://doi.org/10.1007/978-3-540-89689-0_95
8. Hickey, J.M., Di Stefano, P.G., Vasileiou, V.: Fairness by explicability and adversarial SHAP learning. In: Hutter, F., Kersting, K., Lijffijt, J., Valera, I. (eds.) ECML PKDD 2020. LNCS (LNAI), vol. 12459, pp. 174–190. Springer, Cham (2021). https://doi.org/10.1007/978-3-030-67664-3_11
9. Hu, T., et al.: *FairNN* - conjoint learning of fair representations for fair decisions. In: Appice, A., Tsoumakas, G., Manolopoulos, Y., Matwin, S. (eds.) DS 2020. LNCS (LNAI), vol. 12323, pp. 581–595. Springer, Cham (2020). https://doi.org/10.1007/978-3-030-61527-7_38
10. Iosifidis, V., Fetahu, B., Ntoutsi, E.: FAE: a fairness-aware ensemble framework. In: 2019 IEEE Big Data, pp. 1375–1380 (2019)
11. Iosifidis, V., Ntoutsi, E.: AdaFair: cumulative fairness adaptive boosting. In: CIKM 2019, pp. 781–790 (2019)
12. Kang, J., Xie, T., Wu, X., Maciejewski, R., Tong, H.: MultiFair: multi-group fairness in machine learning. arXiv preprint arXiv:2105.11069 (2021)
13. Kearns, M., Neel, S., Roth, A., Wu, Z.S.: Preventing fairness gerrymandering: auditing and learning for subgroup fairness. In: ICML, pp. 2564–2572 (2018)
14. Larson, J., Mattu, S., Kirchner, L., Angwin, J.: How we analyzed the compas recidivism algorithm. ProPublica **9** (2016)
15. Le Quy, T., Roy, A., Iosifidis, V., Zhang, W., Ntoutsi, E.: A survey on datasets for fairness-aware machine learning. Wiley Interdiscip. Rev. Data Mining Knowl. Discov., e1452 (2022)
16. Li, A.H., Bradic, J.: Boosting in the presence of outliers: adaptive classification with nonconvex loss functions. J. Am. Stat. Assoc. **113**(522), 660–674 (2018)
17. Makkonen, T.: Multiple, compoud and intersectional discrimination: bringing the experiences of the most marginalized to the fore (2002)
18. Martinez, N., Bertran, M., Sapiro, G.: Minimax pareto fairness: a multi objective perspective. In: ICML, pp. 6755–6764. PMLR (2020)
19. Morina, G., Oliinyk, V., Waton, J., Marusic, I., Georgatzis, K.: Auditing and achieving intersectional fairness in classification problems. arXiv preprint (2019)

20. Napierala, K., Stefanowski, J.: Types of minority class examples and their influence on learning classifiers from imbalanced data. J. Intell. Inf. Syst. **46**(3), 563–597 (2016)
21. Ntoutsi, E., et al.: Bias in data-driven artificial intelligence systems an introductory survey. Wiley Interdiscip. Rev. Data Mining Knowl. Discov. **10**(3), e1356 (2020)
22. Schapire, R.E.: A brief introduction to boosting. In: Proceedings of the IJCAI (1999)
23. Sun, Y., Kamel, M.S., Wong, A.K., Wang, Y.: Cost-sensitive boosting for classification of imbalanced data. Pattern Recogn. **40**(12), 3358–3378 (2007)
24. Yang, F., Cisse, M., Koyejo, O.O.: Fairness with overlapping groups; a probabilistic perspective. In: Advances in Neural Information Processing Systems 33 (2020)
25. Zafar, M.B., Valera, I., Gomez-Rodriguez, M., Gummadi, K.P.: Fairness constraints: a flexible approach for fair classification. JMLR **20**, 1–42 (2019)

When Correlation Clustering Meets Fairness Constraints

Francesco Gullo[1], Lucio La Cava[2], Domenico Mandaglio[2], and Andrea Tagarelli[2](✉)

[1] UniCredit, Rome, Italy
gullof@acm.org
[2] DIMES Department, University of Calabria, Rende, CS, Italy
{lucio.lacava,d.mandaglio,tagarelli}@dimes.unical.it

Abstract. The study of fairness-related aspects in data analysis is an active field of research, which can be leveraged to understand and control specific types of bias in decision-making systems. A major problem in this context is fair clustering, i.e., grouping data objects that are similar according to a common feature space, while avoiding biasing the clusters against or towards particular types of classes or sensitive features. In this work, we focus on a correlation-clustering method we recently introduced, and experimentally assess its performance in a fairness-aware context. We compare it to state-of-the-art fair-clustering approaches, both in terms of classic clustering quality measures and fairness-related aspects. Experimental evidence on public real datasets has shown that our method yields solutions of higher quality than the competing methods according to classic clustering-validation criteria, without neglecting fairness aspects.

1 Introduction

We live in an era where machine learning is increasingly pervasive in our society. Every day we interact with machine learning systems, even without knowing it, and these acquire more and more decision-making power in our lives. For instance, such systems support, or even replace, decision makers in financial [22], medical [21], or legal [17] domains. Given their delicate role, machine learning systems should guarantee correct functioning and not discriminate those who entrust their decisions. In this context, however, a critical aspect emerges: the data used by such systems are often (intrinsically) biased, resulting from incorrect data collection processes. Thus, it is desirable to avoid machine learning algorithms being affected by, or even amplifying, this bias. For instance, in [16], this refers to removing *disparate impact*, according to which no group of individuals should (even indirectly) be discriminated by a decision-making system.

In this respect, and by focusing on an unsupervised machine learning setting, in this work we tackle the problem of *fair clustering*. This corresponds to clustering a set of data objects such that: (i) analogously to the classic clustering scenario, similar objects are assigned to the same cluster, whereas dissimilar

P. Pascal and D. Ienco (Eds.): DS 2022, LNAI 13601, pp. 302–317, 2022.
https://doi.org/10.1007/978-3-031-18840-4_22

objects are assigned to different clusters, and *(ii)* the clusters are not dominated by a specific type of sensitive data class (e.g., people having the same sex).

Our key assumption is that the above problem can be addressed under a *correlation clustering* framework [7]. Correlation clustering is a well-established tool for partitioning the set of vertices of an input graph into clusters, so as to maximize the similarity of the vertices within the same cluster and minimize the similarity of the vertices in different clusters, according to pairwise vertex weights expressing positive and negative types of co-association. Specifically, following our recent work on correlation clustering [20], here we provide insights into its application to the problem of fair clustering, and we compare it to some state-of-the-art approaches in such a context. Furthermore, albeit we do not aim to provide a comprehensive experimental survey on fair clustering, a by-product of our work is that, to the best of our knowledge, it represents a valuable and unprecedented experimental comparison between approaches of fair clustering.

Our contributions in this work are as follows:

(i) We provide a comparison between state-of-the-art methods in the context of fair clustering, belonging to different approaches;
(ii) We show how, by optimizing aspects of fairness, some methods affect their ability to produce clusters that are qualitatively good according to classic clustering-validation criteria;
(iii) We shed light on the capabilities of our recently proposed algorithm [20] to adapt to a fair clustering scenario. We show that it is able to produce better solutions than the competing methods from a clustering perspective, while still accounting for fairness-related aspects.

The remainder of the paper is organized as follows. Section 2 provides related work on fair clustering. Section 3 describes how the fair clustering problem can be solved through a correlation clustering framework. Section 4 presents our approach to fair correlation clustering. Section 5 and Sect. 6 present experimental methodology, while Sect. 7 discusses our main experimental findings. Section 8 concludes the paper, also providing pointers to future work.

2 Related Work

Although of relatively recent definition, the problem of fairness in clustering has received considerable attention in the literature [13]. With their seminal work, Chierichetti *et al.* [14] were among the first to formalize the notions around fair clustering and the related problem, following the *disparate-impact doctrine* [16]. Their main contribution is a general pre-processing step, i.e., *fairlets decomposition*, to enable traditional algorithms (e.g., k-center and k-median) meeting fairness principles. Following that forerunner work, fairness has become pervasive in the clustering landscape [8,9,23], leading to a fairness-aware declination of numerous traditional clustering formulations, such as k-center [18], k-means [1,24], k-median [6], spectral clustering [19], and hierarchical clustering [2].

The phenomenon of fairness in clustering has also been extended to alternative approaches, such as correlation clustering. In this regard, Ahmadian *et al.* [3] is the first work to leverage the correlation clustering model for the fair clustering task. More specifically, it takes a complete and undirected graph as input, where vertices are assigned a (single) label representing a given protected class attribute (e.g., sex or ethnicity), and the goal is to provide a fair representation of each considered label in the resulting clusters. Recently, Mandaglio *et al.* [20] proposed to model the fair clustering problem of a relational dataset as a correlation clustering instance. Given a set of objects, defined over a set of features, Mandaglio *et al.* build an associated correlation clustering instance by considering the similarity between the tuples. Although Ahmadian *et al.*'s and Mandaglio *et al.*'s approaches aim to cluster different types of data (graphs and tuples, respectively), both approaches reduce the original problem to a correlation clustering instance. However, Mandaglio *et al.*'s formulation is more general than Ahmadian *et al.*'s one, since the former deals with an arbitrary number of labels (or sensitive attributes), while the latter is limited to a single-label setting.

3 Fairness Constraints in Correlation Clustering

3.1 Background on Correlation Clustering

The correlation clustering problem, originally introduced by Bansal *et al.* [7], consists of clustering the set of vertices of a graph whose edges are assigned two nonnegative weights, named positive-type and negative-type weights, respectively. Such weights express the advantage of putting any two connected vertices into the same cluster (positive-type weight) or into separate clusters (negative-type weight). The objective is to partition the vertices so as to either minimize the sum of the negative-type weights between vertices within the same cluster plus the sum of the positive-type weights between vertices in separate clusters (MIN-CC), or maximize the sum of the positive-type weights between vertices within the same cluster plus the sum of the negative-type weights between vertices in separate clusters (MAX-CC). Both the formulations are **NP**-hard [7,25] and they are equivalent in terms of optimality. However, the available (approximation) algorithms for MAX-CC [10,26] are inefficient and poorly usable in practice since they are not able to output more than a fixed number of clusters (i.e., six). Conversely, MIN-CC admits approximation algorithms [4,11] that do not suffer from the limitations of the maximization counterpart. For these reasons, in this work we focus on the minimization formulation of correlation clustering:

Problem 1 (MIN-CC [5]). Given an undirected graph $G = (V, E)$, with vertex set V and edge set $E \subseteq V \times V$, and weights $w^+_{uv}, w^-_{uv} \in \mathbf{R}^+_0$ for all edges $(u, v) \in E$, find a clustering $\mathcal{C} : V \longrightarrow \mathbf{N}^+$ that minimizes:

$$\sum_{(u,v)\in E,\ \mathcal{C}(u)=\mathcal{C}(v)} w^-_{uv} + \sum_{(u,v)\in E,\ \mathcal{C}(u)\neq\mathcal{C}(v)} w^+_{uv}. \tag{1}$$

MIN-CC is **APX**-hard [11], but admits approximation algorithms [5,7,11, 12,27] with guarantees depending on the type of input graph. On general graphs and weights, the best known approximation factor is $\mathcal{O}(\log |V|)$ [11,15], provided by a linear programming approach. Conversely, constant-factor approximation algorithms are possible if the graph is complete and edge weights satisfy the *probability constraint*, i.e., $w_{uv}^+ + w_{uv}^- = 1$ for all $u, v \in V$. Among these, the one which provides the best trade-off between efficiency and theoretical guarantees is the Pivot algorithm [5], which simply picks a random vertex u, builds a cluster as composed of u and all the vertices v such that an edge with $w_{uv}^+ > w_{uv}^-$ exists, and removes that cluster from the graph. The process is repeated until the graph has become empty. This algorithm has $\mathcal{O}(|E|)$ time complexity and it achieves a factor-5 expected guarantee for MIN-CC under the *probability constraint* or if a *global weight bound* holds on the overall edge weights [20].

Next we discuss how a clustering problem with fairness constraints can be profitably solved through a MIN-CC approach.

3.2 Problem Statement

Let $\mathcal{X} = \{X_1, \cdots, X_n\}$ be a set of n objects defined over a set $\mathcal{A}$ of attributes. The latter is assumed to be divided into two sets, $\mathcal{A}^F$ and $\mathcal{A}^{\neg F}$. The $\mathcal{A}^F$ set contains *fairness-aware*, or *sensitive*, attributes such as those identifying sex, race, religion, relationship status in a citizen database and any other attribute over which fairness is to be ensured. $\mathcal{A}^{\neg F}$ denotes the attributes that are relevant to the task of interest, and thus can be regarded as *non-sensitive*. In both cases, we assume that part of the attributes might be numerical, and the others as categorical (binary or multi-value). We use subscripts N and C to distinguish the two types, therefore $\mathcal{A}^F = \mathcal{A}_N^F \cup \mathcal{A}_C^F$ and $\mathcal{A}^{\neg F} = \mathcal{A}_N^{\neg F} \cup \mathcal{A}_C^{\neg F}$.

We consider a clustering task whose goal is to partition the input objects with a twofold objective: (*i*) minimize the inter-cluster similarity according to the non-sensitive attributes $\mathcal{A}^{\neg F}$; (*ii*) minimize the intra-cluster similarity according to the sensitive attributes $\mathcal{A}^F$. The former objective corresponds to the typical clustering objective, since dissimilar objects should belong to different clusters. Pursuing the second objective, instead, would help distribute objects that are similar in terms of sensitive attributes across different clusters, thus fostering the formation of clusters that are equally represented in terms of the sensitive attributes. This is beneficial to ensure that the distribution of groups defined on sensitive attributes within each cluster approximates the distribution across the dataset. Formally, the problem we tackle in this work is:

Problem 2 (FAIR-CC). Given a set of objects $\mathcal{X}$, two subsets of attributes $\mathcal{A}^F$ and $\mathcal{A}^{\neg F}$, and an object similarity function $sim_S(\cdot)$ defined over the subspace S of the attribute set, find a clustering $\mathcal{C}^*$ to minimize:

$$\sum_{u,v \in \mathcal{X},\, \mathcal{C}(u)=\mathcal{C}(v)} sim_{\mathcal{A}^F}(u,v) \quad + \sum_{u,v \in \mathcal{X},\, \mathcal{C}(u)\neq\mathcal{C}(v)} sim_{\mathcal{A}^{\neg F}}(u,v) \tag{2}$$

The objective in Eq. (2) corresponds to solving a complete MIN-CC instance where the set of vertices corresponds to the objects in $\mathcal{X}$ and, for each pair of vertices u and v, the positive-type (resp. negative-type) correlation-clustering weight corresponds to the similarity score between the two vertices according to the non-sensitive (resp. sensitive) attributes.

We remark that the FAIR-CC problem, as stated above, is introduced here for the first time, while in our previous study in [20] we tackled a different problem: given a set of objects defined over sensitive and non-sensitive attributes, find two attribute subsets that lead to pairwise similarity scores satisfying a certain global condition on the correlation-clustering edge weights. The focus in [20] was to show that the global condition can guide the selection of subsets of features that lead to edge weights expressing the best trade-off between an accurate representation of objects vectors (i.e., discarding not too many features), and the way how the weights facilitate the downstream correlation-clustering algorithm performing well, i.e., by making it achieve approximation guarantees [20]. Instead, in this work, the set of attributes, over which the similarity scores are computed, are given as input in the FAIR-CC problem, and hence they are not needed to be discovered. This is also a more realistic scenario for fair clustering, where the set of sensitive attributes is provided by the specific application scenario.

4 Algorithm

The FAIR-CC problem requires a function to measure the similarity between two objects with respect to a set of attributes. Following [20], we quantify the degree of similarity between two objects u and v, according to the set of sensitive and non-sensitive attributes, by means of the following $sim_{\mathcal{A}^{\neg F}}(u, v)$ and $sim_{\mathcal{A}^{F}}(u, v)$ measures, respectively:

$$sim_{\mathcal{A}^{\neg F}}(u,v) := \psi^+ \left(\alpha_N^{\neg F} \cdot sim_{\mathcal{A}_N^{\neg F}}(u,v) + (1-\alpha_N^{\neg F}) \cdot sim_{\mathcal{A}_C^{\neg F}}(u,v)\right), \quad (3)$$

$$sim_{\mathcal{A}^{F}}(u,v) := \psi^- \left(\alpha_N^{F} \cdot sim_{\mathcal{A}_N^{F}}(u,v) + (1-\alpha_N^{F}) \cdot sim_{\mathcal{A}_C^{F}}(u,v)\right), \quad (4)$$

where $\alpha_N^F = |\mathcal{A}_N^F|/(|\mathcal{A}_N^F| + |\mathcal{A}_C^F|)$ and $\alpha_N^{\neg F} = |\mathcal{A}_N^{\neg F}|/(|\mathcal{A}_N^{\neg F}| + |\mathcal{A}_C^{\neg F}|)$ are coefficients to weight similarities proportionally to the number of involved attributes, and $\psi^+ = exp(|\mathcal{A}^F|/(|\mathcal{A}^F|+|\mathcal{A}^{\neg F}|)-1)$ and $\psi^- = exp(|\mathcal{A}^{\neg F}|/(|\mathcal{A}^F|+|\mathcal{A}^{\neg F}|)-1)$ are smoothing factors to penalize correlation-clustering weights that are computed on a small number of attributes. The latter is reasonable as, in a fair clustering task, we usually have fewer sensitive attributes, and it should be avoided that negative-like weights can dominate the positive-like ones. The exponential function enables a mild smoothing, which is desirable.

As FAIR-CC is an instance of MIN-CC, it can be solved by MIN-CC algorithms. Specifically, although it was originally devised for a slightly different problem (as previously explained in Sect. 3), here we borrow the algorithm proposed in [20] and adapt it to solve the FAIR-CC problem. This algorithm,

Algorithm 1. CCBounds [20]

Require: Set of objects $\mathcal{X}$, sensitive attributes $\mathcal{A}^F$, non-sensitive attributes $\mathcal{A}^{\neg F}$, MIN-CC algorithm A

Ensure: Clustering $\mathcal{C}$ of $\mathcal{X}$

1: compute $sim_{\mathcal{A}^{\neg F}}(u,v), sim_{\mathcal{A}^F}(u,v)$, $\forall u,v \in \mathcal{X}$, as in Eqs. (3)–(4)
2: build the instance $I = \langle G = (\mathcal{X}, \mathcal{X} \times \mathcal{X}), \{sim_{\mathcal{A}^{\neg F}}(u,v), sim_{\mathcal{A}^F}(u,v)\}_{u,v \in \mathcal{X}\times\mathcal{X}}\rangle$
3: $\mathcal{C} \leftarrow$ run A on I

dubbed CCBounds[1] and presented in Algorithm 1, consists of building a MIN-CC instance with vertices as the input data objects and edge weights as the similarity scores, and then running a MIN-CC algorithm A on such a MIN-CC instance.

Theoretical Remarks. Let $T_{\mathsf{A}}(\mathcal{X})$ be the running time of the algorithm A on the set of data objects $\mathcal{X}$. CCBounds runs in $\mathcal{O}(|\mathcal{X}|^2|\mathcal{A}|+T_{\mathsf{A}}(\mathcal{X}))$ time complexity since it needs to compute a similarity score, over $\mathcal{A}$ attributes, for each pair of objects in $\mathcal{X}$, and then solve the resulting MIN-CC instance through algorithm A. Also, the space complexity of CCBounds is $\mathcal{O}(|\mathcal{X}|^2)$ for storing the similarity scores in memory. The specific MIN-CC algorithm A used in CCBounds is the one proposed in [4], since it provides (under the probability constraint or the global weight bound stated in [20]) constant-factor approximation guarantee in expectation. Also, taking linear time in the size of the input graph, to the best of our knowledge, it is the most efficient algorithm in the MIN-CC literature. As a result of this choice, the time complexity of CCBounds becomes $\mathcal{O}(|\mathcal{X}|^2|\mathcal{A}|)$.

Another appealing aspect of the fact that FAIR-CC is an instance of MIN-CC is that FAIR-CC inherits the following theoretical result:

Theorem 1 ([20]). *If the condition* $\binom{|\mathcal{X}|}{2}^{-1}\sum_{u,v\in\mathcal{X}}(sim_{\mathcal{A}^{\neg F}}(u,v) + sim_{\mathcal{A}^F}(u,v)) \geq \max_{u,v\in\mathcal{X}} |sim_{\mathcal{A}^{\neg F}}(u,v) - sim_{\mathcal{A}^F}(u,v)|$ *holds on the similarity scores and the oracle A is an α-approximation algorithm for* MIN-CC*, CCBounds is an α-approximation algorithm for* FAIR-CC.

The above theorem provides approximation guarantee on the FAIR-CC objective (cf. Eq. (2)), which combines the cluster quality measure (first summation) and the fairness-related objective (second summation). It is not known how this quality guarantee translates into the single objective, e.g., the fair objective. This is a challenging open question which we defer to future studies.

5 Fairness Evaluation

In this section, we summarize the most-commonly adopted metrics for the evaluation of fairness aspects in clustering. We focus on algorithm-independent measures, i.e., able to generalize across multiple methods, following a *group-level* approach under the *disparate impact doctrine* [16].

[1] https://github.com/Ralyhu/globalCC.

Balance. It is one of the most adopted evaluation metrics for fairness in clustering, initially proposed by Chierichetti *et al.* [14] in a context with one sensitive attribute with two protected groups. It has been successively generalized to m protected groups by Bera *et al.* [8]. According to the latter, the balance of a clustering solution can formally be defined as follows [13]:

$$balance(\mathcal{C}) = \min_{C \in \mathcal{C}, b \in [m]} \min \left\{ R_{C,b}, \frac{1}{R_{C,b}} \right\} \in [0,1], \tag{5}$$

where $R_{C,b}$ is the ratio between the proportion of the objects belonging to a given protected group b in the considered dataset and in a given cluster $C \in \mathcal{C}$.

In such a formulation, the lower and upper bounds of a cluster indicate the fully unbalanced and perfectly balanced scenarios, respectively, where the former indicates the case where all the objects in such a cluster pertain to the same protected group, whereas the latter denotes an equal number of objects from each of the protected groups. Therefore, the higher the balance, the better the obtained solution, in terms of equality. Additionally, the considered generalization allows us to obtain a comprehensive evaluation of the balance of our clustering solutions, as it looks at the dataset context, i.e., it will return high scores provided that the balances of the clustering and the input dataset are comparable.

Average Euclidean Fairness. This metric was introduced by Abraham *et al.* [1] to estimate the unfairness by assessing the deviation between the representation of groups obtained focusing on the sensitive attributes in the whole dataset and the given clustering solution. It expresses the cluster-size weighted average of cluster-level deviations (i.e., Euclidean distances) between two frequency (sensitive) attribute vectors, namely $\mathcal{X}_A$, which is computed over the entire set of objects, and C_A, which is computed for each cluster $C \in \mathcal{C}$, focusing on a sensitive attribute $A \in \mathcal{A}^F$. Formally, it is defined as:

$$AE_A(\mathcal{C}) = \frac{\sum_{C \in \mathcal{C}} |C| \times ED(C_A, \mathcal{X}_A)}{\sum_{C \in \mathcal{C}} |C|}, \tag{6}$$

where ED represents the Euclidean distance between the frequency attribute vectors. Since A can be multi-valued, such a formulation is suited to scenarios where there are multiple protected groups. Also, as this measure is a deviation, smaller values correspond to better solutions.

6 Experimental Methodology

6.1 Competing Methods

In the following, we briefly overview the competing methods we included in our experiments. For each of those methods, we used publicly available code, which we adopted "as-is", i.e., without making any changes or optimizations.

Fair Clustering Through Fairlets [14]**.** This method, here dubbed FAIRLETS, is one of the pillars of fair clustering. It is based on the notion of *fairlets decomposition*, that is a grouping of the input objects into *fairlets*, i.e., minimal subsets of objects that satisfy a given fairness definition, while preserving the clustering objective. Given a good fairlets decomposition, this approach requires traditional clustering algorithms (i.e., k-center or k-median) applied on the centers of the obtained fairlets, to yield the "fair" solutions. FAIRLETS supports two types of fairlets decomposition: an accurate one based on *min cost flow* (MCF), and a more efficient one. We hereinafter refer to those decompositions as *MCF decomposition* and *vanilla decomposition*, respectively. A major limitation of FAIRLETS is that it can handle a single sensitive binary attribute only. We will discuss the impact of such limitations in more detail in Sect. 7.

We involve FAIRLETS in our experimental evaluation by resorting to the unofficial implementation available online.[2]

HST-Based Fair Clustering [6]. This approach, here dubbed HST-FC, focuses on the k-median formulation, and employs a quad-tree decomposition to embed the objects in a tree metric, called *HST*. By leveraging such a tree, HST-FC computes an approximate fairlets decomposition. A fair clustering is ultimately obtained by running k-median algorithms on the produced fairlets. Like FAIRLETS, HST-FC suffers from the limitation that it deals with one binary sensitive attribute only.

In our experiments, we adopt the official implementation made available by the authors of HST-FC.[3]

Fair Correlation Clustering [3]. This method, here dubbed SIGNED, introduces a fairlet-based reduction for the graph clustering scenario with respect to the problem of correlation clustering, leading to the concept of correlation clustering with fairness constraints. Specifically, given a signed graph, i.e., an undirected graph with edges labeled as positive or negative, the algorithm performs a fairlet decomposition (under different fair settings) over the set of vertices. The produced decomposition is used, together with the original graph, to build a reduced (complete and unweighted) correlation clustering instance, where the vertices correspond to the produced fairlets and the sign of the edges between any two fairlets are built according to the majority sign of the edges between vertices within those two fairlets. A clustering on this reduced correlation clustering instance is computed through local-search optimization starting from all singleton clusters, and then expanded into a solution of the original problem. As a fair setting for the fairlets decomposition, we consider the most common case of fair decomposition where clusters are required not to have a sensitive data class. As the SIGNED method requires a signed graph as input, we perform the following preprocessing step to make the relational data compatible with this format. We derive a complete graph whose vertices are the original data objects and an edge (u, v) is labeled as positive with probability $p^+_{uv} = max\{0, sim_{\mathcal{A}^{\neg F}}(u, v) - sim_{\mathcal{A}^F}(u, v)\}$ and as a negative edge with probability $1 - p^+_{uv}$, where the similarity functions are the ones defined in Eqs. (3)–(4). We point out that, although we can adapt the same weighting strategy as CCBounds

[2] https://github.com/guptakhil/fair-clustering-fairlets.

[3] https://github.com/talwagner/fair_clustering.

Table 1. Overview of the datasets involved in our experiments.

	#objs.	*Sensitive* attribute	*Non-sensitive* attribute
Adult	48 842	sex	age, fnlgwt, education_num, capital_gain, hours_per_week
Bank	40 004	marital	age, balance, duration
CreditCard	10 127	sex	customer_age, dependent_count, avg_utilization_ratio, total_relationship_count
Diabetes	101 763	sex	age, time_in_hospital
Student	649	sex	age, study_time, absences

to obtain the edge attributes, we discarded this choice as our experiments showed that it favors the emergence of a degenerated clustering solution (i.e., a single output cluster), due to the strong predominance of positive weights on the edges.

In our evaluation, we use the official implementation made available by the authors of SIGNED.[4]

6.2 Data

We considered five real-world relational datasets, which have been commonly used in the fair clustering literature. The main characteristics of these datasets are summarized in Table 1. As reported in the table, in our evaluation we focused on a smaller subset of the original attributes; note that this is a common practice, which is adopted, among others, by the competing methods outlined above.

Adult.[5] This dataset reports information about the 1994 US Census. For each tuple representing an individual, we considered *age*, *fnlwgt*, *education-num*, *capital-gain* and *hours-per-week* as non-sensitive attributes, and *sex* (i.e., male or female) as a sensitive attribute.

Bank. (See footnote 5) This provides details on phone calls involving direct marketing campaigns of a Portuguese banking institution to assess whether the bank term deposit will be subscribed or not. We considered attributes *age*, *balance* and *duration* as non-sensitive, and *marital status* (i.e., married or not) as sensitive.

CreditCard.[6] This dataset concerns customer credit card services to estimate customer attrition. We considered attributes *customer_age*, *dependent_count*, *avg_utilization_ratio* and *total_relation ship_count* as non-sensitive, and *sex* as sensitive.

Diabetes. (See footnote 5) It reports diabetic patient records, for which we considered *age* and *time_in_hospital* as non-sensitive attributes, and *sex* as a sensitive attribute.

Student. (See footnote 5) This dataset contains student performances for Mathematics and Portuguese language in secondary education of two Portuguese schools. We considered *age*, *study_time* and *absences* as non-sensitive, and *sex* as sensitive.

---

[4] https://github.com/google-research/google-research/tree/master/correlation_clustering.

[5] https://archive.ics.uci.edu/ml/datasets/.

[6] https://www.kaggle.com/sakshigoyal7/credit-card-customers.

6.3 Evaluation Goals

Our evaluation objectives concern both fairness and quality aspects of clustering. In the first case, we use the fairness metrics defined in Sect. 5, which allow us to have a group-wide overview of how a method behaves in terms of fair principles. In the second case, we assess the quality of clustering by means of intra- and inter-clustering similarity, considering both the sensitive and non-sensitive attributes, as described below. Finally, we evaluate running times.

Intra/Inter-Cluster Similarity. As stated in Sect. 3, we take into account the intra-cluster, resp. inter-cluster, similarity among objects to properly distribute them into clusters, either focusing on their sensitive and non-sensitive attributes (cf. Eqs. (3) and (4)). We define the following aggregated scores to have an overall measure of goodness of the clusters:

$$inter(\mathcal{A}^{\neg F}) = \frac{1}{|\Theta|} \sum_{u,v \in \Theta} sim_{\mathcal{A}^{\neg F}}(u,v), \qquad inter(\mathcal{A}^{F}) = \frac{1}{|\Theta|} \sum_{u,v \in \Theta} sim_{\mathcal{A}^{F}}(u,v), \tag{7}$$

$$intra(\mathcal{A}^{\neg F}) = \frac{1}{|\Omega|} \sum_{u,v \in \Omega} sim_{\mathcal{A}^{\neg F}}(u,v), \qquad intra(\mathcal{A}^{F}) = \frac{1}{|\Omega|} \sum_{u,v \in \Omega} sim_{\mathcal{A}^{F}}(u,v), \tag{8}$$

where $\Omega = \{u,v \in \mathcal{X} \,|\, \mathcal{C}(u) = \mathcal{C}(v)\}$, and $\Theta = \{u,v \in \mathcal{X} \,|\, \mathcal{C}(u) \neq \mathcal{C}(v)\}$. In particular, to obtain fair clusters, we need to maximize (resp. minimize) the $inter(\mathcal{A}^F)$, resp. $intra(\mathcal{A}^F)$, scores, so that objects having the same set of *sensitive* attributes will not be clustered together, rather they will be well-distributed across clusters. Conversely, we require to minimize, resp. maximize, the $inter(\mathcal{A}^{\neg F})$, resp. $intra(\mathcal{A}^{\neg F})$, scores, to ensure that objects with the same set of *non-sensitive* attributes will be clustered close with each other and not scattered across different clusters.

Running Times. We measure the running times of CCBounds and the competing methods while executing them on the *Cresco6* cluster.[7]

6.4 Hyper-parameters and Configurations

Data Sampling and Attribute Selection. To test the selected competing methods under different conditions, and run even the most computationally expensive approaches, we adopt the sampling strategy proposed in [14]. Specifically, by sampling (without replacement) we extracted 1k or 10k tuples from the original full set of tuples, by preserving some desired ratio between the protected classes. The details of the sampling strategy used in our experiments are reported in Table 2, where the selected fair attributes and split ratio (i.e., the fraction of tuples pertaining to different sensitive attribute values) are, whenever possible, the same as [14]. Also, both FAIRLETS and HST-FC require two integers p and q as input, whose ratio p/q corresponds to the minimum balance

[7] https://www.eneagrid.enea.it.

Table 2. Configurations and hyper-parameters used in our evaluations w.r.t. different experimental setups. k_{avg} is the avg. number of clusters that were obtained over ten runs of CCBounds, and k corresponds to the parameter value provided to FAIRLETS and HST-FC.

	p, q	Split ratio	k_{avg}	k
Adult-1k	1,2	650/350	3.12	3
Bank-1k	1,2	650/350	3.48	3
Credit-Card-1k	1,6	800/200	5.6	6
Diabetes-1k	1,2	540/460	5.2	5
Student-1k	1,2	266/383	3.88	4
Adult-10k	1,2	6 500/3 500	2.96	3
Bank-10k	1,2	6 500/3 500	3.28	3
Credit-Card-10k	1,6	4 769/5 358	6.32	6
Diabetes-10k	1,2	5 400/4 600	6.44	6
Adult-Full	2,5	32 650/16 192	3.64	4
Bank-Full	2,5	12 790/27 214	3.64	4
Diabetes-Full	1,2	47 055/54 708	OOM	6

required by each clusters, yielded by these algorithms. The configuration of the aforementioned parameters, inspired by [8,14], is reported in Table 2.

We highlight that, as described so far, we focus on a single and binary sensitive attribute to match the minimum requirements that embrace all competing methods. Nonetheless, some approaches (including our CCBounds) can deal with multiple values assigned to a single sensitive attribute.

Number of Clusters. While FAIRLETS and HST-FC require a hyper-parameter k in input, denoting the desired number of output clusters, the same does not apply with the correlation clustering-based approaches. Thus, to create a reasonable comparative environment, we use the (rounded) average number of clusters returned by CCBounds in ten iterations as the k parameter for FAIRLETS and HST-FC. Moreover, we inherit the value k from the nearest subset when the correlation clustering-based approaches run out of memory.

7 Results

Table 3 summarizes the results achieved by CCBounds and the competing methods. With the exception of very high running times and out of memory errors (indicated with NA and OOM, respectively), all reported measurements correspond to averages over 10 runs of the tested algorithms. The similarity values (Eqs. (7)–(8)) were obtained by using Euclidean and Jaccard similarities for numerical and categorical attributes, respectively. Moreover, as for the FAIRLETS method, as previously discussed in Sect. 6.1, we report results only for the vanilla fairlets decomposition, since the min-cost-flow (MCF) counterpart has

Table 3. Summary of results according to the following criteria (columns from left to right): number of clusters, balance score, avg. Euclidean fairness, avg. intra-cluster and inter-cluster similarities according to either the set of selected sensitive attributes or the set of non-sensitive attributes (cf. Table 1), and running time. For each criterion, bold values correspond to the best-performing methods (possibly up to the second decimal point).

		#clust.	balance ↑	AE ↓	$intra(\mathcal{A}^{\neg F})$ ↑	$intra(\mathcal{A}^{F})$ ↓	$inter(\mathcal{A}^{\neg F})$ ↓	$inter(\mathcal{A}^{F})$ ↑	time (s) ↓
Adult-1k	CCBounds	3.12	0.565	**0.007**	**0.685**	0.524	**0.415**	**0.334**	<1
	Fairlets	3	0.805	**0.004**	0.585	**0.319**	0.596	**0.335**	<1
	HST-FC	3	**0.971**	0.01	0.616	0.335	0.599	**0.336**	<1
	Signed	41	0.66	0.03	0.59	**0.32**	0.60	**0.33**	240
Adult-10k	CCBounds	2.96	0.52	0.03	**0.65**	0.43	**0.43**	**0.33**	3.86
	Fairlets	3	0.82	**0.003**	0.60	**0.32**	0.615	**0.33**	<1
	HST-FC	3	**0.98**	**0.006**	0.626	0.336	0.618	**0.336**	3.03
	Signed	NA	NA	NA	NA	NA	NA	NA	>48 h
Adult-Full	CCBounds	3.64	0.56	**0.003**	**0.69**	0.47	**0.42**	0.24	75.5
	Fairlets	4	0.66	0.02	0.59	**0.32**	0.62	**0.34**	**6.5**
	HST-FC	4	**0.96**	**0.008**	0.63	0.34	0.62	**0.34**	72.86
	Signed	NA	NA	NA	NA	NA	NA	NA	>48 h
Bank-1k	CCBounds	3.48	0.565	**0.006**	**0.727**	0.587	**0.441**	**0.369**	<1
	Fairlets	3	0.828	**0.002**	0.606	**0.354**	0.613	**0.364**	<1
	HST-FC	3	**0.968**	**0.007**	0.621	0.365	0.617	**0.365**	<1
	Signed	41	0.7	0.03	0.61	**0.35**	0.63	**0.36**	224
Bank-10k	CCBounds	3.28	0.52	**0.0007**	**0.78**	0.63	**0.45**	**0.36**	4.74
	Fairlets	3	0.7	0.001	0.59	**0.32**	0.63	**0.36**	<1
	HST-FC	3	**0.969**	0.004	0.656	0.365	0.656	**0.365**	3.07
	Signed	NA	NA	NA	NA	NA	NA	NA	>48 h
Bank-Full	CCBounds	3.64	0.55	**0.0004**	**0.72**	0.55	**0.45**	**0.37**	51.1
	Fairlets	4	0.68	0.001	0.62	**0.34**	0.65	0.36	**5.3**
	HST-FC	4	**0.94**	0.008	0.66	0.37	0.66	**0.37**	28
	Signed	NA	NA	NA	NA	NA	NA	NA	>48 h
CreditCard-1k	CCBounds	5.6	0.613	0.127	**0.6**	0.497	**0.46**	0.362	<1
	Fairlets	6	0.4	0.042	0.485	**0.355**	0.486	**0.375**	<1
	HST-FC	6	**0.756**	**0.026**	0.513	0.373	0.481	**0.377**	<1
	Signed	171	0.56	0.1	0.56	0.41	0.49	**0.38**	173
CreditCard-10k	CCBounds	6.32	0.496	0.17	**0.6**	0.46	**0.46**	0.32	4.1
	Fairlets	6	0.94	**0.01**	0.497	**0.34**	0.49	**0.337**	<1
	HST-FC	6	**0.955**	**0.013**	0.52	**0.337**	0.491	**0.337**	2.52
	Signed	NA	NA	NA	NA	NA	NA	NA	>48 h
Diabetes-1k	CCBounds	5.2	0.45	0.33	**0.622**	0.519	**0.512**	0.352	<1
	Fairlets	5	**0.92**	**0.015**	0.537	0.381	0.532	**0.385**	<1
	HST-FC	5	0.872	0.05	0.585	0.386	0.529	**0.386**	<1
	Signed	106	0.85	0.04	0.58	**0.36**	0.54	**0.38**	257
Diabetes-10k	CCBounds	6.44	0.48	0.22	**0.65**	0.54	**0.5**	0.36	4.72
	Fairlets	6	**0.92**	**0.01**	0.53	**0.38**	**0.53**	**0.39**	<1
	HST-FC	6	0.799	0.065	0.59	**0.388**	**0.53**	**0.386**	2.84
	Signed	NA	NA	NA	NA	NA	NA	NA	>48 h
Diabetes-Full	CCBounds	OOM	OOM	OOM	OOM	OOM	OOM	OOM	OOM
	Fairlets	6	**0.93**	**0.01**	OOM	OOM	OOM	OOM	**22.2**
	HST-FC	6	0.81	0.06	OOM	OOM	OOM	OOM	761.2
	Signed	OOM	OOM	OOM	OOM	OOM	OOM	OOM	OOM
Student-1k	CCBounds	3.88	0.51	0.10	**0.625**	0.463	**0.471**	0.224	<1
	Fairlets	4	0.82	**0.013**	0.528	**0.339**	0.543	**0.357**	<1
	HST-FC	4	**0.93**	0.024	0.563	0.357	0.541	**0.358**	<1
	Signed	55	0.82	0.04	0.57	**0.34**	0.55	**0.36**	71

very high running times (more than 7 min on the smallest dataset, i.e., *Student-1k*) and produces solutions that are very similar to the vanilla one (results not shown for the sake of brevity).

As for the balance, we notice that, although CCBounds does not match the high scores obtained by "fairness-native" methods (i.e., FAIRLETS and HST-FC), it is still able to score comparably with its direct competing method, i.e., SIGNED. Exceptions arise in the case of *Student-1k* and *Diabetes-1k*, where CCBounds sets up to lower scores, and for some large datasets, where SIGNED does not terminate in reasonable time, while our CCBounds still obtains good results in reasonable time. The paradigm shifts when we consider small yet heavily unbalanced datasets (i.e., *CreditCard-1k*, with an 80:20 ratio); here, although several competing methods struggle to obtain high scores, CCBounds achieves the second-best balance score. Overall, as the balance obtained by CCBounds in all evaluation scenarios ranges from 0.45 to 0.613, we can conclude that it is able of guaranteeing satisfactory balance scores.

In the case of avg. Euclidean fairness, CCBounds obtains very good scores under different scenarios: it is among the best-performer approaches for the *Adult-1k*, *Adult-Full* and *Bank-1k* datasets, and outperforms all the other methods by an order of magnitude on *Bank-10k* and *Bank-Full*. Conversely, CCBounds is unable to match the best scores obtained by some of the competing methods when focusing on the remaining datasets.

Considering the similarity computed on the sensitive attributes, CCBounds does not achieve the best intra-cluster similarity, meaning that it tends to group a few more objects with the same sensitive attribute value than the other methods. Nevertheless, the inter-cluster similarities are comparable with the other methods, thus indicating that CCBounds is still able to properly separate the objects into clusters, when accounting for the sensitive attribute. Instead, when we focus on the similarity computed on the non-sensitive attributes, CCBounds achieves the best performance in all the considered evaluation scenarios, yielding very high-quality clusters.

Finally, we also investigated on running times, spotting FAIRLETS as the best performer, followed by HST-FC and CCBounds, which both guarantee reasonable running times. Although CCBounds has quadratic time complexity due to pairwise similarity calculations (cf. Sect. 4), we managed to perform in parallel such time-consuming steps. On the contrary, SIGNED requires excessively long execution times, often resulting infeasible in practice, along with an abnormal number of clusters produced, which is particularly large even when considering the smallest *1k* datasets. Overall, it should be noted that, albeit the observed running times should be taken with grain of salt due to the (lack of) code optimizations, major remarks are consistent with the time complexities of the corresponding methods.

Discussion. A number of remarks arise from our experimental evaluation. First, although native fairness-aware approaches are able to produce clustering solutions that optimize fairness notions, we found out that such a capability comes with a cost, as the produced clusters are often far from being qualitatively good.

On the other hand, CCBounds demonstrated itself to be effective and versatile: it was recognized as the best-in-case approach among the tested ones when it comes to find good-quality clusters, while also being able not to excessively penalize aspects related to fairness.

Second, although we unveiled the weakness in quality shown by the native fair-clustering approaches, we nonetheless shed light on how the approaches based on correlation clustering might suffer from computational issues, by being slower than the other methods, and requiring more memory. This is particularly evident with SIGNED, as it is unable to terminate in all datasets having more than 10k tuples, while it is kept under control in CCBounds, which goes down only in the case of *Diabetes-Full* (containing more than 100k tuples, cf. Sect. 6.2), thanks to the numerous optimization adopted under the hood. However, such a dataset makes it difficult to calculate similarities even for traditional and more efficient approaches, despite the computing capabilities at our disposal.

Finally, by wearing the lens of our proposed approach, we can state that it is able to provide performance in terms of fairness-aware metrics that are comparable to its direct competitor (i.e., SIGNED), but, at the same time, it manages to overcome all the state-of-the-art competing methods considered in our assessment, when it comes to generating qualitatively good clusters, anyway preserving aspects of fairness as much as possible.

8 Conclusions

In this paper, we analyzed how a correlation clustering method, called CCBounds, can profitably be used for the problem of fair clustering. Experimental evidence on real data has shown the meaningfulness of the clustering solutions produced by CCBounds, also revealing its ability of yielding clusters of higher quality than the considered competing methods, according to classic clustering-validation criteria, without discarding aspects of fairness.

In the future, we plan to further evaluate the performance of CCBounds under other conditions, e.g., multiple protected values. Also, we aim to investigate on alternative definitions of the similarity functions and push forward the capabilities of CCBounds towards more challenging scenarios, such as embracing multiple sensitive attributes with many values, allowing us to align with more realistic use cases, and strengthen the versatility of the correlation clustering under fairness constraints.

References

1. Abraham, S.S., P, D., Sundaram, S.S.: Fairness in clustering with multiple sensitive attributes. In: Proceedings of the EDBT Conference, pp. 287–298 (2020)
2. Ahmadian, S., et al.: Fair hierarchical clustering. In: Proceedings of the NIPS Conference (2020)
3. Ahmadian, S., Epasto, A., Kumar, R., Mahdian, M.: Fair correlation clustering. In: Proceedings of the AISTATS Conference, pp. 4195–4205 (2020)

4. Ailon, N., Charikar, M., Newman, A.: Aggregating inconsistent information: ranking and clustering. In: Proceedings of the ACM STOC Symposium, pp. 684–693 (2005)
5. Ailon, N., Charikar, M., Newman, A.: Aggregating inconsistent information: ranking and clustering. JACM **55**(5), 23:1–23:27 (2008)
6. Backurs, A., Indyk, P., Onak, K., Schieber, B., Vakilian, A., Wagner, T.: Scalable fair clustering. In: Proceedings of the ICML Conference, pp. 405–413 (2019)
7. Bansal, N., Blum, A., Chawla, S.: Correlation clustering. Mach. Learn. **56**(1), 89–113 (2004)
8. Bera, S.K., Chakrabarty, D., Flores, N., Negahbani, M.: Fair algorithms for clustering. In: Proceedings of the NIPS Conference, pp. 4955–4966 (2019)
9. Bercea, I.O., et al.: On the cost of essentially fair clusterings. In: Proceedings of the APPROX/RANDOM Conference, pp. 18:1–18:22 (2019)
10. Charikar, M., Guruswami, V., Wirth, A.: Clustering with qualitative information. In: Proceedings of the IEEE FOCS Symposium, pp. 524–533 (2003)
11. Charikar, M., Guruswami, V., Wirth, A.: Clustering with qualitative information. JCSS **71**(3), 360–383 (2005)
12. Chawla, S., Makarychev, K., Schramm, T., Yaroslavtsev, G.: Near optimal LP rounding algorithm for correlation clustering on complete and complete k-partite graphs. In: Proceedings of the ACM STOC Symposium, pp. 219–228 (2015)
13. Chhabra, A., Masalkovait-, K., Mohapatra, P.: An overview of fairness in clustering. IEEE Access **9**, 130698–130720 (2021)
14. Chierichetti, F., Kumar, R., Lattanzi, S., Vassilvitskii, S.: Fair clustering through fairlets. In: Proceedings of the NIPS Conference, pp. 5029–5037 (2017)
15. Demaine, E.D., Emanuel, D., Fiat, A., Immorlica, N.: Correlation clustering in general weighted graphs. TCS **361**(2–3), 172–187 (2006)
16. Feldman, M., Friedler, S.A., Moeller, J., Scheidegger, C., Venkatasubramanian, S.: Certifying and removing disparate impact. In: Proceedings of the ACM KDD Conference, pp. 259–268 (2015)
17. Kleinberg, J., Lakkaraju, H., Leskovec, J., Ludwig, J., Mullainathan, S.: Human decisions and machine predictions. Q. J. Econ. **133**(1), 237–293 (2017)
18. Kleindessner, M., Awasthi, P., Morgenstern, J.: Fair k-center clustering for data summarization. In: Proceedings of the ICML Conference, pp. 3448–3457 (2019)
19. Kleindessner, M., Samadi, S., Awasthi, P., Morgenstern, J.: Guarantees for spectral clustering with fairness constraints. In: Proceedings of the ICML Conference, pp. 3458–3467 (2019)
20. Mandaglio, D., Tagarelli, A., Gullo, F.: Correlation clustering with global weight bounds. In: Proceedings of the ECML-PKDD Conference, pp. 499–515 (2021)
21. Martorelli, M., Jayatilake, S.M.D.A.C., Ganegoda, G.U.: Involvement of machine learning tools in healthcare decision making. J. Healthc. Eng. (2021)
22. Mashrur, A., Luo, W., Zaidi, N.A., Robles-Kelly, A.: Machine learning for financial risk management: a survey. IEEE Access **8**, 203203–203223 (2020)
23. Rösner, C., Schmidt, M.: Privacy preserving clustering with constraints. In: Proceedings of the ICALP Colloquim, pp. 96:1–96:14 (2018)
24. Schmidt, M., Schwiegelshohn, C., Sohler, C.: Fair coresets and streaming algorithms for fair k-means. In: Proceedings of the WAOA Workshop, pp. 232–251 (2019)
25. Shamir, R., Sharan, R., Tsur, D.: Cluster graph modification problems. Discret. Appl. Math. **144**(1–2), 173–182 (2004)

26. Swamy, C.: Correlation clustering: maximizing agreements via semidefinite programming. In: Proceedings of the ACM-SIAM SODA Conference, pp. 526–527 (2004)
27. van Zuylen, A., Williamson, D.P.: Deterministic algorithms for rank aggregation and other ranking and clustering problems. In: Proceedings of the WAOA Workshop, pp. 260–273 (2007)

Cooperative Deep Unsupervised Anomaly Detection

Fabrizio Angiulli(✉), Fabio Fassetti, Luca Ferragina, and Rosaria Spada

DIMES, University of Calabria, 87036 Rende (CS), Italy
{f.angiulli,f.fassetti,l.ferragina}@dimes.unical.it

Abstract. In last years deep learning approaches to anomaly detection are becoming very popular. In most of the first methods the paradigm is to train neural networks initially designed for compression (Auto Encoders) or data generation (GANs) and to detect anomalies as a collateral result. Recently new architectures have been introduced in which the expressive power of deep neural networks is associated with objective functions specifically designed for anomaly detection. One of these methods is *Deep-SVDD* which, although created for One-Class classification, has been successfully applied to the (semi-)supervised anomaly detection setting. Technically, *Deep-SVDD* technique forces the deep latent representation of the input data to be enclosed into an hypersphere and labels as anomalies data farthest from its center. In this work we introduce *Deep-UAD*, a neural network approach for unsupervised anomaly detection where, iteratively, a network similar to that of *Deep-SVDD* is alternatively trained with an Auto Encoder and the two networks share some weights in order for each network to improve its training by exploiting the information coming from the other network. The experiments we conducted show that the performances obtained by the proposed method are better than the ones obtained both by deep learning methods and standard shallow algorithms.

1 Introduction

Anomaly detection is a fundamental data mining task whose aim is to isolate samples in a dataset that are suspected of being generated by a distribution different from the rest of the data. The presence of anomalies is due to many reasons like mechanical faults, fraudulent behavior, human errors, instrument error or simply through natural deviations in populations.

Depending on the composition of the dataset, anomaly detection settings can be classified as unsupervised, semi-supervised, and unsupervised [1,14]. In the supervised setting the training data are labeled as normal and abnormal and and the goal is to build a classifier. The difference with standard classification problems is that abnormal data form a rare class. In the semi-supervised setting,

P. Pascal and D. Ienco (Eds.): DS 2022, LNAI 13601, pp. 318–328, 2022.
https://doi.org/10.1007/978-3-031-18840-4_23

the training set is composed by both labelled and unlabelled data. A special case of this setting is the one-class classification when we have a training set composed only by normal class items. In the unsupervised setting the goal is to detect outliers in an input dataset by assigning a score or anomaly degree to each object. Several statistical, data mining and machine learning approaches have been proposed to detect anomalies, namely, statistical-based [11,15], distance-based [6,9,10,24], density-based [12,21], reverse nearest neighbor-based [4,5,19, 25], SVM-based [30,33], and many others [1,14].

In last years deep learning-based methods for anomaly detection [13,17,27] have shown great performances. Auto encoder(AE) based anomaly detection [3,13,20] consists in training an AE to reconstruct a set of examples and then to detect as anomalies those data that show a large reconstruction error. *Variational auto encoders* (VAE) arise as a variant of standard auto encoders designed for *generative* purposes [23]. The key idea of VAEs is to encode each example as a normal distribution over the latent space and regularize the loss by maximizing similarity of these distributions with the standard normal one. Due to similarities to standard auto encoders, VAEs have also been used to detect anomalies. However, it has been noticed that VAEs share with standard AEs the problem that they generalize so well that they can also well reconstruct anomalies [3,7,8,13,22,32]. Generative Adversarial Networks (GAN) [18] are another tool for generative purposes, aiming at learning an unknown distribution by means of an adversarial process involving a discriminator, that outputs the probability for an observation to be generated by the unknown distribution, and a generator, mapping points coming from a standard distribution to points belonging to the unknown one. GANs have also been employed with success to the anomaly detection task [2,16,29,31,34].

Some authors [26,28] have recently observed that all the above mentioned anomaly detection deep learning based methods are not designed to directly discover anomalies, but their main task is data reconstruction (AE and VAE) or data generation (GAN) and anomaly detection is a collateral result. They introduce new methods, called *Deep-SVDD* and *Deep-SAD*, that combine the expressive power of deep neural networks with a loss inspired from SVM-based methods and specifically designed for anomaly detection. These methods are used for one-class and (semi-)supervised settings but we argue that they do not apply very naturally to the unsupervised setting, thus we introduce *Deep-UAD*, a new unsupervised method that deeply modifies the architectures in [26,28]. In particular we build a new training paradigm for the network in [26] that involves an AE which is trained alternatively with the network and with which the network and exchange the information they obtained during the training. This is done by modifying the losses of both the network and the AE. The proposed approach shows sensible improvements in terms of detection performances over both the standard approach in [26,28] and the baseline shallow methods.

The rest of the paper is organized as follows. Section 2 discusses related work with particular emphasis on *Deep-SVDD* and *Deep-SAD*. Section 3 introduces the *Deep-UAD* unsupervised anomaly detection algorithm. Section 4 illustrates experimental results. Finally, Sect. 5 concludes the work.

2 Preliminaries

In this Section we deepen auto encoder and *Deep-SVDD* which are exploited by our technique as basic components and suitably modified to our purposes.

Auto Encoder. An auto encoder (AE) is a neural network architecture successfully employed for anomaly detection [20]. It aims at providing a reconstruction of the input by exploiting a dimensionality reduction step (the *encoder* ϕ_W) followed by a step mapping back from the compressed space (the *latent space*) to the original space (the *decoder* $\psi_{W'}$). Its ability in detecting anomalies depends on the observation that regularities should be better compressed and, hopefully, better reconstructed [20]. The AE loss is $\mathcal{E}(x) = \|x - \hat{x}\|_2^2$, where $\hat{x} = \psi_{W'}(\phi_W(x))$, and coincides with the standard reconstruction error.

One-class SVM. Before discussing *Deep-SVDD* some preliminary notions about One-Class SVM (OC-SVM) [30] are needed. The original OC-SVM method is designed for the one-class setting and has the objective of finding the hyperplane in a feature space that best separates the mapped data from the origin. Given the data $\{x_1, \ldots, x_n\} \subseteq X$, it is defined by the following optimization problem

$$\begin{aligned} \min_{\mathbf{w},\rho,\xi_i} \; & \frac{1}{2}\|\mathbf{w}\|_F^2 - \rho + \frac{1}{\nu n}\sum_{i=1}^{n}\xi_i \\ \text{s. t. } & \langle \phi(x_i), \mathbf{w} \rangle \geq \rho - \xi_i, \\ & \xi_i \geq 0, \quad i = 1, \ldots, n \end{aligned}$$

where ρ is the distance from the origin to the hyperplane $\mathbf{w} \in F$, ξ_i are slack variables and $\nu \in (0, 1]$ is a trade-off hyperparameter. The points in the test set are labelled as normal if they are mapped inside the hyperplane and anomalous if they are mapped outside. Related to OC-SVM, Support Vector Data Descriptor (SVDD) [33] is a method that has the aim of enclosing the input data into a hypersphere of minimum radius. The relative optimization problem is

$$\begin{aligned} \min_{R,\mathbf{c},\xi_i} \; & R^2 + \frac{1}{\nu n}\sum_{i=1}^{n}\xi_i \\ \text{s. t. } & \|\phi(x_i) - \mathbf{c}\| \leq R^2 + \xi_i, \\ & \xi_i \geq 0, \quad i = 1, \ldots, n \end{aligned}$$

where $R > 0$ and $\mathbf{c}$ are the radius and the center of the hypersphere and again ξ_i are slack variables and $\nu \in (0, 1]$ is a trade-off hyperparameter.

Deep-SVDD. In [26], authors apply the same idea expressed in SVDD of enclosing the data into an hypersphere performing the mapping into the feature space with the use of a deep neural network. In particular, let $\phi_W : X \rightarrow F$ be a mapping obtained with a neural network with weights $W = [W_1, \dots, W_L]$ (W_l are the weights relative to the layer $l \in \{1, \dots, L\}$) from the input space $X \subseteq \mathbb{R}^d$ to the output space $F \subseteq \mathbb{R}^k$, with $k < d$. The loss of the network is given by

$$\mathcal{L} = \frac{1}{n}\sum_{i=1}^{n}\|\phi_W(x_i) - \mathbf{c}\|_2^2 + \frac{\lambda}{2}\sum_{l=1}^{L}\|W_l\|_F^2, \tag{1}$$

where the first term forces the network representation $\phi_W(x)$ to stay close to the center $\mathbf{c}$ of the hypersphere and the second term is a weight decay regularizer with hyperparameter $\lambda > 0$. This loss is used in a One-Class anomaly detection setting to map the training set (composed only by normal items) as close as possible to the center $\mathbf{c}$ so that in the testing phase the network is less able to map the anomalies close to $\mathbf{c}$. Because of this, it is defined as anomaly score of the point x the distance of its network representation from the center: $\mathcal{S}(x) = \|\phi_W(x) - \mathbf{c}\|_2^2$.

The center $\mathbf{c}$ is not a trainable parameter and is fixed before the training by means of an AE that is composed so that the encoding part has the same structure as the network ϕ and shares with it the weights W, the structure of the decoding part is symmetric to it and thus the latent space coincides with the space F. The training set is given in input to this AE which is trained with the standard loss and subsequently the center $\mathbf{c}$ is defined as $\mathbf{c} = \frac{1}{n}\sum_{i=1}^{n}\phi_W(x_i)$, that is the mean of the latent representations of all the points in the training set. The same architecture has been applied in [28] for the task of semi-supervised anomaly detection with the following natural adaptation of the loss

$$\mathcal{L} = \frac{1}{n+m}\sum_{i=1}^{n}\|\phi_W(x_i) - \mathbf{c}\|_2^2 + \frac{\eta}{n+m}\sum_{i=1}^{m}\left(\|\phi_W(\tilde{x}_i) - \mathbf{c}\|_2^2\right)^{\tilde{y}_i} + \frac{\lambda}{2}\sum_{l=1}^{L}\|W_l\|_F^2, \tag{2}$$

where $\tilde{x}_i$ are the m labeled data with the relative labels $\tilde{y}_i$ and η is an hyperparameter handling the trade-off between the contributions of labelled and unlabelled data. Let us observe that data labelled as normal ($\tilde{y}_i = +1$) are treated in the usual way which means that they are forced to be mapped close to $\mathbf{c}$ while for the anomalies ($\tilde{y}_i = -1$) the contribution is inverted and they are force to stay as far as possible from $\mathbf{c}$.

It is important to observe that (2) is designed to consider also unlabelled examples. An extreme case occurs when $m = 0$, when all the training data are unlabelled. This scenario is similar to the unsupervised setting but there is a substantial difference: in one case the objective is to detect anomalies in a test set, in the other the anomalies have to be detected among the same data used for the training phase. In this case the losses (2) and (1) coincide, which means that, even if originally the loss (1) has been designed to deal only with normal class items, it can be used in settings that involve the use of unlabeled anomalies in the training phase, thus it can be applicable also to the unsupervised settings.

3 Method

In this Section the technique *Deep-UAD* proposed in this paper is discussed.

A one class based technique, like *Deep-SVDD*, is aimed at building a model for the normal class exploiting input data by assuming they do not contain anomalies and classifying data of a test set. In particular, *Deep-SVDD* tends to map close to the center all the input data and, then, in the unsupervised setting this technique may fail in correctly separating normal and anomalous samples.

Deep-UAD tackles this issue by providing information to the network about the anomaly degree of each sample in order to force the network to approach normal data to the center and to let anomalies far from the center. This is accomplished by exploiting an AE that provides a level of anomaly suspiciousness. Thus, the proposed architecture consists in two components, a neural network *Deep-UAD*$_{NET}$, and an auto encoder *Deep-UAD*$_{AE}$; *Deep-UAD*$_{NET}$ has the same structure of the network of *Deep-SVDD*, thus can be defined by the same mapping function ϕ_W, and it is forced to map the data badly reconstructed by *Deep-UAD*$_{AE}$, namely more suspected to be anomalous, faraway from the center and, conversely, data suggested as normal by *Deep-UAD*$_{AE}$ close to the center. Technically, this is done by introducing this novel loss

$$\mathcal{L}_{\text{NET}} = \frac{1}{n}\sum_{i=1}^{n} \frac{1}{\mathcal{E}\left(x_i\right)} \|\phi_W\left(x_i\right) - \mathbf{c}\|_2^2 + \frac{\lambda}{2}\sum_{l=1}^{L}\|W_l\|_F^2. \tag{3}$$

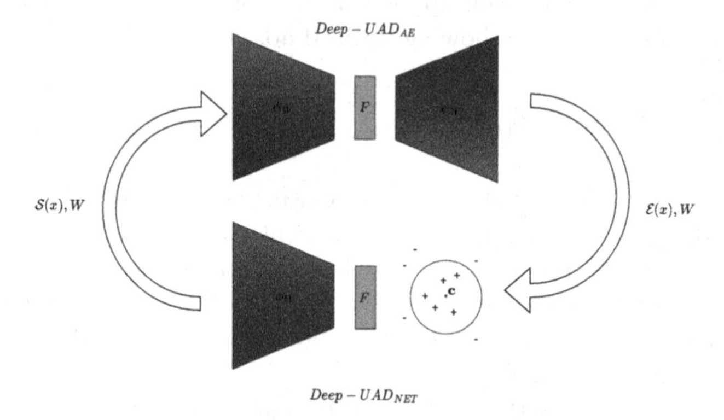

Fig. 1. Diagram of the *Deep-UAD* cooperative strategy: the network *Deep-UAD*$_{NET}$ and the auto encoder *Deep-UAD*$_{AE}$ refine their capabilities to find anomalies by sharing the encoder weights W and passing to each other the information of their own score.

It is inspired by Eq. (1) which is modified by inserting the term $\frac{1}{\mathcal{E}(x_i)}$, directly related to the probability for x_i to be an anomaly according to the AE, and it is used as a weight to control how much is important that the network representation of x_i is mapped close to $\mathbf{c}$. In particular, the smaller is $\mathcal{E}(x_i)$, namely

x_i is probably not an anomaly according to *Deep-UAD*$_{AE}$, the more higher is the weight and thus the network takes more advantage in mapping x_i close to the center; conversely if $\mathcal{E}(x_i)$ is large, x_i is suspected to be an anomaly by the AE, the weight is small and the network has a small advantage in bringing the representation of x_i close to the center.

The strategy of *Deep-UAD* consists in a preliminary phase where the AE, without information by the network *Deep-UAD*$_{NET}$, is trained with standard loss, the center of the hypershpere is computed and the reconstruction error $\mathcal{E}(x_i)$ is evaluated for each sample. Successively, two phases are iteratively executed, during the first one, the network *Deep-UAD*$_{NET}$ is trained with the loss (3) for a certain number of epochs and the score $\mathcal{S}(x_i)$ is calculated, during the second, *Deep-UAD*$_{AE}$ is trained for some epochs with the novel loss

$$\mathcal{L}_{\mathrm{AE}} = \sum_{i=1}^{n} \frac{1}{\mathcal{S}(x_i)} \|x_i - \hat{x_i}\|_2^2. \tag{4}$$

The purpose of $\mathcal{S}(x_i)$ is similar to the one of $\mathcal{E}(x_i)$ in (3), giving a weight to the contribution of each point x_i according to the results obtained by the network. The idea of *Deep-UAD* is that the score obtained from one network improves the training of the other one, the final anomaly score output is $\mathcal{S}(x_i)$.

4 Experimental Results

In this section we report experiments conducted to study the behavior of the proposed method. We focus on three main aspects, namely (*i*) the impact of the dimension of the output space on the performances, (*ii*) the analysis of the cooperative process as the iterations proceed, (*iii*) the comparison with other methods with specific emphasis on *Deep-SVDD*. In our experiments we consider two standard benchmark datasets composed by grayscale images, *MNIST*[1] and *Fashion-MNIST*[2]. They are both composed by 28×28 pixels images divided in 10 classes, thus, in order adapt them for anomaly detection, we adopt a *one-vs-all* policy, i.e. we consider one class as normal and all the others as anomalous. For each class, we create a dataset composed by all the examples of the selected class as normal and s random selected examples from each other class as anomalies.

Sensitivity Analysis on the Dimension K of the Output Space. In this section, our aim is to determine how the dimension of the output space F impacts on the behavior of both our method and the original *Deep-SVDD* algorithm. In order to do this we consider the MNIST dataset in the *one-vs-all* setting and, for each class, we train both models with k varying in the interval $[8, 64]$.

From Fig. 2, in which are reported the results after 5 runs, we can see that for both *Deep-UAD* (in red) and *Deep-SVDD* (in black) the trend is increasing

[1] http://yann.lecun.com/exdb/mnist/.
[2] https://github.com/zalandoresearch/fashion-mnist.

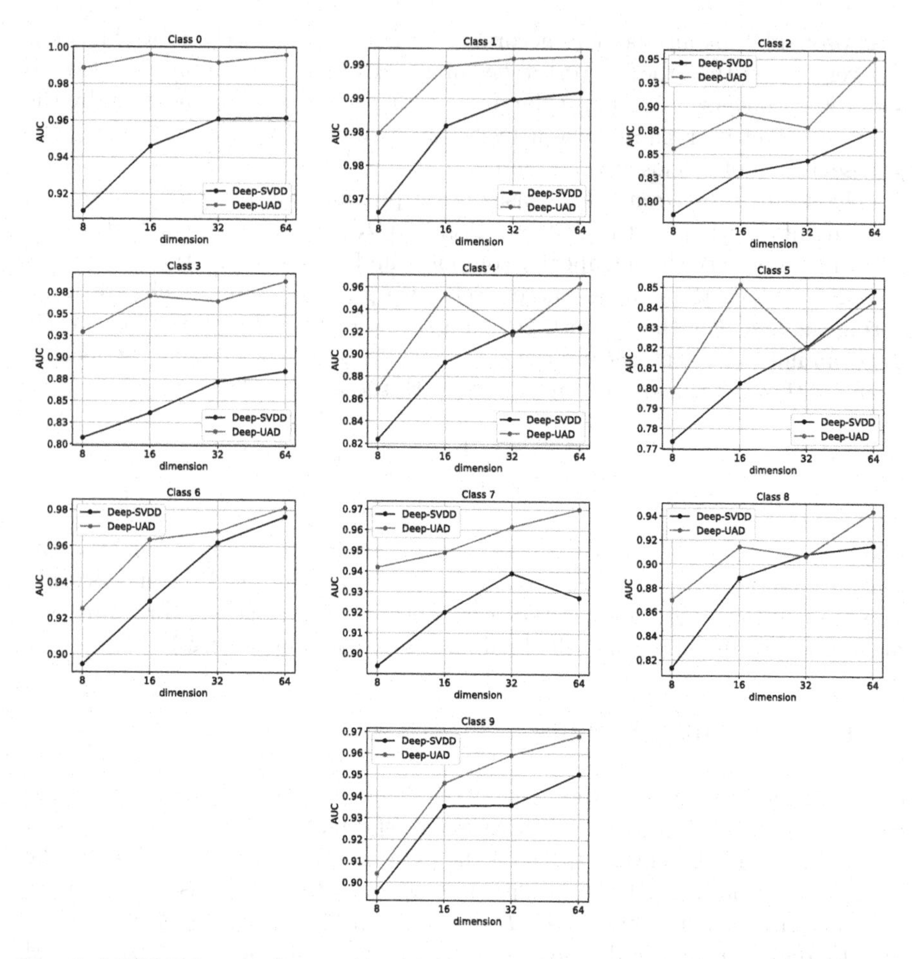

Fig. 2. MNIST dataset (s = 10): AUCs of *Deep-UAD* and *Deep-SVDD* varying the dimension of the final space.

which means that a small dimensional space F is not sufficient, in both cases, to separate the anomalies from the normal examples. Moreover it is important to point out that the performances achieved by our method are better than the ones obtained by *Deep-SVDD* for almost each class and each value of k.

Analysis of the Iterative Process. *Deep-UAD* is based on an iterative process in which the network $Deep\text{-}UAD_{NET}$ and the auto encoder $Deep\text{-}UAD_{AE}$ share information, because of this it is crucial to investigate how the number of iterations affects the performances of both the architectures. We do this by considering MNIST and Fashion-MNIST datasets in the *one-vs-all* setting, performing 5 runs for each class and computing the AUC for each iteration. In each iteration both the network and the AE are trained for 25 epochs.

In Fig. 3 are reported the trends of the two architectures. As we can see, they are always non decreasing, which means that both the architectures are taking advantage of the cooperative strategy. For what concerns *Deep-*UAD_{NET}, which is the one that outputs the score of *Deep-UAD*, the trend becomes substantially stable and constant, sometimes from the very first iteration (as class 0 of MNIST and class *Sandal* of Fashion-MNIST) and other times after a slightly bigger number of iterations (like classes 2 and 7 of MNIST). This means that the parameter of the number of iterations is not hard to fix, since a number around one ten of iterations guarantees always the achievement of a score of *Deep-UAD* close to best possible and an improvement over *Deep-SVDD*.

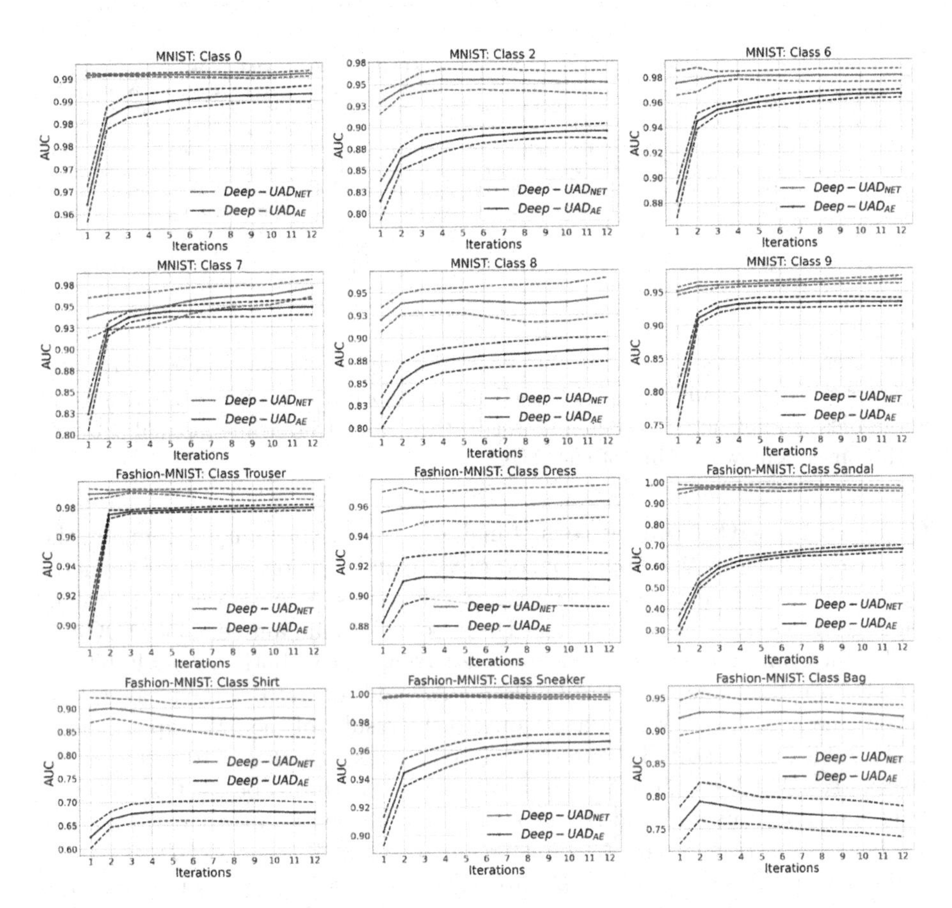

Fig. 3. MNIST and Fashion-MNIST datasets ($s = 10$): AUCs of *Deep-UAD* and AE varying the iterations of the method.

Moreover, *Deep-*UAD_{AE} improves its performances as the iterations proceed. This fact is crucial for the behavior of the whole process, indeed it means that the information provided to *Deep-*UAD_{NET} by *Deep-*UAD_{AE} becomes better at

Table 1. AUC of *Deep-UAD* and competitors on MNIST and Fashion-MNIST with $s = 10$ on the left and $s = 100$ on the right.

MNIST										
	$s = 10$					$s = 100$				
Class	*Deep-SVDD*	*Deep-UAD*	DCAE	IF	KNN	*Deep-SVDD*	*Deep-UAD*	DCAE	IF	KNN
0	.962 ± .018	**.996 ± .001**	.953 ± .008	.951 ± .010	.978 ± .010	.901 ± .015	**.948 ± .006**	.918 ± .005	.868 ± .023	.841 ± .006
1	.991 ± .001	.996 ± .002	.925 ± .013	.994 ± .001	**.998 ± .001**	.974 ± .004	.989 ± .003	.878 ± .014	.991 ± .001	**.996 ± .002**
2	.876 ± .036	**.951 ± .013**	.822 ± .017	.731 ± .024	.891 ± .015	.765 ± .020	**.861 ± .026**	.746 ± .016	.678 ± .028	.703 ± .014
3	.884 ± .014	**.988 ± .006**	.805 ± .026	.804 ± .028	.906 ± .011	.794 ± .027	**.878 ± .026**	.759 ± .027	.762 ± .012	.766 ± .006
4	.924 ± .017	**.964 ± .011**	.772 ± .019	.869 ± .018	.928 ± .008	.827 ± .016	**.918 ± .012**	.698 ± .031	.836 ± .008	.831 ± .007
5	.848 ± .025	.843 ± .037	.728 ± .016	.752 ± .019	**.917 ± .019**	.732 ± .017	**.809 ± .021**	.650 ± .040	.709 ± .020	.765 ± .007
6	.975 ± .007	**.981 ± .005**	.887 ± .019	.894 ± .017	.970 ± .006	.920 ± .014	**.956 ± .007**	.793 ± .018	.848 ± .016	.858 ± .004
7	.927 ± .017	**.970 ± .008**	.853 ± .015	.909 ± .009	.957 ± .007	.875 ± .014	**.946 ± .005**	.781 ± .013	.891 ± .009	.893 ± .003
8	.916 ± .018	**.944 ± .014**	.828 ± .012	.729 ± .021	.857 ± .010	.852 ± .013	**.933 ± .010**	.728 ± .020	.710 ± .014	.731 ± .009
9	.950 ± .009	**.968 ± .014**	.798 ± .019	.872 ± .013	.947 ± .009	.908 ± .011	**.959 ± .005**	.734 ± .012	.858 ± .007	.862 ± .005
Fashion-MNIST										
	$s = 10$					$s = 100$				
Class	*Deep-SVDD*	*Deep-UAD*	DCAE	IF	KNN	*Deep-SVDD*	*Deep-UAD*	DCAE	IF	KNN
0	.868 ± .016	**.929 ± .021**	.793 ± .014	.909 ± .010	.902 ± .010	.770 ± .031	.866 ± .011	.748 ± .012	**.886 ± .008**	.808 ± .006
1	.975 ± .004	**.991 ± .002**	.934 ± .014	.977 ± .003	.987 ± .002	.956 ± .003	**.985 ± .002**	.902 ± .022	.976 ± .002	.967 ± .002
2	.832 ± .016	**.911 ± .013**	.691 ± .021	.873 ± .015	.882 ± .011	.757 ± .013	.828 ± .029	.581 ± .025	**.842 ± .009**	.779 ± .006
3	.923 ± .008	**.963 ± .009**	.898 ± .013	.936 ± .009	.937 ± .011	.867 ± .013	**.955 ± .010**	.858 ± .008	.936 ± .003	.838 ± .004
4	.894 ± .015	**.963 ± .016**	.852 ± .019	.911 ± .014	.888 ± .013	.827 ± .029	**.918 ± .006**	.780 ± .029	.903 ± .005	.807 ± .013
5	.817 ± .025	**.966 ± .013**	.373 ± .041	.928 ± .008	.846 ± .017	.647 ± .041	.801 ± .038	.235 ± .022	**.907 ± .005**	.611 ± .011
6	.756 ± .020	**.874 ± .022**	.619 ± .020	.812 ± .013	.813 ± .012	.691 ± .021	**.781 ± .017**	.537 ± .031	.778 ± .009	.710 ± .005
7	.978 ± .004	**.996 ± .005**	.905 ± .007	.980 ± .007	.979 ± .004	.927 ± .014	**.984 ± .002**	.807 ± .016	.978 ± .002	.915 ± .005
8	.893 ± .015	**.920 ± .018**	.778 ± .016	.886 ± .017	.817 ± .021	.733 ± .025	**.854 ± .017**	.644 ± .025	.822 ± .002	.512 ± .007
9	.980 ± .004	**.991 ± .005**	.970 ± .010	.978 ± .006	.945 ± .013	.917 ± .016	**.983 ± .003**	.912 ± .014	.968 ± .004	.762 ± .014

every iteration, thus *Deep-UAD* succeeds in mapping the anomalies away from the center better than *Deep-SVDD*, when this information is missing and several anomalies are not detected being closer to the center than some normal samples with consequent worsening of the AUC.

Comparison with Competitors. Finally, in this last section, we compare the results *Deep-UAD* with competitors on MNIST and Fashion-MNIST. The methods taken into account are Isolation Forest (IF) and k-Nearest Neighbor as shallow algorithms and *Deep-SVDD* and Deep Convolutional auto encoder (DCAE) as deep learning methods. To ensure a fair comparison, both *Deep-SVDD* and DCAE have the same structure of *Deep-UAD* and for both of them, as well as for our method, we fix $k = 64$ according to the results of the first experiment.

In Table 1 are reported the results for both datasets with $s = 10$ and $s = 100$. We can see that for almost all classes *Deep-UAD* performs better than all considered competitors and in certain cases the differences with their performances are huge. In particular, in the direct comparison with *Deep-SVDD*, the technique that inspires our method, *Deep-UAD* is always winning, meaning that the cooperative work of the network and the AE succeeds in improving the ability of isolating anomalies.

5 Conclusions

In this work is presented *Deep-UAD*, a deep learning approach for unsupervised anomaly detection. It is based on an alternate and cooperative training of an

AE and a neural network aiming at mapping the data close to a fixed center in the output space. Experimental results show that *Deep-UAD* achieves good performances and that the strategy of alternate training brings benefits to both the neural network and the AE improving their capabilities to isolate anomalies.

In the future our main goals are to investigate the application of a cooperative alternate strategy similar to this one to more complex neural architectures, to study possible modifications to the discussed method that may help in improving performances, and to test our algorithm on dataset of different size and nature.

References

1. Aggarwal, C.C.: Outlier Analysis. Springer, Cham (2013). https://doi.org/10.1007/978-3-319-47578-3
2. Akcay, S., Atapour-Abarghouei, A., Breckon, T.P.: GANomaly: semi-supervised anomaly detection via adversarial training (2018)
3. An, J., Cho, S.: Variational autoencoder based anomaly detection using reconstruction probability. Technical Report 3, SNU Data Mining Center (2015)
4. Angiulli, F.: Concentration free outlier detection. In: European Conference on Machine Learning and Knowledge Discovery in Databases, Skopje, Macedonia (2017)
5. Angiulli, F.: CFOF: a concentration free measure for anomaly detection. ACM Trans. Knowl. Discov. Data (TKDD) **14**(1), 4:1-4:53 (2020)
6. Angiulli, F., Basta, S., Pizzuti, C.: Distance-based detection and prediction of outliers. IEEE Trans. Knowl. Data Eng. **2**(18), 145–160 (2006)
7. Angiulli, F., Fassetti, F., Ferragina, L.: Improving deep unsupervised anomaly detection by exploiting VAE latent space distribution. In: Discovery Science (2020)
8. Angiulli, F., Fassetti, F., Ferragina, L.: Latent*Out*: an unsupervised deep anomaly detection approach exploiting latent space distribution. Machine Learning (2022). https://doi.org/10.1007/s10994-022-06153-4
9. Angiulli, F., Pizzuti, C.: Fast outlier detection in large high-dimensional data sets. In: Principles of Data Mining and Knowledge Discovery (PKDD) (2002)
10. Angiulli, F., Pizzuti, C.: Outlier mining in large high-dimensional data sets. IEEE Trans. Knowl. Data Eng. **2**(17), 203–215 (2005)
11. Barnett, V., Lewis, T.: Outliers in Statistical Data. Wiley (1994)
12. Breunig, M.M., Kriegel, H., Ng, R., Sander, J.: LOF: identifying density-based local outliers. In: Proceedings of the International Conference on Managment of Data (SIGMOD) (2000)
13. Chalapathy, R., Chawla, S.: Deep learning for anomaly detection: a survey (2019)
14. Chandola, V., Banerjee, A., Kumar, V.: Anomaly detection: a survey. ACM Comput. Surv. **41**(3), 1–15 (2009)
15. Davies, L., Gather, U.: The identification of multiple outliers. J. Am. Statist. Assoc. **88**, 782–792 (1993)
16. Donahue, J., Krähenbühl, P., Darrell, T.: Adversarial feature learning (2017)
17. Goodfellow, I., Bengio, Y., Courville, A.: Deep Learning. MIT Press, Cambridge (2016)
18. Goodfellow, I., et al.: Generative adversarial nets. In: Advances in Neural Information Processing Systems, vol. 27 (2014)
19. Hautamäki, V., Kärkkäinen, I., Fränti, P.: Outlier detection using k-nearest neighbour graph. In: ICPR, Cambridge, UK (2004)

20. Hawkins, S., He, H., Williams, G., Baxter, R.: Outlier detection using replicator neural networks. In: International Conference on Data Warehousing and Knowledge Discovery (DAWAK), pp. 170–180 (2002)
21. Jin, W., Tung, A., Han, J.: Mining top-n local outliers in large databases. In: Proceedings of the ACM SIGKDD International Conference on Knowledge Discovery and Data Mining (KDD) (2001)
22. Kawachi, Y., Koizumi, Y., Harada, N.: Complementary set variational autoencoder for supervised anomaly detection. In: IEEE International Conference on Acoustics, Speech and Signal Processing (ICASSP), pp. 2366–2370 (2018)
23. Kingma, D.P., Welling, M.: Auto-encoding variational bayes (2013)
24. Knorr, E., Ng, R., Tucakov, V.: Distance-based outlier: algorithms and applications. VLDB J. **8**(3–4), 237–253 (2000)
25. Radovanović, M., Nanopoulos, A., Ivanović, M.: Reverse nearest neighbors in unsupervised distance-based outlier detection. IEEE Trans. Knowl. Data Eng. **27**(5), 1369–1382 (2015)
26. Ruff, L., et al.: Deep one-class classification. In: Proceedings of the 35th ICML, Stockholm, Sweden (2018)
27. Ruff, L., et al.: A unifying review of deep and shallow anomaly detection. Proc. IEEE **109**(5), 756–795 (2021)
28. Ruff, L., et al.: Deep semi-supervised anomaly detection. In: 8th ICLR, Addis Ababa, Ethiopia. OpenReview.net (2020)
29. Schlegl, T., Seeböck, P., Waldstein, S., Langs, G., Schmidt-Erfurth, U.: f-AnoGAN: fast unsupervised anomaly detection with generative adversarial networks. In: Medical Image Analysis 54 (2019)
30. Schölkopf, B., Platt, J.C., Shawe-Taylor, J., Smola, A.J., Williamson, R.C.: Estimating the support of a high-dimensional distribution. Neural Comput. **13**(7), 1443–1471 (2001)
31. Sáinchez-Martín, P., Olmos, P.M., Perez-Cruz, F.: Improved BIGAN training with marginal likelihood equalization (2020)
32. Sun, J., Wang, X., Xiong, N., Shao, J.: Learning sparse representation with variational auto-encoder for anomaly detection. IEEE Access **6**, 33353–33361 (2018)
33. Tax, D.M.J., Duin, R.P.W.: Support vector data description. Mach. Learn. **54**, 45–66 (2004). https://doi.org/10.1023/B:MACH.0000008084.60811.49
34. Zenati, H., Foo, C.S., Lecouat, B., Manek, G., Chandrasekhar, V.R.: Efficient GAN-based anomaly detection (2019)

On the Ranking of Variable Length Discords Through a Hybrid Outlier Detection Approach

Hussein El Khansa(✉), Carmen Gervet, and Audrey Brouillet

Espace-Dev, Univ. Montpellier, IRD, U.Guyane, U.Reunion, Montpellier, France
hussein.elkhansa@ird.fr

Abstract. In this paper we are interested in identifying insightful changes in climate observations series, through outlier detection techniques. Discords are outliers that cover a certain length instead of being a single point in the time series. The choice of the length can be critical, leading to works on computing variable length discords. This increases the number of discords, with potential overlapping, subsumption and reduced insightful results. In this work we introduce a hybrid approach to rank variable length discords and extract the most prominent ones, that can yield more impactful results. We propose a ranking function over extracted variable length discords that accounts for contained point anomalies. We investigate the combination of pattern wise anomaly detection, through the Matrix Profile paradigm, with two different point wise anomaly detectors. We experimented with MAD and PROPHET algorithms based on different concepts to extract point anomalies. We tested our approach on climate observations, representing monthly runoff time series between 1902 and 2005 over the West African region. Experimental results indicate that PROPHET combined with the Matrix Profile method, yields more qualitative rankings, through an extraction of higher values of extreme events within the variable length discords.

Keywords: Matrix profile · Prominent discord discovery · Point and patter outlier detection · Time series · Climate data

1 Introduction

With the advance of computer modeling and data collection in many scientific fields, enormous numbers of data are being generated and collected, to take benefit of the data analyses algorithms, machine learning and neural network are used to study and learn from that it. There are two main types of learning, the fist one is supervised, which utilizes labeled, contrary to the unsupervised learning that does not utilize labeled data. One issue when using supervised learning that labeled data is hard to obtain as it needs expert input, with the rate of data generated and collected it is near impossible to label the data to address

P. Pascal and D. Ienco (Eds.): DS 2022, LNAI 13601, pp. 329–344, 2022.
https://doi.org/10.1007/978-3-031-18840-4_24

this issue, Many approaches are being developed to label data set including text data set [19], image data set [24] and time series [25].

Outlier detection is a field of machine learning to analyze data, covering numerous application domains including fraud detection, insurance [9], medical [33], internet of things [23], cyber security and hydrology [26]. Overall, an outlier can be generally defined as an observation that is significantly dissimilar to other data observation or an observation that does not behave like the expected typical behavior of the other observations. An outlier detection method can be applied to multiple data types such as images, transaction data, sequence data including genomics and time series. There are two main types of anomaly detection, point wise and pattern also called collective anomalies or discords [3].

Discords denote the most unusual time series sub sequences, and are detected using similarity measures that compare sub sequences with each other for a given length. Existing approaches can be either approximate (e.g. HOT-SAX [11], QUICK MOTIF [15], Rare Rule Anomaly [22]) or exact like the matrix profile (MP) paradigm [30], used to detect motifs or discord. The choice of the sub sequence length is critical, set as input parameter and often specified with experts' knowledge which clearly influences the outcome. This observation lead more recently to works on computing variable length discords, over a given length interval (e.g. GraphAn [2], PanMatrix [17]). The increased usage of these approaches has also raised their weaknesses. Indeed, the number of extracted discords is larger, with potential overlapping and subsumption among them, leading to the question of "how can we select the more relevant or actionable ones?". We are not aware of approaches seeking a ranking of variable lengths discords to this date.

In this work, we address this issue in the field of climate data analysis [12]. Climate impact models provide time series based both on historical data and projected data generated from complex simulations of physical processes. The impact models concern many fields, including yields of agricultural crops, biodiversity, runoff of water infiltration. Such models are commonly analyzed using multi-model ensembles means to mitigate uncertainties. With a primary goal of estimating future trends, the use of outlier detection methods has seen limited interest in this field to this date, with some works on hydrological impact models [29,31,31]. Another particularity of these models is the very large scale of the spatio-temporal data series, covering daily or monthly data over 150 years for each spatial point in a given region.

Our objective is to investigate a ranking approach over variable length discords, by integrating two unsupervised anomaly detection frameworks, point anomaly detectors and pattern based anomaly detection using MP. The intuition behind our work, is that a discord will dominate another one, if proportionally to its length, it also contains anomalous points that are greater in numbers and extreme values within the discord sequence. In our work, we compare two types of point outlier detectors, model and predictor based and draw conclusions on their insights to rank variable length discords. We consider respectively the MAD and PROPHET methods. We also define and specify a ranking function over variable length discords. MAD is a general point anomaly detection method, based of how far a point is from the median, while PROPHET is more tailored to time series and detects anomaly points that diverge from the expected values with respect to the projected seasonality and trends derived in the learning phase of PROPHET. Our approach is evaluated on total (surface + subsurface) runoff observations, that reflect the soil water levels.

Our main contributions are: 1) a hybrid approach to rank variable length discords, through the combination of exact discord method with point detectors and a ranking function; and 2) the evaluation through an experimental study over large scale climate impact data, showing the differences among point anomalous detectors and their added value as an effective means to rank variable discords.

This paper is organized as follows. Section 2 gives a background on the concepts and methods we use, Sect. 3 presents our approach. We conclude in Sect. 4.

2 Background

In this section we review existing concepts and approaches we make use of, relative to outlier detection in time series, and review the use of point outliers to extract pattern outliers.

Point outliers over time series can be specified as global outliers when compared to all values in the time series, and local outliers when compared to its neighboring points. Subsequence or pattern outliers in time series, extract collective consecutive points that behave unusually compared to other subsequences of similar length. In such cases a single point in the sequence might or not be also an anomaly on its own.

Methods for point outlier detection can be classified as model outlier detectors or predictor based outlier detectors. Model or statistical outlier detectors derive a probabilistic model which captures the distribution of the time series. If an instance in the time series has a low probability of belonging to the estimated model, it is flagged as an outlier [4]. Examples of model based point outlier detectors are Median Absolute Deviation (MAD) [10] and Minimum Covariance Determinant (MCD) [20].

Predictor based outlier detectors proceed differently. They construct a predictor (like forecast model or a regressor) to learn the normal behavior of the time series, and then exploits the learned information to predict future values. Prediction errors which are predictions that significantly deviate from the true value are labeled as point outliers [14]. Many predictor methods have been developed including ARIMA [21], and SVM [16] and PROPHET [27].

In this work, we will apply one point outlier from each type to extract point outliers within computed variable length discords, as a means to rank those discords. Note that point outliers have also been used to directly segment data series and thus extract discords. We briefly survey these approaches and their suitability for our goal.

2.1 Point Outliers Detectors and Series Segmentation

In [6] the authors proposed IForest-based anomaly detection to detect drift in streaming data, where a fixed sliding window is applied, and in each window the IForest detector is applied to generate the anomaly rate by averaging the instance depth of the forest, if the anomaly rate is smaller than the threshold then their is a drift. Also, in [5] the authors proposed sliding-window convolutional variational autoencoder (SWCVAE) an algorithm to detect point anomalies in robot, by applying sliding windows. Each window is fed to a convolutional variational autoencoder, a trained artificial neural network(ANN), to detect point wise anomaly. The ANN will give each instance in the sliding window an anomaly score, then these scores are averaged to calculate the anomaly score of the discord. If an anomaly score is above the predefined threshold then it is classified as a discord. In [32] the author combines both the result of SAX as a pattern outlier detection and combined with point outlier detection second-order-difference (SOD) and Chebyshev Inequality (CI)-based methods, where discord that does not contain point anomaly detected by SOD and CI are discarded. Those approaches contribute both ranking mechanisms of discords through a common concept, that is the segmentation of the time series and the use of a point outlier detector on each subsequence to give it an anomaly score.

With respect to our goal of comparing variable length discords over large time series, these approaches offer scoring mechanisms for subsequences, but are applied to fixed size windows, and go through the whole data set. This requires the application of outlier detectors over very large sets of sliding windows, without considering variable length windows.

In our approach, we purposely focus on variable length discords, and separate the issue of computing variable length discords, through an exact dedicated approach, with that of exploiting point anomaly detectors for ranking purpose. Part of our goal is to analyse the contribution and added value of point anomaly detectors to rank the discords of variable lengths.

2.2 Matrix Profile, MAD and PROPHET

In this subsection we provide the necessary background on the three methods used in our work, namely the STOMP algorithm to compute Matrix Profiles and extract discords, and the point anomaly detectors MAD and PROPHET.

Matrix Profile Methods. Matrix Profile is a powerful data structure that stores the z-normalized Euclidean distances between each subsequence and its closest neighbor sequence. Two main algorithms have been defined to compute the exact solution of the matrix profile by performing all-pairs-similarity-search on time series, namely STAMP [30] and STOMP [34]. We adopted STOMP for its runtime and scalability over large time series. STOMP computes the matrix profile by calculating the distance profile of every subsequent in time series T and then selecting the minimum value in each distance profile. It uses a similarity search algorithm to compute the matrix profile faster in time complexity of

$O(n^2)$. The produced Matrix can be used to detect discords, where the highest K subsequences in the Matrix Profile are the top K discords, specifying the highest distance of these subsequences with its closest neighbor. The concepts used in this work are recalled here after, relative to discords and the Matrix Profile data structure.

Definition 1. *A time series T is a sequence of real-valued numbers $t_i : T = [t_1, t_2, ..., t_n]$ where n is the length of T.*

Definition 2 (Matrix Profile). *A matrix profile P_m of time series T and given length m is a meta series of the Euclidean distances vector between each subsequence $T_{i,m}$ of given length m where i varies, and its nearest neighbor (closest match) in time series T, together with the corresponding position vector for each closest neighbor associated with $min(D_{i,m})$. We denote it $P_m = [min(D_{1,m}), ..., min(D_{n-m+1,m})]$, where $D_{i,m}(1 \leq i \leq n - m + 1)$ is the distance profile $D_{i,m}$ of time series T for subsequences of length m.*

Definition 3 (Discord). *The discord denoted $\Delta_{j,m}$ is a subsequence $T_{j,m}$ of length m starting at the position j, that has the largest distance to its nearest neighbor. In matrix profile the largest distance corresponds to the maximum distance value in P_m.*

MAD. The Median Absolute Deviation outlier detection method is a point outlier detector that uses the Absolute deviation around the median to measure the distance between a data point and the median [10]. The median is considered a tendency measurement of the data similar to the mean, but unlike the mean it is more robust to detect relevant point outliers in the data, being immune to noise, and to the sample size [13].

Given a time series $T = x_1, x_1, ..., x_i, ..., x_n$ such that i is the time unit, MAD first calculates the median of T, this median is denoted by x̃, and its absolute deviation, by subtracting $\tilde{x}$ from each x in T:

$$MAD = median(|x_i - \tilde{x}|) \tag{1}$$

For each x_i we calculate their modified z-score using the predefined formula [28]:

$$M_i = \frac{0.6745(x_i - \tilde{x})}{MAD} \tag{2}$$

In statistics, the z-score for each x_i measures the difference between the value and the mean how many standard deviations away a value is from the mean in terms of standard deviation. MAD uses the following formula to calculate a z-score:

$$Z - Score = \frac{x_i - \mu}{\sigma} \tag{3}$$

where: μ is the mean value of the data set, σ is the standard deviation of the data set where if the Z-score is 0 then it means the value of x_i is equal to the mean, if z-score is 2 then it is 2 standard deviation $(2 * \sigma)$ away from the mean.

If the absolute value of the modified z-core of x_i is larger than a threshold then is considered as an outlier. [10] proposed the value 3.5 as a threshold, also the default one in MAD.

PROPHET. It is an open source time series forecasting model, that detects outliers based on the data learned through the forecasting component. Its purpose is to forecast future data, and detect outliers, with an uncertainty estimation. Once PROPHET fits the given time series, it generates a data set containing the following variables with an estimation of the uncertainty through a confidence interval for each value. Then for each value in the Time series, it calculates the error and uncertainty based on the confidence interval of each value. If the error of a value is bigger than its uncertainty it will be flagged as an anomaly. PROPHET is based on generalized additive model (GAM), which means that the model is the sum of several components summarized below:

$$y_t = g(t) + s(t) + h(t) + \varepsilon(t) \tag{4}$$

where:

- $g(t)$ denotes the growth function, a piece-wise linear function. The number of linear growths (straight lines), is determined by the number of change points in the time series, that can be entered manually or detected automatically by PROPHET.
$$g(t) = \left(k + \mathbf{a}(t)^\top \boldsymbol{\delta}\right) t + \left(m + \mathbf{a}(t)^\top \boldsymbol{\gamma}\right) \tag{5}$$
where k is the growth rate scalar value, m the offset value, $a(t)$ a binary vector $\in \{1, 0\}$ indicating if there is a growth rate adjustments at time t. $\boldsymbol{\delta}$ is a vector that contains the growth rate value. For instance if there is 10 change point then the vector of $\boldsymbol{\delta}$ will have 10 values.
- $\boldsymbol{\gamma}$ is vector similar too that contains the offset parameter adjustment
- $s(t)$ the seasonality, modeled using Fourier Series as a function of time.
- $h(t)$ holiday seasons parameter, that we will not consider
- $\varepsilon(t)$ the error estimation using a maximum a posteriori estimation.

Not all components are mandatory, and in our case since we will be experimenting on climate impact data. $h(t)$ not considered for instance.

3 Our Approach

Our work is first motivated by the insightful ranking of variable length discords, to extract those that are combined with more eventful changes and potential impact on forecasted data trends. We do so by combining an exact, robust pattern outlier detection approach, with two different point outlier detectors. The exact approach ensures reliable extraction of discords, while the point outlier detectors will contribute to their ranking. We use the Matrix Profile method, STOMP as an exact approach to extract discords, that also offers the qualities of being parameter-free and that does not utilize a similarity or distance threshold. This is important since the point outlier detectors do not have these advantages, and an increase in thresholds tuning would reduce the robustness of the overall results. With respect to the point outlier detectors, we experimented with the generic MAD model approach, and the predictor PROPHET method that also handles trends and seasonality.

Our Algorithm 1 is based on four 4 main phases :

1. (Lines 8–12) Extract variable length discords (Matrix Profile method)
2. (Line 13) Detect point outliers (using PROPHET, MAD) over the time series
3. (Lines 14–20) Score the discords based on the anomalous points they contain
4. (Lines 21–25) Filter and sort the discords with a ranking function

Extract Variable Lengths Discords. We compute the Matrix Profile for variable window sizes j over T, where $4 < j < length(T)/2$. For each window size, we extract the top discord. The output is a list of top discords, one per length.

Detect Point Outliers in the Time Series. This step seeks point wise outliers (through PROPHET and MAD), over the whole time series T. The outcome is a Boolean list denoting for each data in the time series if it is an anomaly.

Scoring the Discords. This step integrates the variable length discords and Boolean anomaly list for scoring purpose. For each discord we count the anomalous points that are included in each discord $\Delta_{i,j}$. Clearly a simple count is not the only element in the scoring, and the discords' length needs to be taken into account. Note that some discords can overlap, thus share anomalous points. To address this issue we define a scoring function as the ratio of the count over each length.

Filtering and Sorting. This procedure filters the discords based on their count, as well as their starting dates, and length. Based on our past work and analysis [7] of subsumption among variable length discords, it is important to compare variable length discords with identical starting dates. The filtering stage groups discords sharing a starting date i and for each group selects the discord with highest ratio, meaning it subsumes the others. Then the discords (of variable length) are sorted in descending order of their ratio value.

Algorithm 1: matrix prophet ensemble

```
   input : Time Series T
   output: list <Discords> SortedList
 1 initialization
 2 int m = length(T)/2
 3 list < Δ > top_discord_list=[ ]                      // List to store top discords
 4 list < Δ > group_list=[ ]          // List to group discords with similar date
 5 list < bool > anomaly_list=[ ]     // Boolean list of anomaly tag per value
 6 list < Δ > sorted_list=[ ]                              // List of ranked discords
 7 list < int > Windows = [4, 5, 6, 7, 8, ..., m];          // List of Window sizes
 8 foreach j in Windows do
 9  |  P_j ← STOMP (T, j)                         // Matrix profile for window size j
10  |  Δ_{i,j} ← max(P_j)                                         // Discord of length j
11  |  top_discord_list ← top_discord_list.add(Δ_{i,j})
12 end
13 anomaly_list ← PROPHET or MAD(T)                     // Store output of detector
14 foreach Δ_{i,j} in top_discord_list do
15  |  start=Δ_{i,j}.i                                // Store the start date of discord
16  |  end=Δ_{i,j}.i + Δ_{i,j}.j                  // Store the end date of the discord
17  |  length=Δ_{i,j}.j                                       //Store the discord length
18  |  int count← count_Anomaly(T,start,end)//Anomaly count per Δ_{i,j}
19  |  Δ_{i,j}.ratio ← count/length
20 end
21 group_list ← groubBy(top_discord_list)   // Group discord list per similar
     starting date
22 foreach group in group_list do
23  |  sorted_list ← add(max(group) ) // Get discord with highest ratio in
    |    each group
24 end
25 sorted_list ← Quicksort(sorted_list,Δ.ratio)
26 return sorted_list
```

3.1 Experimental Comparison of MAD and PROPHET

Climate Data. In our experimental study we used observed monthly runoff data obtained from Global Runoff Reconstruction dataset (GRUN) [8]. In climate and corresponding impact data science, runoff is an impact variable that can be used to quantify flood and drought risks at regional and global scales [1]. It is usually provided in kg/m2/s or in mm/day. The source of the data has a $0.5^\circ \times 0.5^\circ$ spatial grid resolution with a focus on the Sahel region. This data gives indicators on the soil water content over the time period between 1902 and 2005 (i.e. 104 years, 1248 months). This is a standard observation period in historical climate analysis. We spatially average the monthly runoff data over the grid box [5°W-25°E ; 10°N-18°N].

Point Anomaly Detectors. A first comparative study between MAD and PROPHET on the runoff data set is shown in Fig. 1. Those point anomaly detectors use different methods to extract anomalies, and we can see that some anomalous points are shared and many differ. Over the whole data series, MAD extracts a large set of anomalies (green), essentially in the upper part of the data, thus detecting high water levels. On the other hand, PROPHET extracts fewer anomalies, some shared with MAD for the highest values of anomalous water levels, and most importantly detected the drought periods in terms of anomalous points (purple crosses), in 1900–1910 and 1970–1990. This can be explained by the fact that PROPHET through the learning phase was able to identify a normality within seasons, whereas MAD focuses on extremes in the overall data series.

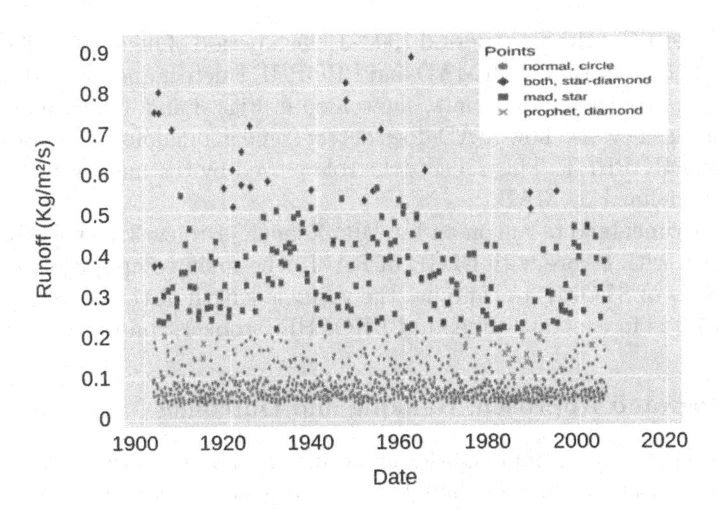

Fig. 1. Anomaly points extracted by MAD and PROPHET: similarities and differences

If we zoom on the period 1981–1984 where 1983 is recorded as one of the driest years in the West African region, this aspect is striking. MAD does not extract any anomalous point, whereas PROPHET detects three anomalies displayed in Fig. 2. By not accounting for, or detecting, seasonality MAD misses important point outliers in such time series. On the other hand, PROPHET takes into account seasonality through the learning phase, and we can see that it extracts anomalies with respect to months which are learnt to be dry and those that should not be. We note that the anomalous points do not necessarily reflect extreme values overall, but anomalous points with respect to the anticipated values (learnt during the forecasting phase).

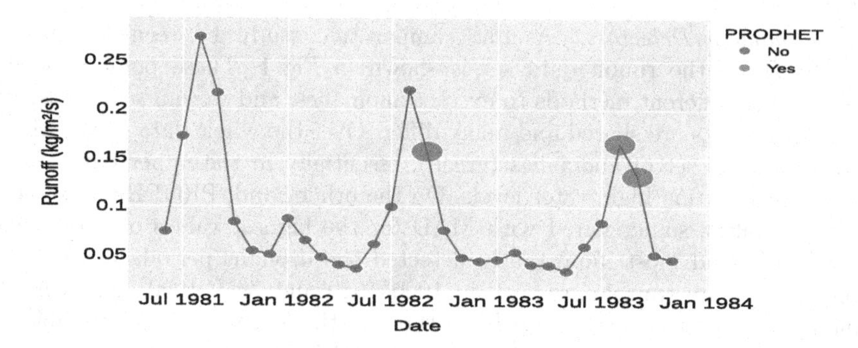

Fig. 2. Anomalous points detected by PROPHET in the subsequence over 1981–1984

We also studied the early period 1902–1905, a period of higher runoff values in average. We can see that both MAD and PROPHET detect anomalies during this period, with some common points, showcased in Figs. 3 and 4. Those figures are interesting as they also how MAD does detect some anomalous points not tagged as such by PROPHET. This reflects the role played by the median function and the z-score defined by MAD.

All experiments were run on an Intel(R) Xeon(R) Bronze 3106 CPU processor at 1.70 GHz with 8 core with 64 GB of RAM. The main computational cost are the N/2 runs of STOMP to compute the variable length discords, taking around 9 min. MAD is in constant time, and PROPHET took 1.5 min.

3.2 Integrated Approach: Ranking and Outcomes

Having compared the commonalities and differences between two point anomaly detectors, we saw that even though they are generic, they can have a great impact on the outliers' detection. This is related to their models and potentially the nature of the data. We now present the results of our methodology. Both

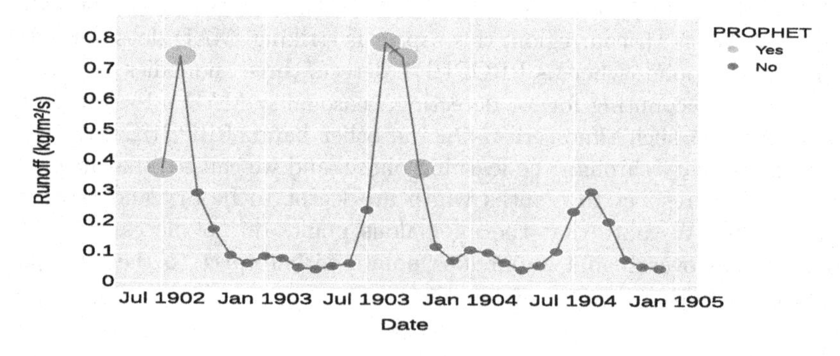

Fig. 3. Anomalous points from PROPHET in the subsequence 07/1902–01/1905

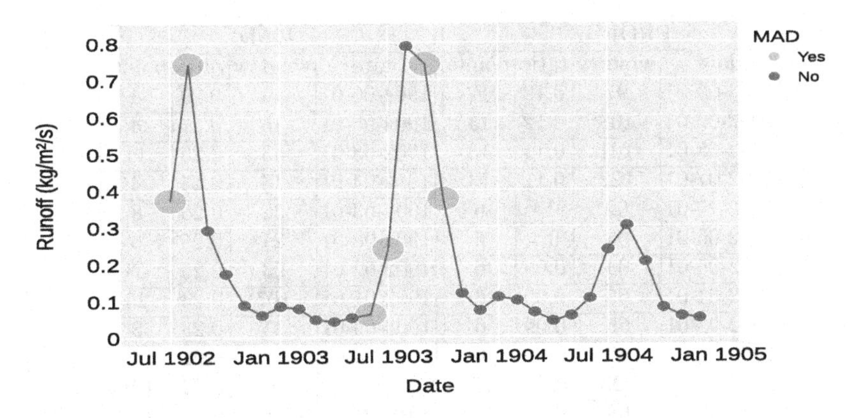

Fig. 4. Anomalous points detected by MAD in the subsequence 07/1902–01/1905

detectors are applied and compared as a means to rank the variable lengths discords produced.

Figure 5 represents respectively for PROPHET and MAD, the top discords ranking with respect to our ratio function. Note that the discords of variable lengths are specified by their starting date, window length and the value of the ratio function. Note that since we are comparing discords of variable lengths, at this stage we only extract the top discord for each given length m, to reduce the combinatorics of our approach. Extracting top k discords per length is part of future work. We recall that the ratio values depend on the length and the number of anomalous points within each discord. In this figure we picked the top 20 discords relative to each detector output ranking. The discords highlighted in yellow, correspond to a filtering, that extracts among those discords the top ones outside a proximity temporal interval of $\pm$ three months. For instance for the year 1982 we have 3 ranked discords starting from July till September with overlapping periods. We kept the one with highest ranking to reduce redundant discord periods. Thus the colored row in the Table 5 represent the top 5 discords resulting from the filtering procedure.

PROPHET				MAD			
date	window	ratio	points	date	window	ratio	points
1983-08-01	97	0.12	12	1904-06-01	14	0.29	4
1982-09-01	108	0.12	13	1904-05-01	15	0.27	4
1982-08-01	121	0.11	13	1902-03-01	43	0.26	11
1982-07-01	122	0.11	13	1904-04-01	16	0.25	4
1902-07-01	62	0.1	6	1904-03-01	32	0.25	8
1902-06-01	63	0.1	6	1904-08-01	24	0.25	6
1902-05-01	64	0.09	6	1902-02-01	93	0.25	23
1902-04-01	65	0.09	6	1902-01-01	107	0.23	25
1902-03-01	66	0.09	6	1903-09-01	35	0.23	8
1902-02-01	67	0.09	6	1927-04-01	623	0.22	140
1984-08-01	123	0.09	11	1927-09-01	617	0.22	138
1902-01-01	68	0.09	6	1912-07-01	591	0.22	132
1984-07-01	125	0.09	11	1927-12-01	614	0.22	137
1984-06-01	129	0.09	11	1912-05-01	592	0.22	132
1984-05-01	130	0.08	11	1927-11-01	615	0.22	137
1984-04-01	131	0.08	11	1912-08-01	589	0.22	131
1983-09-01	143	0.08	12	1927-10-01	616	0.22	137
1903-10-01	12	0.08	1	1904-02-01	18	0.22	4
1983-07-01	158	0.08	13	1903-10-01	36	0.22	8

Fig. 5. Ranking table of top 20 discords

Analysis of the Results:

1. *Point anomalous detectors do matter.* While STOMP produces exact matrix profiles to derive variable length discords, the rankings differ between PROPHET and MAD. By detecting the anomalies in the driest period, where droughts occurred (1982–1984), PROPHET manages to rank top discords in this period. We can see that the variable length are not negligible ranging in this period from 97 months to 123 months. MAD detects mainly the periods with high water levels (1902–1904), yielding high ranking for the corresponding discords. It corresponds to the periods with higher soil water levels and larger inter-annual variability, that comes before the later continuous long-term drying trends observed within Sahel [7]. PROPHET extracted and ranked the 1st, 2nd and 11th discords, that represent the intense droughts that occurred in Sahel, while the 3rd and 19th discords represent the period where the high soil water levels and larger inter-annual variability.
2. *Rankings are insightful.* We can see in the table that the discords vary greatly in length and also that the number of anomalous points extracted in those are not proportional to the length. Thus the integration of exact variable length discords with different anomalous point detectors does bring complementary information and the scoring function is adequate. This information was able to extract the prominent discords over the runoff data that reflected both the driest periods, and periods with highest soil water levels.

3. *Climate-related impact insights.* Those results are compared to basic statistical analysis of runoff data to demonstrate the thematic insights brought by combining anomalous point detection and discord methods (Fig. 6). The first rank of combined PROPHET and discord methods detects an anomalous event/period from August 1983 for 97 months with 12 anomalous points (Table 5, Fig. 2). Following discords 2, 3 and 4 exhibit larger time windows but include this first top discord. This is consistent with well known intense droughts that occurred in Sahel between 1982 and 1987 (Fig. 6a, b), with the most severe drought ever recorded within the African continent in 1983–1984 (e.g. [18]). Moreover, following ranks of combined PROPHET and discords emphasize top discords starting in 1902 for at least 60 months (Table 5). The first ranked of combined MAD and discord methods also detects "anomalies" in 1904 for 14 months, with following ranked ones including this first top ranked (Table 5). These results coincide with large positive anomalies both in annual maximum and annual mean in 1902 and 1903, and with very large year-to-year variability from 1902 to 1907 (Fig. 6a, c). These early 1900s s result also illustrate the well known soil drying trend resulting from a rainfall deficit observed between 1900 and 2013 within the region [8]. PROPHET detects both 1980s s dry events and 1904 abrupt year-to-year changes, and

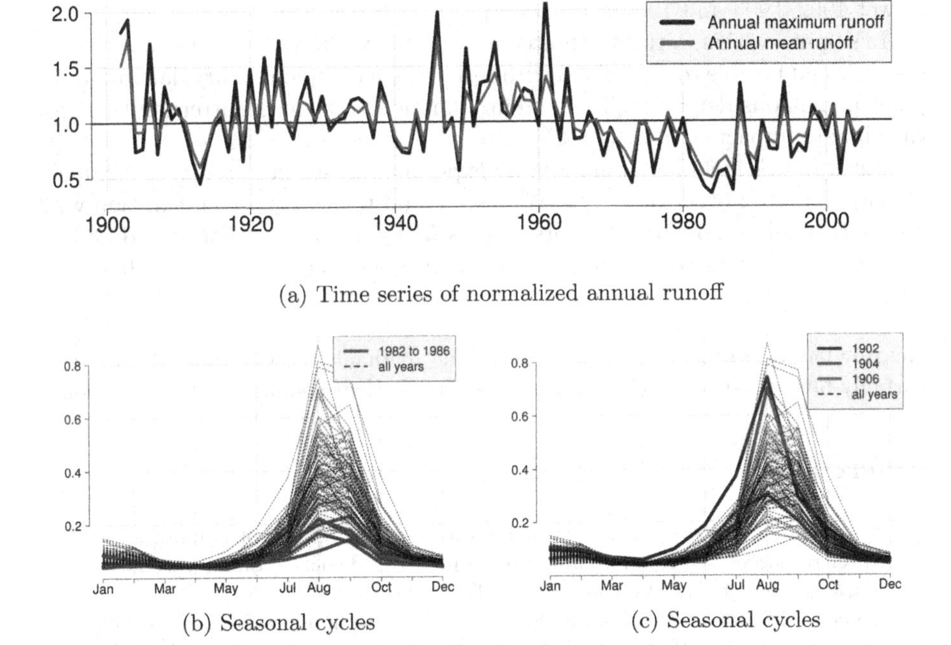

(a) Time series of normalized annual runoff

(b) Seasonal cycles

(c) Seasonal cycles

Fig. 6. (a) Time series of normalized annual maximum and mean runoff (relative to their respective mean climatologies over 1902–2005). **(b)-(c)** Seasonal cycles of absolute runoff values for each year between 1902 and 2005 (in mm/day).

it is important to note that MAD does not detect any 1980 s s dry signal in the first top 20 ranked discords (Table 5).

4 Conclusion and Future Work

In this paper, we proposed a hybrid approach to rank variable length discords that combine an exact pattern wise outlier approach with anomalous point detectors. Our methodology is generic. We chose the Matrix Profile as the pattern outlier detector because it is exact and parameter free, generated variable length discords, and used point wise anomaly detectors to give each discord a score. This score together with the discord length were used to define a ranking function.

We tested two different point anomaly detectors, a model based detector (MAD) and predictor based one (PROPHET). We run experiments on large scale Monthly runoff data of the Sahel region in West Africa from 1902 till 2005. Our results show that PROPHET is more adequate than MAD, as its top 5 ranked discords did capture the actual discords. The top 5 discords extracted with MAD were only able to reflect the periods with highest water level. They also show that PROPHET is suitable both for detecting specific events (i.e. droughts in 1982, 1983, 1984, etc.) and larger timescale anomalies (i.e. dry decade of 1980 s)s). MAD is suitable to detect abrupt inter-annual change (i.e. dry 1904/1905 relative to wet 1902/1903 and 1906 rainy seasons).

In summary, this work contributed a novel approach to exploit variable length discords and be able to rank them with respect to a complementary data analysis, point anomalous detectors. The ranking function confirmed pattern outliers for the climate impact data at hand.

Future work includes running more experimental studies on different impact climate data and other data sets. We also intend to extend our integration with alternative point anomaly detectors such as ARIMA, Random Forest and LSTM, towards a comprehensive approach, given the specificity and impact they have on the extracted point outliers.

Acknowledgements. The authors would like to thank the Occitanie Region, who partially funded this research, and the reviewers for their comments and suggestions.

References

1. Arnell, N.W., Lloyd-Hughes, B.: The global-scale impacts of climate change on water resources and flooding under new climate and socio-economic scenarios. Climatic Change **122**(1–2), 127–140 (2014)
2. Boniol, P., Palpanas, T., Meftah, M., Remy, E.: Graphan: graph-based subsequence anomaly detection. Proceed. VLDB Endow. **13**(12), 2941–2944 (2020)
3. Borges, H., Akbarinia, R., Masseglia, F.: Anomaly detection in time series. In: Hameurlain, A., Tjoa, A.M. (eds.) Transactions on Large-Scale Data- and Knowledge-Centered Systems L. LNCS, vol. 12930, pp. 46–62. Springer, Heidelberg (2021). https://doi.org/10.1007/978-3-662-64553-6_3

4. Chandola, V., Banerjee, A., Kumar, V.: Outlier detection: a survey. ACM Comput. Surv. **14**, 15 (2007)
5. Chen, T., Liu, X., Xia, B., Wang, W., Lai, Y.: Unsupervised anomaly detection of industrial robots using sliding-window convolutional variational autoencoder. IEEE Access **8**, 47072–47081 (2020)
6. Ding, Z., Fei, M.: An anomaly detection approach based on isolation forest algorithm for streaming data using sliding window. IFAC Proceed. Vol. **46**(20), 12–17 (2013)
7. El Khansa, H., Gervet, C., Brouillet, A.: Prominent discord discovery with matrix profile: application to climate data insights. In: Computer Science & Technology Trends, Academy and Industry Research Collaboration Center (AIRCC) (2022)
8. Ghiggi, G., Humphrey, V., Seneviratne, S.I., Gudmundsson, L.: GRUN: an observation-based global gridded runoff dataset from 1902 to 2014. Earth Syst. Sci. Data **11**(4), 1655–1674 (2019)
9. Hansson, A., Cedervall, H.: Insurance fraud detection using unsupervised sequential anomaly detection (2022)
10. Iglewicz, B., Hoaglin, D.C.: How to detect and handle outliers, vol. 16. ASQ Press (1993)
11. Keogh, E., Lin, J., Fu, A.: Hot sax: efficiently finding the most unusual time series subsequence. In: Fifth IEEE International Conference on Data Mining (ICDM2005), p. 8. IEEE (2005)
12. Le Gall, P., Favre, A.-C., Naveau, P., Prieur, C.: Improved regional frequency analysis of rainfall data. Weather Clim. Extremes **36**, 100456 (2022)
13. Leys, C., Ley, C., Klein, O., Bernard, P., Licata, L.: Detecting outliers: do not use standard deviation around the mean, use absolute deviation around the median. J. Exp. Soc. Psychol. **49**(4), 764–766 (2013)
14. Tianyu, Li., et al.: Anomaly scoring for prediction-based anomaly detection in time series. In: 2020 IEEE Aerospace Conference, pp. 1–7. IEEE (2020)
15. Yuhong, Li, Leong, H.U., Yiu, M.L., Gong, Z.: Quick-motif: an efficient and scalable framework for exact motif discovery. In: 2015 IEEE 31st International Conference on Data Engineering, pp. 579–590. IEEE (2015)
16. Ma, J., Perkins, S.: Time-series novelty detection using one-class support vector machines. In: Proceedings of the International Joint Conference on Neural Networks, 2003, vol. 3, pp. 1741–1745. IEEE (2003)
17. Madrid, F., Imani, S., Mercer, R., Zimmerman, Z., Shakibay, N., Keogh, E.: Matrix profile xx: finding and visualizing time series motifs of all lengths using the matrix profile. In: 2019 IEEE International Conference on Big Knowledge (ICBK), pp. 175–182. IEEE (2019)
18. Masih, I., Maskey, S., Mussá, F.E.F., Trambauer, P.: A review of droughts on the African continent: a geospatial and long-term perspective. Hydrol. Earth Syst. Sci. **18**(9), 3635–3649 (2014)
19. Miller, B., Linder, F., Mebane, W.R.: Active learning approaches for labeling text: review and assessment of the performance of active learning approaches. Polit. Anal. **28**(4), 532–551 (2020)
20. Rousseeuw, P.J., Van Driessen, K.: A fast algorithm for the minimum covariance determinant estimator. Technometrics **41**(3), 212–223 (1999)
21. Sanayha, M., Vateekul, P.: Fault detection for circulating water pump using time series forecasting and outlier detection. In: 2017 9th International Conference on Knowledge and Smart Technology (KST), pp. 193–198. IEEE (2017)
22. Senin, P., et al.: Time series anomaly discovery with grammar-based compression. In: EDBT, pp. 481–492 (2015)

23. Sgueglia, A., Sorbo, A.D., Visaggio, C.A., Canfora, G.: A systematic literature review of iot time series anomaly detection solutions. Fut. Gener. Comput. Syst. **134**, 170–186 (2022)
24. Shao, Z., Yang, K., Zhou, W.: Performance evaluation of single-label and multi-label remote sensing image retrieval using a dense labeling dataset. Remote Sensing **10**(6), 964 (2018)
25. Shi, J., Yu, N., Keogh, E., Chen, H.K., Yamashita, K.: Discovering and labeling power system events in synchrophasor data with matrix profile. In: 2019 IEEE Sustainable Power and Energy Conference (iSPEC), pp. 1827–1832. IEEE (2019)
26. Siniosoglou, I., Radoglou-Grammatikis, P., Efstathopoulos, G., Fouliras, P., Sarigiannidis, P.: A unified deep learning anomaly detection and classification approach for smart grid environments. IEEE Trans. Netw. Serv. Manage. **18**(2), 1137–1151 (2021)
27. Taylor, S.J., Letham,. B.: Forecasting at scale. Am. Statist. **72**(1), 37–45 (2018)
28. Wilcox, R.R.: Fundamentals of modern statistical methods: Substantially improving power and accuracy, vol. 249, 2nd edn. Springer (2001). https://doi.org/10.1007/978-1-4419-5525-8
29. Ye, F., Liu, Z., Liu, Q., Wang, Z.: Hydrologic time series anomaly detection based on flink. Mathematical Problems in Engineering (2020)
30. Yeh, C.-C.M., et al.: Matrix profile I: all pairs similarity joins for time series: a unifying view that includes motifs, discords and shapelets. In: 2016 IEEE 16th international conference on data mining (ICDM), pp. 1317–1322. IEEE (2016)
31. Yu, Y., Zhu, Y., Li, S., Wan, D.: Time series outlier detection based on sliding window prediction. Mathematical problems in Engineering (2014)
32. Yue, M.: An integrated anomaly detection method for load forecasting data under cyberattacks. In: 2017 IEEE Power & Energy Society General Meeting, pp. 1–5. IEEE (2017)
33. Zhang, H., Guo, W., Zhang, S., Lu, H., Zhao, X.: Unsupervised Deep Anomaly Detection for Medical Images Using an Improved Adversarial Autoencoder. J. Digit. Imaging, **35**, 153–161 (2021). https://doi.org/10.1007/s10278-021-00558-8
34. Zhu, Y., et al.: Matrix profile II: exploiting a novel algorithm and GPUs to break the one hundred million barrier for time series motifs and joins. In: 2016 IEEE 16th international conference on data mining (ICDM), pp. 739–748. IEEE (2016)

Text, Ontologies and Cross-Modal Learning

TextMatcher: Cross-Attentional Neural Network to Compare Image and Text

Valentina Arrigoni(✉), Luisa Repele, and Dario Marino Saccavino

UniCredit, Milan, Italy
{valentina.arrigoni,luisa.repele,dariomarino.saccavino}@unicredit.eu

Abstract. We study a multimodal-learning problem where, given an image containing a single-line (printed or handwritten) text and a candidate text transcription, the goal is to assess whether the text represented in the image corresponds to the candidate text. This problem, which we dub *text matching*, is primarily motivated by a real industrial application scenario of automated cheque processing, whose goal is to automatically assess whether the information in a bank cheque (e.g., issue date) match the data that have been entered by the customer while depositing the cheque to an automated teller machine (ATM). The problem finds more general application in several other scenarios too, e.g., personal-identity-document processing in user-registration procedures.

We devise a machine-learning model specifically designed for the text-matching problem. The proposed model, termed *TextMatcher*, compares the two inputs by applying a novel cross-attention mechanism over the embedding representations of image and text, and it is trained in an end-to-end fashion on the desired distribution of errors to be detected. We demonstrate the effectiveness of TextMatcher on the automated-cheque-processing use case, where TextMatcher is shown to generalize well to future unseen dates, unlike existing models designed for related problems. We further assess the performance of TextMatcher on different distributions of errors on the public IAM dataset. Results attest that, compared to a naïve model and existing models for related problems, TextMatcher achieves higher performance on a variety of configurations.

Keywords: Multimodal learning · Text recognition · Text matching · Cross attention · Joint embedding learning

1 Introduction

The way we interact with the world concerns stimuli from different senses: images we see, sounds we hear, words we read. All these examples correspond to different *modalities* by which information is presented to us. The same variability can apply to data presented to a machine, such as images, free text, sounds, videos. *Multimodal learning* is an active and challenging research area, whose

P. Pascal and D. Ienco (Eds.): DS 2022, LNAI 13601, pp. 347–362, 2022.
https://doi.org/10.1007/978-3-031-18840-4_25

goal is to build machine-learning models capable of processing and exploiting information from multiple modalities [3]. It includes numerous (classes of) tasks – such as multimodal representation learning, modality translation, multimodal alignment, multimodal fusion, co-learning – and finds application in a wide range of scenarios – such as audio-visual speech recognition, image/video captioning, media description, multimedia retrieval.

In this paper, we introduce the following multimodal-learning task, which we term *text matching*: given an image representing a single line of (printed or handwritten) text and a candidate text transcription, assess whether the text inside the image corresponds to the candidate text.

Applications. The prominent application of the text-matching problem is a real industrial use case of automated cheque processing, which naturally arises in the banking domain. In this context, a customer of a bank deposits a bank cheque to an automated teller machine (ATM). While inserting the cheque into the ATM, the customer is typically required to also type (through the ATM keypad) some information that is written on the cheque, such as issue date, amount, and beneficiary. The match between what is actually written in the cheque and the data entered by the user is a-posteriori verified by back-office operators, who would clearly benefit from a decision support system that has at its core a method to perform this check automatically.

Text matching finds applications in several other real-world scenarios too, in which an image containing text is assigned the (supposedly) corresponding text, and particular kinds of mismatching must be avoided. As an example, softwares for user-registration procedures typically need to collect information regarding personal identity documents. The user is asked to provide an image of her document and also to enter data that are written in the document, such as document identifier, expiration date, and so on. Again, back-office operators later-on check if there is a match between the document and the entered data, and, based on the outcome of the match, they accept or reject the registration.

Challenges. An immediate yet naïve method to solve the text-matching task is to resort to the related well-established problem of *text recognition*, whose goal is, given an image that is assumed to contain text, to recognize and output the text therein [4]. Specifically, the idea would be to use a text-recognition method to extract the text within the input image and then simply compare the extracted text with the candidate text. This is a rather simplistic approach, as it disregards the availability of a candidate text at all. We claim that designing ad-hoc methodologies for text matching, which properly exploit the information of the candidate text and are specifically trained on the desired distribution of non-matching texts, can be more effective. This claim is experimentally confirmed, see Sect. 5 for more details.

Contributions. We tackle the text-matching problem by devising a machine-learning model that is specifically designed for it. The proposed model, dubbed *TextMatcher*, scans the input image horizontally, searching for characters of the candidate text. This is performed by projecting the input image and text into separate *embedding spaces*. Then, a novel *cross-attention mechanism* is employed, which aims to discover local alignments between the characters of the text and the vertical slices of the image. The ultimate similarity score produced by the model is a weighted cosine similarity between features of the characters and features of the slices of the image, where the weights are the computed attention scores. Such a score is eventually used to answer the original yes/no matching question via a thresholding approach.

The model is trained in an end-to-end fashion and, thanks to the cross-attention mechanism, it produces consistent embedding spaces for both image and text, and it is able to successfully specialize to specific distributions of errors. This is desirable because, depending on the application, it can be appropriate to either correct minor typos or enforce the exact spelling of every word.

Summary and Roadmap. To summarize, our main contributions are:

- We study a multimodal-learning problem termed *text matching* (Sect. 3), which finds application in a variety of real scenarios, including an industrial use case of automated cheque processing, peculiar of the banking domain.
- We devise a machine-learning model, termed *TextMatcher*, that is specifically designed for text matching and exploits a novel cross-attention mechanism (Sect. 4).
- We showcase the proposed TextMatcher in the primary application context of automated cheque processing, by carrying out experiments on a real-world (proprietary) dataset of bank cheques provided by UniCredit, a noteworthy pan-European commercial bank (Sect. 5.1).[1]
- We further test the performance of TextMatcher on the popular public IAM dataset [9] (Sect. 5.2). Results on both UniCredit and IAM datasets attest that TextMatcher achieves high accuracy and is capable of properly handling specific distributions of errors. It also consistently outperforms a naïve model and existing text-recognition methods in both those aspects.

Section 2 overviews the related literature. Section 6 concludes the paper.

2 Related Work

The problem we tackle in this work, i.e., text matching, falls into the broad area of multimodal learning. A comprehensive survey of the main challenges, problems, and methods in this area is provided by Baltruvsaitis *et al.* [3]. Referring to the taxonomy reported in that survey, the category that better complies with text matching is the *(implicit) alignment* one, which encompasses multimodal-learning problems whose goal is to identify relationships between sub-elements from different modalities, possibly as an intermediate step for another task.

To the best of our knowledge, the text-matching problem has not been specifically studied in the literature: no ad-hoc method has been designed for it so far. Nevertheless, there exist tasks/methods that share some similarities.In the remainder of this section, we overview such related works.

[1] TextMatcher has been deployed at UniCredit, and it is currently used in production.

Text Recognition. Recognizing text in images has been an active research topic for decades. A plethora of different approaches exist. A prominent state-of-the-art text-recognition model, which we take as a reference in this work, is ASTER [15,16], i.e., an end-to-end neural network that is based on an attentional sequence-to-sequence model to predict a character sequence directly from the input image. For more approaches and details on text recognition, we refer to comprehensive Chen *et al.*'s survey [4].

The main difference between text recognition and our text-matching problem is that the former extracts text from images *without relying on any input candidate text*. A naïve approach to text matching would be to run a text-recognition method on the input image, and using the input candidate text only to check the correspondence with the recognized text. A major limitation of this approach is that it disregards the candidate text at all, thus resulting intuitively less effective than approaches that, like the proposed TextMatcher, are specifically designed for text matching and profitably exploit the candidate text and the given distribution of errors to be recognized. More specifically, the technical strengths of the proposed TextMatcher method over a text-recognition-based approach are:

- While the text-recognition model is trained only on the matching data, TextMatcher is trained with both positive and negative examples, allowing it to better learn the frontier between the two sets whenever it is relevant, for instance when a difference of a single character in a specific position is important for a large portion of data (e.g. "*MR Smith*" vs. "*MS Smith*").
- More importantly, we experimented that the TextMatcher model better generalises to different distributions at inference time thanks to the training through negative matching pairs (see Sect. 5.1).
- If the text-recognition model uses an encoder-decoder architecture (like [16]), the corresponding text-matching model needs only the encoder part, therefore it tends to be faster during inference.

Word Spotting. Given a collection of images representing single words and a query text, word spotting aims at ranking all the images of the word-image collection based on their similarity to the given query [13]. The variant of word spotting where the query is a text string (and not an image), termed *Query-by-String* (QbS), is the one more relevant for our work. Existing approaches to QbS word spotting aim at learning a map from textual representation to image representation. A popular choice for text representation in word spotting is the binary attribute representation referred to as *Pyramidal Histogram of Characters* (PHOC) [1]. Recent works use neural networks in order to learn the mapping from word images to PHOC [6,10,17,18]. For instance, Sudholt and Fink [17] propose the PHOCNet model, which applies a convolutional neural network (CNN) to word images in order to estimate a probability distribution over attributes of the PHOC representation. A more sophisticated approach is proposed by Mhiri *et al.* [10], which learns a mapping from the word images using a CNN and from the input text query with a recurrent neural network (RNN) to a common embedding space, driven by the PHOC representation. On top of the learned embeddings, a matching model is trained to refine the response of the nearest-neighbor queries. Although sharing some similarities with our approach, this method *is not trained end-to-end* together with the matching model, and compares image and text *only after the embedding vectors are produced*, differently from our cross-attention mechanism. More importantly, in general, the word-spotting task is usually designed and evaluated with the assumption that the vocabulary of words *is fixed*, which is not true in our case.

Image-text Matching is another (loosely) related task, whose goal is to measure the semantic similarity between an image and a text [8,11,12,20]. Despite similar in spirit, image-text matching is different from text matching from a conceptual point of view. The fundamental difference is that the input images to image-text matching are *general-purpose* ones, i.e., they are not constrained to represent a (single-line) text. For instance, the goal in image-text matching might be to assess whether an image depicting a dog playing with a ball is well described by the "*A dog is playing with a ball*" text. For this reason, image-text matching considers the semantic content of the image, whereas text matching looks solely at (the syntax of) the text in the image. As a result, image-text matching is typically employed in applications far away from the ones targeted by text matching (e.g., generation of text descriptions from images or image search), and existing approaches to image-text matching cannot (be easily adapted to) work for our text-matching problem. From a methodological point of view, image-text matching and text matching share more similarity, as both the problems can be approached with techniques that involve learning a shared representation for image and text. However, important technical differences still remain. Among the prominent models for image-text matching are the ones proposed in [8,11], which use a cross-attention mechanism to inspect the alignment between image regions and words in the sentence, and [20], which exploits the correlation of semantic roles with positions (those of objects in an image or words in a sentence). The proposed TextMatcher uses attention as well, but, unlike [11,20], it makes a simpler consideration of the horizontal position of a character in the image. Also, while [8,11,20] use pretrained models to generate feature representations for the image regions, our TextMatcher is trained end-to-end, thus being capable of learning the weights of the convolutional layer alongside the attention layer.

3 Text Matching Problem

We tackle a multimodal-learning problem, which we term *text matching* and define as follows: given an image containing a single-line text (printed or handwritten) written horizontally, together with a candidate text transcription, assess whether the text inside the image corresponds to the candidate text. This corresponds to a binary supervised-classification task, in which we are given a dataset of the form $\{\left((I^i, t^i),\ l^i\right) | i = 1, \ldots, n\}$, where I^i and t^i are image and text inputs of the i-th example, and l^i is the corresponding binary label. In particular, we adopt the following convention: an (*image*, *text*) pair is assigned the "1" label if *image* and *text* correspond, and, in this case, the pair is referred to as a *matching pair*. Otherwise, the pair is assigned the "0" label, and it is referred to as a *non-matching pair*. Similarly, we talk about *matching* and *non-matching* texts for a given image. An illustration of the input to text matching is in Fig. 1.

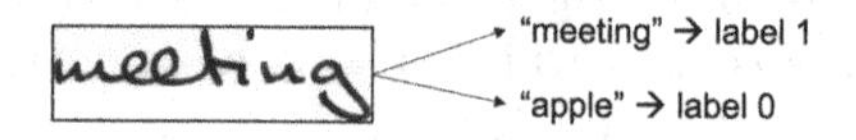

Fig. 1. Text matching as a binary supervised-classification task.

4 Proposed Approach

We propose a model called TextMatcher which directly compares an input image and a candidate text, producing a similarity score. The overall architecture is illustrated in Fig. 2. The image and the text are independently projected as matrices into separate embedding spaces, through *image embedding* and *text embedding* blocks, respectively. These embeddings are then compared with each other through a *cross-attention mechanism*, whose aim is to discover local alignments between the characters of the text and the vertical slices of the image (i.e. rectangular regions obtained by scanning the image along the horizontal axis), and produces in output a similarity score. A key peculiarity of the cross-attention component is that it helps the model specialize to the distribution of specific errors to be recognized. The three blocks in the overall TextMatcher model (i.e., image embedding, text embedding, and cross-attention mechanism) are jointly trained in an end-to-end fashion, via a contrastive loss function. In the following, we describe in detail the various components of TextMatcher.

4.1 Image Embedding

In order to produce the image embedding, the input image is first resized to a fixed dimension, and then it is processed by some convolutional layers, followed by recurrent layers in order to also encode contextual information. The output of this neural network module is the image embedding J of fixed $s_i \times d_i$ dimensionality, where s_i denotes the number of receptive fields, or slices, from the input image, and d_i is the feature dimension. More precisely, in this work we use the encoder block of the ASTER model from [16] to extract the image embedding: the input image is fed into a set of convolutional layers and batch normalization layers, followed by a bidirectional Long Short-Term Memory (LSTM) module. All the weights of the convolutional layers, batch normalization layers, and bidirectional LSTM are jointly learned in the final multimodal task.

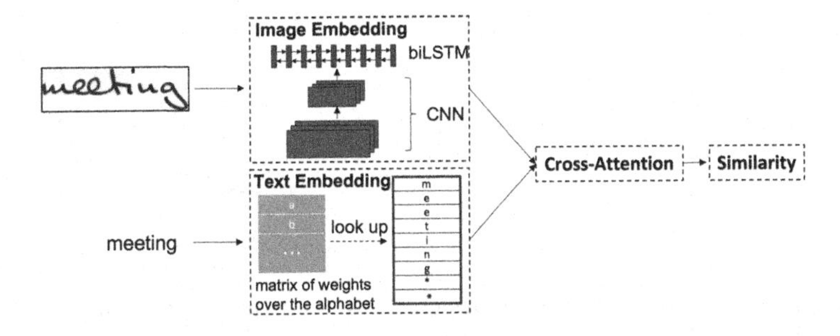

Fig. 2. TextMatcher architecture.

4.2 Text Embedding

As for text embedding, we simply use an embedding matrix over the characters of the alphabet. Let $\mathcal{A}$ be the alphabet at hand, which we assume to include a special character for the padding. The embedding matrix T_{emb} is a learnable matrix of dimension $|\mathcal{A}| \times d_t$. Given a text $c_1c_2 \dots c_l$, we first pad it to a fixed length s_t (or truncate it, if $l > s_t$). Then each character c_i is projected into the embedding space through the embedding matrix T_{emb}. The final embedding of the input text is a matrix T of dimensionality $s_t \times d_t$, whose row i is the row of T_{emb} corresponding to the character c_i, for $i = 1, \dots, s_t$.

4.3 Cross-Attention Mechanism

The attention mechanism was originally devised by Bahdanau *et al.* [2] in the context of encoder-decoder-based machine-translation systems, as a solution to the renowned issue of RNNs of rapid performance degrading as the length of the input sentence increases. Later, this mechanism has been widely employed in other contexts concerning sequential inputs, including natural language processing, computer vision, and speech processing.

The TextMatcher cross-attention component takes as input the J and T embeddings of image and text, and discovers local alignments between them. A similar idea is exploited in the self-attention mechanism of the well-established Transformer architecture [19]. In the latter, the self-attention computes a weighted representation of each token *attending to* the entire sentence. Conversely, in our case a multimodal approach is employed: each character of the input text attends to the vertical slices of the image. Moreover, in [19], the attention scores are used to compute a weighted sum of the value vectors of each token in the sentence, while, in our case, the attention scores are used to compute a weighted sum of cosine similarities between each character and the slices of the image, since our goal is to compute a similarity between image and text.

Specifically, the cross-attention mechanism of TextMatcher is as follows. First of all, in order to inject some positional information, we add independent positional embeddings to both image and text embeddings. The positional embeddings have the same dimension of the corresponding text or image embedding,

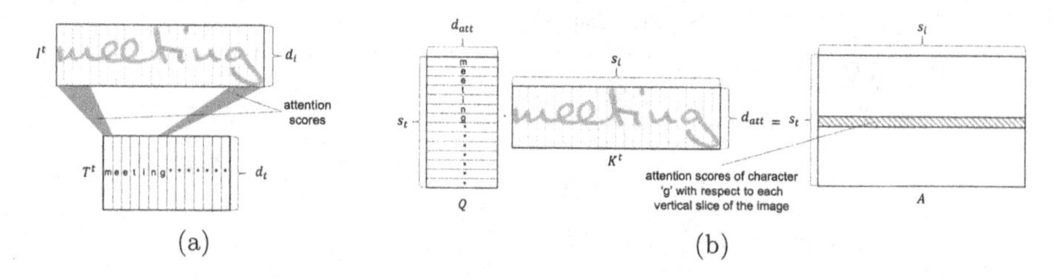

Fig. 3. (a) Visual representation of the cross-attention mechanism employed in the proposed TextMatcher model. (b) Computation of the attention matrix.

and, as such, they can be summed up. Inspired by [19], we use sine and cosine functions of different frequencies, where each dimension of the positional encoding corresponds to a sinusoid. The rationale is that this function would allow the model to easily learn to attend by relative positions. From now on, with a little abuse of notation, we will consider T and J as the text and image embeddings with the addition of the positional embeddings.

Let us consider the perspective of the text: for each character of the text, we want to compute an attention score with respect to each vertical slice of the image embedding, in order to pay more attention to the portion of the image that is expected to contain the corresponding character. The idea of this attention mechanism is depicted in Fig. 3(a). We compute attention scores between the embeddings of the characters and those of the vertical slices of the image by first projecting these vectors into separate embedding spaces of dimension d_{att}, and then computing normalized dot products between all pairs of characters and slices of the image. In particular, we compute *query* vectors of dimension d_{att} for the text and *key* vectors of dimension d_{att} for the image. These vectors are packed together respectively into the query matrix $Q = TQ_t$ and the key matrix $K = JK_i$, where Q_t and K_i are learnable parameters of dimension $d_t \times d_{att}$ and $d_i \times d_{att}$ respectively. The resulting matrices are Q of dimension $s_t \times d_{att}$ and K of dimension $s_i \times d_{att}$. We compute the attention matrix of dimension $s_t \times s_i$ as the dot product between the query Q and the key K, and then we apply a softmax function over the columns of the result, as illustrated in Fig. 3(b):

$$A = \text{softmax}(QK^t, \text{dim} = 1). \tag{1}$$

where the notation dim $= 1$ refers to the computation along the columns. In this way, the i-th row of the attention matrix contains the normalized attention scores of the i-th character of the input text with respect to each vertical slice of the image embedding. Then, the value vectors are used to compute a weighted cosine similarity between characters and vertical slices of the image embedding. First, normalized value matrices are computed for both image and text embeddings:

$$V_{text} = \text{normalize}(TV_t, \text{dim} = 1), \qquad V_{image} = \text{normalize}(JV_i, \text{dim} = 1), \tag{2}$$

with learnable parameters V_t and V_i of dimension $d_t \times d_{att}$ and $d_i \times d_{att}$ respectively. The resulting matrices V_{text} of dimension $s_t \times d_{att}$ and V_{image} of dimension

$s_i \times d_{att}$ are normalized over the columns in order to directly compute cosine similarities as their dot product. The cosine matrix $C = V_{text}V_{image}^t$ has dimension $s_t \times s_i$: the component (i, j) is the cosine similarity between the character at position i and the vertical slice of the image embedding at position j. Then, the cosine matrix is multiplied element-wise with the attention matrix, and a sum over the columns is performed, in order to compute a weighted cosine similarity of each character with respect to each vertical slice of the image embedding:

$$C_{att} = \text{sum}(C \odot A, \text{dim} = 1), \tag{3}$$

where $\odot$ stands for the element-wise multiplication. Finally, the similarities not related to pad characters are summed up, obtaining the final similarity score between the input image and the candidate text: $S_{tm} = \text{sum}(C_{att}[\text{pad} = 1])$.

The S_{tm} score is exploited to ultimately predict the $\hat{l}$ binary label via a thresholding mechanism: given a threshold τ,

$$\hat{l} = \begin{cases} 1, & \text{if } S_{tm} \geq \tau \\ 0, & \text{if } S_{tm} < \tau \end{cases} \tag{4}$$

4.4 Loss

Overall, the TextMatcher model has the following parameters: $W_{encoder}$, T_{emb}, pos_i, pos_t, Q_t, K_i, V_t, V_i, where $W_{encoder}$ contains the weights of the image encoder and pos_t and pos_i are the positional embeddings, possibly carefully initialized and then frozen. Given a dataset of matching and non matching pairs $\{((I^i, t^i),\ l^i) \,|i = 1, \ldots, n\}$, where I^i and t^i are image and text inputs of the i-th example and l^i is the corresponding binary label, the matching network is trained with the following contrastive loss, originally introduced in [7] :

$$L = \alpha l \left(1 - S_{tm}\right)^2 + (1 - l) \max\{m - \left(1 - S_{tm}\right), 0\}^2, \tag{5}$$

where m is the margin and α balances matching and non-matching pairs.

5 Experiments

In this section, we present experiments to empirically assess the performance of our TextMatcher model, on both an industrial use case of automated cheque processing – using a real (proprietary) dataset of bank cheques, and on a more general context – using a popular public dataset of handwritten text (i.e., IAM).

Competitors. We compare our TextMatcher to (i) a text-recognition model adapted to work for text matching, and (ii) a naïve model for text matching.

As for the former, the text transcription $\hat{t}$ produced by a text-recognition model run on an input image I is compared to the input candidate text t so as to produce a similarity score. Specifically, such a similarity score is computed as $S_{tr} = 1 - \frac{\text{Lev}(\hat{t}, t)}{\max\{|\hat{t}|, |t|\}}$, where Lev $(\cdot, \cdot)$ denotes the well-known *Levenshtein* distance between text strings, and $|t|$ is the number of characters in t. The training is performed with the dataset $\{(I^i, t^i) \,|i = 1, \ldots, n\}$, where I^i is the i-th image and t^i the corresponding (correct) text. As a text-recognition model we use the well-established state-of-the-art ASTER [16].

As a naïve text-matching model, we consider a model that computes image and text embeddings separately, and then computes the cosine similarity between their average vectors. The image embedding J of dimension $s_i \times d_i$ and the text embedding T of dimension $s_t \times d_t$ are defined in the same way as in TextMatcher, with the constraint $d_i = d_t$. Then, the average embeddings $\mathbf{T}_{avg} = \text{mean}(T, \text{dim} = 0)$ and $\mathbf{J}_{avg} = \text{mean}(J, \text{dim} = 0)$ are computed, with the convention that rows related to pad characters are not considered in the average of the text embedding. Finally, the output of the model is the cosine similarity between the average image and text $S_n = \frac{\mathbf{T}^t_{avg} \cdot \mathbf{J}_{avg}}{\|\mathbf{T}_{avg}\| \cdot \|\mathbf{J}_{avg}\|}$. The parameters of the convolutional part and the embedding matrix for the text are trained end-to-end in the final multimodal task, using the loss in Sect. 4.4. We term such a naïve text-matching model *NaïveTextMatcher*.

For both competitors, we compute the predicted binary label from the S_{tr} and S_n similarities in the same way as done in TextMatcher (Eq. (4)).

Implementation Details. We resize each grayscale image to 32×256 pixels and normalize pixel values to $[-1.0, 1.0]$. The image embedding part is the encoder of ASTER [16] with a final bidirectional LSTM with 256 hidden dimension, which produces an image embedding of dimension 64×512. The encoder is initialized with the weights of the pretrained model published together with the source code of [16]. For the text embedding we use $d_t = 512$. We add positional embeddings to both image and text embeddings, using the same initialization strategy proposed in [19], and then we freeze them during training. The attention dimension d_{att} is set at 512. The text embedding and the other attention parameters are initialized with the Xavier initialization [5]. The training is performed with Stochastic Gradient Descent (SGD), with 0.9 momentum, using a learning rate equal to 0.005 and batch size 8. As for the loss, we use margin $m = 1$ and $\alpha = 1$ for the experiments on IAM and $\alpha = 0.2$ for the real use case. The maximum number of epochs is 50. NaïveTextMatcher is initialized analogously to TextMatcher.

As for the text-recognition competitor, we use the available source code of ASTER [16].[2] ASTER is initialized with the weights of the publicly available pretrained model. All hyperparameters are set to the default values, except for the batch size set to 64, the height of input images set to 32, and the maximum number of epochs set to 35.

In the automated-cheque-processing use case we use an object-detection model (specifically, the state-of-the-art YoLo [14]) to first extract the part of a cheque containing the field of interest.

[2] https://github.com/ayumiymk/aster.pytorch.

5.1 Industrial Use Case of Automated Cheque Processing

A major application of the text-matching problem is in the context of automated cheque processing. The typical scenario here is that a customer of a bank who deposits a cheque to an ATM is asked to type some information that is written on the cheque, e.g., amount, issue date, beneficiary. Later, back-office operators manually check the correspondence between the information typed by the customer and what is written on the cheque. The main goal of a text-matching solution is to automate such a correspondence verification, in order to help operators perform their manual checks more easily and faster.

We evaluated the proposed TextMatcher in automated cheque processing by using a (proprietary) real dataset provided by UniCredit, a renowned pan-European commercial bank. Although we experimented with other fields too (i.e., amount, beneficiary), here we focus on the verification of the *issue date* of a bank cheque, as it is more challenging and appropriate to showcase the usefulness of text matching. Specifically, as main challenges, a model for matching the date field must be *sensitive to single-digit differences*, and the set of texts available at training time (built from cheques deposited in the past) is *disjoint from the set of texts that will be used at inference time*, as the latter includes *dates that are in the future with respect to the dates observed during training*. In order to handle the future-date issue, we take particular advantage of the proposed text-matching framework, which offers the chance to *inject into the model the desired behaviour through the non-matching texts*. Moreover, the number of non-matching (i.e., negative) samples in the real dataset is very small, therefore we keep these examples for testing purposes and use a synthetic matching dataset for training. In the following, we first describe the strategy for generating synthetic negative samples, and, then, we present the results of the evaluation.

Non-matching Sample Generation. As for the possible difficulties in distinguishing a non-matching text $\tilde{t}$, we observe that: (i) dates $\tilde{t}$ differing from t only for the year are more likely to receive higher similarities if the year in $\tilde{t}$ is present in the training set; and (ii) dates with the same digits as t but in different positions tend to be more challenging, since the position plays an important role. Motivated by these observations, for every matching pair (I, t) (that is originally present in the dataset), we generate a new non-matching pair $(I, \hat{t})$ as follows:

- with probability 0.3, randomly change 1 digit on the **day**;
- with probability 0.3, randomly change 1 digit on the **month**;
- with probability 0.15, randomly change 1 digit out of the last 2 on the **year**;
- with probability 0.15, change the **year** to a different one, chosen among the years represented in the training set;
- with probability 0.1, pick a random date.

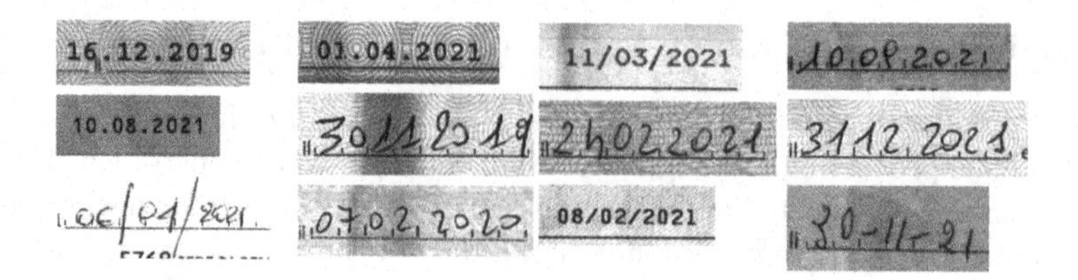

Fig. 4. Examples of images representing issue dates of bank cheques.

Whenever we change one digit in the day or month, with probability 0.5 we sample the replacement from the set $\{0, \ldots, 9\}$, and with probability 0.5 we sample from the set of digits already present in the date.

Dataset Details. We use a real dataset of (matching pairs) of about 50k images, with dates spanning a period from 2018 to 2021. We use dates from the year 2022 to experiment on future dates. We perform an *80-10-10* splitting in training, validation and test sets. Sample images are illustrated in Fig. 4. Note that the available ground truth is in the format *dd/mm/yyyy*, while the exact wording inside the cheque might have the year in two digits (e.g., "21" instead of "2021"), and, in case of day or month lower than 10, the zero can be present or not. For these reasons, we convert the input text to a normalized format of fixed length without separator (*ddmmyy*), where the possible first zero of day and month is encoded in the " $*$ " padding character. For instance the date *02/04/2021* is encoded as the text *$*$2$*$421*. Therefore, the maximum text length is 6 and the alphabet is *0123456789$*$*. To improve the generalization capabilities of the model in handling future dates, we also enlarge the training dataset with 5k examples of synthetic dates with year from 2000 to 2030, generated by concatenating digits from MNIST and adding a background resembling the one of cheques, and with 15k examples of amounts from real cheques with 6 digits and at most 2 zeros.

Results. We report the results of TextMatcher and its competitors on the test set obtained from the splitting of the real dataset of 50k images (*test-50k*), as well as on a real test set of 165k examples of real matching and non-matching examples from cheques deposited in January and February 2022 (*real-2022*). This way, the models are evaluated not only on future unseen dates, but also on *real negative examples.* The rate of negative examples in this real test set is 1.5%, and the handwritten images are 90% of the total. We assess the performance using the false positive (FP) and false negative (FN) rates. These metrics are particularly meaningful for the application scenario at hand, in which the main aim is to have low FP while keeping the FN acceptable (e.g. $FP \leq 2\%$ and $FN \leq 20\%$). We report the results separately for handwritten (FP_h, FN_h) and printed (FP_p, FN_p) images using the optimal threshold that minimizes:

$$10FP + FN, \tag{6}$$

with the constraint $FN \leq 60\%$ to get reasonable results. The results are summarized in Table 1. On the test set with the same distribution as the training

Table 1. Results on automated cheque processing: matching the issue date of a bank cheque. The optimal threshold τ (Eq. (4)) is selected according to Eq. (6) achieved by every method on the validation set.

Dataset	Method	τ	FP_p	FN_p	FP_h	FN_h
test-50k	**TextMatcher**	0.74	0.33	5.96	0.57	8.23
	ASTER	0.10	0.08	4.93	0.00	6.76
	NaïveTextMatcher	0.18	8.71	56.48	11.38	61.84
real-2022	**TextMatcher**	0.74	0.90	**3.55**	0.90	**9.59**
	ASTER	0.10	0.90	49.79	0.00	74.66
	NaïveTextMatcher	0.18	8.80	68.33	7.58	77.26

set (*test-50k*), TextMatcher and ASTER perform comparably. Conversely, in the real test set (*real-2022*), ASTER exhibits a considerable false-negative rate, i.e., 74.66 and 49.79 for the handwritten and printed case, respectively, whereas our TextMatcher generalizes well to unseen future dates as well as to the real distribution of non-matching samples, producing error rates comparable to that obtained on *test-50k*. These results motivated the adoption of TextMatcher in production for automated cheque processing at UniCredit.

5.2 General Applicability of Text Matching (IAM Dataset)

Here we present experiments carried out on the well-known real public IAM handwriting database [9]. This set of experiments aims at investigating the applicability of TextMatcher to more general settings, where the goal is to handle distributions of errors arising from different application scenarios. Moreover, these experiments also give the opportunity to highlight the differences between TextMatcher and the text-recognition models, giving the idea of the kind of application in which TextMatcher can achieve remarkable results.

Non-matching Sample Generation. We consider four ways of injecting errors to generate the negative examples:

- **random**: given a vocabulary V, the text of a non-matching pair is given by a random word of V (e.g., matching text *meeting*, non-matching text *apple*);
- **edit$_1$**: the non-matching text has *Levenshtein* distance equal to 1 from the matching text (e.g., matching text *meeting*, non-matching text *meating*);
- **edit$_{12}$**: the non-matching text has *Levenshtein* distance equal to 1 or 2 (with equal probability) from the matching text;
- **mixed**: the non-matching text is a random word of V with probability $\frac{1}{3}$, has Levenshtein distance equal to 1 with probability $\frac{1}{3}$, or Levenshtein distance equal to 2 with probability $\frac{1}{3}$.

We made 4 datasets containing one non-matching sample for each matching pair.

Dataset Details. The IAM handwriting database [9] consists of 1 539 pages of scanned text from 657 different writers. The database also provides the isolated and labeled words that were extracted from the pages of scanned text using an automatic segmentation scheme (and a-posteriori manually verified). We use the dataset at word level, and consider the available splitting proposed for the *Large Writer Independent Text Line Recognition Task*, in which each writer contributed to one set only. We set the alphabet to *abcdefghijklmnopqrstuvwxyz-'**, and we filter out words with characters outside the alphabet, or words only composed of punctuation marks. Finally, we only retain words with at least 5 characters. The final training, validation and test sets have size 17 550, 4 947, and 4 175, respectively. The maximum word length is 21.

Results. We evaluate the selected models using the confusion matrix and the F1-score. We choose the optimal threshold τ (Eq. (4)) for every method on the validation set, according to the F1-score, and report the performance on the test set. The results presented in Table 2 show that the proposed TextMatcher *outperforms all the competitors in all the configurations.* ASTER was recognized as the best competitor, as expected. In general, for any method, the best performance is achieved on the *random* configuration, whereas $edit_1$ and $edit_{12}$ configurations have the lowest performance, and *mixed* has intermediate performance. This is expected as the higher the similarity between a non-matching text and the corresponding matching text, the higher the difficulty for a model to accomplish the text-matching task, and, hence, the lower the accuracy.

Table 2. Results on the IAM handwriting database, for different configurations of non-matching-sample generation. The optimal threshold τ (Eq. (4)) is selected according to the best F1-score achieved by every method on the validation set.

Configuration	Method	τ	TP	FP	TN	FN	F1
random	**TextMatcher**	0.46	99.21	1.10	98.90	0.79	**99.06**
	ASTER	0.47	97.39	1.05	98.95	2.61	98.15
	NaïveTextMatcher	0.48	90.28	13.49	86.51	9.72	88.61
$edit_1$	**TextMatcher**	0.48	88.91	18.42	81.58	11.09	**85.77**
	ASTER	0.94	73.63	0.00	100.00	26.37	84.81
	NaïveTextMatcher	0.48	90.06	71.02	28.98	9.94	68.99
$edit_{12}$	**TextMatcher**	0.50	89.84	14.23	85.77	10.16	**88.05**
	ASTER	0.94	73.63	0.07	99.93	26.37	84.78
	NaïveTextMatcher	0.46	95.66	75.04	24.96	4.34	70.67
mixed	**TextMatcher**	0.52	92.93	8.07	91.93	7.07	**92.47**
	ASTER	0.95	73.60	0.05	99.95	26.40	84.77
	NaïveTextMatcher	0.48	82.47	35.78	64.22	17.53	75.57

5.3 Discussion

We conclude this section by highlighting some key advantages of our TextMatcher over its most effective competitor, i.e., the text-recognition ASTER model.

The results on IAM show that, in general, TextMatcher achieves better F1-scores than ASTER, in particular for the $edit_{12}$ and *mixed* configurations, where errors of different complexity need to be recognized together. This is related to the distribution of similarities produced by the two models. Indeed, TextMatcher yields a continuous distribution of values, treating different kinds of error similarly. Conversely, ASTER's distribution of similarities is discontinuous, thereby needing different optimal thresholds for different kinds of error.

Moreover, our TextMatcher can be trained on a specific distribution of errors, and even on different distributions at inference time (through the non-matching texts), as for the case of the issue date of bank cheques. This means that the model is flexible yet general enough to handle *any particular application scenario*. Indeed, it can be trained on specific patterns that are known to occur and are perhaps particularly difficult to detect for the application scenario at hand (e.g., distinguish "*facebook ltd*" from "*facebook inc*"). Once trained on the desired negative examples, TextMatcher specializes itself in recognizing these errors, paying more attention to the part of the text that is more relevant for those errors. Conversely, a text-recognition model like ASTER treats all kinds of error in the same way, thus resulting to be less general and versatile.

Finally, our TextMatcher is also more efficient than ASTER at inference time. We tested the CPU inference time of the two trained models for 1 000 random examples taken from the IAM *mixed* configuration: ASTER takes around 0.58 seconds per image on average, whereas TextMatcher takes around 0.07 seconds per image, which corresponds to a 8.75x speed-up.

6 Conclusions

In this paper, we study the task of *text matching*, to assess whether an image containing a single-line text corresponds to a given text transcription, and devise the TextMatcher machine-learning model for this task. The proposed TextMatcher projects image and text into separate embedding spaces, employs a cross-attention mechanism to discover local alignments between those embeddings, and is trained end-to-end on the distribution of errors to be recognized.

We experimentally evaluate TextMatcher on real data, including a proprietary dataset from a real industrial scenario of automated cheque processing, and the popular public IAM dataset. Compared to a naïve model and a state-of-the-art method for the related task of text recognition, TextMatcher proves to be more effective and better suited for handling different distributions of errors.

References

1. Almazán, J., Gordo, A., Fornés, A., Valveny, E.: Word spotting and recognition with embedded attributes. IEEE TPAMI **36**(12), 2552–2566 (2014)
2. Bahdanau, D., Cho, K., Bengio, Y.: Neural machine translation by jointly learning to align and translate. In: ICLR (2015)
3. Baltrušaitis, T., Ahuja, C., Morency, L.P.: Multimodal machine learning: a survey and taxonomy. IEEE TPAMI **41**(2), 423–443 (2018)
4. Chen, X., Jin, L., Zhu, Y., Luo, C., Wang, T.: Text recognition in the wild: a survey. ACM CSUR **54**(2), 42:1-42:35 (2021)

5. Glorot, X., Bengio, Y.: Understanding the difficulty of training deep feedforward neural networks. In: AISTATS, pp. 249–256 (2010)
6. Gómez, L., Rusinol, M., Karatzas, D.: LSDE: levenshtein space deep embedding for query-by-string word spotting. In: ICDAR, pp. 499–504 (2017)
7. Hadsell, R., Chopra, S., LeCun, Y.: Dimensionality reduction by learning an invariant mapping. In: CVPR, pp. 1735–1742 (2006)
8. Lee, K.H., Chen, X., Hua, G., Hu, H., He, X.: Stacked cross attention for image-text matching. In: Ferrari, V., Hebert, M., Sminchisescu, C., Weiss, Y. (eds.) ECCV, pp. 212–228 (2018)
9. Marti, U.V., Bunke, H.: The IAM-database: an English sentence database for offline handwriting recognition. IJDAR **5**(1), 39–46 (2002)
10. Mhiri, M., Desrosiers, C., Cheriet, M.: Word spotting and recognition via a joint deep embedding of image and text. Pattern Recogn. **88**, 312–320 (2019)
11. Qi, X., Zhang, Y., Qi, J., Lu, H.: Self-attention guided representation learning for image-text matching. Neurocomputing **450**, 143–155 (2021)
12. Radford, A., et al.: Learning transferable visual models from natural language supervision. In: ICML, pp. 8748–8763 (2021)
13. Rath, T.M., Manmatha, R.: Word spotting for historical documents. IJDAR **9**(2), 139–152 (2007)
14. Redmon, J., Divvala, S., Girshick, R., Farhadi, A.: You only look once: unified, real-time object detection. In: CVPR, pp. 779–788 (2016)
15. Shi, B., Wang, X., Lyu, P., Yao, C., Bai, X.: Robust scene text recognition with automatic rectification. In: CVPR, pp. 4168–4176 (2016)
16. Shi, B., Yang, M., Wang, X., Lyu, P., Yao, C., Bai, X.: ASTER: an attentional scene text recognizer with flexible rectification. IEEE TPAMI **41**(9), 2035–2048 (2018)
17. Sudholt, S., Fink, G.A.: PHOCNet: a deep convolutional neural network for word spotting in handwritten documents. In: ICFHR, pp. 277–282 (2016)
18. Sudholt, S., Fink, G.A.: Attribute CNNs for word spotting in handwritten documents. IJDAR **21**(3), 199–218 (2018)
19. Vaswani, A., et al.: Attention is all you need. In: NIPS, pp. 5998–6008 (2017)
20. Wang, Y., et al.: Position focused attention network for image-text matching. In: IJCAI, pp. 3792–3798 (2019)

Can Cross-Domain Term Extraction Benefit from Cross-lingual Transfer?

Hanh Thi Hong Tran[1,2,3](✉), Matej Martinc[1], Antoine Doucet[3], and Senja Pollak[2]

[1] Jožef Stefan International Postgraduate School, Jamova cesta 39, 1000 Ljubljana, Slovenia
hanh.usth@gmail.com

[2] Jožef Stefan Institute, Jamova cesta 39, 1000 Ljubljana, Slovenia

[3] University of La Rochelle, 23 Av. Albert Einstein, La Rochelle, France

Abstract. Automatic term extraction (ATE) is a natural language processing task that eases the effort of manually identifying terms from domain-specific corpora by providing a list of candidate terms. In this paper, we experiment with XLM-RoBERTa to evaluate the abilities of cross-lingual and multilingual versus monolingual learning in the cross-domain ATE task. The experiments are conducted on the ACTER corpus covering four domains (Corruption, Wind energy, Equitation, and Heart failure) and three languages (English, French, and Dutch) and on the RSDO5 Slovenian corpus, covering four additional domains (Biomechanics, Chemistry, Veterinary, and Linguistics). Regarding the ACTER test set, the cross-lingual and multilingual models boost the performance in F1-score by up to 5% if the term extraction task excludes the extraction of named entity terms (ANN version) and 3% if including them (NES version) compared to the monolingual setting. By adding an extra Slovenian corpus into the training set, the multilingual model demonstrates a significant improvement in terms of Recall, which, on average, increases by 18% in the ANN version and 13% in the NES version compared with the monolingual setting. Furthermore, our methods defeat state-of-the-art (SOTA) approaches with approximately 2% higher F1-score on average for the ANN version in English and Dutch, and the NES version in French. Regarding the RSDO5 test set, our monolingual approach proves to have consistent performance across all the train-validation-test combinations, achieving an F1-score above 61%. These results are a good indication of the potential in cross-lingual and multilingual language models not only for term extraction but also for other downstream tasks. Our code is publicly available at https://github.com/honghanhh/ate-2022.

Keywords: Term extraction · XLM-RoBERTa · Sequence labeling · Cross-lingual · Cross-domain

P. Pascal and D. Ienco (Eds.): DS 2022, LNAI 13601, pp. 363–378, 2022.
https://doi.org/10.1007/978-3-031-18840-4_26

1 Introduction

Terms are textual expressions that denote concepts in a specific field of expertise. They are beneficial for several terminographical tasks performed by linguists (e.g., construction of specialized term dictionaries [21]). Moreover, terms can also support and improve several complex downstream natural language processing (NLP) tasks, such as topic detection [6], information retrieval [22], machine translation [36], etc. Automatic term extraction (ATE) was born to ease the time and effort needed to manually identify terms from domain-specific corpora.

The TermEval 2020 shared task on monolingual ATE, organized as part of the CompuTerm workshop [30], presented one of the first opportunities to systematically study and compare various ATE systems with the introduction of a new annotated corpus that covers four domains in three languages: The Annotated Corpora for Term Extraction Research (ACTER) dataset [30,31]. While the workshop was an important step forward in systematic comparison, the less-resourced languages (e.g., Slovenian) have not yet been sufficiently explored and remain a research gap. Furthermore, there is still room for improvement in performance and replicability as the open-sourced code is often not available.

Inspired by the success of Transformer-based models in the TermEval 2020 competition [11] and the rise of cross-lingual learning [19], we propose to explore the performance of the multilingual XLM-RoBERTa pretrained model [3] in a multilingual setting, and in a cross-lingual setting, where the model is fine-tuned on several languages and tested on a new unseen language. We model the ATE as a sequence-labeling task. Sequence-labeling approaches have been successfully applied to a range of similar NLP tasks, including Named Entity Recognition [18,34] and Keyword Extraction [16,25]. The experiments are conducted in the cross-domain setting on the ACTER dataset containing texts in four domains (Corruption, Wind energy, Equitation, and Heart failure) with three languages (English, French, and Dutch) and the RSDO5 corpus[1] [12] containing Slovenian texts from four domains (Biomechanics, Chemistry, Veterinary, and Linguistics).

The main contributions of this paper can be summarized as follows:

- We systematically evaluate the performance of XLM-RoBERTa language model on the cross-domain term extraction task on two datasets covering English, French, Dutch, and a less-resourced language, Slovenian.
- We compare the performance of cross- and multilingual toward monolingual approaches to determine the general applicability of multilingual language models for sequence labeling in both rich- and less-resourced languages, for which manually labeled training resources are and are not available.

This paper is organized as follows: Sect. 2 presents the related work. Next, we introduce the dataset, methodology, experimental details as well as evaluation metrics in Sect. 3. The results with error analysis are discussed in Sect. 4 and 5 before we conclude and present future works in Sect. 6.

[1] https://www.clarin.si/repository/xmlui/handle/11356/1470.

2 Related Work

The history of ATE has its beginnings during the 1990s s with research done by Damerau et al. [5], Justeson et al. [14]. ATE systems usually employ the two-step procedure: (1) extracting a list of candidate terms; and (2) determining which candidate terms are correct using supervised or unsupervised techniques. We divide these techniques into the approaches based on (1) term characteristics and (2) machine learning and deep learning.

2.1 Approaches Based on Term Characteristics

Traditional ATE approaches relied on linguistic knowledge and distinctive linguistic aspects of terms to extract possible candidates. Several NLP tools (e.g., tokenization, lemmatization, stemming, chunking, PoS tagging, etc.) are employed to obtain linguistic profiles of term candidates. As a heavily language-dependent approach, the better the quality of the pre-processing tools (e.g., FLAIR [1], Stanza [28]), the better the quality of linguistic ATE methods. More recently, several studies were proposed that preferred the statistical approach toward ATE. The most common statistical approach relies on the assumption that a higher candidate term frequency in a domain-specific corpus implies a higher likelihood that a candidate is an actual term. Some measures relying on this assumption include termhood [35], unithood [4] or C-value [9]. More popular statistical approaches take also into account the frequency of the term internal words compared to the term frequency (e.g., Mutual Information) to identify rare terms and remove frequent words. Many current systems still apply this approach's variation, most commonly in hybrid systems combining linguistic and statistical information [15,29].

2.2 Approaches Based on Machine Learning and Deep Learning

Recently, advances in embeddings and deep neural networks have also influenced the field of term extraction. Several embeddings have been investigated for the task at hand, for example, uni-gram term representations constructed from a combination of local and global vectors [2], non-contextual [37], contextual [17] word embeddings, and the combination of both [10]. The first use of language models for the ATE task is documented in the TermEval 2020 [30] competition on the ACTER dataset, a collection of four domain-specific corpora in three languages (English, French, and Dutch). There, the winning approach on the Dutch corpus used pretrained GloVe word embeddings fed into a BiLSTM-based neural architecture. Meanwhile, the winning approach on the English corpus [11] relied on the extraction of all possible n-gram combinations, which are fed into a BERT binary classifier that determines for each n-gram inside a sentence, whether it is a term or not. Besides, several variations of Transformer-based models have also been investigated (e.g., RoBERTa and CamemBERT have also been used in the TermEval 2020 [11] challenge). Further work inspired by TermEval 2020 includes the HAMLET [32], which proposes a hybrid adaptable machine learning approach that combines the linguistic and statistical clues to detect terms. When it comes to more general related work applicable to ATE task, the research by Kucza et al. [17] was one of the first to propose to model term extraction as a sequence labeling task. Cross-lingual sequence labeling was, on the other hand, explored in Conneau et al. [3] and Lang et al. [19], who take advantage of XLM-RoBERTa, the model we also employ in this work, to compare three cross-lingual approaches, including a binary sequence classifier, a sequence classifier, and a token classifier on several sequence-labeling tasks. Finally, Lang et al. [19] further proposes to use a multilingual encoder-decoder denoising pre-training model called mBART [23] to generate sequences of comma-separated terms from the input. The results demonstrate the capability of multilingual models to outperform monolingual ones in some specific scenarios and the potential of cross-lingual learning.

2.3 Approaches for Slovenian Term Extraction

When it comes to the ATE for Slovenian, and more generally to less-resourced languages, the research is still hindered by the lack of gold standard corpora and limited use of neural methods. The things are nevertheless slowly improving. For example, in recent years, Slovenian KAS corpus was compiled [7]. The release was quickly followed by another corpus designed for term extraction, the RSDO5 corpus that we use in our study [13]. Regarding the employment of ATE models for Slovenian, one of the first approaches was the statistical approach by Vintar et al. [35]. The SOTA was proposed by Ljubevsic et al. [24], where they extract the initial candidate terms using the CollTerm tool [27], a rule-based system employing a complex language-specific set of term patterns (e.g., POS tag,...) from the Slovenian SketchEngine module [8], followed by a machine learning classification approach with features representing statistical term extraction measures. Another recent approach by Repar et al. [29] focuses on term extraction and alignment, where the main novelty is in using an evolutionary algorithm for the alignment of terms. On the other hand, the deep neural approaches have not been explored for Slovenian yet. Another problem is the open-sourced code is often not available for most current benchmark systems, hindering their reproducibility (for Slovenian, only the code from Ljubevsic et al.'s method [24] is available). In our own work [33], we also implemented the Transformers-based sequence labeling approach that we extend in this study in a cross-lingual and multilingual evaluation.

3 Methodology

Section 3.1 presents our chosen datasets with a brief description of the structure, term frequency, and label distribution. We describe the general methodology, experimental setup, and the implementation details in Sects. 3.2 and 3.3. Finally, in Sect. 3.4 we describe the chosen evaluation metrics for the ATE task.

3.1 Dataset

The experiments were conducted on two datasets (ACTER [30] and RSDO5 version 1.1 [12]) containing texts from different languages and domains. The structures of both datasets are presented in Fig. 1.

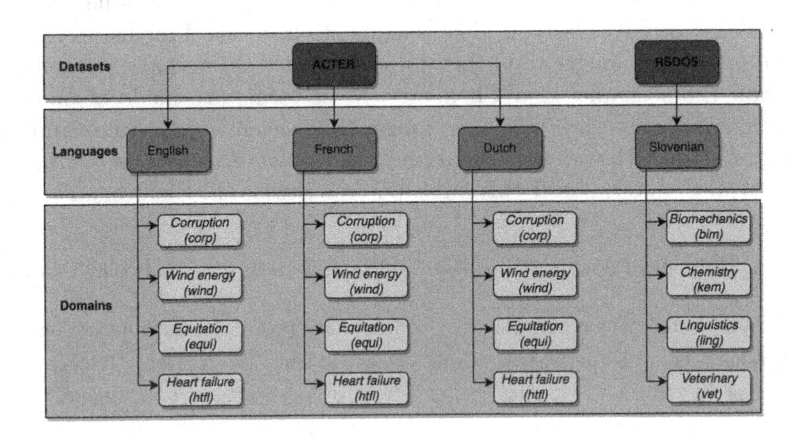

Fig. 1. The structure of RSDO5 and ACTER regarding languages and domains.

The ACTER dataset is a collection of 12 corpora covering four domains (Corruption (corp), Dressage (equi), Wind energy (wind), and Heart failure (htfl)) in three languages (English (en), French (fr) and Dutch (nl)). The dataset has two types of gold standard annotations: one including both terms and named entities (NES), and the other one containing only terms (ANN). Table 1 summarizes the number of documents and unique terms for each domain. Note the discrepancy in size between the Heart failure domain and the other three domains, with the Heart failure domain containing the much more unique terms and documents[2].

Table 1. Number of documents and unique terms in the ACTER dataset.

Languages	Corruption (corp)			Equitation (equi)			Wind energy (wind)			Heart failure (htfl)		
	Docs	Terms		Docs	Terms		Docs	Terms		Docs	Terms	
		ANN	NES		ANN	NES		ANN	NES		ANN	NES
en	19	927	1,173	34	1,155	1,575	5	1,091	1,534	190	2,361	2,585
fr	19	979	1,207	78	961	1,181	2	773	968	210	2,228	2,374
nl	12	1,047	1,295	65	1,393	1,544	8	940	1,245	174	2,074	2,254

The second dataset is the RSDO5 corpus version 1.1 [12] containing texts in Slovenian (sl), a less-resourced Slavic language with rich morphology. Compiled

[2] The detailed description of ACTER can be found in the TermEval competition [30].

during the course of the RSDO[3] national project, the RSDO5 corpus contains 12 documents (including three Ph.D. theses, a scientific book based on a Ph.D. thesis, four graduate level textbooks, and four journal articles) with altogether about 250,000 words collected from diverse sources between 2000 to 2019 covering domains of Biomechanics (bim), Chemistry (kem), Veterinary (vet), and Linguistics (ling). The numbers of documents, tokens, and unique terms per domain are reported in Table 2. The documents from the Linguistics and Veterinary domains are longer (i.e., they have more tokens) and also contain more terms than Biomechanics and Chemistry. Most terms are made of one up to three words and only a few terms are longer than seven words. An example of a long term found in the corpus would be *"stojo po obračanju v nasprotni smeri urinega kazalca"* (stand after turning counterclockwise) in Biomechanics.

Table 2. Number of documents, tokens, and unique terms in the RSDO5 dataset.

Biomechanics (bim)			Chemistry (kem)			Veterinary (vet)			Linguistics (ling)		
Docs	Tokens	Terms	Docs	Tokens	Terms	Docs	Tokens	Terms	Docs	Tokens	Terms
3	61,344	2,319	3	65,012	2,409	3	75,182	4,748	3	109,050	4,601

Furthermore, both datasets contain several nested terms, i.e., a shorter term may appear within a larger term and vice versa. For example, in the RSDO5's Biomechanics domain, the term *"navor"* (torque) appears in terms such as *"sunek navora"* (torque shock), *"zunanji sunek navora"* (external torque shock), and *"izokinetični navor"* (isokinetic torque); in ACTER's English Corruption, term *"confiscation"* appears also in terms such as *"confiscation of corruption proceeds"*, *"confiscation of criminal assets"*, and *"confiscation of the proceeds of crime"*, to mention a few. This makes the labeling harder and the classifier needs to infer from the context whether a specific term is part of a longer term.

3.2 Methodology

We experiment with XLM-RoBERTa, a Transformer-based model pre-trained on 2.5 TB of filtered CommonCrawl data containing 100 languages. We consider ATE as a sequence-labeling task where the model returns a label for each token in a text sequence using the (B-I-O) labeling regime [19,32]. Here, B stands for the beginning word in the term, I stands for the word inside the term, and O stands for the word not part of the term. The terms from a gold standard list are first mapped to the tokens in the raw text and each word inside the text sequence is annotated with one of three labels (see examples in Fig. 2).

The model is first trained to predict a label for each token in the input text sequence (e.g., we model the task as token classification) and then applied to the unseen text (test data). Finally, from the tokens or token sequences labeled as terms, the final candidate term list for the test data is composed.

[3] https://www.cjvt.si/rsdo/en/project/.

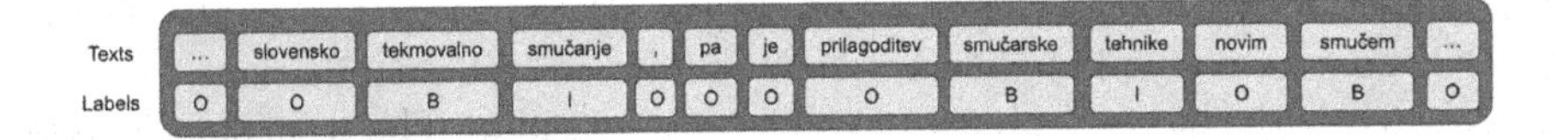

Fig. 2. A sample of our labels in the RSDO5 corpus for term extraction.

We evaluate the cross-domain performance of the model in a monolingual, cross-lingual, and multilingual setting. Altogether, 55 different scenarios are tested. The distinct settings are described below.

1. **Monolingual setup.** We evaluate how well the model performs when there is a language-specific training corpus available and there is a match between the language of the train set and the language of the test set. We fine-tune our model in a single language, which means we train three monolingual models for three languages (English, French, Dutch) and test each model in the same language as well as 12 monolingual models for Slovenian given 12 different combinations of train-validation-test split regarding the domains. This can be considered as a baseline to which we compare other settings.
2. **Cross-lingual setup.** We evaluate the capability of the model to apply the knowledge learned about ATE in one or more languages for ATE in another unseen language. Therefore, we fine-tune the ATE model in one or more languages (e.g., English and Dutch) and test it on another language not appearing in the train set (e.g., French). In this scenario, we, therefore, examine how well the model performs without the language-specific training corpus and how good the knowledge transfer between different languages is.
3. **Multilingual setup.** We fine-tune our model using a.) training datasets from all languages in the ACTER dataset (English, French, and Dutch) or using b.) training datasets from all languages in the ACTER dataset plus the Slovenian training dataset from the RSDO5 corpus, and then apply the model to the test sets of all languages. By doing so, we examine whether adding more data from other languages to the train set that matches the target language improves the predictive performance of the model.

All three settings are applied in a cross-domain evaluation scenario, where we use two domains for training, another domain for validation, and the rest for testing except the multilingual setting with additional Slovenian corpus in the training set where we use two domains from ACTER and all domains from RSDO5 corpus for the training. This way we want to check the generalization capabilities of the model, i.e. whether the knowledge the model obtained on one domain can be applied to the new, unseen domains, which would make the model applicable to arbitrary domains and therefore much more useful. In the ACTER dataset, we use Corruption and Wind energy domains as parts of the training, Equitation domains for validation, and Heart failure domain for testing in order to allow for a direct comparison with other benchmarks sharing the same train-validation-test setting [19], using the same dataset and evaluation setting (predicting on Heart Failure test set) from the related work. Meanwhile, in the RSDO5 corpus, we explore different train-validation-test combinations.

We divide the dataset into train-validation-test splits. The train split is used for fine-tuning the models while the validation split is used to prevent over-fitting during the fine-tuning phase. Finally, the test split is used for evaluation and is excluded during the model training. The model is fine-tuned on the training set to predict the probability for each word in a word sequence whether it is a part of the term (B, I) or not (O). An additional token classification head containing a feed-forward layer with a softmax activation is added on top of each model.

3.3 Implementation Details

We consider ATE as a sequence-labeling task and the models are trained to predict the labels from the (B-I-O) annotation scheme. The distribution across label types and the proportion of (B) and (I) labels in the total number of tokens per domain and per language are presented in Table 3. In the ACTER dataset, the proportion of terms in the texts is the largest for English, followed by French and then Dutch. The proportion of terms increases by from 1% upto 5% when adding NEs into the gold standards. In both datasets, the number of tokens annotated as terms (or parts of the term) only represents up to one-fourth (but usually much less) of the total tokens in the corpus, which means there is a significant imbalance between (B, I) tokens and tokens labeled as not terms (O).

Table 3. Label distribution and proportion of terms appearing per domain.

(a) The ACTER dataset.

Languages	Corruption (corp)				Equitation (equi)				Wind energy (wind)				Heart failure (htfl)			
	B	I	O	% Term	B	I	O	% Term	B	I	O	% Term	B	I	O	% Term
ANN en	4,558	2,200	44,287	13.24	10,745	1,938	46,215	21.53	5,046	3,323	49,873	14.37	9,819	4,504	41,522	25.65
ANN fr	4,461	2,823	51,918	12.30	8,420	2,373	50,487	17.61	5,928	4,405	43,976	19.03	7,165	4,027	43,976	20.29
ANN nl	4,251	1,517	46,730	10.99	10,243	1,509	45,011	20.70	4,174	826	50,642	8.99	8,529	1,391	45,142	18.02
NES en	6,050	3,226	41,769	18.17	11,340	2,377	45,181	23.29	6,040	4,111	48,091	17.43	10,115	4,855	40,875	26.81
NES fr	6,021	3,996	49,185	16.92	8,699	2,632	49,949	18.49	7,356	4,524	53,868	18.07	7,394	4,172	43,602	20.97
NES nl	5,585	2,308	44,605	15.03	10,416	1,625	44,722	21.21	4,708	1,084	49,850	10.41	8,770	1,627	44,665	18.88

(b) The RSDO5 dataset.

Languages	Biomechanics (bim)				Chemistry (kem)				Veterinary (vet)				Linguistics (ling)			
	B	I	O	% Term	B	I	O	% Term	B	I	O	% Term	B	I	O	% Term
sl	7,070	6,835	47,439	22.67	7,614	4,486	52,912	18.61	10,953	6,261	57,968	22.90	12,348	6,079	90,623	16.89

We employ the XLM-RoBERTa token classification model and its "fast" XLM-RoBERTa tokenizer from the Huggingface library[4]. We fine-tune the model for up to 20 epochs (i.e., we employ the early stopping regime) using the learning rate of 2e−05, training and evaluation batch size of 32, and sequence length of 512 tokens, since this hyperparameter configuration performed the best on the validation set. The documents are first split into sentences. Then, the sentences containing more than 512 tokens are truncated, while the sentences with less than

[4] https://huggingface.co/models.

512 tokens are padded with a special *<PAD>* token at the end. During fine-tuning, the model is evaluated on the validation set after each training epoch, and the best-performing model is applied to the test set. The model predicts each word in a word sequence whether it is a part of a term (B, I) or not (O). The sequences identified as terms are extracted from the text and put into a set of all predicted candidate terms. A post-processing step to lowercase all the candidate terms is applied before we compare our derived candidate list with the gold standard.

3.4 Evaluation Metrics

We evaluate the performance of the ATE system by comparing the candidate list extracted on the whole test set level with the manually annotated gold standard of each domain using strictly matching with Precision (P), Recall (R), and F1-score (F1). These evaluation metrics have also been used in the related work, including the TermEval 2020 [11,19,30] and Slovenian benchmark [24]. Therefore, our results are directly comparable to the SOTA methods.

4 Results

In this Section, we determine the predictive power of monolingual, cross-lingual, and multilingual learning in ACTER and RSDO5 test sets as well as compare the results from our proposed approaches to the SOTAs from the related work.

4.1 Prediction on the ACTER Test Set

Table 4 demonstrates the performance of XLM-RoBERTa on the cross-domain sequence-labeling ATE task on the ACTER test set in the monolingual, cross-lingual, and multilingual setting. We group the results according to the test language in the ACTER corpora for better comparison among settings. The results indicate that cross- and multilingual models surpass the performance of the monolingual ones according to all evaluation metrics except for when it comes to the Precision obtained by the French monolingual model on the French test set. Multilingual models tend to outperform cross-lingual ones, except for the cross-lingual model trained in Dutch and applied to the English test set. This multilingual model boosts the F1-score performance by up to 2% in ANN and 1% in the NES task when compared to the second-highest-performing model. By adding the Slovenian corpus with four different domains into the training set, the multilingual model demonstrates a significant improvement in Recall across all test languages, which, on average, increases by 18.17% in ANN and 13.54% in NES test set compared with the monolingual setting.

Table 5 presents a comparison between the best-performing models in this work in terms of F1-score and the benchmark approaches in the ACTER dataset, including the solutions from the winning teams in the competition (TALN-LS2N [11] won on the English and French test set while NLPLab UQAM [20] won on

Table 4. Evaluation on ACTER given Heart failure as test set.

Train language	Enlish test set						French test set						Dutch test set					
	ANN			NES			ANN			NES			ANN			NES		
	P	R	F1	P	R	F1	P	R	F1	P	R	F1	P	R	F1	P	R	F1
en	58.08	48.12	52.63	62.07	52.03	56.61	66.69	47.89	55.75	70.63	53.79	**61.07**	69.23	61.09	64.91	72.95	63.04	67.63
fr	56.94	33.21	41.95	60.01	39.07	47.33	**70.51**	44.43	54.51	**72.41**	48.53	58.11	72.12	51.01	59.76	73.63	55.50	63.29
nl	55.64	56.37	**56.00**	57.60	58.34	**57.97**	66.49	51.48	58.03	67.60	53.16	59.52	70.25	62.15	65.95	73.29	61.49	66.87
en, fr	57.16	51.21	54.02	60.43	51.45	55.58	63.70	52.38	57.49	68.13	52.78	59.48	72.52	61.72	66.69	73.08	63.49	67.95
en, nl	58.00	48.67	52.93	**62.39**	51.33	56.32	65.25	44.17	52.68	68.67	52.36	59.42	69.29	60.17	64.41	74.35	61.71	67.44
fr, nl	**60.84**	46.84	52.93	62.27	50.37	55.69	69.20	48.29	56.88	70.72	49.54	58.26	**75.72**	56.70	64.84	**76.74**	59.58	67.08
en, fr, nl	56.83	53.03	54.86	60.76	52.53	56.35	68.01	50.67	58.07	48.30	65.57	55.63	69.92	64.32	67.00	73.66	62.91	67.86
en, fr, nl, sl	45.88	**66.29**	54.23	48.30	**65.57**	55.63	58.10	**61.62**	**59.81**	59.48	**62.51**	60.96	62.74	**75.51**	**68.54**	63.57	**73.69**	**68.26**

the Dutch test set) and other methods proposed in Rigouts et al. [32] and Lang et al. [19], which are described in Sect. 2. Note that all the approaches from the related work are cross-domain and use the Heart failure domain as the test set and the rest of the data for training or validation. For the ANN task in English and Dutch and the NES task in French, our methods outperform other approaches in terms of F1-score. Despite not surpassing the SOTA in the French ANN task and the other two NES tasks, our method still offers competitive performance being outperformed by the HAMLET approach [32] with a small margin of 0.39% in ANN French, and by the token classifier [19] with about 0.33% in NES English. In terms of multilingual evaluation, we show that in contrast to the findings of Lang et al. [19], adding different languages in general slightly improves the models.

Table 5. F1-score comparison between our results and related work in ACTER.

Methods	English		French		Dutch	
	ANN	NES	ANN	NES	ANN	NES
Winning teams [11]	44.99	46.66	45.94	48.15	18.60	18.70
HAMLET [32]	54.20	55.40	**60.20**	60.80	66.10	66.00
Sequence classifier [19]	x	46.00	x	48.10	x	58.00
NMT [19]	x	55.30	x	57.60	x	59.60
Token classifier [19]	x	**58.30**	x	57.60	x	**69.80**
NMF-based approaches [26]	33.50	33.70	30.90	30.70	30.10	30.30
Our best classifiers	**56.00**	57.97	59.81	**61.07**	**68.54**	68.26

4.2 Evaluation on the RSDO5 Test Set

We also apply monolingual and multilingual cross-domain approaches to the Slovenian RSDO5 dataset. The results grouped by the test domain are presented in Table 6. The monolingual approach, where we use two domains from the RSDO5 corpus for training, validate on the third domain, and test on the last domain, proves to have relatively consistent performance across all the combinations, achieving Precision of more than 62%, Recall of no less than 55%, and F1-score above 61%. The model performs slightly better for the Linguistics and Veterinary domains than for Biomechanics and Chemistry. The difference in the number of terms and length of terms per domain pointed out in Sect. 3.1 might be one of the factors that contribute to this behavior. Moreover, a significant performance boost can be observed for the Linguistics domain when the model is trained in the Chemistry and Veterinary domains, and for the Veterinary domain, when the model is trained in Biomechanics and Linguistics. In these two settings, the model achieves an F1-score of more than 68%.

Table 6. The evaluation of monolingual and multilingual learning in RSDO5.

Validation	Testing	Monolingual setup			Multilingual setup					
					ANN			NES		
		P	R	F1	P	R	F1	P	R	F1
vet	ling	**69.55**	64.05	66.69	67.68	69.55	68.60	67.19	**69.88**	**68.51**
bim	ling	69.48	**73.66**	**71.51**	**69.78**	66.16	67.92	67.81	68.53	68.17
kem	ling	66.20	72.38	69.15	66.50	**71.35**	**68.84**	**67.89**	69.03	68.46
ling	vet	71.06	66.72	68.82	**70.96**	65.27	68.00	69.22	67.40	68.30
kem	vet	**72.66**	65.59	**68.94**	69.75	**68.83**	**69.29**	**70.49**	**67.75**	**69.09**
bim	vet	69.30	**68.07**	68.68	69.77	68.43	69.09	69.26	64.72	66.91
ling	kem	68.67	55.13	61.16	68.26	59.28	63.45	67.54	54.59	60.38
bim	kem	70.14	**60.27**	**64.83**	69.63	**61.19**	**65.14**	**69.25**	52.72	59.86
vet	kem	**70.23**	59.24	64.27	**69.90**	58.41	63.64	67.92	**59.24**	**63.28**
vet	bim	**63.51**	**66.80**	**65.11**	61.14	**64.94**	**62.98**	60.94	66.67	63.68
ling	bim	62.25	65.20	63.69	60.53	63.82	62.13	**62.62**	62.27	62.44
kem	bim	62.35	63.99	63.16	**65.71**	59.16	62.26	61.78	**67.05**	**64.31**

We also explore the performance of multilingual approaches on the RSDO5 test sets. We train the model using the ANN and NES labels from all domains of the ACTER dataset and on two domains from the RSDO5 dataset, validate on the third RSDO5 domain, and test on the last domain. Table 6 demonstrates the comparative performance of the multilingual and the monolingual approaches, which is consistent with the results in the prediction of the ACTER test set.

Furthermore, in Table 7, we present the results from the related work for the RSDO5 dataset [24] in comparison to the proposed monolingual and multilingual approaches. The results from [24]'s method are taken from Hanh et al. [33]. In general, our approach outperforms the approach proposed in Ljubevsic et al. [24] by a large margin on all domains and according to all evaluation metrics, especially when it comes to Recall. Overall, we achieve results roughly twice as high as the approach proposed by Ljubevsic et al. [24] in terms of F1-score for all test domains regarding both monolingual and multilingual learning. We show

Table 7. Comparison between our performance and SOTA in RSDO5 dataset.

Methods	Linguistics			Veterinary			Chemistry			Biomechanics		
	P	R	F1	P	R	F1	P	R	F1	P	R	F1
Monolingual	**69.48**	**73.66**	**71.51**	**72.66**	65.59	68.94	**70.14**	60.27	64.83	**63.51**	66.80	**65.11**
Multilingual	66.50	71.35	68.84	69.75	**68.83**	**69.29**	69.63	**61.19**	**65.14**	61.78	**67.05**	64.31
SOTA [24]	52.20	25.40	34.10	66.90	19.30	29.90	47.80	31.40	37.80	53.80	24.80	33.90

that the multilingual experiments do in several cases improve our monolingual results [33], but this is not systematic.

5 Error Analysis

In order to determine whether the term length affects the models' performance, we calculate Precision and Recall separately for terms of length k = {1,2,3,4, ≥5}. The number of predicted candidate terms (Preds), ground truth (GT), correct predictions (TPs), Precision, and Recall regarding different term lengths k and test domains are presented in Table 8. The results for ACTER's dataset (Table 8a) were obtained by employing the best performing model for a specific language in terms of F1-score on the Heart failure test set. The results for the RSDO5 dataset (Table 8b) were obtained by employing the best-performing model for a specific test domain in F1-score.

Table 8. Performance per term length per domain in each test set.

(a) ACTER test set.

k	Enlish					French					Dutch				
	Preds	GTs	TPs	P	R	Preds	GTs	TPs	P	R	Preds	GTs	TPs	P	R
1	1,009	1,170	639	63.33	54.62	1,153	1,309	829	71.90	63.33	2,005	1,687	1,292	64.44	76.59
2	985	801	501	50.86	62.55	490	620	320	65.31	51.61	661	391	303	45.84	77.49
3	553	377	256	46.29	67.90	163	266	100	61.35	37.59	108	108	55	50.93	50.93
4	163	142	86	52.76	60.56	47	91	24	51.06	26.37	19	35	10	52.63	28.57
≥5	53	95	26	49.06	27.37	13	88	4	30.77	4.55	1	33	1	100.00	3.03

(b) RSDO5 Linguistics test set.

k	Linguistics					Veterinary					Chemistry					Biomechanics				
	Preds	GTs	TPs	P	R	Preds	GTs	TPs	P	R	Preds	GTs	TPs	P	R	Preds	GTs	TPs	P	R
1	2,078	1,728	1,300	62.56	75.23	2,159	2,067	1,472	68.18	71.21	943	890	580	61.51	65.17	1,079	718	22	48.38	72.70
2	2,631	2,404	1,858	70.62	77.29	2,062	2,103	1,448	70.22	68.85	1,073	1,202	768	71.58	63.89	1,153	1,172	822	71.29	70.14
3	322	360	7,191	59.32	53.06	314	446	182	57.96	40.81	164	260	93	56.71	35.77	223	286	124	55.61	43.36
4	57	80	31	54.39	38.75	28	77	10	35.71	12.99	26	46	11	42.31	23.91	26	59	11	42.31	18.64
≥5	12	29	79	75.00	31.03	3	55	2	66.67	3.64	3	11	0	0.00	0.00	11	84	5	45.45	5.95

The models proved to be good at predicting terms containing up to four words for English and Dutch and up to three words for French. The results on the RSDO5 dataset are similar, showing that the models are good at predicting short terms containing up to three words for all four domains of the RSDO5 corpus. The best model applied to the Linguistics test domain also shows relatively good performance when it comes to the prediction of longer terms, achieving 75.00%

Precision and a decent 31.03% Recall for terms with at least five words. Despite the relatively high Precision for prediction of long terms in the Veterinary and Biomechanics test domains, the Recall is pretty low, most likely due to the small amount of longer terms in the dataset on which the models are trained. When it comes to predictions in the Chemistry domain, there are no correct term predictions that consist of more than five words.

6 Conclusion

In summary, we investigated the possibilities of cross- and multilingual learning compared to the monolingual setting in the cross-domain sequence-labeling term extraction given the experiments conducted on multi-domain corpora, namely the ACTER and RSDO5 datasets. We also evaluated the impact of cross- and multilingual models on the ACTER corpora only and by further adding the texts from the Slovenian RSDO5 corpus in the training set. In addition, we examined the cross-lingual effect of rich-resourced training language on less-resourced testing one such as Slovenian. The results demonstrate a promising impact of multilingual and cross-lingual cross-domain learning that outperforms the related works in both datasets, which proves their potential when transferring from the rich- to the less-resourced languages.

However, we believe that there remains room for improvement in the field of supervised term extraction. In the future, we suggest the integration of active learning into our current approach to improve the output of the automated method by dynamical adaptation after human feedback. By learning with humans in the loop, we aim at getting the most information with the least amount of term labels. We will also evaluate the contribution of active learning in reducing the annotation effort and determine the robustness of the incremental active learning framework across different languages and domains.

Acknowledgements. The work was partially supported by the Slovenian Research Agency (ARRS) core research programme Knowledge Technologies (P2-0103), as well as the Ministry of Culture of Republic of Slovenia through project Development of Slovene in Digital Environment (RSDO). The first author was partly funded by Region Nouvelle Acquitaine. This work has also been supported by the TERMITRAD (2020-2019-8510010) project funded by the Nouvelle-Aquitaine Region, France.

References

1. Akbik, A., Bergmann, T., Blythe, D., Rasul, K., Schweter, S., Vollgraf, R.: Flair: an easy-to-use framework for state-of-the-art NLP. In: Proceedings of the 2019 Conference of the North American Chapter of the Association for Computational Linguistics (Demonstrations), pp. 54–59 (2019)
2. Amjadian, E., Inkpen, D., Paribakht, T., Faez, F.: Local-global vectors to improve unigram terminology extraction. In: Proceedings of the 5th International Workshop on Computational Terminology (Computerm2016), pp. 2–11 (2016)

3. Conneau, A., et al.: Unsupervised cross-lingual representation learning at scale. In: ACL (2020)
4. Daille, B., Gaussier, É., Langé, J.M.: Towards automatic extraction of monolingual and bilingual terminology. In: COLING 1994 Volume 1: The 15th International Conference on Computational Linguistics (1994)
5. Damerau, F.J.: Evaluating computer-generated domain-oriented vocabularies. Inf. Process. Manag. **26**(6), 791–801 (1990)
6. ElKishky, A., Song, Y., Wangx, C., Voss, C.R., Han, J.: Scalable topical phrase mining from text corpora. Proc. VLDB Endow. **8**(3), 305–316 (2014)
7. Erjavec, T., Fišer, D., Ljubešić, N.: The KAS corpus of Slovenian academic writing. Lang. Resour. Eval. **55**(2), 551–583 (2021)
8. Fišer, D., Suchomel, V., Jakubícek, M.: Terminology extraction for academic Slovene using sketch engine. In: Tenth Workshop on Recent Advances in Slavonic Natural Language Processing, RASLAN 2016, pp. 135–141 (2016)
9. Frantzi, K.T., Ananiadou, S., Tsujii, J.: The *C-value/NC-value* method of automatic recognition for multi-word terms. In: Nikolaou, C., Stephanidis, C. (eds.) ECDL 1998. LNCS, vol. 1513, pp. 585–604. Springer, Heidelberg (1998). https://doi.org/10.1007/3-540-49653-X_35
10. Gao, Y., Yuan, Yu.: Feature-less end-to-end nested term extraction. In: Tang, J., Kan, M.-Y., Zhao, D., Li, S., Zan, H. (eds.) NLPCC 2019. LNCS (LNAI), vol. 11839, pp. 607–616. Springer, Cham (2019). https://doi.org/10.1007/978-3-030-32236-6_55
11. Hazem, A., Bouhandi, M., Boudin, F., Daille, B.: TermEval 2020: TALN-LS2N system for automatic term extraction. In: Proceedings of the 6th International Workshop on Computational Terminology, pp. 95–100 (2020)
12. Jemec Tomazin, M., Trojar, M., Atelšek, S., Fajfar, T., Erjavec, T., Žagar Karer, M.: Corpus of term-annotated texts RSDO5 1.1 (2021). https://hdl.handle.net/11356/1470, Slovenian language resource repository CLARIN.SI
13. Jemec Tomazin, M., Trojar, M., Žagar, M., Atelšek, S., Fajfar, T., Erjavec, T.: Corpus of term-annotated texts rsdo5 1.0 (2021)
14. Justeson, J.S., Katz, S.M.: Technical terminology: some linguistic properties and an algorithm for identification in text. Nat. Lang. Eng. **1**(1), 9–27 (1995)
15. Kessler, R., Béchet, N., Berio, G.: Extraction of terminology in the field of construction. In: 2019 First International Conference on Digital Data Processing (DDP), pp. 22–26. IEEE (2019)
16. Koloski, B., Pollak, S., Škrlj, B., Martinc, M.: Out of thin air: is zero-shot cross-lingual keyword detection better than unsupervised? arXiv preprint arXiv:2202.06650 (2022)
17. Kucza, M., Niehues, J., Zenkel, T., Waibel, A., Stüker, S.: Term extraction via neural sequence labeling a comparative evaluation of strategies using recurrent neural networks. In: INTERSPEECH, pp. 2072–2076 (2018)
18. Lample, G., Ballesteros, M., Subramanian, S., Kawakami, K., Dyer, C.: Neural architectures for named entity recognition. In: Proceedings of the 2016 Conference of the North American Chapter of the Association for Computational Linguistics: Human Language Technologies, pp. 260–270 (2016)
19. Lang, C., Wachowiak, L., Heinisch, B., Gromann, D.: Transforming term extraction: transformer-based approaches to multilingual term extraction across domains. In: Findings of the Association for Computational Linguistics: ACL-IJCNLP 2021, pp. 3607–3620 (2021)
20. Le, N.T., Sadat, F.: Multilingual automatic term extraction in low-resource domains. In: The International FLAIRS Conference Proceedings, vol. 34 (2021)

21. Le Serrec, A., L'Homme, M.C., Drouin, P., Kraif, O.: Automating the compilation of specialized dictionaries: use and analysis of term extraction and lexical alignment. Terminology. Int. J. Theor. Appl. Issues Spec. Commun. **16**(1), 77–106 (2010)
22. Lingpeng, Y., Donghong, J., Guodong, Z., Yu, N.: Improving retrieval effectiveness by using key terms in top retrieved documents. In: Losada, D.E., Fernández-Luna, J.M. (eds.) ECIR 2005. LNCS, vol. 3408, pp. 169–184. Springer, Heidelberg (2005). https://doi.org/10.1007/978-3-540-31865-1_13
23. Liu, Y., et al.: Multilingual denoising pre-training for neural machine translation. Trans. Assoc. Comput. Linguist. **8**, 726–742 (2020)
24. Ljubešić, N., Fišer, D., Erjavec, T.: KAS-term: extracting slovene terms from doctoral theses via supervised machine learning. In: Ekštein, K. (ed.) TSD 2019. LNCS (LNAI), vol. 11697, pp. 115–126. Springer, Cham (2019). https://doi.org/10.1007/978-3-030-27947-9_10
25. Martinc, M., Škrlj, B., Pollak, S.: TNT-Kid: transformer-based neural tagger for keyword identification. Nat. Lang. Eng. 1–40 (2021). https://doi.org/10.1017/S1351324921000127
26. Nugumanova, A., Akhmed-Zaki, D., Mansurova, M., Baiburin, Y., Maulit, A.: NMF-based approach to automatic term extraction. Expert Syst. Appl. **199**, 117179 (2022)
27. Pinnis, M., et al.: Extracting data from comparable corpora. In: Skadina, I., Gaizauskas, R., Babych, B., Ljubešić, N., Tufiş, D., Vasiljevs, A. (eds.) Using Comparable Corpora for Under-Resourced Areas of Machine Translation. TANLP, pp. 89–139. Springer, Cham (2019). https://doi.org/10.1007/978-3-319-99004-0_4
28. Qi, P., Zhang, Y., Zhang, Y., Bolton, J., Manning, C.D.: Stanza: a Python natural language processing toolkit for many human languages. arXiv preprint arXiv:2003.07082 (2020)
29. Repar, A., Podpečan, V., Vavpetič, A., Lavrač, N., Pollak, S.: TermEnsembler: an ensemble learning approach to bilingual term extraction and alignment. Terminology. Int. J. Theor. Appl. Issues Spec. Commun. **25**(1), 93–120 (2019)
30. Rigouts Terryn, A., Hoste, V., Drouin, P., Lefever, E.: TermEval 2020: shared task on automatic term extraction using the annotated corpora for term extraction research (ACTER) dataset. In: 6th International Workshop on Computational Terminology (COMPUTERM 2020), pp. 85–94. European Language Resources Association (ELRA) (2020)
31. Rigouts Terryn, A., Hoste, V., Lefever, E.: In no uncertain terms: a dataset for monolingual and multilingual automatic term extraction from comparable corpora. Lang. Resour. Eval. **54**(2), 385–418 (2020)
32. Rigouts Terryn, A., Hoste, V., Lefever, E.: HAMLET: hybrid adaptable machine learning approach to extract terminology. Terminology (2021)
33. Tran, H.T.H., Martinc, M., Doucet, A., Pollak, S.: A transformer-based sequence-labeling approach to the Slovenian cross-domain automatic term extraction. In: Submitted to Slovenian Conference on Language Technologies and Digital Humanities (2022, under review)
34. Hanh, T.T.H., Doucet, A., Sidere, N., Moreno, J.G., Pollak, S.: Named entity recognition architecture combining contextual and global features. In: Ke, H.-R., Lee, C.S., Sugiyama, K. (eds.) ICADL 2021. LNCS, vol. 13133, pp. 264–276. Springer, Cham (2021). https://doi.org/10.1007/978-3-030-91669-5_21
35. Vintar, S.: Bilingual term recognition revisited: the bag-of-equivalents term alignment approach and its evaluation. terminology. Int. J. Theor. Appl. Issues Spec. Commun. **16**(2), 141–158 (2010)

36. Wolf, P., Bernardi, U., Federmann, C., Hunsicker, S.: From statistical term extraction to hybrid machine translation. In: Proceedings of the 15th Annual Conference of the European Association for Machine Translation (2011)
37. Zhang, Z., Gao, J., Ciravegna, F.: SEMRE-Rank: improving automatic term extraction by incorporating semantic relatedness with personalised pagerank. ACM Trans. Knowl. Discov. Data (TKDD) **12**(5), 1–41 (2018)

Retrieval-Efficiency Trade-Off of Unsupervised Keyword Extraction

Blaž Škrlj(✉), Boshko Koloski, and Senja Pollak

Jožef Stefan Institute, Ljubljana, Slovenia
blaz.skrlj@ijs.si

Abstract. Efficiently identifying keyphrases that represent a given document is a challenging task. In the last years, plethora of keyword detection approaches were proposed. These approaches can be based on statistical (frequency-based) properties of e.g., tokens, specialized neural language models, or a graph-based structure derived from a given document. The graph-based methods can be computationally amongst the most efficient ones, while maintaining the retrieval performance. One of the main properties, common to graph-based methods, is their immediate conversion of token space into graphs, followed by subsequent processing. In this paper, we explore a novel unsupervised approach which merges parts of a document in sequential form, *prior to* construction of the token graph. Further, by leveraging personalized PageRank, which considers frequencies of such sub-phrases alongside token lengths during node ranking, we demonstrate state-of-the-art retrieval capabilities while being up to two orders of magnitude faster than current state-of-the-art unsupervised detectors such as YAKE and MultiPartiteRank. The proposed method's scalability was also demonstrated by computing keyphrases for a biomedical corpus comprised of 14 million documents in less than a minute.

Keywords: Keyphrase detection · Natural language processing · Text mining

1 Introduction

With the increasing amounts of freely available text-based data sets, methods for efficient keyphrase detection are becoming of high relevance [13]. These methods, given a single or multiple documents, output a ranked list of short phrases (or single tokens), which represents key aspects of the input text. In the recent years, plethora of keyphrase extraction methods were presented; broadly, they can be divided into unsupervised and supervised ones. This paper focuses on unsupervised keyphrase extraction, i.e. the process where no training set of document is needed to learn to estimate keyphrases – they are estimated solely based on statistical/topological properties of a given document. The unsupervised methods can be further divided to the ones which construct a graph based

P. Pascal and D. Ienco (Eds.): DS 2022, LNAI 13601, pp. 379–393, 2022.
https://doi.org/10.1007/978-3-031-18840-4_27

on token co-occurrences and the ones which leverage statistical properties of n-grams [26]. Recently, neural language model-based keyphrase extraction was also proposed [12]. With the abundance of methods, optimization of a single metric becomes less relevant – methods which maximize e.g., F1@k are common. This paper aims to inform the reader that a realm of highly relevant properties beyond simple retrieval performance can be meaningful in practice, and should be the focus of any novel method proposed (including the adaptation of an existing one presented in this paper). The contributions of this paper are multifold (Fig. 1):

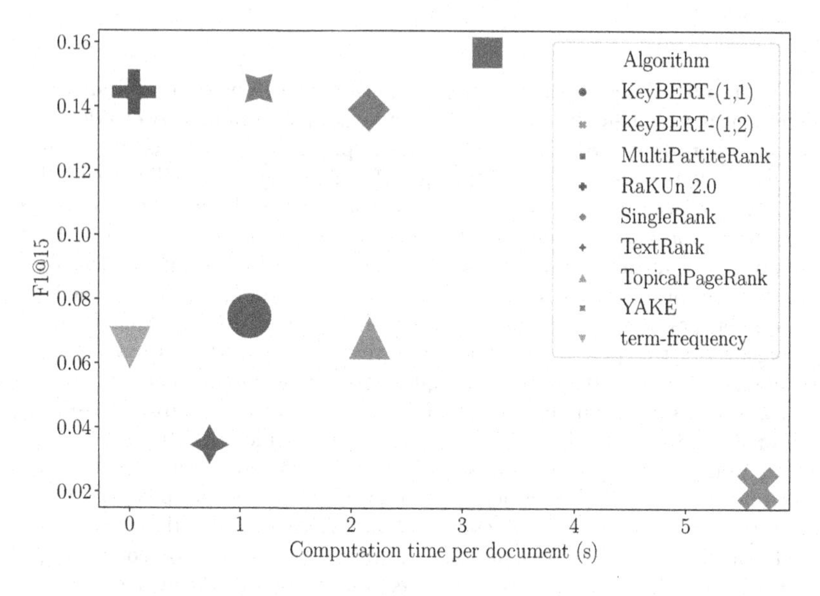

Fig. 1. Performance trade-off (time vs. performance) of keyphrase detection methods averaged across fifteen data sets.

1. We present RaKUn 2.0, a graph-based keyphrase extractor optimized for retrieval-efficiency optimality when considering both retrieval capabilities and performance.
2. A polygon-based visualization suitable for studying and comparing multiple criteria for multiple keyphrase detection algorithms.
3. An extensive benchmark of RaKUn 2.0 against strong baselines (including e.g., the recently introduced KeyBERT).
4. Friedman-Nemenyi-based analysis of average ranks of the algorithms (and their similarity).

2 Selected Related Work

This section contains an overview of the existing keyphrase detection methods, key underlying ideas and possible caveats of different paradigms. This paper focuses exclusively on *unsupervised* keyphrase extraction – the process of transforming an input document D in to a ranked collection of keyphrases, i.e. $K = \{(p, s)_k\}; s_{k+1} \leq s_k$, where k represents the top k hits (detected keyphrases), p a given keyphrase and s a given keyphrase's score. The first branch of approaches are based on text-to-graph transformations, followed by subsequent processing of the obtained graphs. Such methods are able to exploit multilevel structure of a document [6] (MultiPartiteRank), hierarchical structure [29] (SingleRank). An example token graph is shown in Fig. 2. One of the first graph-based methods was TextRank [23], which demonstrated the robustness of graph-based keyphrase detection (and was one of the first to do so). More involved approaches, capable of incorporating topic-level information were also proposed [7] (TopicalPageRank). One of the key issues with graph-based representations is that of node denoising – the process of identifying the relevant space of nodes which are commonly subject to ranking. The graph-based methods are highly dependant on the graph construction approach (based on co-occurrence, syntactic, semantic and similarity information) and node ranking algorithm (e.g. degree, closeness, Page Rank, selectivity, etc.) [3]. A detailed overview of graph-based methods for keyword extraction and various node-ranking measures is provided in [3].

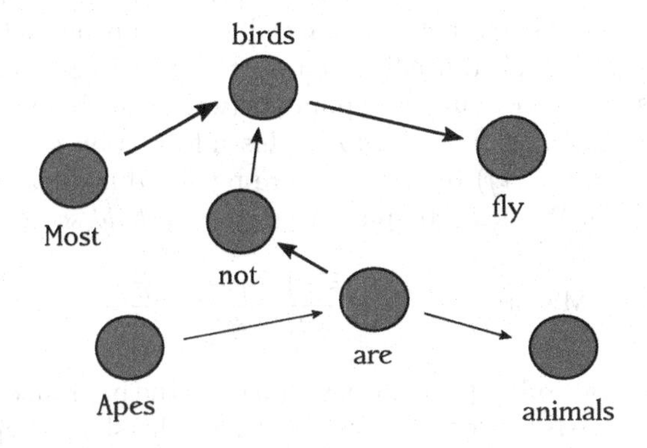

Fig. 2. An example token graph.

Alongside graph-based methods, statistical methods are also actively developed. One of the most recent examples includes YAKE! [8], an approach which considers large amounts of n-gram patterns and scores them so that they represent relevant keyphrases. It operates by extracting statistical features from single documents to select the most important keywords of a text. Keyphrase detection was also considered as a task solvable by considering neural language models [12]. An example of this family of models is AttentionRank [10], which exploits the transformer-based neural language model to extract relevant keywords. A more detailed overview of general keyword detection methods is given in [17].

The discussed approaches seldom focus on metrics beyond retrieval capabilities (e.g., Precision, Recall and F1). One of the purposes of this paper is a comprehensive evaluation of the discussed algorithms with regards to multiple criteria, including computation time and duplication rates (how frequent is a token amongst the space of detected keyphrases).

3 Proposed Algorithm

The proposed approach sources the core idea from the recent paper on meta vertex-based keyphrase detection RaKUn [28]. The considered extension, proposed in this paper is optimized specifically to push the boundary of the retrieval-efficiency front between retrieval performance and retrieval time. We begin with a general overview of the algorithm, followed by theoretical analysis of its complexity (space and time). We refer to the proposed approach as RaKUn 2.0. A high-level overview is shown as Algorithm 1. The main steps include *tokenization, token merging, document graph construction* and *node ranking*. Instead of first constructing (larger) graphs which are subject to node merging into meta vertices, RaKUn 2.0 conducts the merging step at the sequence level, making it more efficient. This step was considered based on an observation that pre-merging tokens in close proximity already offers sufficient results – by considering only tokens close to one another, no specialized metric for string comparison (possibly expensive) was needed, which substantially sped up the detection process. The second idea which substantially sped up the process is related to *bi-gram hashing*. It refers to constructing a mapping between each bi-gram and its count in the document, enabling fast lookup of this information as follows. for each subsequent token pair (t_i, t_j) term counts are retrieved (they are pre-computed during tokenization). We next compute a *merge threshold* score as:

$$\text{MScore} = \frac{|\#t_i - \#b_{ij}| + |\#t_j - \#b_{ij}|}{\#t_i + \#t_j}$$

where t_i and t_j are two subsequent tokens, and b_{ij} is the bi-gram comprised of the two tokens. If MScore is lower than a user-specified threshold (hyperparameter), the merged token is added as a new token to the token space, and term counts of the two individual tokens are diminished by MScore as $\#t_i = \text{MScore} * \#t_i$, i.e., multiplied with the computed score. Values of MScore, lower than one, imply

Algorithm 1: RaKUn 2.0

Data: Input document D, merge factor τ
1 tokens ← tokenizeDocument(D) ‡ `Tokenization.`
2 tokens ← mergeTokens(D) ‡ `Merging.`
3 G ← documentGraph(tokens) ‡ `Weighted graph.`
4 f ← tokenFrequencies(tokens)
5 tokenRanks ← personalizedPR(G,f) ‡ `Ranking.`
6 K ← sort($N(G)$, tokenRanks) ‡ `Sorting.`
7 **return** K;

more emphasis of multi-term keyphrases (individual terms are not as emphasized), and values larger than one imply more individual token keyphrases. Hence, the MScore serves as an intermediary step which *emphasizes* specific tokens during the ranking step.

The token graph G is constructed from the modified list of tokens by considering subsequent, lower-cased tokens as edges. The edge weights are incremented every time a given bi-gram repeats – the transitions between tokens which commonly co-occur are emphasized. The next step is *node ranking*. Here, a real-valued score is assigned to each (pre-merged) token. We consider personalized PageRank algorithm [25], where the personalization vector is constructed based on term counts. This step results in real-valued scores (between 0 and 1) for each token. The final set of scores is obtained by computing an element-wise product between the PageRank scores and token lengths. This step emphasizes longer keyphrases. We traverse the space of scored tokens and remove case-level duplicates (e.g., 'City' and 'city').

The described algorithm for keyphrase detection was conceived with simplicity in mind. This property also resonates with its computational complexity. Let T represent the number of tokens after the merge step (cardinality difference is negligible with regards to the runtime). Both graph construction and merging need one pass across the token sequence ($\mathcal{O}(|T|)$. The computationally most expensive part is computation of personalized PageRank. In theory, PageRank's complexity is $\mathcal{O}(|T| + l)$, where l is the number of links in the constructed token graph. In practice, the obtained graphs are very sparse – only selected bi-grams co-occur. The opposite case, where dense, clique-like graphs would be produced would imply appearance of tokens in highly diverse contexts, which is highly unlikely. The final step requires sorting of tokens based their scores. This yields the final complexity of $\mathcal{O}(|T| \log |T| \cdot l)$. Assuming very sparse graphs (as observed during the experiments), the complexity remains linear with regards to the number of tokens in the token set after the merge step.

4 Evaluation

We next discuss the evaluation procedures used to estimate the performance of individual algorithms, followed by a discussion regarding their comparison. We evaluate each algorithm with regards to three main aspects; retrieval performance, keyword duplication rate and computation time. The retrieval performance was measured as done in the previous work [8]. Precision@k is defined as $\frac{|\text{Gold} \cap \text{k-predicted}|}{k}$. Recall@k is defined as $\frac{|\text{Gold} \cap \text{k-predicted}|}{|\text{Gold}|}$. Precision represents the number of keyphrases retrieved with regards to top k predicted ones, while recall represents the overall retrieval capability. We also computed (macro) F1, which is the harmonic mean of precision and recall, averaged across documents.

The second score is the *duplication rate*. We compute this score as follows; for each detected keyphrase, we first split it to separate tokens (if multi-token keyphrase is considered). For each part, we traverse the space of detected tokens. If there is a match, we increment a duplicate counter, otherwise, we increment

Table 1. Summary of the considered data sets.

Dataset	#Docs	#KW	Mean KW tokens	Mean doc len
wiki20 [22]	20	35.5	2.0	7728.0
fao30 [21]	30	32.2	1.6	4710.3
theses100 [19]	100	6.7	2.0	4813.9
citeulike180 [20]	183	17.4	1.3	4517.9
Nguyen2007 [24]	209	12.0	2.1	4425.6
SemEval2010 [15]	243	15.6	2.2	7093.3
SemEval2017 [2]	493	17.3	2.9	168.3
500N-KPCrowd-v1.1 [18]	500	49.2	1.4	393.9
PubMed [1]	500	14.2	1.9	3880.2
kdd [11]	755	4.1	2.0	74.1
fao780 [21]	779	8.0	1.6	4685.0
Schutz2008 [27]	1231	45.3	1.5	2362.6
www [11]	1330	4.8	1.9	82.0
Inspec [14]	2000	14.1	2.2	112.5
Krapivin2009 [16]	2304	5.3	2.1	7094.1

the non_duplicate counter. The final score is computed as $\frac{\#duplicates}{\#non\_duplicates+1}$, and was observed to be in the interval [0, 1]. The computation time was measured in seconds (for each document). For visualization of retrieval-efficiency tradeoffs with regards to the mentioned scores it makes sense to have uniform meaning of large and small values. Hence, we introduce the following adapted scores which reflect this idea. The retrieval capability already corresponds to e.g., F1 score, meaning that higher values are preferred. We additionally normalize F1 scores to range between 0 and 1 based on the worst-best performing algorithms (on average). This way, an algorithm scored with 0 is the worst-performing one, while the top performing is scored with 1 (see Fig. 8). Similar adaptations were considered for time performance (normalized inverse times) and duplication rates (normalized inverse duplication rates). One of the main results of this paper is a visualization which jointly considers all three aspects. The considered collection of data sets is summarized in Table 1.

The considered baselines are discussed next. The graph-based baselines include MultiPartiteRank [6], SingleRank [29], TextRank [23] and TopicalPageRank [7]. The statistical baseline considered was YAKE [8]. The language model-based baseline is the recent KeyBERT [12]. For all approaches, we considered the default hyperparameter configurations, as we were interested in out-of-the-box performance. We computed, however, two variants of KeyBERT, one which emits single tokens (KeyBERT-(1,1)) and one which permits two term tokens (KeyBERT-(1,2)). Default configuration of KeyBERT variants performed worse than term frequency-based extraction[1], and offered (1,1) adequate performance only when we set the 'maxsum' and 'mmr' flags to 'true'. The stopwords used were the same for all approaches (NLTK's default English stopwords [4]). Other algorithms' implementations were based on the PKE library [5].

5 Results

A summary of algorithm run times (relative to one another) is shown in Fig. 3. As expected, the simplest baseline (term frequency) is up to three orders of magnitude faster than e.g., BERT-based model. The second approach that performs substantially better, while remaining up to two orders of magnitude faster is the proposed RaKUn 2.0. It is closely followed by SingleRank and TopicalPageRank. The duplication levels are shown in Fig. 4. The duplication ablation indicates the highest duplication levels were observed for YAKE, TopicalPageRank and TextRank. MultiPartiteRank and SingleRank had notably lower duplication levels (KeyBERT-(1,1) as well the term frequency (unigram) baseline. The proposed RaKUn 2.0 is at the lower end of the approaches with regards to this score, albeit not being optimal.

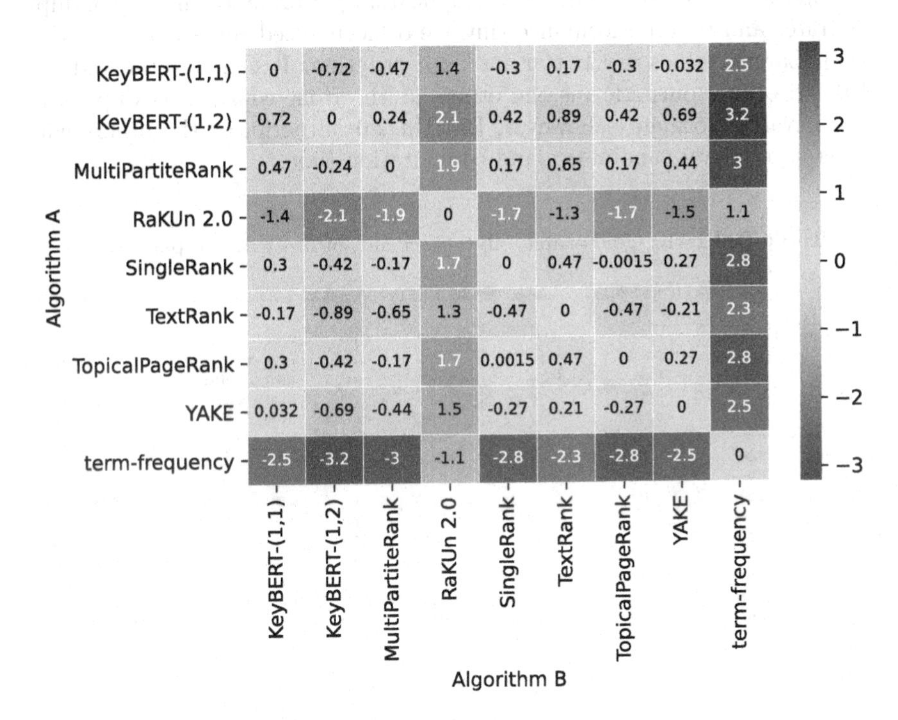

Fig. 3. Pairwise time comparison of average algorithm run times ($\log_{10}(\frac{A}{B})$).

[1] We considered unigrams. Inverse document frequencies were not computed as they require the whole corpus, making them not directly comparable to purely unsupervised methods.

We continue the discussion by presenting the retrieval performance. A systematic investigation of algorithm performance is shown in Fig. 5. The results indicate that on average, MultiPartiteRank is the leading algorithm in the low k scenarios. RaKUn 2.0, however, performs very similarly for up to ten keyphrases, which is one of the most common usecases of such algorithms. A more detailed overview of the scores on the per-data set level is given in Tables 1, 2, 3, 4 and 5. The color codes represent top three performers for each data set (gold = first, silver = second, bronze = third). We additionally conducted rank-based difference significance evaluation [9], where the average algorithm ranks are compared across all data sets. If the algorithms are linked with a red line, they perform very similarly ($p < 0.05$). The diagrams are shown as Figs. 6 and 7. The tests indicate that the difference between the top-performing approaches (MultiPartiteRank, YAKE and RaKUn 2.0) is insignificant. Similar observations can be made based on tabular summaries. Overall, however, we can observe a marginal dominance of RaKUn 2.0 w.r.t. precision. Similar retrieval performance amplifies the purpose of this paper, which transcends the retrieval-only evaluation and incorporates also other properties of either the algorithms or the retrieved space.

In Fig. 8, the selected approaches are compared across the three main evaluation criteria – retrieval performance, duplication performance (inverse of duplication rate) and time performance (inverse of normalized times across all algorithms). Larger values are better for each criterion. It can be observed that MultiPartiteRank outperforms the others at the front considering duplication and retrieval performance, however, RaKUn 2.0 outperforms the others when considering retrieval capabilities and computation time.

Table 2. F1@10 (gold = first, silver = second, bronze = third, per row)

Algorithm Dataset	KeyBERT-(1,1)	KeyBERT-(1,2)	MultiPartiteRank	RaKUn 2.0	SingleRank	TextRank	TopicalPageRank	YAKE	TFreq
500N-KPCrowd-v1.1	0.012	0.012	0.171	0.138	0.164	0.057	0.094	0.127	0.106
Inspec	0.0	0.0	0.22	0.143	0.207	0.126	0.24	0.195	0.041
Krapivin2009	0.051	0.057	0.109	0.097	0.094	0.007	0.02	0.118	0.011
Nguyen2007	0.099	0.058	0.168	0.141	0.152	0.025	0.053	0.188	0.035
PubMed	0.1	0.021	0.087	0.083	0.072	0.002	0.004	0.087	0.036
Schutz2008	0.088	0.023	0.23	0.194	0.219	0.015	0.031	0.15	0.075
SemEval2010	0.071	0.053	0.152	0.139	0.133	0.01	0.023	0.155	0.023
SemEval2017	0.0	0.0	0.216	0.132	0.203	0.122	0.224	0.175	0.056
citeulike180	0.205	0.03	0.172	0.225	0.14	0.004	0.013	0.185	0.097
fao30	0.16	0.027	0.176	0.233	0.161	0.008	0.011	0.15	0.072
fao780	0.116	0.013	0.141	0.138	0.118	0.004	0.009	0.138	0.064
kdd	0.0	0.001	0.107	0.144	0.094	0.058	0.109	0.144	0.056
theses100	0.099	0.017	0.149	0.103	0.128	0.004	0.006	0.093	0.042
wiki20	0.222	0.013	0.186	0.226	0.163	0.0	0.0	0.135	0.021
www	0.0	0.001	0.11	0.113	0.099	0.065	0.109	0.129	0.062

Table 3. Precision@10 (gold = first, silver = second, bronze = third, per row)

Algorithm Dataset	KeyBERT-(1,1)	KeyBERT-(1,2)	MultiPartiteRank	RaKUn 2.0	SingleRank	TextRank	TopicalPageRank	YAKE	TFreq
500N-KPCrowd-v1.1	0.046	0.037	0.38	0.323	0.36	0.129	0.173	0.262	0.192
Inspec	0.0	0.0	0.174	0.112	0.165	0.101	0.189	0.152	0.032
Krapivin2009	0.037	0.04	0.079	0.069	0.068	0.005	0.014	0.084	0.008
Nguyen2007	0.09	0.05	0.151	0.124	0.138	0.022	0.048	0.166	0.032
PubMed	0.065	0.013	0.057	0.055	0.047	0.001	0.003	0.057	0.023
Schutz2008	0.193	0.047	0.504	0.433	0.48	0.029	0.065	0.329	0.163
SemEval2010	0.075	0.055	0.159	0.146	0.14	0.011	0.024	0.162	0.023
SemEval2017	0.0	0.001	0.293	0.184	0.278	0.169	0.3	0.235	0.077
citeulike180	0.208	0.03	0.172	0.228	0.14	0.003	0.012	0.183	0.097
fao30	0.183	0.033	0.21	0.28	0.19	0.01	0.013	0.18	0.087
fao780	0.075	0.008	0.092	0.09	0.077	0.002	0.006	0.089	0.041
kdd	0.0	0.001	0.064	0.087	0.056	0.036	0.065	0.085	0.034
theses100	0.064	0.011	0.098	0.068	0.084	0.002	0.004	0.06	0.027
wiki20	0.19	0.01	0.155	0.19	0.135	0.0	0.0	0.12	0.02
www	0.0	0.001	0.066	0.068	0.06	0.04	0.065	0.076	0.037

Table 4. Recall@10 (gold = first, silver = second, bronze = third, per row)

Algorithm Dataset	KeyBERT-(1,1)	KeyBERT-(1,2)	MultiPartiteRank	RaKUn 2.0	SingleRank	TextRank	TopicalPageRank	YAKE	TFreq
500N-KPCrowd-v1.1	0.007	0.007	0.144	0.119	0.139	0.041	0.087	0.129	0.113
Inspec	0.0	0.0	0.356	0.233	0.331	0.194	0.388	0.326	0.07
Krapivin2009	0.097	0.119	0.212	0.187	0.182	0.014	0.041	0.236	0.021
Nguyen2007	0.135	0.087	0.23	0.216	0.205	0.036	0.078	0.279	0.05
PubMed	0.27	0.065	0.223	0.209	0.181	0.009	0.013	0.239	0.096
Schutz2008	0.063	0.017	0.161	0.133	0.153	0.011	0.022	0.104	0.053
SemEval2010	0.071	0.054	0.152	0.139	0.133	0.01	0.024	0.156	0.023
SemEval2017	0.0	0.0	0.183	0.11	0.17	0.101	0.189	0.15	0.046
citeulike180	0.221	0.034	0.187	0.242	0.151	0.005	0.016	0.205	0.104
fao30	0.149	0.023	0.159	0.21	0.147	0.007	0.01	0.134	0.065
fao780	0.335	0.036	0.396	0.39	0.321	0.01	0.025	0.39	0.193
kdd	0.0	0.003	0.384	0.514	0.346	0.188	0.398	0.562	0.194
theses100	0.266	0.045	0.389	0.262	0.326	0.019	0.02	0.254	0.116
wiki20	0.294	0.017	0.251	0.297	0.221	0.0	0.0	0.166	0.023
www	0.0	0.004	0.393	0.412	0.352	0.221	0.394	0.502	0.237

Table 5. Retrieval time (s). (gold = first, silver = second, bronze = third, per row)

Algorithm Dataset	KeyBERT-(1,1)	KeyBERT-(1,2)	MultiPartiteRank	RaKUn 2.0	SingleRank	TextRank	TopicalPageRank	YAKE	TFreq
500N-KPCrowd-v1.1	0.422	0.763	0.477	0.009	0.454	0.417	1.552	0.699	0.0
Inspec	0.202	0.337	0.399	0.006	0.4	0.394	1.554	0.74	0.0
Krapivin2009	1.561	11.282	6.144	0.08	4.332	1.111	2.852	1.423	0.007
Nguyen2007	1.333	6.318	3.528	0.05	2.45	0.87	2.505	1.304	0.005
PubMed	1.237	4.865	2.823	0.046	1.945	0.766	2.312	1.249	0.004
Schutz2008	1.58	6.926	3.993	0.038	2.675	0.772	2.428	1.236	0.004
SemEval2010	1.675	12.479	6.135	0.076	4.213	1.117	2.707	1.378	0.008
SemEval2017	0.213	0.431	0.403	0.007	0.395	0.389	1.596	0.947	0.0
citeulike180	1.556	8.006	3.937	0.051	2.411	0.812	2.446	1.322	0.005
fao30	1.528	7.599	4.665	0.056	2.793	0.877	2.47	1.573	0.005
fao780	1.531	7.806	5.284	0.056	3.111	0.838	2.53	1.479	0.005
kdd	0.153	0.268	0.394	0.006	0.394	0.383	1.371	0.549	0.0
theses100	1.52	7.069	4.05	0.053	2.603	0.811	2.293	1.644	0.004
wiki20	1.598	10.453	5.674	0.066	3.8	0.952	2.586	1.429	0.006
www	0.152	0.269	0.39	0.006	0.396	0.396	1.283	0.552	0.0

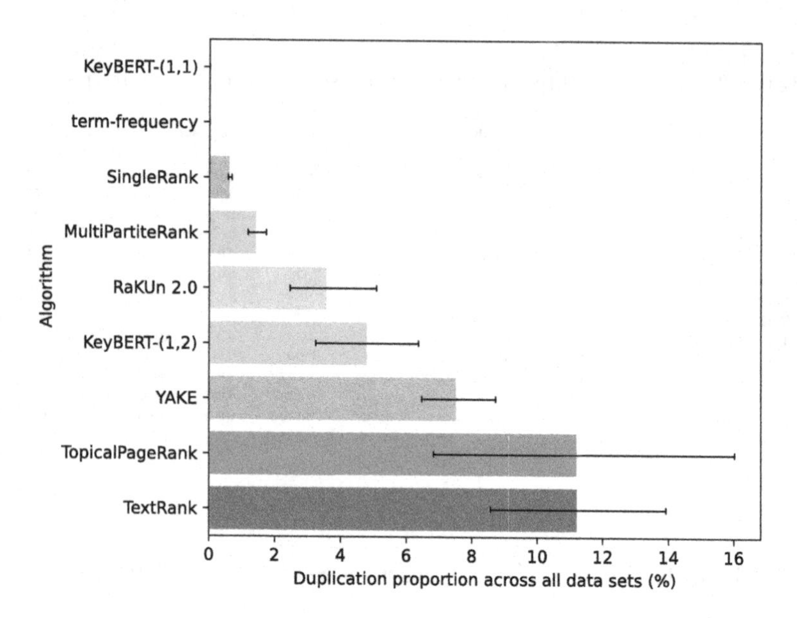

Fig. 4. Duplication levels for different algorithms.

5.1 Scaling to 14M Documents

A direct way of testing the complexity bounds stated in the methods section was to attempt and run RaKUn 2.0 directly on the collection of approximately 14 million biomedical articles – the MeDAL corpus [30][2]. The corpus was parsed into a list of documents and fed into the default configuration of RaKUn 2.0. The computation took approximately forty seconds (including text reading) on a virtual machine with 12 cores and 32GB of RAM. The list of top ten keyphrases is shown as Table 6.

The top keyphrases correspond to rather general biological terms, which are some of the main topics related to the considered documents. The results were obtained by maintaining the merge_threshold hyperparameter set to one – single term keyphrases can be obtained if this threshold is lowered. For example, if set to 0.5, the top three keyphrases are 'activity', 'concentration' and 'enzyme'.

[2] https://www.reddit.com/r/MachineLearning/comments/jx63fd/r_a_14m_articles_dataset_for_medical_nlp/.

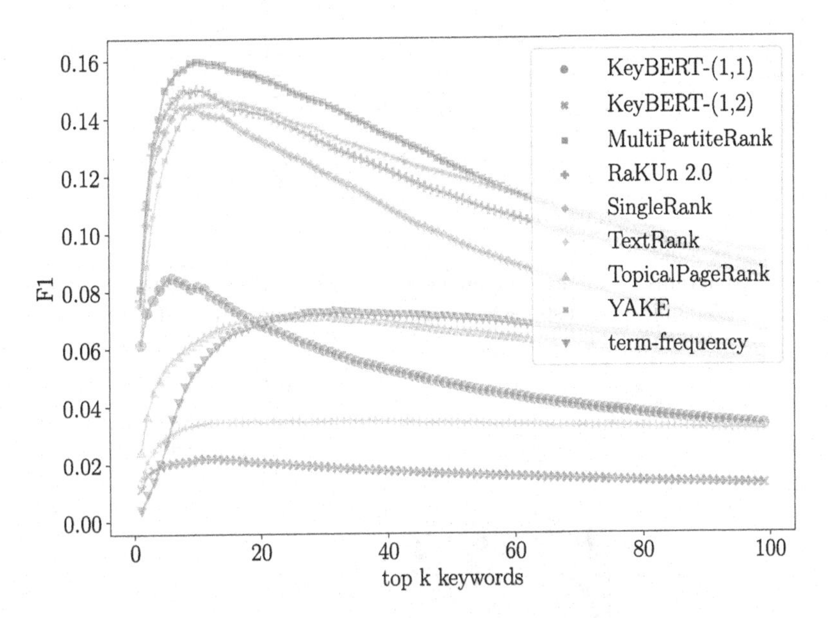

Fig. 5. F1 score for different top k keyphrases, averaged across all data sets.

Table 6. 14M articles summarized as top ten keyphrases.

Keyphrase	Score
presence	0.02041868080426608
molecular weights	0.01313742352650019
glutamine synthetase	0.01081927396059080
growth hormone	0.01081481738381907
arterial blood	0.00973761662559790
investigated	0.00926714499542069
rate constant	0.00904369510973679
blood flow	0.00899499866920862
molecular weight	0.00865807865159297
sodium dodecyl	0.00865611530561878

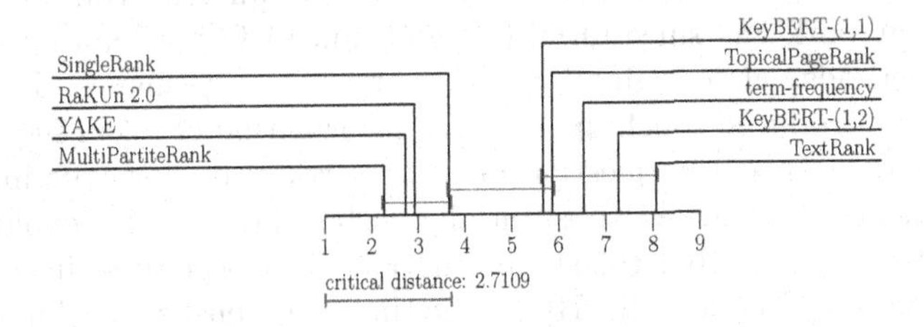

Fig. 6. Critical difference diagram - F1@15. RaKUn 2.0's performance is (statistically) comparable to the recent state-of-the-art approaches.

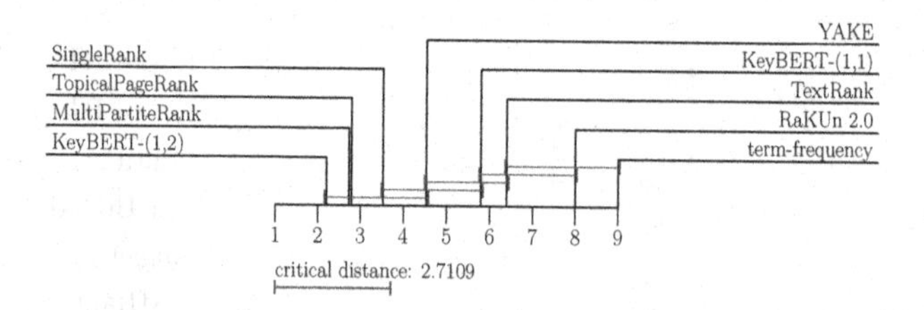

Fig. 7. CD diagrams – time per document. Higher ranks indicate faster compute time. RaKUn 2.0 is significantly faster when compared to other state-of-the-art methods.

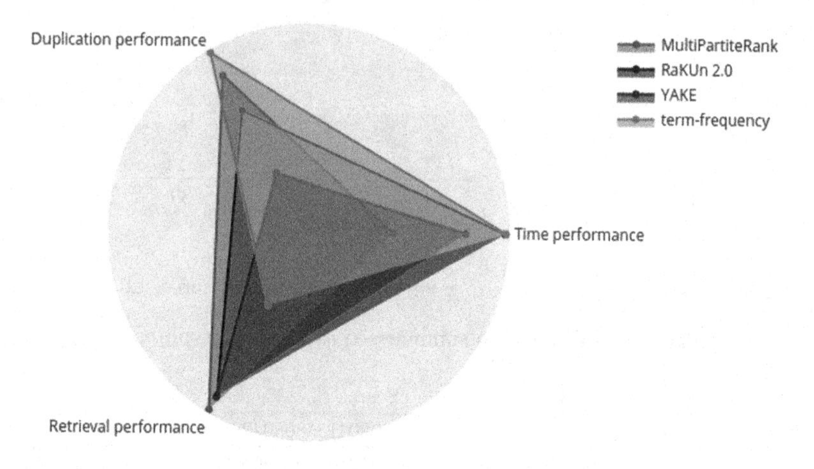

Fig. 8. A visualization comparing best and worst-performing approaches with regards to three different criteria relevant in practice. Note that the scores are relative with regards to the considered methods' performances.

6 Discussion and Conclusions

In this paper we presented an approach to unsupervised keyphrase detection, aimed specifically at pushing the limits of computation time and retrieval performance. The main contributions of this paper are an algorithm for keyphrase detection that performs substantially (significantly) faster than current state-of-the-art methods, while maintaining the retrieval performance. The algorithmic novelties introduced touch upon the transformation of token sequences into graphs, and re-address the question of meta vertices by constructing them at the sequence level, which is substantially faster. Further, by exploiting personalized PageRank, global token information is incorporated into keyphrase ranking alongside token lengths. By conducting an extensive benchmark against established baselines, this paper presents an evaluation which incorporates both retrieval capabilities, but further details into computation time and duplication rates amongst the retrieved keyphrases.

Analysis of keyphrase detection algorithms with regards to multiple evaluation criteria is becoming of higher relevance, as many low-latency applications cannot afford expensive detection phase. To our knowledge, this paper is similarly one of the first to evaluate the performance based on critical difference diagrams, exactly assessing the significance of observed differences (in time and retrieval performance).

Further work includes exploration of lower-level implementations of top-performing approaches, alongside their parts that could be subject to parallelism. A potentially interesting endeavor would also include background knowledge (as graphs), possibly enabling detection of keywords beyond the ones found in a given document, while remaining unsupervised.

7 Replicability

The RaKUN 2.0 algorithm is available as a simple-to-use Python library available at https://github.com/SkBlaz/rakun2.

Acknowledgements. The work was supported by the Slovenian Research Agency (ARRS) core research programme Knowledge Technologies (P2-0103), and projects Computer-assisted multilingual news discourse analysis with contextual embeddings (J6-2581) and Quantitative and qualitative analysis of the unregulated corporate financial reporting (J5-2554). The work was also supported by the Ministry of Culture of Republic of Slovenia through project Development of Slovene in Digital Environment (RSDO).

References

1. Aronson, A.R., et al.: The NLM indexing initiative. In: Proceedings of the AMIA Symposium, p. 17. American Medical Informatics Association (2000)
2. Augenstein, I., Das, M., Riedel, S., Vikraman, L., McCallum, A.: SemEval 2017 task 10: ScienceIE - extracting keyphrases and relations from scientific publications. In: Proceedings of the 11th International Workshop on Semantic Evaluation (SemEval-2017), pp. 546–555. Association for Computational Linguistics, Vancouver (2017). https://doi.org/10.18653/v1/S17-2091, https://aclanthology.org/S17-2091
3. Beliga, S., Meštrović, A., Martincic-Ipsic, S.: An overview of graph-based keyword extraction methods and approaches. J. Inf. Organ. Sci. **39**, 1–20 (2015)
4. Bird, S., Klein, E., Loper, E.: Natural Language Processing with Python: Analyzing Text with the Natural Language Toolkit. O'Reilly Media, Inc., Sebastopol (2009)
5. Boudin, F.: PKE: an open source python-based keyphrase extraction toolkit. In: Proceedings of COLING 2016, the 26th International Conference on Computational Linguistics: System Demonstrations, Osaka, Japan, pp. 69–73, December 2016. https://aclweb.org/anthology/C16-2015
6. Boudin, F.: Unsupervised keyphrase extraction with multipartite graphs. In: Proceedings of the 2018 Conference of the North American Chapter of the Association for Computational Linguistics: Human Language Technologies, Volume 2 (Short Papers), pp. 667–672. Association for Computational Linguistics, New

Orleans (2018). https://doi.org/10.18653/v1/N18-2105, https://aclanthology.org/N18-2105
7. Bougouin, A., Boudin, F., Daille, B.: TopicRank: graph-based topic ranking for keyphrase extraction. In: Proceedings of the Sixth International Joint Conference on Natural Language Processing, pp. 543–551. Asian Federation of Natural Language Processing, Nagoya (2013). https://aclanthology.org/I13-1062
8. Campos, R., Mangaravite, V., Pasquali, A., Jorge, A., Nunes, C., Jatowt, A.: Yake! keyword extraction from single documents using multiple local features. Inf. Sci. **509**, 257–289 (2020). https://doi.org/10.1016/j.ins.2019.09.013
9. Demšar, J.: Statistical comparisons of classifiers over multiple data sets. J. Mach. Learn. Res. **7**(1), 1–30 (2006). https://jmlr.org/papers/v7/demsar06a.html
10. Ding, H., Luo, X.: AttentionRank: unsupervised keyphrase extraction using self and cross attentions. In: Proceedings of the 2021 Conference on Empirical Methods in Natural Language Processing, pp. 1919–1928. Association for Computational Linguistics, Online and Punta Cana, Dominican Republic, November 2021. https://doi.org/10.18653/v1/2021.emnlp-main.146, https://aclanthology.org/2021.emnlp-main.146
11. Gollapalli, S.D., Caragea, C.: Extracting keyphrases from research papers using citation networks. In: Brodley, C.E., Stone, P. (eds.) Proceedings of the Twenty-Eighth AAAI Conference on Artificial Intelligence, 27–31 July 2014, Québec City, Québec, Canada, pp. 1629–1635. AAAI Press (2014). https://www.aaai.org/ocs/index.php/AAAI/AAAI14/paper/view/8662
12. Grootendorst, M.: KeyBERT: minimal keyword extraction with BERT (2020). https://doi.org/10.5281/zenodo.4461265
13. Hasan, K.S., Ng, V.: Automatic keyphrase extraction: a survey of the state of the art. In: Proceedings of the 52nd Annual Meeting of the Association for Computational Linguistics (Volume 1: Long Papers), pp. 1262–1273. Association for Computational Linguistics, Baltimore (2014). https://doi.org/10.3115/v1/P14-1119, https://aclanthology.org/P14-1119
14. Hulth, A.: Improved automatic keyword extraction given more linguistic knowledge. In: Proceedings of the 2003 Conference on Empirical Methods in Natural Language Processing, pp. 216–223 (2003). https://aclanthology.org/W03-1028
15. Kim, S.N., Medelyan, O., Kan, M.Y., Baldwin, T.: SemEval-2010 task 5: automatic keyphrase extraction from scientific articles. In: Proceedings of the 5th International Workshop on Semantic Evaluation, pp. 21–26. Association for Computational Linguistics, Uppsala (2010). https://aclanthology.org/S10-1004
16. Krapivin, M., Autaeu, A., Marchese, M.: Large dataset for keyphrases extraction (2009)
17. Kumar, T., Mahrishi, M., Meena, G.: A comprehensive review of recent automatic speech summarization and keyword identification techniques. Artif. Intell. Ind. Appl. 111–126 (2022)
18. Marujo, L., Viveiros, M., da Silva Neto, J.P.: Keyphrase cloud generation of broadcast news (2013)
19. Medelyan, O.: Human-competitive automatic topic indexing. Ph.D. thesis, The University of Waikato (2009)
20. Medelyan, O., Frank, E., Witten, I.H.: Human-competitive tagging using automatic keyphrase extraction. In: Proceedings of the 2009 Conference on Empirical Methods in Natural Language Processing, pp. 1318–1327. Association for Computational Linguistics, Singapore (2009). https://aclanthology.org/D09-1137

21. Medelyan, O., Witten, I.H.: Domain-independent automatic keyphrase indexing with small training sets. arXiv preprint abs/10.1002 (2010). https://arxiv.org/abs/10.1002
22. Medelyan, O., Witten, I.H., Milne, D.: Topic indexing with Wikipedia. In: Proceedings of the AAAI WikiAI Workshop, vol. 1, pp. 19–24 (2008)
23. Mihalcea, R., Tarau, P.: TextRank: bringing order into text. In: Proceedings of the 2004 Conference on Empirical Methods in Natural Language Processing, pp. 404–411. Association for Computational Linguistics, Barcelona (2004). https://aclanthology.org/W04-3252
24. Nguyen, T.D., Kan, M.-Y.: Keyphrase extraction in scientific publications. In: Goh, D.H.-L., Cao, T.H., Sølvberg, I.T., Rasmussen, E. (eds.) ICADL 2007. LNCS, vol. 4822, pp. 317–326. Springer, Heidelberg (2007). https://doi.org/10.1007/978-3-540-77094-7_41
25. Page, L., Brin, S., Motwani, R., Winograd, T.: The pagerank citation ranking: bringing order to the web. Technical report 1999-66, Stanford InfoLab (1999). https://ilpubs.stanford.edu:8090/422/, previous number = SIDL-WP-1999-0120
26. Papagiannopoulou, E., Tsoumakas, G.: A review of keyphrase extraction. Wiley Interdisc. Rev. Data Min. Knowl. Discov. **10**(2), e1339 (2020)
27. Schutz, A.T., et al.: Keyphrase extraction from single documents in the open domain exploiting linguistic and statistical methods. M. App. Sc thesis (2008)
28. Škrlj, B., Repar, A., Pollak, S.: *RaKUn*: *Ra*nk-based *K*eyword extraction via *Un*supervised learning and meta vertex aggregation. In: Martín-Vide, C., Purver, M., Pollak, S. (eds.) SLSP 2019. LNCS (LNAI), vol. 11816, pp. 311–323. Springer, Cham (2019). https://doi.org/10.1007/978-3-030-31372-2_26
29. Wan, X., Xiao, J.: CollabRank: towards a collaborative approach to single-document keyphrase extraction. In: Proceedings of the 22nd International Conference on Computational Linguistics (COLING 2008), pp. 969–976. COLING 2008 Organizing Committee, Manchester, UK (2008). https://aclanthology.org/C08-1122
30. Wen, Z., Lu, X.H., Reddy, S.: MeDAL: medical abbreviation disambiguation dataset for natural language understanding pretraining. In: Proceedings of the 3rd Clinical Natural Language Processing Workshop, pp. 130–135. Association for Computational Linguistics, Online (2020). https://doi.org/10.18653/v1/2020.clinicalnlp-1.15, https://aclanthology.org/2020.clinicalnlp-1.15

A Fuzzy OWL Ontologies Embedding for Complex Ontology Alignments

Houda Akremi[1](✉), Mouhamed Gaith Ayadi[2], and Sami Zghal[1,3]

[1] Faculty of Sciences of Tunis, LIPAH -LR11ES14, University of Tunis El Manar, Tunis 2092, Tunisia
houda.akremi@fst.utm.tn

[2] ISG, BESTMOD-LR99ES04, University of Tunis, Bardo 2000, Tunisia
mouhamed.gaith.ayadi@gmail.com

[3] Faculté des Sciences Juridiques, Économiques et de Gestion de Jendouba, Université de Jendouba, Campus Universitaire, Jendouba 8189, Tunisia
sami.zghal@fsjegj.rnu.tn

Abstract. The semantic heterogeneity concern in the information integration can be handled by applying ontology alignment. The purpose of the ontology alignment procedure is to locate concepts that are semantically identical in two ontologies. But, one of these alignments' downsides is the lack of expressiveness and uncertainties, which can be accounted by using fuzzy complex alignments. To address this issue, the use of an effective strategy, consisting of two parts, is applied. We proceeded by establishing of a fuzzification approach that enables a semantic representation of both crisp and fuzzy data. The next step was to model fuzzy OWL 2 ontologies in vector space by a semantic embedding-based ontology matching technique and compute their similarity scores to determine the correlation levels. Then, it is reinforced by a stable marriage-based alignment extraction algorithm to establish a high-quality matching. Our proposed alignment scheme has been validated and reviewed on the benchmark tracks supplied by the Ontology Alignment Evaluation Initiative (OAEI). Experimental findings demonstrated the effectiveness of our matching method.

Keywords: Ontology · Ontology fuzzification · Complex alignment · Ontology embedding

1 Introduction

On the Semantic Web, the issue of managing heterogeneity among multiple information resources is becoming increasingly challenging. Ontology alignment (also known by Ontology matching) is a critical task for managing semantic heterogeneity. The majority of the provided ontology matching approaches are conventional [12], which seek correspondences between related entities in various ontologies. This discipline is structured by two ' paradigms: simple and complex matching. Simple methods [34] are restricted to matching single entities and they are insufficiently expressive to surmount ontological conceptual heterogeneity [33].

P. Pascal and D. Ienco (Eds.): DS 2022, LNAI 13601, pp. 394–404, 2022.
https://doi.org/10.1007/978-3-031-18840-4_28

However, complex ontology alignment methods enable the mapping of a concept from the first ontology to numerous concepts from the second ontology [28]. These approaches can provide correlations that better express the links between entities in various ontologies. To deal with that, experts have investigated numerous ontology alignment methodologies and created a variety of semi-automatic and automated ontology matching systems [30–32]. Yet, present complex ontology matching techniques have lot of limitations, such as the matcher's inadequate ontology similarity computation and ontology mapping findings, etc. Moreover, ontologies are still significantly constrained in terms of expressing information and knowledge in the real world. So, ontologies fuzzification could be a technique for modeling incorrect knowledge. In particular, fuzzy ontological representations can more accurately describe practical knowledge in reference to a specific topic. Fuzzy knowledge tends to boost the decision-making process' clarity and effectiveness. To overcome these shortcomings, we established a fuzzification approach that allows the semantic representation of both crisp and fuzzy data. Then, a new complex alignment approach of ontologies is provided. In this regard, we suggest putting in place a framework to generate semantic embeddings for OWL 2 ontologies. The plan is to map entities in vector space and to compute their similarity values later to ensure the matching task. Furthermore, a stable marriage-based ontology extraction approach is provided to increase alignment quality. The output alignment attempts to entirely encompass the two ontologies' shared topic.

The remainder of the paper is organized as follows. Section 2 outlines the related work. Section 3 focuses on the proposed approach. Section 4 overviews an evaluation of our approach. Finally, Sect. 5 provides the major conclusions and discusses its perspectives.

2 Related Work

In this section, we first go over the fundamentals of complex ontology alignments and related applications (2.1), and then we go over for the fuzzification of the ontological representation (2.2). Finally, we go over for various ontology embedding approaches (2.3).

2.1 Complex Alignments

The procedure of establishing an alignment A for a pair of ontologies O_1 and O_2 is referred as ontology matching [33]. A is directional which means to find an alignment between a source ontology O_1 and a target ontology O_2, denoted $A_{O_1 \longrightarrow O_2}$. So, an alignment is a set of correspondences among entities belonging to the matched ontologies with various cardinalities: $1:1$ (one-to-one), $1:m$ (one-to-many), $n:1$ (many-to-one) or $n:m$ (manyto-many). A correspondence is a triple (e_1, e_2, r):

- e_1 and e_2 are entities, e.g., classes and properties of the first and the second ontology, respectively;

- r is a relation, e.g., equivalence ($=$), more general ($\sqsupseteq$), disjointness ($\perp$), holding between e_1 and e_2.

The correspondence (e_1, e_2, r) claims that the relation r holds among the ontology entities e_1 and e_2. According to that, we explore two sorts of correspondences: simple and complex. In fact, a complex alignment includes at least one complex correspondence. Complex alignment creation seems to be more challenging than simple alignment creation. Certainly, the alignment space, which reflects the set of all possible correspondences across ontologies, is wider in complex matching generation compared to simple matching generation. Simple alignments are not as expressive as complex alignments. Works employing such complex alignments have been suggested for a variety of tasks including ontology evolution [11], data translation [4] and ontology merging [5]. As a start, the AROA system (Association Rule-based Ontology Alignment) [24] provides a collection of matching conditions for detecting matching patterns. Then, KAOM (Knowledge-Aware Ontology Matching) [13] uses several matching methods used to identify transformation function correspondences and logical relation correspondences. Likewise, the CANARD framework [27] is concerned with linear Competency Questions for Alignment. Finally, The AML (AgreementMakerLight) [7] is a mechanism applying lexical matching methods for correlating ontologies. With this variety of proposed models, researchers frequently ought to employ several methodologies to raise the efficiency of the matching task. It is clear that it is possible to compute semantic similarity via OWL 2 ontologies embeddings which increases the matcher's accuracy.

2.2 Ontologies Fuzzification

Fuzzy logic is currently experiencing an alternate challenge, which is affecting the Semantic Web viewpoint. A fuzzy ontology enables for the appropriate knowledge to be mapped out in an unambiguous ontology. According to Li et al. [17], on the principle of fuzzy representation, the following ontological concepts are homogeneous to regular recognition:

- **Fuzzy concept:** refer to concepts with no sufficient certainty;
- **Fuzzy roles:** explain the fuzzy connection between concept instances;
- **Fuzzy data types:** allow for improper perceptions of attribute values throughout the fuzzification process.

Zhai et al. [15] sought to clarify the fuzzification of the ontological representation in a fuzzy ontological scheme. Then, Bouaziz et al. [2] conducted a nearly identical pilot investigation, focusing on the variation between these fuzzy aspects. Zekri et al. [36] provided a particular fuzzy ontological description of Alzheimer's disease, known as AlzFuzzyOnto. El-Sappagh et al. [6] report the advancement of diabetes detection research through "CBRDiabOnto". Gomez-Romero et al. [8] proposed an adaptive fuzzy analysis-related expansion integrating semantic BIMs (building information models) that provide support for imperfect knowledge categorization. A procedural methodology known as FODM was introduced by Li et al. [17] for fuzzy ontological structure. The fact that neither of the present fuzzy ontological methodological structures addresses the negative effects of utilizing the current crisp structural features throughout the determination process also has to be brought up. In light of that, this study presents a sketch of a fuzzy ontological structure characterized by two specific details: *(i)* utilizing the existing crisp ontological representation and *(ii)* recognizing fuzzy ontological features. Therefore, the use of a knowledge ontological structure is the first step in our fuzzification technique. The next steps are the acquisition of the detection of the components of a fuzzy ontological structure and their explanation.

2.3 Ontology Embedding

There are several methods in the literature that conduct (RDF) knowledge graphs embeddings, which mainly concentrate on the data instances embedding [20,23,29]. But, embeddings benefit from data instances, and the ontology's knowledge is often neglected. In perspective of concept embeddings, significant attempts have been made to utilize word embeddings to correlate the lexical information with a vector representation. This method has typically been used for ontology matching assignments [16,21] . The basic drawback of this method is the neglect of the rich semantics of the ontologies especially with domain-specific terminology. Onto2Vec [25] and OPA2Vec [26] are special systems for ontology embedding. Starting from these methods limitations, the framework OWL2Vec* [3] is implemented to arrange: *(i)* the generation of significant vectors for small-medium ontologies, *(ii)* the embeddings noise due to OWL constructs, *(iii)* the generation of similar embeddings using Word2Vec. This method is followed and exploited in our work to develop an ontology embedding matching system reinforced by a stable marriage-based ontology extraction algorithm, which we will detail later.

3 The Fuzzy Complex Ontology Alignment Framework

This study expands a fuzzy ontology matcher that computes the similarity values of two fuzzy entities based on a semantic OWL embedding model. This framework is made up of four major components: **Parsing component**, **Fuzzification component**, **Embedding component**, and **Alignment component**.

3.1 Parsing Component

The parsing (or pretreatment) process is vital for complex alignment ontologies. It is carried out with the help of the OWL API[1]. This phase turns the studied ontologies into a format suitable for the remaining treatments. The aim is to treat all of the available information in both loaded OWL files, whose each entity is defined by all of its properties. A linguistic pretreatment process precedes

[1] http://owlapi.sourceforge.net/.

this procedure, including cleaning empty words, removing special characters, and lemmatizing. This process gets the ontologies ready for the fuzzification procedure.

3.2 Fuzzification Component

In this part, we clearly outline each step of the fuzzification methodology to indicate ambiguous ontological entities, by injecting fuzzy aspects. It is defined by three phases. : crisp ontology analysis, ontology fuzzification and consistency checking.

Crisp Ontology Analysis. The purpose of this phase is to pinpoint and to gather the imprecise ontological entities (concepts, relations, axioms, etc.) in the crisp domain ontology that can be fuzzified according to a linguistic variable's particular value. Linguistic variables are the terminology applied to describe a scenario, a phenomena, or a procedure, like as temperature, age, etc. We took advantage of the lexical database WordNet[2] to form the fuzzy entities.

Ontology Fuzzification. This step is about the fuzzification of the crisp domain ontology. We consider adopting the Fuzzy OWL2 language to enable the description of fuzzy data-types, fuzzy modified concepts, weighted concepts, weighted sum concepts, fuzzy nominals, fuzzy modifiers, fuzzy modified roles, data-types, and fuzzy axioms. In the ontology, these components are expressed as classes, relations, and individuals, accordingly.The latter has seven sub-steps, which are as follows:

- **Crisp ontology annotation:** Both logics of Zadeh [35] and that of Lukasiewicz [18] are commonly employed for the annotation.
- **Concepts fuzzification:** The fuzzy linguistic values are used to assess the fuzzy concepts via a fuzzy OWL2 annotation.A degree of truth among 0 and 1 can be used to characterize the fuzzy concept.
- **Object properties fuzzification:** Object properties (roles) can really be turned into fuzzy abstract roles depending on a set of requirements including someValuesFrom, allValuesFrom, minCardinality and maxCardinality [37]. A new role is adapted to accommodate the fuzzy object attributes.
- **Data-types fuzzification:** A data-type attribute specified as an OWL class instance connects individuals to data values.To enhance imprecision, a basic OWL2 datatype can be translated into fuzzy data-types via a fuzzy membership functions.
- **Fuzzy modifiers and fuzzy modified data-types:** A fuzzy data-type is one that connects individuals to data values with the respect of its features and constraints [37]. Fuzzy modifiers can be used to adjust the membership degree of fuzzy data-types.

[2] https://wordnet.princeton.edu/.

- **Fuzzy modified data-type properties:** The attributes of fuzzy modified data-type are expressed as fuzzy modified data-type.
- **Axioms fuzzification:** The links among individuals, concepts, properties and relations are specified by the fuzzy axioms. Information about the fuzzy ontology structure is described using the fuzzy axioms [37].

Consistency Checking. The syntactic-level assessment is characterized by consistency checking via FuzzyDL reasoner to determine whether the fuzzy ontology is coherent and free of errors. It is a DL reasoning engine that supports fuzzy logic reasoning to examine the fuzzy domain ontology.This process gets the fuzzy ontologies ready for the embedding procedure.

3.3 Embedding Component

The purpose of this assignment is to determine semantic embeddings. Fundamentally, this method begins with the projection of the ontology into a graph. Then, multiple techniques are employed to walk the ontology graph. After that, a corpus of sentences is produced depending on walking patterns. Finally, concept embeddings will be generated from that corpus. To recap, the actual embedding structure is comprised of three major modules:

- **Ontology projection.** To project the ontology into a RDF graph, a simplified model applied by Agibetov et al. [1] is employed and followed. By analogy, the graph nodes indicate the ontology concepts and the edges are labeled with potential correlations between those concepts.
- **Walk strategy.** A set of mechanisms to walk the ontology graph were provided, based on an inspired versions of RDF2Vec [23] and node2vec [9] which consist in the ontology projection as an input and the weighted edges integration for the walks. This new strategy permitted the generation of sentences containing the concept URI and/or concept labels. This solution operates well with massive ontologies and offers semantic similarity even for structures that are identical.
- **Concept embeddings.** The walk techniques are adaptable, allowing the production of several sorts of sentences corpora that result to concept embeddings with various characteristics based on Word2Vec [19] and FastText [14].

3.4 Alignment Component

After the semantic embedding procedure, each entity is expressed in the form of a vector in the vector space. All classes and properties in both ontologies are covered and then, we computed the similarity values using two types of metrics (cosine and linguistic similarity), which it has a great impact on the outcomes of the ontology matching. As previously stated, the cosine similarity metric, as specified below, has been used to compute the similarity of the two entities.

$$Cosine \quad Similarity(V_{w1}, V_{w2}) = \frac{V_{w1}.V_{w2}}{V_{w1}.V_{w2}} \tag{1}$$

where V_{w1} and V_{w2} are, respectively, the vectors of two words $w1$ and $w2$ and V_{w1} and V_{w2}, respectively, denote their norms. The more similar they are, the closer the outcome is to 1.

The two metrics generate two similarity matrices, and an aggregation approach is required to combine the two matrices into one matrix. The higher of two similarity values is considered as the final similarity value, which serves to verify the alignment's accuracy. To improve mapping outcomes, a stable marriage-based ontology extraction technique [10,22] with a thresholding method is adapted. It enables for the selection of suitable mappings. The process is ended when all values in the similarity matrix are less than or equal to the threshold (0.5). So, a similarity of less than 0.5 is considered unreliable. These steps enable the creation of stable mappings.

4 Evaluation

In this section, the effectiveness of our approach is evaluated through a series of tests applying the benchmark track released by the Ontology Alignment Evaluation Initiative (OAEI). The OAEI provided an innovative ontology alignment evaluation report track regarding Complex alignments[3]. We employed a real-world dataset as a potential complex alignment benchmark from the GeoLink project, supported by the U.S. National Science Foundation's EarthCube program. It consists of two ontologies: the GeoLink Base Ontology (GBO) and the GeoLink Modular Ontology (GMO). The matching of the two ontologies was created in collaboration with domain experts from a number of geoscience research institutions. Table 1 exposes the number of classes and properties in both ontologies. Additional details are available in [38]. We have computed to adopt it for our ontology matching method.

Table 1. The number of classes, object properties, and data properties in both GeoLink ontologies.

Ontology	Classes	Object properties	Data properties
GeoLink base ontology	40	149	49
GeoLink modular ontology	156	124	46

The following are the usual definitions of ontology alignment metrics for evaluating the quality of ontology matching findings:

$$Recall = \frac{Correct\_found\_correspondences}{All\_possible\_correspondences} \quad (2)$$

$$Precision = \frac{Correct\_found\_correspondences}{All\_found\_correspondences} \quad (3)$$

[3] http://oaei.ontologymatching.org/2018/complex/index.html.

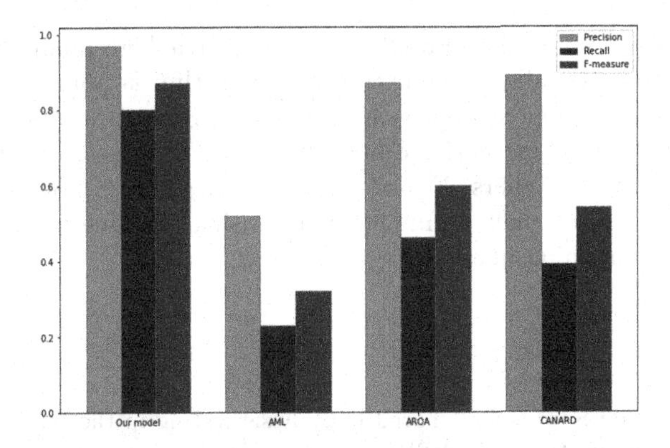

Fig. 1. Performance Comparison with state-of-the-art methods

Table 2. Performance Comparison on GeoLink Benchmark

Evaluation metrics	Our model	AML	AROA	CANARD
Precision	97.00%	52.00%	87.00%	89.00%
Recall	80.00%	23.00%	46.00%	39.00%
F-measure	87.68%	32.00%	60.00%	54.00%

$$F\_measure = \frac{2 * Recall * Precision}{Recall + Precision} \tag{4}$$

where precision and recall reflect the accuracy and sufficiency of the matching findings, respectively, and F-measure allows to adjust them with balance. Figure 1 depicts the precision, recall, and F-measure computed for various state-of-the-art complex alignment methods compared with our model.

It is presented also in Table 2. In fact, our alignment model provides encouragement results. Our model achieves 80.52% F-measure for the simple $(1:1)$ correspondences. Moreover, it provides 87.68% F-measure for complex $(1:n)$ correspondences. According to the performance comparison, only our model and the CANARD system can create almost perfect complex matching. To summarize, the proposed method's success is demonstrated by comparison with other matchers.

5 Conclusion

In this paper, we have proposed a new ontology complex alignment method. To that purpose, this study initially uses an innovative fuzzification methodology that performs in three steps to convert crisp domain ontology into a fuzzy one. Then, this study uses ontologies semantic embedding to model entities in vector space. It computes the similarity values through using cosine similarity metric and the linguistic similarity. Finally, to establish a high-quality matching, a stable marriage-based alignment extraction algorithm is applied. Studies have revealed that our alignment approach provides significant results. Experimental findings show that our method produces better matching outcomes than state-of-the-art ontology matchers. Future work includes the exploration of a more improved ontology complex alignment model using machine learning paradigm to improve the matching quality.

References

1. Asan, A., et al.: Supporting shared hypothesis testing in the biomedical domain. J Biomed. Seman. **9**(1), 9 (2018)
2. Bouaziz, R., Ghorbel, H., Bahri, A.: Fuzzy ontologies model for semantic web. In: The Second International Conference on Information and Knowledge Management, eKNow, Maorten, Netherlands Antilles (2010)
3. Chen, J., Hu, P., Jiménez-Ruiz, E., Holter, O., Antonyrajah, D., Horrocks, I.: Owl2vec*: embedding of owl ontologies. Mach. Learn. **110**, 1813–1845 (2021)
4. David, J., Euzenat, J., Scharffe, F., Trojahn dos Santos, C.: The alignment API 4.0. Semant. Web **2**(1), 3–10 (2011)
5. Dou, D., Qin, H., Lependu, P.: Ontograte: towards automatic integration for relational databases and the semantic web through an ontology-based framework. Int. J. Semant. Comput. **4**(1), 123–151 (2010)
6. El-Sappagh, S., Elmogy, M., Riad, A.: A fuzzy-ontology-oriented case-based reasoning framework for semantic diabetes diagnosis. Artif. Intell. Med. **65**(3), 179–208 (2015)
7. Faria, D., Pesquita, C., Santos, E., Cruz, I.F., Couto, F.M.: Agreement maker light results for OAEI 2013. In: Proceedings of the 8th International Conference on Ontology Matching, vol. 1111, pp. 101–108. CEUR-WS.org, Aachen, DEU (2013)
8. Gomez-Romero, J., Bobillo, F., Ros, M., Molina-Solana, M., Ruiz, M., Martín-Bautista, M.: A fuzzy extension of the semantic building information model. Autom. Constr. **57**, 202–212 (2015)
9. Grover, A., Leskovec, J.: node2vec: Scalable feature learning for networks. CoRR (2016)
10. Gusfield, D., Irving, R.W.: The Stable Marriage Problem: Structure and Algorithms. Foundations of Computing (2013)
11. Hartung, M., Groß, A., Rahm, E.: Conto-diff: generation of complex evolution mappings for life science ontologies. J. Biomed. Inform. **46**(1), 15–32 (2013)
12. Jain, P., Hitzler, P., Sheth, A.P., Verma, K., Yeh, P.Z.: Ontology alignment for linked open data. In: Patel-Schneider, P.F., Pan, Y., Hitzler, P., Mika, P., Zhang, L., Pan, J.Z., Horrocks, I., Glimm, B. (eds.) ISWC 2010. LNCS, vol. 6496, pp. 402–417. Springer, Heidelberg (2010). https://doi.org/10.1007/978-3-642-17746-0_26
13. Jiang, S., Lowd, D., Kafle, S., Dou, D.: Ontology matching with knowledge rules. In: Hameurlain, A., Küng, J., Wagner, R., Chen, Q. (eds.) Transactions on Large-Scale Data- and Knowledge-Centered Systems XXVIII. LNCS, vol. 9940, pp. 75–95. Springer, Heidelberg (2016). https://doi.org/10.1007/978-3-662-53455-7_4
14. Joulin, A., Grave, E., Bojanowski, P., Douze, M., Jégou, H., Mikolov, T.: Fasttext.zip: Compressing text classification models. CoRR (2016)

15. Jun, Z., Yiduo, L., Jiatao, J., Yi, Y.: Fuzzy Ontology Models Based on Fuzzy Linguistic Variable for Knowledge Management and Information Retrieval. In: Proceedings of Intelligent Information Processing, pp. 58–67. Beijing, China (2008)
16. Kolyvakis, P., Kalousis, A., Kiritsis, D.: DeepAlignment: Unsupervised ontology matching with refined word vectors. In: Proceedings of the 2018 Conference of the North American Chapter of the Association for Computational Linguistics: Human Language Technologies, Volume 1), pp. 787–798. Association for Computational Linguistics, New Orleans, Louisiana (2018)
17. Li, G., Yan, L., Ma, Z.: An approach for approximate subgraph matching in fuzzy RDF graph. Fuzzy Sets Syst. 376, (2019)
18. Lukasiewicz, T., Straccia, U.: Managing uncertainty and vagueness in description logics for the semantic web. J. Web Semant. **6**(4), 291–308 (2008)
19. Mikolov, T., Sutskever, I., Chen, K., Corrado, G., Dean, J.: Distributed representations of words and phrases and their compositionality. CoRR (2013)
20. Moon, C., Jones, P., Samatova, N.F.: Learning entity type embeddings for knowledge graph completion. In: Proceedings of the 2017 ACM on Conference on Information and Knowledge Management, pp. 2215–2218. Association for Computing Machinery, New York (2017)
21. Nkisi-Orji, I., Wiratunga, N., Massie, S., Hui, K.-Y., Heaven, R.: Ontology alignment based on word embedding and random forest classification. In: Berlingerio, M., Bonchi, F., Gärtner, T., Hurley, N., Ifrim, G. (eds.) ECML PKDD 2018. LNCS (LNAI), vol. 11051, pp. 557–572. Springer, Cham (2019). https://doi.org/10.1007/978-3-030-10925-7_34
22. Ouali, I., Ghozzi, F., Taktak, R., Hadj Sassi, M.S.: Ontology alignment using stable matching. Procedia Comput. Sci. **159**, 746–755 (2019), knowledge-Based and Intelligent Information & Engineering Systems: Proceedings of the 23rd International Conference KES2019
23. Ristoski, P., Rosati, J., Noia, T.D., Leone, R.D., Paulheim, H.: Rdf2vec: RDF graph embeddings and their applications. Semant. Web **10**, 721–752 (2019)
24. Ritze, D., Meilicke, C., Šváb Zamazal, O., Stuckenschmidt, H.: A pattern-based ontology matching approach for detecting complex correspondences, vol. 551, pp. 25–36 (2009)
25. Smaili, F.Z., Gao, X., Hoehndorf, R.: Onto2Vec: joint vector-based representation of biological entities and their ontology-based annotations. Bioinformatics **34**(13), i52–i60 (2018)
26. Smaili, F.Z., Gao, X., Hoehndorf, R.: OPA2Vec: combining formal and informal content of biomedical ontologies to improve similarity-based prediction. Bioinformatics **35**(12), 2133–2140 (2018)
27. Thiéblin, E., Haemmerlé, O., Trojahn dos Santos, C.: Complex matching based on competency questions for alignment: a first sketch. In: 13th International Workshop on Ontology Matching co-located with the 17th International Semantic Web Conference (OM@ISWC 2018), Monterey, United States, pp. 66–70 (2018)
28. Todorov, K., Hudelot, C., Popescu, A., Geibel, P.: Fuzzy ontology alignment using background knowledge. Int. J. Uncertain. Fuzziness Knowl. Based Syst. **22**(1), 75–112 (2014)
29. Wang, Q., Mao, Z., Wang, B., Guo, L.: Knowledge graph embedding: a survey of approaches and applications. IEEE Trans. Knowl. Data Eng. **29**(12), 2724–2743 (2017)
30. Xue, X., Wang, H., Zhang, J., Zhang, J., Chen, D.: An automatic biomedical ontology meta-matching technique. J. Netw. Intell. **4**(3), 109–113 (2019)

31. Xue, X., Wang, Y.: Optimizing ontology alignments through a memetic algorithm using both matchfmeasure and unanimous improvement ratio. Artif. Intell. **223**, 65–81 (2015)
32. Xue, X., Wang, Y.: Using memetic algorithm for instance coreference resolution. IEEE Trans. Knowl. Data Eng. **28**(2), 580–591 (2016)
33. Xue, X., Yao, X.: Interactive ontology matching based on partial reference alignment. Appl. Soft Comput. **72**, 355–370 (2018)
34. Xue, X., Zhang, J.: Matching large-scale biomedical ontologies with central concept based partitioning algorithm and adaptive compact evolutionary algorithm. Appl. Soft Comput. **106**, 107343 (2021)
35. Zadeh., L.A.: A fuzzy-algorithmic approach to the definition of complex or imprecise concepts. Intl. J. Man Mach. Stud. 8(3), 249–291 (1976)
36. Zekri, F., Turki, E., Bouaziz, R.: Alzfuzzyonto : Une ontologie floue pour l'aide à la décision dans le domaine de la maladie d'alzheimer. In: Actes du 18ème Congrès INFORSID, pp. 83–98. Biarritz, France (2015)
37. Zhang, F., Cheng, J., Ma, Z.: A survey on fuzzy ontologies for the semantic web. Knowl. Eng. Rev. **31**(3), 278–321 (2016)
38. Zhou, L., Cheatham, M., Krisnadhi, A., Hitzler, P.: A complex alignment benchmark: geolink dataset. In: Vrandečić, D., et al. (eds.) ISWC 2018. LNCS, vol. 11137, pp. 273–288. Springer, Cham (2018). https://doi.org/10.1007/978-3-030-00668-6_17

Optimization and Network Analysis

Optimal Decoding of Hidden Markov Models with Consistency Constraints

Alexandre Dubray[1(✉)], Guillaume Derval[2], Siegfried Nijssen[1], and Pierre Schaus[1]

[1] Institute of Information and Communication Technologies, Electronics and Applied Mathematics (ICTEAM), Uclouvain, Louvain-la-Neuve, Belgium
{alexandre.dubray,siegfried.nijssen,pierre.schaus}@uclouvain.be
[2] Department of Electrical Engineering and Computer Science, ULiège, Liège, Belgium
gderval@uliege.be

Abstract. Hidden Markov Models (HMM) are interpretable statistical models that specify distributions over sequences of symbols by assuming these symbols are generated from hidden states. Once learned, these models can be used to determine the most likely sequence of hidden states for unseen observable sequences. This is done in practice by solving the shortest path problem in a layered directed acyclic graph using dynamic programming. In some applications, although the hidden states are unknown, we argue that it is known that some observable elements must be generated from the same hidden state. Finding the most likely hidden state in this contrained setting is however a hard problem. We propose a number of alternative approaches for this problem: an Integer Programming (IP), Dynamic Programming (DP), a Branch and Bound (B&B) and a Cost Function Network (CFN) approach. Our experiments show that the DP approach does not scale well; B&B scales better for a small number of constraints imposed on many elements and CFNs are the most robust approach when many smaller constraints are imposed. Finally, we show that the addition of consistency constraints indeed allows to better recover the correct hidden states.

Keywords: Hidden markov model · Constrained viterbi · Branch and bound · Cost function networks

1 Introduction

Hidden Markov Models (HMM) are a class of probabilistic models in which it is assumed that symbols in sequences are generated independently from each other, from hidden states. For a sequence of observed data, it is assumed that there is a sequence of hidden states that generated it with a given probability; determining the hidden states that generated the symbols is here useful in understanding the data. HMMs have been used in various real-world applications such as protein structure prediction [15], trajectory mining [16], speech recognition [10] or human activity recognition [6,9]. The decoding problem in HMMs

P. Pascal and D. Ienco (Eds.): DS 2022, LNAI 13601, pp. 407–417, 2022.
https://doi.org/10.1007/978-3-031-18840-4_29

is to find the most likely sequence of hidden states, for an observed sequence, and is usually solved by the Viterbi algorithm [19], which has a polynomial run time. The decoding problem in HMMs can be reduced to solving the shortest path problem in a layered directed acyclic graph (DAG). Since in such graphs the shortest and longest path problems are equivalent, and the applications in Sect. 5 are concerned with HMMs, we will refer to this problem as the most likely path problem in the rest of this paper. However the presented methods also work for layered DAGs not associated with HMMs.

In this work we argue that in many applications, a better decoding can be found by exploiting background knowledge stating that symbols in a given sequence must have been generated from the same hidden state. Such connections between sequences are not taken into account in classical HMM decoding, in which multiple sequences are decoded independently. However, in practice such background knowledge exists. For example, in part-of-speech tagging, it is likely that within one sentence, multiple occurrences of the same uncommon word must be given the same tag. Another application can be found in the analysis of traffic data, where we consider a truck state assignment problem as an example. In this task, constraints are imposed stating that trucks in the same area at the same time must be labeled identically. Finally, in human activity recognition problems, natural consistency constraints also arise when activities are registered near to each other (e.g., same room, same sensor). To take into account the background knowledge that symbols in the sequence must originate from the same state, the Viterbi algorithm cannot be used anymore.

The rest of this paper is organized as follows. The decoding problem under constraints is presented in Sect. 3. Then, three of the four approaches for solving the problem are presented in more detail: Dynamic Programming, a Branch and Bound and a Cost Function Network approach. These methods are compared in Sect. 5 as well as the benefit of the consistency constraints. We conclude in Sect. 6.

2 Related Work

As we will see, the decoding problem can be seen as a problem of finding the most likely path in a DAG under logical constraints between nodes or groups of nodes. This problem has been studied in multiple contexts. In the case of HMMs, and more generally conditional random fields, Roth et al. solved the decoding problem using Integer Programming and proposed constraints useful for the semantic role labeling problem [11]. With a focus on the alignment of biological sequences, Christiansen et al. proposed a constrained version of HMMs [3]. They implemented various constraints in the PRISM language [13], but no consistency constraint between sequence elements.

We will show in this work that finding the most likely sequence of hidden states can be expressed as a weighted constraint satisfaction problem, also known as a Cost Function Network (CFN). In a CFN, the goal is to find an assignment to discrete decision variables such that a sum of functions defined on these variables is optimized while respecting defined constraints. In this work we will rely on dedicated solvers for CFNs, such as Toulbar2 [4,8].

In [17], for finding longest paths in a general DAG, the logical constraints are represented in a Binary Decision Diagram (BDD) and a dynamic program, taking into account the BDD nodes, is designed to find the optimal solution. In [20], consistency constraints are imposed between words to improve logical reasoning from sentences in natural language. They use Dual Decomposition [12] to solve the problem, which solves a Lagrangian relaxation of the problem; in contrast to our approach, however, this approach does not guarantee finding the optimal solution.

3 Problem Definition

In this section we formalize the problem of finding the most likely path in a layered DAG under consistency constraints. Solving this problem allows to also solve the HMM decoding problem. We first introduce the notation as well as the notions of layer and consistency constraints in a DAG, then express the problem of finding the most likely path in it.

3.1 Most Likely Path in a Layered DAG with Consistency Constraints

We define the HMM decoding problem over labeled Directed Acyclic Graphs (DAGs). Let $G = (V, E)$ be a graph with V the set of nodes and E the set of edges. Each node $v \in V$ has a label, from a set $\mathcal{L}$, denoted l_v and V is divided into T layers $L_1, \ldots, L_T$ such that $V = \bigcup_{j=1,\ldots,T} L_j$ and $L_i \cap L_j = \emptyset \; \forall i \neq j$. In each layer, no two nodes have the same label. Thus, when clear from the context, a node can be identified by its label. We denote by $e = (l, l', t) \in E$ an edge from the node with label l at layer L_t to the node with label l' at layer L_{t+1} $(1 \leq t < T)$ with weight w_e, where weights can be both positive and negative. In the HMM decoding problem, each layer has the same number of nodes representing the hidden states. An example of such a graph is shown in Fig. 1.

A path in G from L_1 to L_T selects one node per layer and can be identified by the sequence of node labels on the path. More formally, let $P = \langle P_1, \ldots, P_T \rangle \in \mathcal{L}^T$ be a path from L_1 to L_T such that $P_i \in L_i$. The cost of P, is the sum of the weights of the arcs in the path: $\sum_{t=1}^{T-1} w_{(P_t, P_{t+1}, t)}$.

A consistency constraint is specified in our work by identifying a set of layers for which the same label must be selected in each layer of the path. More formally, $C = \{C_1, \ldots, C_k\}$ are k consistency constraints with $C_i = \{c_1^i, \ldots, c_{k_i}^i\} \subseteq \{1, \ldots, T\}$ and $C_i \cap C_j = \emptyset$ for $i \neq j$. The set of all constrained layers is denoted $L_C = \bigcup_{i=1}^{k} C_i$. We also define a vector $c \in \{0, \ldots, k\}^T$ that gives for each layer the index of its constraint or 0 if the layer is unconstrained. For example, in Fig. 1 we have $c = \langle 0, 0, 1, 0, 2, 0, 2, 0, 1, 0, 0 \rangle$. A path P is said to be consistent

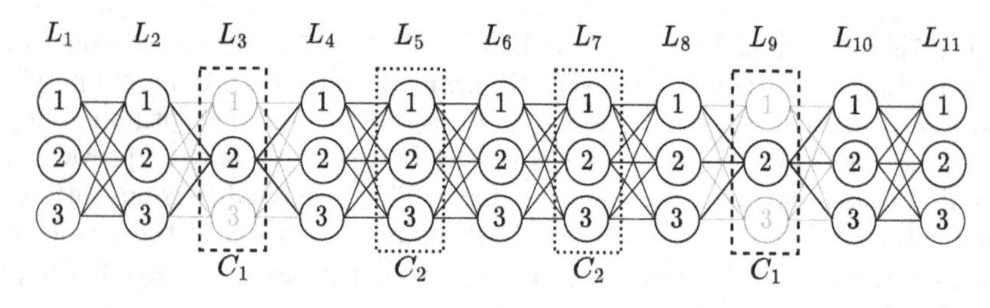

Fig. 1. Example of a layered DAG for the decoding problem in a HMM with three hidden states and two consistency constraints. The edges in the DAG are oriented from left to right and the labels on the nodes represent the hidden states. In this example, the constraint C_1 has node 2 assigned to it, hence the other nodes are faded.

if all the consistency constraints are respected. The problem of finding the most likely consistent path is thus formalized as follows:

$$P^\star = \arg\max_{P \in \mathcal{L}^T} \sum_{t=1}^{T-1} w_{(P_t, P_{t+1}, t)} \tag{1}$$

$$\text{s.t. } P_{c_1^i} = \ldots = P_{c_{k_i}^i} \qquad \forall C_i \in C \tag{2}$$

The importance of this problem to HMM decoding is that an instance of this problem, including the DAG and its weights, can be constructed for a specific HMM decoding problem on a sequence of symbols. Note that the problem defined by Eq. (1)–(2) is NP-hard, as we showed in a technical report [5].

4 Solving the Problem

In this section, three of the four approaches to solve the problem defined by Eqs. (1)–(2) are explained. We omit the IP formulation as it is very similar to the one presented in [11] but with equality constraints. First a Dynamic Programming approach (DP) is introduced, followed by a Branch and Bound (B&B) method and finally a model based on cost-function networks (CFN) is presented.

4.1 Dynamic Programming

For solving the unconstrained problem, the Viterbi algorithm [19] is the classical dynamic programming approach. The recurrence relation computes the value of the most likely path from L_1 to a node $i \in L_t$ from layer L_{t-1} and stores it in a $T \times |\mathcal{L}|$ table. The entries of the table are computed as follows:

$$V[t,i] = \begin{cases} 0 & \text{if } t = 1 \\ \max_{j \in L_{t-1}} V[t-1,j] + w_{(j,i,t-1)} & \text{otherwise} \end{cases} \tag{3}$$

and the value of the most likely path is given by $\max_{i \in L_T} V[T,i]$.

This equation uses the fact that the graph is organized into layers and a path ending at layer L_t always comes from layer L_{t-1}. Thus, the most likely path to a node $i \in L_t$ is one of the most likely paths to a node in L_{t-1} plus the edge to i. However, when adding consistency constraints, this equation does not work anymore because it does not take into account consistency. We resolve this by adding assignments of labels to consistency constraints in the DP; by assigning a label to a constraint we assign the same label to all layers in the constraint. Let $P_C = \langle P_{C_1}, \ldots, P_{C_k} \rangle \in (\mathcal{L} \cup \{\bot\})^k$ be an assignment of labels to the consistency constraints with $P_{C_i} = \bot$ if no label is assigned to C_i. Then if $P_{C_i} \neq \bot$, the path from L_1 to L_T must pass through P_{C_i} for every layer L_t with $t \in C_i$. We define the assignment operator $P_C|_{j,l}$ which assigns l to P_{C_j}. The values of the most likely paths can now be stored in a $T \times |\mathcal{L}| \times k^{|\mathcal{L}|}$ table, taking into account the possible assignments of labels to the constraints. The entries in the table are computed as follows:

$$V[t,i,P_C] = \begin{cases} 0 & \text{if } t = 1 \\ \max\limits_{l \in L_{t-1}} V[t-1,l,P_C] + w_{(l,i,t-1)} & \text{if } L_{t-1} \notin L_C \\ V[t-1,P_{C_{c[t]}},P_C] + w_{(P_{C_{c[t]}},i,t-1)} & \text{if } L_{t-1} \in L_C \wedge P_{C_{c[t]}} \neq \bot \\ \max\limits_{l \in L_{t-1}} V[t-1,l,P_C|_{c[t],l}] + w_{(l,i,t-1)} & \text{if } L_{t-1} \in L_C \wedge P_{C_{c[t]}} = \bot \end{cases} \tag{4}$$

The first two cases of Eq. (4) are the same as Eq. (3) because there are no constraints to consider. However when the layer L_{t-1} is constrained, there are two situations. If there is a choice for this constraint in P_C, then in order to be consistent with P_C, the path must pass by it. In that case there is no need to consider the other nodes in the layer. However, when there is not yet a node assigned to this constraint, then every node $j \in L_{t-1}$ must be considered to compute the most likely path. In this case, the P_C vector is updated to reflect the choice made.

4.2 Branch and Bound

Let $P_C \in (\mathcal{L} \cup \{\bot\})^k$ be, as for the DP, a vector of node assignments for the consistency constraint. The search starts from the vector $\langle \bot, ..., \bot \rangle$. The idea of this method is to branch on the P_{C_i} values and to compute the most likely path from L_1 to L_T while being consistent with P_C. Initially some P_{C_i} are unassigned; as long as the constraint is unassigned, we ignore the constraint and the cost is an upper bound on the optimal solution in the branch. An example is shown on Fig. 1 where there are two consistency constraints and $P_C = \langle 2, \bot \rangle$. The most likely path from L_1 to L_{11} can be seen as the most likely path from L_1 to L_3, then L_3 to L_9 and finally from L_9 to L_{11}. As long as $P_{C_2} = \bot$, we ignore constraint C_2 and an upper bound on the most likely path is obtained.

In practice, a $T \times |\mathcal{L}|$ array, denoted V, is used to store the values of the most likely paths from the layers in L_c to the other layers. At the root of the search tree, the V array is filled with Eq. (3) since there are no consistency constraints imposed. When a value P_{C_i} is assigned, the whole table does not need to be recomputed. Let us look at Fig. 1 as an example. When the search assigns $P_{C_1} = 2$, the layers constrained by C_1 act as new source layers. The computed values, in V, for layers L_1 to L_3 still represent the values a recursive equation computes for the most likely path from L_1 to L_3 and thus, need not be recomputed. Let us assume now that the search assigns $P_{C_2} = 1$. The values in V for L_1 to L_5 and L_9 to L_{11} are still valid, and only the values from L_6 to L_8 need to be updated.

Notice that when all the edges have a negative weight, as for the HMM decoding problem, then the values in the V array can be computed between consistency constraints, even if not assigned. In the example in Fig. 1, the consequence is that when P_{C_1} is set to 2, the values from L_3 are only computed until L_5 and not L_9. Since the edges only have negative weights, this gives a less tight upper bound on the optimal solution, but is faster to compute.

4.3 Cost Function Networks

In Cost Function Networks (CFNs) [4], a set of functions is defined, each of which maps a subset of the assignment in P_c to a cost. The goal is to find an assignment of P_C such that the sum of the function's cost (evaluated on the assignment) is minimal. We model our problem in CFNs by dividing the graph into segments between successive constrained layers; a function is defined on each of these segments. These functions map the choice for the constrained layers at the start and end of the segment (i.e., a partial assignment of P_C) to the value of the most likely path on the segment, consistent with P_C. For a full assignment of P_C, the sum of the most likely path on the segments gives the value of the most likely path in the full graph.

More formally, let $L_t, L_{t'}$ such that $c[t] \neq c[t']$ and $\nexists\, t'' : t < t'' < t' \wedge c[t] \neq c[t'']$ be two successive constrained layers of different consistency constraints. A function $f_{t,t'} : \mathcal{L} \times \mathcal{L} \mapsto \mathbb{R}$ is defined on the segment between L_t and $L_{t'}$, mapping each choice of $c[t]$ and $c[t']$ to the value of the most likely path from L_t to $L_{t'}$ consistent with the choices. For every node $u \in L_t$ and $v \in L_{t'}$, a simple dynamic program finds the value of the most likely path between u and v and stores it in a $\mathcal{L} \times \mathcal{L}$ table. Since the table fully defines the function, $f_{t,t'}$ is used to refer to the function as well as its table of values and $f_{t,t'}(P_C)$ refers to the value associated with the choices for L_t and $L_{t'}$ in P_C.

Let $L_{t_1}, \ldots, L_{t_m}$ be all the constrained layers. Without loss of generality, we assume that they are sorted in chronological order so that $t_1 < t_2, t_2 < t_3, \ldots$ Let $F = \{f_{start}, f_{end}, f_{t_1,t_2}, \ldots, f_{t_{m-1},t_m}\}$ be the functions, as defined above, for each segment of the graph and two additional special functions. The $f_{start} : \mathcal{L} \mapsto \mathbb{R}$ function maps, for each choice for $c[t_1]$, the value of the most likely path from L_1 to L_{t_1}. The function f_{end} is defined in the same way for the layer L_{t_m} to L_T. The value of a path P_C, which we wish to optimize, is then given by

$$f_{start}(P_C) + f_{end}(P_C) + \sum_{i=1}^{m-1} f_{t_i,t_{i+1}}(P_C). \tag{5}$$

Table 1. Execution time in seconds of the methods in function of the proportion of constraints in the model on the truck state assignment problem. The entries for the CFN method represent the time needed to compute the functions plus the optimization time by Toulbar2.

Proportion of constraints	0.00	0.10	0.20	0.30	0.40	0.50	0.60	0.70	0.80	0.90	1.00
DP	6.00	360.10	324.40	299.30	278.60	255.70	232.70	211.90	192.20	173.90	157.80
IP	588.20	841.67	911.33	926.60	927.20	1007.20	1001.25	1169.67	1061.67	1141.00	1047.00
B&B	**2.63**	**3.59**	**3.59**	**3.55**	**3.43**	**3.42**	**3.34**	**3.31**	**3.27**	**3.21**	**2.93**
CFN	–	35.94	34.28	32.43	30.92	28.80	26.85	25.15	23.00	20.80	18.93

Dedicated solvers for CFN are designed to find the assignment to P_C such that the value of Eq. (5) is minimal. From this optimal assignment we can easily recover the solution using a dynamic program.

5 Experimental Results

In this section we analyze the run time of the methods presented in Sect. 4 on two different HMM applications with different characteristics in terms of sequence lengths and number of consistency constraints. We finish this section by analyzing, on a third application, the impact of the consistency constraints on the output of the decoding problem. The IP is solved with the Gurobi solver [7] and for the B&B we use the variation of the algorithm that supports only negative weights, as we experiment only with HMMs and it gives, in our experiments, the best results.[1]. For the CFN method the Toulbar2 solver [14] is used.

Truck Trajectory Mining. HMMs have been used to identify activity stops in truck trajectories [16]. Four hidden states represent if the truck is driving, in a traffic jam, resting or doing work-related actions. In this context, it is natural to assume that trucks located in similar geographical areas do the same activity. Four consistency constraints are created based on the type of point (stop or driving) in some geographical areas (e.g., rest areas, highways).

We experiment on a data set of trajectories of trucks described in [1], which contains roughly 6 million data points (and thus as many layers in the graph). We successively kept a given percentage of each constraint in order to evaluate the impact of the constraints size on the run time.

Table 1 shows the run time of the methods with different proportions of states included in constraints, where constraints are larger if they involve more states. The run time of the DP and IP methods both increases with the size of the constraints. For the DP method, more choices must be propagated through the recursion while in the IP model there are more linear constraints. The run time of the B&B method is stable with the constraint size. The size of the constraints only impacts the computation of the V array in each node of the search tree. As the whole array still needs to be computed in order to have

[1] The source code and the data sets can be found at https://github.com/AlexandreDubray/consistent-viterbi.

Table 2. Run time in seconds of the methods in function of the number of consistency constraints for the POS tagging problem. Timeout has been set to 1 h and is indicated by T.O. while out of memory errors are indicated by O.O.M.

Dataset	conll2000					treebank					brown				
Number of constraints	25 9104					100 676					1 161 192				
Number of layers	2	3	4	5	6	2	3	4	5	6	2	3	4	5	6
DP	153.2	128.7	O.O.M	O.O.M	O.O.M	56.0	1132.0	O.O.M	O.O.M	O.O.M	1047.4	O.O.M	O.O.M	O.O.M	O.O.M
IP	97.6	137.6	86.0	128.1	130.6	34.6	33.4	32.1	32.2	47.6	485.3	480.9	790.1	O.O.M	O.O.M
B&B	**19.6**	**30.75**	290.25	T.O.	T.O.	14.95	66.94	116.93	777.78	T.O	**81.66**	1054.34	T.O.	T.O.	T.O.
CFN	52.33	72.56	**72.22**	**72.59**	**71.88**	**7.64**	**29.58**	**31.2**	**29.8**	**29.93**	294.71	**301.91**	**293.4**	**294.31**	**297.66**

a feasible solution, the impact is limited. Finally, the run time of the CFN method decreases with the size of the constraints. In that case, the run time is dominated by the computation of the local functions F. Once computed, Toulbar2 is able to find the optimal solutions in a few milliseconds. With more constraints, the segments are shorter and thus faster to compute, which makes the overall approach faster.

Overall the B&B method is the fastest on this data set because there are few constraints and few choices per constraint. Thus even if the CFN approach is much better than the DP and IP, the time needed to compute the local functions F makes it slower than B&B.

Part of Speech Tagging. The goal of this application is to assign to each word of a sentence, or text, a part of speech (POS) tag. The NLTK Python package [2] provides data sets of sentences with annotated POS. We experiment on three data sets with the 12 universal POS tags and consistency constraints are imposed on layers with the same POS tag.

Table 2 shows the run time of the methods in function of the number of consistency constraints. First, let us note that only the CFN method is able to solve the problem for all numbers of constraints on all data sets. The B&B and DP methods both time-out or reach a memory limit quickly as the number of constraints increases. For the DP method, with more constraints, the number of constraint choices to propagate increases exponentially. For the B&B the search space becomes too large and the upper bound is not strong enough to prune large part of the search space to make the approach tractable. The IP methods can handle more constraints but, on the brown data set, which is larger, the amount of memory needed to model the problem is too large. For these three methods, the run time increases with the number of constraints which is expected.

On the contrary, the run time of the CFN is stable with the number of constraints and the method is the most efficient for these data sets. Adding new constraints has little impact on the time needed to compute F since it is done by computing the values between successive constrained layers (i.e., all layers of the graph are processed $|\mathcal{L}|^2$ times). In addition to that, Toulbar2 is very efficient at finding the optimal solution, in few milliseconds. Hence the total run time of the CFN method is stable with the number of constraints.

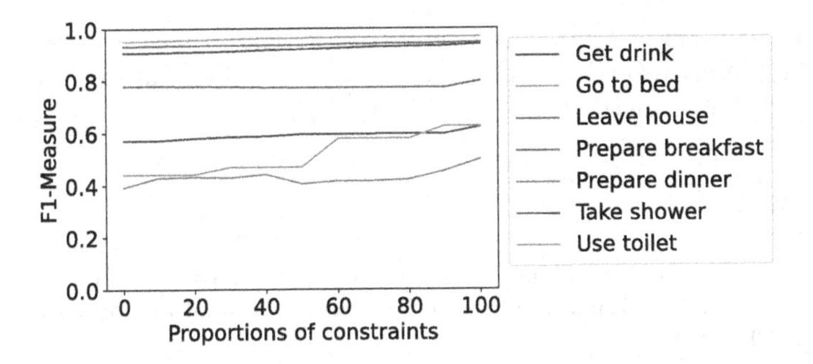

Fig. 2. F1-Measure in function of the proportion of constraints for each activity

Human Activity Recognition. Finally, in this section we analyze the impact of the consistency constraints on the output of the decoding problem, using a real-world data set for Human Activity Recognition (HAR). In HAR the goal is to find which activities a person is doing based on inputs from sensors which can be placed on the person (e.g., a smartwatch) or in their environment (e.g., light sensors in the house). We use the annotated data sets as described in [18] for this experiment. These data sets provide the activities (based on the activation of sensors in their house) made by three persons for multiple days.

The F1-Measure per activity is shown in Fig. 2 for one of the houses (the results are similar for the other houses). The F1-Measure was computed, for a proportion of the constraints, following the same methodology as in [9]. It can be seen that the activities are better recovered as the proportion of constraints increases. The biggest impact is on the activities that are not well recovered using a classical decoding algorithm (e.g. "Go to bed", "Prepare dinner"). The activities that have a high F1 measure when there are no constraints also benefit from the constraints, but in a less marked way.

6 Conclusions and Future Work

In many applications using Hidden Markov Models, consistency constraints between sequences can be found but are not used in the classical decoding algorithm. In this work, we formalized this problem as finding the most likely path in a layered directed acyclic graph with consistency constraints on the layers of the graph. We proposed an Integer Programming (IP), a Dynamic Program, a Branch and Bound (B&B) and a Cost Function Network method to solve the problem. We showed that Branch and Bound scales better for a few large constraints, while the CFN is better for many smaller constraints. Finally, our experiments on a real-world human activity recognition data set showed the benefit of consistency constraints.

In this work we focused on consistency constraints, imposing that the same node is selected between different layers. However in some applications, it might be acceptable to have sets of nodes that can appear together in the layers of a consistency constraint (e.g., a non-activity stops and a rest stop, in the Truck Trajectory Mining problem). The impact of additional logical constraints on the Branch and Bound method could also be investigated.

References

1. Adam, A., Finance, O., Thomas, I.: Monitoring trucks to reveal belgian geographical structures and dynamics: From GPS traces to spatial interactions. J. Transp. Geogr. **91**, 102977 (2021)
2. Bird, S., Klein, E., Loper, E.: Natural Language Processing with Python: Analyzing Text with the Natural Language Toolkit. O'Reilly Media, Inc. (2009)
3. Christiansen, H., Have, C.T., Lassen, O.T., Petit, M.: Inference with constrained hidden markov models in prism. Theory Pract. Logic Program. 10, (2010)
4. Cooper, M.C., De Givry, S., Sánchez, M., Schiex, T., Zytnicki, M., Werner, T.: Soft arc consistency revisited. Artif. Intell. **174**, 449-478 (2010)
5. Dubray, A., Derval, G., Nijssen, S., Schaus, P.: On the complexity of the shortest path problem in a layered directed acyclic graph with consistency constraints (2022). 2078.1/264677
6. Fallmann, S., Kropf, J.: Human activity recognition of continuous data using hidden markov models and the aspect of including discrete data. In: UIC, pp.121–126 (2016)
7. Gurobi Optimization, LLC: Gurobi Optimizer Reference Manual (2022). https://www.gurobi.com
8. Hurley, B., et al.: Multi-language evaluation of exact solvers in graphical model discrete optimization. Constraints **21**(3), 413–434 (2016). https://doi.org/10.1007/s10601-016-9245-y
9. Kabir, M.H., Hoque, M.R., Thapa, K., Yang, S.H.: Two-layer hidden markov model for human activity recognition in home environments. Int. J. Distrib. Sens. Netw. IJDSN. **2016**, 1–12 (2016)
10. Rabiner, L.R.: A tutorial on hidden markov models and selected applications in speech recognition. In: Proceedings of the IEEE (1989)
11. Roth, D., Yih, W.T.: Integer linear programming inference for conditional random fields. In: ICML (2005)
12. Rush, A.M., Sontag, D., Collins, M., Jaakkola, T.: On dual decomposition and linear programming relaxations for natural language processing (2010)
13. Sato, T., Kameya, Y.: Prism: a language for symbolic-statistical modeling. In: IJCAI (1997)
14. Schiex, T., de Givry, S., Sanchez, M.: Toulbar2-an open source weighted constraint satisfaction solver (2006). https://toulbar2.github.io/toulbar2
15. Sonnhammer, E.L., et al.: A hidden markov model for predicting transmembrane helices in protein sequences. In: ISMB (1998)
16. Taghavi, M., Irannezhad, E., Prato, C.G.: Identifying truck stops from a large stream of GPS data via a hidden markov chain model. In: ITCS (2019)
17. Takeuchi, F., Nishino, M., Yasuda, N., Akiba, T., Minato, S.I., Nagata, M.: BDD-constrained a* search: a fast method for solving constrained shortest-path problems. IEICE Trans. Inform. Syst. **10**(12), 2945–2952 (2017)
18. Van Kasteren, T., Noulas, A., Englebienne, G., Kröse, B.: Accurate activity recognition in a home setting. In: UbiComp, pp. 1-9,(2008)

19. Viterbi, A.: Error bounds for convolutional codes and an asymptotically optimum decoding algorithm. IEEE Trans. Inform. Theo. **13**(2), 260–269 (1967)
20. Yoshikawa, M., Mineshima, K., Noji, H., Bekki, D.: Consistent CCG parsing over multiple sentences for improved logical reasoning. arXiv preprint (2018)

Semi-parametric Approach to Random Forests for High-Dimensional Bayesian Optimisation

Vladimir Kuzmanovski[1,3,4(✉)] and Jaakko Hollmén[1,2]

[1] Department of Computer Science, Aalto University, Espoo, Finland
vladimir.kuzmanovski@aalto.fi
[2] Department of Computer and Systems Sciences, Stockholm University, Stockholm, Sweden
jaakko.hollmen@dsv.su.se
[3] Smart City Center of Excellence, Tallinn University of Technology, Tallin, Estonia
[4] Department of Knowledge Technologies, Jožef Stefan Institute, Ljubljana, Slovenia

Abstract. Calibration of simulation models and hyperparameter optimisation of machine learning and deep learning methods are computationally demanding optimisation problems, for which many state-of-the-art optimisation methods are adopted and applied in various studies. However, their performances come to a test when the parameter optimisation problems exhibit high-dimensional spaces and expensive evaluation of models' or methods' settings. Population-based (evolutionary) methods work well for the former but not suitable for expensive evaluation functions. On the opposite, Bayesian optimisation eliminates the necessity of frequent simulations to find the global optima. However, the computational demand rises significantly as the number of parameters increases. Bayesian optimisation with random forests has overcome issues of its state-of-the-art counterparts. Still, due to the non-parametric output, it fails to utilise the capabilities of available acquisition functions. We propose a semi-parametric approach to overcome such limitations to random forests by identifying a mixture of parametric components in their outcomes. The proposed approach is evaluated empirically on four optimisation benchmark functions with varying dimensionality, confirming the improvement in guiding the search process. Finally, in terms of running time, it scales linearly with respect to the dimensionality of the search space.

1 Introduction

Models or algorithms built for generalising observed processes or phenomena require a specific configuration level when adapting to a new problem. Among the most widely known such issues are calibration of simulation models and hyperparameter optimisation (HPO) for machine learning (ML) and deep learning (DL) methods [16,19,37]. In both cases, the accuracy of the outcome is highly dependent on (hyper)parameter settings, a selection of which may pose a hard computational (optimisation) problem and its automation pose a great optimisation challenge [10]. The problem features high-dimensional and complex configuration

P. Pascal and D. Ienco (Eds.): DS 2022, LNAI 13601, pp. 418–428, 2022.
https://doi.org/10.1007/978-3-031-18840-4_30

(search) spaces with an intractable likelihood (unavailable gradient) of underlying loss functions, and expensive model evaluations [12]. Such settings require methods that have consistent performance, effectively use of parallel resources, and are characterised by scalability, robustness and flexibility [11].

Constrained by the limiting factors, various state-of-the-art optimisation methods, including primarily black-box gradient-free meta-heuristics, like Grid and Random search [12] and population-based methods [15,26,27], are becoming inefficient. On the opposite, the Bayesian optimisation (BO) eliminates the necessity to simulate a large sample set for finding the global optima by sequential sampling design. Hence, it is a valuable method for the HPO and calibration processes [12,18,30].

The BO [31] is an optimisation method that seeks the global optima through sequential sampling design and approximation of underlying likelihood function with a surrogate model over a surrogate (response) surface. The surrogate surface is defined by a parameter space and discrepancy between observed and simulated (predicted) outputs. The sequential sampling design follows an iterative approach, through which new samples (parameters' values) that maximise the expected (acquisition) utility are acquired. State-of-the-art performances of the BO are achieved using Gaussian processes (GPs) regression prior [31], as applied in various domains, such as population genetics [18], spreading of pathogens [22], atomic structure of materials [36,38], as well as cosmology [21].

However, the GP prior, and hence the BO, have limited applicability in settings of high-dimensional data, constituting a bottleneck for their broader adoption in settings of complex parameter spaces [11,28], as well as their adapted counterparts using dimensionality reduction [5], or synthetic parametric likelihoods introduced [1,28,32] that don't circumvent the obstacle of the evaluation cost. Therefore, replacing the GP prior with more robust regression surrogate is considered in previous studies, leading to improved performances [12,37]. Widely adopted alternatives are random forests (RF) regression [6,16,28] and Bayesian neural networks [33,34], with the latter being computationally more expensive.

The RF is limited to a non-probabilistic output and previous study adopted it by empirical (mean and variance) [16] or quantile statistics over predictions of base models [28]. As such, these adoptions cannot detect multi-modal posteriors, which are highly probable in predictions by RF base models, because the RF tends to increase the variance between the base models.

This study aims to examine the potential of the semi-parametric approach to RF regression for overcoming its limitation in the context of BO. We propose a semi-parametric approach to the RF outputs by modeling the base predictions as a mixture of Gaussian components. Each component is then evaluated with the acquisition function, where the obtained utilities are linearly combined using weights from the mixture model.

The performance of the proposed approach are empirically examined over a set of benchmark functions for high-dimensional optimisation: Levy [20], Schwefel [29], Ackley [3] and Griewank [23] function. The achieved performances are compared with the random search and the state-of-the-art BO methods, with GP (where applicable) and empirical RF.

The novelty of our work represents an extension to the BO with RF non-parametric surrogate model, enabling a semi-parametric output. The semi-parametric output improves the applicability of the RF as a surrogate model in the context of the BO by estimating the outputs' uncertainty over homogeneous clustered sub-spaces derived from the divide-and-conquer approach of the RF regression. In addition to the novelty, our work contributes to:

- formalising and implementing a BO framework, using R programming language, with ability for new surrogate models and benchmark functions;
- evaluating performances of the state-of-the-art BO methods with GP and RF surrogate models, over benchmark problems with varying dimensionality.

The rest of the manuscript is organised as follows. In Sect. 2, we present the background of the BO and our contributions to the proposed approach. In Sect. 3, the experimental design is presented, followed by results. Finally, summary and conclusions are presented in Sect. 5.

2 Materials and Methods

Simulation models and ML (DL) methods represent generative processes that generate a hypothesis, trying to fit the given observed data of the modelled phenomenon. The generative process is driven by selecting (hyper)parameter values of the underlying models [14]. Formally, the parameter selection corresponds to statistical inference of a finite number of parameters $\theta \in \mathbb{R}^d$ of a model or method from a set of observations Y_o:

$$p(\theta|Y_o) = \frac{p(Y_o|\theta) \cdot p(\theta)}{p(Y_o)}, \tag{1}$$

where $p(\theta)$ encodes our prior beliefs on the distribution of parameter values and $p(Y_o|\theta)$ represents the likelihood of the observations, derived from the known function $\mathcal{L}(\theta)$. Since the analytical form of $\mathcal{L}(\theta)$ is unknown in the underlying challenge, we use the notation $L(\theta)$ that need to be approximated over a set of N samples - $\tilde{L}^N(\theta)$. The notation is simplified if the marginal distribution $p(Y_o)$ is omitted because it does not depend on θ, $p(\theta|Y_o) \propto L(\theta) \cdot p(\theta)$. The $L(\theta)$ is approximated over finite sample set ($\tilde{L}^N(\theta)$) and it is reconstructed as the number of samples increases, i.e., $\lim_{N\to\infty} \tilde{L}^N(\theta) = L(\theta)$.

The approximation ($\tilde{L}^N(\theta)$) of the likelihood function ($L(\theta)$) can be performed in parametric or non-parametric manner [14,31]. We focus on the latter, with utilisation of a surrogate regression using random forests.

Sequential sampling design feature an *acquisition function* $A(\theta)$, whereby $s \in \mathbb{R}$ generated samples are credited with an utility. The BO enriches the evidence with evaluated $k \leq s$ samples that maximise the utility. A wide range of acquisition function are developed [17], but for the purpose of this study, we adopt the *expected improvement* (*EI*) [25]:

$$EI(\theta|\mu, \sigma, f^*) = \sigma(\theta)[z\Phi(z) + \phi(z)]; \quad z = \frac{f^* - \mu(\theta)}{\sigma(\theta)}, \tag{2}$$

where $\sigma(\theta)$ and $\mu(\theta)$ are statistics of the inferred posterior distribution (under GP posterior distribution they represent functions), f^* is the most optimal output, i.e., active optima discovered, and Φ and ϕ are probability density and cumulative distribution function in terms of the standard normal distribution, respectively. The expected improvement $EI(\theta) = 0$ if $\sigma(\theta) = 0$. The analogy behind (2) reveals the exploration-exploitation trade-off that favours larger uncertainty proximal to the known optimal region(s).

Random forests (RF) [6], the adopted surrogate, is an ensemble method composed of C regression trees. Each regression tree is built over a subspace of the parameter space, designed by random subsets of both the features (dimensions) and bootstrap samples. Therefore, given a dataset, each regression tree predicts the target for a specific region in the defined space. The prediction of the ensemble, on the other hand, is an aggregation (average) of the outcomes of all C tree base predictors:

$$\mathcal{RF}(\theta|\Theta, Y) = \frac{1}{C}\sum_{i=1}^{C} \tau_i; \quad \tau_i = T_i(\theta|\Theta_i, Y_i), \tag{3}$$

where Θ_i and Y_i are training dataset of i-th regression tree T_i that provide a prediction τ_i, while Θ and Y global training dataset.

The RF method has small number of hyper-parameters that can significantly influence the outcome and has shown excellent robustness over high-dimensional data, which limits the bias of the overall predictions by maximising the variance between base predictors [13]. However, in the context of the BO, RF models lack: (i) uncertainties quantification of predictions (non-probabilistic output); and (ii) predicting a value outside of the observed range. Thus, as a standalone surrogate model, the RF greatly affects the efficiency of a probabilistic acquisition function $A(\theta)$ (e.g., EI) in acquiring new promising samples.

We propose an extension to the previous works with the RF as a surrogate by a semi-parametric estimates of the prediction uncertainties. The semi-parametric approach constructs a mixture of parametric components, i.e., Gaussian Mixture model [8,35] and estimates the uncertainty of the predictions over homogeneously clustered sub-spaces (base predictors) that reduce the variance within the identified components.

Gaussian Mixture Model (GMM) [8,35] is a semi-parametric density function composed of weighted sum of M parametric components, where each component m_i follows Gaussian distribution with a mean μ_i and standard deviation σ_i:

$$p(x) = \sum_{i=1}^{M} w_i\, \mathcal{N}(x|\mu_i, \sigma_i); \quad \sum_{i=1}^{M} w_i = 1, \; w_i > 0. \tag{4}$$

where $x \in \mathbb{R}$ and in the context of this study, corresponds to a base prediction (τ_j) of a decision tree (T_j) in a $\mathcal{RF}$.

The proposed extension to RF named as *random forests with semi-parametric output* (RFw/SPO) estimates the acquisition utility from a sample, i.e., parameter values θ, as a linear combination of component-wise acquisition utilities from a mixture of Gaussian components (η_i), over base prediction (τ_j) of a RF model:

$$\widetilde{EI}(\theta|\mu_{1..M}, \sigma_{1..M}, f^*) = \sum_{i=1}^{M} w_i \; EI(\theta|\mu_i, \sigma_i, f^*). \tag{5}$$

The rationale behind the proposed extension to the RF is based on the ideas from [6], where the sampling variance of an RF is shown to be governed by the variance v of the base predictors and their correlation ρ, leading to the ensemble variance being $v * \rho$. In settings of high-dimensional and sparsely sampled functions, the correlation between base predictors is hardly observable. It may even lead to sparse space divisions during the divide-and-conquer approach. When aggregated over the whole set of base predictors, predictions of such sparse subspaces might heavily underestimate the ensemble variance. Therefore, by identifying distant homogeneous components that concentrate around their expected values with reduced within-component variance, the probabilistic acquisition function will have more detailed estimates of the prediction uncertainties during the derivation of samples' utility. This approach allows for the identification of potential gaps by estimating a multi-modal density function, unlike the empirical uncertainty estimation used in [16].

The proposed method and BO with GP and RF as a surrogate are implemented in a custom framework, using R programming language[1]. The GMM models are fitted using expectation-maximization (EM) algorithm [9] for maximum likelihood estimation. The implementation expects a specification of three hyperparameters for managing the optimisation process. Namely, it requires setting up a sample size at the initial (n_i) and iteration sampling (n_t), as well as the maximum number of iterations (max_t).

3 Experimental Design

To examine the proposed method's properties, we investigate its performance over four benchmark optimisation functions: Levy, Schwefel, Ackley and Griewank (Fig. 1); and compare them against random search (RS) [4], and BO with GP (BO-GP) [31] and RF (BO-RF) surrogates [16]. All benchmark functions are continuous with varying intrinsic dimensionality.

Ackley function [3] is a non-convex function, with multi-modal surface (Fig. 1(a)) defined on n-dimensional space:

$$f(\mathbf{x}) = -a \; exp\left(-b\sqrt{\frac{1}{n}\sum_{i=1}^{n} x_i^2}\right) - exp\left(\frac{1}{n}\sum_{i=1}^{n} cos(cx_i)\right) + a + e, \tag{6}$$

where the a, b and c are constants, with default values being $a = 20$, $b = 0.2$ and $c = 2\pi$. Common input domain is $x_i \in [-32, 32]$. The function surface has many local optima and a single global optimum $f(\hat{x}) = 0$ at $\hat{x} = (0, 0, \ldots, 0)$.

[1] Code: https://tinyurl.com/2xtsaaut.

Griewank function [23] is a non-convex function (Fig. 1(b)) defined on n-dimensional space:

$$f(\mathbf{x}) = 1 + \sum_{i=1}^{n} \frac{x_i^2}{4000} - \prod_{i=1}^{n} cos\left(\frac{x_i}{\sqrt{i}}\right). \tag{7}$$

This function has many local optima, equidistantly spread across all dimensions, and a single global optimum $f(\hat{x}) = 0$ at $\hat{x} = (0, 0, \ldots, 0)$. Frequently used input domain is $x_i \in [-600, 600]$.

Levy function [20] is a non-convex function, with multi-modal surface (Fig. 1(c)) defined on n-dimensional space:

$$f(\mathbf{x}) = sin^2(\pi w_1) + \sum_{i=1}^{n} -1(w_i - 1)^2(1 + 10sin^2(\pi w_i + 1))$$
$$+ (w_n - 1)^2(1 + sin^2(2\pi w_n)); \quad w_i = 1 + \frac{x_i - 1}{4}, i = 1, \ldots, n. \tag{8}$$

This function has a single optimum (minimum) $f(\hat{x}) = 0$ at $\hat{x} = (1, 1, \ldots, 1)$, and it is usually evaluated over input domain of $x_i \in [-10, 10]$.

Schwefel function [29] is a non-convex and multi-modal function (Fig. 1(d)) that can be defined on space with arbitrary (n) dimensions:

$$f(\mathbf{x}) = 418.9829d - \sum_{i=1}^{n} x_i sin\left(\sqrt{|x_i|}\right). \tag{9}$$

This function has a single global optimum $f(\hat{x}) = 0$ at the edge of the space, i.e., $\hat{x} = (420.9687, \ldots, 420.9687)$, where the input domain is $x_i \in [-500, 500]$.

The methods' performances are empirically evaluated by a set of optimisation tasks that are repeatedly performed for each benchmark function across four dimensions (10, 50, 100, and 500). Each task is set to terminate after 100 iterations and is repeated ten times, across which performances are summarised.

The BO is configured with the same set of hyperparameters for the defined optimisation tasks. Initial sample size (n_i) is set to 50 and iteration sample

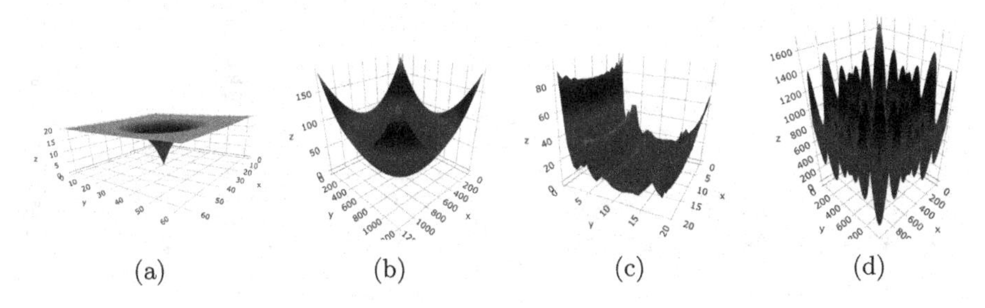

(a) (b) (c) (d)

Fig. 1. Visualisation of 2-D benchmark optimisation functions: (a) Ackley, (b) Griewank, (c) Levy, and (d) Schwefel.

size (n_t) to 200. Surrogate models are mainly used with the default parameter settings, except for the number of trees in an RF ensemble, set to 500 for lower dimensions (10 and 50) and 1000 for the 50- and 100-dimensional problems.

4 Results and Discussion

We evaluate the performances of the applied method(s) with regards to their *optimisation curve*, *resulted global optimum*, and *running time*.

Overall, the proposed method BO-RFw/SPO tends to outperform the BO-RF while lagging behind the BO-GP method. The former is observed across most of the benchmark problems with higher dimensionalities, except for the Schwefel function, where no significant difference is observed. The lag of BO-RFw/SPO behind the BO-GP is observed across multiple functions and lower dimensionality. RS is performing worst in all cases (Fig. 2).

The performances of our method show closer behaviour to the BO-GP than to BO-RF, among problems with the lowest dimension. This is explained by unimodal distributions of the RF predicted values (due to the low dimensionality) and the property of the SPO to avoid possible outliers by maximising the likelihood of a Gaussian component and identifying the modes. Thus, with a single component, it closely resembles the GP outcome.

Another observation is that the BO-RFw/SPO iteratively improves by gathering new samples that guide the search process, unlike the BO-RF, which frequently has a good start, but lack improvement during the later iterations. This is explained by the fact that the acquired samples from SPO are more informative for the region of interest and thus learns more accurate surrogate models through the iterations. This is observed in settings with higher dimensions, mainly due to the increased modality of the density function of the base predictions.

Regarding the global optima, on average, BO-RFw/SPO get closer to the known optimum when compared to the BO-RF, in particular in high-dimensional problems (Fig. 3 (right)). Worth noting is the case of the Schwefel function, where

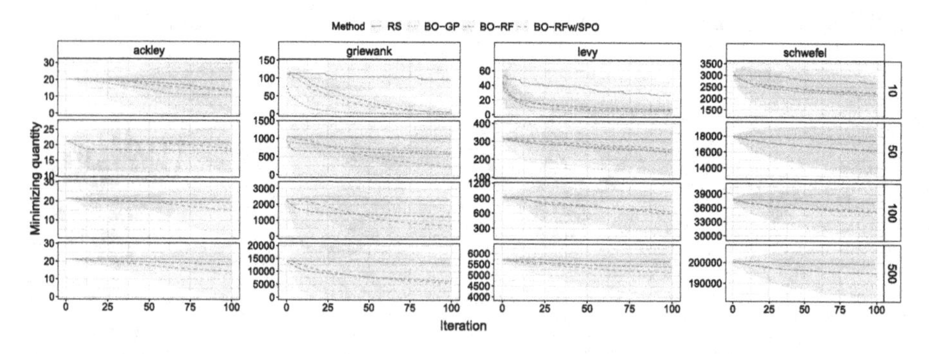

Fig. 2. Optimisation curves given per benchmark function and dimensionality of the problem, summarised over all repetitions.

none of the methods can make a significant step toward the global optimum (Fig. 3 (right)). It is the case due to the location of the global optimum at the edge of the function surface and the sampling method used.

The observed running time confirms the expected behaviours, primarily due to the fixed stopping criteria (100 iterations) of the implemented algorithms. Figure 3 (left) summarises the running time, from where we can confirm that the RF surrogate model is the fastest, followed by the proposed SPO approach, which has added complexity of maximising the likelihood of identified mixture of Gaussian components. Finally, for lower-dimensional problems, the BO-RFw/SPO is comparable to the BO-GP.

5 Summary and Conclusions

We have introduced a semi-parametric approach to the random forests surrogate model with Gaussian mixture models as an extension to Bayesian optimisation for high-dimensional problems. This extends previous work, which has been based on regression with Gaussian processes as a surrogate model.

In our study, we test the performance of our proposed approach and other established methods in four different benchmark optimisation problems. We vary the dimension of the problem to investigate the feasibility and performance of the methods. We note that the Gaussian process-based method is computationally demanding in higher dimensions, while our proposed method works well across all dimensions in presented scenarios. Furthermore, the quality of the solutions of the semi-parametric approach resembles the one from the Gaussian process as a surrogate model. It outperforms both the random search and random forests surrogate model that uses an empirical estimation of the model variance.

The performances of the proposed approach are achieved without compromising the computational complexity and scalability of the method across higher-dimensional problems. On average, the running time of the semi-parametric approach is higher than the running time of random forests by a constant factor.

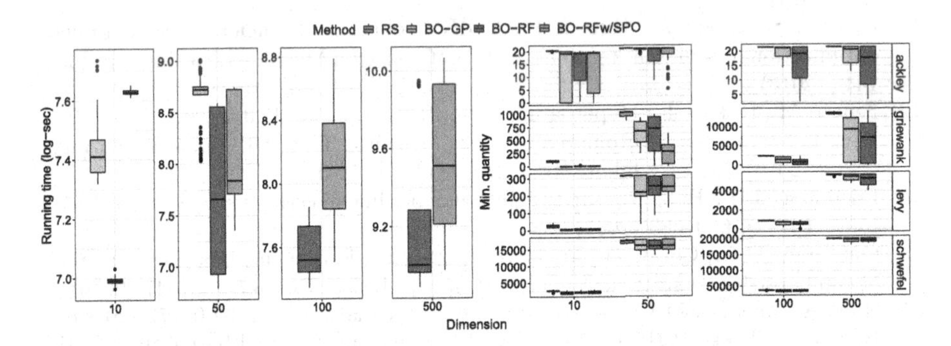

Fig. 3. Performance summary in terms of running time (in log-seconds) (**left**) and resulting global optima per function (**right**) shown across different dimensions.

Therefore, it is confirmed that our approach that transforms a non-parametric surrogate model (random forests) into semi-parametric output is advantageous for Bayesian optimisation in high-dimensional problems.

In further work, we intend to examine the effects of the random forests' hyper-parameters over the search pace through the function space and other approaches to generating new samples at each iteration of the Bayesian optimisation.

Acknowledgments. This work is supported by the European Commission through the H2020 project Finest Twins (grant No. 856602).

References

1. An, Z., Nott, D.J., Drovandi, C.: Robust Bayesian synthetic likelihood via a semi-parametric approach. Stat. Comput. **30**(3), 543–557 (2020)
2. Aushev, A., Pesonen, H., Heinonen, M., Corander, J., Kaski, S.: Likelihood-free inference with deep Gaussian processes. arXiv preprint arXiv:2006.10571 (2020)
3. Bäck, T.: Evolutionary Algorithms in Theory and Practice: Evolution Strategies, Evolutionary Programming, Genetic Algorithms. Oxford University Press, USA (1996)
4. Bergstra, J., Bengio, Y.: Random search for hyper-parameter optimization. J. Mach. Learn. Res. **13**(2), 281–305 (2012)
5. Blum, M., Nunes, M., Prangle, D., et al.: Comparative review of dimension reduction methods in approximate Bayesian computation. Stat. Sci. **28**(2), 189–208 (2013)
6. Breiman, L.: Random forests. Mach. Learn. **45**(1), 5–32 (2001)
7. Chen, B., Castro, R., Krause, A.: Joint optimization and variable selection of high-dimensional Gaussian processes. arXiv preprint arXiv:1206.6396 (2012)
8. Day, N.: Estimating the components of a mixture of normal components. Biometrika **56**(3), 463–474 (1969)
9. Dempster, A.P., Laird, N.M., Rubin, D.B.: Maximum likelihood from incomplete data via the EM algorithm. J. R. Stat. Soc. Ser. B (Methodological) **39**(1), 1–38 (1977)
10. Eggensperger, K., et al.: Towards an empirical foundation for assessing Bayesian optimization of hyperparameters. In: NIPS Workshop on BO in Theory and Practice (2013)
11. Falkner, S., Klein, A., Hutter, F.: BOHB: robust and efficient hyperparameter optimization at scale. In: International Conference on Machine Learning, pp. 1437–1446 (2018)
12. Feurer, M., Hutter, F.: Hyperparameter optimization. In: Hutter, F., Kotthoff, L., Vanschoren, J. (eds.) Automated Machine Learning. TSSCML, pp. 3–33. Springer, Cham (2019). https://doi.org/10.1007/978-3-030-05318-5_1
13. Friedman, J.H., Hall, P.: On bagging and nonlinear estimation. J. Stat. Plann. Infer. **137**(3), 669–683 (2007)
14. Gutmann, M.U., Corander, J.: Bayesian optimization for likelihood-free inference of simulator-based statistical models. J. Mach. Learn. Res. **17**(1), 1–47 (2016)
15. Hansen, N.: The CMA evolution strategy: a tutorial. arXiv:1604.00772 (2016)
16. Hutter, F., Hoos, H.H., Leyton-Brown, K.: Sequential model-based optimization for general algorithm configuration. In: Coello, C.A.C. (ed.) LION 2011. LNCS, vol. 6683, pp. 507–523. Springer, Heidelberg (2011). https://doi.org/10.1007/978-3-642-25566-3_40

17. Järvenpää, M., Gutmann, M.U., Pleska, A., Vehtari, A., Marttinen, P., et al.: Efficient acquisition rules for model-based approximate Bayesian computation. Bayesian Anal. **14**(2), 595–622 (2019)
18. Järvenpää, M., Gutmann, M.U., Vehtari, A., Marttinen, P., et al.: Gaussian process modelling in approximate Bayesian computation to estimate horizontal gene transfer in bacteria. Ann. Appl. Stat. **12**(4), 2228–2251 (2018)
19. Kuzmanovski, V., Hollmén, J.: Composite surrogate for likelihood-free bayesian optimisation in high-dimensional settings of activity-based transportation models. In: Abreu, P.H., Rodrigues, P.P., Fernández, A., Gama, J. (eds.) IDA 2021. LNCS, vol. 12695, pp. 171–183. Springer, Cham (2021). https://doi.org/10.1007/978-3-030-74251-5_14
20. Laguna, M., Marti, R.: Experimental testing of advanced scatter search designs for global optimization of multimodal functions. J. Glob. Optim. **33**(2), 235–255 (2005)
21. Leclercq, F.: Bayesian optimization for likelihood-free cosmological inference. Phy. Rev. D **98**(6) (2018)
22. Lintusaari, J., Gutmann, M., Dutta, R., Kaski, S., Corander, J.: Fundamentals and recent developments in approximate Bayesian computation. Syst. Biol. **66**, e66–e82 (2017)
23. Locatelli, M. A Note on the Griewank Test Function. J. Glob. Optim. **25**, 160–174 (2003). https://doi.org/10.1023/A:1021956306041
24. Meinshausen, N.: Quantile regression forests. JMLR **7**, 983–999 (2006)
25. Mockus, J.: On Bayesian Methods for Seeking the Extremum. In: Marchuk, G.I. (eds.) Optimization Techniques IFIP Technical Conference. LNCS. Springer, Heidelberg (1975). https://doi.org/10.1007/978-3-662-38527-2_55
26. Oh, S., Seshadri, R., Azevedo, C., Ben-Akiva, M.E.: Demand calibration of multimodal microscopic traffic simulation using weighted discrete SPSA. Transp. Res. Rec. **2673**(5), 503–514 (2019)
27. Petrik, O., Adnan, M., Basak, K., Ben-Akiva, M.: Uncertainty analysis of an activity-based microsimulation model for Singapore. Future. Gener. Comput. Sys. **110**, 350–363 (2018)
28. Raynal, L., Marin, J., Pudlo, P., Ribatet, M., Robert, C., Estoup, A.: ABC random forests for Bayesian parameter inference. Bioinformatics **35**(10), 1720–1728 (2019)
29. Schwefel, H.P.: Numerical Optimization of Computer Models. Wiley (1981)
30. Sha, D., Ozbay, K., Ding, Y.: Applying Bayesian optimization for calibration of transportation simulation models. Transp. Res. Rec. **2674**, 036119812093625 (2020)
31. Shahriari, B., Swersky, K., Wang, Z., Adams, R.P., De Freitas, N.: Taking the human out of the loop: a review of Bayesian optimization. Proc. IEEE **104**(1), 148–175 (2015)
32. Sisson, S.A., Fan, Y., Beaumont, M.: Handbook of Approximate Bayesian Computation. CRC Press (2018)
33. Snoek, J., Rippel, O., Swersky, K., Kiros, R., Satish, N., Sundaram, N., Patwary, M., Prabhat, M., Adams, R.: Scalable Bayesian optimization using deep neural networks. In: International Conference on Machine Learning, pp. 2171–2180 (2015)
34. Springenberg, J.T., Klein, A., Falkner, S., Hutter, F.: Bayesian optimization with robust Bayesian neural networks. In: Proceedings of the 30th International Conference on Neural Information Processing Systems, pp. 4141–4149 (2016)
35. Titterington, D., Smith, A., Makov, U.: Statistical Analysis of Finite Mixture Distributions. Series in Probability and Mathematical Statistics. Wiley (1985)

36. Todorović, M., Gutmann, M., Corander, J., Rinke, P.: Bayesian inference of atomistic structure in functional materials. NPJ Comput. Mater. **5**(1), 35 (2019)
37. Yu, T., Zhu, H.: Hyper-parameter optimization: a review of algorithms and applications. arXiv preprint arXiv:2003.05689 (2020)
38. Zhang, Y., Apley, D.W., Chen, W.: Bayesian optimization for materials design with mixed quantitative and qualitative variables. Sci. Rep. **10**(1), 4924 (2020)

A Clustering-Inspired Quality Measure for Exceptional Preferences Mining—Design Choices and Consequences

Ruben Franciscus Adrianus Verhaegh, Jacco Johannes Egbert Kiezebrink, Frank Nusteling, Arnaud Wander André Rio, Márton Bendegúz Bendicsek, Wouter Duivesteijn(✉), and Rianne Margaretha Schouten(✉)

Eindhoven University of Technology, Eindhoven, the Netherlands
{r.f.a.verhaegh,j.j.e.kiezebrink,f.nusteling,a.w.a.rio, m.b.bendicsek}@student.tue.nl, {w.duivesteijn,r.m.schouten}@tue.nl

Abstract. Exceptional Preferences Mining (EPM) combines the research fields of Preference Learning and Exceptional Model Mining. It is a local pattern mining task, where we try to find coherent subgroups of the dataset featuring unusual preferences between a fixed set of labels. We introduce a new quality measure for Exceptional Preferences Mining, inspired by concepts from Clustering. On top of that, we draw conclusions on two design choices that must necessarily be made whenever one defines a quality measure for any version of Exceptional Model Mining: on the one hand, exceptional behavior is easily (spuriously) found in tiny subgroups, so what is the best way to compensate for that; on the other hand, when gauging exceptionality of a subgroup's behavior, what does one use as reference for the normal behavior? We find that the choice of correction factor not only influences the subgroup size but it also effects the presumed exceptionality of found subgroups. The entropy function allows for detecting exceptional subgroups of a meaningful size, both when a candidate subgroup is evaluated against its complement and against the entire dataset.

Keywords: Exceptional preferences mining · Label ranking · Exceptional model mining · Preference learning · Pattern mining

1 Introduction

Exceptional Preferences Mining (EPM) [15,16] combines the two research fields of Preference Learning (PL) [5] and Exceptional Model Mining (EMM) [4,12]. In PL, rather than predicting the relevance of individual labels for records of the dataset, the focus lies on learning whether a record of the dataset prefers a label over another. Hence, PL is mostly concerned with analyzing how labels relate to each other, rather than the individual expression of a single label. A subfield

P. Pascal and D. Ienco (Eds.): DS 2022, LNAI 13601, pp. 429–444, 2022.
https://doi.org/10.1007/978-3-031-18840-4_31

of PL is Label Ranking (LR) [2,17], where one tries to learn a preference order (ranking) on a set of labels. This is the part of PL that is of specific concern to EPM. The other research field, EMM, seeks interesting subgroups of the dataset. A subgroup is interesting if it satisfies two properties. On the one hand, subgroups must be *interpretable*: we must be able to define them in terms of few conditions on attributes of the dataset, so that we can understand and build real-life policies on them. On the other hand, subgroups must be *exceptional*: a few columns of the dataset are split off to form the target space, over which we build a model, and subgroups are interesting if their behavior in this target space is unusual. For instance, when analyzing sequential data in target space, Markov chains can capture behavior in a subgroup, and one could assess exceptionality of Markov model parameters to gauge the quality of a subgroup [18]. Within EPM, the exceptionality of a label ranking becomes the target concept of an EMM run: we find subgroups displaying unusual rankings of a set of labels.

Existing EPM quality measures [15,16] gauge exceptionality of the label ranking within a subgroup on three separate levels of granularity (discussed in more detail in Sect. 3), but they all share one trait: they only assess whether records of the subgroup behave exceptionally, but not whether there is consistency behind the measured exceptionalities. These measures neglect that exceptionality of behavior might be achieved by lumping together disparate, heterogeneous kinds of behavior (cf. [1] for a similar argument in Subgroup Discovery, correcting for dispersion). In this paper, we propose a quality measure for EPM that not only captures exceptional behavior, but additionally encourages subgroups to have homogeneous target distributions. More specifically, we propose a quality measure for EPM based on the principles of clustering, where one optimizes for low within-cluster and high between-cluster distance. Comparably, our proposed quality measure assigns a high quality value to subgroups with preference relations that are dissimilar compared to records outside the subgroup but simultaneously very similar across records inside the subgroup.

When developing a quality measure for EMM (and hence also for EPM), two design choices must be made. On the one hand, exceptional behavior is easily (spuriously) found in tiny subgroups, so one must incorporate a component in the quality measure to promote non-tiny subgroups. Typical solutions are using the entropy of the subgroup/complement split, the size of the subgroup, or the square root thereof. On the other hand, exceptional behavior cannot exist in a vacuum: behavior can only be exceptional w.r.t. a reference behavior. Typical choices are using the behavior on the entire dataset as normal behavior, or using the behavior on the subgroup's complement as reference. Crucially, for both these design choices, very little evidence exists on what the right choice would be. In this paper, we show that the choice of correction factor not only influences the subgroup size but it also effects the presumed exceptionality of found subgroups, and we further demonstrate differences in outcomes under different reference behaviors in the context of EPM.

1.1 Main Contributions

The main contributions of this paper are:

1. a new quality measure for EPM that allows for the finding of exceptional and coherent subgroups in both descriptive and target space;
2. an exploration of the effect of subgroup size correction functions on the exceptionality of the found subgroups;
3. a demonstration of how outcomes differ depending on whether a subgroup is evaluated against the global model or against its complement.

2 Preliminaries

Exceptional Preferences Mining (EPM) [15,16] is a mix of Preference Learning (PL) [5] on the one hand and Exceptional Model Mining (EMM) [4,12] on the other hand. It combines the task of "learning to rank" [5, p. 3] with the task of identifying subgroups in a dataset that behave exceptionally. Specifically, EPM focuses on Label Ranking (LR) [2,17], a type of problem in PL that aims to map instances to rankings over a predefined set of labels, or classes. One can consider LR to be a variant of the conventional classification problem, but instead of assigning a case to a specific class, LR aims to assign a complete order of labels.

Assume a dataset Ω, which is a bag of N records $r \in \Omega$ of the form

$$r = (a_1, \ldots, a_k, t_1, \ldots, t_\ell)$$

where k and ℓ are positive integers. Target attributes $t_1, \ldots, t_\ell$ contain values associated with ℓ unique labels or classes from the set $\mathcal{L} = \{\lambda_1, \ldots, \lambda_\ell\}$. Thus, t_1 contains values associated with label λ_1, t_2 contains values associated with label λ_2, etcetera. The exact meaning of the values depends on the application domain. For instance, in a classification problem, t_v can be the probability that a record r belongs to class $\lambda_v \in \mathcal{L}$. Alternatively, in Sect. 6 of this paper, we analyze the Dutch parliament elections in 2021 and consider record $r \in \Omega$ to be a municipality; attributes $t_1, \ldots, t_\ell$ contain the number of votes for ℓ distinct political parties.

2.1 Order Relations

We are interested in the ordering of the political parties by the number of votes. The idea is to construct an ordering of the associated labels such that label λ_v precedes λ_w when $t_v > t_w$, $v \neq w$ and $1 \leq v, w \leq \ell$. Here, we consider total order relations $\succ$ on $\mathcal{L}$, which means that label λ_v cannot have the same position as λ_w. In other words, the ordering is a ranking and $\lambda_v \succ \lambda_w$ not only means that λ_v precedes λ_w but also that it is preferred over λ_w. Depending on the application, the user can decide what total order should be assigned to labels with equal values. In the case of Dutch elections, political parties with an equal number of votes will be ranked based on their position on the voting list.

Formally, a total order $\succ$ is a permutation π of the set $\{1, 2, \ldots, \ell\}$ such that $\pi(v)$ is the position of label λ_v in the order. For instance, if we consider the total order $\lambda_4 \succ \lambda_1 \succ \lambda_3 \succ \lambda_2$ for $\ell = 4$, $\pi = (2, 4, 3, 1)$.

Table 1. Example toy datasets: the shared descriptor space, and separate target spaces for SD, EMM, and EPM.

Attribute name	a_1	a_2	a_3	a_4	...	a_k
Meaning	Name	Legs	Swims?	Flies?	...	Fluffy?
r^1	Cat	4	no	no	...	a bit
r^2	Fish	0	yes	no	...	no
r^3	Owl	2	no	yes	...	no
r^4	Sheep	4	no	no	...	very yes
r^5	Snail	0	no	no	...	no

(a) Descriptor space

t_1
Friendly
no
yes
no
yes
yes

(b) SD

t_1	t_2	...	t_m
Length	Weight	...	Life span
46	4 500	...	15
10	227	...	12
41	1 585	...	8
1 500	95 000	...	11
2	6	...	6

(c) EMM

$\pi(1)$	$\pi(2)$	$\pi(3)$
Grass rank	Bread rank	Meat rank
2	3	1
2	1	3
2	3	1
1	2	3
1	2	3

(d) EPM ($\ell = 3$)

2.2 Local Pattern Mining Methods: SD, EMM, and EPM

In the setting of both LR and EPM, preferences on $\mathcal{L}$ are associated with particular (groups of) dataset records through a set of features or attributes. In EMM and EPM terms, these features are *descriptive* attributes, or descriptors. Attributes $a_1, \ldots, a_k$ are these descriptors. The task of Local Pattern Mining methods [7,13] is to find subgroups of the dataset, defined as a conjunction of conditions on a few descriptors. Subgroup Discovery (SD) [8,10,20] seeks subgroups displaying an unusual distribution of a single target attribute, Exceptional Model Mining (EMM) [4,12] seeks subgroups displaying an unusual interaction between multiple target attributes, and Exceptional Preferences Mining (EPM) [15,16] seeks subgroups where this interaction is exceptional preference relations. Hence, EMM can be seen as the multitarget generalization of SD, and EPM can be seen as a specific instantiation of the generic EMM framework.

Table 1 displays a toy dataset of some animals in a zoo. SD, EMM, and EPM all share the descriptor space of Table 1a; any target space from Tables 1b, c, and d can be appended. Combining Tables 1a and b, SD would find that the subgroup "flies? = no" has a 75% share of "friendly = yes", while this share is 60% in the overall population. Combining Tables 1a and c, EMM would find subgroups with an unusual interaction between the m targets (for example, exceptional regression coefficients of length and weight while predicting life span, when using the EMM model class from [3]). Combining Tables 1a and d, EPM would find that the subgroup "Legs $\leq$ 1" always ranks meat last.

2.3 Definitions

The task of EPM is to identify subgroups in the dataset with exceptional preferences. The subgroups are defined by a description over the collective domain of descriptive attributes. Formally, a description is a function $D : \mathcal{A} \mapsto \{0, 1\}$, and a record r^i is covered by description D if and only if $D(a_1^i, ..., a_k^i) = 1$.

Definition 1. The subgroup corresponding to a description D is the bag of records $S_D \in \Omega$ that D covers: $S_D = \{r^i \in \Omega \mid D(a_1^i, \ldots, a_k^i) = 1\}$.

We denote the number of records in a subgroup S with n. Every subgroup has a complement $S^C = \Omega \setminus S$ which contains all $n^C = N - n$ records not in S. Whether a subgroup has exceptional preferences is evaluated with a *quality measure* (QM):

Definition 2. Given a description language $\mathcal{D}$ governing which subgroups can be formulated on a given dataset, a quality measure is a function $\varphi : \mathcal{D} \mapsto \mathbb{R}$.

The goal is to find the top-q subgroups with the highest quality value. It is practically impossible to investigate all candidate subgroups exhaustively since the number of candidates scales exponentially with the number of descriptive attributes. Therefore, we perform a heuristically guided search called beam search. We will further discuss beam search in Sect. 4.1 while discussing the time complexity of our approach.

3 Related Work

Local pattern mining methods have been used to understand preference relations. For instance, the Olympic ranking of countries has been studied [14] with SD. Casting German federal Bundestag election vote shares (and vote share changes between subsequent elections) within regions as preference relations, a traditional SD analysis can be performed by averaging across the ℓ parties [6]. Instead, we are interested in finding subgroups with an unusual interaction between ℓ target attributes (and therefore consider our approach to be EMM).

Existing EPM QMs [15,16] are based on preference matrices (PM). A PM $\in \{-1, 0, 1\}^{\ell \times \ell}$ is a square matrix that for each pair of labels λ_v, λ_w in a ranking π evaluates whether they precede (1) or succeed (-1) each other ($\forall v, w \in \{1, \ldots, \ell\}$). PMs of individual records can be averaged, which allows for the comparison of matrix M^D, the PM for the entire dataset, with M^S, the PM for the subgroup. Denoting the difference between matrices M^D and M^S with L^S, [15,16] propose three quality measures, for exceptionality on three distinct levels of behavioral granularity:

$$\varphi_{\text{norm}} = \sqrt{n/N} \cdot \sqrt{\sum_{v=1}^{\ell}\sum_{w=1}^{\ell} L^S(v,w)^2} \tag{1}$$

$$\varphi_{\text{labelwise}} = \sqrt{n/N} \cdot \max_{v=1,\dots,\ell} \frac{1}{(\ell-1)} \sum_{w=1}^{\ell} L^S(v,w) \tag{2}$$

$$\varphi_{\text{pairwise}} = \sqrt{n/N} \cdot \max_{v,w=1,\dots,\ell} L^S(v,w) \tag{3}$$

The first QM, φ_{norm}, takes the Frobenius norm of L^S to search for preference deviations that occur spread out across the entire difference matrix. Zooming in, $\varphi_{\text{labelwise}}$ evaluates whether there is one particular label λ_v that ranks substantially different in the subgroup, ignoring interactions between other labels. Zooming in even further, $\varphi_{\text{pairwise}}$ studies pairwise preferences [9], evaluating whether any pair of labels interacts unusually in the subgroup. All three QMs compare the PM of the subgroup with the PM of the entire dataset, and share the choice for subgroup size correction factor:

$$\xi_{\text{sqrt}} = \sqrt{n/N}. \tag{4}$$

In developing our quality measure we will borrow principles from clustering. EMM is a local pattern mining technique whereas clustering is a global analysis task, partitioning all records into homogeneous clusters. In EMM, subgroups have an interpretable description, and records may be assigned to any number of subgroups. Methods on the crossroads of local and global pattern mining have been proposed, such as Predictive Clustering Rules (PCR) [21], SD with a classification rule learning algorithm (CN2-SD) [11], Cluster Grouping (CG) [22] and Multi-Response Subgroup Discovery (MR-SD) [19]. Although our quality measure is inspired by principles in clustering, our method is a purely local one.

4 Proposed Method: A Clustering-Based Quality Measure

We propose to perform EPM using the following clustering-based quality measure. Given a subgroup S and its complement S^C, let π^i denote the ranking of labels in the ith record of S, and let π^j denote the ranking in the jth record of S^C, where $1 \leq i \leq n$ and $1 \leq j \leq n^C$. We seek subgroups of records with exceptional label preferences. Those subgroups should have rankings dissimilar from the rankings in its complement. We define this notion of inter-subgroup distance as

$$\alpha_{\text{compl}} = \frac{1}{n \cdot n^C} \cdot \sum_{i=1}^{n}\sum_{j=1}^{n^C} d(\pi^i, \pi^j), \tag{5}$$

where $d(\cdot,\cdot)$ is some distance metric between the two rankings.

In addition, we want the cases in the subgroup to have similar rankings (i.e. to have small distance to one another), because coherent and homogeneous subgroups are 1) easier to interpret and 2) more practically relevant than heterogeneous subgroups. We define this notion of intra-subgroup distance as

$$\beta = \frac{1}{n \cdot (n-1)} \cdot \sum_{h=1}^{n} \sum_{i=1}^{n} d(\pi^h, \pi^i). \tag{6}$$

Next, we divide the inter-subgroup distance α by the intra-subgroup distance β, which results in the following quality measure,

$$\varphi_{\text{clus}} = \frac{\alpha}{\beta + 1} \cdot \xi, \tag{7}$$

where ξ is a function that corrects for the subgroup size. We add 1 to the denominator to account for perfect homogeneous subgroups (where $\beta = 0$).

Quality measure φ_{clus} is expected to give a high value when the subgroup is homogeneous (β small), when the subgroup's rankings are different from those in its complement (α large) or, ideally, both. Hence, our proposed quality measure is generic. Simultaneously, the distance function $d(\cdot, \cdot)$ can be specified by the user, which allows for searching subgroups with specific ranking deviations.

If one is interested in comparing a subgroup with the average ranking in the entire dataset, Eq. (5) can easily be adapted as follows,

$$\alpha_{\text{average}} = \frac{1}{n} \sum_{i=1}^{n} d(\pi^i, \pi^D), \tag{8}$$

where π^D is the label ranking when all N data records are taken into account. In this scenario, β does not change: we still aim to find coherent subgroups with exceptional label rankings; only the reference behavior has changed.

4.1 Time Complexity

To traverse the space of candidate subgroups, we apply beam search, a commonly used algorithm that is flexible in handling descriptive attributes of binary, categorical, and/or numerical type [4]. The algorithm performs a level-wise search of d levels, where the first level evaluates candidate subgroups with descriptions based on 1 descriptor and each subsequent level refines the descriptions of the top-w subgroups. The time complexity of beam search for EMM [4] is given by

$$\mathcal{O}(dwkE(c + \mathcal{M}(N, \ell) + \log(wq))), \tag{9}$$

where E is the worst-case number of categories (binary and numerical attributes are refined faster), c refers to the complexity of comparing the model in the subgroup against another model, $\mathcal{M}(N, \ell)$ is the cost of learning a model on N records and ℓ targets and d, w, k, and q are as described before (cf. [4, Section 4.2.1] for more details).

To evaluate the exceptionality of a candidate subgroup with quality measure φ_{clus}, α_{compl} requires $n \cdot n^C$ comparisons, α_{average} requires n comparisons and β

requires $n \cdot (n-1)$ comparisons. The time complexity of one comparison depends on the number of target attributes ℓ. That means that the time complexity of calculating φ_{clus} scales quadratically: $\mathcal{O}(\ell(n \cdot (n-1) + n \cdot n^C)) = \mathcal{O}(N^2 \cdot \ell)$ (since n and n^C are both $\mathcal{O}(N)$). The effect of c is already incorporated here.

The original EPM QMs [15,16] have a different time complexity. Calculating a PM costs $\mathcal{O}(\ell^2)$ per record; an average PM over n records then has a complexity of $\mathcal{O}(n\ell^2)$. In Sect. 5, we will further analyze these run times with synthetic data. Section 6 evaluates the performance of our proposed quality measure on real-world data.

4.2 Qualitative Differences Between φ_{clus} and Existing QMs

The added value of the quality measure is that it finds interesting results based on the distance between the sum of the permutations of the subgroup and the complement of the subgroup. Therefore, this quality measure should excel in finding those subgroups where the general ranking of the target variables differs greatly. Where previous work uses a general mean norm quality measure to find subgroups for label ranking [6], φ_{clus} seems intuitively very similar to a norm-based quality measure. It is different in that it tries to find subgroups based on the deviation from the overall mean of the permutations of the labels.

Existing work introduces different approaches in order to find subgroups for label ranking. Within EPM, preference matrices [16] are used; beyond pattern mining, a meta learning technique to reduce label ranking to binary classification was proposed [2]. Both these papers rely on preference matrices: label ranks are transformed to an interval $[0, 1]$ by averaging preferences of label pairs, thus accumulating them to matrices, one representing the dataset (M_D) and another representing the subgroup (M_S) [16]. Our algorithm, on the other hand, calculates the average distance of a label in the subgroup compared to those within the subgroup S (β) and to those in the complement subgroup S^C (α). The quality measures presented in the studied literature all have clear use cases as mentioned in Sect. 3, while our measure aims to be more generic.

The approach of the quality measure created in this paper is different from all above-mentioned ones, thus could yield different interesting results. Besides this, φ_{clus} should be robust with respect to variations in dataset metacharacteristics that theoretically ought not to negatively affect the outcome of an Exceptional Preferences Mining run. More specifically, the number of rows in the dataset will likely not influence the quality measure as the similarities are normalized. An increase in the number of target variables will likely make finding subgroups more stable: more target variables will reduce the opportunity for sudden peaks in the distance function. The number of descriptive attributes in a dataset almost always affects local pattern mining techniques such as EPM: an increase in the number of descriptors exponentially inflates the search space, making interesting subgroups harder to find. The expectation is that this will be no different for this quality measure.

5 Synthetic Data Experiment

We generate data with $N \in \{100, 500, 1000\}$ records. Each of these records can be described by $k \in \{2, 8, 32\}$ binary descriptors, which are independently sampled from a binomial distribution $a_h \sim \text{Bin}(N, p)$ with $p = 0.5$ for all $1 \leq h \leq k$. For the sake of simplicity and consistency, we let the true subgroup cover records where $a_1 = 1 \wedge a_2 = 1$, resulting in subgroups with size $n = \frac{1}{4}N$.

Each record has a ranking based on $\ell \in \{2, 8, 32\}$ target attributes. Since we want our synthetic data to resemble a real-world scenario as much as possible, we first analyze the average ranking of the $\ell = 37$ political parties in the real-world dataset (see Sect. 6). There, a party with rank $v+1$ has about 0.7 times as many votes as the party with rank v, for all $1 \leq v \leq \ell$. The variance of the number of votes over the records had an average ratio with the number of votes of 0.03. For the synthetic dataset, we therefore draw ℓ target attributes from a normal distribution $t_v \sim \mathcal{N}(\mu_v, \sigma_v^2)$ with mean $\mu_v = 0.7^{(v-1)}$ and variance $\sigma_v^2 = 0.03\mu_v$. Given the number of votes per party, a ranking π per record is obtained as per Sect. 2. Because of random sampling, $\pi(v)$ may or may not have value v, but on average the ranking in the entire dataset will be $\pi^D = (1, 2, \ldots, \ell)$.

We experiment with three types of subgroups (N.B.: every dataset contains one true subgroup, whose type is a simulation parameter):

reversed: we invert the values of the target attributes; the values of t_1 are swapped with the values of t_ℓ, the values of t_2 are swapped with $t_{\ell-1}$, etc. On average, this will result in $\pi_{\text{rev}} = (\ell, \ell-1, \ldots, 2, 1)$.

pairwise-swapped: we swap each consecutive pair of attributes; the values of t_1 are swapped with the values of t_2, the values of t_3 are swapped with t_4, etc. This will result in $\pi_{\text{pair}} = (2, 1, 4, 3, \ldots, \ell, \ell-1)$ for even values of ℓ.

last-to-first: here, no matter the values of attribute t_ℓ, we put λ_ℓ at rank 1, resulting in $\pi_{\text{ltf}} = (2, 3, \ldots, \ell-1, \ell, 1)$.

Note that because we generate the entire dataset first, and then replace the target values of the records covered by the true subgroup definition, $\pi^{SG^C} = \pi^D$.

We evaluate the performance of φ_{clus} with $\alpha = \alpha_{\text{compl}}$, and experiment with three types of subgroup size corrections: ξ_{sqrt} as given in Eq. (4), the entropy function as proposed for EMM by [12],

$$\xi_{\text{entropy}} = -\frac{n}{N}\log\frac{n}{N} - \frac{n^C}{N}\log\frac{n^C}{N}, \tag{10}$$

and no correction: $\xi_{\text{none}} = 1$. The three ways of correction are chosen such that they have opposite objectives: ξ_{sqrt} prefers larger subgroups, ξ_{entropy} prefers a 50/50 split of the dataset and ξ_{none} guides the search towards small subgroups. We run beam search with $w = 20$, $d = 3$ and evaluate whether or not the true subgroup is present in the top-q subgroups with $q = 10$. For every combination of parameters, we run the experiment $nreps = 5$ times and report the proportion of true subgroups not appearing in the top-q result list, the average rank of the true subgroups in that result list and the average run time. See https://github.com/bendicsekb/data_mining_election for all source code and results.

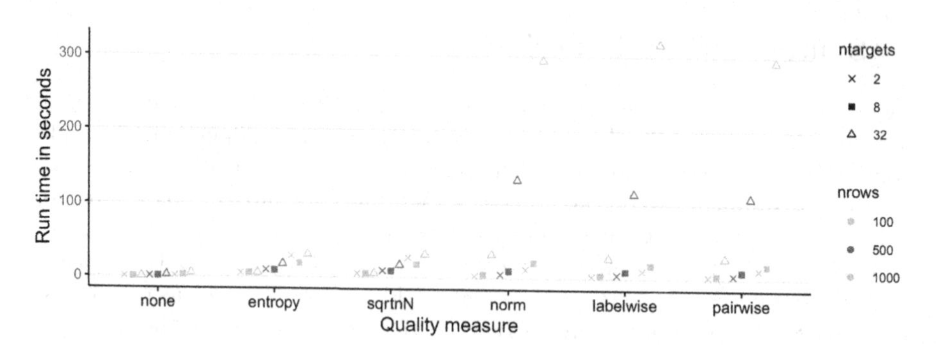

Fig. 1. Run time in seconds of the beam search algorithm for six quality measures, varying dataset size and varying number of target attributes. (Color figure online)

5.1 Results

For the reversed subgroup type and the last-to-first subgroup type, all six QMs ($\varphi_{\text{norm}}, \varphi_{\text{labelwise}}, \varphi_{\text{pairwise}}$, and φ_{clus} with three variants of ξ) find the ground truth subgroup in 100% of the cases at the first position in the result list. Quality measure φ_{clus} finds the pairwise-swapped subgroup type under all simulation conditions when $\xi = \xi_{\text{none}}$ (when no subgroup size correction is applied). For ξ_{entropy} and ξ_{sqrt}, the true subgroup cannot be found when both k and ℓ are 32. Like φ_{clus} with ξ_{none}, quality measures φ_{norm} and $\varphi_{\text{labelwise}}$ find the pairwise-swapped subgroup under all simulation conditions. When $N = 100$, $\varphi_{\text{labelwise}}$ has difficulty when the number of descriptors k is too large relative to the number of targets ℓ (8 vs. 2, 32 vs. 8 and 32 vs. 32). The problem disappears when N increases to 500 or 1000.

Figure 1 presents the run times in seconds for all six quality measures for the reversed subgroup type. Conclusions are similar for the pairwise-swapped and last-to-first subgroup type. As discussed in Sect. 4.1, we see that φ_{clus} scales with the number of rows N while the EPM QMs scale with the number of targets ℓ.

6 Real-World Data Experiment

We analyze data from the 2021 Dutch general election, publicly available at https://www.verkiezingsuitslagen.nl/verkiezingen/detail/TK20210317. The dataset contains the number of votes for $\ell = 37$ political parties in $N = 351$ municipalities. We add information about socio-economic characteristics of the municipalities such as the number of citizens, gender balance, age distribution, migration background, number of companies, how many ducks go for slaughter (proxy for rurality), total road length, and much more. That dataset is made available by Statistics Netherlands (https://opendata.cbs.nl/statline/portal.html?_la=nl&_catalog=CBS&tableId=70072ned&_theme=237). Consequently, we have $k = 83$ numerical descriptors.

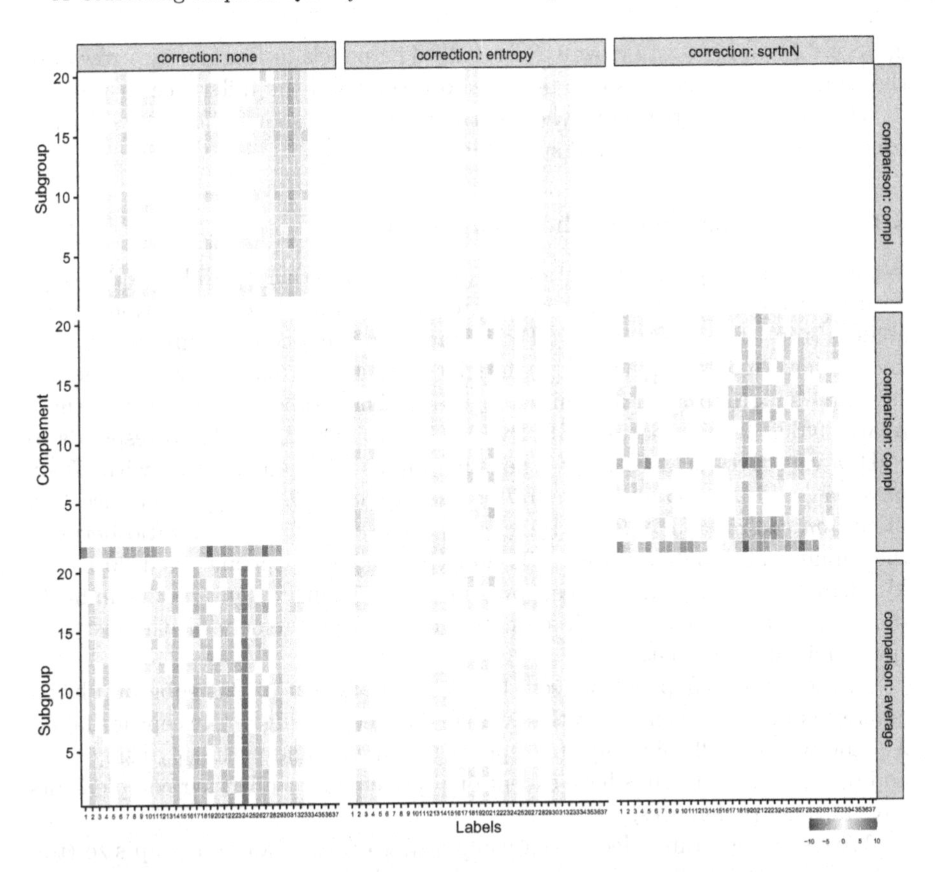

Fig. 2. Difference in rank between a group and the average dataset ranking π^D, obtained with quality measure φ_{clus} (red: higher rank in group, blue: lower rank in group, white: no difference). (Color figure online)

The goal is to find coherent subgroups of municipalities with an exceptional ranking of political parties. We evaluate the performance of φ_{clus} in six scenarios, combining the two comparisons with reference models α_{compl} and α_{average} with three types of subgroup size correction ξ_{none}, ξ_{entropy}, and ξ_{sqrt}. As distance function $d(\cdot,\cdot)$ within φ_{clus} we employ the Euclidean norm. The beam search is run with parameters $w = 30$, $d = 3$, and $q = 20$, and minimum subgroup size constraint of $c_{\text{size}} = 10\%$.

Figure 2 presents results. Each of the 9 panels shows the difference in label ranking between a subgroup and the average dataset ranking π^D ($q = 20$ by $\ell = 37$). Here, red indicates that a political party has a higher rank in the subgroup (more votes, moved to the left), blue represents a lower rank in the subgroup (fewer votes, moved to the right) and white means that there is no difference. The three columns in Fig. 2 correspond to the three types of subgroup size correction. The three rows correspond to reference models α_{compl} (row 1 and

2, Sect. 6.1) and α_{average} (row 3, Sect. 6.2). The panels in the first two rows use the same beam search results, but the top row gives the difference in ranking for the discovered subgroups whereas the second row shows the results for the complements of the discovered subgroups.

6.1 Comparing Against the Complement

When no subgroup size correction is applied, comparing candidate subgroups with their complements results in subgroups with quite some exceptional preferences (top left panel in Fig. 2). For instance, in the second subgroup, label λ_5 and λ_6 have moved to the right, while label λ_7 and λ_8 have moved to the left (the first subgroup is an all-white subgroup). These labels correspond to three relatively left-leaning parties (SP, PvdA, GroenLinks). Label λ_8 corresponds to FvD, a very right-leaning party. The subgroup covers municipalities with *Green pressure* $\geq$ *37.2%* $\wedge$ *Surinam migration background* $\geq$ *0.8%* $\wedge$ *Any-non western migration background* $\geq$ *9.6%*. Here, *green pressure* refers to the ratio between the number of people aged 0 to 20 and the number of people aged 20 to 65. Municipalities with a high green pressure skew younger. Our results indicate that younger citizens, or their parents, vote more extremely on the electoral spectrum than older citizens.

In the center left panel, we see that the complement of this subgroup has a label ranking that is similar to the average dataset ranking π^D, except for labels λ_{30} and λ_{31}. For all subgroups (except for the first) found with φ_{clus} using ξ_{none} and α_{compl}, the subgroups have exceptional preferences while their complements have average preferences.

We see an opposite effect when using ξ_{sqrt} to correct for subgroup size (top right panel). Here, all $q = 20$ subgroups do not deviate from the average dataset ranking. However, the complements of these subgroups show very exceptional preferences relations (center right panel). For some of these complements, a Christian party (CU) has obtained fewer votes (see blue color for λ_{10}). That happens for instance in the complement of subgroup 4, which is described by *Dutch background* $\leq$ *92%*. Apparently, in municipalities where the percentage of citizens with a Dutch background is > 92% (the complement), people tend to vote less for this particular Christian party.

Finding contrasting results for two opposite types of subgroup size correction gives us more insight in the performance of φ_{clus} specifically and quality measures in EMM in general. Remember that φ_{clus} is designed to generate exceptional and homogeneous subgroups. Then, when the algorithm divides the dataset into two groups and compares one group with the other (e.g., the complement), it rewards the more homogeneous group and chooses that to be the subgroup.

Using ξ_{sqrt} as a correction factor and giving preference to larger subgroups, the algorithm will select subgroups with records that are close to the dataset norm; it is likely that there are more records with average ranking behavior than there are records with similar non-average behavior. Using ξ_{none}, we find exceptional subgroups. Unfortunately, the consequence is that the subgroups are very small and have a size just larger than the minimum constraint.

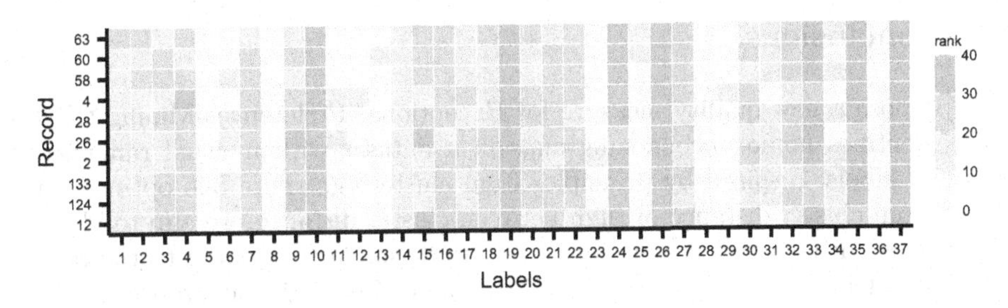

Fig. 3. Label ranking of 10 random records in the best-scoring subgroup found with φ_{clus} when comparing with the average dataset ranking and using no correction factor. (Color figure online)

The entropy function gives results that are in-between (see center top panel in Fig. 2). Although φ_{clus} still tends to select homogeneous subgroups, the preference for subgroups that contain half the number of records generates subgroups that do deviate from the norm, albeit for only a few labels. The complements of these subgroups have more exceptional preferences.

6.2 Comparing Against the Average Dataset Ranking

The bottom row of Fig. 2 presents results for φ_{clus} using α_{average} instead of α_{compl}. A clear effect of that can be seen when we use ξ_{none}. Then, the subgroups have label rankings that deviate much more from the average ranking than when we evaluate a candidate subgroup against its complement (compare left top panel with left bottom panel). In this scenario, the inter-subgroup distance as given by α_{average} is larger than α_{compl}. At the same time, the subgroups are coherent in target space and have small β, as can be seen for subgroup 1 in Fig. 3 where we present the label ranking of 10 random records in the subgroup. Although some fluctuations and differences between records exist, in general the records have similar rankings. Unfortunately, these results do not carry over to the scenario where we prefer larger subgroups (ξ_{sqrt}). Then, even though the subgroups have average ranking behavior, the intra-subgroup distance β dominates the quality value (bottom right panel in Fig. 2).

Like before, ξ_{entropy} finds an in-between solution and presents subgroups with exceptional preferences while making sure that the subgroups have a meaningful size (bottom center panel). Interestingly, the results are similar to those in the center panel, where we evaluate the complements of subgroups that are found under α_{compl}. Apparently, when using ξ_{entropy} in φ_{clus}, the reference group does not matter as much and similar exceptional subgroups are found. The difference is that these exceptional subgroups are not selected when we evaluate them against their complements, because the latter are more homogeneous and will therefore have a higher quality value.

7 Conclusions

We propose a new quality measure for Exceptional Preferences Mining (EPM) that identifies homogeneous subgroups in a dataset with unusual rankings of a set of labels. Inspired by principles from clustering, where one optimizes for low within-cluster distance or high between-cluster distance, we aim to identify subgroups with preference relations that are dissimilar compared to the rest of the dataset but very similar compared to records inside the subgroup.

As synthetic data experiments show (cf. Fig. 1), the time complexity of our quality measure scales with the number of dataset records, as opposed to existing quality measures for EPM that scale with the number of labels. Runtimes are roughly equivalent when the number of targets is 2 or 8, but our QM is substantially faster when this number is 32.

When developing a quality measure for EMM (and hence also for EPM), a correction for subgroup size should be included in order to steer the search away from tiny subgroups. Furthermore, one has to choose the reference behavior: is a candidate subgroup compared with its complement or with the average behavior? We investigate these scenarios for a real-world dataset with information about the voting behavior of municipalities in the Netherlands (cf. Fig. 2). If we compare candidate subgroups S with their complements S^C, we find that a size correction that prefers larger subgroups results in a search where the intra-subgroup distance dominates. Interestingly, exceptional ranking behavior is happening in this result set, but on the complements of the subgroups that EPM reports, which themselves display consistent but unexceptional behavior. In contrast, when we do not apply a correction for subgroup size, EPM reports subgroups are exceptional and homogeneous, but they only barely pass the minimum support constraint. To find subgroups of substantial size that also display exceptional behavior themselves, the entropy function gives the best results.

Comparing candidate subgroups with the average dataset Ω delivers subgroups with very exceptional preferences, especially when there is no correction for subgroup size. Then, the inter-subgroup distance will increase while exceptional subgroups are still coherent and homogeneous. When the entropy function is used, we again find subgroups with exceptional preferences of meaningful size. Comparing with Ω instead of S^C leads to comparable results, but the exceptional behavior is more often encompassed by the subgroups resulted by EPM instead of hidden away in their complements.

References

1. Boley, M., Goldsmith, B.R., Ghiringhelli, L.M., Vreeken, J.: Identifying consistent statements about numerical data with dispersion-corrected subgroup discovery. Data Min. Knowl. Discov. **31**(5), 1391–1418 (2017)
2. Cheng, W., Henzgen, S., Hüllermeier, E.: Labelwise versus pairwise decomposition in label ranking. In: Proceedings of the 15th LWA Workshops: KDML, IR and FGWM, pp. 129–136 (2013)

3. Duivesteijn, W., Feelders, A., Knobbe, A.J.: Different slopes for different folks: mining for exceptional regression models with Cook's distance. In: Proceedings of the 18th ACM SIGKDD International Conference on Knowledge Discovery and Data Mining (KDD 2012), pp. 868–876 (2012)
4. Duivesteijn, W., Feelders, A.J., Knobbe, A.: Exceptional model mining – supervised descriptive local pattern mining with complex target concepts. Data Min. Knowl. Disc. **30**(1), 47–98 (2016)
5. Fürnkranz, J., Hüllermeier, E.: Preference learning: an introduction. In: Fürnkranz, J., Hüllermeier, E. (eds.) Preference Learning, pp. 1–17. Springer, Heidelberg (2010). https://doi.org/10.1007/978-3-642-14125-6_1
6. Grosskreutz, H., Boley, M., Krause-Traudes, M.: Subgroup discovery for election analysis: a case study in descriptive data mining. In: Proceedings of the 13th International Conference on Discovery Science (DS 2010), pp. 57–71 (2010)
7. Hand, D.J., Adams, N.M., Bolton, R.J. (eds.): Pattern Detection and Discovery. LNCS (LNAI), vol. 2447. Springer, Heidelberg (2002). https://doi.org/10.1007/3-540-45728-3
8. Herrera, F., Carmona, C.J., González, P., Del Jesus, M.J.: An overview on subgroup discovery: foundations and applications. Knowl. Inf. Syst. **29**(3), 495–525 (2011)
9. Hüllermeier, E., Fürnkranz, J., Cheng, W., Brinker, K.: Label ranking by learning pairwise preferences. Artif. Intell. **172**(16–17), 1897–1916 (2008)
10. Klösgen, W.: Explora: a multipattern and multistrategy discovery assistant. In: Advances in Knowledge Discovery and Data Mining, pp. 249–271 (1996)
11. Lavrač, N., Kavšek, B., Flach, P., Todorovski, L.: Subgroup discovery with CN2-SD. J. Mach. Learn. Res. **5**, 153–188 (2004)
12. Leman, D., Feelders, A., Knobbe, A.: Exceptional model mining. In: Proceedings of the Joint European Conference on Machine Learning and Knowledge Discovery in Databases (ECMLPKDD 2008), pp. 1–16 (2008)
13. Morik, K., Boulicaut, J.-F., Siebes, A. (eds.): Local Pattern Detection. LNCS (LNAI), vol. 3539. Springer, Heidelberg (2005). https://doi.org/10.1007/b137601
14. Pieters, B.F., Knobbe, A., Džeroski, S.: Subgroup discovery in ranked data, with an application to gene set enrichment. In: Proceedings of the Preference Learning Workshop at Joint European Conference on Machine Learning and Knowledge Discovery in Databases (ECMLPKDD 2010), pp. 1–18 (2010)
15. de Sá, C.R., Duivesteijn, W., Azevedo, P.J., Jorge, A.M., Soares, C., Knobbe, A.J.: Discovering a taste for the unusual: exceptional models for preference mining. Mach. Learn. **107**(11), 1775–1807 (2018)
16. de Sá, C.R., Duivesteijn, W., Soares, C., Knobbe, A.: Exceptional preferences mining. In: Proceedings of the 19th International Conference on Discovery Science (DS 2016), pp. 3–18 (2016)
17. de Sá, C.R., Soares, C., Knobbe, A.: Entropy-based discretization methods for ranking data. Inf. Sci. **329**, 921–936 (2016)
18. Schouten, R.M., Bueno, M.L., Duivesteijn, W., Pechenizkiy, M.: Mining sequences with exceptional transition behaviour of varying order using quality measures based on information-theoretic scoring functions. Data Min. Knowl. Disc. **36**, 379–413 (2022)
19. Umek, L., Zupan, B.: Subgroup discovery in data sets with multi-dimensional responses. Intell. Data Anal. **15**(4), 533–549 (2011)
20. Wrobel, S.: An algorithm for multi-relational discovery of subgroups. In: Proceedings of PKDD, pp. 78–87 (1997)

21. Ženko, B., Džeroski, S., Struyf, J.: Learning predictive clustering rules. In: Proceedings of the International Workshop on Knowledge Discovery in Inductive Databases, pp. 234–250 (2005)
22. Zimmermann, A., De Raedt, L.: Cluster-grouping: from subgroup discovery to clustering. Mach. Learn. **77**(1), 125–159 (2009)

Recurrent Segmentation Meets Block Models in Temporal Networks

Chamalee Wickrama Arachchi(✉) and Nikolaj Tatti

HIIT, University of Helsinki, Helsinki, Finland
{chamalee.wickramaarachch,nikolaj.tatti}@helsinki.fi

Abstract. A popular approach to model interactions is to represent them as a network with nodes being the agents and the interactions being the edges. Interactions are often timestamped, which leads to having timestamped edges. Many real-world temporal networks have a recurrent or possibly cyclic behaviour. For example, social network activity may be heightened during certain hours of day. In this paper, our main interest is to model recurrent activity in such temporal networks. As a starting point we use stochastic block model, a popular choice for modelling static networks, where nodes are split into R groups. We extend this model to temporal networks by modelling the edges with a Poisson process. We make the parameters of the process dependent on time by segmenting the time line into K segments. To enforce the recurring activity we require that only $H < K$ different set of parameters can be used, that is, several, not necessarily consecutive, segments must share their parameters. We prove that the searching for optimal blocks and segmentation is an **NP**-hard problem. Consequently, we split the problem into 3 subproblems where we optimize blocks, model parameters, and segmentation in turn while keeping the remaining structures fixed. We propose an iterative algorithm that requires $\mathcal{O}\left(KHm + Rn + R^2H\right)$ time per iteration, where n and m are the number of nodes and edges in the network. We demonstrate experimentally that the number of required iterations is typically low, the algorithm is able to discover the ground truth from synthetic datasets, and show that certain real-world networks exhibit recurrent behaviour as the likelihood does not deteriorate when H is lowered.

1 Introduction

A popular approach to model interactions between set of agents is to represent them as a network with nodes being the agents and the interactions being the edges. Naturally, many interactions in real-world datasets have a timestamp, in which case the edges in networks also have timestamps. Consequently, developing methodology for temporal networks has gained attention in data mining literature [17].

Many temporal phenomena have recurrent or possibly cyclic behaviour. For example, social network activity may be heightened during certain hours of day.

P. Pascal and D. Ienco (Eds.): DS 2022, LNAI 13601, pp. 445–459, 2022.
https://doi.org/10.1007/978-3-031-18840-4_32

Our main interest is to model recurrent activity in temporal networks. As a starting point we use stochastic block model, a popular choice for modelling static networks. We can immediately extend this model to temporal networks, for example, by modelling the edges with a Poisson process. Furthermore, Corneli et al. [6] modelled the network by also segmenting the timeline and modelled each segment with a separate Poisson process.

To model the recurrent activity we can either model it explicitly, for example, by modelling explicitly cyclic activity, or we can use more flexible approach where we look for segmentation but restrict the number of distinct parameters. Such notion was proposed by Gionis and Mannila [10] in the context of segmenting sequences of real valued vectors.

In this paper we extend the model proposed by Corneli et al. [6] using the ideas proposed by Gionis and Mannila [10]. More formally, we consider the following problem: given a temporal graph with n nodes and m edges, we are looking to partition the nodes into R groups and segment the timeline into K segments that are grouped into H levels. Note that a single level may contain non-consecutive segments. An edge $e = (u, v)$ is then modelled with a Poisson process with a parameter λ_{ijh}, where i and j are the groups of u and v, and h is the level of the segment containing e.

To obtain good solutions we rely on an iterative method by splitting the problem into three subproblems: (i) optimize blocks while keeping the remaining parameters fixed, (ii) optimize model parameters Λ while keeping the blocks and the segmentation fixed, (iii) optimize the segmentation while keeping the remaining parameters fixed. We approach the first subproblem by iteratively optimizing block assignment of each node while maintaining the remaining nodes fixed. We show that such single round can be done in $\mathcal{O}\left(m + Rn + R^2H + K\right)$ time, where n is the number of nodes and m is the number of edges. Fortunately, the second subproblem is trivial since there is an analytic solution for optimal parameters, and we can obtain the solution in $\mathcal{O}\left(m + R^2H + K\right)$ time. Finally, we show that we can find the optimal segmentation with a dynamic program. Using a stock dynamic program leads to a computational complexity of $\mathcal{O}\left(m^2KH\right)$. Fortunately, we show that we can speed up the computation by using a SMAWK algorithm [2], leading to a computational complexity of $\mathcal{O}\left(mKH + HR^2\right)$.

In summary, we extend a model by Corneli et al. [6] to have recurring segments. We prove that the main problem is **NP**-hard as well as several related optimization problems where we fix a subset of parameters. Navigating around these **NP**-hard problems we propose an iterative algorithm where a single iteration requires $\mathcal{O}\left(KHm + Rn + R^2H\right)$ time, a linear time in edges and nodes.

The rest of the paper is organized as follows. First we introduce preliminary notation, the model, and the optimization problem in Sect. 2. We then proceed to describe the iterative algorithm in Sect. 3. We present the related work in Sect. 4. Finally, we present our experiments in Sect. 5 and conclude the paper with discussion in Sect. 6. The proofs are provided in Appendix[1].

[1] The appendix is available at https://arxiv.org/abs/2205.09862.

2 Preliminary Notation and Problem Definition

Assume a *temporal graph* $G = (V, E)$, where V is a set of nodes and E is a set of edges, where each edge is tuple (u, v, t) with $u, v \in V$ and t being the timestamp. We will use $n = |V|$ to denote the number of nodes and $m = |E|$ the number of edges. For simplicity, we assume that we do not have self-loops, though the models can be adjusted for such case. We write $t(e)$ to mean the timestamp of the edge e. We also write $N(u)$ to denote all the edges adjacent to a node $u \in V$.

Perhaps the simplest way to model a graph (with no temporal information) is with Erdos-Renyi model, where each edge is sampled independently from a Bernoulli probability parameterized with q. Let us consider two natural extensions of this model. The first extension is a block model, where nodes are divided into k blocks, and an edge (u, v) are modelled with a Bernoulli probability parameterized with q_{ij}, where i is the block of u and j is the block of v. Given a graph, the optimization problem is to cluster nodes into blocks so that the likelihood of the model is optimized. For the sake of variability we will use the words *block* and *group* interchangeably.

A convenient way of modelling events in temporal data is using Poisson process: Assume that you have observed c events with timestamps $t_1, \ldots, t_c$ in a time interval T of length Δ. The log-likelihood of observing these events at these exact times is equal to $c \log \lambda - \lambda \Delta$, where λ is a model parameter. Note that the log-likelihood does not depend on the individual timestamps.

If we were to extend the block model to temporal networks, the log-likelihood of c edges occurring between the nodes u and v in a time interval is equal to $c \log \lambda_{ij} - \lambda_{ij} \Delta$, where λ_{ij} is the Poisson process parameter and i is the block of u and j is the block of v. Note that λ_{ij} does not depend on the time, so discovering optimal blocks is very similar to discovering blocks in a static model.

A natural extension of this model, proposed by Corneli et al. [6], is to make the parameters depend on time. Here, we partition the model into k segments and assign different set of λs to each segment.

More formally, we define a *time interval* T to be a continuous interval either containing the starting point $T = [t_1, t_2]$ or excluding the starting point $T = (t_1, t_2]$. In both cases, we define the duration as $\Delta(T) = t_2 - t_1$.

Given a time interval T, let us define

$$c(u, v, T) = |\{e = (u, v, t) \in E \mid t \in T\}|$$

to be the number of edges between u and v in T.

The log-likelihood of Poisson model for nodes u, v and a time interval T is

$$\ell(u, v, T, \lambda) = c(u, v, T) \log \lambda - \lambda \Delta(T).$$

We extend the log-likelihood between the two sets of nodes U and W, by writing

$$\ell(U, W, T, \lambda) = \sum_{u, w \in U \times W} \ell(u, w, T, \lambda),$$

where $U \times W$ is a set of all node pairs $\{u, w\}$ with $u \in U$ and $w \in W$ and $u \neq v$. We consider $\{u, w\}$ and $\{w, u\}$ the same, so only one of these pairs is visited.

Given a time interval $D = [a, b]$, a K-segmentation $\mathcal{T} = T_1, \ldots, T_K$ is a sequence of K time intervals, such that $T_1 = [a, t_1], T_2 = (t_1, t_2], \ldots T_i = (t_{i-1}, t_i], \ldots$, and $T_K = (t_{K-1}, b]$. For notational simplicity, we require that the boundaries t_i must collide with the timestamps of individual edges. We also assume that D covers the edges. If D is not specified, then it is set to be the smallest interval covering the edges.

Given a K-segmentation, a partition of nodes $\mathcal{P} = P_1, \ldots, P_R$ into R groups, and a set of $KR(R+1)/2$ parameters $\Lambda = \{\lambda_{ijk}\}^2$, the log-likelihood is equal to

$$\ell(\mathcal{P}, \mathcal{T}, \Lambda) = \sum_{i=1}^{R} \sum_{j=i}^{R} \sum_{k=1}^{K} \ell(P_i, P_j, T_k, \lambda_{ijk}).$$

This leads immediately to the problem considered by Corneli et al. [6].

Problem 1. ((K, R) model). Given a temporal graph G, a time interval D, integers R and K, find a node partition with R groups, a K-segmentation, and a set of parameters Λ so that $\ell(\mathcal{P}, \mathcal{T}, \Lambda)$ is maximized.

We should point out that for fixed $\mathcal{P}$ and $\mathcal{T}$, the optimal Λ is equal to

$$\lambda_{ijk} = \frac{c(P_i, P_j, T_k)}{|P_i \times P_j| \Delta(T_k)}.$$

In this paper we consider an extension of (K, R) model. Many temporal network exhibit cyclic or repeating behaviour. Here, we allow network to have K segments but we also limit the number of distinct parameters to be at most $H \leq K$. In other words, we are forcing that certain segments *share* their parameters. We do not know beforehand which segments should share the parameters.

We can express this constraint more formally by introducing a mapping $g : [K] \rightarrow [H]$ that maps a segment index to its matching parameters. We can now define the likelihood as follows: given a K-segmentation, a partition of nodes $\mathcal{P} = P_1, \ldots, P_R$ into R groups, a mapping $g : [K] \rightarrow [H]$, and a set of $HR(R+1)/2$ parameters $\Lambda = \{\lambda_{ijh}\}$, the log-likelihood is equal to

$$\ell(\mathcal{P}, \mathcal{T}, g, \Lambda) = \sum_{i=1}^{R} \sum_{j=i}^{R} \sum_{k=1}^{K} \ell\left(P_i, P_j, T_k, \lambda_{ijg(k)}\right).$$

We will refer to g as *level mapping.*

This leads to the following optimization problem.

Problem 2. ((K, H, R) model). Given a temporal graph G, a time interval D, integers R, H, and K, find a node partition with R groups, a K-segmentation, a level mapping $g : [K] \rightarrow [H]$, and parameters Λ maximizing $\ell(\mathcal{P}, \mathcal{T}, g, \Lambda)$.

[2] For notational simplicity we will equate λ_{ijh} and λ_{jih}.

Algorithm 1: Main loop of the algorithm

```
1 P ← random groups; Λ ← random values;
2 T, g ← FindSegments(P, Λ);
3 Λ ← UpdateLambda(P, T, g);
4 while convergence do
5     P ← FindGroups(P, Λ, T, g);
6     Λ ← UpdateLambda(P, T, g);
7     T, g ← FindSegments(P, Λ);
8     Λ ← UpdateLambda(P, T, g);
```

3 Fast Algorithm for Obtaining Good Model

In this section we will introduce an iterative, fast approach for obtaining a good model. The computational complexity of one iteration is $\mathcal{O}\left(KHm + Rn + R^2H\right)$, which is linear in both the nodes and edges.

3.1 Iterative Approach

Unfortunately, finding optimal solution for our problem is **NP**-hard.

Proposition 1. *Problem 2 is* ***NP****-hard, even for* $H = K = 1$ *and* $R = 2$.

Consequently, we resort to a natural heuristic approach, where we optimize certain parameters while keeping the remaining parameters fixed.

We split the original problem into 3 subproblems as shown in Algorithm 1. First, we find good groups, then update Λ, and then optimize segmentation, followed by yet another update of Λ.

When initializing, we select groups $\mathcal{P}$ and parameters Λ randomly, then proceed to find optimal segmentation, followed by optimizing Λ.

Next we will explain each step in details.

3.2 Finding Groups

Our first step is to update groups $\mathcal{P}$ while maintaining the remaining parameters fixed. Unfortunately, finding the optimal solution for this problem is **NP**-hard.

Proposition 2. *Finding optimal partition* $\mathcal{P}$ *for fixed* Λ, $\mathcal{T}$ *and* g *is* ***NP****-hard, even for* $H = K = 1$ *and* $R = 2$.

Due to the previous proposition, we perform a simple greedy optimization where each node is individually reassigned to the optimal group while maintaining the remaining nodes fixed.

We should point out that there are more sophisticated approaches, for example based on SDP relaxations, see a survey by Abbe [1]. However, we resort to a simple greedy optimization due to its speed.

Algorithm 2: Algorithm FINDGROUPS($\mathcal{P}, \Lambda$) for finding groups for a fixed segmentation $\mathcal{T}$, g and parameters Λ

1 $p(v) \leftarrow$ group index of v;
2 $s(e) \leftarrow$ segment index of e;
3 $d[h] \leftarrow \sum_{g(k)=h} \Delta(T_k)$;
4 **foreach** $v \in V$ **do**
5 $\quad b \leftarrow p(v)$;
6 $\quad c[j, h] \leftarrow$ array c_{jh} as defined in Proposition 3;
7 $\quad$ **foreach** $a = 1, \ldots R$ **do**
8 $\quad\quad x[a] \leftarrow \sum_{h=1}^{H} \lambda_{bah} d[h] + \sum_{j=1}^{R} c[j, h] \log \lambda_{ajh} - |P_j| \lambda_{ajh} d[h]$;
9 $\quad p(v) \leftarrow \arg\max_a x[a]$ (update $\mathcal{P}$ also);
10 **return** $\mathcal{P}$;

A naive implementation of computing the log-likelihood gain for a single node may require $\Theta(m)$ steps, which would lead in $\Theta(nm)$ time as we need to test every node. Luckily, we can speed-up the computation using the following straightforward proposition.

Proposition 3. *Let $\mathcal{P}$ be the partition of nodes, Λ set of parameters, and $\mathcal{T}$ and g the segmentation and the level mapping. Let $\mathcal{S}_h = \{T_k \in \mathcal{T} \mid h = g(k)\}$ be the segments using the hth level.*

Let u be a node, and let P_b be the set such that $u \in P_b$. Select P_a, and let $\mathcal{P}'$ be the partition where u has been moved from P_b to P_a. Then

$$\ell(\mathcal{P}', \mathcal{T}, g, \Lambda) - \ell(\mathcal{P}, \mathcal{T}, g, \Lambda) = Z + \sum_{h=1}^{H} \lambda_{bah} t_h + \sum_{j=1}^{R} c_{jh} \log \lambda_{ajh} - |P_j| \lambda_{ajh} t_h,$$

where Z is a constant, not depending on a, $t_h = \Delta(\mathcal{S}_h)$ is the total duration of the segments using the hth level and $c_{jh} = c(u, P_j, \mathcal{S}_h)$, is the number of edges between u and P_j in the segments using the hth level.

The proposition leads to the pseudo-code given in Algorithm 2. The algorithm computes an array c and then uses Proposition 3 to compute the gain for each swap, and consequently to find the optimal gain.

Computing the array requires iterating over the adjacent edges, leading to $\mathcal{O}(|N(v)|)$ time, and computing the gains requires $\mathcal{O}(R^2 H)$ time. Consequently, the computational complexity for FINDGROUPS is $\mathcal{O}(m + R^2 Hn + K)$.

The running time can be further optimized by modifying Line 8. There are at most $2m$ non-zero $c[i, j]$ entries (across all $v \in V$), consequently we can speed up the computation of a second term by ignoring the zero entries in $c[i, j]$. In addition, for each a, the remaining terms

$$\sum_{h=1}^{H} \lambda_{bah} d[h] + \sum_{j=1}^{R} |P_j| \lambda_{ajh} d[h]$$

can be precomputed in $\mathcal{O}(RH)$ time and maintained in $\mathcal{O}(1)$ time. This leads to a running time of $\mathcal{O}\left(m + Rn + R^2H + K\right)$.

3.3 Updating Poisson Process Parameters

Our next step is to update Λ while maintaining the rest of the parameters fixed. This refers to UPDATELAMBDA in Algorithm 1. Fortunately, this step is straightforward as the optimal parameters are equal to

$$\lambda_{ijh} = \frac{c\left(P_i, P_j, \mathcal{S}_h\right)}{\left|P_i \times P_j\right| \Delta\left(\mathcal{S}_h\right)},$$

where $\mathcal{S}_h = \{T_k \in \mathcal{T} \mid h = g(k)\}$ are the segments using the hth level. Updating the parameters requires $\mathcal{O}\left(m + R^2H + K\right)$ time.

In practice, we would like to avoid having $\lambda = 0$ as this forbids any edges occurring in the segment, and we may get stuck in a local maximum. We approach this by shifting λ slightly by using

$$\lambda_{ijh} = \frac{c\left(P_i, P_j, \mathcal{S}_h\right) + \theta}{\left|P_i \times P_j\right| \Delta\left(\mathcal{S}_h\right) + \eta},$$

where θ and η are user parameters.

3.4 Finding Segmentation

Our final step is to update the segmentation $\mathcal{T}$ and the level mapping g, while keeping Λ and $\mathcal{P}$ fixed. Luckily, we can solve this subproblem in linear time.

Note that we need to keep Λ fixed, as otherwise the problem is **NP**-hard.

Proposition 4. *Finding optimal Λ, $\mathcal{T}$ and g for fixed $\mathcal{P}$ is* ***NP****-hard.*

On the other hand, if we fix Λ, then we can solve the optimization problem with a dynamic program. To be more specific, assume that the edges in E are ordered, and write $o[e, k]$ to be the log-likelihood of k-segmentation covering the edges prior and including e. Given two edges $s, e \in E$, let $y(s, e; h)$ be the log-likelihood of a segment $(t(s), t(e)]$ using the hth level of parameters, $\lambda_{\cdot\cdot h}$. If s occurs after e we set y to be $-\infty$. Then the identity

$$o[e, k] = \max_h \max_s y(s, e; h) + o[s, k-1]$$

leads to a dynamic program.

Using an off-the-shelf approach by Bellman [5] leads to a computational complexity of $\mathcal{O}\left(m^2KH\right)$, assuming that we can evaluate $y(s, e; h)$ in constant time.

However, we can speed-up the dynamic program by using the SMAWK algorithm [2]. Given a function $x(i, j)$, where $i, j = 1, \ldots, m$, SMAWK computes $z(j) = \arg\max_i x(i, j)$ in $\mathcal{O}(m)$ time, under two assumptions. The first assumption is that we can evaluate x in constant time. The second assumption is that x is *totally monotone.* We say that x is totally monotone, if $x(i_2, j_1) > x(i_1, j_1)$, then $x(i_2, j_2) \geq x(i_1, j_2)$ for any $i_1 < i_2$ and $j_1 < j_2$.

We have the immediate proposition.

Proposition 5. *Fix h. Then the function $x(s, e) = y(s, e; h) + o[s, k-1]$ is totally monotone.*

Our last step is to compute x in constant time. This can be done by first precomputing $f[e, h]$, the log-likelihood of a segment starting from the epoch and ending at $t(e)$ using the hth level. The log-likelihood of a segment is then $y(s, e; h) = f[e, h] - f[s, h]$, which we can compute in constant time.

Algorithm 3: Algorithm FindSegments($\mathcal{P}, \Lambda$) for finding optimal segmentation for fixed groups $\mathcal{P}$ and parameters Λ

1 $t_{min} \leftarrow \min \{t \mid (u, v, t) \in E\}$;
2 $f[e, h] \leftarrow$ log-likelihood of a segment $[t_{min}, t(e)]$ using parameters $\lambda_{\cdot\cdot h}$;
3 **foreach** $e \in E$ **do** $o[e, 1] \leftarrow \max_h f[e, h]$
4 **foreach** $k = 2, \ldots, K$ **do**
5 $\quad x(s, e; h) \leftarrow o[s, k-1] + f[e, h] - f[s, h]$;
6 $\quad$ **foreach** $h = 1, \ldots, H$ **do**
7 $\quad\quad z[e, h] \leftarrow \arg\max_s x(s, e; h)$ for each $e \in E$ (use SMAWK);
8 $\quad o[e, k] \leftarrow \max_h x(z[e, h], e; h)$ for each $e \in E$;
9 $\quad r[e, k] \leftarrow \arg\max_h x(z[e, h], e; h)$;
10 $\quad q[e, k] \leftarrow z[e, r[e, k]]$;
11 Use r and q to recover the optimal segmentation $(T_1, \ldots, T_K)$ and the level mapping g ;
12 **return** $(T_1, \ldots, T_K)$, g;

The pseudo-code for finding the segmentation is given in Algorithm 3. A more detailed version of the pseudo-code is given in Appendix. Here, we first precompute $f[e, h]$. We then solve segmentation with a dynamic program by maintaining 3 arrays: $o[e, k]$ is the log-likelihood of k-segmentation covering the edges up to e, $q[e, k]$ is the starting point of the last segment responsible for $o[e, k]$, and $r[e, k]$ is the level of the last segment responsible for $o[e, k]$.

In the inner loop we use SMAWK to find optimal starting points. Note that we have to do this for each h, and only then select the optimal h for each segment. Note that we do define x on Line 5 but we do not compute its values. Instead this function is given to SMAWK and is evaluated in a lazy fashion.

Once we have constructed the arrays, we can recursively recover the optimal segmentation and the level mapping from q and r, respectively.

FindSegments runs in $\mathcal{O}\left(mKH + HR^2\right)$ time since we need to call SMAWK $\mathcal{O}(HK)$ times.

We were able to use SMAWK because the optimization criterion turned out to be totally monotone. This was possibly only because we fixed Λ. The notion of using SMAWK to speed up a dynamic program with totally monotone scores was proposed by Galil and Park [9]. Fleischer et al. [7], Hassin and Tamir [14] used this approach to solve dynamic program segmenting monotonic one-dimensional sequences with L_1 cost.

We fixed Λ because Proposition 4 states that the optimization problem for $H < K$ cannot be solved in polynomial time if we optimize $\mathcal{T}$, g, and Λ at the same time. Proposition 4 is the main reason why we cannot use directly the ideas proposed by Corneli et al. [6] as the authors use the dynamic program to find $\mathcal{T}$ and Λ at the same time.

However, if $K = H$, then the problem is solvable with a dynamic program but requires $\mathcal{O}\left(Km^2R^2\right)$ time. However, if we consider the optimization problem as a minimization problem and shift the cost with a constant so that it is always positive, then using algorithms by Guha et al. [26], Tatti [11] we can obtain $(1+\epsilon)$-approximation with $\mathcal{O}\left(K^3 \log K \log m + K^3\epsilon^{-2} \log m\right)$ number of cost evaluations. Finding the optimal parameters and computing the cost of a single segment can be done in $\mathcal{O}\left(R^2\right)$ time with $\mathcal{O}\left(R^2 + m\right)$ time for precomputing. This leads to a total time of $\mathcal{O}\left(R^2(K^3 \log K \log m + K^3\epsilon^{-2} \log m) + m\right)$ for the special case of $K = H$.

4 Related Work

The closest related work is the paper by Corneli et al. [6] which can be viewed as a special case of our approach by requiring $K = H$, in other words, while the Poisson process may depend on time they do not take into account any recurrent behaviour. Having $K = H$ simplifies the optimization problem somewhat. While the general problem still remains difficult, we can now solve the segmentation $\mathcal{T}$ *and* the parameters Λ simultaneously using a dynamic program as was done by Corneli et al. [6]. In our problem we are forced to fix Λ while solving the segmentation problem. Interestingly enough, this gives us an advantage in computational time: we only need $\mathcal{O}\left(KHm + HR^2\right)$ time to find the optimal segmentation while the optimizing $\mathcal{T}$ and Λ simultaneously requires $\mathcal{O}\left(R^2Km^2\right)$ time. On the other hand, by fixing Λ we may have a higher chance of getting stuck in a local maximum.

The other closely related work is by Gionis and Mannila [10], where the authors propose a segmentation with shared centroids. Here, the input is a sequence of real valued vectors and the segmentation cost is either L_2 or L_1 distance. Note that there is no notion of groups $\mathcal{P}$, the authors are only interested in finding a segmentation with recurrent sources. The authors propose several approximation algorithms as well as an iterative method. The approximation algorithms rely specifically on the underlying cost, in this case L_1 or L_2 distance, and cannot be used in our case. Interestingly enough, the proposed iterative method did not use SMAWK optimization, so it is possible to use the optimization described in Sect. 3 to speed up the iterative method proposed by Gionis and Mannila [10].

In this paper, we used stochastic block model (see [3,16], for example) as a starting point and extend it to temporal networks with recurrent sources. Several past works have extended stochastic block models to temporal networks: Matias and Miele [29], Yang et al. [21] proposed an approach where the nodes can change block memberships over time. In a similar fashion, Xu and Hero [27] proposed a model where the adjacency matrix snapshots are generated with a logistic function whose latent parameters evolve over time. The main difference with our approach is that in these models the group memberships of nodes are changing while in our case we keep the memberships constant and update the probabilities of the nodes. Moreover, these methods are based on graph snapshots while we work with temporal edges. In another related work, Matias et al. [22] modelled interactions using Poisson processes conditioned by stochastic block model. Their approach was to estimate the intensities non-parametrically through histograms or kernels while we model intensities with recurring segments. For a survey on stochastic block models, including extensions to temporal settings, we refer the reader to a survey by Lee and Wilkinson [19].

Stochastic block models group similar nodes together; here similarity means that nodes in the same group have the similar probabilities connecting to nodes from other group. A similar notion but a different optimization criterion was proposed by Arockiasamy et al. [4]. Moreover, Henderson et al. [15] proposed a method where nodes with similar neighborhoods are discovered.

In this paper we modelled the recurrency by forcing the segments to share their parameters. An alternative approach to discover recurrency is to look explicitly for recurrent patterns [8,12,13,20,23,28]. We should point out that these works are not design to work with graphs; instead they work with event sequences. We leave adapting this methodology for temporal networks as an interesting future line of work.

Using segmentation to find evolving structures in networks have been proposed in the past: Kostakis et al. [18] introduced a method where a temporal network is segmented into k segments with $h < k$ summaries. A summary is a graph, and the cost of an individual segment is the difference between the summary and the snapshots in the segment. Moreover, Rozenshtein et al. [25] proposed discovering dense subgraphs in individual segments.

5 Experimental Evaluation

The goal in this section is to experimentally evaluate our algorithm. Towards that end, we first test how well the algorithm discovers the ground truth using synthetic datasets. Next we study the performance of the algorithm on real-world temporal datasets in terms of running time and likelihood. We compare our results to the following baselines: the running times are compared to a naive implementation where we do not utilize SMAWK algorithm, and the likelihoods are compared to the likelihoods of the (R, K) model.

We implemented the algorithm in Python[3] and performed the experiments using a 2.4 GHz Intel Core i5 processor and 16 GB RAM.

Synthetic Datasets: To test our algorithm, we generated 5 temporal networks with known groups and known parameters Λ which we use as a ground truth. To generate data, we first chose a set of nodes V, number of groups R, number

[3] The source code is available at https://version.helsinki.fi/dacs/.

Table 1. Dataset characteristics and results from the experiments. Here, n is the number of nodes, m is the number of edges, R is the number of groups, K is the number of segments, H is the number of levels, LL_1 is the normalized log-likelihood for the ground truth, G is the Rand index, LL_2 is the discovered normalized log-likelihood, I is the number of iterations, and CT is the computational time in seconds.

Dataset	n	m	L	K	H	LL_1	R	LL_2	I	CT
Synthetic-1	50	76 332	2	2	2	0.95	1	0.94	2	2.81 s
Synthetic-2	30	95 889	3	3	3	0.95	1	0.94	3	5.36 s
Synthetic-3	20	65 056	3	3	3	0.97	1	0.97	3	3.91 s
Synthetic-4	60	537 501	3	4	3	0.94	1	0.93	3	23.13 s
Synthetic-5	10	33 475	2	10	5	0.91	1	0.91	4	10.27 s
Email-Eu-1	309	61 046	3	10	7			0.89	12	188 s
Email-Eu-2	162	46 772	4	8	7			0.87	9	177 s
MathOverflow	21 688	107 581	2	3	2			0.91	20	263 s
CollegeMsg	1 899	59 835	3	8	5			0.87	19	662 s
MOOC	7 047	411 749	2	3	2			0.81	6	208 s
Bitcoin	3 783	24 186	3	10	10			0.91	7	115 s
Santander	735	33 116	3	7	5			0.94	20	60 s

of segments K, and number of levels H. Next we assumed that each node has an equal probability of being chosen for any group. Based on this assumption, the group memberships were selected at random.

We then randomly generated Λ from a uniform distribution. More specifically, we generated H distinct values for each pair of groups and map them to each segment. Note that, we need to ensure that each distinct level is assigned to at least one segment. To guarantee this, we first deterministically assigned the set of H levels to first H segments and the remaining $(K - H)$ segments are mapped by randomly selecting $(K - H)$ elements from H level set.

Given the group memberships and their related Λ, we then generated a sequence of timestamps with a Poisson process for each pair of nodes. The sizes of all synthetic datasets are given in Table 1.

Real-World Datasets: We used 7 publicly available temporal datasets. *Email-Eu-1* and *Email-Eu-2* are collaboration networks between researchers in a European research institution.[4] *Math Overflow* contains user interactions in Math Overflow web site while answering to the questions.[4] *CollegeMsg* is an online message network at the University of California, Irvine.[4] *MOOC* contains actions by users of a popular MOOC platform.[4] *Bitcoin* contains member rating interactions in a bitcoin trading platform.[4] *Santander* contains station-to-station links

[4] http://snap.stanford.edu.

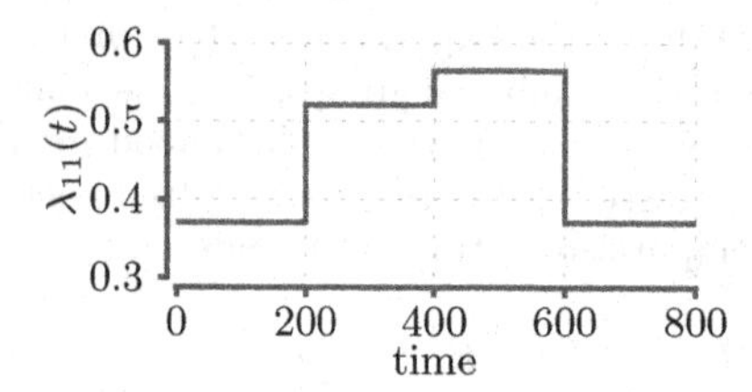

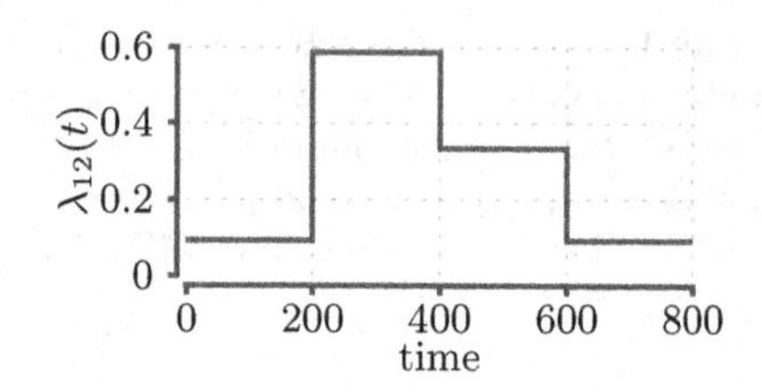

Fig. 1. Discovered parameters $\lambda_{11}(t)$, $\lambda_{12}(t)$ for the *Synthetic-4* dataset. Parameter $\lambda_{12}(t)$ implies the Poisson process parameter between group 1 and group 2 as a function of time.

that occurred on Sep 9, 2015 from the Santander bikes hires in London.[5] The sizes of these networks are given in Table 1.

Results for Synthetic Datasets: To evaluate the accuracy of our algorithm, we compare the set of discovered groups with the ground truth groups. Here, our algorithm found the ground truth: in Table 1 we can see that Rand index Rand [24] (column G) is equal to 1,

Next we compare the log-likelihood values from true models against the log-likelihoods of discovered models. To evaluate the log-likelihoods, we normalize the log-likelihood, that is we computed $\ell(\mathcal{P}, \mathcal{T}, g, \Lambda)/\ell(\mathcal{P}', \mathcal{T}', g', \Lambda')$, where $\mathcal{P}', \mathcal{T}', g', \Lambda'$ is a model with a single group and a single segment. Since all our log-likelihood values were negative, the *normalized log-likelihood* values were between 0 and 1, and *smaller* values are better.

As demonstrated in column LL_1 and column LL_2 of Table 1, we obtained similar normalized log-likelihood values when compared to the normalized log-likelihood of the ground truth. The obtained normalized log-likelihood values were all slightly better than the log-likelihoods of the generated models, that is, our solution is as good as the ground truth.

An example of the discovered parameters, λ_{11} and λ_{12}, for *Synthetic-4* dataset are shown in Fig. 1. The discovered parameters matched closely to the generated parameters with the biggest absolute difference being 0.002 for *Synthetic-4*. The figures for other values and other synthetic datasets are similar.

Computational Time: Next we consider the computational time of our algorithm. We varied the parameters R, K, and H for each dataset. The model parameters and computational times are given in Table 1. From the last column CT, we see that the running times are reasonable despite using inefficient Python libraries: for example we were able to compute the model for *MOOC* dataset, with over 400 000 edges, under four minutes. This implies that the algorithm scales well for large networks. This is further supported by a low number of iterations, column I in Table 1.

Next we study the computational time as a function of m, number of edges.

[5] https://cycling.data.tfl.gov.uk.

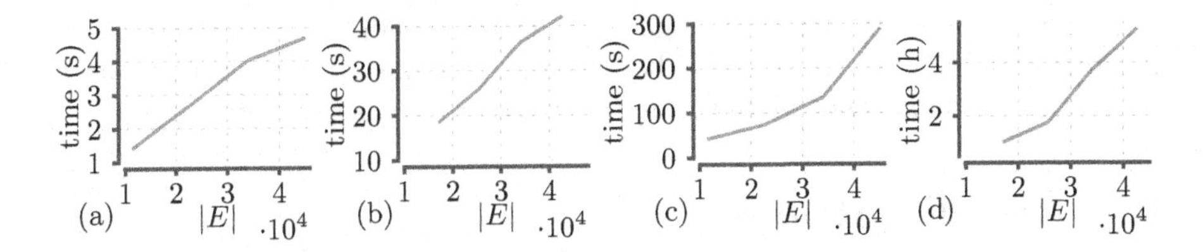

Fig. 2. Computational time as a function of number of temporal edges ($|E|$) for *Synthetic-large* (a, c) and *Santander-large* (b, d). This experiment was done with $R = 3$, $K = 5$, and $H = 3$ using SMAWK algorithm (a–b) and naive dynamic programming (c–d). The times are in seconds in (a–c) and in hours in (d).

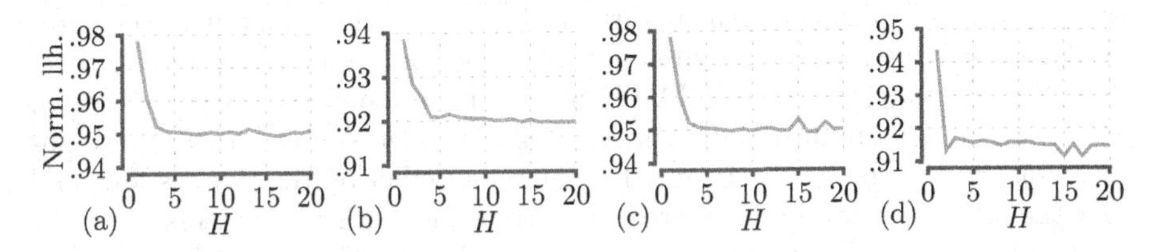

Fig. 3. Normalized log-likelihood as a function of number of levels (H) for the *Santander* dataset (a), *bitcoin* dataset (b), *Synthetic-5* dataset (c), and *Email-Eu-1* dataset (d). This experiment is done for $R = 2$, $K = 20$, and $H = 1, \ldots, 20$.

We first prepared 4 datasets with different number of edges from a real-world dataset; *Santander-large*. To vary the number of edges, we uniformly sampled edges without replacement. We sampled like a .4, .6, .8, and 1 fraction of edges.

Next we created 4 different *Synthetic-large* dataset with 30 nodes, 3 segments with unique λ values but with different number of edges. To do that, we gradually increase the number of Poisson samples we generated for each segment.

From the results in Fig. 2 we see that generally computational time increases as $|E|$ increases. For instance, a set of 17 072 edges accounts for 18.46s whereas a set of 34 143 edges accounts for 36.36s w.r.t *Santander-large*. Thus a linear trend w.r.t $|E|$ is evident via this experiment.

To emphasize the importance of SMAWK, we replaced it with a stock solver of the dynamic program, and repeat the experiment. We observe in Fig. 2 that computational time has increased drastically when stock dynamic program algorithm is used. For example, a set of 34 143 edges required 3.7h for *Santander-large* dataset but only 36.36s when SMAWK is used.

Likelihood vs Number of Levels: Our next experiment is to study how normalized log-likelihood behaves upon the choices of H. We conducted this experiment for $K = 20$ and vary the number of levels (H) from $H = 1$ to $H = 20$. The results for the *Santander*, *Bitcoin*, *Synthetic-5*, and *Email-Eu-1* dataset are shown in Fig. 3. From the results we see that generally normalized log-likelihood decreases as H increases. That is due to the fact that higher the H levels, there exists a higher degree of freedom in terms of optimizing the likelihood. Note that if $H = K$, then our model corresponds to the model studied by Corneli et al. [6]. Interestingly enough, the log-likelihood values plateau for values of $H \ll K$ suggesting that existence of recurring segments in the displayed datasets.

6 Concluding Remarks

In this paper we introduced a problem of finding recurrent sources in temporal network: we introduced stochastic block model with recurrent segments.

We showed that finding optimal blocks and recurrent segmentation was an **NP**-hard problem. Therefore, to find good solutions we introduced an iterative algorithm by considering 3 subproblems, where we optimize blocks, model parameters, and segmentation in turn while keeping the remaining structures fixed. We demonstrate how each subproblem can be optimized in $\mathcal{O}(m)$ time. Here, the key step is to use SMAWK algorithm for solving the segmentation. This leads to a computational complexity of $\mathcal{O}\left(KHm + Rn + R^2H\right)$ for a single iteration. We show experimentally that the number of iterations is low, and that the algorithm can find the ground truth using synthetic datasets.

The paper introduces several interesting directions: Gionis and Mannila [10] considered several approximation algorithms but they cannot be applied directly for our problem because our optimization function is different. Adopting these algorithms in order to obtain an approximation guarantee is an interesting challenge. We used a simple heuristic to optimize the groups. We chose this approach due to its computational complexity. Experimenting with more sophisticated but slower methods for discovering block models, such as methods discussed in [1], provides a fruitful line of future work.

Acknowledgements. This research is supported by the Academy of Finland projects MALSOME (343045).

References

1. Abbe, E.: Community detection and stochastic block models: recent developments. JMLR **18**(1), 6446–6531 (2017)
2. Aggarwal, A., Klawe, M., Moran, S., Shor, P., Wilber, R.: Geometric applications of a matrix-searching algorithm. Algorithmica **2**(1–4), 195–208 (1987)
3. Anderson, C.J., Wasserman, S., Faust, K.: Building stochastic blockmodels. Soc. Netw. **14**(1), 137–161 (1992)
4. Arockiasamy, A., Gionis, A., Tatti, N.: A combinatorial approach to role discovery. In: ICDM, pp. 787–792 (2016)
5. Bellman, R.: On the approximation of curves by line segments using dynamic programming. Commun. ACM **4**(6), 284–284 (1961)
6. Corneli, M., Latouche, P., Rossi, F.: Multiple change points detection and clustering in dynamic networks. Stat. Comput. **28**(5), 989–1007 (2018)
7. Fleischer, R., Golin, M.J., Zhang, Y.: Online maintenance of k-medians and k-covers on a line. Algorithmica **45**(4), 549–567 (2006)

8. Galbrun, E., Cellier, P., Tatti, N., Termier, A., Crémilleux, B.: Mining periodic patterns with a MDL criterion. In: ECML PKDD, pp. 535–551 (2019)
9. Galil, Z., Park, K.: A linear-time algorithm for concave one-dimensional dynamic programming. IPL **33**(6), 309–311 (1990)
10. Gionis, A., Mannila, H.: Finding recurrent sources in sequences. In: RECOMB, pp. 123–130 (2003)
11. Guha, S., Koudas, N., Shim, K.: Approximation and streaming algorithms for histogram construction problems. TODS **31**(1), 396–438 (2006)
12. Han, J., Dong, G., Yin, Y.: Efficient mining of partial periodic patterns in time series database. In: ICDE, pp. 106–115 (1999)
13. Han, J., Gong, W., Yin, Y.: Mining segment-wise periodic patterns in time-related databases. In: KDD (1998)
14. Hassin, R., Tamir, A.: Improved complexity bounds for location problems on the real line. Oper. Res. Lett. **10**(7), 395–402 (1991)
15. Henderson, K., et al.: RolX: structural role extraction & mining in large graphs. In: KDD, pp. 1231–1239 (2012)
16. Holland, P.W., Laskey, K.B., Leinhardt, S.: Stochastic blockmodels: first steps. Soc. Netw. **5**(2), 109–137 (1983)
17. Holme, P., Saramäki, J.: Temporal networks. Phys. Rep. **519**(3), 97–125 (2012)
18. Kostakis, O., Tatti, N., Gionis, A.: Discovering recurring activity in temporal networks. DMKD **31**(6), 1840–1871 (2017)
19. Lee, C., Wilkinson, D.J.: A review of stochastic block models and extensions for graph clustering. Appl. Netw. Sci. **4**(122), 1–50 (2019)
20. Ma, S., Hellerstein, J.: Mining partially periodic event patterns with unknown periods. In: ICDE, pp. 205–214 (2001)
21. Matias, C., Miele, V.: Statistical clustering of temporal networks through a dynamic stochastic block model. J. Roy. Stat. Soc. Seri. B (Stat. Methodol.) **79**(4), 1119–1141 (2017)
22. Matias, C., Rebafka, T., Villers, F.: Estimation and clustering in a semiparametric poisson process stochastic block model for longitudinal networks. Biometrika **105**(3), 665–680 (2018)
23. Ozden, B., Ramaswamy, S., Silberschatz, A.: Cyclic association rules. In: ICDE, pp. 412–421 (1998)
24. Rand, W.M.: Objective criteria for the evaluation of clustering methods. J. Am. Stat. Assoc. **66**(336), 846–850 (1971)
25. Rozenshtein, P., Bonchi, F., Gionis, A., Sozio, M., Tatti, N.: Finding events in temporal networks: segmentation meets densest subgraph discovery. KAIS **62**(4), 1611–1639 (2020)
26. Tatti, N.: Strongly polynomial efficient approximation scheme for segmentation. Inf. Process. Lett. **142**, 1–8 (2019)
27. Xu, K.S., Hero, A.O.: Dynamic stochastic blockmodels for time-evolving social networks. JSTSP **8**(4), 552–562 (2014)
28. Yang, J., Wang, W., Yu, P.: Mining asynchronous periodic patterns in time series data. TKDE **15**(3), 613–628 (2003)
29. Yang, T., Chi, Y., Zhu, S., Gong, Y., Jin, R.: Detecting communities and their evolutions in dynamic social networks–a Bayesian approach. Mach. Learn. **82**, 157–189 (2011)

Community Detection in Edge-Labeled Graphs

Iiro Kumpulainen(✉) and Nikolaj Tatti

HIIT, University of Helsinki, Helsinki, Finland
{iiro.kumpulainen,nikolaj.tatti}@helsinki.fi

Abstract. Finding dense communities in networks is a widely-used tool for analysis in graph mining. A popular choice for finding such communities is to find subgraphs with a high average degree. While useful, interpreting such subgraphs may be difficult. On the other hand, many real-world networks have additional information, and we are specifically interested in networks that have labels on edges. In this paper, we study finding dense subgraphs that can be explained with the labels on edges. More specifically, we are looking for a set of labels so that the induced subgraph has a high average degree. There are many ways to induce a subgraph from a set of labels, and we study two cases: First, we study conjunctive-induced dense subgraphs, where the subgraph edges need to have all labels. Secondly, we study disjunctive-induced dense subgraphs, where the subgraph edges need to have at least one label. We show that both problems are **NP**-hard. Because of the hardness, we resort to greedy heuristics. We show that we can implement the greedy search efficiently: the respective running times for finding conjunctive-induced and disjunctive-induced dense subgraphs are in $\mathcal{O}(p \log k)$ and $\mathcal{O}(p \log^2 k)$, where p is the number of edge-label pairs and k is the number of labels. Our experimental evaluation demonstrates that we can find the ground truth in synthetic graphs and that we can find interpretable subgraphs from real-world networks.

1 Introduction

Finding dense communities in networks is a common tool for analyzing networks with potential applications in diverse domains, such as bioinformatics [9,13], finance [8], social media [2], or web graph analysis [9].

While useful on their own, analyzing dense structures without any additional explanation may be difficult and may limit its impact. Consequently, several works have been proposed for searching dense subgraphs that also can be explained using available additional information [10,17]. Here the authors were looking for communities that are both dense and that can be explained using labels on nodes, see Sect. 5 for a more detailed discussion.

In this paper, we consider finding dense subgraphs in networks with labeled edges. More formally, we are looking for a label set that induces a dense subgraph. As a measure of density, a subgraph (W, F) will use $|F|/|W|$, the ratio of

P. Pascal and D. Ienco (Eds.): DS 2022, LNAI 13601, pp. 460–475, 2022.
https://doi.org/10.1007/978-3-031-18840-4_33

edges over the nodes, a popular choice for measuring the density of a subgraph. Note that optimizing this measure is equivalent to optimizing the average degree of nodes in W. Finding the densest subgraph—with no label constraints—can be done in polynomial time [11] and can be 2-approximated in linear time [7]. Unfortunately, additional requirements on the labels will make solving the optimization problem exactly computationally intractable.

We consider two cases: conjunctive-induced and disjunctive-induced dense subgraphs. In the former, the induced subgraph consists of all the edges that have the given label set. In the latter, the induced subgraph consists of all the edges that have at least one label common with the label set.

We show that both problems are **NP**-hard, which forces us to resort to heuristics. We propose a greedy algorithm for both problems: we start with an empty label and keep adding the best possible label until no additions are possible. We then return the best observed induced subgraph.

The computational bottleneck of the greedy method is selecting a new label. If done naively, evaluating a single label candidate requires enumerating over all the edges. Since this needs to be done for every candidate during every addition, the running time is $\mathcal{O}(p|L|)$, where $|L|$ is the number of labels and p is the number of edge-label pairs. By keeping certain counters we can speed up the running time. We show that conjunctive-induced graphs can be discovered in $\mathcal{O}(p \log |L|)$ time using a balanced search tree, and that disjunctive-induced graphs can be discovered in $\mathcal{O}(p \log^2 |L|)$ time with the aid of an algorithm originally used to maintain convex hulls.

The remainder of the paper is organized as follows. In Sect. 2 we introduce the notation and formalize the optimization problem. In Sects. 3–4 we present our algorithms. Section 5 is devoted to the related work. Finally, we present the experimental evaluation in Sect. 6 and conclude with a discussion in Sect. 7.

2 Preliminary Notation and Problem Definition

In this section, we first describe the common notation and then introduce the formal definition of our problem.

Assume that we are given an *edge-labeled graph*, that is, a tuple $G = (V, E, lab)$, where V is the set of vertices, $E \subseteq \{(x, y) \mid (x, y) \in V^2, x \neq y\}$ is the set of undirected edges, and $lab : E \rightarrow 2^L$ is a function that maps each edge $e \in E$ to the set of labels $lab(e)$. Here L is the set of all possible labels.

Given a label $\ell \in L$, let us write $E(\ell)$ to be the edges having the label ℓ. In addition, let us write $V(\ell)$ to be the nodes adjacent to $E(\ell)$.

Our goal is to search for dense regions of graphs that can be explained using the labels. In other words, we are looking for a set of labels that induce a dense graph. More formally, we define an *inducing* function to be a function f that maps two sets of labels to a binary number. An example of such a function could be $f(A; B) = [B \subseteq A]$ which returns 1 if and only if B is a subset of A.

Given a set of labels $B \subseteq L$, an inducing function f, and a graph G, we define the *label-induced subgraph* $H = G(f, B)$ as $(V(B), E(B), lab)$, where

$$E(B) = \{e \in E \mid f(lab(e); B) = 1\}$$

is the subset of edges that satisfy f, and $V(B)$ is the set of vertices that are adjacent to $E(B)$.

Given a graph G with vertices V and edges E, we measure the *density* of the graph $d(G)$ as the number of edges divided by the number of vertices: $d(G) = \frac{|E|}{|V|}$.

We are now ready to state our generic problem.

Problem 1. (LD). Let $G = (V, E, lab)$ be an edge-labeled graph over a set of labels L with multiple labels being possible for each edge. Assume an inducing function f. Find a set of labels L^* such that the density $d(H)$ of the label-induced subgraph $H = G(f, L^*)$ is maximized.

We consider two special cases of LD. Firstly, let us define $f_{AND}(A; B) = [B \subseteq A]$, that is, the induced edges need to contain every label in B. We will denote the problem LD paired with $f_{AND}()$ as LDAND. Secondly, we define $f_{OR}(A; B) = [B \cap A \neq \emptyset]$, that is, the induced edges need to have one common label with B. Then, we denote the corresponding problem as LDOR.

3 Finding Dense Conjunctive-Induced Graphs

In this section, we focus on LDAND, that is, finding conjunctive-induced graphs that are dense. We will first prove that LDAND is **NP**-hard.

Theorem 1. LDAND *is* ***NP****-hard.*

Proof. We will prove the claim by reducing 3ExactCover to the densest subgraph problem. In 3ExactCover we are given a set X and a family $\mathcal{C}$ of subsets of size 3 over X and asked if there is a disjoint subset of $\mathcal{C}$ whose union is X.

Assume that we are given a set X and a family $\mathcal{C} = \{C_1, \ldots, C_N\}$ of N subsets. We set labels to be $L = \{1, \ldots, N\}$. The vertices V contain N vertices $y_1, \ldots, y_N$, and an additional vertex z. We connect each y_i to z, labeled with $L \setminus \{i\}$. For each overlapping C_i and C_j, we introduce $4N$ additional vertices and $2N$ edges, each edge connecting two *unique nodes*, and labeled as $L \setminus \{i, j\}$.

We claim that for $|X| \geq 5$, 3ExactCover has a solution if and only if there is an induced graph H with $d(H) \geq |X|/(|X| + 3)$.

Assume that we are given a set of labels $A \subset L$. Let $B = L \setminus A$. Let k be the number of set pairs in B that are overlapping, that is,

$$k = |\{\{i, j\} \mid i, j \in B, C_i \cap C_j \neq \emptyset\}|.$$

Then the density of the corresponding graph $H = G(f_{AND}(), A)$ is equal to

$$d(H) = \frac{|B| + 2Nk}{|B| + 1 + 4Nk} \quad .$$

Assume that $k > 0$. Since $|B| \leq N$, we can bound the density with

$$d(H) = \frac{|B| + 2Nk}{|B| + 1 + 4Nk} \leq \frac{N + 2Nk}{N + 1 + 4Nk} < \frac{N + 2Nk}{N + 4Nk} \leq \frac{3}{5} \quad .$$

Assume that $k = 0$. Then the density is equal to $|B|/(|B| + 1)$. Let $\mathcal{U} = \{C_i \mid i \in B\}$. Since $\mathcal{U}$ is disjoint, $3|B| \leq |X|$ and the equality holds if and only if $\mathcal{U}$ covers X.

Assume that there is a subgraph $H = G(f_{AND}(), A)$ with $d(H) \geq |X|/(|X|+3)$. Since we assume that $|X| \geq 5$, we have $d(H) \geq 5/8 > 3/5$, and the preceding discussion shows that the sets corresponding to A form an exact cover of X.

On the other hand, if there is an exact cover in $\mathcal{C}$, then $d(G(f_{AND}(), A)) = |X|/(|X| + 3)$, where A is the set of labels corresponding to the cover. This shows that maximizing the density of the label-induced subgraph is an **NP**-hard problem. □

The **NP**-hardness forces us to resort to heuristics. Here, we use the algorithm for 2-approximating dense subgraphs [7] as a starting point. The algorithm iteratively removes a node with the smallest degree, and returns the best solution among the observed subgraphs. We propose a similar greedy algorithm, where we greedily add the best possible label, and repeat until the induced subgraph is empty. We then select the best observed labels as the output.

To avoid enumerating over the edges every time we look for a new label, we maintain several counters. Let A be the current set of labels. For each label, we maintain the number of nodes n_k and edges m_k of the candidate graph, that is, $n_k = |V(A \cup \{k\})|$ and $m_k = |E(A \cup \{k\})|$. We store the densities m_k/n_k in a balanced search tree (for example, a red-black tree), which allows us to obtain the largest element quickly. Once we update set A, we also update the counters and update the search tree. Maintaining the node counts n_k requires us to maintain the counters $r_{v,k}$, number of edges labeled as k adjacent to v: once the counter reduces to 0, we reduce n_k by 1. The pseudo-code of the algorithm is given in Algorithm 1.

We conclude with an analysis of the computational complexity of GreedyAnd.

Theorem 2. GreedyAnd *runs in* $\mathcal{O}(p \log |L| + |V| + |E|)$ *time, where* p *is the number of edge-label pairs* $p = |\{(e, k) \mid e \in E, k \in lab(e)\}|$.

Proof. Initializing counters in GreedyAnd can be done in $\mathcal{O}(|V| + |E| + |L|)$ time while initializing the tree can be done in $\mathcal{O}(|L| \log |L|)$ time.

Let us consider the inner for-loop. Since an edge is deleted once it is processed, the inner for-loop is executed at most p times during the search. Since this is the only way the counters get updated, the tree T is updated p times, each update requiring $\mathcal{O}(\log |L|)$ time.

The outer loop is executed at most $|L|$ times. During each round, selecting and removing the label requires $\mathcal{O}(\log |L|)$ time.

Algorithm 1: GREEDYAND, greedy search for the conjunctive-induced dense subgraphs

```
1  n_ℓ ← |V(ℓ)|, for each label ℓ ∈ L;
2  m_ℓ ← |E(ℓ)|, for each label ℓ ∈ L;
3  r_{v,ℓ} ← |{e ∈ E(ℓ) | e is adjacent to v}|, for each vertex v and label ℓ;
4  T ← labels sorted by the density values m_k/n_k (e.g., in a red-black tree);
5  A_0 ← ∅ and i ← 0;
6  while there are labels do
7      pick and remove label k that has the maximum density in T;
8      A_{i+1} ← A_i ∪ {k};
9      for each edge e without label k do
10         for each label ℓ of edge e = (u, v) do
11             m_ℓ ← m_ℓ − 1;
12             r_{v,ℓ} ← r_{v,ℓ} − 1; r_{u,ℓ} ← r_{u,ℓ} − 1;
13             if r_{v,ℓ} = 0 then n_ℓ ← n_ℓ − 1 if r_{u,ℓ} = 0 then n_ℓ ← n_ℓ − 1
14         remove edge e;
15     update T for all labels ℓ with changed values of m_ℓ or n_ℓ;
16     i ← i + 1;
17 return the set of labels A_i that yields the highest density;
```

In summary, the algorithm requires

$$\mathcal{O}\left(|V| + |E| + |L| + |L| \log |L| + p \log |L|\right) \subseteq \mathcal{O}\left(|V| + |E| + p \log |L|\right)$$

time, completing the proof. □

4 Finding Dense Disjunctive-Induced Graphs

In this section, we focus on LDOR, that is, finding disjunctive-induced graphs that are dense. We will first prove that LDOR is **NP**-hard.

Theorem 3. LDOR *is* ***NP****-hard.*

Proof. We will prove the claim by reducing 3EXACTCOVER to the densest subgraph problem. In 3EXACTCOVER we are given a set X and a family $\mathcal{C}$ of subsets of size 3 over X and asked if there is a disjoint subset of $\mathcal{C}$ whose union is X.

Assume that we are given a set X and a family $\mathcal{C} = \{C_1, \ldots, C_N\}$ of N subsets. The vertices V consists of the set X, N additional vertices $y_1, \ldots, y_N$, and 2 more vertices $Z = z_1, z_2$. We have N labels, $L = \{1, \ldots, N\}$.

Next, we define the edges E. Connect each $x \in X$ to Z, and label the edges with labels $\{i \mid x \in C_i\}$. Then for each C_i, we connect z_1 to y_i, labeled with i.

We claim that 3EXACTCOVER has a solution if and only if the optimal label-induced graph has the density of $7|X|/(6 + 4|X|)$.

Given a non-empty set of labels $A \subseteq L$, the density of the corresponding graph H is equal to $g(k, |A|)$, where $g(s,t) = \frac{2s+t}{2+s+t}$, and k is the size of the union of sets in $\mathcal{C}$ corresponding to A.

Note that since $k \geq 3$, we have $2k > 2 + k$. Thus, $\partial \log g / \partial t = 1/(2k + t) - 1/(2 + k + t) < 0$, and consequently $g(k,t) > g(k,t')$ when $t < t'$.

Since each set in $\mathcal{C}$ is of size 3, we have $|A| \geq k/3$. Thus,

$$g(k, |A|) \leq g(k, k/3) = \frac{7k}{6 + 4k} \leq \frac{7|X|}{6 + 4|X|},$$

where the equalities hold if and only if $k = |X|$ and $3|A| = k$, that is, A corresponds to an exact cover of X. □

Similar to LDand, we resort to a greedy search to find good subgraphs: We start with an empty label set, and iteratively add the best possible label. Once done, we return the best observed label set.

However, we maintain a different set of counters as compared to GreedyAnd. The reason for having different counters is to avoid a significantly higher number of updates: the inner loop would need to go over the edge-label pairs that are *not* present in the graph. More formally, we maintain values n and m representing the number of nodes and edges in the subgraph induced by the current set of labels, say A. We also maintain n_k and m_k, the number of *additional* nodes and edges if k is added to A. At each iteration, we select the label optimizing $\frac{m+m_k}{n+n_k}$. We will discuss the selection process later. Once the label is selected, we update the counters m_k and n_k. To maintain n_k properly, we keep track of what nodes are already in $V(A)$, using an indicator r_v with $r_v = 1$ if $v \in V(A)$. The pseudo-code for the algorithm is given in Algorithm 2.

During each iteration, we need to select the label maximizing $\frac{m+m_k}{n+n_k}$. We cannot use priority queues any longer since n and m change every iteration. However, we can speed up the selection using a convex hull, a classic concept from computational geometry, see for example, [14]. First, let us formally define a lower-right convex hull.

Definition 1. Given a set of points $X = \{(x_i, y_i)\}$ in a plane, we define a *lower-right convex hull* $H = \mathit{hull}(H)$ to be a subset of X such that $q = (x_q, y_q) \in X$ is *not* in X if and only if there is a point $r = (x_r, y_r) \in H$ such that $x_q \leq x_r$ and $y_q \geq y_r$, or if there are two points $p, r \in H$ such that q is above or at the segment joining q and r.

If we were to plot X on a plane, then $\mathit{hull}(X)$ is the lower-right portion of the complete convex hull, that is, a set of points in X that form a convex polygon containing X. For notational simplicity, we will refer to $\mathit{hull}(X)$ as the convex hull. Note that if we order the points in $\mathit{hull}(X)$ by their x-coordinates, then the y-coordinates and the slopes of the intermediate segments are also increasing.

We will first argue that we only need to search the convex hull when looking for the optimal label.

Algorithm 2: GreedyOr, greedy search for the disjunctive-induced dense subgraphs

1 $n \leftarrow 0$; $m \leftarrow 0$;
2 $n_\ell \leftarrow |V(\ell)|$, for each label $\ell \in L$;
3 $m_\ell \leftarrow |E(\ell)|$, for each label $\ell \in L$;
4 $S_v \leftarrow \{\ell \in L \mid$ there is an edge with label ℓ adjacent to $v\}$;
5 $r_v \leftarrow 0$, for each vertex v;
6 $A_0 \leftarrow \emptyset$ and $i \leftarrow 0$;
7 **while** *there are labels* **do**
8 pick and remove label k that yields the maximum density $\frac{m+m_k}{n+n_k}$;
9 $A_{i+1} \leftarrow A_i \cup \{k\}$;
10 **for** *each edge* $e = (u, v)$ *with label* k **do**
11 **for** *each label* ℓ *of edge* $e = (u, v)$ **do** $m_\ell \leftarrow m_\ell - 1$ $m \leftarrow m + 1$;
12 **if** $r_v = 0$ **then**
13 **for** *each label* ℓ *in* S_v **do** $n_\ell \leftarrow n_\ell - 1$ $n \leftarrow n + 1$;
14 **if** $r_u = 0$ **then**
15 **for** *each label* ℓ *in* S_u **do** $n_\ell \leftarrow n_\ell - 1$ $n \leftarrow n + 1$;
16 $r_v \leftarrow 1$; $r_u \leftarrow 1$;
17 remove edge e;
18 $i \leftarrow i + 1$;
19 **return** the set of labels A_i that yields the highest density;

Theorem 4. *Let X be a set of positive points (m_i, n_i), and let $H = hull(X)$ be the convex hull. Select $m, n \geq 0$. Then* $\max_{p \in X} \frac{m+m_i}{n+n_i} = \max_{p \in H} \frac{m+m_i}{n+n_i}$.

Proof. Let $k = (m_k, n_k)$ be the optimal point in X. Assume that $k \notin H$. Assume that there is a point $q = (m_q, n_q)$ in H such that $m_q \geq m_k$ and $n_q \leq n_k$. Then $\frac{m+m_k}{n+n_k} \leq \frac{m+m_q}{n+n_q}$, so the point q is also optimal.

Assume there is no such point q. Then, the x-coordinate of point k falls between two consecutive points p and q in H, that is, $m_p < m_k < m_q$. Then k must be above the segment between p and q as otherwise, k would also be a part H. Therefore, the slope for the segment between p and k must be greater than the slope of the segment between p and q, and the slope for the segment between k and q must be smaller,

$$\frac{n_q - n_k}{m_q - m_k} \leq \frac{n_q - n_p}{m_q - m_p} \leq \frac{n_k - n_p}{m_k - m_p}. \tag{1}$$

Furthermore, since $k \notin H$, we must have $n_k > n_p$. By assumption, we also have $n_k < n_q$. In summary, we have $n_p < n_k < n_q$ and $m_p < m_k < m_q$, which means that the slopes in Eq. 1 are all positive. By taking the reciprocals this then gives,

$$\frac{m_q - m_k}{n_q - n_k} \geq \frac{m_q - m_p}{n_q - n_p} \geq \frac{m_k - m_p}{n_k - n_p}. \tag{2}$$

Denote then the objective value at point k by $c = \frac{m+m_k}{n+n_k}$. Let $x_1 = c(n + n_p) - m$. Then, the optimality of k implies $\frac{m+x_1}{n+n_p} = c \geq \frac{m+m_p}{n+n_p}$, which means $x_1 \geq m_p$. The definition of c leads to $m = c(n + n_k) - m_k$, which in turns leads to $x_1 = c(n_p - n_k) + m_k$. Solving for c we get $c = \frac{m_k - x_1}{n_k - n_p}$. Substituting $x_1 \geq m_p$ yields $c \leq \frac{m_k - m_p}{n_k - n_p}$, using Eq. 2 then yields $c \leq \frac{m_q - m_k}{n_q - n_k}$.

Next, let $x_2 = c(n_q - n_k) + m_k$ which means that $c = \frac{x_2 - m_k}{n_q - n_k}$. Now since $c \leq \frac{m_q - m_k}{n_q - n_k}$ we must have $x_2 \leq m_q$. Since $m_k = c(n + n_k) - m$, we also have $x_2 = c(n_q + n) - m$, yielding $c = \frac{m+x_2}{n+n_q} \leq \frac{m+m_q}{n+n_q}$, thus q is also optimal. □

Theorem 4 states that we need to only consider the convex hull H of the set $\{(m_i, n_i)\}$ when searching for the optimal new label. Note that H does not depend on n or m. Moreover, we can use the algorithm by Overmars and Van Leeuwen [16] to maintain H as n_k and m_k are updated in $\mathcal{O}\left(\log^2 |L|\right)$ time per update. We will see that the number of needed updates is bounded by the number of edge-label pairs.

However, the convex hull can be as large as the original set, so our goal is to avoid enumerating over the whole set. To this end, we design a binary search strategy over the hull. We will first introduce two quantities used in our search.

Definition 2. Given two points $p, q \in hull\,(X)$, we define the inverse slope as $s(p,q) = \frac{m_q - m_p}{n_q - n_p}$ and the bias term as $b(p,q) = \frac{m_q n_p - m_p n_q}{n_q - n_p}$.

First, let us prove that both s and b are monotonically decreasing.

Lemma 1. *Let p, q, and r be three consecutive points in $hull\,(X)$. Then we have $n \times s(q,r) + b(q,r) \leq n \times s(p,q) + b(p,q)$, for any $n \geq 0$.*

Proof. The slope for the segment between p and q is less than or equal to the slope for the segment between q and r. Inversing the slopes leads to

$$s(q,r) = \frac{m_r - m_q}{n_r - n_q} \leq \frac{m_q - m_p}{n_q - n_p} = s(p,q).$$

By cross-multiplying, adding $m_q n_q - m_q n_p - m_q n_r + \frac{m_q n_p n_r}{n_q}$ to both sides, multiplying by $\frac{n_q}{(n_r - n_q)(n_q - n_p)}$, and simplifying, we get

$$b(q,r) = \frac{m_r n_q - m_q n_r}{n_r - n_q} \leq \frac{m_q n_p - m_p n_q}{n_q - n_p} = b(p,q).$$

Combining the two equations proves the claim. □

Next, we show the key necessary condition for the optimal point.

Lemma 2. *Let p, q, and r be 3 consecutive points in $hull\,(X)$. Select $n, m \geq 0$. If q optimizes $\frac{m_q + m}{n_q + n}$, then $n \times s(q,r) + b(q,r) \leq m \leq n \times s(p,q) + b(p,q)$.*

Proof. Since q is optimal, we have $\frac{m+m_p}{n+n_p} \le \frac{m+m_q}{n+n_q}$. Solving for m gives us $m \le n\frac{m_q-m_p}{n_q-n_p} + \frac{m_q n_p - m_p n_q}{n_q-n_p} = n \times s(p,q) + b(p,q)$. Similarly, due to optimality, $\frac{m+m_r}{n+n_r} \le \frac{m+m_q}{n+n_q}$, and solving for m leads to $m \ge n \times s(p,q) + b(p,q)$, proving the claim. □

The two lemmas allow us to use binary search as follows. Given two consecutive points p and q we test whether $m \le n \times s(p,q) + b(p,q)$. If true, then the optimal label is q or to the right of q, if false, the optimal point is to the left of q. To perform the binary search, we can use directly the structure maintained by the algorithm by Overmars and Van Leeuwen [16] since it stores the current convex hull in a balanced search tree. Moreover, the algorithm allows evaluating any function based on the neighboring points. Specifically, we can maintain s and b. In summary, we can find the optimal label in $\mathcal{O}(\log |L|)$ time.

Our next result formalizes the above discussion.

Theorem 5. GREEDYOR *runs in* $\mathcal{O}\left(p \log^2 |L| + |V| + |E|\right)$ *time, where* p *is the number of edge-label pairs* $p = |\{(e,k) \mid e \in E, k \in lab(e)\}|$.

Proof. The proof is similar to the proof of Theorem 2, except we have replaced a search tree with the convex hull structure by Overmars and Van Leeuwen [16]. The inner for-loops are evaluated at most $\mathcal{O}(p)$ times since an edge or a node is visited only once, and $\sum_v |S_v| \in \mathcal{O}(p)$. Maintaining the hull requires $\mathcal{O}\left(\log^2 |L|\right)$ time, and there are at most $\mathcal{O}(p)$ such updates. Searching for an optimal label requires $\mathcal{O}(\log |L|)$ time, and there are at most $|L|$ such searches. □

We should point out that a faster algorithm by Brodal and Jacob [5] maintains the convex hull in $\mathcal{O}(\log |L|)$ time. However, this algorithm does not provide a search tree structure that we can use to search for the optimal addition.

5 Related Work

A closely related work to our method is an approach proposed by Galbrun et al. [10]. Here the authors search for multiple dense subgraphs that can be explained by conjunction on (or the majority of) the *node* labels. The authors propose a greedy algorithm for finding such communities. Interestingly enough, the authors do not show that the underlying problem is **NP**-hard—although we conjecture that this is indeed the case—instead, they show that the subproblem arising from the greedy approach is an **NP**-hard problem.

Another closely related work is an approach proposed by Pool et al. [17], where the authors search for dense subgraphs that can be explained by queries on the *nodes*. The quality of the subgraphs is a ratio S/C, where S measures the goodness of a subgraph using the edges within the subgraph as well as the cross-edges, and C measures the complexity of the query.

The major difference between our work and the aforementioned work is that our method uses labels on the edges. While conceptually a small difference, this distinction leads to different algorithms and different analyses of those algorithms. Moreover, we cannot apply directly the previously discussed methods to networks that only have labels on edges.

An appealing property of finding subgraphs that maximize $|E(W)|/|W|$, or equivalently an average degree, is that we can find the optimal solution in polynomial time [11]. Furthermore, we can 2-approximate the graph with a simple linear algorithm [7]. The algorithm iteratively removes the node with the smallest degree and then selects the best available graph. This algorithm is essentially the same as the algorithm used to discover k-cores, subgraphs that have the *minimum* degree of at least k. The connection between the k-cores and dense subgraphs is further explored by Tatti [20], where the dense subgraphs are extended to create an increasingly dense structure. A variant of a quality measure was proposed by Tsourakakis [21], where the quality of the subgraph is the ratio of triangles over the vertices. In another variant by Bonchi et al. [4], the edges were replaced with paths of at most length k. Finding such structures in labeled graphs poses an interesting line of future work.

While finding dense subgraphs is polynomial, finding cliques is an **NP**-hard problem with a very strong inapproximability bound [12]. Finding cliques may be impractical as they do not allow any absent edges. To relax the requirement, Abello et al. [1] and Uno [22] proposed searching for quasi-cliques, that is subgraphs with a high proportion of edges, $|E(W)|/\binom{|W|}{2}$. Another relaxation of cliques is k-plex where k absent edges are allowed for a vertex [18]. Finding k-plexes remain an **NP**-hard problem [3]. Alternatively, we can relax the definition by considering n-cliques, where vertices must be connected with an n-path [6], or n-clans where we also require that the diameter of the graph is n [15]. Since 1-clique (and 1-clan) is a clique, these problems remain computationally intractable.

6 Experimental Evaluation

In this section, we describe our experimental evaluation of the GreedyAnd and GreedyOr algorithms. First, we observe how the algorithms behave on synthetic data with increasing randomness. Then we apply the algorithms to real-world datasets and analyze the results.

We implement our algorithms in Python and the source code is available online[1]. Since the number of labels in our experiments was not exceedingly large, we did not use the speed up using convex hulls when implementing disjunctive-induced graphs. Instead, we search for the optimal label from scratch leading to a running time of $\mathcal{O}\left(|L|^2 + p\right)$.

Experiments with Synthetic Data: We evaluate the greedy algorithms on synthetic graphs of 200 vertices and 50 labels. We select 5 of the labels as target labels and construct graphs for the conjunctive and disjunctive cases such that selecting the subgraph induced by these 5 labels gives the best density. We then

[1] https://version.helsinki.fi/dacs.

add random noise to the network by introducing a noise parameter ϵ, which controls the probability of randomly adding and removing edges as well as adding new labels to the edges.

For the conjunctive case, we create five disjoint cliques of 10 vertices such that all edges on the kth clique have all except the kth of the target labels. Finally, we add one more 20 vertex clique that has all of the target labels. Since each of the smaller cliques is missing one of the target labels, selecting the conjunction of all of them yields the densest subgraph as the clique of 20 vertices.

Given the noise parameter ϵ, we then add noise by having each of the edges in the cliques removed with probability ϵ, as well as having any other edges added between any pair of vertices with probability ϵ. Finally, for each of the edges in the cliques we add any of the other labels with probability ϵ each, except for adding the remaining target labels to edges in the cliques.

For the disjunctive case, we have created one clique with 40 vertices. The edges in the clique are split into five sets, such that each set of edges gets one of the target labels. Now, selecting the disjunction of the five target labels induces the clique as the subgraph and results in the highest density.

We then add noise by adding removing edges from the clique and adding new edges between any other pair of vertices with probability ϵ. In addition, each edge gains any of the other labels also with probability ϵ.

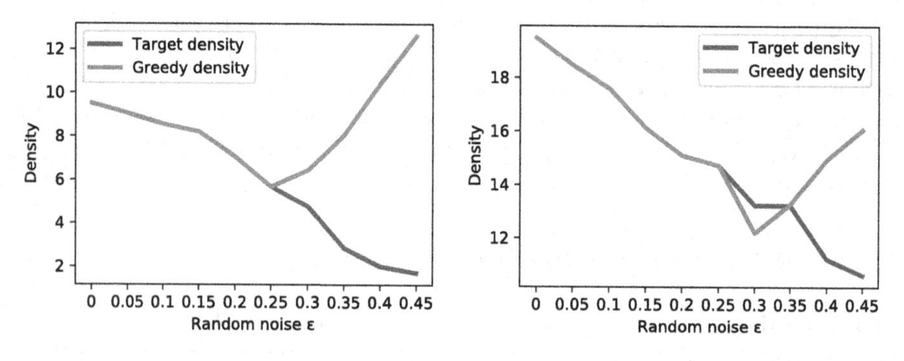

Fig. 1. Density of the subgraph induced by the target labels and the subgraph induced by the labels chosen by the greedy algorithms as a function of noise ϵ in the network. The results for GreedyAnd algorithm are on the left and for GreedyOr on the right.

We repeat the experiments with increasing values of ϵ and compare the density of the subgraph induced by the target labels to the density of the subgraph induced by the labels of the greedy algorithms. The results are shown in Fig. 1.

In both cases, the greedy algorithms correctly find the target labels for small values of ϵ. After $\epsilon > 0.25$ for GreedyAnd and after $\epsilon > 0.35$ for GreedyOr, the algorithms start to find other sets of labels, which yield higher densities than the target labels as many of the edges in the target clique have been removed and other edges have been added. However, at $\epsilon = 0.30$, the GreedyOr returns a suboptimal solution that yields a slightly lower density than the target labels.

Experiments with Real-World Datasets: We test the greedy algorithms by running experiments on four real-world datasets. The first dataset is the Enron Email Dataset[2], which consists of publicly available emails from employees of a former company called Enron Corporation. We collect the emails in sent mail folders and construct a graph where new edges are added between the sender and the recipients of each email. Each edge has labels consisting of the stemmed words in the email's title, with stop words and words including numbers removed.

The second dataset consists of high energy physics theory publications (HEP-TH) from the years 1992 to 2003. The data was originally released in KDD Cup[3] but we use a preprocessed version of the data available in GitHub[4] We create the network by adding authors as vertices, and edges between any two authors are added if they share at least two publications. The edges between authors are then given labels which consist of the preprocessed words in the titles of the shared articles between the two authors.

The third dataset consists of publications from the DBLP[5] dataset [19]. From this dataset, we chose publications from ECMLPKDD, ICDM, KDD, NIPS, SDM, and WWW conferences. The network is constructed in the same way as for the HEP-TH data, with authors as vertices, two or more shared publications as edges, and preprocessed and filtered words from the titles as labels.

The fourth and final dataset consists of the latest 10000 tweets collected from Twitter API[6] with the hashtag #metoo by the 27th of May, 23:59 UTC. We create the network by having users as vertices with an edge between a pair of users if one of them has retweeted or responded to one of the other's tweets. The labels on the edge are then any hashtags in the retweets or response tweets between the two users.

We construct the networks by filtering out labels that appear in less than 0.1% of the edges in the Enron and Twitter datasets, or labels that occur in less than 0.5% of the papers in the case of the HEP-TH and DBLP datasets. The sizes, label counts, and densities of the resulting graphs are shown in Table 1.

We run the greedy algorithms on each of these graphs, and compare the results against the densest subgraph ignoring the labels (DENSE). We report the statistics for the label-induced subgraphs and the densest subgraphs in Table 2.

For each of the datasets, both algorithms find label-induced subgraphs with higher densities than in the original graphs. In most cases, the restriction of constructing label-induced subgraphs results in clearly lower densities compared to the densest label-ignorant subgraphs. Interestingly, for the DBLP dataset GREEDYAND finds a label-induced subgraph with a very high density that is close to the density of the densest subgraph ignoring the labels. The running times are practical: the algorithm processes networks with 100 000 edge label pairs in seconds.

---

[2] https://www.cs.cmu.edu/~./enron/.
[3] https://www.cs.cornell.edu/projects/kddcup/datasets.html.
[4] https://github.com/chriskal96/physics-theory-citation-network.
[5] https://www.aminer.org/citation.
[6] https://developer.twitter.com/en/docs/twitter-api.

For Enron and HEP-TH datasets, the GreedyOr returns large sets of labels resulting in large subgraphs, whereas the GreedyAnd algorithm selects only a few labels with smaller induced subgraphs in each case. For the Twitter dataset, both greedy algorithms select only one label, which induces a small subgraph with a notably higher density than the original graph.

Table 1. Basic characteristics of the networks: number or vertices $|V|$, number or edges $|E|$, number of labels $|L|$, number of edge-label pairs p, and the density $d = |E|/|V|$.

Dataset	$\|V\|$	$\|E\|$	$\|L\|$	p	d
Enron	11 024	18 072	2 604	361 000	1.64
HEP-TH	4 738	7 767	240	78 078	1.64
DBLP	10 550	16 811	268	16 0850	1.60
Twitter	7 973	9 314	248	19 849	1.17

Table 2. Statistics for the resulting subgraphs for the greedy algorithms and the label-ignorant densest subgraph algorithm. For the label-induced subgraphs, we have the number of vertices n, the number of edges m, the size of the best set of labels $|A|$, density d, and running time t in seconds. For the densest subgraph, we show the number of vertices n and density d.

Dataset	GreedyAnd					GreedyOr					Dense	
	n	m	$\|A\|$	d	t	n	m	$\|A\|$	d	t	n	d
Enron	18	31	2	1.72	10.43	1 233	2 711	193	2.2	25.02	85	11.35
HEP-TH	7	14	4	2	1.96	3 284	5 588	40	1.7	5.74	58	3.81
DBLP	25	300	3	12	4.06	243	538	1	2.21	1.74	44	12.52
Twitter	12	31	1	2.58	0.77	12	31	1	2.58	1.89	19	3.37

Case Study: We analyze the communities for the Twitter and DBLP datasets by repeatedly running the GreedyAnd algorithm for these graphs. After running the algorithm, we exclude the edges from the output edge-induced subgraph, and run the algorithm again on the remaining graph. The first 8 resulting sets of labels, as well as densities and sizes for the induced subgraphs, are shown in Table 3.

For the DBLP graph, the algorithm finds a group of 25 authors that have each written at least two papers together with a shared topic, as well as other relatively large groups of authors whose edges form almost perfect cliques. The labels representing stemmed words can be used to interpret the topics of publications for these groups of authors having tight collaboration.

For the Twitter data of #metoo tweets, the densest label-induced subgraphs are formed by mostly looking at individual hashtags. This detects groups of people tweeting about #MeTooASE referring to the French Me Too movement for foster children, as well as groups closely discussing other topics in the context of the Me Too movement such as live streaming or the recent trial between Johnny Depp and Amber Heard.

Table 3. Label sets with corresponding subgraph densities and sizes were selected by running the GreedyAnd algorithm ten times on the graphs for DBLP and Twitter datasets. The labels are stemmed words from publication titles for DBLP, and tweet hashtags for Twitter data. The densities are not monotonically decreasing as the greedy algorithm does not always find the optimal solution.

DBLP				Twitter			
d	labels	n	m	d	labels	n	m
12.0	novel, rate, techniqu	25	300	2.58	metooase	12	31
10.74	identif, combin, process	23	247	1.88	streamer	16	30
6.2	forecast, experi, use	15	93	1.75	anubhavmohanty	16	28
6.0	heterogen, manag, stream, use	13	78	1.71	victimservices	7	12
2.0	heterogen, segment	5	10	1.83	causette, lfi	6	11
3.13	heterogen, manag, use, dynam	8	25	1.63	istandwithjohnny	8	13
2.5	heterogen, sourc, toward	6	15	1.43	rupertmurdock	7	10
2.5	heterogen, construct, dimension, network	6	15	1.25	marilynmanson	8	10

7 Concluding Remarks

In this paper, we considered the problem of finding dense subgraphs that are induced by labels on the edges. More specifically, we considered two cases: conjunctive-induced dense subgraphs, where the edges need to contain the given label set, and disjunctive-induced dense subgraphs, where the edges need to have only one label in common. As a measure of quality, we used the average degree of a subgraph. We showed that both problems are **NP**-hard, and we proposed a greedy heuristic to find dense induced subgraphs. By maintaining suitable counters we were able to find subgraphs in quasi-linear time: $\mathcal{O}(p \log |L|)$ for conjunctive-induced graphs and $\mathcal{O}\left(p \log^2 |L|\right)$ for disjunctive-induced graphs. We then demonstrated that the algorithms are practical, they can find ground truth in synthetic datasets, and find interpretable results from real-world networks.

While this paper focused on the conjunctive and disjunctive cases, future work could explore other ways to induce graphs from a label set and design efficient algorithms for such tasks. Another direction for future work is to relax the requirement that every edge/node must be induced from labels. Instead, we can allow some deviation from this requirement but then penalize the deviations appropriately when assessing the quality of the subgraph.

Acknowledgements. This research is supported by the Academy of Finland projects MALSOME (343045).

References

1. Abello, J., Resende, M.G.C., Sudarsky, S.: Massive quasi-clique detection. In: Rajsbaum, S. (ed.) LATIN 2002. LNCS, vol. 2286, pp. 598–612. Springer, Heidelberg (2002). https://doi.org/10.1007/3-540-45995-2_51
2. Angel, A., Koudas, N., Sarkas, N., Srivastava, D., Svendsen, M., Tirthapura, S.: Dense subgraph maintenance under streaming edge weight updates for real-time story identification. VLDB J. **23**(2), 175–199 (2014)
3. Balasundaram, B., Butenko, S., Hicks, I.V.: Clique relaxations in social network analysis: the maximum k-plex problem. Oper. Res. **59**(1), 133–142 (2011)
4. Bonchi, F., Khan, A., Severini, L.: Distance-generalized core decomposition. In: SIGMOD, pp. 1006–1023 (2019)
5. Brodal, G.S., Jacob, R.: Dynamic planar convex hull. In: FOCS, pp. 617–626 (2002)
6. Bron, C., Kerbosch, J.: Algorithm 457: finding all cliques of an undirected graph. Commun. ACM **16**(9), 575–577 (1973)
7. Charikar, M.: Greedy approximation algorithms for finding dense components in a graph. In: Jansen, K., Khuller, S. (eds.) APPROX 2000. LNCS, vol. 1913, pp. 84–95. Springer, Heidelberg (2000). https://doi.org/10.1007/3-540-44436-X_10
8. Du, X., Jin, R., Ding, L., Lee, V.E., Thornton Jr, J.H.: Migration motif: a spatial-temporal pattern mining approach for financial markets. In: KDD, pp. 1135–1144 (2009)
9. Fratkin, E., Naughton, B.T., Brutlag, D.L., Batzoglou, S.: Motifcut: regulatory motifs finding with maximum density subgraphs. Bioinformatics **22**(14), e150–e157 (2006)
10. Galbrun, E., Gionis, A., Tatti, N.: Overlapping community detection in labeled graphs. DMKD **28**(5), 1586–1610 (2014)
11. Goldberg, A.V.: Finding a maximum density subgraph. University of California Berkeley Technical report (1984)
12. Håstad, J.: Clique is hard to approximate within $n^{1-\epsilon}$. In: FOCS, pp. 627–636 (1996)
13. Langston, M.A., et al.: A combinatorial approach to the analysis of differential gene expression data. In: Shoemaker, J.S., Lin, S.M. (eds.) Methods of Microarray Data Analysis, pp. 223–238. Springer, Boston (2005). https://doi.org/10.1007/0-387-23077-7_17
14. Li, F., Klette, R.: Convex hulls in the plane. In: Li, F., Klette, R. (eds.) Euclidean Shortest Paths: Exact or Approximate Algorithms, pp. 93–125. Springer, London (2011). https://doi.org/10.1007/978-1-4471-2256-2_4
15. Mokken, R.J.: Cliques, clubs and clans. Qual. Quant. **13**(2), 161–173 (1979)
16. Overmars, M.H., Van Leeuwen, J.: Maintenance of configurations in the plane. J. Comput. Syst. Sci. **23**(2), 166–204 (1981)
17. Pool, S., Bonchi, F., van Leeuwen, M.: Description-driven community detection. TIST **5**(2), 1–28 (2014)
18. Seidman, S.B.: Network structure and minimum degree. Soc. Netw. **5**(3), 269–287 (1983)
19. Tang, J., Zhang, J., Yao, L., Li, J., Zhang, L., Su, Z.: Arnetminer: extraction and mining of academic social networks. In: KDD, pp. 990–998 (2008)

20. Tatti, N.: Density-friendly graph decomposition. TKDD **13**(5), 1–29 (2019)
21. Tsourakakis, C.E.: The k-clique densest subgraph problem. In: WWW, pp. 1122–1132 (2015)
22. Uno, T.: An efficient algorithm for solving pseudo clique enumeration problem. Algorithmica **56**(1), 3–16 (2010)

A Fast Heuristic for Computing Geodesic Closures in Large Networks

Florian Seiffarth[1(✉)], Tamás Horváth[1,2,3], and Stefan Wrobel[1,2,3]

[1] Department of Computer Science, University of Bonn, Bonn, Germany
{seiffarth,horvath,wrobel}@cs.uni-bonn.de
[2] Fraunhofer IAIS, Schloss Birlinghoven, Sankt Augustin, Germany
[3] Fraunhofer Center for Machine Learning, Sankt Augustin, Germany

Abstract. Motivated by the increasing interest in applications of graph geodesic convexity in machine learning and data mining, we present a heuristic for approximating the geodesic convex hull of node sets in large networks. It generates a small set of (almost) maximal outerplanar spanning subgraphs for the input graph, computes the geodesic closure in each of these graphs, and regards a node as an element of the convex hull if it belongs to the closed sets for at least a user specified number of outerplanar graphs. Our heuristic algorithm runs in time linear in the number of edges of the input graph, i.e., it is faster with one order of magnitude than the standard algorithm computing the closure exactly. Its performance is evaluated empirically by approximating convexity based core-periphery decomposition of networks. Our experimental results with large real-world networks show that for most networks, the proposed heuristic was able to produce close approximations significantly faster than the standard algorithm computing the exact convex hulls. For example, while our algorithm calculated an approximate core-periphery decomposition in 5 h or less for networks with more than 20 million edges, the standard algorithm did not terminate within 50 days.

Keywords: Geodesic closure · Outerplanar graphs · Convex cores

1 Introduction

In recent years, there has been a growing interest in applications of *geodesic convexity* in graphs (see, e.g., [14]). Besides other fields of computer science (e.g., genome rearrangement problems [6]), this concept has been utilized successfully also in *machine learning* and *data mining*. Examples include exact *cluster* recovery with queries [4], vertex classification in batch [2,15,17] and active *learning* [18], or *mining* complex networks [19,20]. Regarding this latter application, a *new* type of network decomposition [3] based on geodesic convexity has been proposed in [19]. More precisely, a broad class of real-world networks can be decomposed into a dense convex *core*, "surrounded" by a sparse

P. Pascal and D. Ienco (Eds.): DS 2022, LNAI 13601, pp. 476–490, 2022.
https://doi.org/10.1007/978-3-031-18840-4_34

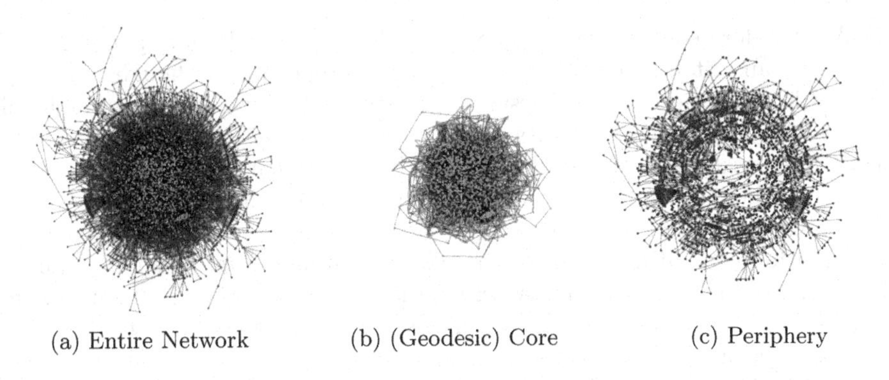

(a) Entire Network (b) (Geodesic) Core (c) Periphery

Fig. 1. (a) CA-GrQc network [12], (b) its (geodesic) core, (c) its periphery.

non-convex *periphery*[1] (see Fig. 1 for a relatively small example). The results in [19–21] clearly demonstrate that such a *core-periphery* decomposition provides new useful insights into the network's structure. For example, mainly the nodes in the core govern the degree distribution of the entire network or they have higher clustering coefficients and smaller geodesic distances to each other than the periphery nodes. A further nice property of core-periphery decomposition is that it is *not* characteristic to all network types, enabling a distinction between different types of networks.

This and other applications of geodesic convexity rely on computing the *geodesic convex hull*, also referred to as the *closure* of a set of vertices. Given a graph G and a set X of vertices of G, the closure of X is the *smallest* set C of vertices of G which contains X as well as *all* vertices of *all* shortest paths with both endpoints in C. Such a smallest set always exists, it is unique, and can be computed in $O(nm)$ time with a standard algorithm (cf. [14]), where n (resp. m) denotes the number of vertices (resp. edges) of G. Since $m = O\left(n^2\right)$ in the worst case, all approaches relying on computing geodesic convex hulls become practically *infeasible* for *large* networks.

To overcome this problem, we give up the demand for *correctness* and propose a *novel* approach to calculate only an *approximation* of the closure of X. More precisely, our heuristic is based on the following main steps: Generate a set of spanning subgraphs of G independently at random, compute the closure of X in these subgraphs separately, and regard a vertex of G as an element of the convex hull of X iff it belongs to the closure of X in at least a user specified percentage of the number of spanning subgraphs. The main question for this scheme is the choice of the class of the spanning subgraphs. Regarding forests, a closer look at the problem as well as our empirical results reveal that already for graphs that are structurally very close to forests, a poor approximation performance can be obtained in this way. This is because spanning forests may drastically distort shortest paths.

[1] The concepts of convex and non-convex node sets are used in an opposite way in [19,20], by speaking about non-convex cores and convex peripheries. This is because of a slightly different definition of convexity: While the authors in [19,20] look at the geodesic closure of induced subgraphs, we consider the whole structure.

We therefore consider the class of *outerplanar graphs* [5] for spanning subgraphs because it has several nice algorithmic properties. Although it is only slightly beyond the class of forests in the structural hierarchy, our empirical results with *large real-world* networks show that a close approximation of the geodesic convex hull can be obtained with outerplanar spanning subgraphs. Regarding the first step of the proposed heuristic, a maximal outerplanar spanning subgraph can be generated in $O(m)$ time [8]. For the second step we present an algorithm computing the (geodesic) convex hull in an *outerplanar* graph G. This can be done in time $O(nf)$, where f, the *face number* of G, is the maximum number of interior faces over the biconnected components of G. Our algorithm is linear in n in *practice* because f is typically *negligible* w.r.t. n. For example, in case of outerplanar spanning subgraphs of Erdős-Rényi random graphs with around 1,000,000 edges, the average face number was consistently less than 80.

We report experimental results with *large* real-world networks. They clearly show that their cores can be approximated closely (with a Jaccard similarity between 82 and 99%) with this scheme in *feasible* time, using only 100 spanning outerplanar subgraphs. Since we are not aware of any other approach approximating geodesic convex hulls in graphs, we compared the runtime results obtained by our heuristic with those of the standard algorithm mentioned above. In particular, in case of networks with more than 20 million (and up to 117 million) edges, the approximate decomposition could be computed in 5 h or less with our algorithm. In contrast, the computation of the exact core-periphery decomposition with the standard algorithm had to be aborted after 50 days. Because of the close approximation, the approximate cores inherit several properties of the exact ones. For example, their degree distributions were consistently close to those of the exact ones.

The rest of the paper is organized as follows. In Sect. 2 we collect the necessary notions and fix the notation. Section 3 contains the description of the algorithm computing the closure of a set of vertices in outerplanar graphs. In Sect. 4 we empirically evaluate our approach. Finally, in Sect. 5 we formulate some questions for further research. For space limitations, proofs are omitted in this short version. They can be found in the technical report [16].

2 Notions and Notation

For basic notions in graph theory, we refer to some standard textbook (see, e.g., [7]). The set V of vertices (resp. E of edges) of a graph $G = (V, E)$ is denoted by $V(G)$ (resp. $E(G)$). By graphs we always mean finite *undirected* and *unweighted* graphs without loops and parallel edges that are *connected* and denote $|V(G)|$ and $|E(G)|$ by n and m, respectively.

Given a graph G, the function $I : V \times V \to 2^V$, called the *geodesic interval*, maps (u, v) to the union of the sets of vertices on all shortest paths between u and v. A set $X \subseteq V(G)$ is *(geodesically) convex* or *closed* if $I(u, v) \subseteq X$ for all $u, v \in X$. For all $X \subseteq V(G)$, there exists a unique smallest closed set $X' \supseteq X$, called the *convex hull* or *closure* of X. Furthermore, the function ρ_G mapping the subsets of $V(G)$ to their convex hulls is a *closure operator*, that is, it is *extensive* (i.e., $X \subseteq \rho_G(X)$), *monotone* (i.e., $X \subseteq Y$ implies $\rho_G(X) \subseteq \rho_G(Y)$), and *idempotent* (i.e., $\rho_G(\rho_G(X)) = \rho_G(X)$) for all $X, Y \subseteq V(G)$. We omit G from ρ_G if it is clear from the context. For a graph G and $X \subseteq V(G)$, $\rho(X)$ can be computed with the *standard* algorithm relying on the *single-source shortest path* (SSSP) problem (see, e.g., [14]). That is, iterate over *all* elements $u \in \rho(X)$, starting with an arbitrary element of X, as follows: Let $X' \supseteq X$ be the set of elements in $\rho(X)$ that have already been generated before we process the next element u. Then add $Y = \bigcup_{v \in X'} I(u, v)$ to X', where Y can be calculated by solving the SSSP problem (for unweighted graphs) from u to all elements of X'. After all elements in X' have been processed, we have $X' = \rho(X)$. It is a folklore result that the SSSP problem can be solved with *breadth-first search* (BFS) in $O(n + m)$ time, implying that $\rho(X)$ can be computed in $O(nm)$ time.

A graph G is *outerplanar* [5] if it can be embedded in $\mathbb{R}^2$ in a way that no two edges cross each other (except possibly in their endpoints) and there exists a point $P \in \mathbb{R}^2$ such that each vertex of G can be reached from P by a simple curve that does not cross any of the edges. Removing all points and curves from the plane corresponding to the vertices and edges of G, respectively, we obtain a set of connected "pieces" of the plane, called *faces*. Since G is finite, all faces are bounded except for one, the *outer* face. The bounded faces are called *interior* faces. The *face number* of a biconnected outerplanar graph is the number of its interior faces; the face number of an outerplanar graph G, denoted $\Phi(G)$, is the maximum of the face numbers over its biconnected components.

Let G be an outerplanar graph. All biconnected components, called *blocks* of G consist of a unique Hamiltonian cycle and a possibly empty set of (non-crossing) diagonals. Edges not belonging to blocks are called *bridges*. The *block and bridge tree* (BB-tree) $\widetilde{G}$ of G is defined as follows [10]: For each block B of G, (i) introduce a new vertex, called *block vertex* v_B, (ii) remove all edges belonging to B, and (iii) for every vertex v of B, connect v with v_B by an edge if v is adjacent to a bridge or to another biconnected component of G; otherwise remove v. It holds that $\widetilde{G}$ is a (free) tree and can be computed in $O(n)$ time.

While $\rho(X)$ can be computed in $O(nm)$ time for arbitrary graphs, Theorem 1 gives rise to a faster algorithm for outerplanar graphs

Theorem 1 ([1]). *Let G be an outerplanar graph. Then for all $X \subseteq V(G)$, $\rho(X) = \bigcup_{u,v \in X} I(u, v)$.*

Thus, in case of outerplanar graphs, it suffices to perform a BFS only from the elements of X, resulting in the following corollary, by noting that $m = O(n)$ in case of outerplanar graphs:

Corollary 1. *Let G and X be as in Theorem 1. Then $\rho(X)$ can be solved in time $O(m|X|) = O(n|X|)$.*

3 The Heuristic

As mentioned earlier, our heuristic for approximating the closure of a set $X \subseteq V(G)$ for an arbitrary graph G consists of the following main steps:

(i) Generate s (inclusion) maximal outerplanar spanning subgraphs $G_1, \ldots .G_s$ of G independently at random, for some $s > 0$ integer. Each outerplanar spanning subgraph can be generated in $O(m)$ time, where $m = |E(G)|$ [8].[2] The number s of spanning subgraphs can be regarded as a constant (e.g., it was set to 100 in our experiments, independently of the networks' size). Thus, the total time of this step is *linear* in m in *practice.*

(ii) For all outerplanar graphs G_i generated in step (i), calculate the closure $\rho_{G_i}(X)$. Corollary 1 implies that $\rho_{G_i}(X)$ can be computed in $O(n|X|)$ time. Below we give a more sophisticated algorithm. Its complexity is $O(nf)$, where f is the face number of G_i. Thus, its complexity is *independent* of the cardinality of X, which makes our algorithm *superior* to the standard one in terms of runtime. Since $f = O(n)$, it does not improve the *theoretical* worst-case complexity of the standard algorithm. Still, it has two important advantages over the standard algorithm. The first one is *practical*: Our experiments with various graphs clearly show that the face number of spanning outerplanar graphs is *negligible*, compared to their size (i.e., number of vertices). The second one is of *theoretical* interest: Allowing only at most c faces per biconnected components in the spanning outerplanar graphs for some *constant* c, our algorithm runs in guaranteed *linear* time. The importance of this property is that already a few number of diagonals in the biconnected components result in a substantial improvement of the approximation performance, as demonstrated experimentally in the next section.

(iii) Finally, a vertex $u \in V(G)$ is regarded as an element of $\rho_G(X)$ iff there is a set $S \subseteq \{G_1, \ldots, G_s\}$ with $|S| \geq t$ for some $0 < t \leq s$ integer such that $u \in \rho_{G'}(X)$ for all $G' \in S$.

The rest of this section is devoted to step (ii) above. More precisely, we deal with the following problem for *outerplanar* graphs:

Problem 1. Given a graph G and $X \subseteq V(G)$, *compute* $\rho(X)$.

The algorithm solving Problem 1 for outerplanar graphs is given in Algorithm 1 (see, also, Figs. 2, 3, 4 for a running example). We assume that G is *connected*, by noting that all results can easily be generalized to disconnected outerplanar graphs as well. Algorithm 1 first calculates the BB-tree $\tilde{G}$ for the input outerplanar graph G and then stores the input vertex set X and the set of block nodes of $\tilde{G}$ in the variables X_0 and Y, respectively (lines 1–3). In line 4, it computes the set C_1 of block nodes representing the blocks of G that have at least one vertex in X_0. In a similar way, C_2 contains the set of nodes of $\tilde{G}$ that belong to X_0 (cf. line 5) (see Fig. 2 for an example).

The closure of $C_1 \cup C_2$ in $\tilde{G}$ is calculated in C (line 6) and the union of X_0 and the set of vertices in C that belong to $V(G)$ is stored in X_1 (line 7) (see

[2] To the best of our knowledge, there exists no (simple) algorithmic realization of this result stated in [8] (see, also, the discussion in Sect. 3.5 in [11]). In [16] we present a *fast* and *easy to implement* algorithm computing an *almost* inclusion *maximal* outerplanar spanning subgraph in $O(m)$ time.

Algorithm 1: OUTERPLANAR GRAPHS: CLOSURE

Input: outerplanar graph G and $X \subseteq V(G)$
Output: $\rho(X)$
1 construct the BB-tree $\widetilde{G}$ for G;
2 $X_0 \leftarrow X$;
3 $Y \leftarrow$ set of block nodes of $\widetilde{G}$;
4 $C_1 = \{v_B \in Y : V(B) \cap X_0 \neq \emptyset\}$;
5 $C_2 \leftarrow V(\widetilde{G}) \cap X_0$;
6 $C \leftarrow \tau(\widetilde{G}, C_1 \cup C_2),$;
7 $X_1 \leftarrow X_0 \cup (C \cap V(G)),\ i \leftarrow 1$;
8 **foreach** $v_B \in Y \cap C$ **do**
9 | **if** $|V(B) \cap X_i| > 1$ **then**
10 | | $X_{i+1} \leftarrow X_i \cup \beta(B, V(B) \cap X_i)$;
11 | | $i \leftarrow i + 1$;
12 **return** X_i;

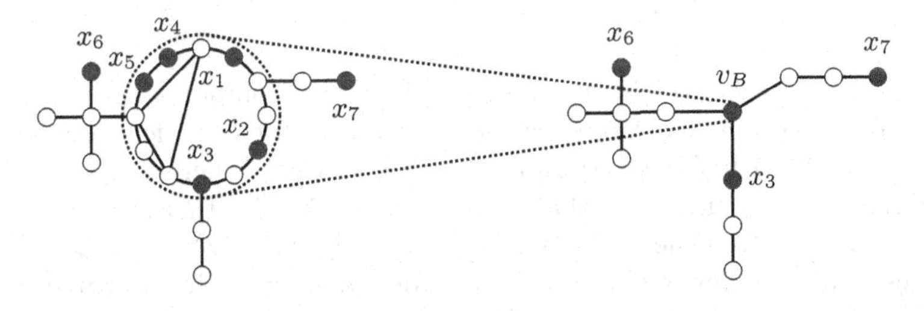

Fig. 2. *Left:* Outerplanar graph G and input set $X = X_0 = \{x_1, \ldots, x_7\} \subseteq V(G)$ in blue, *Right:* BB-tree $\tilde{G}$ constructed from G (for the biconnected outerplanar component on the left hand side, see dotted circle, a new node v_B is added). The sets $C_1 = \{v_B\}$ and $C_2 = \{x_3, x_6, x_7\}$ are marked in blue (Lines 1–5 of Algorithm 1).

Fig. 3 for an example). Note that at this point of the algorithm we have $v \in X_1 \subseteq \rho_G(X)$ for all $v \in \rho_G(X)$ that do not belong to a biconnected component of G. Furthermore, for all $v \in \rho_G(X) \setminus X_1$, v is on a shortest path in one of the blocks and with both endpoints in X. Accordingly, in loop 8–11, the algorithm takes all block nodes v_B of $\tilde{G}$ that belong to the closed set C, computes the closure of the set of vertices of the corresponding block B over B that are known to be closed (i.e., belong to X_i), updates the set of already known closed vertices in X_{i+1}, and increments the loop variable i. At the end, it returns the set X_i.

It remains to discuss the functions τ and β (cf. lines 6 and 10). Regarding τ (see Algorithm 2), it computes the closure of a set of nodes of a tree. It iteratively removes all leaves of T that are not in X and returns the set of all nodes of T at the end that have not been deleted (see Fig. 2 (right) and Fig. 3 (left)). The proof of the following lemma is straightforward:

Lemma 1. *For any tree T with n nodes and for any $X \subseteq V(T)$, Algorithm 2 returns $\rho_T(X)$ in $O(n)$ time.*

Fig. 3. *Left:* Output of Algorithm 2 applied to the BB-tree $\tilde{G}$ from Fig. 2 (nodes in $X_1 \setminus X_0$ are marked in red). *Right:* Biconnected outerplanar graph B corresponding to v_B, nodes in $X_1 \cap V(B)$ which are not in X_0 are marked in red. (Color figure online)

Algorithm 2: FUNCTION τ

Input: tree T and $X \subseteq V(T)$
Output: $\rho_T(X)$
1 **while** $\exists v \in V(T) \setminus X$ *with* $d(v) \leq 1$ **do**
2 | remove v from T;
3 **return** $V(T)$;

Regarding β (see Algorithm 4 and Fig. 4), which computes the closure over biconnected outerplanar graphs, we first show that for any biconnected outerplanar graph B with $f = \Phi(B)$ and for any $X \subseteq V(B)$, there is a set $G_X \subseteq X$ of cardinality *linear* in f such that $\rho_B(G_X) = \rho_B(X)$. Furthermore, G_X can be constructed in *linear* time as follows (see, also, Algorithm 3): Initialize G_X with $\emptyset$ (cf. line 1) and process all interior faces F of B one by one in an arbitrary order as follows: If F has no vertex from X then disregard F; o/w choose an arbitrary vertex w from $X' = V(F) \cap X$. For that w, calculate the furthest vertex $u \in X'$ and the furthest vertex $v \in (X' \setminus \rho_F(\{u, w\})) \cup \{w\}$, and add u and v to G_X (cf. lines 6 and 7 of Algorithm 3). Note that $\rho_F(\{u, w\})) = V(F)$ if $d(u, w) = \ell/2$, where ℓ is the (cycle) length of F; o/w it is the set of vertices of the (unique) shortest path between u and w. If w does not lie on a shortest path between u and v (cf. line 8), then add w to G_X as well. Note that u and v can be equal to w. Hence, we add at least one and at most three vertices of X' to G_X for F.

To illustrate the above steps in our running example, consider the biconnected outerplanar graph B and the set $X \subseteq V(B)$ marked with color blue in Fig. 4. A generator set G_X computed for the input set marked with blue in Fig. 4 (left) is given in Fig. 4 (middle). It contains four vertices marked with black. In case of the largest face of G, suppose we first select $w \in X$. For w, we first add u and then v to G_X by Algorithm 3 (see Fig. 4 (left) for u, v, and w); w is not added because it is on a shortest path between u and v. The closure $\rho(X) = \rho(G_X)$ is given in Fig. 4 (right).

We have the following result about Algorithm 3:

Lemma 2. *Let B be a biconnected outerplanar graph with $f = \Phi(B)$. Then for all $X \subseteq V(B)$, Algorithm 3 computes a set $G_X \subseteq X$ in $O(n)$ time such that $\rho_B(G_X) = \rho_B(X)$ and $|G_X| = O(f)$.*

Algorithm 3: FUNCTION GENERATORSET

```
Input: biconnected outerplanar graph B, X ⊆ V(B)
Output: G_X ⊆ X such that ρ_B(G_X) = ρ_B(X)
1  G_X ← ∅                          // G_X ⊆ X: generator set for ρ_B(X) ;
2  forall the interior faces F of B do
3  |   X' ← V(F) ∩ X;
4  |   if |X'| > 0 then
5  |   |   select an arbitrary vertex w from X';
6  |   |   add u = arg max_{x∈X'} d(x, w) to G_X ;
7  |   |   add v = arg max_{x∈(X'\ρ_F({u,w}))∪{w}} d(x, w) to G_X ;
8  |   |   if w ∉ ρ_F({u, v}) then
9  |   |   |   add w to G_X ;
10 return G_X;
```

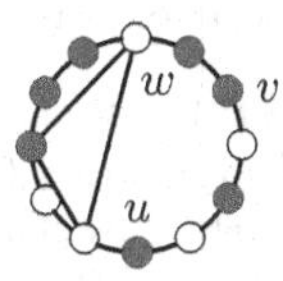

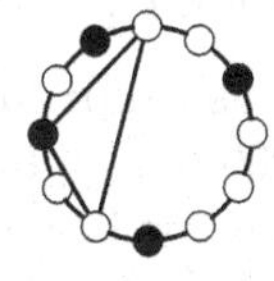
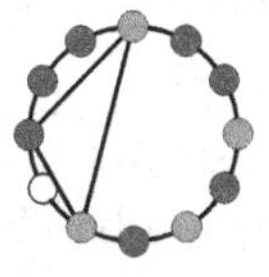

Fig. 4. *Left:* Biconnected outerplanar graph B with $X \subseteq V(B)$ in blue (cf. Fig. 3 (right)), *Middle:* generator set $G_X \subseteq X$ in black, *Right:* $\rho(X) = \rho(G_X)$, newly added nodes in red. (Color figure online)

We are ready to present Algorithm 4 computing the closure of a set of vertices over a biconnected outerplanar graph (see line 10 in Algorithm 1). The input of Algorithm 4 consists of a biconnected outerplanar graph B and a set $X \subseteq V(B)$. Using Algorithm 3, it first computes a generator set G_X for B and X and, utilizing the results in Corollary 1, computes $\rho_B(X) = \rho_B(G_X)$ in time $O(|V(B)| \cdot |G_X|)$.

Lemma 3. *Let B, f, and X be as in Lemma 2. Then Algorithm 4 computes $\rho_B(X)$ correctly and in $O(|V(B)|f)$ time.*

Using Lemmas 1–3, one can show the following result:

Theorem 2. *Algorithm 1 solves Problem 1 for outerplanar graphs correctly and in $O(nf)$ time, where $f = \Phi(G)$.*

4 Experimental Results

Our experiments presented in this section are concerned with some basic properties of the proposed heuristic. First, we compare the runtime of our outerplanar closure algorithm (Algorithm 1) to that of the naïve algorithm for outerplanar graphs (see Sect. 2). Second, using *large real-world* networks [12], we empirically

Algorithm 4: FUNCTION β

Input: biconnected outerplanar graph B, $X \subseteq V(B)$
Output: $\rho_B(X)$

1 $G_X \leftarrow$ GENERATORSET(B, X);
2 **return** $\rho_B(G_X)$;

evaluate the approximation performance of our heuristic on the *core-periphery* decomposition [19] problem. For the implementation[3] we used the C++-library SNAP 6.0 [13]. All experiments were conducted on a machine with AMD Ryzen 9 3900X and 64 GB RAM.

4.1 Datasets

The following synthetic and real world data sets are used in our experiments:

ERDŐS-RÉNYI. This dataset contains Erdős-Rényi connected random graphs [9] with 10 different sizes from $n = 1{,}000$ to $n = 10{,}000$, with a step size of 1,000 and with edge probabilities ranging from $p = 0.006$ to $p = 0.02$, with step size 0.002. Below $p = 0.006$, the graphs were too sparse for our purpose. For $n = 10{,}000$ and $p = 0.02$, the graphs contain around 1,000,000 edges. For all configurations of (n, p), 100 *connected* Erdős-Rényi random graphs have been generated.

LARGE REAL-WORLD NETWORKS. This dataset contains 15 real-world networks from [12] (see Table 2). In case of disconnected graphs, only their largest connected components were considered.

4.2 Computing Closures in Outerplanar Graphs

In this section we empirically evaluate Algorithm 1 on synthetically generated data. More precisely, we first generate an outerplanar spanning subgraph G at random for each graph in the ERDŐS-RÉNYI dataset. To compare our algorithm with the standard one based on SSSP, for each outerplanar graph G we construct a graph G' with the same number of nodes and edges, but with the difference that G' is not necessarily outerplanar. That is, for each G we first generate a random spanning tree T of G and construct then a possibly non-outerplanar graph G' via adding $m-n+1$ random edges to T. Thus, G and G' have the same number of vertices and edges. Fig. 5 (left) shows the average runtime needed to calculate the closures on G and G' for a random subset of 1% of the vertices. (C1) is the naïve closure algorithm for the outerplanar graphs G using the result of Corollary 1 (i.e., it calculates the shortest paths between all pairs of input vertices). (C2) is our Algorithm 1 and (CGraph) is the naïve closure algorithm for the arbitrary graphs G'. Recall that the complexity of (CGraph) is $O(nm)$, where $m = O(n)$ by construction, it is $O(n|X|)$ for (C1), where $|X| = n/100$,

[3] The code is available at https://github.com/fseiffarth/GCoreApproximation.

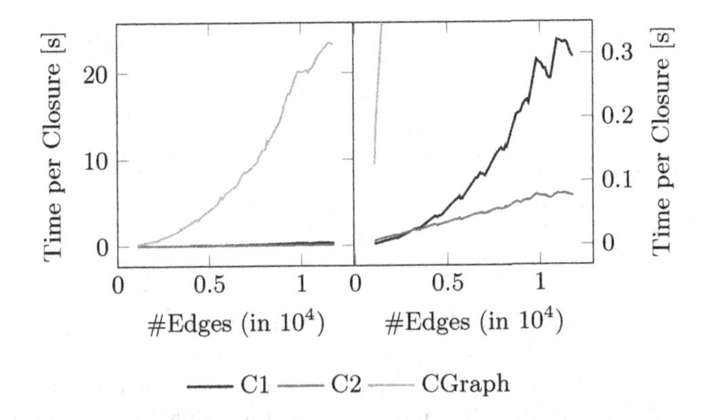

Fig. 5. (left) Closure runtimes for outerplanar graphs with the naïve alg. (C1) and with Algorithm 1 (C2) and for arbitrary graphs (CGraph), with the same number of nodes and edges. The generator set is a random subset of 1% of the vertices. (right) Runtime scaled down for (C1) and (C2).

Table 1. Average order and face number of the spanning outerplanar subgraphs of the ERDŐS-RÉNYI random graphs with fixed size of $n = 10^4$. The properties are averaged over 100 samples.

Edge Prob.	#Edges	Avg. #Output Edges	Avg. Face Number
0.008	399,960	11,077.61 (± 19.76)	76.11 (± 25.38)
0.012	599,940	11,342.36 (± 23.01)	70.52 (± 16.36)
0.016	799,920	11,561.69 (± 25.60)	71.77 (± 19.26)
0.020	999,900	11,755.85 (± 27.71)	65.95 (± 14.14)

and $O(nf)$ for our algorithm (C2), which is independent of $|X|$. The results are in accordance with these complexities. In particular, the closure computation on the arbitrary graphs G' is slower by a factor up to 300 than on the outerplanar graphs G with (C1) and (C2) (see left of Fig. 5). The right part of Fig. 5 is scaled down for (C1) and (C2). It clearly shows that (C2) (i.e., Algorithm 1) is much faster in practice than the naïve algorithm (C1). In particular, (C2) seems to be the only out of the three algorithms which scales *linearly* with the number of edges. This indicates that the face number f in the time complexity $O(nf)$ is *negligible* in practice.

This observation is supported by Table 1. It reports the average *face number* of the generated spanning outerplanar subgraphs for the graphs with $n = 10^4$ vertices in the ERDŐS-RÉNYI dataset. Somewhat surprisingly, the average face number does *not* increase with the density. Figure 6 shows the average face number as a function of the number of vertices (left) and the number of edges of the input graphs (right), where the colors represent different edge probabilities. The results indicate that in practice, the face number seems to be *sublinear* in the graph size for fixed density (in our experiments, it was always less than 80),

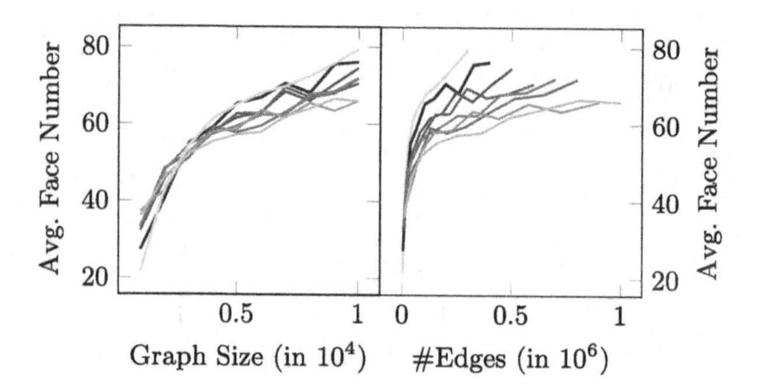

Fig. 6. Face numbers for spanning outerplanar subgraphs generated for the Erdős-Rényi dataset. (left) Average face number against the graph size. (right) Average face number against input edge number (different colors depict different edge probabilities).

justifying the *better* runtime of our closure computation algorithm (see Fig. 5 (right)).

4.3 Core Approximation in Real-World Networks

Finally, applying the heuristic described above, we present experiments concerning the approximation of *cores* in *large real-world* networks. Similar to [19], the core $\mathcal{C}$ of a graph G is defined by $\bigcap_{j=1}^{i} C_j$, where i is the smallest integer satisfying $\bigcap_{j=1}^{i} C_j = \bigcap_{j=1}^{i+1} C_j$ and $C_j = \rho(X_j)$ is the closure of $X_j \subseteq V(G)$ containing $l > 0$ vertices selected independently and uniformly at random. Note that this definition is not deterministic, but our experiments and those in [19] show that if a core exists, then it is *stable*. That is, the choice of X_j and especially l does not affect the core if l is sufficiently large. In particular, as a compromise between runtime and stability w.r.t. random effects, we choose $l = 10$. For each of the networks in Table 2, the fixed point was reached after $i = 3$ iterations.

We used 15 networks from [12] in our experiments. The size (n) and order (m) of some of them are more than 1,000 times larger than those in [19]. Table 2 contains the size of the exact cores and the runtime of computing them. While the *exact* core of the 3 largest networks could not be computed within 50 days with the standard algorithm sketched in Sect. 2, our algorithm produced the *approximate* cores in 5h for these large networks; in less than 40min for all other graphs.

For the approximation, for each large network we generated $s = 100$ spanning outerplanar subgraphs and calculated the closure of l randomly chosen vertices on each of these outerplanar graphs with Algorithm 1. Given the 100 closed sets in the outerplanar subgraphs obtained in this way, a vertex $v \in G$ was regarded as closed iff it was contained in at least t out of the $s = 100$ closed sets. The approximate core $\tilde{\mathcal{C}}$ was then calculated in the same iterative way as the exact one, but with the approximate closed sets. We compared exact and approximate cores with each other using Jaccard similarity. The first value in

Table 2. Large real-world networks from [12] with number of vertices (n), number of edges (m), density, number of vertices and edges in the core, time to calculate the exact core in seconds (or n.a. if it was not possible within 50 days), size of the approximated core, time to calculate the approximated core, and the Jaccard similarities of the exact and approximated cores obtained by grid search over l (number of random nodes in the generator set) and t (absolute frequency threshold for considering a node to be an element of the approximate core), and for $l = 5, t = 1$ in brackets (values of at least 0.9 in bold). The networks are sorted by nm.

Graph	Size n	#Edges m	Density	Size Core	#Edges Core	Time [s] Exact	Approx Core	Time [s] Approx	Jaccard sim best ($l = 5$, $t = 1$)
com-Orkut	3,072,441	117,185,083	2.5e−05	n.a	n.a	n.a	2,915,420	1.8e+04	n.a
soc-LiveJournal1	4,843,953	43,362,750	3.7e−06	n.a	n.a	n.a	3,018,149	8.7e+03	n.a
soc-pokec-relationships	1,632,803	22,301,964	1.7e−05	n.a	n.a	n.a	1,390,297	6.5e+03	n.a
com-youtube.ungraph	1,134,890	2,987,624	4.6e−06	390,825	2,169,158	8.9e+05	338,654	2.2e+03	0.82 (0.73)
com-dblp.ungraph	317,080	1,049,866	2.1e−05	90,077	438,265	7.0e+04	92,833	5.3e+02	**0.92** (0.87)
com-amazon.ungraph	334,863	925,872	1.7e−05	216,109	643,075	2.2e+05	231,618	5.2e+02	0.88 (0.87)
Slashdot0902	82,168	582,533	1.7e−04	48,718	514,338	1.4e+04	45,558	1.6e+02	**0.92** (0.68)
Cit-HepPh	34,401	420,828	7.1e−04	32,111	417,050	6.1e+03	32,309	9.6e+01	**0.99** **(0.97)**
Cit-HepTh	27,400	352,059	9.4e−04	24,832	347,918	3.5e+03	25,049	7.7e+01	**0.98** **(0.98)**
CA-AstroPh	17,903	197,031	1.2e−03	9,487	142,943	6.4e+02	9,522	3.0e+01	**0.95** **(0.93)**
CA-CondMat	21,363	91,342	4.0e−04	8,603	49,682	4.0e+02	8,761	3.5e+01	**0.94** **(0.90)**
CA-HepPh	11,204	117,649	1.9e−03	4,825	63,548	1.8e+02	4,804	1.8e+01	**0.93** **(0.91)**
Wiki-Vote	7,066	100,736	4.0e−03	4,579	98,026	1.3e+02	4,452	1.5e+01	**0.97** (0.78)
CA-HepTh	8,638	24,827	6.7e−04	3,605	14,161	4.6e+01	3,669	1.2e+01	**0.96** **(0.93)**
CA-GrQc	4,158	13,428	1.6e−03	1,336	5,036	7.0e+00	1,380	6.0e+00	**0.92** (0.88)

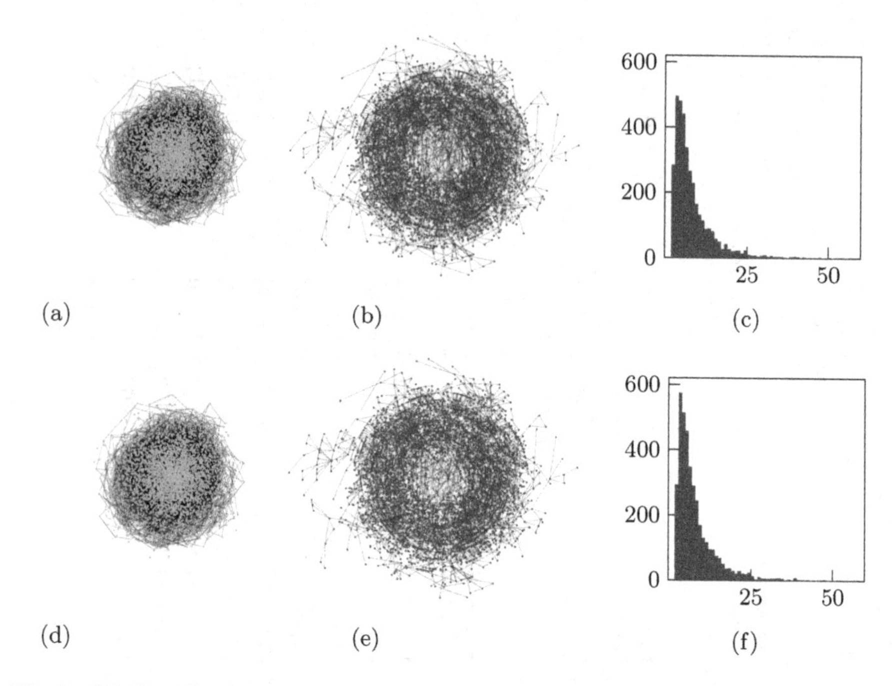

Fig. 7. CA-HepTh network, its exact (a) core, (b) periphery, (c) degree distribution of the core and its approximated (d) core, (e) periphery, (f) degree distribution of the approx. core.

the last column of Table 2 denotes the *best* Jaccard similarity achieved via grid search over $l \in \{5, \ldots, 2000\}$ and $t \in \{1, \ldots, 10\}$. We stress that using higher values of l has *no* impact on the time complexity of our algorithm, as it depends on n and the face number only (cf. Sect. 3). The second value (in brackets) denotes the Jaccard similarity for the approximate core obtained for $l = 5$ and $t = 1$.

For 12 out of the 15 graphs, we obtained a Jaccard similarity of around 0.8 or more; for 9 even at least 0.9. As an example, in Fig. 7 we show the exact core and periphery of the CA-HepTh network (see (a) and (b)) and their approximations (see (d) and (e)) for $l = 5$ and $t = 1$ (see, also, Table 2). We also plot the degree distribution of the exact core (see (c)) and that of the approximate one (see (f)) obtained for these values. One can see that the two distributions are fairly similar to each other, by noting that the Jaccard similarity obtained for $l = 5$ and $t = 1$ was 0.93 (see Table 2). A similar behavior could be observed for the other networks as well.

5 Concluding Remarks

Our experimental results clearly demonstrate that the presence of *cyclic* edges in the spanning subgraphs is essential for a close approximation of the geodesic convex hull. Thus, it is natural to ask whether further graph classes *beyond* forests can also be considered for spanning subgraphs. Such a graph class should fulfill at least two properties: (i) A (potentially maximal) spanning subgraph from this class could be generated in time *linear* in the order of the input graph and (ii) for the graphs in this class, the preclosure of any set of vertices should be its closure at the same time (cf. Theorem 1 in Sect. 2). This second condition indicates that the graphs in the class should be $K_{2,3}$-free (w.r.t. forbidden minor). A somewhat related question is if we can find an alternative simple algorithmic realization of the result stated in [8] for the problem of finding a *maximal* spanning outerplanar graph, preserving at the same time the linear runtime (in the number of edges).[4]

Although our primary focus in this work was on an effective approximation of geodesic convex hulls in large graphs, the results of Sect. 4.3 raise some interesting questions towards large real-world networks. For example, we are investigating whether it is possible to approximate the set of nodes with the highest *betweenness centrality* in large networks by that in their approximate cores? Furthermore, we are going to study if it is possible to estimate the quality of our approximation without knowing the exact core in advance by looking at the *variance* in the outputs of several approximation runs.

Our empirical results concerning core approximation in large real-world networks have been obtained for relatively small sets of generator elements (parameter l) and for low frequency thresholds (parameter t). The choice of these two parameters seem crucial for a close approximation (see Table 2). The related question is how to choose them, especially in case of large networks? *Sampling* seems a natural way, the question is whether it is possible to utilize the structure of the network at hand during sampling? Last but not least, it would be interesting to *systematically* study further types of random as well as large real-world networks for their core-periphery decomposition.

Acknowledgements. This research has been funded by the Federal Ministry of Education and Research of Germany and the state of North-Rhine Westphalia as part of the Lamarr-Institute for Machine Learning and Artificial Intelligence, LAMARR22C. The authors gratefully acknowledge this support.

References

1. Allgeier, B.: Structure and properties of maximal outerplanar graphs. Ph.D. thesis (2009)
2. de Araújo, P.H.M., Campêlo, M.B., Corrêa, R.C., Labbé, M.: The geodesic classification problem on graphs. In: Proceedings of the 10th Latin and American Algorithms, Graphs and Optimization Symposium, LAGOS. Electronic Notes in Theoretical Computer Science, vol. 346, pp. 65–76. Elsevier (2019)

[4] In the long version of this paper we present a linear time algorithm for generating an *almost* maximal spanning outerplanar graph that is easy to implement. Our empirical results show that only at most 0.03% of the edges were missing for maximality.

3. Borgatti, S., Everett, M.: Models of core/periphery structures. Soc. Netw. **21**, 375–395 (1999)
4. Bressan, M., Cesa-Bianchi, N., Lattanzi, S., Paudice, A.: Exact recovery of clusters in finite metric spaces using oracle queries. In: COLT 2021. Proceedings of Machine Learning Research, vol. 134, pp. 775–803. PMLR (2021)
5. Chartrand, G., Harary, F.: Planar permutation graphs. Annales de l'I.H.P. Probabilités et statistiques **3**(4), 433–438 (1967)
6. Cunha, L., Protti, F.: Closure of genomic sets: applications of graph convexity to genome rearrangement problems. Electron. Notes Discrete Math. **69**, 285–292 (2018)
7. Diestel, R.: Graph Theory. GTM, vol. 173. Springer, Heidelberg (2017). https://doi.org/10.1007/978-3-662-53622-3
8. Djidjev, H.N.: A linear algorithm for the maximal planar subgraph problem. In: Akl, S.G., Dehne, F., Sack, J.-R., Santoro, N. (eds.) WADS 1995. LNCS, vol. 955, pp. 369–380. Springer, Heidelberg (1995). https://doi.org/10.1007/3-540-60220-8_77
9. Erdős, P., Rényi, A.: On random graphs. Publicationes Mathematicae **6**, 290–297 (1959)
10. Horváth, T., Ramon, J., Wrobel, S.: Frequent subgraph mining in outerplanar graphs. Data Min. Knowl. Discov. **21**(3), 472–508 (2010)
11. Leipert, S.: Level planarity testing and embedding in linear time. Ph.D. thesis (1998)
12. Leskovec, J., Krevl, A.: SNAP Datasets: Stanford large network dataset collection, June 2014. http://snap.stanford.edu/data
13. Leskovec, J., Sosič, R.: Snap: a general-purpose network analysis and graph-mining library. ACM TIST **8**(1), 1 (2016)
14. Pelayo, I.M.: Geodesic Convexity in Graphs. Springer, New York (2013). https://doi.org/10.1007/978-1-4614-8699-2
15. Seiffarth, F., Horváth, T., Wrobel, S.: Maximal closed set and half-space separations in finite closure systems. In: Brefeld, U., Fromont, E., Hotho, A., Knobbe, A., Maathuis, M., Robardet, C. (eds.) ECML PKDD 2019. LNCS (LNAI), vol. 11906, pp. 21–37. Springer, Cham (2020). https://doi.org/10.1007/978-3-030-46150-8_2
16. Seiffarth, F., Horváth, T., Wrobel, S.: A fast heuristic for computing geodesic cores in large networks (2022). arXiv:2206.07350
17. Stadtländer, E., Horváth, T., Wrobel, S.: Learning weakly convex sets in metric spaces. In: Oliver, N., Pérez-Cruz, F., Kramer, S., Read, J., Lozano, J.A. (eds.) ECML PKDD 2021. LNCS (LNAI), vol. 12976, pp. 200–216. Springer, Cham (2021). https://doi.org/10.1007/978-3-030-86520-7_13
18. Thiessen, M., Gärtner, T.: Active learning of convex halfspaces on graphs. In: Advances in Neural Information Processing Systems (2021)
19. Tilen, M., Šubelj, L.: Convexity in complex networks. Netw. Sci. **6**(2), 176–203 (2018)
20. Šubelj, L., Fiala, D., Ciglaric, T., Kronegger, L.: Convexity in scientific collaboration networks. J. Informet. **13**(1), 10–31 (2019)
21. Šubelj, L.: Convex skeletons of complex networks. J. R. Soc. Interface **15**(145), 20180422 (2018)

Explainability and Interpretability

JUICE: JUstIfied Counterfactual Explanations

Alejandro Kuratomi(✉), Ioanna Miliou, Zed Lee, Tony Lindgren, and Panagiotis Papapetrou

Department of Computer and Systems Sciences, Stockholm University, Borgarfjordsgatan 12, 16455 Kista, Sweden
{alejandro.kuratomi,ioanna.miliou,zed.lee,tony,panagiotis}@dsv.su.se

Abstract. Complex, highly accurate machine learning algorithms support decision-making processes with large and intricate datasets. However, these models have low explainability. Counterfactual explanation is a technique that tries to find a set of feature changes on a given instance to modify the models prediction output from an undesired to a desired class. To obtain better explanations, it is crucial to generate faithful counterfactuals, supported by and connected to observations and the knowledge constructed on them. In this study, we propose a novel counterfactual generation algorithm that provides faithfulness by *justification*, which may increase developers and users trust in the explanations by supporting the counterfactuals with a known observation. The proposed algorithm guarantees justification for mixed-features spaces and we show it performs similarly with respect to state-of-the-art algorithms across other metrics such as proximity, sparsity, and feasibility. Finally, we introduce the first model-agnostic algorithm to verify counterfactual justification in mixed-features spaces.

Keywords: Machine learning · Interpretability · Counterfactuals · Faithfulness · Justification · Mixed-features space

1 Introduction

Highly accurate, complex machine learning (ML) models are becoming increasingly ubiquitous in different applications, including high-stakes decisions (medical diagnoses [7], loan applications [2], recidivism prediction [5]). As their utilization increases, so are the interpretability requirements to enable user understanding and trust [13,21]. Consider, for example, a complex ML model built on a large medical dataset that accurately predicts the risk of future severe disease occurrence. The most valuable knowledge for doctors, given a patient with a predicted high risk of severe disease development, would be the reasons for this undesired outcome and how to prevent it [7,13].

One way to convey such knowledge is to define a counterfactual (CF) explanation, which aims to explain the output of a complex ML model on an instance of

P. Pascal and D. Ienco (Eds.): DS 2022, LNAI 13601, pp. 493–508, 2022.
https://doi.org/10.1007/978-3-031-18840-4_35

interest in a model-agnostic manner [10,13,21]. CF explanations are paramount because they provide feature relevance and potential actions to achieve a desired class [10,21]. Given a classifier f, an instance $\mathbf{x}$ with predicted class label $\mathbf{c}$, defined over a set of variables $\boldsymbol{v}$, the aim of a CF explanation is to answer a fundamental question: *How should the variables $\boldsymbol{v}$ of $\boldsymbol{x}$ change to obtain class label $\mathbf{c}$' instead of $\mathbf{c}$?* [1]. A CF explanation $\mathbf{x}$' is structured in the following manner: *Outcome $\mathbf{c}$ occurs because variables $\boldsymbol{v}$ in $\boldsymbol{x}$ have values $\boldsymbol{v}_0$. If $\boldsymbol{x}$ is instead altered to $\boldsymbol{x}$' with values $\boldsymbol{v} = \boldsymbol{v}_1$, then outcome $\mathbf{c}$' occurs.*

The CF may be obtained through different algorithms which optimize diverse metrics, but there is currently no consensus on the best metric to optimize for better explanation quality [13,18]. This study focuses on generating CFs that satisfy desiderata such as proximity, sparsity, feasibility, and, most importantly, faithfulness through their connection to observations.

Intuitively, a desired CF metric is *proximity*, referring to the distance between the CF and the instance of interest. The shortest path is preferred in terms of actionability, (the shortest way to move from $\mathbf{x}$ to $\mathbf{x}$') [11,19,21]. While CF proximity is optimized by prioritizing distance, it may ignore other relevant CF desiderata, such as *sparsity* [13]. Minimizing distance may change several features. In a medical scenario, doctors and patients would prefer *sparse* explanations [3,12,21], i.e., the smallest number of feature changes to prevent a disease diagnosis. A sparse CF may prove a more actionable target [12,21].

Additionally, proximity or sparsity do not guarantee *feasibility* [21]. If CF values are not physically possible, the derived CF explanation loses credibility. Furthermore, immutable features like gender or ethnicity, should remain unchanged, while others could change, but only in one direction, like age. Feasibility considers feature *plausibility* (physically possible values), *mutability* (whether values may change), and *directionality* (the possible direction of the changes).

Lastly, CF *faithfulness* [8,10,16,18,21], indicates whether the generated CF is supported by observations. This metric has not been prioritized in traditional CF generation and has recently gained attention to facilitate user trust in the models [8,16]. A faithful CF is connected to and supported by observations and the knowledge constructed on them, i.e., in a medical scenario, the doctors may have a guarantee that the CF is not a spurious example, and that the recommended treatments are justified by the data and may be trusted [10,16,21].

One way to attain CF faithfulness is through likelihood [4,8,16]. Pawelczyk et al. find likely CFs with respect to the observations as part of an autoencoder cost function [16]. CF faithfulness may also be achieved by guaranteeing *justification*. As proposed by Laugel et al. [10], justification is a property shown by CFs that are *connected* to a correctly classified CF observation through a path where no decision boundary is met. Thus, a more complete CF explanation $\mathbf{x}$' can be defined: *Outcome $\mathbf{c}$ occurs because variables $\boldsymbol{v}$ in $\boldsymbol{x}$ have values $\boldsymbol{v}_0$. If $\boldsymbol{x}$ is instead altered to $\boldsymbol{x}$' with values $\boldsymbol{v} = \boldsymbol{v}_1$, then outcome $\mathbf{c}$' occurs, which is justified by the observation $\mathbf{e}$.* An example is shown in Fig. 1 [10], where, although point $\mathbf{b}$ is the closest CF to $\mathbf{x}$, a further CF may be preferred (point $\mathbf{a}$) if it preserves a connection to an observation (point $\mathbf{e}$). Alas, to the best of our

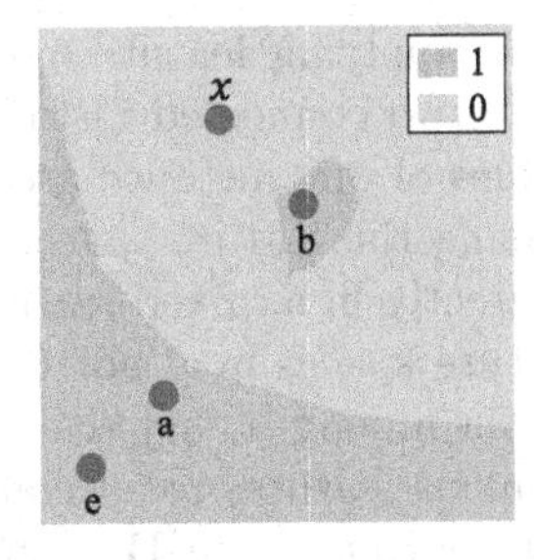

Fig. 1. x is the instance of interest, b is a close CF, but it is not justified by observed data as a is through observation e (highlighted in green outline).

knowledge, there is no algorithm that *generates* justified CFs while preserving other CF desiderata such as proximity, sparsity, and feasibility in mixed-features spaces [20]. Most importantly, justification has been previously approximated through ϵ-justification (a DBSCAN process that verifies the connection between synthetic points in a ball of radius ϵ) on continuous feature spaces [10,16], which cannot be applied directly in a mixed-features scenario, i.e., the CF justification verification in spaces with binary, categorical, ordinal, and continuous features, to the best of our knowledge, has not yet been established [20].

Therefore, we hereby present two algorithmic solutions for the problems described above: (1) the JUstIfied Counterfactual Explanations (JUICE) generation algorithm, and (2) the mixed-features justification verification algorithm. The main contributions of this paper are described below:

1. **JUICE:** a novel algorithm that generates justified CFs by using an exhaustive feature search in a mixed-features space, where feature value plausibility, mutability, and directionality are considered. JUICE attains competitive performance against state-of-the-art CF generation algorithms in terms of proximity and sparsity, providing complete CF explanations (see Sect. 3.1).
2. **Mixed-features justification verification:** a novel algorithm that empirically verifies whether any CF is justified or not by a CF observation traversing paths among any type of features (see Sect. 3.2).
3. **Empirical evaluation and benchmark:** JUICE generates justified CFs while providing as-good performance with respect to proximity, sparsity, feasibility, and run time when compared to other state-of-the-art CF algorithms. The CF justification of the latter algorithms is evaluated using the mixed-features verification algorithm (see Sect. 4).

2 The CF Justification Problem

Justification has been defined on continuous feature spaces by Laugel et al. [10], requiring a *connection* in space between the CF and a correctly classified observation. A *connection* is a path in space that connects two same-labeled instances, according to a classifier f, by applying feature operations, where no decision boundary of f is met, i.e., the two connected instances lie in the same classification region. All feature values of one instance are transformed to the feature values of the other by traversing the feature space. This definition of justification [10] cannot be applied directly in a mixed-features scenario, given that the connection in continuous feature spaces may not be feasible for other types of features. Hence, the connections among binary (categorical features are one-hot encoded), ordinal, and continuous features require additional considerations.

For binary features, a *binary connection* **B** is shown in Fig. 2a. Each green arrow represents an operation in the binary space to connect point $\boldsymbol{x'} = (0, 0)$ to point $\mathbf{e} = (1, 1)$. These operations are called *binary feature flips.* The binary path $B_B = [1, 2]$ is followed, meaning that first feature 1 is flipped, followed by feature 2, and points $\boldsymbol{x'}$, $(1, 0)$, and $\mathbf{e}$ belong to class 1. Therefore, this binary path guarantees a binary connection $\mathbf{B}(\mathbf{x'}, \mathbf{e})$ between $\boldsymbol{x'}$ and $\mathbf{e}$, by converting all binary features. Meanwhile, the red path B_A does not guarantee a binary connection, because $f((0, 1)) = 0 \neq f(\mathbf{e})$.

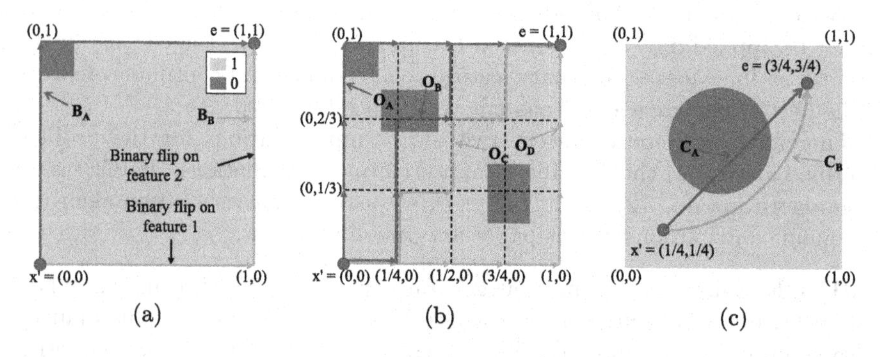

Fig. 2. Examples of different paths connecting CF **x'** and observation **e** in different spaces. Figure 2a: Binary paths in a binary space. Figure 2b: Ordinal paths in an ordinal space. Figure 2c: Continuous paths in a continuous space.

For ordinal features, an *ordinal connection* **O** is shown in Fig. 2b. Each green arrow represents an operation in the ordinal feature space to connect point $\boldsymbol{x'} = (0, 0)$ to point $\mathbf{e} = (1, 1)$. These operations are called *ordinal jumps*, which only visit plausible adjacent values following the feature values order. The ordinal path $O_C = [1, 2, 1, 2, 1, 2, 1]$ is followed, meaning that first feature 1 is changed, then feature 2, then feature 1, and so on, always in the direction $\mathbf{x'} \rightarrow \mathbf{e}$ and each point along this path belongs to class 1. Therefore, this ordinal path guarantees ordinal connection $\mathbf{O}(\mathbf{x'}, \mathbf{e})$ between $\boldsymbol{x'}$ and $\mathbf{e}$, by converting all ordinal features. Similarly, the ordinal path $O_D = [1, 1, 1, 1, 2, 2, 2]$ also guarantees ordinal connection. Meanwhile, none of the red paths O_A and O_B guarantee ordinal connections because $f((0, 1)) = 0 \neq f(\mathbf{e})$ and $f((\frac{1}{4}, \frac{2}{3})) = 0 \neq f(\mathbf{e})$.

For continuous features, a *continuous connection* **C** is shown in Fig. 2c. Each green arrow represents an operation in the continuous space to connect points

$\boldsymbol{x}' = (\frac{1}{4}, \frac{1}{4})$ and $\mathbf{e} = (\frac{3}{4}, \frac{3}{4})$. These operations are called *continuous jumps*, which traverse the continuous space. Each of the points along path C_B belongs to class 1, therefore, it guarantees continuous connection $\mathbf{C}(\mathbf{x}', \mathbf{e})$ between $\boldsymbol{x}'$ and $\mathbf{e}$. Meanwhile, the red path C_A does not, for at least a point v, $f(v) = 0 \neq f(\mathbf{e})$.

Let $\mathcal{X}$ be a data space, $X \in \mathcal{X}$ be a set of N labeled data examples, and $\mathcal{Y}$ the set of their corresponding class labels, such that $x_i \in X$ is of class $y_i \in \mathcal{Y}$, $\forall i \in \{1, \ldots, N\}$. Hence, the general justification problem is (based on [10]):

Problem 1. (General Justification). *Given a classifier $f : \mathcal{X} \to \mathcal{Y}$ trained on a mixed-features dataset X, a counterfactual example $\boldsymbol{x}' \in \mathcal{X}$ is generally justified by an observation $\mathbf{e} \in X$, if $f(\boldsymbol{x}') = f(\mathbf{e})$, and if there exists a binary, an ordinal and a continuous connection $\boldsymbol{B}$, $\boldsymbol{O}$, $\boldsymbol{C}$, whenever these types of features exist, between $\boldsymbol{x}'$ and $\mathbf{e}$ such that no decision boundary of f is met.*

There is no algorithm that generates justified counterfactuals for mixed-features spaces. We define the problem of Justified Counterfactual Generation:

Problem 2. (Justified Counterfactual Generation). *Given a classifier $f : \mathcal{X} \to \mathcal{Y}$ trained on a mixed-features dataset X, a cost function $L(\cdot)$, and all the possible counterfactuals z, a justified counterfactual example $\boldsymbol{x}' \in \mathcal{X}$ can be found by solving the following problem:*

$$\mathbf{x}' = \arg\min_{z}\{L(x, z) | f(x) \neq f(z) \wedge \exists\, \mathbf{e} \in X : \mathbf{B}(z, \mathbf{e}) \wedge \mathbf{O}(z, \mathbf{e}) \wedge \mathbf{C}(z, \mathbf{e})\}, \quad (1)$$

where $L(\cdot)$ can be the L2-norm or the L0-norm (see Sec. 3.1).

3 Proposed Algorithms

We hereby present the JUICE algorithm, which tackles Problem 2, and the mixed-features justification verification algorithm, which solves Problem 1. These algorithms may be found at the GitHub repo[1].

3.1 JUstIfied Counterfactual Explanations (JUICE)

We define the label of the CF instance as $CF_{label} = f(x')$. JUICE first finds a correctly classified CF observation, then navigates from it towards the instance of interest in this order: binary features, then ordinal features, and finally continuous features, improving proximity and sparsity. For each point v along the paths, it verifies $f(v) = CF_{label}$ and returns the closest point v to the instance of interest.

JUICE is presented in Algorithm 1. Lines 2 and 3 find the nearest neighbor CF and calculate the gradient in the $CF_{NN} \to x$ direction. Then, lines 4–6 find the binary, ordinal, and continuous feature indices where the instance of interest and the CF_{NN} differ. Lines 7–9 store the possible permutations (or paths) for

[1] https://github.com/alku7660/JUICE.

each of these three sets of indices. Lines 10–12 try to navigate through these paths in the binary subspace, from the CF_{NN} towards the instance of interest. This navigation stops only when either: (1) a path led to the exact same values as the instance of interest with all points along the path having the CF label, or (2) all binary paths were checked. Lines 13–15 and 16–18 do the same with ordinal and continuous features, respectively. The $TryNavigationThrough()$ function generates changes from the CF_{JUICE} to the instance of interest based on the binary, ordinal, or continuous paths, verifying $f(v) = CF_{label}$ for every admissible and changed v example along the paths, guaranteeing justification.

Algorithm 1: JUstified Counterfactual Explanations Pseudoalgorithm

input : x: instance of interest, x_{label}: x label, f: model, *type*: feature type, T: Train subset, *priority*: proximity or sparsity
output: CF_{JUICE}: JUICE CF, $justifier$ justifier example

1 $CF_{NN} \leftarrow$ NN(x, T, x_{label})
2 $CF_{JUICE} \leftarrow CF_{NN}$
3 gradient $\leftarrow x - CF_{NN}$
4 binaryIndices $\leftarrow$ **where**$(gradient \neq 0 \wedge type = binary)$
5 ordinalIndices $\leftarrow$ **where**$(gradient \neq 0 \wedge type = ordinal)$
6 continuousIndices $\leftarrow$ **where**$(gradient \neq 0 \wedge type = continuous)$
7 binaryIndicesPermutations $\leftarrow$ **permute**$(binaryIndices)$
8 ordinalIndicesPermutations $\leftarrow$ **permute**$(ordinalIndices)$
9 continuousIndicesPermutations $\leftarrow$ **permute**$(continuousIndices)$
10 **for** $binaryPath \in binaryIndicesPermutations$ **do**
11   **if TryNavigationThrough**$(CF_{JUICE}, binaryPath, f, priority)$ **then**
12   **Update**(CF_{JUICE})
13 **for** $ordinalPath \in ordinalIndicesPermutations$ **do**
14   **if TryNavigationThrough**$(CF_{JUICE}, ordinalPath, f, priority)$ **then**
15   **Update**(CF_{JUICE})
16 **for** $continuousPath \in continuousIndicesPermutations$ **do**
17   **if TryNavigationThrough**$(CF_{JUICE}, continuousPath, f, priority)$ **then**
18   **Update**(CF_{JUICE})
19 **return** CF_{JUICE}, CF_{NN}

The *priority* parameter is either *proximity* or *sparsity*. The proximity version of JUICE (*JUICEP*) minimizes the L-2 norm and the sparsity version (*JUICES*) minimizes the L-0 norm. JUICES finds the closest CF that has one single feature change (CF_{JUICES}). If JUICES fails, it ignores the single-feature restriction and optimizes for proximity to try to find the closest connected CF (CF_{JUICEP}) to the instance of interest which is connected to CF_{NN}. In binary features, flips in the binary paths always lead to binary values. In the case of one-hot encoded categorical variables, the sum of their values for every flip should equal 1. In ordinal features, changes are done in plausible values following the order described by the ordinal path. In the continuous features, the jumps are done using 1-percentual changes of the *gradient* value.

3.2 Mixed-features Justification Verification Algorithm

We propose the mixed-features justification verification algorithm. To the best of our knowledge, this is the first model-agnostic algorithm for CF justification evaluation in mixed-features spaces. Given a CF, the mixed-features justification verification algorithm finds the nearest correctly classified CF observation, T_{NN}. It then tries to find connections between the CF and T_{NN} through the different feature types, whenever applicable, since some datasets may not have all types of features. In the presence of binary, ordinal and continuous features, it first attempts a binary connection between the given CF and T_{NN}. If successful, it tries to find an ordinal connection. Finally, if successful, it performs an ϵ-justification search process [10] in the continuous-feature subspace, to find a continuous connection to *any* observation having the same binary and ordinal feature values as those of the T_{NN}. This works because, if the binary and ordinal connections with the T_{NN} are verified, then *any other observation with the same binary and ordinal feature values as those in the* T_{NN} *also has a binary and ordinal connection to the CF*. Hence, any observation with these values (not only the T_{NN}) may be continuously connected and verifies the CF general justification. If a feature type is missing, the algorithm skips to the next type.

The Mixed-Features justification verification is presented in Alg. 2. Line 2 finds the T_{NN}. If T_{NN} is the same as the given CF, then the CF is justified by itself and returns it. Otherwise it proceeds in lines 6 and 7 to calculate the gradient and perform the binary connection search. If there is a binary connection, the ordinal connection is searched in line 9, and if it exists, the continuous connection is searched. For the binary and ordinal connections, the search process (lines 7 and 9) stops whenever: (1) the values of the CF binary and ordinal features are equal to those of T_{NN} and the CF_{label} is preserved, or (2) the list of paths is fully checked and no successful binary or ordinal connections between T_{NN} and the CF were found, in which case, the CF is not justified. If the continuous connection search is executed, then the ϵ-justification search process is carried, as in [10,16]. A set of synthetic instances of size n is generated, for which the binary and ordinal features have the same values as the T_{NN}, and the continuous features are randomly and evenly spaced in the continuous subspace. Then, these instances are mixed with the observations, creating a dense continuous subspace. Finally, using a ball of radius r ($ball_r$), the algorithm jumps between examples v_{ball_r} inside the $ball_r$, verifying $f(v_{ball_r}) = CF_{label}$. Each evaluated example is then used as the center of the $ball_r$ to search for other instances until a CF observation is reached. If a CF observation lies in any $ball_r$, then the CF is ϵ-justified by it or continuously connected to it, and this observation is the justifier example.

3.3 Complexity

JUICE and the mixed-feature justification verification algorithms are possible solutions to the justified CF search and the justification verification problems. The computationally expensive solution in the original features space should join the binary, ordinal and continuous features and generate all permutations. This would allow the creation of paths, such as O_B and O_C in Fig. 2b, which are currently unchecked. The complexity of such solution would be $O((b+c+o+k)! \cdot (b+c+\sum_{i=1}^{o} m_i + kp))$, where b, c, o, and k are the number of binary, categorical, ordinal, and continuous features respectively. Additionally, p is the number of possible values of continuous features (in this case, $p = 100$) and m_i is the number of possible values of the ordinal feature $i, \forall i \in o$. In comparison JUICE has a lower complexity of $O(b! \cdot b + c! \cdot c + o! \cdot \sum_{i=1}^{o} m_i + k! \cdot kp)$.

Algorithm 2: Mixed-Features Justification Pseudoalgorithm

input : CF, CF_{label}, T: train subset, f: model, n: examples to generate, r: continuous subspace search radius.
output: $justifier$: justifier example.

```
1  justifier ← ∅
2  T_NN ← NN(CF, T, CF_label)
3  if T_NN = CF then
4  │  justifier ← CF
5  else
6  │  grad ← T_NN − CF
7  │  binConnect ← binaryConnect(CF, T_NN, grad, f, CF_label)
8  │  if binConnect then
9  │  │  ordConnect ← ordinalConnect(CF, T_NN, grad, f, CF_label)
10 │  if ordConnect then
11 │  │  conConnect, justifier ← contConnect(CF, T_NN, grad, f, CF_label, n, r)
12 return justifier
```

4 Empirical Evaluation

We present here the datasets used, the CF algorithms, the performance metrics used to evaluate JUICE and the baseline methods and the results obtained.

4.1 Datasets

8 binary classification datasets are used for the evaluation of JUICE and the other CF methods. Two datasets (1 and 2) are synthetic with a mixed-features space. Datasets 3, 5, 6, 7, and 8 can be found at the UCI Machine Learning website. For every feature, the properties of mutability, directionality, and plausible

values are manually set. The processing pipeline implemented for publicly available datasets follows the preprocessing guidelines of state-of-the-art research, as shown in [6]. The details of the generation of datasets 1 and 2, the features of all datasets, and the dataset pipelines are included in the GitHub repo.

1. **Disease**: A synthetic dataset with the task of predicting whether a person will have a severe disease or not, with 7 features.
2. **Athlete**: A synthetic dataset with the task of predicting whether an Olympic athlete will win a medal or not, with 6 features.
3. **Ionosphere**: The ionosphere dataset that originally contains 34 continuous features. Only 8 are selected based on feature importance from an RF model.
4. **Compass**: The Propublica dataset aimed at predicting criminal recidivism based on past personal records. Available at the Propublica website[2].
5. **Credit**: The prediction of credit card clients dataset, with 24 features.
6. **Adult**: The adult income prediction dataset, with 14 features.
7. **German**: The Statlog german credit dataset that classifies clients as good or bad credit risks, with 20 features.
8. **Heart**: The heart disease dataset, with 75 features.

4.2 CF Generation Methods

A total of 9 state-of-the-art CF generation methods are implemented and briefly described below. For the GS, DiCE, FACE, and CCHVAE algorithms, the CARLA framework, a python framework for different CF generation algorithms presented by Pawelczyk et al. [15], is adapted for the respective datasets:

1. **Nearest Neighbor Tweaking (NN)**: Selects the closest training observation carrying the CF label as the CF [11].
2. **Actionable Feature Tweaking (FT)**: Selects the feature value changes based on tree-ensembles value thresholds for majority vote class flips [19].
3. **Random Forest Similarity Tweaking (RT)**: Combines the FT and NN methodology to provide a training justified CF in tree-ensembles. It trains a random forest and lists the training example frequencies in the leaf nodes where the instance of interest lies. The most frequent example is the CF [11].
4. **Minimum Observable (MO)**: Searches the dataset for the closest CF [22].
5. **Growing Spheres (GS)**: Grows spheres around the instance of interest until they reach the closest decision boundary to find a nearby CF [9].
6. **Diverse Counterfactual Explanations (DiCE)**: Outputs a set of diverse CFs for user selection, based on the idea that, for a given user, a single CF may not work because the user may not be able to implement the changes suggested by it. Therefore, DiCE provides diverse actionable CFs [14].
7. **Feasible and Actionable Counterfactual Explanations (FACE)**: Outputs a CF that considers the plausibility of the actions to reach it, where higher density paths in the dataset are more likely and plausible [17].

[2] https://www.propublica.org/datastore/dataset/compas-recidivism-risk-score-data-and-analysis.

8. **Model-Agnostic Counterfactual Explanations (MACE)**: A multi-objective algorithm that uses Satisfiability Modulo Theories (SMT) to achieve sparsity, proximity, and feasibility through logical formulae [6].
9. **Counterfactual Conditional Heterogeneous Autoencoder (CCH-VAE)**: uses a variational autoencoder to generate CFs in the latent space in likely data regions, according to the dataset distribution. It uses ϵ-justification to measure justification in the continuous latent space [16].

4.3 Performance Metrics

Five CF performance metrics are used to evaluate JUICE CFs and benchmark them against the state-of-the-art CF generation algorithms:

1. **Justification**: True - if the CF is connected to a correctly classified CF observation through the mixed-features justification verification, False - otherwise.
2. **Feasibility**: The intersection of mutability, directionality, and plausibility:

$$F_{CF} = (mutable_{CF} = 1) \wedge (direction_{CF} = 1) \wedge (plausible_{CF} = 1), \quad (2)$$

   F_{CF} outputs 1 if the obtained CF satisfies simultaneously all three conditions, otherwise it outputs 0.
3. **Sparsity**: 1 minus the ratio of changed features $n_{changed}$:

$$S_{CF} = \begin{cases} 1 - n_{changed}/n & \text{if}\, n_{changed} > 1 \\ 1 & \text{if}\, n_{changed} = 1 \end{cases}$$

4. **Proximity**: the euclidean distance between the CF and the instance of interest.
5. **Time**: Run time in seconds of the CF generation method.

4.4 Results and Discussion

This section presents the results of the empirical evaluation. First, the black-box models used for each dataset and their F1 score performance is shown. Then, the main results of the CF evaluation metrics are presented and discussed.

Black-box models and available test samples. We considered Decision Tree, Support Vector Machines, Random Forest (RF), and Multi-Layer Perceptron (MLP) as the black-box models. A grid search was performed to select the best-performing model. For the datasets used in this study, RF and MLP yielded the best performance and are used as the predictor f. Table 1 shows the selected models for each dataset with their F1 score. Some of the CF generation methods require a specific ML classifier, i.e., the RT and the FT algorithms require an RF. In these cases, an additional model was trained, achieving a similar classification performance. Finally, the number of instances of interest per dataset was obtained after balancing the datasets and considering only instances with undesired predicted classes. The number of instances of interest per dataset is reported in the last row of Table 1. Further details are found in the GitHub repo.

Table 1. Black-box models performance and available test samples.

Dataset	Disease	Compass	Adult	Heart	German	Credit	Ionosphere	Athlete
Model	RF	RF	RF	RF	MLP	MLP	MLP	MLP
F1	0.74	0.65	0.83	0.83	0.70	0.72	0.96	0.75
Instances	41	49	39	23	30	46	27	42

CF generation methods. The 5 metrics results are shown in Fig. 3, aggregated for all available instances of interest specified in Table 1. The proposed methods are the light and dark green bars, JUICEP (proximity version) and JUICES (sparsity version), respectively. Note that lower is better for Proximity and Time.

Regarding justification, JUICE always provides justified CFs, matched by NN and RT. This was expected since the latter select observations as CFs (justified by themselves). Notice that the justification property strongly depends on the dataset. Additionally, JUICES and JUICEP provide the highest feasibility, together with NN and MO. RT CFs are not always feasible because they may not comply with mutability or directionality.

In terms of sparsity and proximity, MACE is the best algorithm. Nevertheless, JUICES and JUICEP provide good performance, occupying the third and fifth places in sparsity and the fourth and fifth in proximity. This is close to the performance of DiCE (second) in sparsity and GS and RT (second and third) in proximity. DiCE is outstanding in sparsity because the CFs have a diverse, smaller number of changes. However, DiCE, as well as CCHVAE, do not perform well in terms of proximity because they prioritize likelihood as a proxy for faithfulness. Likely CFs imply a nearby CF closer to the center of the data distribution, which may increase the distances to the instance of interest [16].

The MACE optimization process takes the longest time for all datasets, in some cases being 1 or 2 orders of magnitude higher than the next best performing method. The JUICE algorithms ranks eighth and ninth in run time, a result which was expected due to the complexity that was previously discussed in Sect. 3.3. However, this is a good trade-off for guaranteeing justified CFs and attaining competitive performance in terms of the other explanation metrics.

To make the comparison easier, the Nemenyi test is run, and the CD plots are shown in Fig. 4. In terms of justification, even though NN, RT, and JUICE methods provide 100% justification, they are not statistically different from CCHVAE, but they are with respect to MACE, which performed best in proximity and sparsity. However, JUICE methods performed as good as MACE in both of these metrics. In terms of feasibility, NN, MO, MACE, and JUICE methods are the best, always providing feasible CFs, making them statistically better than CCHVAE and FACE, which are methods focused on the likelihood or path-adherent CFs. Regarding run time, MACE is by far the least favored (last place) but not statistically different from FT, JUICE methods, FACE, and CCHVAE.

These results indicate that JUICE derived methods are as good as the best performing methods in traditional CF metrics, such as proximity and sparsity,

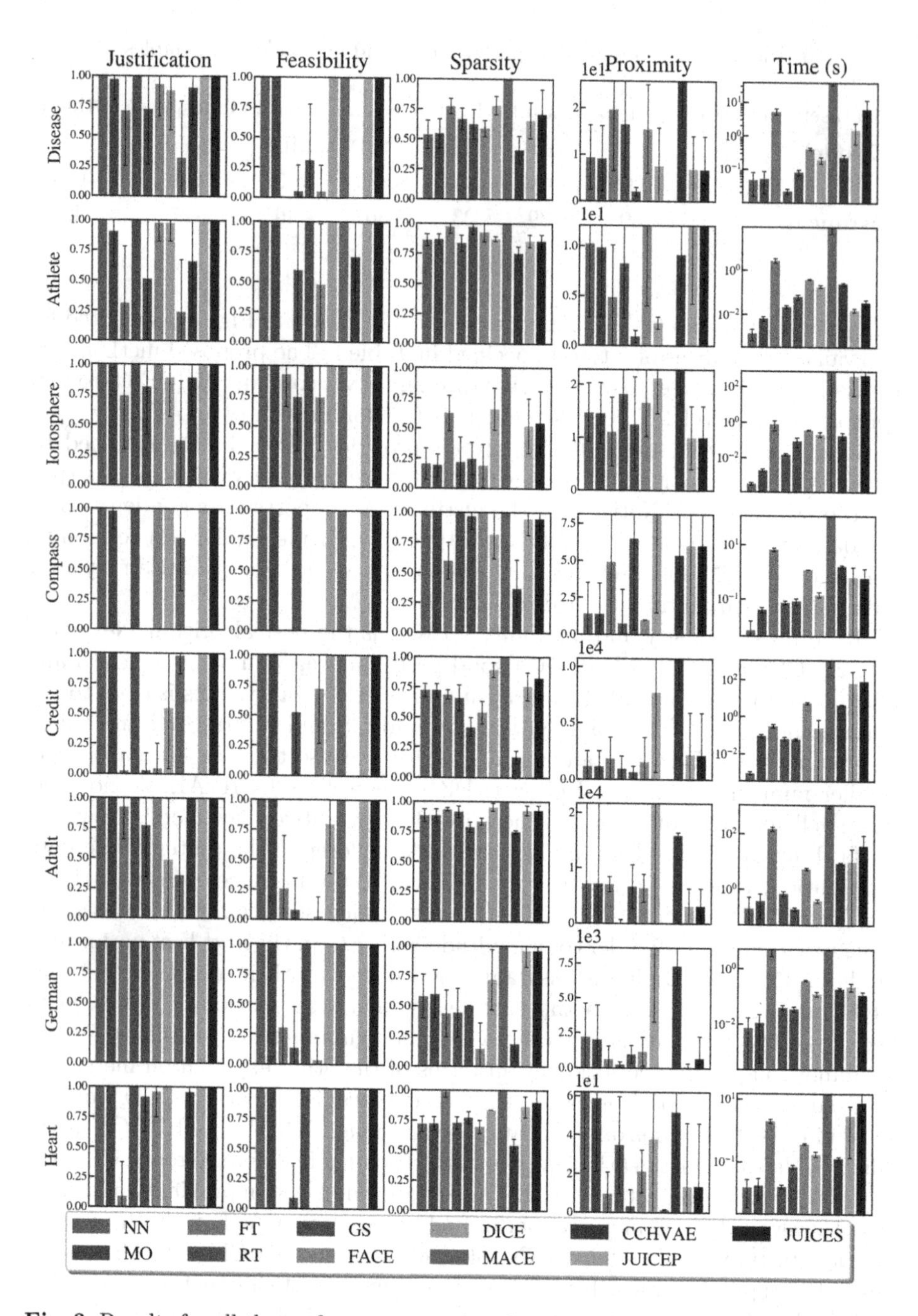

Fig. 3. Results for all the performance metrics, datasets, and methods. Rows of figures correspond to different datasets, while columns to different metrics. Each bar in the plots corresponds to one method. Note that the Time (s) plot (right column of bar plots) has a logarithmic scale in the y-axis (the lower the better, as in the proximity metric).

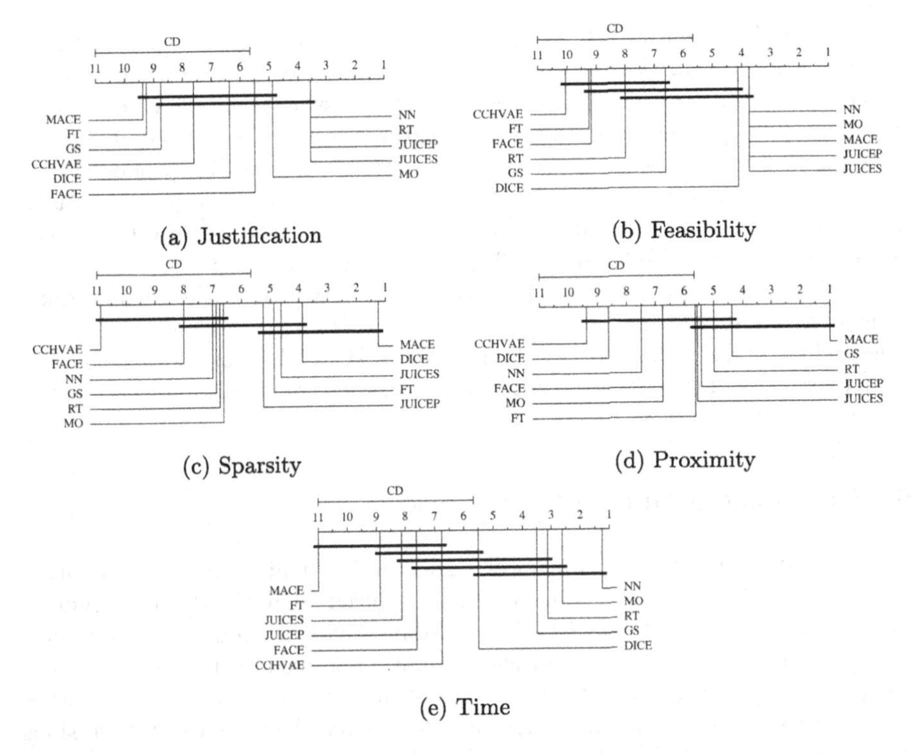

(a) Justification (b) Feasibility (c) Sparsity (d) Proximity (e) Time

Fig. 4. Different CD plots (Nemenyi tests) for the 5 performance metrics.

are the best in feasibility and justification, and performed slightly worse in relation to others in the run time metric, which is for CFs not as critical as proximity or faithfulness. Additional datasets may be included to strengthen the tests.

5 Related Work

There are a few intuitive approaches for faithful CFs, such as NN and RT [11]. Since the obtained CFs belong to the observed dataset, they are justified. However, they have two main problems: (1) proximity and sparsity are restricted to the closest observations (closer or sparser CFs may exist), and (2) there could be a mismatch between the model-predicted label and ground-truth label. Additionally, RT (as well as FT) only work in tree-ensemble models.

Likelihood also leads to faithfulness [8]. Dandl et al. [4] propose MOC to obtain likely CFs by analyzing the k-nearest observations without considering justification. Moreover, CCHVAE, introduced by Pawelczyk et al. [16], attempts to improve ϵ-connectedness in the continuous latent space. However, there is no guarantee of justification (even with high levels of CF likelihood). The method also suggests a trade-off between faithfulness and CF actionability difficulty, which is related to the CF proximity.

The GS algorithm introduced in Laugel et al. [9] is similar to the justification verification algorithm hereby proposed, but it does not take into account either the existence of multiple paths from one classification region to another (it only applies a greedy selection of the features to be changed one at a time) or the mixed-features scenario. Additionally, Mothilal et al. [14] propose DiCE that preserves the distance measures used in the original space to allow easier interpretation. However, both GS and DiCE may obtain unfaithful CFs, since neither considers CF likelihood or CF justification.

Furthermore, Poyiadzi et al. [17] propose the FACE algorithm, which takes actionability a step further and considers the problem of possible paths to reach a desired CF. Finally, Karimi et al. [6] propose MACE, which provides optimality in sparsity, proximity, and feasibility. However, these models do not necessarily guarantee justification.

6 Conclusions and Future Work

In this study, we proposed two novel algorithms that aim towards generating and verifying justified CFs. Firstly, we introduced JUICE, which guarantees the *justification* property on CFs in accordance with a trained model f on a mixed-features dataset. The empirical evaluation showed that JUICE achieves comparative performance with respect to state-of-the-art algorithms in other desiderata, such as proximity, sparsity, and feasibility. We presented two versions of JUICE, namely, JUICEP and JUICES, that prioritize proximity or sparsity based on the focus of the user. Additionally, we introduced the first model-agnostic justification verification algorithm to evaluate the justification of any CF obtained with respect to a given trained model f in a mixed-features scenario.

Future work should consider scalability and multi-connection to clusters of observations and their distance to the CF, combining both justification and likelihood as a stronger measure of CF faithfulness. Moreover, additional metrics, such as diversity, robustness, and fairness, should be considered, which are becoming increasingly important when providing actionable plans for end users. Finally, a natural extension is to adapt JUICE to image and temporal datasets to increase the utility of the algorithm and the power of the statistical tests.

References

1. Bobek, S., Nalepa, G.J.: Explainability in knowledge discovery from data streams. In: 2019 First International Conference on Societal Automation (SA), pp. 1–4. IEEE (2019)
2. Boer, N., Deutch, D., Frost, N., Milo, T.: Just in time: personal temporal insights for altering model decisions. In: 2019 IEEE 35th International Conference on Data Engineering (ICDE), pp. 1988–1991. IEEE (2019)
3. Byrne, R.M.: Counterfactuals in explainable artificial intelligence (XAI): evidence from human reasoning. In: IJCAI, pp. 6276–6282 (2019)

4. Dandl, S., Molnar, C., Binder, M., Bischl, B.: Multi-objective counterfactual explanations. In: Bäck, T. (ed.) PPSN 2020. LNCS, vol. 12269, pp. 448–469. Springer, Cham (2020). https://doi.org/10.1007/978-3-030-58112-1_31
5. Dodge, J., Liao, Q.V., Zhang, Y., Bellamy, R.K., Dugan, C.: Explaining models: an empirical study of how explanations impact fairness judgment. In: Proceedings of the 24th International Conference on Intelligent User Interfaces, pp. 275–285 (2019)
6. Karimi, A.H., Barthe, G., Balle, B., Valera, I.: Model-agnostic counterfactual explanations for consequential decisions. In: International Conference on Artificial Intelligence and Statistics, pp. 895–905. PMLR (2020)
7. Kyrimi, E., Neves, M.R., McLachlan, S., Neil, M., Marsh, W., Fenton, N.: Medical idioms for clinical Bayesian network development. J. Biomed. Inform. **108**, 103495 (2020)
8. Laugel, T., Lesot, M.J., Marsala, C., Detyniecki, M.: Issues with post-hoc counterfactual explanations: a discussion. arXiv preprint arXiv:1906.04774 (2019)
9. Laugel, T., Lesot, M.J., Marsala, C., Renard, X., Detyniecki, M.: Inverse classification for comparison-based interpretability in machine learning. arXiv preprint arXiv:1712.08443 (2017)
10. Laugel, T., Lesot, M.-J., Marsala, C., Renard, X., Detyniecki, M.: Unjustified classification regions and counterfactual explanations in machine learning. In: Brefeld, U., Fromont, E., Hotho, A., Knobbe, A., Maathuis, M., Robardet, C. (eds.) ECML PKDD 2019. LNCS (LNAI), vol. 11907, pp. 37–54. Springer, Cham (2020). https://doi.org/10.1007/978-3-030-46147-8_3
11. Lindgren, T., Papapetrou, P., Samsten, I., Asker, L.: Example-based feature tweaking using random forests. In: 2019 IEEE 20th International Conference on Information Reuse and Integration for Data Science (IRI), pp. 53–60. IEEE (2019)
12. Miller, T.: Explanation in artificial intelligence: insights from the social sciences. Artif. Intell. **267**, 1–38 (2019)
13. Molnar, C.: Interpretable machine learning: a guide for making black-box models explainable (2021). https://christophm.github.io/interpretable-ml-book/limo.html
14. Mothilal, R.K., Sharma, A., Tan, C.: Explaining machine learning classifiers through diverse counterfactual explanations. In: Proceedings of the 2020 Conference on Fairness, Accountability, and Transparency, pp. 607–617 (2020)
15. Pawelczyk, M., Bielawski, S., Heuvel, J.v.d., Richter, T., Kasneci, G.: CARLA: a python library to benchmark algorithmic recourse and counterfactual explanation algorithms. arXiv preprint arXiv:2108.00783 (2021)
16. Pawelczyk, M., Broelemann, K., Kasneci, G.: Learning model-agnostic counterfactual explanations for tabular data. In: Proceedings of The Web Conference 2020, pp. 3126–3132 (2020)
17. Poyiadzi, R., Sokol, K., Santos-Rodriguez, R., De Bie, T., Flach, P.: Face: feasible and actionable counterfactual explanations. In: Proceedings of the AAAI/ACM Conference on AI, Ethics, and Society, pp. 344–350 (2020)
18. Rudin, C.: Stop explaining black box machine learning models for high stakes decisions and use interpretable models instead. Nat. Mach. Intell. **1**(5), 206–215 (2019)
19. Tolomei, G., Silvestri, F., Haines, A., Lalmas, M.: Interpretable predictions of tree-based ensembles via actionable feature tweaking. In: Proceedings of the 23rd ACM SIGKDD International Conference on Knowledge Discovery and Data Mining, pp. 465–474 (2017)

20. Verma, S., Dickerson, J., Hines, K.: Counterfactual explanations for machine learning: a review. arXiv:2010.10596 (2020)
21. Wachter, S., Mittelstadt, B., Russell, C.: Counterfactual explanations without opening the black box: automated decisions and the GDPR. Harv. JL Tech. **31**, 841 (2017)
22. Wexler, J., Pushkarna, M., Bolukbasi, T., Wattenberg, M., Viégas, F., Wilson, J.: The what-if tool: interactive probing of machine learning models. IEEE Trans. Vis. Comput. Graph. **26**(1), 56–65 (2019)

Explaining Siamese Networks in Few-Shot Learning for Audio Data

Andrea Fedele, Riccardo Guidotti(✉), and Dino Pedreschi

University of Pisa, Pisa, Italy
a.fedele7@studenti.unipi.it, {riccardo.guidotti,dino.pedreschi}@unipi.it

Abstract. Machine learning models are not able to generalize correctly when queried on samples belonging to class distributions that were never seen during training. This is a critical issue, since real world applications might need to quickly adapt without the necessity of re-training. To overcome these limitations, few-shot learning frameworks have been proposed and their applicability has been studied widely for computer vision tasks. Siamese Networks learn pairs similarity in form of a metric that can be easily extended on new unseen classes. Unfortunately, the downside of such systems is the lack of explainability. We propose a method to explain the outcomes of Siamese Networks in the context of few-shot learning for audio data. This objective is pursued through a local perturbation-based approach that evaluates segments-weighted-average contributions to the final outcome considering the interplay between different areas of the audio spectrogram. Qualitative and quantitative results demonstrate that our method is able to show common intra-class characteristics and erroneous reliance on silent sections.

Keywords: Explainable AI · Siamese Networks · Audio Data

1 Introduction

In recent years, Artificial Intelligence (AI) significantly sped up its pace thanks to the availability of large datasets, the emergence of powerful computing devices, and the development of sophisticated algorithms [10]. Machine Learning (ML) models have been widely employed, achieving significant results in different fields such as computer vision and audio signal processing. Despite their thriving achievements, traditional ML systems rely on learning from big and large-scale datasets, while real-world applications typically involve constraints that might lead to a limited fuel of samples. More importantly, current techniques fail when asked to generalize rapidly from few samples only. On the contrary, humans are capable of learning by quickly generalizing on their prior knowledge. For example, if a child is presented with a few pictures of a person he has never seen before, he will still be able to match and identify the right individual among a reasonable number of pictures portraying different subjects.

To overcome these limitations, recent studies proposed *few-shot learning* (FSL) frameworks where a ML model must learn to classify unseen classes given

P. Pascal and D. Ienco (Eds.): DS 2022, LNAI 13601, pp. 509–524, 2022.
https://doi.org/10.1007/978-3-031-18840-4_36

only few samples per class [23]. FSL methods typically consider a *C-way k-shot* classification task where C represents the number of classes the model is asked to classify from, while k is the number of labeled samples per each of these classes. Such parameters define the *support set* dimension, which is an auxiliary set of data the classifier is provided with to be guided in its verdict. One-shot learning is a special case of FSL where exactly one instance of the target class is present in the training set, while zero-shot learning aims to predict the correct class of a given instance without being previously exposed to such class. Different algorithms have been proposed to tackle few-shot metric-learning [13,19,20,22]. Siamese Networks (SNs) [13] make use of two identical sub-networks that map the inputs into an embedding space where a distance function is later employed to calculate the distance between such embedded representations. Based on such distance, a similarity score is finally computed for the given inputs. The capability of SNs intrigued scientists to verify their robustness on audio inputs. In [12], the authors demonstrate that SNs work well on audio inputs, producing good results when asked to generalize on unseen classes.

A big issue of SNs is the lack of explainability. Understanding the reason why a model takes a specific decision is hugely important to developers, organizations, and end-users. In recent years, researchers examined the eXplainable Artificial Intelligence (XAI) topic from various perspectives [3,9]. A possible characterization of XAI techniques distinguishes between *gradient-based* and *perturbation-based* approaches [3,9]. While both aim to understand the contribution that each input feature has on a specific outcome, they solve the problem differently. Gradient-based approaches estimate feature contribution by means of forward and backward propagation passes throughout the network. On the other hand, perturbation-based methods perturb the input and measure the output changes with respect to the original input. Different XAI methods have found very good feedbacks in the research community, but few techniques have been presented to interpret and explain SNs. Moreover, to the best of our knowledge, none of the available techniques has ever been tested on audio inputs processed by SNs in the context of *C-way one-shot learning.*

In this paper, we propose a local perturbation-based method to explain Siamese Networks in the context of *C-way one-shot* learning on audio input data. Our *SIamese Networks EXplainer* (SINEX) seeks to expose the discriminative features for a ML model that is asked to quickly learn and generalize like humans do. We want our explainers to answer questions such as: *What is the model listening to when it correctly matches the class of two audio it has never heard before? Why is a given recording more similar to a specific audio than it is to others? Why is the model miss-classifying a given audio?* SINEX uses a perturbation approach evaluating *segment-weighted-average* contribution values to the final outcome. A *coalation-based* variant of SINEX (SINEXC) is also presented. SINEXC uses a similar approach that also considers the interplay between different areas of the input as a whole. The contribution values can then be visualized as heatmaps to have a intuitive and user-friendly idea of the SN behavior. We employ SINEX to explain SNs reaching state-of-the-art performance on both 1-

second and 5-seconds long recordings in the context of *5-way one-shot* learning. Our results illustrate that class-homogeneous datasets seem to lead to robust networks, while class-heterogeneous ones result in more inaccurate classification behaviors. SINEX also brings to light an erroneous reliance on silenced areas that, in some cases, is the cause of miss-classification errors.

The paper is organized as follows. Related works are reviewed in Sect. 2. In Sect. 3 we formalize the problem we face, while in Sect. 4 we illustrate our proposals. Section 5 reports the experimental results and our main findings. Finally, Sect. 6 summarizes our contributions and open research directions.

2 Related Works

In this section we illustrate the few existing works to explain SNs. In [26], SNs are employed in a query-by-vocal-imitation retrieval system. The last section of [26] focuses on visualizing and *sonifying* the input patterns that maximize the activation of certain neurons, using the activation maximization approach. By visualizing the patterns that maximize the activation of random neurons from each layer, the authors suggest that it is possible to have a glimpse of what the network believes is important. Unfortunately, the fact that only random neurons are considered might cause an isolated effect w.r.t. the overall job of the SN. Also, it is not possible to comprehend whether the returned features have a positive or negative impact on the outcomes. A similar approach is described in [1], where the system is developed in the context of one-shot learning to identify bird species. The authors visualize how audio spectrograms [7] are decomposed by each layer and consider the last one as the explanation layer. They conclude that the most distinctive feature is the distribution of the signal's energy in species-dependent frequency bands. Both in [26] and [1], the explanation is derived only from the convolutional encoders, bypassing the SN core similarity scoring layer.

In [21], similarly to [8,14], a special auto-encoder is used as core to the explanation algorithm. First, the encoder is trained to reconstruct the input instances of the training set starting from the embedded representation given by the SN's built-in embedder. Then, the decoder is trained to reconstruct the original input given the hidden representation. Once the auto-encoder is trained, a pair of inputs to explain is given as input both to the SN and the auto-encoder. The vectors resulting from the SN's encoders get perturbed on what the authors refer to as *important features*. Such features are chosen considering the smallest distance if the two inputs are semantically close, while the biggest distance is considered otherwise. The resulting perturbed vectors are submitted to the decoder, which maps them back to the original input space. The embedded vectors get randomly perturbed, and the mean contribution value of each feature is measured as the difference between its value in the reconstructed input after perturbation and its value in the reconstructed input without perturbation. Limitations of this approach are the large number of tuning parameters, including the choice of the number of important features, and the need for a large number of data to train the auto-encoder. Moreover, this algorithm requires to train an

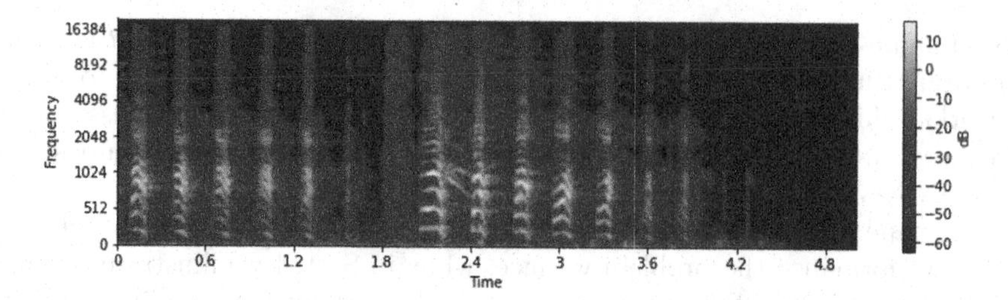

Fig. 1. Log-mel spectrogram of an audio containing a person laugh. Darker areas represents lower dBs, i.e., silence, lighter pixels represent sounds audible by human hear.

additional ML model that needs access to the training set. In [5], another post-hoc explanation approach for SNs is proposed. The authors argue that unseen instances in the support set might make existing perturbation-based XAI methods over-sensitive to irrelevant perturbations. To control such possible superfluous variations, the authors find global invariant salient features for individual objects using self-supervision. Then, they formulate an optimization problem to adapt the global salient features to explain a SN prediction for an input pair. Results on tabular and graph data show that their explanations robustly respects the self-learned invariance. Unfortunately, also these approaches [5,21] do not account for the similarity scoring layer but only operate on the data embedded representation. Lastly, we mention a different approach explored in [24], where a *class-to-class SN* (C2C-SN) is trained to learn patterns of both similarity and difference between classes. The authors demonstrate the use of C2C-SNs for explanation purposes by means of prototypical case finding and contrastive cases. Differently from our work, [24] does not explore *C-way k-shot* learning tasks, and it does not query the model on unseen classes.

The explanation approach we propose differs from the ones employed in [1,5,26] since they use *gradient-based* methods that tend to focus on the encoder part of the SNs alone. These works limit their exploration to the last convolutional layer of the network, not considering the SN architecture as a whole. Also, differently from [21], we do not train any additional model. Moreover, our explanation method does not need access to the training data, and it can be deployed directly at prediction time. The main goal of our project is to explain SNs in their entirety, focusing on the layer responsible for the final similarity score. Moreover, our proposal addresses the problem of explaining SNs in the context of *C-way one-shot* learning which is not tackled by any work in the literature.

3 Problem Formulation

An audio dataset $D = \langle X, y \rangle$ in the FSL setting is composed by a set of n tracks $X = \{x_1, \ldots, x_n\}$ such that it is always possible to represent each track x_i as a matrix in $\mathbb{R}^{p \times q}$ where each x_i is the spectrogram [1,7] of the ith track in the dataset, q is the length of the track and p the number of frequencies observed. Thus, $x_{i,j,k}$ indicates the intensity value at time j of the kth frequency for the ith track. The intensity is typically expressed in terms of decibel (dB). An example of spectrogram is shown in Fig. 1. The vector $y = \{y_1, \ldots, y_n\}$ indicates the class y_i associated to x_i with $y_i \in [0, \ldots, L-1]$ and L is the total number of classes. Typically, L is a high number if compared to traditional multi-class problems. L is greater or equal than 50 in our experiments. In the context of *C-way one-shot learning* where we pose our analysis, the support set S is composed by C different tracks where each belongs to a different class and only appears once in S. We highlight that typically $C \ll L$. We refer the interested reader to [13,23] for further details on how SNs are trained.

In such learning framework, a SN is a deep learning model f that takes as input a support (or reference) set $S = \langle\{s_1, \ldots, s_C\}, \{y_1, \ldots, y_C\}\rangle$ containing tracks $s_i \in \mathbb{R}^{p \times q}$ and their corresponding class labels y_i, a query instance x for which the associated class label is unknown, and predicts the class label y_i for x by comparing x with each $s_i \in S$ with respect to a learned similarity function sim [13] to finally select the highest of the obtained scores, i.e.,

$$y_i = f(x, S) = \underset{\forall s_i, y_i \in S}{\arg\max}\ sim(x, s_i)$$

The problem we aim to address in this paper is defined as follows. Given a pretrained SN f, our objective is to define a post-hoc local explanation method g such that, taken as input f, S and a query instance x, g returns an explanation E that unveil what f listened to in order to assign the class y_i to x. More formally, our objective is to define a function g such that $E = g(f, x, S)$. We formalize the explanation E as a set of heatmaps for each track in S, i.e., $E = \{h_1, \ldots, h_C\}$ where h_i is the heatmap of the track $s_i, y_i \in S$, and the value $h_{i,j,k}$ indicates the importance/saliency at time j of the kth frequency for the ith support track.

4 Siamese Network Explainer

In this section, we present our local post-hoc SIamese Networks EXplainer SINEX. SINEX implements the explanation function g w.r.t. the problem formulation above. The architecture of the SNs used for audio spectrograms comprehends two main parts: two identical convolution-based encoding sub-networks, and the final layer computing the distance between the encoded vectors to generate their similarity score. Employing a *gradient-based* explanation technique on the isolated sub-networks would only highlight how they work when reducing inputs in different convolution stages, but it would bypass the final layer's impact on the similarity score prediction. Differently from the existing proposal in the literature (see Sect. 2), we propose to explain the SN outcome by generating an explanation based on the layers contributing to the final similarity score. Hence, we decided to follow a *perturbation-based* approach measuring the outcome similarity prediction after different input perturbations.

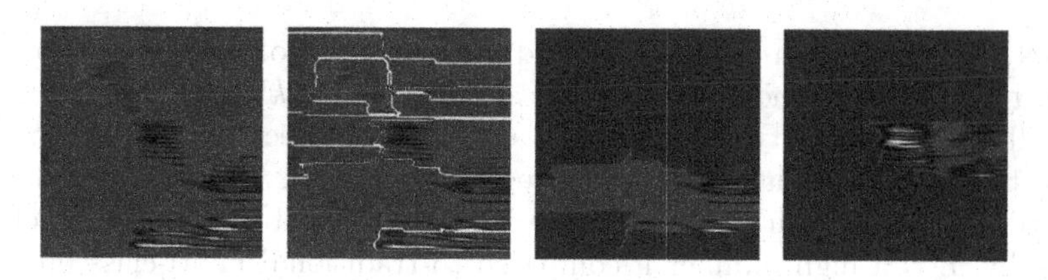

Fig. 2. Perturbation procedure overview. From left to right: support set sample spectrogram, segmentation on 12 regions, perturbation on the 3rd and 5th segment.

The first aspect to consider is how to perturb a spectrogram. As a first attempt, we inspected a window-occlusion-based approach [25]. The main limitation of occluding with a fixed size shape is that the features contribution values might vary significantly for different window sizes. Preliminary experiments with scrolling windows of shapes from 1×1 to 50×50 showed that both smaller and bigger window dimensions tend to result in contribution values equally important across the whole spectrum, without being able to discriminate between different portions of it. Defining the right window size might become a hard task since the same window might have various results even on different instances of the same class label. Even worst, using fixed-length segments to divide the input w.r.t. the time or the frequency axes individually would imply a direct bond between specific class instances and their segmentation. Thus, relying deeply on time segmentation alone would mean assuming that every sound event starts and end at the same moment in time in a given recording. In real-world applications, it is unimaginable to assume that the actual sound event occurs at the exact same time under different recording takes.

A spectrogram can be intended as an image due to its matrix structure, but it should not be confused as one. In fact, it is crucial to expand it by means of an additional channel dimension so that the spectrogram can be used as "image-like" input data with common convolution-based architectures. Therefore, we propose to segment audio data by means of techniques typically used for image inputs. Examples are the Felzenszwalb approach [6] which computes the Felsenszwalb's graph-based image segmentation using a minimum spanning tree-based clustering on the image grid, and SLIC [2] which, instead, segments images using k-Means clustering. Unfortunately, the single-fake-channeled nature of audio inputs deriving from such augmentation procedure precludes the possibility of experimenting with other segmentation algorithms. For this reason, Felzenszwalb and SLIC are the algorithms we decide to test since they can operate on grey-scale images.

Considering the *C-way one-shot* setting, we have to specify what we want to perturb between the query instance x and the instances in the support set S to get the explanation. Since x is classified w.r.t. the instances in S, we decided to segment and perturb the instances $s_i \in S$ and analyze how the estimation of the similarity between x and s_i changes when parts of s_i are hidden to the SN.

Algorithm 1: SINEX(f, x, S)

```
   Input  : f - Siamese Network, x - instance to explain, S - support set
   Param  : seg - segmentation algorithm, γ min nbr. segments
   Output: E - explanation
1  E ← ∅;                                          // init. explanation
2  for s_i ∈ S do                                  // for each support sample
3  |   v ← sim(x, s_i);                            // calculate similarity
4  |   R ← seg(s_i, γ);                            // apply segmentation
5  |   h_i ← 0^{p×q};                              // init. sample contribution
6  |   for r_j ∈ R do                              // for each segment
7  |   |   z_i ← (s_i[¬r_j] ← 0);                  // perturb sample
8  |   |   u ← sim(x, z_i);                        // calculate similarity
9  |   |   δ ← v − u;                              // compute segment similarity delta
10 |   |   h_ij ← δ/|r_i|;                         // weight and update contribution
11 |   E ← E ∪ {h_i};                              // add heatmap to explanation
12 return E;
```

We leave as a future study the analysis of our method in the opposite setting where the segmentation and perturbation are applied on the query instance x.

Algorithm 1, illustrates SINEX as our proposal for implementing the function g to explain a SN f w.r.t. a query instance x and a support set S. For every sample of the support set (lines 2–11), SINEX measures the SN similarity v between the query input x and the support sample s_i before perturbation (line 3). Then, it segments the support set sample using the segmentation algorithm *seg* (line 4), and obtains a set of segments R containing at least γ segments. For each segment r_j its contribution is computed as follows and stored in the saliency map h_i that models the importance of the various areas of support sample s_i. Saliency maps are initialized with a null contribution, i.e., with a matrix with value 0 and having the same dimensionality of the support sample spectrogram (line 5). First, the support set sample s_i is perturbed (line 7) by obscuring/silencing everything except the region r_j. With the notation $s_i[\neg r_j] \leftarrow 0$ we are indicating the fact that, the support sample s_i is taking the value 0 in all regions except r_j. The value 0 in Algorithm 1 line 7 symbolizes a default value. Since in our setting we are dealing with audio inputs, obscuring a segment means setting its value to -80, as this is the smallest value in the dB scale. After that, the new similarity outcome u is computed, pairing the query sample x with the perturbed version of the support set spectrogram z_i (line 8). The difference δ between the starting similarity score v and the one resulting after-perturbation v is finally computed (line 9). Such difference is then weighted w.r.t. the current segment size $|r_j|$ and updated in the corresponding saliency map h_i (line 10). Examples of the perturbation procedure on a sample instance are shown in Fig. 2.

This version of SINEX can suffer from some known problems of perturbation-based methods. First of all, when we perturb instances, we might generate *out-of-distribution* (OOD) data point. Since well-trained ML models generalize correctly on new samples as long as they belong to a known distribution, we have no guarantee on the significance of the similarity measures w.r.t. implausible instances. A solution to this problem is to re-train the model on a dataset that includes the perturbed data points, accepting the compromise of the addi-

Algorithm 2: SINEXC(f, x, S)

Input : f - Siamese Network, x - instance to explain, S - support set
Param : seg - segmentation algorithm, γ min nbr. segments, α - per-segment coalitions, β - per-coalition active segments
Output: E - explanation

```
1  E ← ∅;                                    // init. explanation
2  for s_i ∈ S do                            // for each support sample
3      v ← sim(x, s_i);                      // calculate similarity
4      R ← seg(s_i, γ);                      // apply segmentation
5      h_i ← 0^{p×q};                        // init. sample contribution
6      for r_j ∈ R do                        // for each segment
7          Π ← perturb(s_i, α, β);           // make α coalition with β active segments
8          ū ← 0;                            // sums of segment similarities
9          for π_k ∈ Π do                    // for each coalition
10             z_i ← (s_i[¬r_j ∧ ¬π_k] ← 0); // perturb sample
11             ū ← ū + sim(x, z_i);          // calculate similarity
12         δ ← v − ū/|Π|;                    // compute segment similarity delta
13         h_ij ← δ/|r_i|;                   // weight and update contribution
14     E ← E ∪ {h_i};                        // add heatmap to explanation
15 return E;
```

tional time resources needed. Secondly, we have to consider the *isolated effect* such techniques might lead to. Measuring the prediction changes of singular segment perturbations might help us understand how that segment is contributing to the final outcome, but it will disregard it completely from the interplay it has with the remaining input areas. We take into account both problems as follows. Considering the FSL context and the fact that we evaluate the model on new classes, we assume SNs robustness to OODs since unseen sample might be considered OOD themselves. Despite this benefit, the nature of audio inputs might still play an important role when a spectrogram from an unknown distribution is compared with a perturbed sample belonging to that same distribution. Our proposal to mitigate these downsides is to consider how a specific segment contributes to the final outcome by considering its *average prediction* value. Inspired by SHAP [15], we want such value to consider the interplay between the segment in analysis and the remaining others. However, differently from *SHAP*, our context does not allow us to compute "baseline values" w.r.t. the training set data. Since our goal is to design an explanation method that is independent from its training data, we extend SINEX as described in Algorithm 2 that reports the SINEXC version, i.e., SINEX with Coalitions.

In SINEXC, to measure the interplay between each segment and the remaining areas of the spectrogram, we introduce the parameter α to control the number of per-segment coalitions to generate, and the parameter β to control the overall number of segments that must remain active in each coalition. Through this approach, we take into account not only how a specific segment influences the outcome but also how its interplay with other areas of the spectrogram impacts the final similarity score. For each coalition (lines 9–11, Alg. 2), we store the sum of the similarity scores obtained querying the model in $\bar{u}$. Then we compute the segment contribution δ (line 12, Alg. 2) as the difference between the original similarity score v and the mean similarity value $\bar{u}/|\Pi|$ obtained w.r.t. the $|\Pi|$

coalitions. Finally, the *segment-average prediction* value is weighted according to the segment size (line 13, Alg. 2). This process is iterated over each pair of a C-way one-shot batch keeping fixed the query sample and applying such coalition-based methodology to each s_i element of the support set S.

As a result of the run of SINEX and SINEXC, the explanation E contains the saliency maps $\{h_1, \ldots, h_C\}$ for each support set sample $\{s_1, \ldots, s_C\}$ indicating the contribution values per each segment within its own spectrogram segmentation. Each saliency map can be visualized as a heatmap having the same size as the input spectrogram. Therefore, we can visualize such heatmaps showing how each segment is influencing - either positively or negatively - the final similarity score outcome. Finally, to completely target the problem of explaining the *C-way one-shot* classification task as a whole, for visualization purposes, we employ a common representation scale where the contribution values are considered in terms of their absolute values and normalized in the scale $[-T, +T]$ where T is the maximum absolute value among all the contributions in E.

5 Experiments

We report here the experiments carried out to validate SINEX and SINEXC. We implemented SINEX and its coalition variation SINEXC in Python[1]. After some preliminary tests using Felzenszwalb and SLIC segmentation algorithms[2], we decided to select the Felzenszwalb approach for the experiments reported in the following using with the setting: $scale$ = 50 and $sigma$ = 1.5. The minimum number of segments γ parameter has been analyzed in the ranges $[50, 1000]$ and $[100, 1000]$ with a step of 50 and 100 for SINEX and SINEXC, respectively. SINEXC additional parameters setting was set as $\alpha = 200$ and $\beta = 0.15$[3].

Datasets and Siamese Networks Models. We experimented with two datasets[4]: AudioMNIST and ESC-50. *AudioMNIST* is composed of 30k recordings of spoken digits (0–9) in English. Each digit is repeated 50 times for each of the 60 different speakers, which are divided into 12 females and 48 males. Due to the high number of available speakers, we used this dataset to pursue a *speaker recognition* task creating three disjoint sets: the training set was composed of 50 classes, while the remaining 10 were divided equally between validation and test set. The *ESC-50* dataset is a collection of 2k annotated 5-second audio clips divided into 50 different classes and 40 repetitions per class. In this case, we decided to pursue an *environmental audio classification* task. We split the dataset to have 40 training class, 5 validating and 5 testing ones. We pre-processed both datasets with `librosa` to extract their log-mel spectrogram representation[5] which is commonly used for audio classification tasks [11,17].

---

[1] Code available at https://github.com/andreafedele/SINEX.

[2] https://scikit-image.org/.

[3] Higher α increases the execution time without revealing variation in the segment contribution values. Higher β tends to consider all segment as equally important the more β is closer to 0.5.

[4] https://github.com/soerenab/AudioMNIST, https://github.com/karolpiczak/ESC-50.

[5] `librosa`: https://librosa.org/. AudioMNIST tracks were down-sampled to 41kHz and zero-padded to have vectors of equal length. Then, each track was converted using an FFT window size of

Table 1. 5-way one-shot mean test accuracy scores.

	AudioMNIST						ESC-50					
Class	04	56	55	27	46	Avg	Glass Breaking	Church Bells	Frog	Laughing	Door Wood Creaks	Avg
Accuracy	.93	.91	.88	.82	.78	.86	.99	.93	.91	.82	.71	.87

In line with [1,12,13,26], we adopted the following structure for the architecture of the SNs for both datasets. The overall architecture is formed by two convolution-based encoders, followed by a distance layer and a final output scoring layer. The two encoders share the same architecture and weights, which are updated simultaneously during training so to result in identical embedding subnetworks. Each convolution-based encoder is composed of three 2D-convolution blocks with 64, 32, 12 filters, and kernel size 5×5, 5×5, 3×3, respectively. Each convolution layer is followed by a max-pooling layer. Finally, a fully connected layer of 4096 units takes the input back to a 1-dimension form. The square distance operates as a distance layer between the two encoded inputs, and a final fully connected layer composed of only 1 unit calculates the similarity score by means of a sigmoid activation function. Both SNs were trained using a binary cross-entropy loss function. The maximum number of training epochs was set to 5000, with the model performance being evaluated every 100 epochs on 300 random 5-way 1-shot tasks. For both datasets, the training procedure stopped because no 5-way 1-shot mean accuracy improvement was recorded after 10 consecutive evaluation runs. The SN working on AudioMNIST reached its best mean 5-way one-shot accuracy after 900 training epochs with a value of 0.83, while the SN using the ESC-50 dataset needed 1900 epochs to obtain a score of 0.86. Further details on the SNs architecture and the training procedure are accessible in the technical report available in the repository.

We highlight that the SNs for both datasets are validated and tested on sets composed of unseen classes[6]. We report in Table 1 the mean 5-way one-shot accuracy for each class of both datasets. Results show that the SN on ESC-50 reaches accuracy higher than 90% on 3 out of 5 classes, but the *Door Wood Creaks* class lowers the overall result with a mean accuracy of 71%. Such value is smaller than the smallest result in AudioMNIST, which scores accuracy values that seems to be better distributed among the 5 classes. We highlight that these performance are better than state-of-the-art SNs for the ESC-50 dataset [12].

Qualitative Evaluation. Fig. 3 and 5 show examples of explanation returned by SINEXC for AudioMNIST and ESC-50, respectively. The visualization color of our explanations is affected by a diverging light blue to pink colormap inspired by [15]. Light blue areas represent segments of negative influence on the SNs

4096, hop length of 197 samples, and 224 mel-bands. Such a process leads to spectrograms of sizes equal to 224×224. For ESC-50, the sampling rate was 44.1kHz, and no down-sampling was applied. We converted tracks of ESC-50 using a FFT window size of 2048, hop length of 512 samples, and 128 mel-bands. The dimensions of the resulting spectrogram is 128×431.

[6] Not even one sample of the test classes was ever seen during training, leading to a situation which is commonly referred as zero-shot learning. Despite this, in this paper, we use the term C-way one-shot learning to indicate that the additional support set is composed of a singular sample for each of the C classes. We are therefore asking the model to zero-shot the right classification of an unseen query class by providing exactly one other sample of that same class in the support set.

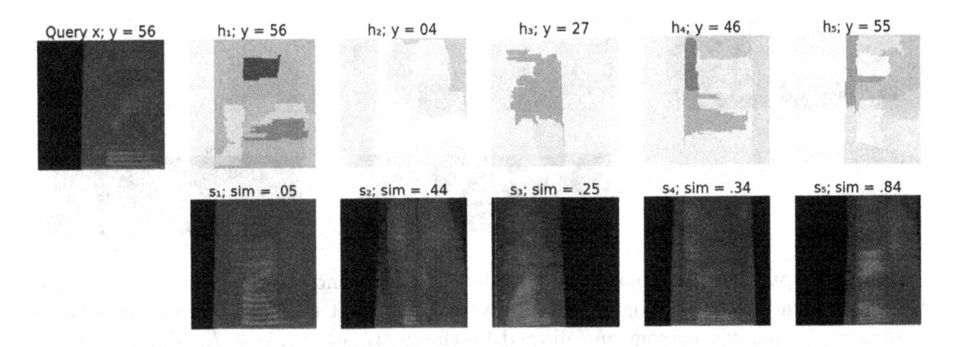

Fig. 3. SINEXC explanation with Felzenszwalb and $\gamma = 900$ of a correct 5-way one-shot classification on speaker 56 (female) for AudioMNIST. From left to right, the top line shows the query sample x spectrogram followed by the heatmaps $h_i \in E$ related to their $s_i \in S$ spectrograms. Each s_i is shown in the bottom line right below its heatmap.

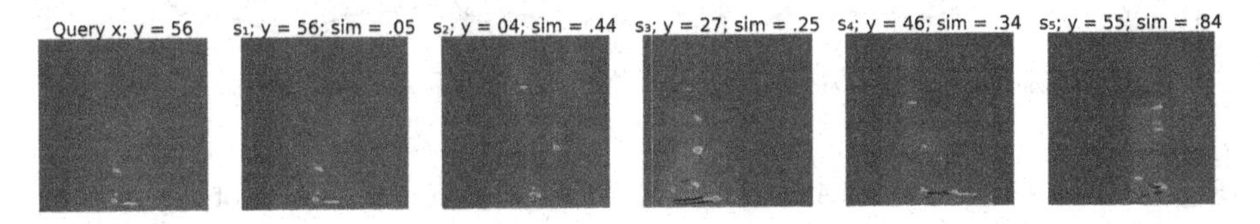

Fig. 4. GRAD-CAM heatmaps highlighting important pixels for the SNs encoders. Dark-violet areas indicate non influential pixels, red colors indicate influential ones.

similarity outcome, while pink portions indicate segments that positively affect the similarity score. White areas are instead neutral to the SNs classification process. In Fig. 5 we only show heatmaps for the two support set samples scoring the highest similarity scores to respect writing space constraints. The same AudioMNIST query sample x and support set S data used to generate the SINEXC explanation in Fig. 3, were also inspected with the Gradient-weighted Class Activation (GRAD-CAM) technique [18] in Fig. 4. We remind the reader that GRAD-CAM only focuses on the encoder part of the SNs architecture, limiting its exploration to the last convolutional layer of the encoder itself.

SINEXC explanations on AudioMNIST illustrate that the correct classification of female speaker recordings is mainly due to medium-high frequency segments, while their miss-classification depends primarily on segments that reside at the very bottom of the frequency range. A symmetrical behavior is observed for male speakers' audios: a correct classification is usually based on lower frequency values, while incorrect classifications are generally due to segments higher in the frequency spectrum. Reliance on silent-areas is also exposed by SINEXC on AudioMNIST. Experiments with GRAD-CAM show that encoders typically focus on the same frequency areas between different samples of a given class. Generally, heatmaps are located in spoken areas of the spectrogram, and they rarely show up in silent areas. Differently from SINEX and SINEXC, for GRAD-CAM we observe few important pixels which span vertically across the whole spectrum, highlighting the fundamental frequencies and their harmonics as verified in [4].

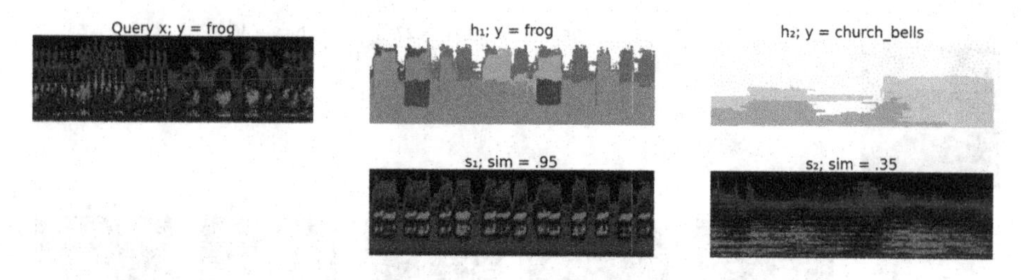

Fig. 5. Excerpt of SINEX explanation with Felzenszwalb and $\gamma = 900$ of a correct 5-way one-shot classification on frog class for ESC-50. From left to right, the top line shows the query sample x spectrogram followed by the heatmaps $h_1, h_2 \in E$ related to their $s_1, s_2 \in S$ spectrograms. Each spectrogram is in the bottom line below its heatmap.

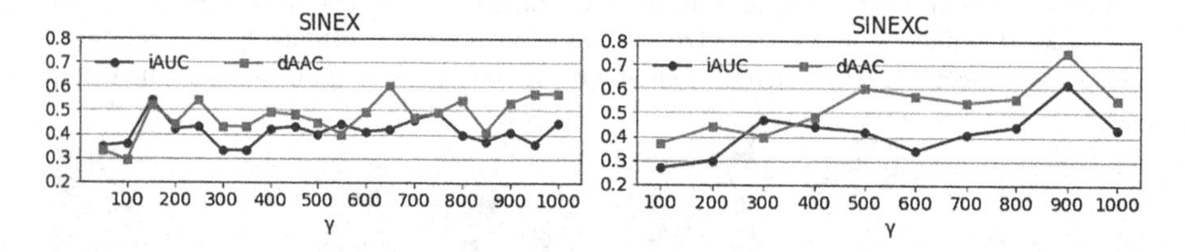

Fig. 6. Mean $iAUC$ and $dAAC$ on AudioMNIST varying min number of segments γ.

The application of SINEXC on ESC-50 extracts different insights. Analyzing correct and incorrect classifications for *Laughing* and *Door Wood Creaks* classes, we explored the SN inability to discriminate on medium-high frequencies between these classes. Experiments also led us to think that the decay between sound events plays a big role, especially if such events are repeated frequently in the overall recording. Correct classification of the *Frog* label happens when the frog croaks are well separated, while correct *Church Bells* classification happens if a delay is present and clear between the distinct sound events.

The applications of SINEXC unveil a strong dependence on the recordings domain. AudioMNIST is a class-homogeneous dataset, while ESC-50 presents a much higher class-heterogeneousness due to the different sources and recording environments. Such collection brings to spectrograms of very different morphology, despite them being labeled as belonging to the same class. These insights were possible thanks to the explanations performed through SINEXC, while GRAD-CAM does not allow to perform such reasoning.

Quantitative Evaluation. To assess the qualitative significance of the explanations found by SINEX and SINEXC, we followed the methodology described in [16] computing the insertion/deletion scores by increasingly inserting/deleting the most influential pixels returned by our explainers from an empty/full spectrogram, respectively. In particular, we expect the insertion curve to rapidly increase after only a small percentage of pixels are inserted, resulting in a large insertion-area-under-curve $iAUC$. In a dual manner, we expect the deletion curve to decrease rapidly after only a few pixels removal operations, therefore, resulting

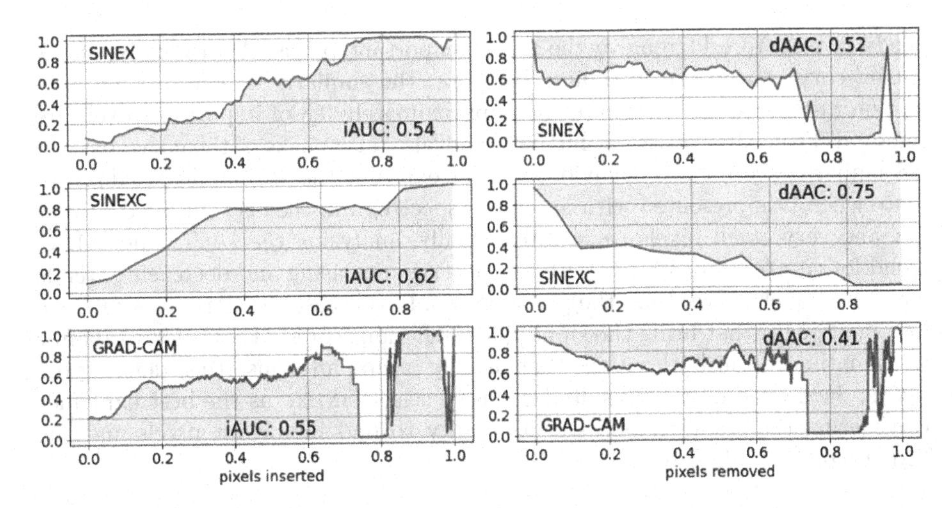

Fig. 7. *iAUC* and *dAAC* curves for SINEX (1st row), SINEXC (2nd row), and GRAD-CAM (3rd row) vary the percentage of pixels inserted/deleted.

in a large deletion-area-above-curve *dAAC*. For both measures, the closer to one, the better the result. Values close to one would imply that an explainer is capable of finding important pixels for the classification process [16]. We conducted this experiment on AudioMNIST due to execution time reasons[7]. However, preliminary results on ESC-50 show the same results presented in the following.

For SINEX and SINEXC, we carried out 500 experiments, dividing them into 20 different *iAUC* and *dAAC* evaluations per each of the Felzenszwal tested values $\gamma \in [100, 1000]$ with a step of 100. For each γ, we run 25 experiments dividing them into 5 per each of the test classes, measuring the overall *iAUC* and *dAAC* mean values. Figure 6 shows the resulting *iAUC* and *dAAC* mean values. Recalling that we want such values to be simultaneously as closer to one as possible, these experiments show that the best configuration is $\gamma = 150$ for SINEX and $\gamma = 900$ for SINEXC. On such configurations, SINEX results in values of $iAUC = 0.54$ and $dAAC = 0.52$, while SINEXC reaches scores of $iAUC = 0.62$ and $dAAC = 0.75$. With these settings, SINEX requires on average 0.02 minutes to return an explanation, while SINEXC needs about 0.70 minutes.

We used the same evaluation measures to objectively and quantitatively compare SINEX and SINEXC against GRAD-CAM. For SINEX and SINEXC we adopted the best parameters configuration described above. Figure 7 show the average *iAUC* and *dAAC* trends for the three explanation methods. SINEXC curves, with $iAUC = 0.62$ and $dAAC = 0.75$, (2nd row) show that by removing the smallest amount of 0.1 of important pixels, the SN similarity score decrease rapidly from 1.0 to 0.4. Similarly, the addition of 0.4 of important pixels brings the similarity score outcomes to values close to 0.8. To reach the same results, SINEX (1st row)

[7] Generally, experiments on ESC-50 require double the time needed by AudioMNIST. Traditional window occlusion-based method requires 15 times the time needed by SINEXC on AudioMNIST.

needs instead to add/remove the 0.7 of important pixels. Additionally, SINEX *dAAC* curve unveils an intriguing behavior: the similarity mean outcome prediction results values close to 0 when approximately 0.8 of important pixels are removed. From that point on, further removals cause peaks of the curve back to 0.8 similarity score. Such behavior might be exposing the SN inability to discriminate when it is presented with an empty spectrogram that is composed of very few and very small pixels ($\gamma = 150$). Finally, analyzing the *iAUC* and *dAAC* trend for GRAD-CAM (3rd row), we can observe increasing and decreasing trends which are much slower than the previous two. In this case, similarly to SINEX, we find final peaks that bring the similarity prediction scores back to $0.8 - 1.0$ after a dramatic drop in both curves that usually occurs after inserting/deleting 0.7 pixels. Such experimental results lead us to select SINEXC as the best explainer considering the trade-off between its ability to find important pixels and the time needed to calculate them.

6 Conclusion

We have presented SINEX, an explanation method for Siamese Networks that process audio inputs in the context of C-way k-shot learning. By using a local perturbation-based approach on the spectrogram morphology, SINEX is able to highlight important areas for the SN classification process, both towards large and small similarity scores. Experiments proved the utility of SINEX and its coalition variation SINEXC, as well as their superiority with respect to existing approaches employed to fulfill the same task. Future research directions are related to refining the explanations, both on the front-end visualization and to the method itself. Concerning the front-end visualization, our idea is to combine the visual communicative effect with the soundification of the most important audio segments. Through soundification, our approach would be able to play to human users the parts SNs focus on to derive a certain similarity measure. We did not apply soundification at this stage due to the concise duration of the tracks of the datasets analyzed. Furthermore, we aim to investigate deeper which are the samples to perturb among those in the support set. Also, we would like to experiment with SINEX on other data types. The approach as it is, is applicable on single-channel images and it can be extended on RGB-images after minor adaptation of the segmentation algorithm. This opens for experiments in any domain that utilizes images. Finally, we plan to conduct an extrinsic interpretability evaluation of SINEX explanation through a human decision-making task driven by its explanations. In this way, we could objectively evaluate the goodness of these explanations.

Acknowledgment. This work has been partially supported by the European Community Horizon 2020 programme under the funding schemes: H2020-INFRAIA-2019–1: Research Infrastructure G.A. 871042 *SoBigData++*, G.A. 952026 *HumanE-AI Net*, ERC-2018-ADG G.A. 834756 *XAI: Science and technology for the eXplanation of AI decision making*, G.A. 952215 *TAILOR*, CHIST-ERA grant *CHIST-ERA-19-XAI-010*.

References

1. Acconcjaioco, M., et al.: One-shot learning for acoustic identification of bird species in non-stationary environments. In: International Conference on Pattern Recognition (ICPR), pp. 755–762. IEEE (2020)
2. Achanta, R., et al.: SLIC superpixels compared to state-of-the-art superpixel methods. IEEE Trans. Pattern Anal. Mach. Intell. **34**(11), 2274–2282 (2012)
3. Adadi, A., et al.: Peeking inside the black-box: a survey on explainable artificial intelligence (XAI). IEEE Access **6**, 52138–52160 (2018)
4. Becker, S., et al.: Interpreting and explaining deep neural networks for classification of audio signals. CoRR arXiv:abs/1807.03418 (2018)
5. Chen, C., et al.: Self-learn to explain siamese networks robustly. In: International Conference on Data Mining (ICDM), pp. 1018–1023. IEEE (2021)
6. Felzenszwalb, P.F., et al.: Efficient graph-based image segmentation. Int. J. Comput. Vis. **59**(2), 167–181 (2004)
7. Flanagan, J.L.: Speech Analysis Synthesis and Perception, vol. 3. Springer Science & Business Media (2013)
8. Guidotti, R., Monreale, A., Matwin, S., Pedreschi, D.: Black box explanation by learning image exemplars in the latent feature space. In: Brefeld, U., Fromont, E., Hotho, A., Knobbe, A., Maathuis, M., Robardet, C. (eds.) ECML PKDD 2019. LNCS (LNAI), vol. 11906, pp. 189–205. Springer, Cham (2020). https://doi.org/10.1007/978-3-030-46150-8_12
9. Guidotti, R., et al.: A survey of methods for explaining black box models. ACM Comput. Surv. **51**(5), 93:1-93:42 (2019)
10. Haenlein, M., et al.: A brief history of artificial intelligence: on the past, present, and future of artificial intelligence. Cal. Manag. Rev. **61**(4), 5–14 (2019)
11. Hershey, S., et al.: CNN architectures for large-scale audio classification. In: International Conference on Acoustics, Speech and Signal Processing (ICASSP), pp. 131–135. IEEE (2017)
12. Honka, T.: One-shot learning with siamese networks for environmental audio (2019)
13. Koch, G., et al.: Siamese neural networks for one-shot image recognition. In: ICML Deep Learning Workshop, vol. 2. Lille (2015)
14. Van Looveren, A., Klaise, J.: Interpretable counterfactual explanations guided by prototypes. In: Oliver, N., Pérez-Cruz, F., Kramer, S., Read, J., Lozano, J.A. (eds.) ECML PKDD 2021. LNCS (LNAI), vol. 12976, pp. 650–665. Springer, Cham (2021). https://doi.org/10.1007/978-3-030-86520-7_40
15. Lundberg, S.M., et al.: A unified approach to interpreting model predictions. In: NIPS, pp. 4765–4774 (2017)
16. Petsiuk, V., et al.: RISE: randomized input sampling for explanation of black-box models. In: BMVC, p. 151. BMVA Press (2018)
17. Piczak, K.J.: Environmental sound classification with convolutional neural networks. In: Machine Learning for Signal Processing (MLSP), pp. 1–6. IEEE (2015)
18. Selvaraju, R.R., et al.: Grad-CAM: visual explanations from deep networks via gradient-based localization. Int. J. Comput. Vis. **128**(2), 336–359 (2020)
19. Snell, J., et al.: Prototypical networks for few-shot learning. In: Adv. Neural. Inf. Process. Syst. **30**, (2017)
20. Sung, F., et al.: Learning to compare: relation network for few-shot learning. In: Conference on Computer Vision and Pattern Recognition (CVPR), pp. 1199–1208. Computer Vision Foundation (2018)

21. Utkin, L.V., et al.: Explanation of siamese neural networks for weakly supervised learning. Comput. Informat. **39**(6), 1172–1202 (2020)
22. Vinyals, O., et al.: Matching networks for one shot learning. In: Advances in Neural Information Processing Systems 29 (2016)
23. Wang, Y., et al.: Generalizing from a few examples: a survey on few-shot learning. ACM Comput. Surv. **53**(3), 63:1-63:34 (2020)
24. Ye, X., Leake, D., Huibregtse, W., Dalkilic, M.: Applying class-to-class siamese networks to explain classifications with supportive and contrastive cases. In: Watson, I., Weber, R. (eds.) ICCBR 2020. LNCS (LNAI), vol. 12311, pp. 245–260. Springer, Cham (2020). https://doi.org/10.1007/978-3-030-58342-2_16
25. Zeiler, M.D., Fergus, R.: Visualizing and understanding convolutional networks. In: Fleet, D., Pajdla, T., Schiele, B., Tuytelaars, T. (eds.) ECCV 2014. LNCS, vol. 8689, pp. 818–833. Springer, Cham (2014). https://doi.org/10.1007/978-3-319-10590-1_53
26. Zhang, Y., et al.: Siamese style convolutional neural networks for sound search by vocal imitation. IEEE ACM Trans. Audio Speech Lang. Proc. **27**(2), 429–441 (2019)

Interpretable Latent Space to Enable Counterfactual Explanations

Francesco Bodria[1], Riccardo Guidotti[2](✉), Fosca Giannotti[1], and Dino Pedreschi[2]

[1] Scuola Normale Superiore, Piazza dei Cavalieri, 7, Pisa, Italy
{francesco.bodria,fosca.giannotti}@sns.it

[2] University of Pisa, Largo Bruno Pontecorvo, 3, Pisa, Italy
{riccardo.guidotti,dino.pedreschi}@unipi.it

Abstract. Many dimensionality reduction methods have been introduced to map a data space into one with fewer features and enhance machine learning models' capabilities. This reduced space, called latent space, holds properties that allow researchers to understand the data better and produce better models. This work proposes an interpretable latent space that preserves the similarity of data points and supports a new way of learning a classification model that allows prediction and explanation through counterfactual examples. We demonstrate with extensive experiments the effectiveness of the latent space with respect to different metrics in comparison with several competitors, as well as the quality of the achieved counterfactual explanations.

1 Introduction

The booming research in eXplainable AI (XAI) of recent years has focused mainly on post-hoc explanation [12,30], or how to add a transparency layer on top of an opaque machine learning model [16]. Post-hoc explanation methods have several shortcomings, including robustness and trustworthiness of the explanations [8,13]. An emerging, more ambitious objective is to define novel Machine Learning (ML) methodologies to construct models that are *transparent-by-design*, i.e., models that natively deliver accurate classifications together with trustworthy explanations [31].

In this paper we propose a new approach to perform classification, named *ILS* for *Interpretable Latent Space*. ILS foresees the simultaneous construction of an Interpretable Latent Space and of a classifier trained on such latent space in the training phase. The latent space is built to obtain classification and retrieve explanations simultaneously. ILS uses a similarity loss to transform data from the real space to the latent space using a linear model. Then, from this latent space, a counterfactual explanation is extracted. We show how this approach enables a new use of the learned model such that, when applied to an instance x, besides the prediction, it also returns a *counterfactual example*, i.e., another instance x' with minimal changes to the features of x that is classified differently.

P. Pascal and D. Ienco (Eds.): DS 2022, LNAI 13601, pp. 525–540, 2022.
https://doi.org/10.1007/978-3-031-18840-4_37

For example, consider a ML model for credit approval and an application x classified as "declined". Our proposal also returns a counterfactual instance [11] illustrating the minimal changes to x needed to obtain credit approval, e.g., *(Hours Per Week → 53.5, occupation → Prof Specialty)*.

Thus, the main contribution of this paper is twofold. First, an interpretable latent space is defined based on a linear encoding of the original data space. Second, we show how the newly defined interpretable latent space properties allow us to find a counterfactual example. We extensively evaluate our proposal with various tabular and image datasets. First, we observe that the interpretable latent space actually preserves similarities better than other approaches in the literature [3,29]. Second, we assess qualitatively and quantitatively the counterfactuals provided by our method compared to others.

The rest of the paper is organized as follows. In Sect. 2, we analyze the existing works about latent space creation. After that, Sect. 3 illustrate the proposed methodology. Then, Sect. 4 reports the experimental results comparing our proposal against state-of-the-art methods. Conclusions and future research directions are discussed in Sect. 5.

2 Related Works

Latent Space Models. When it comes to compressing high dimensional data into a lower dimensional space, several options are available: dimensionality reduction methods that create features with linear combinations of original ones, and generative models that incorporate non-linear relationships. Principle Component Analysis (PCA) [3] is the most famous dimensionality reduction technique. It is defined as an orthogonal linear transformation that maps the data into a new coordinate system, reducing variance. Uniform Manifold Approximation and Projection (UMAP) [23] uses graph layout algorithms to arrange data in low-dimensional space and preserve the local structure of the data. TriMap [5] tries to balance the importance of the local and the global structure of points in the created latent space using triplet constraints of the kind: "point i is closer to point j than point k." Generative models are the most recent approaches to the problem of latent space creation. The most common architecture is called AutoEncoder [26], and it is composed of: an encoder that encodes the data into a reduced representation called latent space and a decoder that decodes the data from the latent space back to its original space. Both the encoder and the decoder are Neural Networks, and the training is performed by minimizing the reconstruction loss between the original data and the one generated by the decoder. Several variations have been proposed for years to improve data representation in the latent space. Variational Auto-Encoders (VAE) [29] improve the representation in the latent space by adding a constraint on the encoding network that forces it to generate latent vectors that follow a Gaussian unit distribution. The problem with autoencoders is the amount of parameters not trivial to select during training.

Latent Space Applications. Latent spaces are largely used in several domains ranging from healthcare [1,24,32], social network analysis [10,15] and anomaly

Algorithm 1: $ILS(x, X, Y, K, f)$

Input : x - instance to classify and explain X - training data, Y - labels, K - list of latent space dimensions, f - classifier training function
Output: y - classification, x' - counterfactual explanation

Train(X, Y, f, K):
1 $\mathcal{M} \leftarrow LearnBestLatentSpace(X, f, K)$; //find best latent space
2 $Z \leftarrow \mathcal{M}(X)$; //turn training data into latent space
3 $b \leftarrow f(Z, Y)$; //train classifier on the latent space
4 **return** $b, \mathcal{M}$;

Predict and Explain$(x, b, Z, \mathcal{M})$:
5 $z \leftarrow \mathcal{M}(x)$; //get latent representation
6 $\hat{y} \leftarrow b(z)$; //apply prediction
7 $x' \leftarrow GetCounterfactual(z, b, Z, \mathcal{M})$ //get counterfactual explanation
8 **return** $\hat{y}, x'$;

detection [6,28]. Several studies have demonstrated that the model's latent space can hold better clustering performances than the original feature space [25,27, 39]. Other works have shown that classifiers benefit from latent spaces [40,41].

Interpretation of the Latent Space. The latent space is always created using a generative model, and post-hoc analysis is performed to give insights into the dimension created. The most popular technique is to apply vector transformation in the latent space in order to control the generation of the data [2,33,38]. Style transfer is another way to use the latent space, the idea is to transfer the style of a data to another one with minimal changes [14]. Recent approaches have tried to explore the possibility of building a transparent latent space. The exploration is done by moving into the latent space and using the transparent proprieties to analyze the reconstructed data [17,37].

3 Methodology

We introduce here the Interpretable Latent Space (ILS) method, and we show how it is able to return an explanation in the form of counterfactual instances besides the classification outcome. The idea of ILS is to create an interpretable latent space in which the position of a point can be explained exactly in terms of input characteristics. We claim that our space is interpretable since the linear mapping of the input features can be represented in the latent space in the form of vectors (Fig. 1 (left)). Transparency can be exploited to obtain the counterfactual explanation of a classifier trained in such a space. Like most classification methods, ILS has a *train phase* and a *predict phase*. A novelty is that the latter is indeed a *predict & explain phase*.

The whole procedure, illustrated in Algorithm 1, starts by learning the latent space model $\mathcal{M}$ from the training set X, and with reference to a varying number of latent dimensions k chosen among the list of dimensions K (line 1). With $\mathcal{M}$, we indicate a model able to turn the input dataset $X \in \mathbb{R}^n$ into a latent version $Z \in \mathbb{R}^k$. Details about the latent space learning are discussed and formalized in Sect. 3.1. Then, $\mathcal{M}$ is applied to X to obtain the latent representation of the dataset $Z \in \mathbb{R}^k$ (line 2). Finally, a classifier b is trained on Z through

Algorithm 2: *LearnBestLatentSpace*(X, K, f)

```
   Input  : X - training data, K - list of latent space dimensions, f - classifier training function
   Output: M - Trained model
1  M ← ∅;                                                   //empty best transformation model
2  s ← 0;                                                   //init. best model score
3  for k ∈ K do
4  |   M' ← LearnLatentSpace(X, k);                         //learn latent space with k dimensions
5  |   Z ← M'(X);                                           //get latent representation
6  |   b ← f(Z, Y)                                          //train classifier on Z
7  |   s' ← evaluate(b(Z), Y);                              //evaluate classification performance
8  |   if s < s' then
9  |   |   M ← M';                       //take best transformer w.r.t classifier performance
10 return M;
```

the training function f (line 3). After the latent space training, the predict & explain procedure works as follows. Given an instance x, x is turned into its latent representation z, and the classifier is applied to obtain its prediction (lines 4–5). Then, the prediction $\hat{y}$ is explained by exploiting the interpretability of the learned space, returning a counterfactual instance x'. Details on how the counterfactual is constructed are given in Sect. 3.2.

3.1 Interpretable Latent Space Learning

ILS is based on a linear transformation that enables the transparent mapping between the input and latent features. The idea is to create/learn a latent space by combining a similarity loss analogous to the one utilized in t-SNE [22] with the mapping reasoning of PCA: the data are mapped into the space based on the similarity between them. In addition to PCA and t-SNE, latent space is also created using black-box predictions, augmenting data information. Our objective is that *similar instances in the original input space should be close also in the latent space that we are trying to build.* Linear models have been proven in recent years [21] to be the best methodology to produce explanations, in the sense that it is possible to isolate the contribution of each feature to the prediction. In line with these insights, we propose a procedure to build an interpretable latent space using a linear mapping $\mathcal{M}$ that transforms the input space X of dimension n into a latent space of dimension k, i.e., $Z = \mathcal{M}(X)$ such that $z_j = w_0x_0 + w_1x_1 + \cdots + w_ix_i + \cdots + w_nx_n$, where w are the weights of the model $\mathcal{M}$, x is an instance belonging to the input space $\mathbb{R}^n$, z its transformation to the latent space $\mathbb{R}^k$, and k is the number of latent dimensions. In the literature, the k parameter is challenging to select and is usually provided heuristically. Hence, the objective of *LearnLatentSpace*, described by Algorithm 3, is to find the "best" weights w for the linear model $\mathcal{M}$, given a specific dimension k.

LearnLatentSpace starts by initializing the model $\mathcal{M}$ with random weights (line 3) and fitting it to the data X. Thus, we use gradient optimization [18] to minimize an unsupervised loss that encourages similarity between near points. We adopt the similarity probability loss introduced in [22], with a different purpose: instead of using the similarity loss for visualizing a space in two dimensions, we use it to create a new data space that enjoys the requested similarity property.

Algorithm 3: *LearnLatentSpace*(X, k)

Input : X - training data, k - latent space dimension
Output: $\mathcal{M}$ - trained transformation model

```
1  i ← 0;                                    //init. iteration index
2  L_i ← ∞;                                  //initialize loss
3  M ← init();                               //initialize model weights
4  S_X ← PairwiseSimilarity(X);              //original similarity matrix
5  while L_{i-1} > L_i do                    //until the loss decreases
6  |   Z ← M(X)                              //get latent representation
7  |   S_Z ← PairwiseSimilarity(Z);          //latent similarity matrix
8  |   L_i ← KLD(S_X, S_Z);                  //compute Kullback-Leibler Divergence loss
9  |   M ← update(M, L_i);                   //Update the model using backpropagation
10 |   i ← i + 1
11 return M;
```

More in detail, the similarity of a point x_j to a point x_i is the probability that x_i would pick x_j as its neighbor if neighbors were picked in proportion to their probability density under a Gaussian distribution centered at x_i. Formally, the probability is given by

$$PairwiseSimilarity = \frac{\exp\left(-||x_i - x_j||^2/2\sigma_i^2\right)}{\sum_{k \neq i} \exp\left(-||x_i - x_k||^2/2\sigma_i^2\right)}$$

where x_i and x_j are the two points, and σ is the variance of the Gaussian. Algorithm 3 computes the similarity probability of any two points x_i and x_j in the input space (lines 4 and 7). The more two points are similar, the higher this value. From a computational point of view, the issue with using this similarity loss is that it requires calculating the similarity between every pair of instances. However, since the fit is performed by gradient optimization, it is possible to divide the data into fixed-sized batches and compute the similarity matrix separately for any small batch of points. This operation can be done only once for the input data, but it needs to be repeated every time for the latent space since the position of the point in Z changes after every iteration. The two matrices must have similar distributions to enjoy the previously described similarity property. Therefore, the final loss function that we minimize is the Kullback-Leibler divergence [19] (line 8) between the matrices S_X and S_Z:

$$L(S_X, S_Z) = KL(S_X \| S_Z)$$

where S_X and S_Z are the similarity matrices computed respectively on the input space X and the latent space Z. This loss is back-propagated to update the weights w_i of the model $\mathcal{M}$ until convergence (line 9)[1].

The function *LearnBestLatentSpace* described by Algorithm 2 is designed to select the best space by varying the dimension of the latent space k. After initializing an empty model and setting its score to zero (lines 1–2), the following procedure is repeated for each dimension k (cycle for 3–10). Given k, it learns $\mathcal{M}'$, mapping the input space to a k-dimensional latent space (line 4). Next, X

[1] For the convergence problem, we used the early stopping technique.

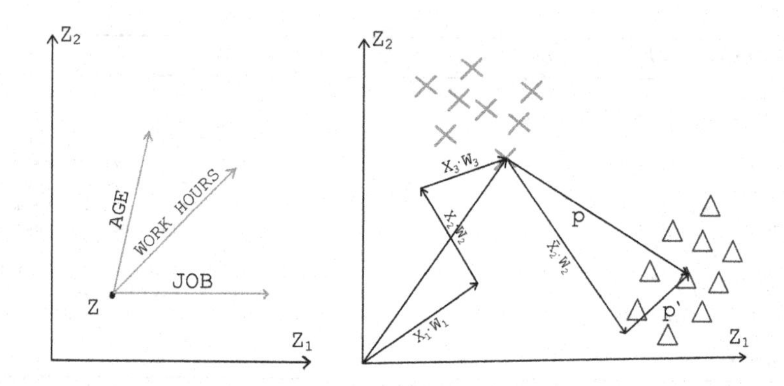

Fig. 1. Scheme of the vector model used for creating explanations. Left: representation of the input features in the latent space. Right: we illustrate a step in the ILS algorithm to modify the input features based on the position of the latent space, as explained in Sect. 3.2. The best update found by ILS is $(\bar{X}_2)$, p is the projector to the different class center, p' is the new projector for the next step.

is mapped into its latent representation Z according to M', and a classifier b is trained on Z (line 6). The performance of the classifier is used to assess the goodness of $\mathcal{M}'$ (line 8). Finally, the latent space with the smallest size returning the highest classification performance is returned by *LearnBestLatentSpace* (lines 8–9). The *LearnBestLatentSpace* procedure is costly because it trains the classifier for every latent space dimension. We highlight that k is a crucial parameter, as its value can determine the goodness of the latent space.

3.2 Counterfactual Explanations

This section describes the methodology employed by ILS to extract counterfactual explanations exploiting the interpretability of the latent space. We refer here to binary classification, but the approach easily extends to multi-class classifiers.

Let $x = \{x_1, x_2, \ldots, x_n\}$ be an input data point for which we want to provide a prediction and its explanation. The first step of *GetCounterfactual*, illustrated in Algorithm 4 and used by ILS in line 7 of Algorithm 1, is to find the best counterfactual explanation by choosing the direction in the latent space. Given z as the latent representation of x (line 1), *GetCounterfactual* computes the position of the nearest centroid of the points in the latent space with opposite predictions of z's (line 2). This is realized using a clustering algorithm on those points in the latent space with opposite predictions with respect to z and by taking the centroid c of the cluster nearest to z (line 3). Taking the nearest sample is insufficient since we could move towards a single sample that could be wrongly classified. The direction to move in the latent space to change the outcome for z is expressed in line 7 by the projector $p = c - z$. The goal is to find the best feature x_i such that its new value $\bar{x}'_i$ moves the candidate counterfactual $\bar{x}$ towards the desired prediction. In particular, each input feature x_i is responsible for a direction of movement in the latent space as shown in the example in Fig. 1

Algorithm 4: *GetCounterfactual*(x, b, Z, M)

Input : x - instance to classify and explain, b - classifier, Z - latent training set, M - latent transformation model,
Output: x' - counterfactual explanation

1 $z \leftarrow M(x)$; // get latent representation
2 $C \leftarrow Clustering(z' | z' \in Z_{\neq})$; // centroids with different prediction
3 $i \leftarrow \arg\min_i d(C_i, z)$ // find the centroid
4 $c \leftarrow C_i$
5 $\bar{x} \leftarrow x$; // init. counterfactual
6 **while** $b(M(x)) = b(M(\bar{x}))$ **do**
7 $\quad p \leftarrow c - z$ // Find the vector projecting in the centroid direction
8 $\quad u \leftarrow \emptyset$; // possible updates
9 $\quad$**for** $i \in [1, n]$ **do**
10 $\quad\quad x'_i \leftarrow Equation2(\bar{x}, i, p, M)$ //calculate update for feature i
11 $\quad\quad u_i \leftarrow x'_i$; //store update
12 $\quad i \leftarrow \arg\min_{i \in [1,n]} \{d_{euclidean}(u_i, c)\}$ // find best update
13 $\quad \bar{x}_i \leftarrow x'_i$ // apply the best update
14 **return** $\bar{x}$;

(left)). This is repeated (lines 6–15) until the prediction for $\bar{x}$, the counterfactual candidate, is different from the prediction of the instance under analysis, i.e., until $b(\mathcal{M}(x)) \neq b(\mathcal{M}(\bar{x}))$.

The goal of ILS is to find the new value of x_i such that the projection p' of the instance point x' is perpendicular to the feature direction (Fig. 1 (right)). More formally, this translate in $p' \cdot (W_i x'_i) = 0$ (Equation (1)), where $p' = c - z'$, z' is the position in the latent space by modifying the input feature i, and W_i is the vector of the ith weight of the model $\mathcal{M}$ with dimension k.

$$0 = p'(W_i x'_i) = x'_i \left(\sum_{j=1}^{k} p'_j w_{ji} \right) = x'_i \sum_{j=1}^{k} \left(c_j - z'_j \right) w_{ji} = \sum_{j=1}^{k} \left(c_j - \sum_{l=1}^{n} x'_l w_{jl} \right) w_{ji} \tag{1}$$

$$\rightarrow x'_i = \frac{\sum_{j=1}^{k} c_j w_{ji} - \sum_{j=1}^{k} \left(\sum_{l \neq i}^{n} x_l w_{jl} \right) w_{ji}}{\sum_j w_{ji}^2} \tag{2}$$

By substituting the value of p', we obtain Equation (1). This equality would be valid only if the scalar product of p' and W_i part would be 0. Then, ILS substitutes the value of z' and extracts the x'_i value from the summation corresponding to the modification needed. The rest of the steps retrieves x'_i.

Going back to Algorithm 4, by applying the formula of Equation (2), ILS finds all the possible modifications of the ith feature of x (line 10). Then, it selects the update x'_i that, if applied to $\bar{x}$ brings it to be more similar to c than the other possible updates analyzed (line 11). At this stage, the update is applied

Table 1. Datasets statistics.

Dataset	Credit	Adult	Cover	Clean1	Clean2	Isolet	Madelon	Sonar	Soybean	Anneal	Mnist	Fashion
Instances	1,000	48,842	581,012	476	6,598	7,797	2,600	208	683	898	70,000	70,000
Features	59	7	54	166	166	617	500	60	35	38	784	784
Class values	2	2	7	2	2	26	2	2	19	6	10	10

to the ith feature (line 11). The procedure is iterated until a different prediction is obtained for $\bar{x}$. We underline that it is not said that the features to update are different in every iteration. Indeed, the best feature i to be updated may be the same that was already modified some iteration ago. This is due to the fact that the position in the latent space is changing at every iteration, and it is necessary for some refinement of the modification done before.

4 Experiments

We conducted two types of experiments. The first type of experiment is aimed to verify the goodness of the latent space created and compare it with other literature approaches. The second type of experiment is aimed at validating the counterfactual explanations produced.

Datasets. We ran experiments on a selection of twelve small and medium-sized datasets widely referenced for classification tasks and publicly available. Table 1 shows summary statistics on the datasets[2].

4.1 Latent Space Evaluation

First, we evaluate the quality of the latent space created by ILS. In line with [36], we used different methods, datasets, and metrics described in the following.

Competitors. We compared ILS against two categories of algorithms: autoencoders and dimensionality reduction methods. Since ILS is a hybrid approach of these two categories, we decided to include both in the experiments. The methods tested are PCA [3], UMAP [23], TMAP [5], and VAE [29], described in Sect. 2. For ILS, we used the Adam optimizer [18] with a learning rate of 1e-3 and a batch size of 4096. For the VAE, we trained it using early stopping of 5 for a maximum of 1000 epochs using the Adam optimizer, a learning rate of 1e-4, and a batch size of 4096. We decided to use three hidden layers with dimensions equal to the number of input features divided by 2. For PCA, UMAP and TRIMAP, we used the standard parameters. We highlight that we did not compare against t-SNE as its main goal is to define a 2d space for visualization purposes rather than a latent space to perform further mining. Also, t-SNE is rarely employed for latent dimensions higher than 3.

[2] ILS code; UCI and pytorch datasets; PCA, UMAP, and TMAP methods links.

Table 2. Space quality metrics. The best scores are in bold.

	Random triplet accuracy					Outlier preservation				
Name	ILS	VAE	PCA	UMAP	TMAP	ILS	VAE	PCA	UMAP	TMAP
`Credit`	**.9525**	.6731	**.9525**	.6958	.6803	**.0000**	**.0000**	**.0000**	**.0000**	.0109
`Adult`	**.9560**	.7218	.9309	.7318	.6093	.0309	.0946	.0394	**.0031**	.0111
`Cover`	**.9838**	.7191	.9740	.7369	.6863	**.0013**	.0018	.0026	.0643	.1905
`Clean1`	.9862	.7730	**.9868**	.8069	.8132	.0220	.0063	.0031	**.0000**	.0145
`Clean2`	.9861	.8371	**.9895**	.6949	.7761	.0079	**.0007**	.0052	.0622	.1207
`Isolet`	**.9669**	.7972	.9572	.7498	.7912	.0021	.0013	**.0002**	.0153	.0510
`Madelon`	**.7738**	.5197	.7052	.5977	.6246	.0115	.0011	**.0000**	**.0000**	.4038
`Sonar`	.9511	.7928	**.9885**	.7813	.7180	**.0000**	**.0000**	**.0000**	**.0000**	.0153
`Soybean`	**.9654**	.7807	.9479	.7685	.7733	.0306	.0284	**.0197**	.0349	.0243
`Anneal`	**.9927**	.7348	.9880	.7441	.7537	.0033	**.0000**	.0017	.0216	.1464
`Mnist`	**.9425**	.7375	.9130	.6278	.5993	.0012	**.0010**	.0011	.0044	.0195
`Fashion`	**.9734**	.7888	.9598	.7365	.7772	.0020	.0031	**.0016**	.0074	.0343
Wins	9	0	4	0	0	3	5	6	5	0

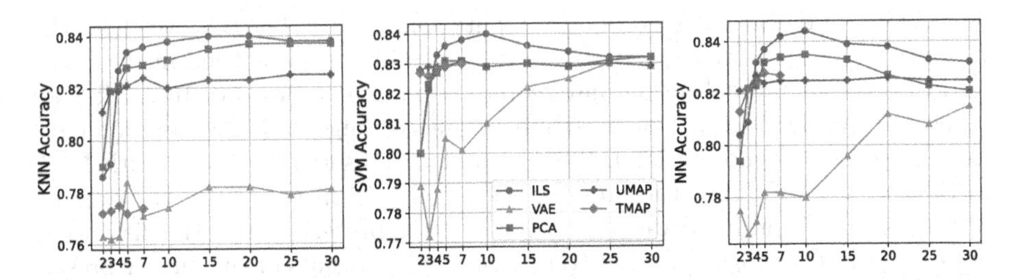

Fig. 2. Accuracy of classifiers on `adult` varying the number of latent dimensions k.

Metrics. We considered two types of evaluation metrics: *space quality metrics* and *accuracy metrics*. Space quality metrics verify different desired proprieties of the space, while accuracy metrics measure the performance of classification models trained on the latent space. To measure the relative positioning of neighborhoods, we sample observations and compute the Random Triplet Accuracy [36], which is the percentage of triplets whose relative distance order is preserved in the high and low-dimensional spaces; the closer to 1, the better. Also, we measure the outliers preservation: we want an outlier in the input space to remain an outlier in the latent space. We used the Local Outlier Algorithm (LOF) [9] to measure which points are labeled outlier or inlier in space. We ran the algorithm in both spaces to check for changes. The percentage of the changes gives the final score. We called this metric *Outlier Preservation*: the lower, the better. The classification quality of the latent space is measured using three classification models. A K-Nearest Neighbours (KNN) [35] classifier, a SVM [35] and a Neural Network (NN). For each method, we partition the embedding into five folds, each time using four folds as the training data and the remaining fold to evaluate accuracy. The metric is denoted as accuracy: the closer to 1, the better.

Results. We trained every algorithm on different latent space dimensions and evaluated the metrics for every dimension. We tested the following latent dimensions $K = \{2, 3, 4, 5, 7, 10, 15, 20, 25, 30\}$ to covert most of the possible

Table 3. Accuracy metrics. The best scores are in bold. Uncertainty is on the third decimal.

	KNN accuracy					SVM accuracy					NN accuracy				
Name	ILS	VAE	PCA	UMAP	TMAP	ILS	VAE	PCA	UMAP	TMAP	ILS	VAE	PCA	UMAP	TMAP
Credit	**.749**	.704	.745	.715	.701	**.743**	.710	.742	.710	.704	**.742**	.736	.706	.713	.699
Sdult	**.840**	.784	.837	.825	.775	**.840**	.830	.832	.831	.830	**.844**	.815	.835	.827	.828
Cover	.719	.605	**.722**	.671	.536	.930	.921	**.938**	.897	.891	.836	.772	**.877**	.801	.776
Clean1	**.878**	.610	.877	.770	.830	**.843**	.793	.840	.840	.833	**.905**	.588	.880	.685	.799
Clean2	.934	.857	**.955**	.924	.883	**.965**	.961	.962	.955	.949	.972	.899	**.988**	.933	.929
Isolet	.803	.579	.826	**.850**	.748	**.879**	.515	.875	.864	.853	**.936**	.651	.928	.834	.853
Madelon	.722	.537	**.795**	.614	.581	.847	.542	**.884**	.605	.676	.753	.521	**.870**	.579	.668
Sonar	**.826**	.663	.813	.791	.791	**.878**	.791	.842	.871	.806	**.835**	.748	.813	.769	.756
Soybean	**.888**	.536	.873	.884	.873	.897	.827	.886	**.902**	.899	.891	.580	**.895**	.847	.884
Anneal	**.963**	.769	.958	.917	.910	**.985**	.938	.977	.953	.948	**.987**	.769	.978	.889	.907
Mnist	**.977**	.953	.928	.969	.743	.973	.973	.974	.972	**.975**	.973	.972	.973	.969	**.976**
Fashion	.813	**.830**	.822	.811	.612	.851	.841	**.855**	.827	.828	**.867**	.866	.860	.815	.840
Wins	7	1	3	1	0	7	0	3	1	1	7	0	4	0	1

dimensions while not exaggerate on computational times. For ILS, we chose the variance in the similarity loss $\sigma = 1$, since our data are normalized in the range $[-1, 1]$; different normalization may require a different value of σ. In Table 2 we report the best latent space dimension results according to space evaluation metrics. Table 2 (left) shows that ILS is the best to preserve distances, with PCA as the second best. The other approaches largely fail in preserving the original distances. This is probably due to the fact that the preservation of the distance is not explicitly minimized. Table 2 (right), shows the results of the outlier preservation metric. We do not have a clear winner with respect to this score. TMAP is significantly the worst approach[3].

In Fig. 2 we report the scores of the classification models with varying latent dimension k for the `adult` dataset. Other datasets have similar behavior. Overall, increasing the latent dimensions leads to better results, although there is a sort of "magic" dimension for every dataset at which the improvement is saturated. This supports the approaches taken by most papers in literature where a fixed latent dimension is used for all the experiments. Still, it is unclear how to find this dimension without trying them all, and ILS is not an exception. Table 3 shows KNN, SVM, and NN accuracy. For KNN, ILS produces a better latent space for most of the datasets. For SVM accuracy, we observe similar results. We notice that VAE does not perform well for tabular data, but it recovers on images. UMAP is generally better than TMAP, while PCA is the second best approach.

4.2 Explanations Evaluation

In this section, we discuss the creation of counterfactual explanations through the interpretable latent space. As a classifier, we use KNN, but the same process can be directly applied to any classifier.

[3] TMAP crashed for $k > 10$ due to the exponential computational cost.

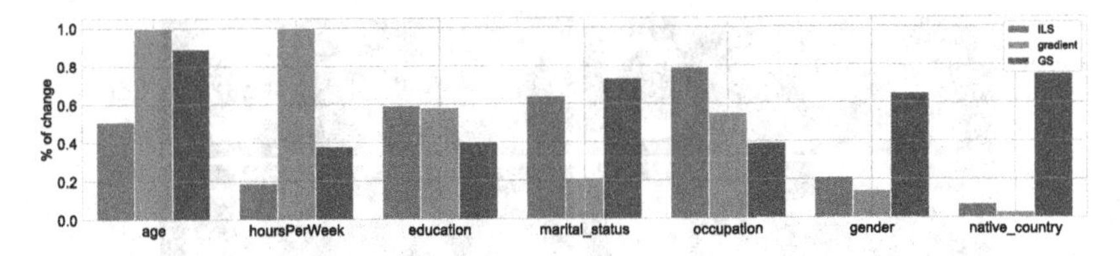

Fig. 3. Histogram of the percentage of changes in the features among the three methods. GD and GSG algorithms focus more on the first two features that are the continuous one for the adult dataset.

Table 4. Example of counterfactuals produced by ILS, Gradient, and GSG for adult.

		Age	Hours	Education	Married	Occupation	Gender	Country
$y_1 = 0$	x_1	**−0.315**	**0.000**	**3**	**1**	**0**	**1**	**1**
$\hat{y}_1 = 0$	ILS	−0.315	0.000	3	0	1	1	1
	GD	0.051	0.463	2	1	1	1	1
	GSG	−0.231	−0.106	3	0	0	1	1
$y_3 = 1$	x_2	**−0.123**	**0.000**	**3**	**0**	**3**	**1**	**1**
$\hat{y}_3 = 0$	ILS	−0.123	0.000	3	0	1	1	1
	GD	−0.168	−0.044	3	0	3	1	1
	GSG	−0.121	0.001	3	0	3	1	1
$y_4 = 0$	x_3	**−0.315**	**0.000**	**1**	**0**	**2**	**1**	**1**
$\hat{y}_4 = 1$	ILS	−0.315	0.000	1	1	2	1	1
	GD	−0.634	−0.363	1	0	2	1	1
	GSG	−0.321	0.006	1	0	2	1	1

Competitors. We compare our proposal against two model-agnostic methods that return counterfactual explanations differently. We selected these algorithms because, similarly to ILS, they are among the few model and data agnostic approaches. As a first comparison method, we search for a counterfactual of a sample x by minimizing the distance between the sample x and the centroid c of the opposite class, following the gradient descent (GD). After every gradient iteration, we modify the instance and check the prediction. We stop iterating as soon as the prediction changes. The other method is called Growing Spheres Generation [20] (GSG). The GSG procedure relies on a generative approach, growing a sphere of synthetic instances around x to find the closest counterfactual x'. Given x, GSG ignores the direction of the closest classification boundary. Indeed, GSG generates candidate counterfactuals randomly in all directions of the feature space until the decision boundary of the classifier is crossed and the closest counterfactual to x is retrieved. We selected these two approaches among the many available ones [7,34] since they are two popular agnostic approaches to search for counterfactual explanations regardless of model and data type.

Qualitative Evaluation. We report in Tables 4 and Fig. 4 the counterfactual explanations returned by ILS, GD and GSG for the adult and mnist dataset, respectively. We highlight that, for adult, the features age and hoursPerWeek are continuous, while the others are discrete. In datasets with mixed continuous

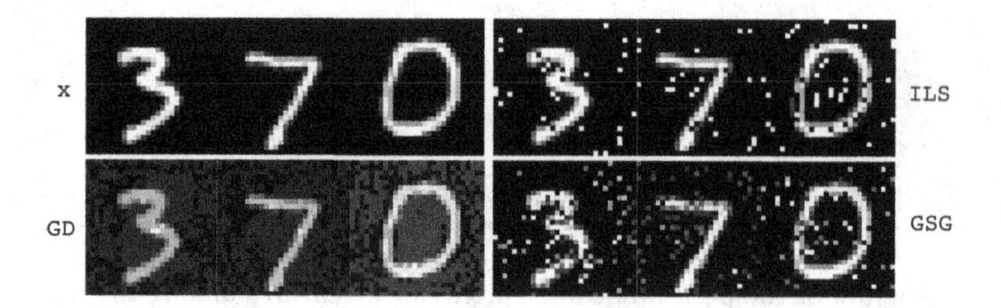

Fig. 4. Example of counterfactuals produced by ILS, GD, and GSG for the `mnist` dataset. The counterfactual classes target are 8, 9, and 8 from left to right.

Table 5. Counterfactual explanations evaluation in terms of distance and plausibility. ILS returns counterfactuals with minimal changing of input features while retaining a good result also in the distance metrics.

	d_{dist}			d_{count}			*impl*			% Success			Run time		
	ILS	GD	GSG	ILS	GD	GSG	ILS	GD	GSG	ILS	GD	GSG	ILS	GD	GSG
Adult	2.74	1.15	**0.54**	**0.43**	0.50	0.60	**0.14**	0.32	0.42	**1.00**	**1.00**	0.99	**0.01**	0.10	1.92
Credit	2.94	2.86	**1.40**	**0.11**	0.99	0.39	2.62	1.84	**1.40**	**1.00**	**1.00**	0.98	**0.03**	0.20	1.75
Clean1	4.16	4.09	**1.16**	**0.09**	1.00	0.19	3.99	4.01	**1.16**	**1.00**	**1.00**	0.72	**0.04**	0.26	1.33
Clean2	4.09	3.32	**0.45**	**0.06**	1.00	0.18	3.73	3.31	**0.45**	**1.00**	**1.00**	0.11	**0.21**	0.51	14.2
Madelon	1.96	**1.02**	-	**0.004**	1.00	-	1.95	**1.02**	-	**1.00**	**1.00**	0.00	**0.08**	0.07	-
Mnist	4.65	**3.11**	8.13	**0.03**	1.00	0.17	4.65	**3.11**	5.30	**1.00**	**1.00**	1.00	**0.18**	0.24	12.6

and discrete values, we observe that GD and GSG methods tend to focus more on the continuous features to change the prediction. For example, for x_1 of Table 4, ILS produces a counterfactual by modifying the marital status and the occupation of the person, while GD and GSG also modify the age and the hours per week. Another example that highlights this behavior is given by x_3, where ILS produces a counterfactual by only modifying the marital status while GD and GSG change the first two continuous features again. To further highlight this, we generated counterfactuals for the whole data in the `adult` dataset and checked which features were modified by the method.

In Fig. 3 is reported for the percentage of modifications of each input feature. We observe that GSG and GD change the first two features (age and hoursPerWeek) considerably more times than ILS because it is easier in the real space to modify continuous variables than categorical ones to obtain the desired effect. For image datasets, such as the `mnist` examples illustrated in Fig. 4, all methods resemble adversarial attacks [4] where only a few pixels are modified, and the counterfactuals found are very far from real samples in the dataset. The counterfactuals found by GD are very confused, and the modifications from the original image look like random background noise. On the other hand, ILS and GSG capture a small set of pixels that modify the attributed class.

Metrics. The quality of the found counterfactuals is evaluated with different metrics. We have chosen datasets with binary classification and a dataset on images for comparison with different data types. For `mnist`, since it is a multi-class dataset, we follow the approach proposed in [20] and search counterfactuals for a selected class. In two different fashions, we decided to measure the proximity between x and its counterfactual $\bar{x}$. The first one, named dis_{dist}, is the average Euclidean distance between x and the counterfactual $\bar{x}$. The second measure computed, dis_{count}, quantifies the average number of features changed between a counterfactual $\bar{x}$ and x.

$$dis_{dist} = \frac{1}{|X|} \sum_{x \in X} d(x, \bar{x}) \qquad dis_{count} = \frac{1}{|X|m} \sum_{x \in X} \sum_{i=1}^{m} \mathbb{1}_{\bar{x}_i \neq x_i}$$

where $\mathbb{1}$ returns 1 if *cond* is true, 0 otherwise, and m is the number of features. Also, we measured the *implausibility* of the generated counterfactuals in terms of how close a counterfactual $\bar{x}$ is to the reference population X. It is the average distance of $\bar{x}$ from the closest instance in the X. The lower, the better.

$$impl = \frac{1}{|X|} \sum_{x \in X} \min_{\hat{x} \in X} d(\bar{x}, \hat{x})$$

We used the test set as reference population X. Finally, we computed the success rate of the algorithm to produce a counterfactual instance.

Results . The results are presented in Table 5. By modifying the instance in the latent space, our ILS method searches for counterfactuals focusing on fewer features than other methods. The counterfactuals found by ILS are more human understandable since human reasoning often involves modifying only one feature at a time. ILS as GD has a success rate of producing a counterfactual of 100% in contrast to GSG, which is not always successful. In particular, for the dataset `clean2` the success rate of GSG is lower than 10% and for `madelon` GSG completely fails. ILS is the faster method among the three to return counterfactuals. Since ILS can select the right feature to modify to change the prediction, it is faster than GSG, which has to generate many points and call the classifier for each of them to obtain a prediction. All three approaches fail to find plausible counterfactuals for images, but ILS modifies the fewest number of pixels.

5 Conclusion

In this paper, we introduce ILS, a new way of performing classification that foresees the construction of an Interpretable Latent Space for simultaneously classifying and explaining. We have compared our methods against several approaches producing similar types of latent space, demonstrating that our proposal improves with respect to the state-of-the-art because, besides interpretability, we have observed superior performance on different metrics assessing the quality of the latent space and the accuracy of classification models built on it. Also, the transparent nature of the transformation enables the interpretation of the point's position in the latent space in terms of vectors facilitating the explanations of any classifier methods built on it. We have shown how a counterfactual explanation can be produced using linear transformations and its effectiveness compared to state-of-the-art explainers.

Future directions involve studying the k parameter and experimenting with neuroevolution or other types of AutoML. Other future research directions include studying more sophisticated counterfactual explanations that may be enabled by the properties of the vector space, including intentional representations and interactive, visual exploration of the counterfactual explanations by the user beyond the first one. Moreover, by leveraging the fact that the trained ILS model is linear, it is possible to extract the saliency values for a given sample or the whole model from the weights. Additionally, the possibility of using the dataset labels to produce a better latent space should be explored.

Acknowledgment. This work has been partially supported by the European Community Horizon 2020 program under the funding schemes: H2020-INFRAIA-2019-1: Research Infrastructure GA 871042 *SoBigData++*, G.A. 952026 *HumanE-AI Net*, ERC-2018-ADG GA 834756 *XAI: Science and technology for the eXplanation of AI decision making*, G.A. 952215 *TAILOR*.

References

1. Abati, D., et al.: Latent space autoregression for novelty detection. In: Conference on Computer Vision and Pattern Recognition (CVPR), pp. 481–490. Computer Vision Foundation/IEEE (2019)
2. Abdal, R., et al.: Image2stylegan: how to embed images into the stylegan latent space. In: International Conference on Computer Vision (ICCV), pp. 4431–4440. IEEE (2019)
3. Abdi, H., Williams, L.J.: Principal component analysis. Wiley Interdisc. Rev. Comput. Stat. **2**(4), 433–459 (2010)
4. Akhtar, N., et al.: Threat of adversarial attacks on deep learning in computer vision: survey II. CoRR arXiv:2108.00401 (2021)
5. Amid, E., Warmuth, M.K.: Trimap: large-scale dimensionality reduction using triplets. CoRR arXiv:1910.00204 (2019)
6. Angiulli, F., Fassetti, F., Ferragina, L.: Improving deep unsupervised anomaly detection by exploiting VAE latent space distribution. In: Appice, A., Tsoumakas, G., Manolopoulos, Y., Matwin, S. (eds.) DS 2020. LNCS (LNAI), vol. 12323, pp. 596–611. Springer, Cham (2020). https://doi.org/10.1007/978-3-030-61527-7_39
7. Artelt, A., Hammer, B.: On the computation of counterfactual explanations - a survey. CoRR arXiv:1911.07749 (2019)
8. Bodria, F., et al.: Benchmarking and survey of explanation methods for black box models. CoRR arXiv:2102.13076 (2021)
9. Breunig, M.M., et al.: LOF: identifying density-based local outliers. In: SIGMOD Conference, pp. 93–104. ACM (2000)
10. Grover, A., Leskovec, J.: node2vec: scalable feature learning for networks. In: Knowledge Discovery and Data Mining (KDD), pp. 855–864. ACM (2016)
11. Guidotti, R.: Counterfactual explanations and how to find them: literature review and benchmarking. In: Data Mining and Knowledge Discovery (DAMI), pp. 1–55 (2022)

12. Guidotti, R., et al.: Factual and counterfactual explanations for black box decision making. IEEE Intell. Syst. **34**(6), 14–23 (2019)
13. Guidotti, R., et al.: A survey of methods for explaining black box models. ACM Comput. Surv. **51**(5), 93:1–93:42 (2019)
14. Guo, W., Diab, M.T.: Modeling sentences in the latent space. In: Association for Computational Linguistics (ACL), vol. 1, pp. 864–872. The Association for Computer Linguistics (2012)
15. Hoff, P.D., et al.: Latent space approaches to social network analysis. J. Am. Stat. Assoc. **97**(460), 1090–1098 (2002)
16. Kim, B., et al.: Examples are not enough, learn to criticize! criticism for interpretability. In: Neural Information Processing Systems (NIPS), pp. 2280–2288 (2016)
17. Kim, J., Cho, S.: Explainable prediction of electric energy demand using a deep autoencoder with interpretable latent space. Expert Syst. Appl. **186**, 115842 (2021)
18. Kingma, D.P., et al.: Adam: a method for stochastic optimization. In: ICLR (2015)
19. Kullback, S., Leibler, R.A.: On information and sufficiency. Ann. Math. Stat. **22**(1), 79–86 (1951)
20. Laugel, T., Lesot, M.-J., Marsala, C., Renard, X., Detyniecki, M.: Comparison-based inverse classification for interpretability in machine learning. In: Medina, J., et al. (eds.) IPMU 2018. CCIS, vol. 853, pp. 100–111. Springer, Cham (2018). https://doi.org/10.1007/978-3-319-91473-2_9
21. Lundberg, S.M., et al.: A unified approach to interpreting model predictions. In: Neural Information Processing Systems (NIPS), pp. 4765–4774 (2017)
22. Van der Maaten, L., Hinton, G.: Visualizing data using t-sne. J. Mach. Learn. Res. **9**(11) (2008)
23. McInnes, L., Healy, J.: UMAP: uniform manifold approximation and projection for dimension reduction. CoRR arXiv:1802.03426 (2018)
24. Medrano-Gracia, P., et al.: Atlas-based anatomical modeling and analysis of heart disease. Drug Discov. Today Dis. Model. **14**, 33–39 (2014)
25. Mukherjee, S., et al.: Clustergan: latent space clustering in generative adversarial networks. In: AAAI, pp. 4610–4617. AAAI Press (2019)
26. Ng, A., et al.: Sparse autoencoder. CS294A Lect. Notes **72**(2011), 1–19 (2011)
27. Peng, X., et al.: Structured autoencoders for subspace clustering. IEEE Trans. Image Process. **27**(10), 5076–5086 (2018)
28. Pol, A.A., et al.: Anomaly detection with conditional variational autoencoders. CoRR arXiv:2010.05531 (2020)
29. Pu, Y., et al.: Variational autoencoder for deep learning of images, labels and captions. Adv. Neural Inf. process. Syst. **29** (2016)
30. Ribeiro, M.T., et al.: "why should I trust you": explaining the predictions of any classifier. In: Knowledge Discovery and Data Mining (KDD). ACM (2016)
31. Rudin, C.: Stop explaining black box machine learning models for high stakes decisions and use interpretable models instead. Nat. Mach. Intell. **1**(5), 206–215 (2019)
32. Schreyer, M., et al.: Detection of accounting anomalies in the latent space using adversarial autoencoder neural networks. CoRR arXiv:1908.00734 (2019)
33. Spinner, T., et al.: Towards an interpretable latent space: an intuitive comparison of autoencoders with variational autoencoders. In: IEEE (2018)
34. Stepin, I., et al.: A survey of contrastive and counterfactual explanation generation methods for explainable artificial intelligence. IEEE Access **9**, 11974–12001 (2021)
35. Tan, P., et al.: Introduction to Data Mining, 2nd edn. Pearson, Boston (2019)

36. Wang, Y., et al.: Understanding how dimension reduction tools work: an empirical approach to deciphering t-sne, umap, trimap, and pacmap for data visualization. J. Mach. Learn. Res. **22**, 201:1–201:73 (2021)
37. Winant, D., Schreurs, J., Suykens, J.A.K.: Latent space exploration using generative kernel PCA. In: Bogaerts, B., Bontempi, G., Geurts, P., Harley, N., Lebichot, B., Lenaerts, T., Louppe, G. (eds.) BNAIC/BENELEARN -2019. CCIS, vol. 1196, pp. 70–82. Springer, Cham (2020). https://doi.org/10.1007/978-3-030-65154-1_5
38. Wu, J., et al.: Learning a probabilistic latent space of object shapes via 3D generative-adversarial modeling. In: Neural Information Processing Systems (NIPS), pp. 82–90 (2016)
39. Yang, B., et al.: Towards k-means-friendly spaces: simultaneous deep learning and clustering. In: International Conference on Machine Learning (ICML), vol. 70, pp. 3861–3870. PMLR (2017)
40. Yeh, C., et al.: Learning deep latent space for multi-label classification. In: AAAI. AAAI Press (2017)
41. Zhang, L., et al.: LSDT: latent sparse domain transfer learning for visual adaptation. IEEE Trans. Image Process. **25**(3) 1177–1191 (2016)

Shapley Chains: Extending Shapley Values to Classifier Chains

Célia Wafa Ayad[1,2(✉)], Thomas Bonnier[2], Benjamin Bosch[2], and Jesse Read[1]

[1] LIX, École Polytechnique, Institut Polytechnique de Paris, Palaiseau, France
wafa.ayad@polytechnique.edu
[2] Société Générale, Paris, France

Abstract. In spite of increased attention on explainable machine learning models, explaining multi-output predictions has not yet been extensively addressed. Methods that use Shapley values to attribute feature contributions to the decision making are one of the most popular approaches to explain local individual and global predictions. By considering each output separately in multi-output tasks, these methods fail to provide complete feature explanations. We propose Shapley Chains to overcome this issue by including label interdependencies in the explanation design process. Shapley Chains assigns Shapley values as feature importance scores in multi-output classification using classifier chains, by separating the direct and indirect influence of these feature scores. Compared to existing methods, this approach allows to attribute a more complete feature contribution to the predictions of multi-output classification tasks. We provide a mechanism to distribute the hidden contributions of the outputs with respect to a given chaining order of these outputs. Moreover, we show how our approach can reveal indirect feature contributions missed by existing approaches. Shapley Chains helps to emphasize the real learning factors in multi-output applications and allows a better understanding of the flow of information through output interdependencies in synthetic and real-world datasets.

Keywords: Machine learning explainability · Classifier chains · Multi-output classification · Shapley values

1 Introduction

A multi-output model predicts several outputs from one input. This is an important learning problem for decision-making involving multiple factors and complex criteria in the real-world scenarios, such as in healthcare, the prediction of multiple diseases for individual patients. Classifier chains [8] is one such approach for multi-output classification, taking output dependencies into account by connecting individual base classifiers, one for each output. The order of output nodes and the choice of the base classifiers are two parameters yielding different predictions thus different explanations for the given classifier chain.

P. Pascal and D. Ienco (Eds.): DS 2022, LNAI 13601, pp. 541–555, 2022.
https://doi.org/10.1007/978-3-031-18840-4_38

To address the lack of transparency in existing machine learning models, solutions such as SHAP [5], LIME [9], DEEPLIFT [11] and Integrated Gradients [12] have been proposed. Using Shapley values [10] is one approach to attribute feature importance in machine learning. The framework SHAP [5] provides Shapely values used to explain model predictions, by computing feature marginal contributions to all subsets of features. This theoretically well founded approach provides instance-level explanations and a global interpretation of model predictions by combining these local (instance-level) explanations.

However, these methods are not suitable for multi-output configurations, especially when these outputs are interdependent. In addition, the SHAP framework provides separate feature importance scores only for independent multi-output classifiers. By assuming the independence of outputs, one ignores the indirect connections between features and outputs, which leads to assigning incomplete feature contributions, thus an inaccurate explanation of the predictions.

Figure 1 is a graphical representation of a classifier chain: patients with two conditions, obesity (Y_{OB}) and psoriasis (Y_{PSO}), given four features: genetic components (X_{GC}), environmental factors (X_{EF}), physical activity (X_{PA}) and eating habits (X_{EH}). From a clinical point of view, all factors X are associated with both conditions Y, obesity and psoriasis. However, since obesity is a strong feature for predicting psoriasis [4] (indeed, a motivating factor for using such a model is that predictive accuracy can be improved by incorporating outputs as features), it may mask the effects of other features. Namely, X_{PA} and X_{EH} will be found by methods as SHAP applied to each output separately to have zero contribution towards predicting Y_{PSO}, and one might interpret that psoriasis is mainly affected by factors which cannot be modified by the patient (environment and genetics). The *indirect* effects (physical activity and eating habits) will not be detected or explained.

We propose Shapley Chains to address this limitation of incomplete attribution of feature importance in multi-output classification tasks by taking into account the relationships between outputs and distributing their importance among the features with respect to a given order of these outputs. Calculating the Shapley values of outputs helps to better understand the importance of the chaining that connects these outputs and to visualize this relationship impact on the prediction of subsequent outputs in the chain. For these subsequent outputs, the computation of the Shapley values of the associated outputs shows the indirect influence of some features through the chain, which is generally not intuitive and missed by existing work. Our method will successfully explain these *indirect* effects. By attributing importance to the features X_{PA} and X_{EH}, Shapley Chains will help doctors to emphasize the importance of eating healthy and practicing physical activities in order to prevent and better cure psoriasis instead of blaming only genetics and exterior environmental factors.

This paper addresses the problem of attributing feature contributions in multi-output classification tasks with classifier chains when outputs are interdependent. Our contribution in this paper is resumed to:

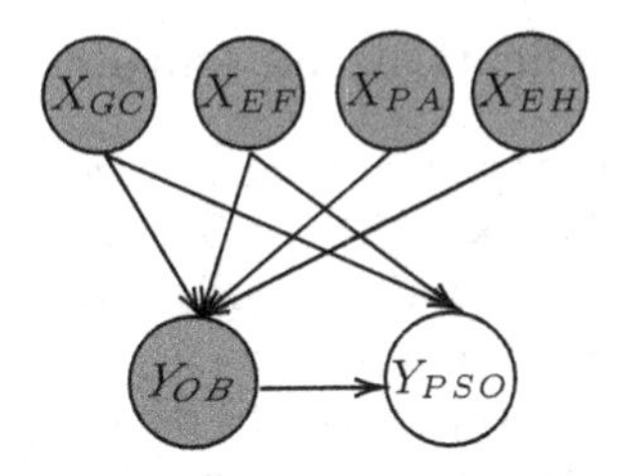

Fig. 1. An example of a multi-output task: predicting Y-outputs from X-features. A classifier chain uses the first output Y_{OB} as an additional feature to predict the second output Y_{PSO}.

- We propose Shapley Chains, a novel post-hoc model agnostic explainability method designed for multi-output classification task using classifier chains.
- Shapley Chains attribute feature importance to all features that directly or indirectly contribute to the prediction of a given output, by tracking all the related outputs in the given chain order.
- Compared to existing methods, we show a more complete distribution of feature importance scores in multi-output synthetic and real-world datasets.

We devote Sect. 2 to a background and related work. In Sect. 3, we detail our proposed method Shapley Chains. Finally in Sect. 4, we run experiments on a synthetic and real-world datasets. The results of our method compared to SHAP values applied to independent classifiers are then discussed.

2 Background and Related Work

In this section we review multi-output classification, output dependencies, classifier chains and Shapley values to serve as a background for the rest of this paper. The notation we used is summarized in the next table (Table 1).

Table 1. Notation.

Notation	Meaning
$\mathbf{x}$	A given instance vector
$\mathbf{y}$	A given output vector
x_i	The ith feature of instance $\mathbf{x}$
y_j	The jth output
X	The feature space of x_i
Y	The output space of y_j
n	The number of features for each instance $\mathbf{x}$
m	The number of outputs

2.1 Multi-output Classification and Output Dependencies

A multi-output classifier H is a mapping function that for a given instance $\mathbf{x}=\{x_1, x_2, ..., x_n\}$, such that $\mathbf{x} \in X$, it learns a vector of base classifiers $\mathsf{H}(\mathbf{x}) = h_1(\mathbf{x}), h_2(\mathbf{x}), ..., h_m(\mathbf{x})$ and returns a vector of predicted values $\mathbf{y} = \{y_1, y_2, ..., y_m\}$, with $y_j \in \{0, 1\}$ and $\mathbf{y} \in Y$.

In real-world applications, outputs can be dependent or independent. Designing classifiers that incorporate these output dependencies makes it possible to better represent the relationships in the data (between outputs, therefore between features and outputs). There are two types of output dependencies wrt subsequent outputs; namely marginal independencies, $P(\mathbf{y}) = \prod_{j=1}^{m} P(y_j)$, and conditional output dependencies:

$$P(\mathbf{y}|\mathbf{x}) = \prod_{j=1}^{m} P(y_j|X, y_1, ..., y_{j-1}) \quad (1)$$

In this article, we focus on output conditional dependencies. The nature of the relationship between features and outputs and between outputs is not restricted to causality. Therefore, no prior knowledge of the causal graph is necessary. This specific subject is partially covered in Shapley Flow [13], which is designed for single-output tasks.

2.2 Classifier Chains

A classifier chain is one multi-output method that learns m classifiers (one classifier for each output, also referred as base classifier). All the classifiers are linked in a chain. The chaining method passes output information between classifiers, allowing this method to take into account output dependencies [7] when learning a given output in the chaining.

This method is exactly an expression of Eq. 1, if expressed according to the chain rule of probability (i.e., Fig. 2 as a probabilistic graphical model representation). That is one reason why conditional dependencies are interesting in this context. However, a classifier chain is not faithful to a 'proper' inference procedure, and rather takes a greedy approach to inference, plugging in predictions as observations; and proceeds much as a forward pass across a neural network. This creates some ambiguity between how much effect is gained from probabilistic dependence (as a probabilistic graphical model would) and feature effect (as one encounters via the latent layers of deep learning). Although discussion has been ongoing e.g., [7,8], there is not yet a consistent understanding in practice of what role a prediction plays as a feature to another label. By propagating output contributions among the features, Shapley Chains helps to clarify these prediction roles, and confirm which outputs are interdependent using the Shapley value described in the next section.

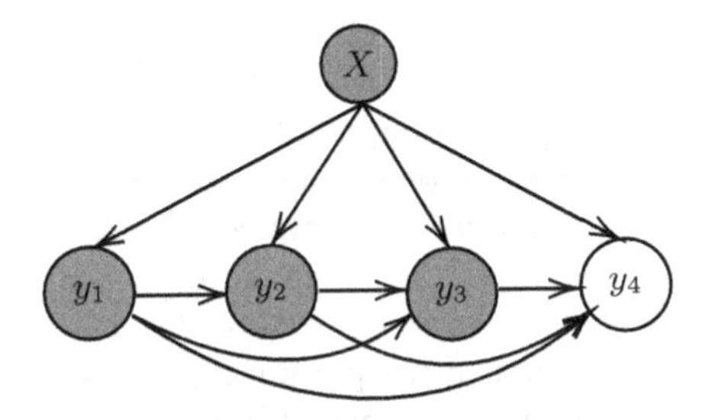

Fig. 2. One example of a classifier chain structure.

2.3 Shapley Values

The Shapley value expresses the contribution of feature x_i, to predict output y_j as a weighted sum:

$$\phi_{y_j} x_i = \sum_{S \subseteq X \setminus \{i\}} \frac{|S|!\ (|X| - |S| - 1)!}{|X|!} \left[f_x\ (S \cup \{i\}) - f_x\ (S) \right] \quad (2)$$

where $S \subseteq X$, and f_x is the value function that defines each feature's contribution to each subset S. It computes each feature's average added value to each combination of features when making a prediction for instance $\mathbf{x}$.

Additivity is one axiom of a fair attribution mechanism that is satisfied by the Shapley value. It finds a good interpretation in multi-output classification. Consider two prediction tasks $(X,\ f)$, $(X,\ g)$ composed of the same set of features. We create a coalition prediction task $(X, f+g)$ by adding the two previous prediction tasks in the following way: $(f+g)(S) = f(S) + g(S)$ for all $S \subseteq X$. The additivity axiom states that the allocation of the prediction $(X, f+g)$ will be equal to the sum of the allocations of the two original prediction tasks. One should note that in this definition, we assume that the two prediction tasks are completely independent meaning that feature contributions to one prediction has no effect on the second one, which is not always the case because in real-world applications tasks are more often interdependent. One approach we propose is to use classifier chains because it permits to represent these relationships by introducing different chaining orders of these outputs. The overall feature Shapley values for a classifier chain can be calculated by marginalizing over all possible output chain structures. $\forall c \in \mathsf{C}$, the Shapley value of x_i in Eq. 2 can be written as follows:

$$\phi_{y_j x_i} = \frac{1}{|\mathsf{C}|} \sum_{c \subseteq \mathsf{C}} \phi_{y_j^c} x_i \quad (3)$$

with $\phi_{y_j^c}$ being the contribution of feature x_i to the prediction of y_j with respect to the given chaining order c. For the matter of simplicity, we use ϕ_{y_j} to refer to $\phi_{y_j^c}$ in the rest of this paper. We report feature contribution for each chain structure independently to show the impact of different chaining orders and the marginalization over these orders in Sect. 4.1.

2.4 Related Work

The explainability of machine learning is an active research topic in the recent years. Several contributions have been made to explain single-output models and predictions. Inspecting feature importance scores of existing models is an intuitive approach that has served for many studies. These feature importance scores are either derived directly from feature weights in a linear regression for instance, or learned from feature permutations based on the decrease in model performance. Other more complex methods like LIME [9] learn a surrogate model locally (around a given instance) in order to explain the predictions of the initial model with simple and interpretable models like decision trees. On the other hand, DeepLift [11], Integrated gradient [12] and LRP [6] are some neural network specific methods proposed to explain deep neural networks.

The SHAP framework is one popular method attributing Shapley values as feature contributions. It provides a wide range of model-specific and model-agnostic explainers. Researchers have also proposed other Shapley value inspired methods incorporating feature interactions in the explanation process. For example, asymmetric Shapley values [3] incorporates causal knowledge into model explanations. This method attributes importance scores to features that do not directly participate in the prediction process (confounders), but fails to capture all direct feature contribution. On the other hand, on manifold Shapley values [2] focus on better representing the out of coalition feature values but provides misleading interpretation of feature contributions. Wang et al. [13] have proposed Shapley Flow, providing both direct and indirect feature contributions when a causal graph is provided. Resuming feature interactions to causality and assuming the causal graph is provided and accurate are two downsides of this method. These methods significantly contributed to advancing the explainability of machine learning models but none of them have tackled multi-output problems, more specifically when outputs are interdependent. Shapley Chains addresses this limitation.

3 Proposed Method: Shapley Chains

In this section, we introduce our approach to compute direct and indirect feature Shapley values for a classifier chain model. Note that our proposed method is model-agnostic, meaning that our computations do not depend directly on the chosen base learner used by the classifier chain.

We want to compute feature contributions to the prediction of each output $y_j \in Y$ for each instance $\mathbf{x}$. For example, Fig. 3 shows the direct and indirect contributions of x_i to predict output y_4 given in Fig. 2. In the next two sections, we detail the computations of the Shapley value of each feature to predict each output. We refer to these Shapley values as direct and indirect feature contributions.

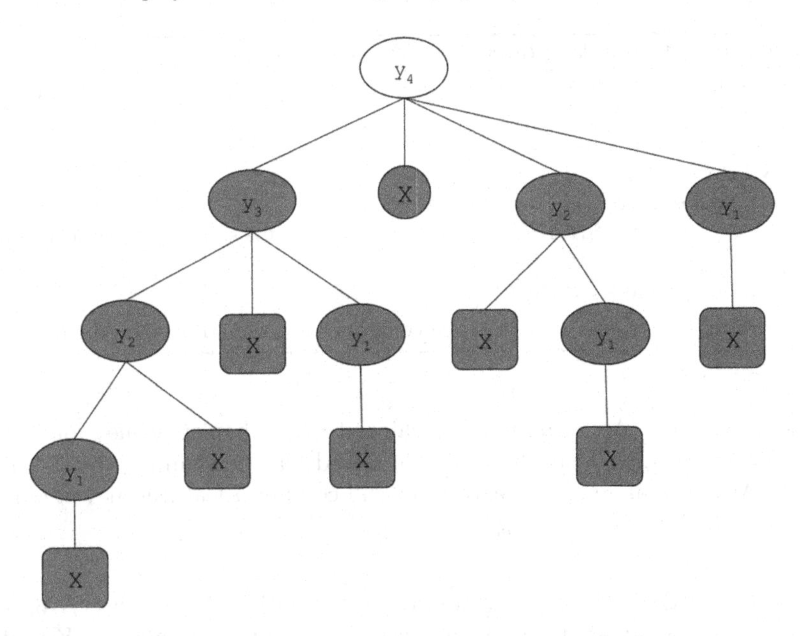

Fig. 3. Representation of direct and indirect contributions for a dataset with 4 outputs (y_1, y_2, y_3 and y_4). For example: the 4th output y_4 has 7 indirect Shapley values (7 paths ending with square leave) and one direct Shapley value (one path ending with a circle leaf).

Direct Contributions. The direct contributions are computed for features and outputs as in Eq. 2. Consider again the example of patients with the two conditions: psoriasis and obesity. For both Y_{OB} and Y_{PSO}, we use the framework SHAP in order to compute the Shapley value of each feature: X_{GC}, X_{EF}, X_{PA} and X_{EH}. This will attribute non zero Shapley values to X_{GC} and X_{EF} to predict Y_{OB} and Y_{PSO} separately. On the other hand, X_{EF} and X_{PA} will have non-zero Shapley values to predict Y_{OB} and zero values for the prediction of Y_{PSO}. The classifier chain method will add Y_{OB} to the feature set to predict Y_{PSO}. By running the SHAP framework on this new set, Y_{OB} will have a non zero Shapley value because it is dependent to Y_{PSO}. This Shapley value will be attributed to the features that are correlated to Y_{OB}. The attribution mechanism of direct feature (and output) contributions can be generalized to the classifier H with m base classifiers as shown in Algorithm 1.

For the first output y_1, we calculate the Shapley value of each feature according to Eq. 2, as done in the SHAP framework. This marginal value of all possible subsets to which the feature can be associated to is the feature's contribution to predict the first output y_1. For the second output y_2, we append the predictions y_1 made by the first classifier h_1 to the features set, and we train a second classifier h_2 to learn the second output y_2. We again use the SHAP framework to assign Shapley values to features and the first output y_1. Here, the feature set includes the first prediction. We perform the same steps for each

Algorithm 1. Computing direct feature contributions

```
1: procedure DICONTRIBUTION(X, Y, H)   ▷ features, outputs, classifier chain model
2:     i = j = 0
3:     Φ=[]
4:     while j < len(Y) do
5:         while i < len(X) do
6:             Φ_{y_j} x_i ← SHAP(X, y_j, H)   ▷ Shapley values of inputs wrt each output
7:             append y_j to X
8:             append Φ_{y_j} x_i to Φ
9:     return Φ   ▷ Φ contains features and outputs Shapley values
```

remaining output. At each step, we calculate the Shapley values for features and previous predicted outputs that are linked via the chaining to the current output. At the final step, the feature set will contain n features and m outputs: $X = \{x_1, x_2, ..., x_n, y_1, y_2, ..., y_m\}$.

Indirect Contributions. The indirect contribution $\Phi_{indirect} y_j(x_i)$ of x_i to predict y_j is the weighted sum of the direct contributions of all $y_k \in Y$ that are chained to y_j. $\Phi_{indirect} y_j(x_i)$ is computed according to the Eq. 4.

$$\Phi_{indirect} y_j(x_i) = \sum_{k=1}^{j-1} \Phi y_j(y_k) \cdot Z_k(x_i) \tag{4}$$

where $j > 1$ and the function $Z_k(x_i)$ computes the weight vector for all paths from output y_k down to x_i. For $k > 1$ and $Z_1(x_i) = W(y_1, x_i)$, $Z_k(x_i)$ is recursively computed as follows:

$$Z_k(x_i) = \sum_{l=1}^{k-1} W(y_k, y_{k-l}) \cdot Z_{k-l}(x_i) + W(y_k, x_i) \tag{5}$$

where $W(y_k, y_{k-l})$ is the corresponding weight of y_{k-l} to predict the next output y_k (the direct contribution of y_{k-l} to predict y_k. And, $W(y_k, x_i)$ is the weight of x_i to predict y_k (the direct contribution of x_i to predict y_k). The weights $W(y_k, y_{k-l})$ and $W(y_k, x_i)$ are calculated according to:

$$W(y_k, .) = \frac{|\Phi y_k(.)|}{\left(\sum_{q=1}^{n} |\Phi y_k(x_q)| + \sum_{p<k} |\Phi y_k(y_p)|\right)} \tag{6}$$

where $\Phi y_k(x_q)$ is the direct contribution, as in Eq. 2; of each feature x_q to predict y_k). $p < k$ means the output p is chained to the output j forming a directed acyclic graph illustrated in Fig. 2.

For instance, in order to have a complete fair distribution of feature importance for the prediction of Y_{PSO}, we compute the indirect Shapley values of the features X_{PA} and X_{EH}. We do so by distributing the direct Shapley value of

Y_{OB} computed previously to the four features. By the distribution operation, we mean the multiplication of the direct Shapley value of each feature by the direct Shapley value of Y_{OB}, divided by the sum of the shapley values of all features to predict the same output (here Y_{OB}).

We generalize this mechanism in Algorithm 2 of calculating indirect Shapley values to the chain structure in Fig. 2.3. The first output y_1 has always zero indirect Shapley values because there is no output that precedes it in the chaining. Thus, for the rest of this section, we compute feature indirect contributions for $y_j \in \{y_2, y_3, ..., y_m\}$. For each output y_j, there exists one direct path to the features thus one direct feature contributions and $2^j - 1$ indirect paths for each feature.

Algorithm 2. Computing feature indirect contributions

```
1: procedure INCONTRIBUTION(X, Y, Φ)      ▷ inputs, outputs, Shapley values of
   features and outputs
2:     i = j = 0
3:     while j < len(Y) do
4:         while i < len(X) do
5:             compute W(y_k, y_{k-l}) and W(y_k, x_i) in Eq. 6
6:             compute Z_k(x_i) in Eq. 5
7:     return Φ_indirect y_j(x_i) in Eq. 4     ▷ returning indirect feature contributions.
```

One should notice that for the matter of the simplicity of understanding, we take the absolute value in Eq. 6. Thus, all the contributions will be positive. These absolute values can be replaced by the raw Shapley values in order to keep the positive or negative sign of feature contributions. Keeping the sign helps to understand if the feature penalizes or is in favor of the prediction.

4 Experiments

In order to assess the importance of the features that is attributed by our proposed framework[1] to explain their contributions to predict multiple outputs with a classifier chain, we run experiments on both synthetic and real-world datasets: a *xor* data that we describe next, and the Adult Income dataset from the UCI repository [1]. Here, we rely on human explanation to validate our results.

4.1 Synthetic Data

To demonstrate our work, we first run experiments on a multi-output synthetic dataset containing two features (x_1 and x_2) and three outputs (*and*, *or* and *xor*) corresponding to the logical operations of the same names performed on x_1 and x_2. We split this dataset to 80% for the training and 20% for the test of our classifier.

Next, we construct a classifier chain with the chaining order illustrated in Fig. 4. We use a logistic regression as the base learner. Our method is model

[1] https://github.com/cwayad/shapleychains.

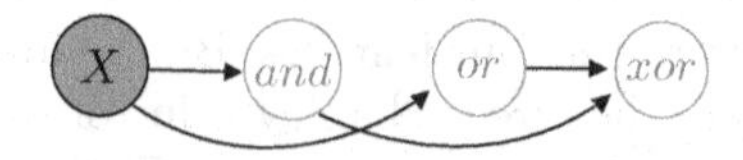

Fig. 4. The classifier chain structure for *xor* data. X is the set of features x_1 and x_2. *and*, *or* and *xor* are the outputs for which we want to compute direct and indirect Shapley values.

agnostic meaning that it can be applied to a classifier chain with any other base learners. The use of the logistic regression as the base learner to predict *xor* is justified by the accuracy that this model achieves compared to other classifiers like decision trees. The classifier chain is trained on the train set using x_1 and x_2 to predict *and* and *or* separately. Then, we append these two predicted outputs to the features set in order to predict *xor*. Here, the order in which we predict *and* and *or* does not change our method's behavior.

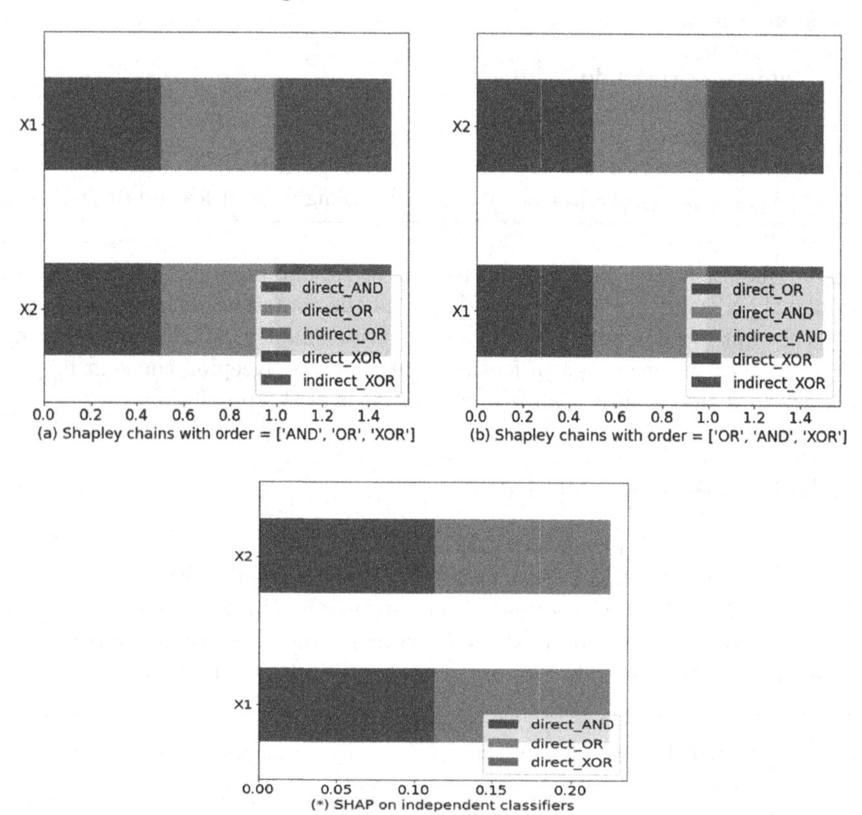

Fig. 5. A comparison of SHAP applied on independent classifiers and Shapley Chains. From the left to the right. (*a*) and (*b*) Normalized direct and indirect feature contributions made by Shapley Chains to predict *and*, *or* and *xor* for chain orders [*and*, *or*, *xor*] and [*or*, *and*, *xor*]. (∗) SHAP assigns contributions to x_1 and x_2 only to predict *and* and *or* outputs and completely misses their contributions to predict *xor*. Absent colors refer to null Shapley values.

To explain the influence of x_1 and x_2 on the prediction of xor, we compared the application of the framework SHAP on each classifier independently and Shapley Chains on the trained classifier chain. We report our analysis on the test data. The results of the comparison shown in Fig. 5 indicate that the output chaining propagates the contributions of x_1 and x_2 to predict xor via *and* and *or*. Specifically, Fig. 5(a) and Fig. 5(b) illustrate that our method detects the indirect contributions of x_1 and x_2 (indirect_xor) to predict xor thanks to the chaining of *and* and *or* to xor implemented with the classifier chain model, which tracks down all feature contributions through the chaining of outputs. Furthermore, Fig. 5(a) and Fig. 5(b) confirm that predicting *or* before *and* or vice versa does not affect the feature contributions attribution, which confirms the chain structure for this data. On the other hand, these contributions of x_1 and x_2 are completely neglected by the SHAP framework on independent classifiers (Fig. 5($*$)).

Impact of the Chaining Order on the Classifier Chain Explainability. In order to measure the impact of the chaining order on the explainability of our classifier chain model with Shapley Chains, we performed analysis on the $3! = 6$ possible output chaining orders in the synthetic dataset (scenarios (a) and (b) in Fig. 5 and scenarios (c), (d), (e) and (f) in Fig. 6).

The information known to the classifier chain when training each output changes depending on the order of these outputs. For instance, in scenarios a and b (Fig. 5), we first learn the two outputs *and* and *or* using x_1 and x_2 features. xor is then predicted using *and* and *or*. Here, in both scenarios, both features x_1 and x_2 contribute indirectly (through *and* and *or*) to predict xor. Meanwhile in the scenario c (or d), the model relies on *and*(or *or*), x_1 and x_2 to predict xor. We observe that x_1 and x_2 have direct and indirect contributions, meaning that the classifier chain relies partially on these two features to predict xor (direct contributions of x_1 and x_2), and on *and* (indirect contributions of x_1 and x_2 via *and*). The last two scenarios e and f show no contribution of x_1 and x_2 to predict xor, which is explained by the fact that using only these two features, the model can not predict xor without having the information about the dependencies of xor to *and* and *or*.

These results show that the chain order of *and*, *or* and xor outputs has an important role in the explainability of the classifier chain, because feeding different inputs to the classifier chain yields different predictions, thus different Shapley values are attributed to the features. x_1 and x_2 importance scores can either be derived from a direct inference of xor output only if there is additional information on output dependencies (for example *and* is linked to xor) or by extracting it from the chain that links *and* and *or* to xor. In the absence of all output dependencies of *and* or *or* to xor, the model completely ignores the importance of features x_1 and x_2 in the prediction of xor.

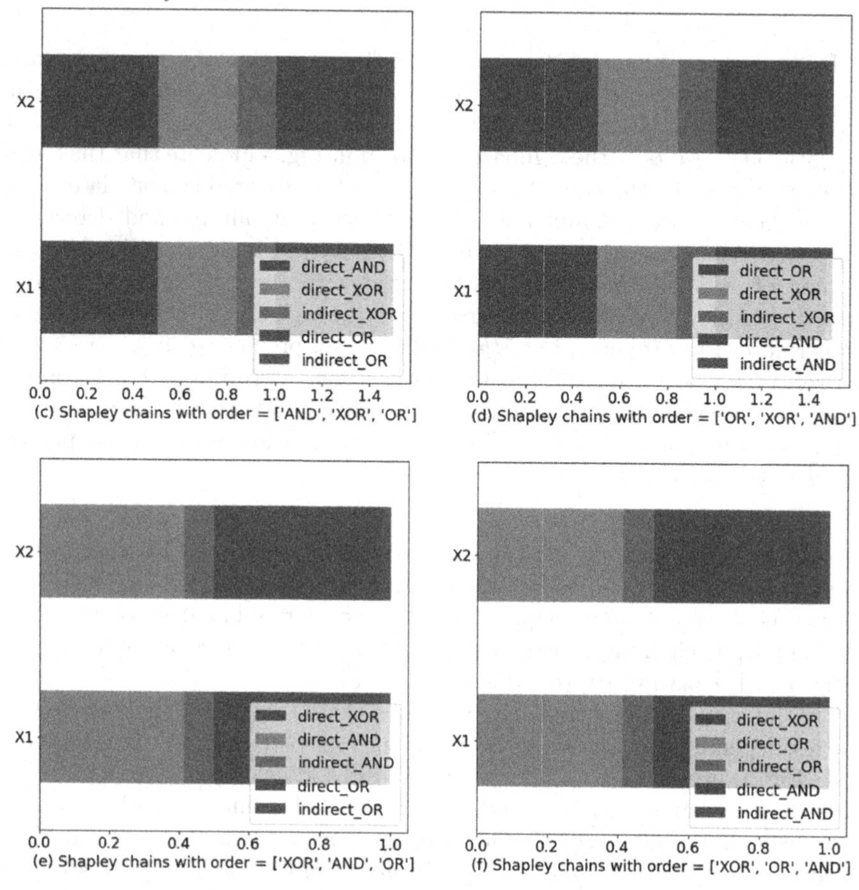

Fig. 6. Possible output chaining orders for *xor* data. Normalized total feature contributions (direct and indirect Shapley values) for c, d, e and f.

4.2 Explaining Adult Income with Shapley Chains

We run Shapley Chains on the UCI Adult Income dataset. This dataset contains over 32500 instances with 15 features. We first discretize *workclass*, *marital status* and *relationship* characteristics. We remove *race*, *education* and *native country* and normalize the dataset with the min/max normalizer. Next, we split it into two subsets, using 80% for the training and the remaining 20% for testing. We evaluated the hamming loss of a classifier chain with different base learners and we kept the best base classifier, the logistic regression in this case.

In order to explain feature contributions to the predictions of the three outputs *sex*, *occupation* and *income*, we compared the results of Shapley Chains against classic Shapley values applied on separate logistic regression classifiers for different chain orders. Figure 7 shows graphical representation of normalized and stacked feature contributions when applying Shapley Chains on our dataset (Fig. 7(a)), and stacked feature contributions from independent logistic regression classifiers (Fig. 7(b)). In both cases, the magnitude of the feature contributions is greater in Shapley Chains compared to independent Shapley values, which confirms our initial hypothesis of some contributions are missed by SHAP framework, and these contributions can be detected when we take into account output dependencies. For example, the number of hours worked in a week (*hours.per.week*) has a more important indirect contribution to predict individual's *occupation* than a direct contribution. This is explained by the fact that *sex* is related to *occupation*, and this relationship is propagated to the features by Shapley Chains. *relationship* is another example of Shapley Chains detecting indirect feature contributions to predict *occupation*. Furthermore, feature rankings are different in Shapley Chains. For example, the ranking of *capital.gain* comes in the fourth position (before *workclass*) using SHAP applied to independent classifiers. In our method, this feature's ranking is always less important (according to different chaining orders) than *workclass* to predict *sex*, *occupation* and *income* which makes more sens to us.

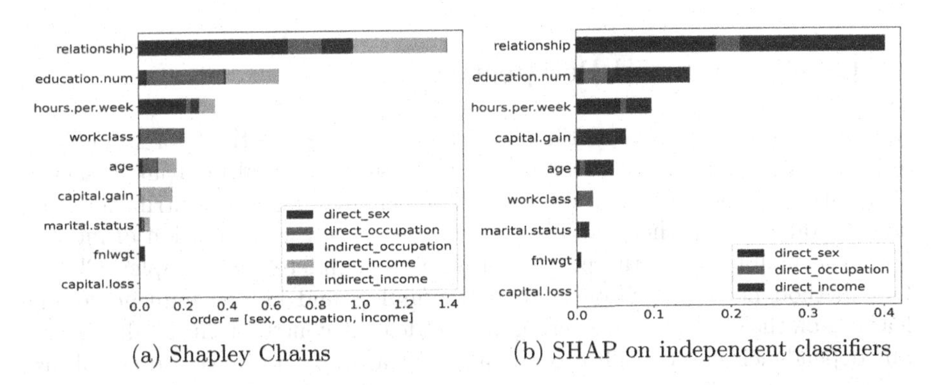

(a) Shapley Chains (b) SHAP on independent classifiers

Fig. 7. (a) Direct and indirect Shapley values on Adult Income data: we normalize and stack each feature's direct and indirect contributions to each output. *sex* has only direct contributions because it is the first output we predict in this chain order. (b) Stacked Shapley values of independent classifiers on Adult Income data.

We also tested the impact of different chain orders of these three outputs on the feature importance attribution. Figure 8 illustrates three different chaining orders. Each different order allows each classifier to use different prior knowledge to learn these outputs. For example in Fig. 8(b), we first predict *income* and *sex* and we use this information to predict *occupation*. Intuitively, *occupation* is correlated to individual's *sex* and *income*. The classifier chain uses this information provided to the third classifier to predict *occupation*. Here, Shapley Chains attribute more importance to the factors that predict both *income* and *sex*, when predicting *occupation*. Shapley Chains preserves the order of feature importance scores across all the chaining orders in general, but the magnitude of each feature's importance differs from one chain to another. This is due to the prior knowledge that is fed into the classifier when learning each output. In addition, these feature importance scores are always more important in Shapley Chains compared to Shapley values of independent classifiers for all chain orders.

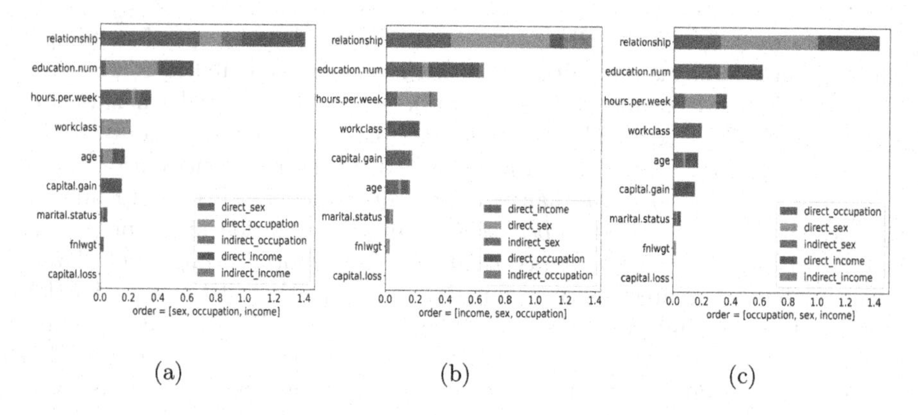

Fig. 8. Stacked direct and indirect feature contributions for 3 different chain structures over Adult Income data.

5 Conclusions and Perspectives

In this paper, we presented Shapley Chains, a novel method for calculating feature importance scores based on Shapley values for multi-output classification with a classifier chain. We defined direct and indirect contribution and demonstrated on synthetic and real-world data how the attribution of indirect feature contribution to the prediction is more complete with Shapley Chains. Our method helps practitioners to better understand hidden influence of the features on the outputs by detecting indirect feature contributions hidden in output dependencies. Although the rankings of feature importance are not always different from independent feature importance scores, the magnitude of these scores is always important in Shapley Chains, which is more important to look at in applications that are sensitive to the magnitude of these importance scores rather than their rankings. By extending the Shapley value to feature importance attribution for classifier chains, we make use of output interdependencies that is implemented in classifier chains in order to represent the real learning factors of a multi-output classification task.

To extend this work, Shapley Chains could be evaluated on multi-output regression tasks. Exploring the relationship's type between the outputs, and studying whether Shapley Chains preserves all these relationships when attributing feature contributions is another open question of our work.

References

1. Dua, D., Graff, C.: UCI machine learning repository (2017). http://archive.ics.uci.edu/ml
2. Frye, C., de Mijolla, D., Begley, T., Cowton, L., Stanley, M., Feige, I.: Shapley explainability on the data manifold (2021)

3. Frye, C., Rowat, C., Feige, I.: Asymmetric shapley values: incorporating causal knowledge into model-agnostic explainability (2021)
4. Jensen, P., Skov, L.: Psoriasis and obesity. Dermatology **232**(6), 633–639 (2016)
5. Lundberg, S., Lee, S.I.: A Unified Approach to Interpreting Model Predictions (2017)
6. Montavon, G., Binder, A., Lapuschkin, S., Samek, W., Müller, K.-R.: Layer-wise relevance propagation: an overview. In: Samek, W., Montavon, G., Vedaldi, A., Hansen, L.K., Müller, K.-R. (eds.) Explainable AI: Interpreting, Explaining and Visualizing Deep Learning. LNCS (LNAI), vol. 11700, pp. 193–209. Springer, Cham (2019). https://doi.org/10.1007/978-3-030-28954-6_10
7. Read, J., Pfahringer, B., Holmes, G., Frank, E.: Classifier chains for multi-label classification. Mach. Learn. **85**(3), 333–359 (2011)
8. Read, J., Pfahringer, B., Holmes, G., Frank, E.: Classifier chains: a review and perspectives. J. Artif. Intell. Res. **70**, 683–718 (2021)
9. Ribeiro, M.T., Singh, S., Guestrin, C.: "Why Should I Trust You?": Explaining the Predictions of Any Classifier (2016)
10. Rozemberczki, B., Sarkar, R.: The Shapley Value of Classifiers in Ensemble Games (2021)
11. Shrikumar, A., Greenside, P., Kundaje, A.: Learning Important Features Through Propagating Activation Differences (2019)
12. Sundararajan, M., Taly, A., Yan, Q.: Axiomatic Attribution for Deep Networks (2017)
13. Wang, J., Wiens, J., Lundberg, S.: Shapley flow: a graph-based approach to interpreting model predictions. In: Proceedings of The 24th International Conference on Artificial Intelligence and Statistics, pp. 721–729. PMLR (2021)

Explaining Crash Predictions on Multivariate Time Series Data

Francesco Spinnato[1], Riccardo Guidotti[2(✉)], Mirco Nanni[3], Daniele Maccagnola[4], Giulia Paciello[4], and Antonio Bencini Farina[4]

[1] Scuola Normale Superiore, Pisa, Italy
francesco.spinnato@sns.it
[2] University of Pisa, Pisa, Italy
riccardo.guidotti@unipi.it
[3] ISTI-CNR, Pisa, Italy
mirco.nanni@isti.cnr.it
[4] Generali Italia, Mogliano Veneto, Italy
{daniele.maccagnola,giulia.paciello,antonio.bencini}@generali.com

Abstract. In Assicurazioni Generali, an automatic decision-making model is used to check real-time multivariate time series and alert if a car crash happened. In such a way, a Generali operator can call the customer to provide first assistance. The high sensitivity of the model used, combined with the fact that the model is not interpretable, might cause the operator to call customers even though a car crash did not happen but only due to a harsh deviation or the fact that the road is bumpy. Our goal is to tackle the problem of interpretability for car crash prediction and propose an eXplainable Artificial Intelligence (XAI) workflow that allows gaining insights regarding the logic behind the deep learning predictive model adopted by Generali. We reach our goal by building an interpretable alternative to the current obscure model that also reduces the training data usage and the prediction time.

Keywords: Multivariate time series · Crash prediction · Explainability · Interpretable machine learning · Car insurance · Case study

1 Introduction

Crash Data Recorders (CDR) have been increasingly used inside cars to monitor safety measures and record impact speeds [23]. Through machine learning models, these data sources that can be exploited by insurance companies to monitor and improve customer service quality [22]. We collaborate with Assicurazioni Generali to detect car crashes, relying on their multivariate time series data and their Artificial Intelligence (AI) system, based on a deep learning model. Generali is one of the largest global insurance companies and is developing an automatic AI-based decision-making system to provide first assistance to its customers. Generali records speed and acceleration as multivariate time series from

P. Pascal and D. Ienco (Eds.): DS 2022, LNAI 13601, pp. 556–566, 2022.
https://doi.org/10.1007/978-3-031-18840-4_39

each insured customer's car. Such data is used to train a deep Convolutional Neural Network (CNN) that enables Generali to warn human operators of possible car crashes. In turn, the operator will physically call the customer to check if something bad happened and to know which kind of assistance is required.

Two weaknesses are currently present. First, the high sensitivity of the AI system might cause unnecessary and harassing calls. Second, the AI system is based on a deep learning model that is inherently not interpretable. An interpretability layer is helpful for numerous reasons, and eXplainable Artificial Intelligence (XAI) can help build user trust toward more transparent AI decisions. XAI is a branch of AI that focuses on allowing humans to comprehend the decisions of complex black-box models used by AI systems [3]. Thus, the objective of this work is to tackle the problem of interpretability for car crash prediction, proposing a pipeline that allows to gain insights regarding the logic behind the CNN and build a more transparent predictive model on a multivariate time series dataset. We highlight that XAI for multivariate time series is still an underexplored topic.

The literature on crash prediction studies car accidents from various perspectives. We mainly distinguish between *real-time* and *long-term* crash prediction. Long-term crash prediction is a relatively little explored area, with a few works based on movement statistics and mining models [4,13,20]. A large part of the literature focuses on real-time crash prediction by analyzing areas and condition of collision [15,19] through internal and external sensor recording mobility features [5,12] or physiological parameters [1]. The interested reader can find in [9] a survey analyzing the key problems. We believe our analysis is more closely related to real-time crash prediction. However, in our scenario, the decision system is used after the crash, unlike all the works mentioned above, where classifiers are used prior to the crash. Also, a limitation of all the models used in the aforementioned approaches is that they are not interpretable.

Since we aim to explain a neural network, our interest is focused on post-hoc model-specific XAI methods [3]. Many approaches are based on Grad-CAM [16] and can analyze the gradients of CNNs to understand the most important features in the time series. However, to the best of our knowledge, none of them was tested on large multivariate time series with signals of different lengths. Furthermore, they can only be applied on CNNs, which would limit the proposed framework's expandability. For these reasons, we make use of GradientExplainer [10] which can deal with any deep learning model and with signals of different lengths, ensuring fast feature-based explanations. We use GradientExplainer to distinguish the areas of attention in the time series. Then, we reduce the dataset dimensionality and train subsequence-based surrogate trees to imitate the prediction of the CNN in a more interpretable way. Finally, we train the best performing surrogate as a possible interpretable predictor to be used as a replacement of the CNN. Preliminary results show that the proposed XAI workflow is promising *(i)* in terms of efficiency, as it reduces the data usage and cuts the prediction time, and *(ii)* in terms of effectiveness, providing an interpretability layer that helps operators better understand the prediction of the AI system.

2 Problem Description and Explanation Methodologies

Generali collects high-dimensional time series data through CDR. This data is transferred in real-time to insurance company servers, and it is processed through an automatic AI decision-making system. If the AI system signals the presence of a possible crash, the operator calls the customers to check if there is a need for first assistance. Generali aims at solving two criticalities in the current approach. First, *reduce false positives* by increasing the model precision in order to avoid unnecessary calls. Second, *increase the interpretability of the automatic decision-making system* to help the operator choose the right course of action.

In the following, we present the background necessary to comprehend each step. Formally, we define a multivariate time series as follows:

Definition 1 (Multivariate Time Series). A multivariate *time series* $X = \{x_{1,1}, \dots x_{j,k}, \dots, x_{m,d}\} \in \mathbb{R}^{m \times d}$ is an ordered set of m real-valued observations, each having $d > 1$ dimensions (or signals/channels).

A Time Series Classification (TSC) dataset is defined as follows:

Definition 2 (TSC Dataset). A *time series classification dataset* $\mathcal{D} = (\mathcal{X}, \mathbf{y})$ is a set of n time series, $\mathcal{X} = \{x_{1,1,1}, \dots x_{i,j,k}, \dots, x_{n,m,d}\} \in \mathbb{R}^{n \times m \times d}$, with a vector of assigned labels (or classes), $\mathbf{y} = \{y_1, y_2, \dots, y_n\} \in \{0, 1\}^n$.

Observation (i, j, k) of a dataset is denoted by $x_{i,j,k}$: i denotes the ith multivariate time series in the dataset, j denotes the jth time-step, k denotes the kth signal of the time series. Hence, we define Time Series Classification (TSC) as:

Definition 3 (TSC). Given a TSC dataset $\mathcal{D}$, *Time Series Classification* is the task of training a function f from the space of possible inputs $\mathcal{X}$ to a probability distribution over the class variable values in $\mathbf{y}$.

Dataset. The car crash dataset ($\mathcal{D}$) provided by Generali contains $n = 81{,}173$ instances. Each instance is a multivariate time series composed of four signals ($d = 4$), namely the *acceleration* of the car for the x, y, z axes, and the *speed* of the car. The acceleration signals of the car contain $m_1 = 2{,}490$ observations for each axis, and they are, in turn, a concatenation of two signals sampled at different frequencies. The speed signal is a recording containing $m_2 = 41$ time-steps. Each multivariate time series X_i is labelled either as $y_i = 1$ when it is a crash or as $y_i = 0$ when it is a no-crash. Crashes are rarer than no-crashes and represent only about the 6% of the dataset. From a classification perspective, the main critical issues in dealing with this dataset are the presence of signals with a big difference in length, i.e., speed and acceleration, and the heavy label unbalance. An example of a no-crash instance is depicted in Fig. 1. The dataset is split by Generali into 50% training set, 25% validation set, and 25% test set.

Predictive Model. Generali implements the predictive model with a Convolutional Neural Network (CNN) [6][1]. On the test set, the CNN has an accuracy

[1] Details can not be disclosed due to company policies.

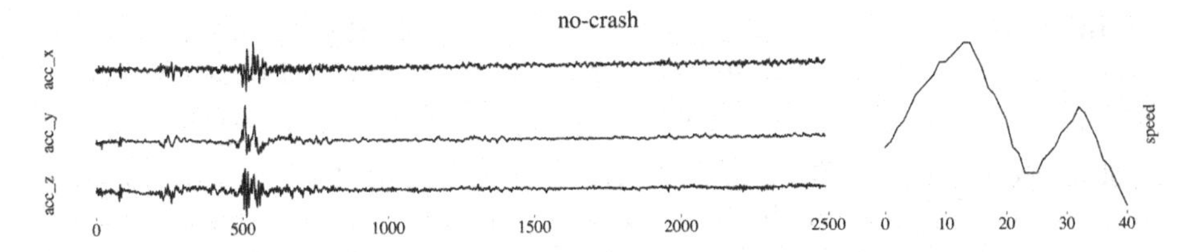

Fig. 1. Multivariate time series sample of a no-crash instance: *left* - acceleration times series for the three axes, *right* - speed time series.

of 0.961 and a precision of 0.701. Precision is paramount in this setting because every false positive causes unnecessary calls by human operators. Also, due to the CNN architecture, the predictive model is a black-box [3]. Thus, the challenges raised by Generali are *(i)* gaining insights regarding the logic behind the prediction of the provided CNN, without making any modification to the data or the model, to understand which parts of the data and which patterns of the multivariate time series are more important for the classification, *(ii)* building an efficient and effective, interpretable alternative for the CNN model, possibly reducing false positives and optimizing data usage. Also, Generali requires high efficiency at test time, to minimize response time when dealing with unseen instances.

Explanation Methodologies and Workflow. We meet these challenges through a set of XAI methodologies organized as the following workflow:

1. **Analyze the attention** of the CNN, inspecting its gradients, discovering the parts of the time series more relevant for the prediction (Sect. 3);
2. **Build interpretable subsequence-based surrogates** of the CNN to imitate its output, focusing on the parts previously highlighted (Sect. 4);
3. **Train an interpretable subsequence-based model** on the real labels, as a replacement of the CNN, still focusing only on the parts highlighted in the first step while also leveraging the results of the second step (Sect. 5).

3 Gradient-Based Explainer

To inspect the CNN we use GradientExplainer, a gradient-based interpretability approach [17]. GradientExplainer returns an explanation in the form of a saliency map, highlighting the contribution of each time-step for the classification [3].

Definition 4 (Saliency Map)**.** Given a time series X, a *saliency map* $\Phi = \{\phi_{j,k} \mid \forall j \in [1, m], k \in [1, d]\}$ contains a score $\phi_{j,k}$ for every observation $x_{j,k}$ of X.

This saliency map is returned in terms of approximated SHAP values [10], obtained by computing the expectations of gradients, sampling reference values from a background dataset. To compute the SHAP values, X is perturbed using a matrix $Z' \in \{0,1\}^{m \times d}$ to decide which values to keep and which values to replace in X. Each input observation receives a positive or negative SHAP value, depending on its contribution to the model output. Formally:

Definition 5 (Additive Feature Attribution). An additive feature attribution method g has an explanation model that is a linear function of binary variables, $g(Z') = \phi_0 + \sum_{j=1}^{m} \sum_{k=1}^{d} \phi_{j,k} z'_{j,k}$, where $z'_{j,k} \in \{0,1\}$ and $\phi_{j,k} \in \mathbb{R}$.

In other words, given a time series X, the explanation model g tries to transparently approximate the prediction of a black-box classifier f in the local neighborhood of X, i.e., $g(Z') \approx f(X)$. In our binary classification setting, for each time-step j and each signal k, a $\phi_{j,k}$ SHAP value close to 0 indicates that the point $x_{j,k}$ is almost irrelevant for the classification; a positive value indicates a contribution towards the class crash, while a negative value indicates a contribution towards the class no-crash. ϕ_0 is the base value, i.e., the default classification output for an "empty" time series. This kind of explanation is *local*, meaning that it can be used to shed light on the decision of the CNN for single time series. However, by retrieving such explanations for each instance of the dataset and aggregating the SHAP values at different granularity, we can gain an overview of the global behavior of the model. Despite being simple operations, to the best of our knowledge, *the SHAP values aggregations for multivariate time series classification detailed in the following is a novel contribution in XAI.*

Signal Importance. The signal-wise aggregation for each multivariate time series is obtained by summing the SHAP values of each signal. Thus, for every prediction, we can understand the impact of the different dimensions for every instance in the dataset. These sums are collected in $\Phi^{\text{signal}} \in \mathbb{R}^{n \times d}$ such that $\Phi^{\text{signal}}_{i,k}$ is defined as: $\Phi^{\text{signal}}_{i,k} = \sum_{j=1}^{m} \phi_{i,j,k}$ with $i \in [1,n]$, $k \in [1,d]$. In our setting, Φ^{signal} is a matrix with n rows and $d = 4$ columns, corresponding to the sums of the SHAP values of all time-steps for each time series in $\mathcal{X}$ for each signal, i.e., the accelerations on the x, y, z *axes* and the *speed*. These sums, not reported due to space limits, tell us that *the most relevant dimension is the acceleration on the x-axis*, for which the SHAP values deviate the most from 0. This suggests a higher degree of contribution both for crash and no-crash time series. *The acceleration on the y and z-axis and the speed seem to be less impactful.* However, their contributions are not irrelevant, even if smaller w.r.t. the *x-axis*.

Time-Step Importance. A more fine-grained insight can be obtained by averaging the SHAP values by time-step. Similarly to the previous point, we collected these averages in a matrix $\Phi^{\text{point}} \in \mathbb{R}^{m \times d}$ such that $\Phi^{\text{point}}_{j,k}$ is defined as: $\Phi^{\text{point}}_{j,k} = \frac{1}{n} \sum_{i=1}^{n} \phi_{i,j,k}$ with $j \in [1,m]$, $k \in [1,d]$. In this case, for each time-step of each signal, we can see its average contribution. In the barplot in Fig. 2, we present such averages, stacking the accelerations on the left-hand side plots for better readability. We notice that, *the main area of attention for acceleration is around the* 500th *time-step, independently of the class.* Moreover, regarding no-crash instances, there is also a minor contribution around the 100th time step, with a high peak of positive SHAP values, corresponding to the concatenation of signals with different frequencies. This concatenation seems to nudge the CNN towards the class crash, even in no-crash instances. This event signals a possible defect of the CNN provided by Generali. The second part of the time series seems

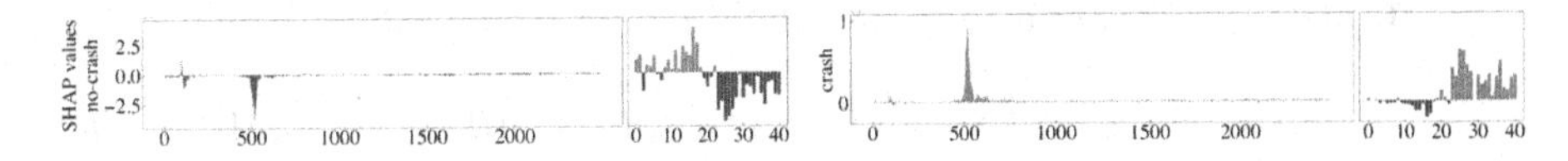

Fig. 2. Mean of SHAP values aggregated pointwise. *left* - speed and acceleration for no-crash instances. *right* - speed and acceleration for crash instances. Red and blue colors highlight positive and negative SHAP values. (Color figure online)

more relevant for the speed signal, especially for the class crash. *These insights are extremely important as they suggest that only a very small part of the data is relevant for the classification.* In the following, we use this information to optimize data usage, while we will continue to consider the four dimensions.

4 Subsequence-Based Surrogates

Subsequences are one of the most common ways to build interpretable models in the time series domain. Formally:

Definition 6 (Subsequence). Given a single signal $\mathbf{x} = \{x_1, \ldots, x_m\}$ of the multivariate time series X, a *subsequence* $\mathbf{s} = \{x_j, \ldots, x_{j+l-1}\}$ of length l is an ordered sequence of values such that $1 \leq j \leq m - l + 1$.

Subsequences can be real-valued, like shapelets [21], or can be symbolic, like SAX-based subsequences [7]. Subsequence extraction is computationally expensive, and therefore, using the insights previously gained, *we trim the x, y, z acceleration signals taking only the observations between the* 400th *and* 800th *time-step*. In this way, the issue earlier raised regarding the concatenation of the two signals sampled at different frequencies, is also avoided. On the other hand, we leave the speed signal as it is. We denote this filtered version of $\mathcal{X}$ with $\mathcal{X}'$.

Shapelet-Based Subsequences. We run the Shapelet Transform [8] with the Learning-Shapelets (LTS) algorithm [2]. Once the most discriminative shapelets are found, the dataset is transformed into a simplified representation.

Definition 7 (Shapelet Transform). Given a time series dataset $\mathcal{X}$ and a set S containing h shapelets, the *Shapelet Transform*, σ, converts $\mathcal{X} \in \mathbb{R}^{n \times m \times d}$ into a real-valued matrix $T \in \mathbb{R}^{n \times h}$, obtained by taking the minimum Euclidean distance between each time series in $\mathcal{X}$, and each shapelet in S, via a sliding-window.

SAX-Based Subsequences. Symbolic Aggregate approXimation (SAX) [7] transforms time series into strings. We perform subsequences extraction with MR-SEQL [14], which greedily selects the most discriminative symbolic subsequences. Thus, the dataset is transformed into an interpretable representation having as features the extracted subsequences and as values 0 or 1 depending on the absence or presence of subsequences. Formally:

Table 1. Surrogates performance *(higher is better, best values in bold).*

	SAX DT	SAX RF	SAX XGB	SAX LGB	SAX CAT	SHP DT	SHP RF	SHP XGB	SHP LGB	SHP CAT
Accuracy	0.954	0.966	0.975	0.974	**0.976**	0.950	0.950	0.951	0.951	0.951
Precision	0.621	**0.912**	0.862	0.855	0.862	0.639	0.641	0.624	0.623	0.623
Recall	0.567	0.475	0.681	0.678	**0.697**	0.360	0.361	0.435	0.438	0.437
Fscore	0.593	0.625	0.761	0.756	**0.771**	0.460	0.462	0.513	0.514	0.514

Definition 8 (Symbolic Subsequence Transform). Given a dataset $\mathcal{X}$ and a set S containing h symbolic subsequences, the *Symbolic Subsequence Transform*, σ, converts $\mathcal{X} \in \mathbb{R}^{n \times m \times d}$ into a binary-valued matrix $T \in \{0,1\}^{n \times h}$, obtained by checking if each subsequence in S is contained or not in each time series in $\mathcal{X}$.

Tree-Based Global Surrogates. Regardless of the way subsequences are extracted, the tabular representation $T' = \sigma(\mathcal{X}')$ can be paired with any classification model, with the advantage of an interpretable input [8]. We independently extract the subsequences from each signal of the multivariate time series[2], and we concatenate the resulting transformed datasets column-wise. As classification models, we adopt tree-based classifiers [18] because they simultaneously offer good performance and provide partial interpretability by granting the possibility to access the feature importance. We train the following five models[3]: a standard Decision Tree (DT) as baseline, Random Forest (RF), XGBoost (XGB), LightGBM (LGB) and CatBoost (CAT). We highlight that *we train these classifiers as global surrogates*, i.e., not on the original dataset labels $\mathbf{y}$, but on the prediction of the CNN, $\mathbf{y}_{\text{CNN}} = \text{CNN}(\mathcal{X})$. To guarantee a high level of generalization, the surrogate models are trained on the prediction of the CNN on the validation set.

We measure the performance on the test set and report the results in Table 1. All the metrics are computed w.r.t. the prediction of the CNN, i.e., given a model M: $metric = eval(\mathbf{y}_{\text{CNN}}, \mathbf{y}_{\text{M}})$ with $\mathbf{y}_{\text{M}} = \text{M}(\sigma(\mathcal{X}'))$. In general, all SAX-based methods outperform their shapelet-based counterpart. They are quite successful in imitating the output of the CNN, using a fraction of the input data. Specifically, SAX-LGB, SAX-XGB, SAX-CAT and SAX-RF perform better than SAX-DT. SAX-RF achieves the highest precision; however, it falls behind in all other performance metrics. From these results, it is clear that overall the best model is SAX-CAT.

[2] The subsequence extraction is performed using the default implementation parameters for `MR-SEQL` and the heuristic proposed in [2] for `LTS`.

[3] All the models are trained using the default library implementation parameters: `Scikit-learn` for DT, RF, `XGBoost` for XGB, `LightGBM` for LGB, `CatBoost` for CAT.

5 Subsequence-Based Classifier

The surrogates trained in the previous section are *interpretable complements* of the black-box, explaining its prediction using subsequences. Depending on the goal of the analysis, these models can also be used as *interpretable replacements* of the black-box. For this purpose, SAX-CAT is chosen as a candidate replacement of the CNN. As for the surrogates, SAX-CAT is trained on the transformed version of the trimmed set, $T' = \sigma(\mathcal{X}')$, which only contains around 17% of the initial observations. SAX-CAT is trained on the original training set labels $\mathbf{y}$.

Table 2. Models performance.

	Accuracy	Precision	Recall	Fscore	roc-auc	Runtime
CNN	0.961	0.701	0.660	0.680	0.924	784 ± 69.5
SAX-CAT	0.958	0.760	0.471	0.582	0.911	218 ± 23.9

In Table 2 we benchmark SAX-CAT and CNN on the test set, comparing $\mathbf{y}_{\text{CNN}}$ and $\mathbf{y}_{\text{SAX-CAT}}$ with $\mathbf{y}$. Results show a comparable performance in terms of accuracy, with a substantial improvement in precision. Besides, there is a degradation terms of recall and f-score. However, the main purpose of Generali was to reduce the number of false positives, giving less weight to false negatives. On training time, the most expensive computation for SAX-CAT is the extraction of discriminative subsequences, which takes about 7 h. However, this search has to be performed only once. On the other hand, the training of the CNN takes less than an hour. On test time, the average runtimes to classify and explain an unseen test instance are 784 ms $\pm$ 69.5 ms for the CNN and 218 ms $\pm$ 23.9 ms for SAX-CAT. The runtime of SAX-CAT includes the subsequence transform, the classification and explanation in terms of features importance, which is attached to the SAX-based features. The most significant advantage of using tree-based approaches is that their predictions can be efficiently interpreted using SHAP's TreeExplainer [11]. The local explanations can be aggregated to have a general global overview of the logic behind the model, i.e., in our setting, we can understand which subsequences are more relevant for classifying crash and no-crash instances.

The global explanation plot is presented in Fig. 3 (top). This summary global plot depicts the SHAP values for all the instances and subsequences in the dataset and sorts them by the overall impact on the model prediction. Specifically, on the y-axis are presented the top-10 subsequences, sorted from the most influential (top) to the least influential (bottom). Each point on each row in the plot corresponds to one multivariate time series. Points colored in orange indicate that the corresponding subsequence is contained in the multivariate time series, while points colored in green represent a not-contained subsequence. The SHAP values are plotted on the x-axis, showing the contribution of feature values toward the classification. In general, this plot helps visualize if the

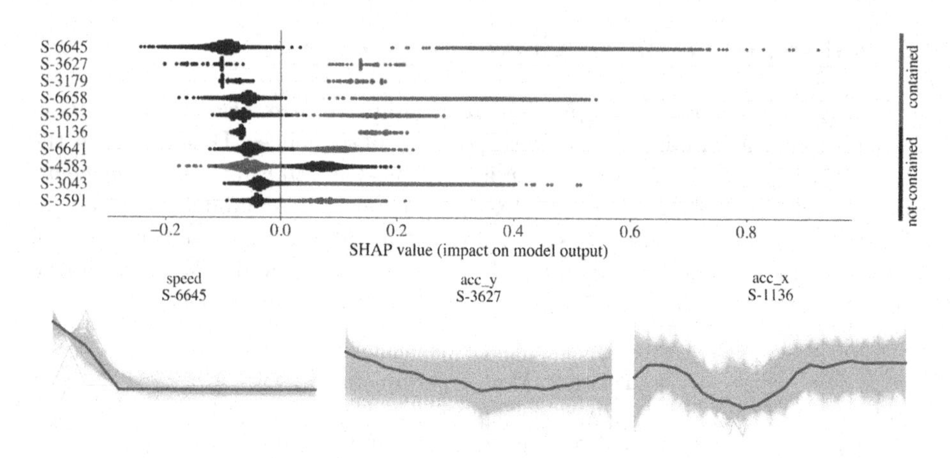

Fig. 3. SHAP values summary plot (*top*); sample of relevant subsequences (*bottom*).

presence/absence of subsequences contributes to the no-crash (negative SHAP values) or crash (positive SHAP values) class at a global level. In our case, the most influential subsequence is *S-6645*, belonging to the *speed signal*, which represents a decrement of car speed followed by a stop. As expected, the presence of *S-6645* contributes to the class crash, while its absence is an indicator of the class no-crash.

Three example subsequences are presented in Fig. 3 (bottom). Given that they are symbolic subsequences, they can assume slightly different shapes in the dataset; therefore, a representative subsequence is presented in orange, while the other subsequences are presented in a faded gray. The representative is computed as a medoid. Given their higher frequency, subsequences belonging to the acceleration axes are inherently harder to interpret. However, they could still be of interest to domain experts. In fact, the importance of these subsequences for the model output suggests that the car's jerk (or jolt), i.e., the rate at which the car's acceleration changes w.r.t. time, is probably relevant to the classification. For example, the most important subsequence for the *y-axis* is *S-3627*, and presents a decrement in acceleration, contributing to the class crash. *S-1136* is instead most relevant for the *x-axis* and presents a decrement followed by an increment in acceleration, also contributing to the class crash.

Fig. 4. SHAP values explanation for sample in Fig. 1 obtained from SAX-CAT.

In Fig. 4 we report the local explanation for the instance depicted in Fig. 1. Features contributing to the class crash are in red, while features contributing to the class no-crash are in blue. This plot, combined with the knowledge of the shape of the subsequences, can help domain experts to understand the logic behind the prediction of the tree-based approach. In this case, the presence of *S-6641* and *S-6699* pushes the prediction toward the class crash, while the presence of *S-1641* and the absence of *S-6645* contribute towards no-crash. As expected, contributions towards no-crash are greater in magnitude w.r.t. contributions towards crash, given that the instance is a no-crash.

6 Conclusion

We have presented a workflow to tackle explainability in the domain of crash prediction and multivariate time series. We have observed that, with respect to the CNN adopted by Generali, the predictive performance of the proposed SAX-CAT is comparable in terms of accuracy, and achieves higher precision while being interpretable, which were the goals of this work. In addition, SAX-CAT is three times faster than the CNN in making the prediction and uses only 17% of the original data. In future research directions, we would like to improve the performance of SAX-CAT, increasing the crash recall by further fine-tuning the subsequence-based models and improving the interpretability of the framework through prototypical and counterfactual instances.

Acknowledgment. This work has been partially supported by the European Community Horizon 2020 programme under the funding schemes: G.A. 871042 *SoBigData++*, G.A. 952026 *HumanE AI Net*, and G.A. 834756 *XAI*.

References

1. Ba, Y., et al.: Crash prediction with behavioral and physiological features for advanced vehicle collision avoidance system. TR_C **74**, 22–33 (2017)
2. Grabocka, J., et al.: Learning time-series shapelets. In: KDD. ACM (2014)
3. Guidotti, R., et al.: A survey of methods for explaining black box models. ACM Comput. Surv. **51**(5), 1–42 (2019)
4. Guidotti, R., et al.: Crash prediction and risk assessment with individual mobility networks. In: MDM. IEEE (2020)
5. Kweon, Y.J., et al.: Development of crash prediction models with individual vehicular data. TR_C **19**(6), 1353–1363 (2011)
6. LeCun, Y., et al.: Gradient-based learning applied to document recognition. Proc. IEEE **86**(11), 2278–2324 (1998)
7. Lin, J., et al.: Experiencing SAX: a novel symbolic representation of time series. Data Min. Knowl. Discov. **15**, 107–144 (2007)
8. Lines, J., et al.: A shapelet transform for time series classification. In: KDD, KDD 2012, pp. 289–297. ACM, New York (2012)
9. Lord, D., et al.: The statistical analysis of crash-frequency data: a review and assessment of methodological alternatives. TR_A **44**(5), 291–305 (2010)

10. Lundberg, S.M., et al.: A unified approach to interpreting model predictions. In: NIPS, pp. 4768–4777 (2017)
11. Lundberg, S.M., et al.: From local explanations to global understanding with explainable AI for trees. Nat. Mach. Intell. **2**(1), 56–67 (2020)
12. Mannering, F.L., et al.: Analytic methods in accident research: methodological frontier and future directions. Anal. Methods Accid. Res. **1**, 1–22 (2014)
13. Nanni, M., et al.: City indicators for geographical transfer learning: an application to crash prediction. GeoInformatica 1–32 (2022)
14. Nguyen, T.L., et al.: Interpretable time series classification using linear models and multi-resolution symbolic representations. DAMI **33**(4), 1183–1222 (2019)
15. Salim, F.D., et al.: Collision pattern modeling and real-time collision detection at road intersections. In: ITSC, pp. 161–166. IEEE (2007)
16. Selvaraju, R.R., et al.: Grad-CAM: visual explanations from deep networks via gradient-based localization. In: ICCV, pp. 618–626 (2017)
17. Sundararajan, M., et al.: Axiomatic attribution for deep networks. In: ICML. Proceedings of Machine Learning Research, vol. 70, pp. 3319–3328. PMLR (2017)
18. Tan, P.N.: Introduction to Data Mining. Pearson Education India (2018)
19. Wang, J., et al.: Real-time driving danger level prediction (2010)
20. Wang, Y., et al.: ML methods for driving risk. In: EM-GIS. ACM (2017)
21. Ye, L., et al.: Time series shapelets: a new primitive for data mining (2009)
22. Zantalis, F., et al.: A review of machine learning and IoT in smart transportation. Future Internet **11**(4), 94 (2019)
23. Ziebinski, A., et al.: Review of advanced driver assistance systems (ADAS) (2017)

Author Index

Printed in the United States
by Baker & Taylor Publisher Services